Inverse Modeling

An introduction to the theory and methods of inverse problems and data assimilation

Inverse Modeling

An introduction to the theory and methods of inverse problems and data assimilation

Gen Nakamura and Roland Potthast

Hokkaido University, Japan
Inha University, Korea
University of Reading, United Kingdom
German Meteorological Service (DWD)

IOP Publishing, Bristol, UK

Media content for this book is available from http://iopscience.iop.org/book/978-0-7503-1218-9/page/about-the-book.

ISBN 978-0-7503-1218-9 (ebook)
ISBN 978-0-7503-1219-6 (print)
ISBN 978-0-7503-1220-2 (mobi)

DOI 10.1088/978-0-7503-1218-9

Version: 20151201

IOP Expanding Physics
ISSN 2053-2563 (online)
ISSN 2054-7315 (print)

British Library Cataloguing-in-Publication Data: A catalogue record for this book is available from the British Library.

Published by IOP Publishing, wholly owned by The Institute of Physics, London

IOP Publishing, Temple Circus, Temple Way, Bristol, BS1 6HG, UK

US Office: IOP Publishing, Inc., 190 North Independence Mall West, Suite 601, Philadelphia, PA 19106, USA

Inverse modeling is *universal* in the sciences and economics.
How could you do without it?

Mathematics combines the *beauty of the mind*
with the *practice of life*.

It is joy to use *simple principles*
to carry out *complex tasks*.

It is extremely powerful to use a *balance*
of *generalization* and *specialization*,
to step back and see the *basic principles*
and then focus on the *detail* of particular problems.

Science is and has always been
an interplay between *many ideas*,
between *many minds*,
between *many humans.*

Science is and has always been
the quest to *understand creation*,
—the work of a *great God*,
and our Saviour in *Christ Jesus*.

Contents

Preface xiii

Acknowledgements xv

Author biographies xvi

1 Introduction **1-1**

1.1 A tour through theory and applications 1-2
1.2 Types of inverse problems 1-14
1.2.1 The general inverse problem 1-15
1.2.2 Source problems 1-16
1.2.3 Scattering from obstacles 1-17
1.2.4 Dynamical systems inversion 1-19
1.2.5 Spectral inverse problems 1-21
Bibliography 1-22

2 Functional analytic tools **2-1**

2.1 Normed spaces, elementary topology and compactness 2-1
2.1.1 Norms, convergence and the equivalence of norms 2-1
2.1.2 Open and closed sets, Cauchy sequences and completeness 2-5
2.1.3 Compact and relatively compact sets 2-7
2.2 Hilbert spaces, orthogonal systems and Fourier expansion 2-10
2.2.1 Scalar products and orthonormal systems 2-10
2.2.2 Best approximations and Fourier expansion 2-13
2.3 Bounded operators, Neumann series and compactness 2-19
2.3.1 Bounded and linear operators 2-19
2.3.2 The solution of equations of the second kind and the Neumann series 2-25
2.3.3 Compact operators and integral operators 2-27
2.3.4 The solution of equations of the second kind and Riesz theory 2-32
2.4 Adjoint operators, eigenvalues and singular values 2-33
2.4.1 Riesz representation theorem and adjoint operators 2-33
2.4.2 Weak compactness of Hilbert spaces 2-36
2.4.3 Eigenvalues, spectrum and the spectral radius of an operator 2-38
2.4.4 Spectral theorem for compact self-adjoint operators 2-40
2.4.5 Singular value decomposition 2-46
2.5 Lax–Milgram and weak solutions to boundary value problems 2-48

2.6	The Fréchet derivative and calculus in normed spaces	2-50
	Bibliography	2-55
3	**Approaches to regularization**	**3-1**
3.1	Classical regularization methods	3-1
	3.1.1 Ill-posed problems	3-1
	3.1.2 Regularization schemes	3-2
	3.1.3 Spectral damping	3-4
	3.1.4 Tikhonov regularization and spectral cut-off	3-7
	3.1.5 The minimum norm solution and its properties	3-10
	3.1.6 Methods for choosing the regularization parameter	3-14
3.2	The Moore–Penrose pseudo-inverse and Tikhonov regularization	3-20
3.3	Iterative approaches to inverse problems	3-22
	3.3.1 Newton and quasi-Newton methods	3-23
	3.3.2 The gradient or Landweber method	3-25
	3.3.3 Stopping rules and convergence order	3-31
	Bibliography	3-34
4	**A stochastic view of inverse problems**	**4-1**
4.1	Stochastic estimators based on ensembles and particles	4-1
4.2	Bayesian methods	4-4
4.3	Markov chain Monte Carlo methods	4-6
4.4	Metropolis–Hastings and Gibbs sampler	4-11
4.5	Basic stochastic concepts	4-14
	Bibliography	4-19
5	**Dynamical systems inversion and data assimilation**	**5-1**
5.1	Set-up for data assimilation	5-3
5.2	Three-dimensional variational data assimilation (3D-VAR)	5-5
5.3	Four-dimensional variational data assimilation (4D-VAR)	5-8
	5.3.1 Classical 4D-VAR	5-8
	5.3.2 Ensemble-Based 4D-VAR	5-13
5.4	The Kalman filter and Kalman smoother	5-16
5.5	Ensemble Kalman filters (EnKFs)	5-22
5.6	Particle filters and nonlinear Bayesian data assimilation	5-29
	Bibliography	5-33

6	**Programming of numerical algorithms and useful tools**	**6-1**
6.1	MATLAB or OCTAVE programming: the butterfly	6-1
6.2	Data assimilation made simple	6-4
6.3	Ensemble data assimilation in a nutshell	6-8
6.4	An integral equation of the first kind, regularization and atmospheric radiance retrievals	6-9
6.5	Integro-differential equations and neural fields	6-12
6.6	Image processing operators	6-15
	Bibliography	6-18
7	**Neural field inversion and kernel reconstruction**	**7-1**
7.1	Simulating neural fields	7-4
7.2	Integral kernel reconstruction	7-8
7.3	A collocation method for kernel reconstruction	7-17
7.4	Traveling neural pulses and homogeneous kernels	7-20
7.5	Bi-orthogonal basis functions and integral operator inversion	7-23
7.6	Dimensional reduction and localization	7-26
	Bibliography	7-30
8	**Simulation of waves and fields**	**8-1**
8.1	Potentials and potential operators	8-1
8.2	Simulation of wave scattering	8-11
8.3	The far field and the far field operator	8-15
8.4	Reciprocity relations	8-21
8.5	The Lax–Phillips method to calculate scattered waves	8-23
	Bibliography	8-26
9	**Nonlinear operators**	**9-1**
9.1	Domain derivatives for boundary integral operators	9-1
9.2	Domain derivatives for boundary value problems	9-8
9.3	Alternative approaches to domain derivatives	9-11
	9.3.1 The variational approach	9-11
	9.3.2 Implicit function theorem approach	9-18
9.4	Gradient and Newton methods for inverse scattering	9-21
9.5	Differentiating dynamical systems: tangent linear models	9-27
	Bibliography	9-30

10 Analysis: uniqueness, stability and convergence questions **10-1**

10.1 Uniqueness of inverse problems 10-3

10.2 Uniqueness and stability for inverse obstacle scattering 10-4

10.3 Discrete versus continuous problems 10-7

10.4 Relation between inverse scattering and inverse boundary value problems 10-9

10.5 Stability of cycled data assimilation 10-14

10.6 Review of convergence concepts for inverse problems 10-18

10.6.1 Convergence concepts in stochastics and in data assimilation 10-19

10.6.2 Convergence concepts for reconstruction methods in inverse scattering 10-21

Bibliography 10-24

11 Source reconstruction and magnetic tomography **11-1**

11.1 Current simulation 11-2

11.1.1 Currents based on the conductivity problem 11-2

11.1.2 Simulation via the finite integration technique 11-4

11.2 The Biot–Savart operator and magnetic tomography 11-8

11.2.1 Uniqueness and non-uniqueness results 11-12

11.2.2 Reducing the ill-posedness of the reconstruction by using appropriate subspaces 11-16

11.3 Parameter estimation in dynamic magnetic tomography 11-25

11.4 Classification methods for inverse problems 11-28

Bibliography 11-32

12 Field reconstruction techniques **12-1**

12.1 Series expansion methods 12-2

12.1.1 Fourier–Hankel series for field representation 12-2

12.1.2 Field reconstruction via exponential functions with an imaginary argument 12-6

12.2 Fourier plane-wave methods 12-10

12.3 The potential or Kirsch–Kress method 12-12

12.4 The point source method 12-20

12.5 Duality and equivalence for the potential method and the point source method 12-27

Bibliography 12-29

13 Sampling methods **13-1**

13.1 Orthogonality or direct sampling 13-2
13.2 The linear sampling method of Colton and Kirsch 13-4
13.3 Kirsch's factorization method 13-10
Bibliography 13-17

14 Probe methods **14-1**

14.1 The SSM 14-2
14.1.1 Basic ideas and principles 14-2
14.1.2 The needle scheme for probe methods 14-7
14.1.3 Domain sampling for probe methods 14-9
14.1.4 The contraction scheme for probe methods 14-10
14.1.5 Convergence analysis for the SSM 14-13
14.2 The probing method for near field data by Ikehata 14-16
14.2.1 Basic idea and principles 14-17
14.2.2 Convergence and equivalence of the probe and SSM 14-20
14.3 The multi-wave no-response and range test of Schulz and Potthast 14-21
14.4 Equivalence results 14-26
14.4.1 Equivalence of SSM and the no-response test 14-27
14.4.2 Equivalence of the no-response test and the range test 14-29
14.5 The multi-wave enclosure method of Ikehata 14-31
Bibliography 14-38

15 Analytic continuation tests **15-1**

15.1 The range test 15-1
15.2 The no-response test of Luke–Potthast 15-6
15.3 Duality and equivalence for the range test and no-response test 15-10
15.4 Ikehata's enclosure method 15-11
15.4.1 Oscillating-decaying solutions 15-13
15.4.2 Identification of the singular points 15-18
Bibliography 15-20

16 Dynamical sampling and probe methods **16-1**

16.1 Linear sampling method for identifying cavities in a heat conductor 16-2
16.1.1 Tools and theoretical foundation 16-4
16.1.2 Property of potential 16-14
16.1.3 The jump relations of $\mathcal{K}^*$ 16-16

16.2 Nakamura's dynamical probe method 16-17

16.2.1 Inverse boundary value problem for heat conductors with inclusions 16-17

16.2.2 Tools and theoretical foundation 16-18

16.2.3 Proof of theorem 16.2.6 16-20

16.2.4 Existence of Runge's approximation functions 16-24

16.3 The time-domain probe method 16-26

16.4 The BC method of Belishev for the wave equation 16-29

Bibliography 16-35

17 Targeted observations and meta-inverse problems 17-1

17.1 A framework for meta-inverse problems 17-1

17.2 Framework adaption or zoom 17-7

17.3 Inverse source problems 17-8

Bibliography 17-13

Appendix A A-1

Preface

Simulation of natural phenomena has been in the focus of science now for many centuries. The simulations are based on observations and scientific reasoning, which leads to the qualitative and quantitative description of natural or industrial processes.

When you try to bring *measurement data or observations* into simulations, you come to the fields of *inverse problems* and *data assimilation*. The key task of inverse problems is to infer knowledge about the structure of some system from measurements. Usually it is based on equations which model the underlying physical, chemical or biological processes, and it reconstructs sources or structural information. When the processes are *dynamical*, an important part of any realistic simulation is the evaluation of *initial states* in addition to *structural information* and underlying *parameter functions*. Data assimilation algorithms calculate initial states on which forecasts can be based.

Today, inverse problems are universal in the sciences. Basically each and every field of research or technology has its own inverse problems which are to be solved. Different fields have their own terminology, ranging from *medical imaging* to *tomography*, from *seismic exploration* to *inverse scattering* and *radar*, from *neuroscience* to *nondestructive testing*. At the same time, all these fields share many questions and phenomena, such that there is a high need for a joint approach. Here, functional analysis provides a very adequate framework with the language of normed spaces and Hilbert spaces – which has been highly successful in the world of physics now for 100 years.

Data assimilation has grown from geophysical and meteorological communities, where the determination of initial states has been of crucial importance for applications from the very beginning, since *forecasts* are possible only when reliable estimators for the initial conditions are available. Today, operational centers for *numerical weather prediction* (NWP) and climate projection are using top-500 supercomputers to serve our societies day by day[1]. At the same time, many further scientific communities discover the need for *data assimilation*, since they are moving from static estimates and qualitative science to quantitative forecasting for technological or medical phenomena.

Inverse Problems and Data Assimilation touch quite different *mathematical communities*. Inversion algorithms have first been suggested in the framework of mathematical analysis, partial differential equations and applied mathematics. It links into optimization as well as analysis in function spaces. Data Assimilation has been strongly linked to the stochastic communities, estimation theory and filtering. At the same time, it has been influenced by linear algebra and deterministic optimization. Today, we observe some convergence of these fields driven by the

[1] For example the Met Office in the UK, the European Center for Medium Range Weather Prediction (ECMWF) at Reading, UK, or the German Meteorological Service (DWD) at Frankfurt/Offenbach, Germany.

need to communicate and understand each other when working on joint applications.

Clearly, one single book cannot introduce the whole of inverse problems and data assimilation today – the selection we present might appear biased to some colleagues. We feel that it provides a broad generic selection, which will be useful for many different related problems as well. Also, different scientists work on the background of their particular field and its language. Our book aims to be introductory in many of its parts, with some deeper sections and chapters. Our goal is to reach a broad range of scientists from mathematics and applications. To this end we have included a lot of material about functional analysis which we found to be helpful for our own research over time, and also basic material about stochastic estimation theory and Markov Chain Monte Carlo methods.

We have added about 170 scripts in OCTAVE or MATLAB to the book (available for download **here**), which serve as a first step towards simulations and more sophisticated inversion or data assimilation algorithms. Extended versions of these scripts have been successfully applied to real data for practical problems such as acoustic sound analysis, magnetic tomography for fuel cells and cognitive neuroscience.

Our book has the claim to be full of insight which is highly relevant to practical applications. The second author works with algorithms in an operational environment, which is linked to many millions of users worldwide, ranging from about 40 other states to the flight control in central Europe and German federal weather warnings for all parts of society. This work is linked to dozens of different inverse problems, integrating them all into core data assimilation methods.

We are convinced that both the insight given by mathematical analysis as well as the small-scale and large-scale numerical tests are important for applications. Development and progress today is only possible when theory and practice come together.

Acknowledgements

The authors would like to acknowledge the strong and continuing support of their families during the past years, without which their research and joint activity would not have been possible.

Further, thanks belong to our colleagues and teachers Rainer Kress, David Colton, Gunther Uhlmann, Andreas Kirsch, Masahiro Yamamoto, Peter Monk, Hyeonbae Kang, Simon Chandler-Wilde and Jin Cheng, who have strongly supported our work over time.

The authors are grateful to the funding organizations, which have financed part of our work, including the German Science Foundation, EPSRC in the UK, the Leverhulme Trust, the German Meteorological Service (DWD), JSPS in Japan and NRF in Korea.

Last but not least we would like to express our thanks to our God, the creator of this world in Christ Jesus, for fascinating science, and for his deep grace and continuing guidance.

Author biographies

Gen Nakamura

Gen Nakamura received his Bachelor's Degree 1971 from International Christian University, his Master's Degree 1974 from the Graduate School of Tokyo Metropolitan University and Doctor of Science Degree 1977 from the Graduate School of Tokyo Metropolitan University.

He worked as Assistant Professor 1977-1991 at Josai University, 1981–1982 Lecturer of MIT, 1982–1983 Japan-US Exchange Fellow, 1989 Visiting Scientist of Brown University, 1991–1994 Professor of Josai University, 1994–1997 Professor of Science University of Tokyo, 1997–2001 Professor of Gunma University, 2001–2012 Professor of the Graduate School of Science, Hokkaido University, 2012–2013 Specially appointed Professor of the Graduate School of Science, Hokkaido University, 2013–2015 Professor of Inha University, since 2015 Harim Professor (special professor) Inha University, 2001- Visiting Professor of South East University. Professor Nakamura has served 2006–2009 as Program Officer of Japan Science for the Promotion of Science, 2009–2013 Board Member of Mathematical Society of Japan, 2011–2015 Expert Committee Member of Research Institute of Mathematical Science, Kyoto University and since 2011 Committee Member Based on the Fields of Specialties of Science Council of Japan.

He received several prizes, for example the Autumn Prize of Mathematical Society of Japan, 2000 and the Prize for Science and Technology by the Minister of Education, Culture, Sports, Science and Technology of Japan, 2009. He is Editor of Eurasian Journal of Mathematical and Computer Applications and Member of the steering committee of Eurasian Association for Inverse Problems.

Professor Nakamura has published 170 publications recorded in the MathScinet data base.

Roland Potthast

Roland Potthast received his Diplom 1993, a PhD in Mathematics 1994 and a Habilitation Degree in Mathematics 1999 at the University of Göttingen, Germany. Postdoctoral positions led him 1996–1998 to the University of Delaware, USA. After the Habilitation, Roland worked in a Consultancy Company in Cologne, Germany, 1999–2000, and at the Brunel University in London 2000–2001. From 2001 to 2006 he was leader of a *Young Researcher Group* on Inverse Problems in Göttingen, funded by the Volkswagen Foundation. He received an extraordinary professorship at the University of Göttingen in 2004. As a part-time activity, he was Owner and CEO of an

IT-company based in Göttingen for seven years 2005–2012, providing internet-based services to large-scale European companies.

Roland moved to the UK in 2006, with a lectureship at the University of Reading 2006, with promotions to Reader (Associate Professor) 2008 and Full Professor in 2010. Roland has been visiting professor at the University of Rennes, France, 2007–2009 and at the Research Center Jülich, Germany, 2009 and had a call to a W3 professorship in Munich 2010. Since 2010 he holds a post as 'Director and Professor' (B1) at the German *Federal Ministry of Transport and digital Infrastructure (BMVI),* leading the *Data Assimilation Department* of the German Weather Service (DWD) in Frankfurt/Offenbach, with a part-time involvement as full professor at the University of Reading, UK. He is supervising a group of about 25 scientists and coordinates a network of about 45 researchers in data assimilation and inverse problems working in collaboration with DWD.

Roland received several prizes and awards, among them 1994 the best PhD award of the German Mathematical Society, four years of full-time funding by the German Science foundation DFG and five years by the Volkswagen Foundation; in the UK an EPSRC Springboard fellowship in 2007 and a Leverhulme research fellowship in 2008, a 'Bridging-the-Gaps Award' by EPSRC in 2009–12; and the *Pichorides Distinguished Lectureship* of the University of Crete, Greece, in 2015.

Professor Potthast has published more than 65 mathematical research papers, which received more than 1000 citations, a book on inverse scattering theory 2001 and a book on neural field theory 2014.

IOP Publishing

Inverse Modeling

An introduction to the theory and methods of inverse problems and data assimilation

Gen Nakamura and Roland Potthast

Chapter 1

Introduction

The goal of this book is to provide an introduction to the *techniques, tools and methods* for *inverse problems* and *data assimilation*. It is written at the interface between mathematics and applications, and is designed for students, and researchers and developers in mathematics, physics, engineering, acoustics, electromagnetics, meteorology, biology, environmental and other applied sciences. Our challenge is to be accessible to a wide range of readers without compromising the mathematical insight and content.

The authors are rooted in mathematical analysis, but have worked on a wide range of applications from applied acoustics and magnetic source reconstruction via cognitive neuroscience to weather prediction and environmental remote sensing. These applications will find their way into the demonstration of concepts and tools in the different chapters of this book.

We will give an overview and introduce the reader to evolving techniques and results, covering important *basic concepts* and *methods*. We are not only interested in *formulating and practically testing* schemes, which is usually the main road for many applications, but in *understanding* the properties of concepts and methods. This includes the study of *convergence, consistency and stability* as well as *duality results* and *equivalence statements*.

Mathematical analysis and functional analysis provide a rich set of concepts and tools to study inverse problems and data assimilation problems and algorithms. Often, for people from applications these methods are not easily accessible, since the concepts and main tools are scattered within the literature. To this end we have included chapter 2 on *functional analysis* and an examination of *basic probability theory* in section 4.5, giving a brief and concise summary of the main results which we consider to be indispensable for a proper understanding of inversion and data assimilation methods.

doi:10.1088/978-0-7503-1218-9ch1

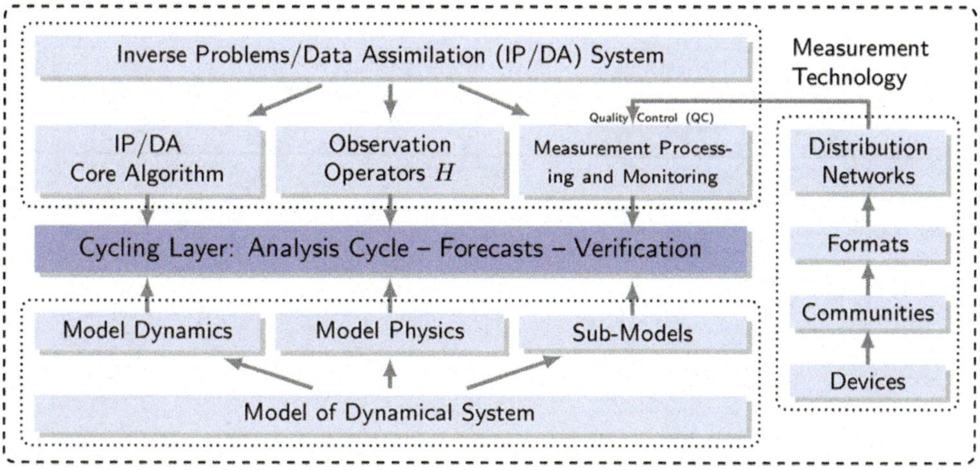

Figure 1.1. The different components of inverse problems and data assimilation systems.

1.1 A tour through theory and applications

The goal of this introductory section is to provide a short tour through both theoretical elements and different applications, discussing their role and indicating in what way inverse problems and data assimilation techniques are needed for applications. For each topic we briefly discuss what *types of inverse problems* arise from the application and what *types of questions* we usually have when we study inverse problems.

We will observe that *inverse problems* live between a quite *general* mathematical theory and a strong and wide range of very *special* applications. Many techniques for inverse problems or data assimilation are *universal* and can be described in a general framework, for example by a dynamical system on a normed space. But they are also *specific*: inverse problems are linked to particular application areas and it is of crucial importance to know the application and the properties of solutions, their settings and their environment.

Figure 1.1 shows the components of inverse problems or data assimilation systems. Usually, the time evolution of some natural quantity is modeled by a *dynamical system*. This can, e.g., be the state of the planetary atmosphere, or the state of the brain in neurodynamics. The modeling includes dynamical equations, often some particular type of physics and a range of sub-models, indicated by the three boxes in the fourth row of figure 1.1.

Different communities use different notation; the state is denoted by either x or by φ[1]. The model M maps φ_k at time t_k into φ_{k+1} at time t_{k+1}, where $k \in$ is a natural number.

[1] Note that when the *functional analytic* notation is used, then x denotes a point in space and $\varphi(x)$ is the function φ evaluated at x. In stochastics or *data assimilation* x is a random variable which can be a full function or a discretized version of the function. Since different communities have a strong inclination to keep their standard notation, here we will flexibly use both of them in different parts of our presentation.

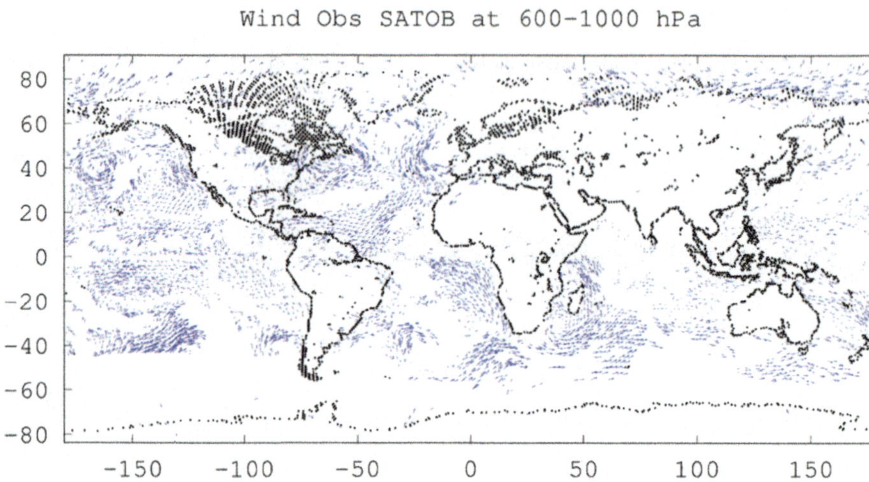

Figure 1.2. Wind measurements given by AMVs, a remote sensing technique for wind reconstruction from satellite images of clouds.

Given a system state φ, we now carry out some type of *measurement*. For measurements we use the letters y or f. Measurements can evaluate the function φ at some given points x in space, or some quite general function of φ. Measurements can consist of an integrated value of φ, leading to versions of *tomographic inverse problems*. We call the mapping H of the state φ onto the observation f the *observation operator*. Observation operators can be complex, for example the *atmospheric motion vectors* (AMVs) displayed in figure 1.2 are generated from sequential images of clouds taken by satellites. In this case, the observation of the atmospheric motion is generated using a *reconstruction*. Subsequently, the winds could be directly fed into the core algorithms of the data assimilation system with the purpose of determining or correcting the state of the full atmosphere with all its variables.

A typical example is the measurement of *infrared* or *microwave radiation* by the satellite instruments `IASI`[2], or `CrIS`[3] (infrared), or `ATMS`[4] (microwave). Waves of many frequencies are measured on top of an atmospheric column, for example `IASI` measures 8461 channels (or frequencies) and CrIS measures 1400 frequencies. The waves arise from different parts of an atmospheric column and their propagation is modeled within a *radiative transfer model* such as `RTTOV`[5], which is used to build the observation operator H mapping the state φ onto the measured data.

As visualized in figure 1.1, modeling of the observation operator is an important part of the inversion or data assimilation process. Usually, measurements need some

[2] For IASI see http://www.eumetsat.int/website/home/Satellites/.

[3] For CrIS see http://npp.gsfc.nasa.gov/cris.html.

[4] For ATMS see http://npp.gsfc.nasa.gov/atms.html.

[5] See the pages of the Numerical Weather Prediction Satellite Application Facility (NWP-SAF) https://nwpsaf.eu/.

- *pre-processing*,
- *quality control* and
- *monitoring*.

Measurements are made on the basis of some *measurement technology* such as microwave sounding carried out by the ATMS instrument. Engineers build devices (see figure 1.1, right column), based on the science carried out in a corresponding *community*, for example the Tiros Operational Vertical Sounder (TOVS) community[6]. *Formats* of data and transmission are of crucial importance, in particular when measurements are exchanged on an international level. In meteorology, for example, the World Meteorological Organization[7] defines the BUFR format[8]. Today, distribution networks such as the *Global Telecommunication System*[9] in weather, climate and environmental sciences are highly developed, and several dozen communities are involved in collecting the data which is employed day-to-day by centers for *numerical weather prediction* (NWP), and *climate monitoring and projection*.

For this book, we will take the *measurement data* with their particular properties as given. For the observations, we will restrict our attention to *generic situations*, where we can carry out the *simulations* in a simple way to focus on the formulation and analysis of inversion methods. *Quality control* as in figure 1.1 (second row) usually tests outliers, trends and biases of data. For our study, we will assume that our data has already been processed by some *quality control component* of a full system and we can focus on the *inversion*, i.e. on the *extraction of information* from quality controlled measurement data.

The core inversion or assimilation is carried out on the basis of the observations, the observation operators, and the underlying state space and its dynamical system. Usually, the assimilation and model runs are cycled, which is called the *analysis cycle*. This means that some initial state x_0 (or φ_0) at time t_0 is used. Then,

$$\begin{aligned}&\text{(S1) at every time } t_{k-1} \text{ for } k \in \mathbb{N} \text{ the model is run forward in time,}\\&\quad\text{providing a first guess } x_k^{(b)} \left(\text{or } \varphi_k^{(b)}\right) \text{ for the state at time } t_k.\end{aligned} \tag{1.1.1}$$

$$\begin{aligned}&\text{(S2) Then, the data } y_k \left(\text{or } f_k\right) \text{ at time } t_k \text{ and the first guess are used to}\\&\quad\text{calculate a reconstruction } x^{(a)} \left(\text{or } \varphi_k^{(b)}\right), \text{ which we call \textit{the analysis}.}\end{aligned} \tag{1.1.2}$$

This analysis is used as initial data for another model run from t_k to t_{k+1} and the steps S1 and S2 are cycled. In figure 1.1 this cycling layer is shown in the third row.

[6] There are regular conferences held, for example, by the International TOVS working group, see https://cimss.ssec.wisc.edu/itwg/index.html.

[7] See https://www.wmo.int/pages/index_en.html.

[8] BUFR stands for Binary Universal Form for the Representation of meteorological data, see https://www.wmo.int/pages/prog/gcos/documents/gruanmanuals/ECMWF/bufr_user_guide.pdf.

[9] See https://www.wmo.int/pages/prog/www/TEM/index_en.html.

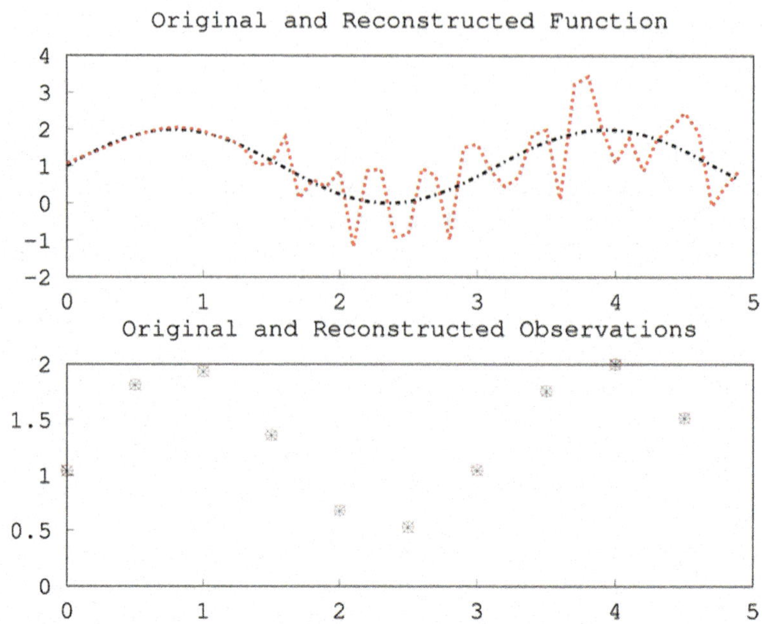

Figure 1.3. We demonstrate the ill-posedness of an integral equation with a *Gaussian kernel*. Here, the measurement is a sampled version of a transformed function, where the transform is a Gaussian smoothing. Inversion without regularization leads to severe instabilities, such that in general naive reconstruction does not lead to satisfactory results.

Classical *inverse problem* techniques provide a rich set of tools and insight into the core inversion step, which searches for a state φ which contains both the prior knowledge given by $x^{(b)}$ or $\varphi^{(b)}$ and matches the data f in the sense of

$$H(\varphi) = f \tag{1.1.3}$$

in some observation space Y. Usually, since $f = f^{\delta}$ contains measurement error, we are not searching for an exact match, but for an appropriate approximation to this equation. In our diagram in figure 1.1 the classical inverse problem is located in the top dotted box containing rows one and two, carrying out the *inversion* based on the measurement data and the state space. In this book, we will study the inverse problem from two different perspectives:

1. A *classical inverse problem* is given by the inversion task for an equation of the type (1.1.3), where a state or structure φ needs to be reconstructed based on given data f.
2. A *dynamical inverse problem* or *data assimilation problem* is given by the repeated reconstruction task of the type (1.1.3) when cycling as in steps (S1) and (S2) is carried out.

The dynamical problem can be based on methods for the solution of the classical inverse problem in each assimilation step. We can, for example, calculate a solution $\varphi_{a,k}$ of equation (1.1.3) with data f_k at time t_k in each assimilation step and then assimilate this reconstructed state into the dynamical system. Often, such reconstructions can be carried out in limited parts of the larger system, for example in an atmospheric column only. In this case it is called *retrieval* by the corresponding community. Alternatively, we can employ the full observation operator H (e.g. on many atmospheric columns in parallel) within some data assimilation framework and reconstruct the state φ_k at time t_k by one of the data assimilation algorithms such as *three-dimensional variational assimilation* (3D-VAR, section 5.2) or the ensemble Kalman filter (EnKF, section 5.5).

Clearly, different observation operators H and the choice of the set-up which is reflected in the state space X and the observation space Y lead to a variety of different inversion tasks. Both inverse problems and data assimilation share what is called *ill-posedness*, i.e. the reconstruction can depend unstably on the data. We will introduce the corresponding terminology in chapter 3. Dynamical inversion or data assimilation leads to *further research questions*; we will, for example, study the *instability* of cycled inverse problems in section 10.5. Here, we briefly introduce the reader to the content of the upcoming chapters.

Chapter 1: Introduction. After the introduction of section 1.1, the basic mathematical framework for different types of inverse problems and data assimilation is reviewed in section 1.2. We present the *general inverse problem* in section 1.2.1, basically understanding it as a solution to a nonlinear equation in a Banach space environment. The *inverse source problem* is described in section 1.2.2. *Inverse scattering problems* are characterized in section 1.2.3. When inversion is employed in the framework of *dynamical systems*, we come to *dynamical systems inversion* in section 1.2.4. This is the core task of *data assimilation*, but we need to keep in mind that for data assimilation we usually integrate a whole range of inverse problem techniques at different levels of the assimilation process. Finally, *spectral inverse problems* are discussed in section 1.2.5.

Chapter 2: Functional analytic tools. Our second chapter contains a mini-course on functional analytic tools. Working with a broad range of applied scientists, we have learned that it is crucial to have the main definitions and properties of these concepts and tools at hand when you study particular algorithms for inversion.

We describe the Banach space and Hilbert space settings, which we use to carry out the analysis of inverse problems and data assimilation, in sections 2.1 and 2.2. This includes an important tool for best approximation which we will employ again and again in later sections, as well as on Fourier series in a Hilbert space environment. The Fourier series expansion will be a key tool both for general regularization theory as well as for many specific applications, e.g. for neural field inversion.

In section 2.3 we study operators and operator equations. To fully understand the *ill-posedness* of most inverse or data assimilation problems, this section is of key importance—in particular for data assimilation with remote sensing measurements. Eigenvalues, singular values and more general spectral theory for operators in Hilbert spaces are introduced in section 2.4. We note that analysis based on spectral decompositions and singular vectors has today penetrated basically every aspect of both forward modeling as well as inverse problems and data assimilation.

Chapter 3: Regularization. As a key goal of an *inverse problem* we usually try to infer knowledge from given measurement data y in a space Y, which are linked to an unknown state or parameter function φ in a space X. We call the operator which maps a state x onto the particular observation (or sets of observations) the *observation operator* $H : X \to Y$. Often, H is *nonlinear*, there is more than one solution, in general solutions do not exist and even if a solution x exists, it depends in an unstable way on y. In any of these cases we call the problem or the operator equation *ill-posed*, compare for example the reconstruction shown in Figure 1.3. Ill-posed problems need special approaches for their stable solution, which are called *regularization*. Classical regularization approaches are introduced and analyzed in section 3.1. In section 3.3 we study iterative approaches to inverse problems and investigate the regularizing properties of *stopping rules*.

Chapter 4: Stochastic view. Today, many applications mix tools and concepts from classical applied mathematics and from stochastics. Many parts of deterministic algorithms can be seen as a version or part of a stochastic algorithm, and vice versa. An important example is the famous *Tikhonov regularization*, which in stochastics is the *maximum likelihood estimator* for a *Bayes' posterior distribution* in the case of *Gaussian densities*. It can also be seen as one step of the *Kalman filter*, see section 5.4.

We provide an introduction into the stochastic viewpoint on inverse problems in chapter 4. We first study *stochastic estimators* based on ensembles or particles in section 4.1. This theory is the background to *ensemble methods* in data assimilation. Then, we introduce *Bayesian methods* in section 4.2. The *Bayesian view* will always be complementary to any deterministic regularization method which we study or use. The construction of ensembles which sample some given distribution based on *Markov chains* is described in section 4.3. Different ways to realize such chains are introduced in section 4.4, where we introduce the *Metropolis–Hastings sampler* and the *Gibbs sampler*. An example is shown in figure 1.4, where we sample some bimodal distribution by a Metropolis-Hastings sampler.

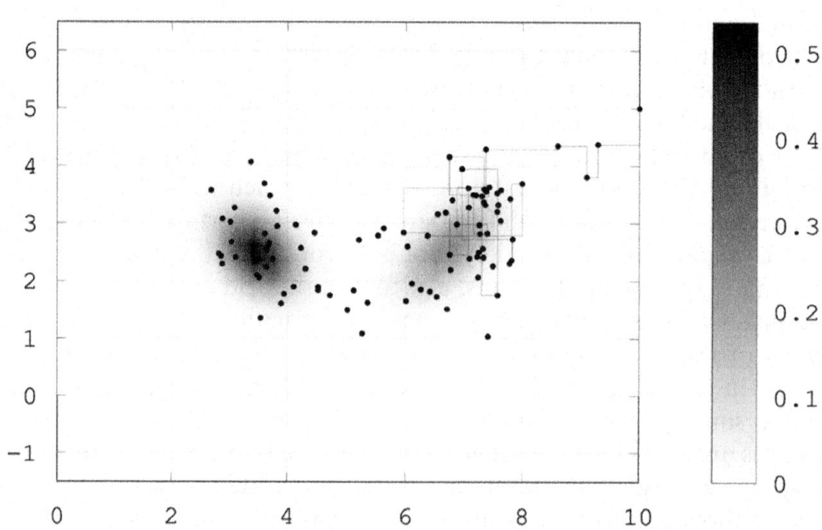

Figure 1.4. The sampling of some Bayes' posterior density using the Metropolis–Hastings algorithm within a Markov chain Monte Carlo method.

Chapter 5: Core data assimilation. Data assimilation denotes the use of measurement data to estimate initial conditions and to determine unknown parameter functions of *dynamical systems*. The theory has grown from important applications such as NWP and *oceanography*, but it is a general research area today, which basically touches upon all applications where *modeling* and *forecasting* of phenomena changing in time is needed.

We will describe the general set-up in section 5.1. A variational approach to data assimilation is described in sections 5.2 and 5.3, where 3D-VAR basically consists in a state estimation by Tikhonov regularization at different subsequent time steps of the dynamical system under consideration, and *four-dimensional variational assimilation* (4D-VAR) fits whole trajectories over a time window. Usually, such a state estimate is repeated with some frequency over time, which is known as *cycling*. We study the stability of cycled systems in section 10.5.

Both 3D-VAR and 4D-VAR do not fully use the knowledge we gain about the dependencies between different variables throughout the subsequent assimilation steps. The *Kalman filter* which we introduce in section 5.4 fills this gap, updating the covariances in an appropriate way in each time step. An ensemble-based version of the Kalman filter is the EnKF, which we describe in section 5.5. Finally, methods which deal with *nonlinear* data assimilation are described in section 5.6, including various variants of *particle filters* and *ensemble particle filters*.

```
a = 0;              % set interval left boundary
b = 5;              % set interval right boundary
N = 50;             % number of quadrature points
N2 = 10;            % number of measurement points
h = (b-a)/N;        % distance of quadrature points
h2 = (b-a)/N2;      % distance of measurement points
%
x  = a:h:(b-h);             % vector of quadrature points
x2 = a:h2:(b-h2);           % vector of quadrature points
xmat = repmat(x,N2,1);      % matrix of repeated x
ymat = repmat(x2',1,N);     % matrix of repeated x2
dmat = (xmat-ymat).^2;      % matrix of differences^2
%
sigma = 2;                  % parameter for Gaussian
k = exp(-sigma*dmat)*h;     % define Gaussian kernel
phitrue = (1 + sin(2*x))'; % true function
ftrue = k*phitrue;            % true observations
% we add 0.1% noise to the true observations
f = k*phitrue + 0.001*(rand(N2,1)-0.5);
%
alpha = 1e-12;    % regularization parameter
% regularized solution using Tikhonov regularization
phi_alpha = inv(alpha*eye(N,N)+k'*k)*k'*f;
```

Figure 1.5. Programming code for approximately solving an integral equation in OCTAVE, the case $\alpha = 0$ is visualized in figure 1.3.

Chapter 6: Programming. A broad range of programming languages and high-level programming systems are available today for the implementation of scientific algorithms.

For efficient code development, it is very popular to first create easy versions of algorithms in a high-level system, before implementation in more classical programming languages or GPU-based libraries can be approached. For the study of algorithms we consider it to be crucial to gain practical experience on an easy level. To this end we provide many codes for this introductory book which allow the reader to test methods and to play around with parameters and examples. In chapter 6 we describe basic elements of our approach based on `OCTAVE` or `MATLAB`®. In particular, `OCTAVE`[10] is freely available and easily installed on either `Linux` or `Windows`® computers. As an example, we show the code for an ill-posed integral equation in figure 1.5.

Chapter 7: Neural field inversion. The modeling of *cognitive processes* is one of the great challenges of our time. With the strong growth of *medical imaging techniques*, our insight into brain processes has grown considerably. This leads to a broad collection of inverse and data assimilation tasks in neuroscience and medicine. Here, we concentrate on continuous so-called *neural field* or *neural mass* models for the description of neural activity and its time dynamics.

We introduce the Cowan–Wilson and Amari *neural field equation* in section 7.1, studying the existence and properties of solutions based on fixed-point theorem. The theory can be seen as a continuous version of and an analytical approach to *neural networks*. A key ingredient of this model is the *connectivity kernel* $w(y, x)$, which describes the influence of activities at some point x to activities at some point y in space. Then, the *inverse neural kernel problem* is studied and solved in sections 7.2–7.6. This is the basis for *dynamic cognitive modeling* [1], where the realization of cognitive activity in a neural field dynamical system is investigated.

Chapter 8: Waves. *Inverse* problems for acoustic or electromagnetic waves usually need good knowledge and advanced methods to solve the direct or *forward* problem as well, see for example the field shown in figure 1.6—either just for the purpose of simulations to calculate *simulated* measurements, or as an integral part of the inversion algorithm.

We introduce *integral equation methods* for the solution of scattering problems in chapter 8, solving the basic *boundary value problems* of acoustic scattering theory. They will also be the basis for our numerical testing—such that we provide `OCTAVE` codes for a range of generic two-dimensional examples. First, *potentials* and *potential operators* are introduced in section 8.1. We describe the simulation of the scattering process in section 8.2. The behavior of the scattered field $u^s(x)$ when $|x|$ is large is expressed by the *far field*, which we describe in section 8.3. Reciprocity relations describe important properties of either far field patterns or scattered fields, respectively. These relations are derived in section 8.4. Finally, we will introduce the *Lax–Phillips method* to calculate scattered waves in section 8.5.

[10] We used `OCTAVE` version 3.8.1 or higher.

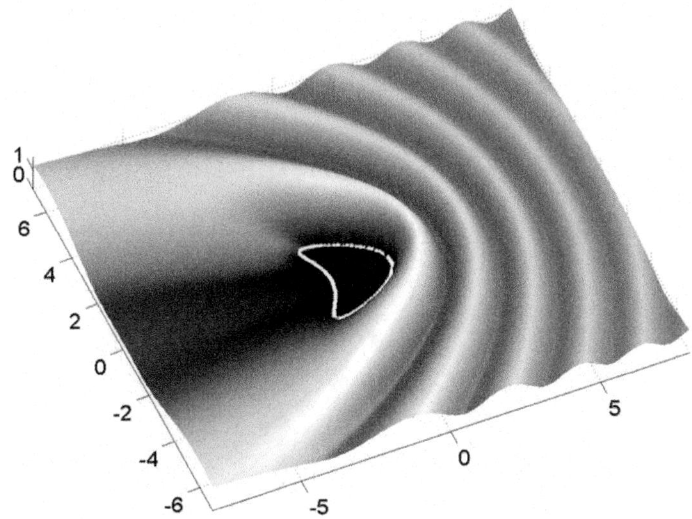

Figure 1.6. Scattering of some acoustic wave from an obstacle. We display the modulus of the total field calculated by boundary integral equations.

Chapter 9: Nonlinearity. Many important inverse problems are *nonlinear* in nature. Differentiation is an important technique for inversion and data assimilation. Since the unknown quantity is usually a real-valued function or even a set of functions, for a thorough study it is natural to work in Banach spaces or Hilbert spaces. We summarize the calculus of *Fréchet derivatives* in section 2.6. When we search the shape of unknown objects, boundary integral equations are an important tool to model direct and inverse wave scattering problems. We derive the differentiability of *boundary integral operators* with respect to variations of the boundary in section 9.1. The differentiability of the corresponding *boundary value problems* is studied in section 9.2. Alternative approaches to prove the Fréchet differentiability of these problems are discussed in section 9.3. Newton's method and the gradient method for an inverse obstacle scattering problem are carried out in section 9.4. In our final section 9.5 we study the Fréchet differentiability of dynamical models with respect to their initial states. It establishes the so-called *tangent linear model* and its adjoint. This differentiability and the calculation of the derivative is a key component of the 4D-VAR method introduced in section 5.3.

Chapter 10: Uniqueness and stability. The study of uniqueness, stability and convergence both from a theoretical as well as from a practical viewpoint is crucial for inverse problems. Here, we will present a selection of ideas. We first discuss uniqueness concepts in section 10.1. In particular, the connection between uniqueness and stability is discussed in more detail and unconventional concepts such as ϵ-uniqueness and ϵ-stability are treated. We then study the uniqueness of the inverse obstacle scattering problem in section 10.2. This uniqueness result has inspired many probing and sampling methods which will be presented in subsequent chapters. We will study the stability of data assimilation methods with ill-posed observation

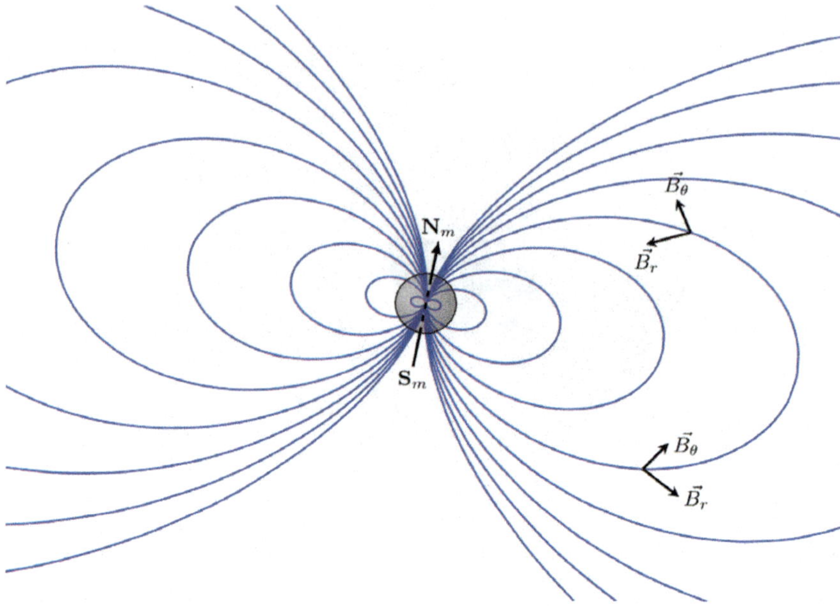

Figure 1.7. An electric current generates magnetic fields. The goal of magnetic tomography is to reconstruct the sources, i.e. the *current distribution* when the measured magnetic field is a superposition of the magnetic fields generated by a continuous superposition of such dipoles.

operators in section 10.5. Finally, we provide a review of convergence concepts for inverse problems in section 10.6.

Chapter 11: Magnetic tomography. Magnetic tomography is concerned with the reconstruction of currents from their magnetic field. Figure 1.7 visualizes such a magnetic field generated by a current element. This is of importance both in medical applications, where in *magnetoencephalography* magnetic fields are employed to reconstruct the activity within the brain based on highly sensitive sensors, the SQUIDS. It is also of importance for engineering applications, for example to monitor and study the current distribution within fuel cells. Here, we will provide basic insight which applies to both areas. We describe the simulation of a current distribution in section 11.1. This includes, in section 11.1.1, the study of the *conductivity problem* which is the basis of important inverse problems in *electric impedance tomography* or *electric resistance tomography*, where the first term is usually used in medical applications, the second term in geophysics. The finite integration approach to simulate such currents is introduced in section 11.1.2.

The problem of magnetic tomography is then studied in section 11.2. We study the *uniqueness* and *non-uniqueness* of magnetic tomography in section 11.2.1, and describe its stability analysis and methods to reduce its ill-posedness in section 11.2.2. We use the inverse source problem of magnetic tomography to present two further areas of research. First, we carry out data assimilation in combination with a *parameter estimation* for the dynamic magnetic tomography problem in section 11.3. Second, we study *classification problems* in the framework of inverse and ill-posed problems. We will show that the ill-posedness of some problem carries over to the classification problem, with the consequence that in general we need to be very careful with the naive application of classification methods to data generated by a compact or ill-posed observation operator.

Chapter 12: Field reconstruction. The reconstruction of unknown fields in acoustics, electromagnetics or fluid dynamics is one of the basic tasks of the field of inverse problems. We present several approaches, based on the Fourier–Hankel expansion in section 12.1, on Fourier-plane-wave expansions in section 12.2, on potentials following the method of Kirsch and Kress in section 12.3 and Green's theorem and the approximation of point sources in section 12.4. The duality and equivalence of the Kirsch–Kress method and the *point source method* will be shown in section 12.5. The method applies to problems in acoustics, electromagnetics or fluid dynamics.

Chapter 13: Sampling methods. Sampling methods are close relatives of *probing methods*, although usually their set-up and convergence analysis is quite different from the methods described in chapter 14. Sampling methods formulate an indicator function based on the measured far field patterns or scattered fields, which provides insight and knowledge about some unknown scatterer. We will start with the *orthogonality sampling* method which constructs a stable weighted superposition of the measurements to find the unknown scatterers in section 13.1. Then, the *linear sampling method* of Colton–Kirsch is the topic of section 13.2. The *factorization method* of Kirsch, which we present in section 13.3, is a modified version of *linear sampling*, but opens a whole further area of study with its factorization tools. We present the application of the linear sampling method to an *inverse heat conduction problem* in section 16.1.

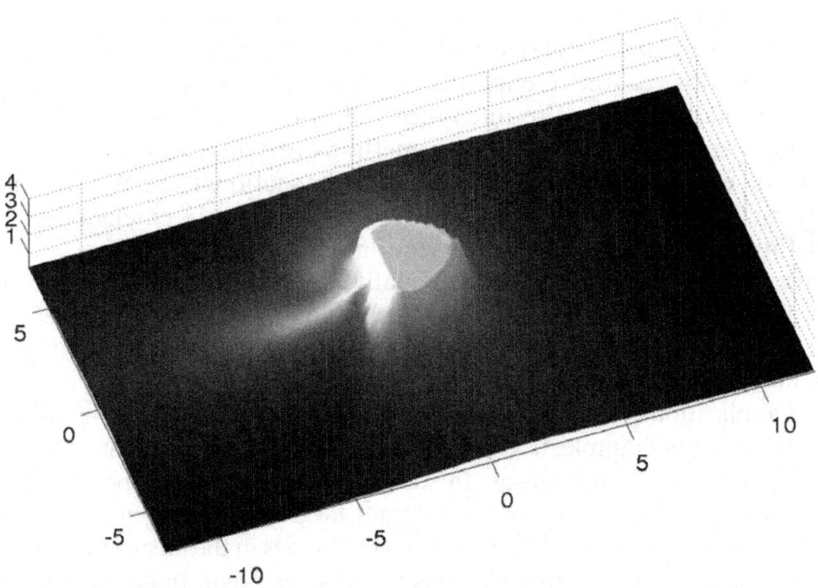

Figure 1.8. We display the *indicator function* of the singular sources method to detect the shape of an unknown object.

Chapter 14: Probing methods. It is a very basic idea to probe some unknown space using waves to acquire knowledge about an inaccessible region. Using a set of incident waves and measuring the corresponding scattered fields, probing methods construct virtual probes, reconstruct the corresponding scattered fields and use these to find the *location* and *shapes* of unknown objects. We describe the *singular sources method* in section 14.1. In its simplest version it uses a point source and the reconstruction of the scattered field in the source point as a virtual probe to reconstruct objects and their properties. Its indicator function is displayed in figure 1.8. Related *probing methods* for the near field following Ikehata and Nakamura are introduced in section 14.2, also based on probing with particular singular sources.

Many of the analytic continuation methods have a multi-wave version which carries out some type of probing. For example, the *multi-wave no-response test* described by Schulz and Potthast is presented in section 14.3 and the *multi-wave enclosure method* by Ikehata is described in section 14.5.

There are quite different ways in which *probing* can be carried out. We use the term *point sampling* if some particular field which has a singularity in a point z in space is employed and some *indicator function* $\mu(z)$ is sampled for a choice of points z to infer information about the unknown scatterer. Other methods are based on *domain sampling* which we will introduce in section 14.1.3, where the indicator function is calculated for whole *test domains* G, and the unknown shape is found as the intersection of test domains with particular bounds on the indicator function. Further approaches use a contracting sequence of test domains, such as the *contraction scheme* in section 14.1.4.

Chapter 15: Analytic continuation tests. One key line of research in recent years has been the study of the range of operators as a tool to solve inverse problems. For example, many *sampling methods*, which will be our topic in section 13, are based on this principle. A related task is to investigate the extensibility of some field as a solution to the Helmholtz or Maxwell's equations. This leads to *analytic continuation tests*. We introduce the *range test* in section 15.1 and the *no-response test* in section 15.2. These methods are dual to each other and can be seen to be equivalent, which we will show in section 15.3. We will introduce a further method which test extensibility or approximate extensibility of some solution for the *enclosure method* in section 15.4.

Chapter 17: Meta-inverse problems. Usually *inverse problems* are based on *direct problems* in the sense that the inversion tries to find or reconstruct some parameter function p or state x which is part of the direct problem. Measurements y of some quantity are the input and the task is to find p or x.

In most applications we have the freedom to choose parts of the set-up of the whole problem. We are, for example, free to choose additional measurements. Or we are free to alter the location of some sensor. In some problems we might be free to alter the location of sources. This leads to the important task of *improving the reconstructions*, the data assimilation process and the forecast by changes in the measurement process or further parameters. We call this task a *meta-inverse problem*. In data assimilation the area of *targeted measurements* and *measurement-to-analysis sensitivity* or *measurement-to-forecast-sensitivity* are parts of this area of research.

When ill-posed observation operators are involved, the search for an improved or optimal measurement set-up needs to take care of the ill-posedness and the regularization algorithms which are part of the solution of the inverse problem. We will introduce a general *framework* for solving *meta-inverse problems* in section 17.1. It is a concept for taking care of all ingredients of the task with proper care. We also show that the adaptation of the set-up of an inverse problem can have a significant effect on the quality of reconstructions. A method of *framework adaption* or *zoom* is formulated in section 17.2. We apply it to an inverse source problem in section 17.3.

Appendix. Our appendix will collect notations and basic integration formulas and formulas of vector calculus which are helpful for many of our derivations and examples.

Further literature. Our introduction aims to be as self-contained as possible, but we rely on the work of colleagues in many places. Science is and has always been an interplay between many ideas, between many minds, between many humans. There is a fast-growing literature, both on data assimilation as well as on inverse problems.

Please take advantage of the books on *inverse problems* of our highly appreciated colleagues, in particular Kirsch [2], Engl, Hanke and Neubauer [3], Colton and Kress [4], Cakoni and Colton [5], Kirsch and Grinberg [6], Potthast [7], Bertero and Boccaci [8], Aster *et al* [9], Muller and Siltanen [10], Groetsch [11], Isakov [12], Lavrent'ev, Romanov and Shishat-skii [13], Ammari and Kang [14], Natterer [15], Sharafutdinov [16], Klibanov and Timonov [17], Beilina and Klibanov [18], Kaltenbacher, Neubauer and Scherzer [19], Schuster, Kaltenbacher, Hofmann and Kazimierski [20], Hansen [21], Kabanikhin [22], Wang, Yagola and Yang [23], Tikhonov, Goncharsky and Stepanov [24], Bal, Finch, Kuchment, Schotland, Stefanov and Uhlmann [25], and Kuchment [26], see also Scherzer [27].

The *stochastic aspect* is treated for example in Biegler *et al* [28], and Kaipio and Somersalo [29].

Introductions to *data assimilation* include recent books and reviews by Leeuwen and Reich [30], Law, Stuart and Zygalakis [31], Andersen [32], Aster *et al* [9], Freitag and Potthast [33], and Kaipio and Somersalo [29]. We also refer to more classical texts by Kalnay [34, 35], Lewis *et al* [36] and Daley [37]. Techniques with applications in climate and weather are treated in Cullen [38], Blayo *et al* [39], Enting [40], Wunsch [41, 42] and Peng [43]. For satellite-based remote sensing we refer, for example, to [44–47].

A comprehensive introduction to *neural field theory* is given by Coombes, beim Graben, Potthast and Wright [48], with many contributions by key researchers in the field.

1.2 Types of inverse problems

We now introduce a selection of the key types of inverse problems to further introduce the typical set-up and notation.

1.2.1 The general inverse problem

The most general view on an inverse problem is that we have some particular system we are modeling and some data we are measuring. Usually, the system state is described by a variable x in a state space X and the measurements by a variable y in some observation space Y. The modeling of the observation provides a mapping H of x onto y, we have already introduced this as the *observation operator*. Then, the *inverse* problem is to reconstruct x when y is given. In this sense the inverse problem is just the task of the inversion of some (in general nonlinear) equation $H(x) = y$.

We need X to be a linear space equipped with some measure of distance of its elements. Here, complete normed spaces (Banach spaces, see definition 2.1.14) or complete spaces with a scalar product (Hilbert spaces, see section 2.2.1) are an adequate environment for most of our arguments, since they include $\mathbb{R}^n$ equipped with maximum or ℓ^2-norms, weighted ℓ^2-spaces as well as most function spaces such as bounded continuous functions $BC(\Omega)$ on some domain Ω or the space of square-integrable functions $L^2(\Omega)$ on Ω. We are now prepared to set the general framework.

Definition 1.2.1 (General inverse problem). *Consider a subset U of a Banach space X, another Banach space Y and a mapping $H : U \to Y$. The inverse problem consists of the solution of the equation*

$$H(x) = y, \tag{1.2.1}$$

which by setting $F(x) := H(x) - y$ can be written in the form

$$F(x) = 0. \tag{1.2.2}$$

The setting of definition 1.2.1 is very general and applies to almost all inverse problems. Even in this general setting we can formulate properties and solution procedures such as iterative methods (see sections 3.3 and 9.4). With more specific properties and settings we will be able to formulate better and more specific solutions. In the following sections, we will study inverse problems on several levels, usually summarized into particular *configurations*.

Theorem 1.2.2 (Least squares formulation). *The general inverse problem is equivalent to the minimization of the functional*

$$J(x) = \|H(x) - y\|^2, \quad x \in U. \tag{1.2.3}$$

When the Banach space Y is some L^2-space, then this is a classical least squares *problem.*

In principle, we can approach the general inverse problem by solving a minimization problem and we can apply the whole range of optimization methods to solve an inverse problem. When x is a solution to some partial differential equation (PDE), we speak of *PDE-constraint optimization*, see for example [49]. It is a very popular area of current research.

However, in many cases solving the full optimization problem is not efficient or not a realistic approach for the particular problem under consideration. The minimization problem (1.2.3) does not provide insight into the structure, instability and non-uniqueness of the task. Often, the combination of *generic* (general) approaches with *specific* tools has proven to be of large value to develop efficient and flexible solution methods.

1.2.2 Source problems

Many natural processes have sources, for example acoustic sources which generate sound, heat sources, the sources of light or other electromagnetic radiation. The *source problem* is to reconstruct the location and strength of such sources. For example, figure 1.9 displays the radiation measured by a polar orbiting satellite arising in different heights of each atmospheric column.

Often, source problems can be modeled by knowing how some quantity is transported from a point y in space to a point x in space. If we have a source strength of size $\varphi(y)$ at the point y and assume that stronger sources create a stronger field with linear dependence on the source strength, then this yields a field $\Phi(x, y)\varphi(y)$ at points x with some propagation function $\Phi(x, y)$ usually depending on x and y. The actual measurements $u(x)$ on some set Λ are given by the sum over all sources multiplied by $\Phi(x, y)$. If we use a continuous distribution of sources in some domain D, this is modeled by the *volume integral operator*

$$(V\varphi)(x) = \int_D \Phi(x, y)\varphi(y)\, dy, \quad x \in \Lambda. \tag{1.2.4}$$

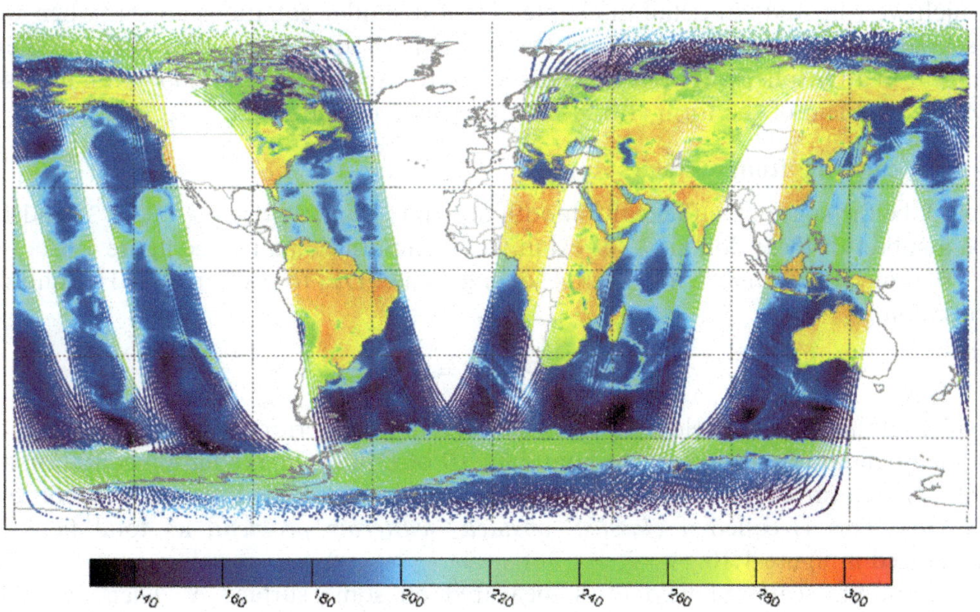

Figure 1.9. The measurement of infrared radiation emitted by the atmospheric columns as measured by a polar orbiting satellite. The reconstruction of the temperatures constitutes a classical inverse source problem in each column.

Here, the operator (1.2.4) is a *linear integral operator*. For example, for time-harmonic acoustic scattering the kernel Φ is given by the *free-space fundamental solution* to the *Helmholtz equation*, i.e.

$$\Phi(x, y) = \frac{1}{4\pi} \frac{e^{i\kappa|x-y|}}{|x - y|}, \quad x \neq y \in \mathbb{R}^3.$$

Source reconstruction aims to determine φ on D given measurements of $V\varphi$ on Λ.

Definition 1.2.3 (Linear source reconstruction problem). *Let X be some Banach space of source densities φ on D and Y be a Banach space of functions on Λ and assume that V is a linear integral operator of the form (1.2.4) which maps X into Y. Then, the* linear source reconstruction problem *is to solve*

$$V\varphi = y \tag{1.2.5}$$

given some measurements $y \in Y$.

Clearly, the linear source reconstruction problem of definition 1.2.3 can be seen as a special linear version of the general inverse problem 1.2.1. But with the definition (11.2.4) we have a very special structure, such that (1.2.5) is a linear integral equation *of the first kind.* If the kernel is continuous or weakly singular, it is well known that the equation is compact and thus *ill-posed* in the sense of definition 3.1.1. Then, the theory which we will work out in section 3.1 applies.

A very generic example of a source problem is given by *magnetic tomography*, which we will study in detail in chapter 11. But similar equations appear when we seek *sources of particles* which are transported by some given flow field, as they appear in atmospheric aerosol or volcanic ash transport. *Deconvolution* is an inverse source problem of the form (1.2.5) which appears in optics.

1.2.3 Scattering from obstacles

Let X be the space of sufficiently smooth domains D in $\mathbb{R}^m$ for $m \in \{1, 2, 3\}$ and U be a subset of these domains, for example C^2 smooth domains, see figure 1.6.

We can consider incident waves u^i solving some PDE, for example the Helmholtz equation

$$\Delta u + \kappa^2 u = 0,$$

such that the scattered field u^s satisfies the PDE in the exterior of the domain $D \in U$, some boundary condition on the boundary ∂D of D and some radiation condition at infinity (see for example definition 8.2.1 with the Sommerfeld radiation condition (8.2.6)). This is called a *classical obstacle scattering problem*, its total field is visualized in figure 1.10.

Often, the scattered field u^s is measured on some surface Λ. Then, we can reformulate the simulation of the scattering problem into an equation

$$F(D, u^i) = u^s|_\Lambda. \tag{1.2.6}$$

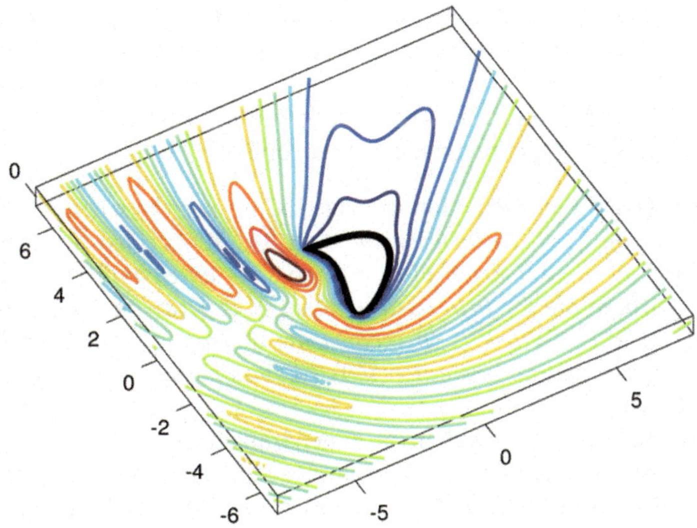

Figure 1.10. A contour plot of the total field $u = u^{\mathrm{i}} + u^{\mathrm{s}}$ for scattering of the obstacle shown as a black curve from an incident plane wave coming from the bottom left. The scripts for the direct problem are presented in section 8.2, the scripts for inversion in chapters 9 and 12–15.

When measurements are carried out for different incident waves, we consider $u^{\mathrm{i}} = u_j^{\mathrm{i}}$ for $j \in J$ with some index set J. Here, J might be a countable set $J = \{1, 2, 3, \ldots\}$ or the set of all directions $d \in \mathbb{S}$ on the unit sphere.

For the above obstacle scattering problem we immediately obtain two specific options for the reconstruction of D. We can decompose the task into two subtasks:

(a) First reconstruct the scattered field u^{s} in the exterior of the domain D.

(b) Then reconstruct the domain D as the set of points where the boundary condition is satisfied.

The particular setting allows the development of specific methods, which are generally known as *decomposition methods*. We summarize the setting which allows decomposition methods into the following general configuration.

Definition 1.2.4 (Scattering problem configuration). *Let X be the space of sufficiently smooth domains D in $\mathbb{R}^m$ and U be a subset of these domains. We can consider incident waves u_j^{i} for $j \in J$ solving some* PDE. *With the PDE in the exterior of the domain $D \in U$, some* boundary condition *on the boundary ∂D of D and some* radiation condition *at infinity we assume that the scattered field u_j^{s}, $j \in J$, which satisfies the* PDE, boundary condition and radiation condition, *is uniquely determined.*

Often, the asymptotic behavior of the field $u^{\mathrm{s}}(x)$ for $|x| \to \infty$ is used to define a far field pattern u^{∞} *in the general form*

$$u^{\mathrm{s}}(x) = \Phi(x)\{u^{\infty}(\hat{x}) + O(|x|^{-\tau})\}, \quad \hat{x} = \frac{x}{|x|}, \quad x \in \mathbb{R}^m \tag{1.2.7}$$

with some spherical wave $\Phi(x)$ *and a constant* $\tau > 0$. *We speak of the* scattering configuration with a far field pattern.

For the scattering configuration we can formulate two canonical inverse problems: the field and the shape reconstruction problem.

Definition 1.2.5 (Field reconstruction problem). *Consider the scattering configuration described in definition 1.2.4. Given* u^s *on some set* $\Lambda \subset \mathbb{R}^m$ *reconstruct* u^s *in the exterior of D and reconstruct the extension of* u^s *into possible extensibility sets in* $\mathbb{R}^m$.

In the setting where the far field pattern u^∞ of u^s *is given on some subset* $\Lambda \subset \mathbb{S}$ *of the unit sphere in* $\mathbb{R}^m$, *the task is to reconstruct* u^s *from its far field pattern in the exterior of D or to reconstruct the extension of* u^s *into possible extensibility sets in* $\mathbb{R}^m$.

The field reconstruction problem is a generalization of the classical problem to extend a complex function into its Riemannean surface. In more general terms the problem is to calculate a function from its values on some surface or set Λ. Clearly, this can only be carried out if the function has particular properties, for example being a solution to some PDE which has a *unique continuation property*, i.e. its continuation into a larger set from values in a smaller set is uniquely determined.

Definition 1.2.6 (Shape reconstruction problem). *Consider the scattering configuration described in definition 1.2.4. Given* u_j^s *on some set* $\Lambda \subset \mathbb{R}^m$ *for* $j \in J$ *reconstruct the shape* ∂D *of the domain D. Alternatively, we might use the far field patterns* u_j^∞, $j \in J$, *given on* $\Lambda \subset \mathbb{S}$ *for reconstructions. In this case we speak of the* shape reconstruction problem from the far field patterns.

We will discuss several different approaches to these problems, in particular *iterative methods* in chapter 9, *field reconstruction techniques* in chapter 12, *sampling schemes* in chapter 13, *probing methods* in chapter 14 and *analytic continuation tests* in chapter 15.

1.2.4 Dynamical systems inversion

Consider a *dynamical system* of the form

$$\dot{x} = F(t, x), \quad t \geqslant t_0, \tag{1.2.8}$$

with initial condition

$$x(t_0) = x_0, \tag{1.2.9}$$

with initial state $x_0 \in \mathbb{R}^n$ for some $n \in \mathbb{N}$ and $F : \mathbb{R} \times \mathbb{R}^n \to \mathbb{R}$. As an example we display the trajectory of the Lorenz 63 model in figure 1.11.

Often, the initial conditions of such a system are not known. Then, measurements of some parts or function $H(x)$ of the system state x are used to determine the initial conditions and with the initial conditions via (1.2.8) and (1.2.9) the full *state* of the dynamical system. The mapping H is known as the *observation operator*.

Definition 1.2.7 (Dynamical system state estimation). *If measurements*

$$y_1, \dots, y_n \in \mathbb{R}^m \tag{1.2.10}$$

of a function $H(x)$ of the state $x = x(t) \in \mathbb{R}^n$ are carried out at points $t_0 < t_1 < t_2 < \dots, t_n$ in time, the task of the state reconstruction *problem is to find initial conditions $x_0 = x(t_0)$ such that the solution of (1.2.8) and (1.2.9) satisfies*

$$H\big(x(t_j)\big) = y_j, \quad j = 1, \dots, n. \tag{1.2.11}$$

Often, knowledge from previous times is used for the problem solution at later time steps. When we carry out such problems *sequentially* over time in a *cycled* way as described in (1.1.1) and (1.1.2) in section 1.1, we call it a *data assimilation problem*.

Often, for a dynamical system of the form (1.2.8) and (1.2.9) the mapping F contains unknown parameters or parameter functions p. In the easiest case the initial condition x_0 is known, but F is in some class parametrized by $p \in \mathcal{P}$ in general both problems are coupled.

Definition 1.2.8 (Parameter estimation). *If measurements*

$$y_1, \dots, y_n \in \mathbb{R}^m \tag{1.2.12}$$

of a function $H(x)$ of the state $x = x(t) \in \mathbb{R}^n$ are given at points $t_0 < t_1 < t_2 < \dots, t_n$ in time, the task of the parameter estimation *problem is to find parameters*

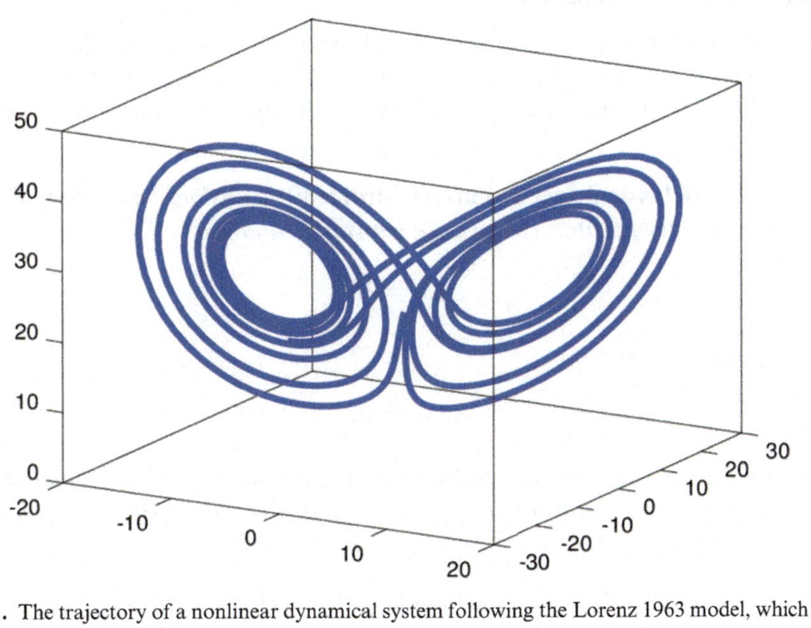

Figure 1.11. The trajectory of a nonlinear dynamical system following the Lorenz 1963 model, which is used as a basic example for data assimilation techniques in the chapters 5, 6 and 9.5.

$p \in \mathcal{P}$ and initial conditions $x_0 = x(t_0)$ such that the solution of (1.2.8) and (1.2.9) satisfies

$$H\big(x[p](t_j)\big) = y_j, \quad j = 1, \ldots, n. \tag{1.2.13}$$

Parameter estimation has a long history in engineering applications and stochastics. It is a standard approach to understanding parameter estimation as a special data assimilation problem, where the state is *augmented* by adding the parameters, i.e. we define a new state

$$\tilde{x} := \begin{pmatrix} x \\ p \end{pmatrix} \in \mathbb{R}^{n+s} \tag{1.2.14}$$

when the parameter vector $p \in \mathcal{P}$ has dimension s. A more general way is to work with a state space X, an observation space Y and a parameter space Z, such that the augmented state $\tilde{x}$ is an element of $X \times Z$. When we employ a constant dynamics $\dot{p} = 0$ and define $H(\tilde{x}) := H(x)$, the parameter estimation problem becomes a classical data assimilation problem for $\tilde{x}$ defined in definition 1.2.7.

We will introduce and analyze methods for *data assimilation* in chapter 5, with numerical examples in sections 6.2, 6.3 and 9.5, and parameter estimation in combination with data assimilation in the framework of magnetic tomography in section 11.3.

1.2.5 Spectral inverse problems

Many variants of spectral inverse problems have been investigated in the history of inverse problems. The classical task is often formulated by the phrase *can you hear the shape of a drum*? In mathematical terms, this leads to a spectral inverse problem as follows.

Definition 1.2.9 (Classical spectral inverse problem). *Consider some domain $D \subset \mathbb{R}^n$, a partial differential operator L and an eigenvalue equation*

$$Lu = \lambda u \tag{1.2.15}$$

on D with boundary condition

$$Bu = 0, \tag{1.2.16}$$

where B might denote the trace operator mapping a function onto its boundary values or it might map u onto its normal derivative $\partial u / \partial \nu$ on ∂D. Given the eigenvalues $(\lambda_n)_{n \in \mathbb{N}}$ of L with boundary condition (1.2.16), find the shape D.

This book will not consider spectral inverse problems further, but directs the reader to the comprehensive work of colleagues such as [50, 51].

Bibliography

[1] Graben P B and Potthast R 2009 Inverse problems in dynamic cognitive modelling *Chaos* **19** 015103

[2] Kirsch A 1996 *An Introduction to the Mathematical Theory of Inverse Problems* (*Applied Mathematical Sciences* vol 120) (New York: Springer)

[3] Engl H W, Hanke M and Neubauer A 1996 *Regularization of Inverse Problems* (*Mathematics and its Applications* vol 375) (Dordrecht: Kluwer Academic)

[4] Colton D and Kress R 1992 *Inverse Acoustic and Electromagnetic Scattering Theory* (*Applied Mathematical Sciences* vol 93) 2nd edn (Berlin: Springer)

[5] Cakoni F and Colton D 2006 *Qualitative Methods in Inverse Scattering Theory* (Berlin: Springer)

[6] Kirsch A and Grinberg N 2008 *The Factorization Method for Inverse Problems* (Oxford: Oxford University Press)

[7] Potthast R 2001 *Point Sources and Multipoles in Inverse Scattering Theory* (*Chapman and Hall/CRC Research Notes in Mathematics* vol 127) (Boca Raton, FL: CRC)

[8] Bertero M and Boccacci P 1998 *Introduction to Inverse Problems in Imaging* (Boca Raton, FL: CRC)

[9] Aster R C, Borchers B and Thurber C H 2013 *Parameter Estimation and Inverse Problems* (Boston, MA: Academic)

[10] Muller J and Siltanen S 2012 *Linear and Nonlinear Inverse Problems with Practical Applications* (Philadelphia, PA: Society for Industrial and Applied Mathematics)

[11] Groetsch C W 1993 *Inverse Problems in the Mathematical Sciences* (Braunschweig: Vieweg)

[12] Isakov V 1998 *Inverse Problems for Partial Differential Equations* (*Springer Series in Applied Mathematical Science* vol 127) (Berlin: Springer)

[13] Lavrent'ev M, Romanov V and Shishat-skii S 1986 *Ill-posed Problems of Mathematical Physics and Analysis* (Providence, RI: American Mathematical Society)

[14] Ammari H and Kang H 2007 *Polarization and Moment Tensors* (*Applied Mathematical Sciences* vol 162) (New York: Springer)

[15] Natterer F 2001 *The Mathematics of Computerized Tomography* (*Classics in Applied Mathematics* vol 23) (Philadelphia, PA: Society for Industrial and Applied Mathematics)

[16] Sharafutdinov V A 1994 *Integral Geometry of Tensor Fields* (*Inverse and Ill-posed Problems Series*) (Utrecht: VSP)

[17] Klibanov M V and Timonov A 2004 *Carleman Estimates for Coefficient Inverse Problems and Numerical Applications* (Utrecht: VSP)

[18] Beilina L and Klibanov M V 2012 *Approximate Global Convergence and Adaptivity for Coefficient Inverse Problems* (New York: Springer)

[19] Kaltenbacher B, Neubauer A and Scherzer O 2008 *Iterative Regularization Methods for Nonlinear Ill-Posed Problems* (Berlin: De Gruyter)

[20] Schuster T, Kaltenbacher B, Hofmann B and Kazimierski K S 2012 *Regularization Methods in Banach Spaces* (Berlin: De Gruyter)

[21] Hansen P C 2010 *Discrete Inverse Problems–Insight and Algorithms* (Philadelphia, PA: Society for Industrial and Applied Mathematics)

[22] Kabanikhin S I 2011 *Inverse and Ill-posed Problems* (Berlin: De Gruyter)

[23] Wang Y, Yagola A G and Yang C (ed) 2010 *Optimization and Regularization for Computational Inverse Problems and Applications* (Heidelberg: Springer)

[24] Tikhonov A N, Goncharsky A and Stepanov A V V 1995 *Numerical Methods for the Solution of Ill-Posed Problems* (Dordrecht: Springer)

[25] Guillaume B, Finch D, Kuchment P, Schotland J, Stefanov P and Uhlmann G 2011 *Tomography and Inverse Transport Theory* (Providence, RI: American Mathematical Society)

[26] Kuchment P 2014 *The Radon Transform and Medical Imaging* (Cambridge: Cambridge University Press)

[27] Scherzer O (ed) 2011 *Handbook of Mathematical Methods in Imaging* (Dordrecht: Springer)

[28] Biegler L *et al* 2011 *Large-Scale Inverse Problems and Quantification of Uncertainty (Wiley Series in Computational Statistics)* (New York: Wiley)

[29] Kaipio J and Somersalo E 2005 *Statistical and Computational Inverse Problems* (*Applied Mathematical Sciences* vol 160) (New York: Springer)

[30] van Leeuwen P J, Cheng Y and Reich S 2015 *Nonlinear Data Assimilation* (Cham: Springer)

[31] Law K, Stuart A and Zygalakis K 2015 *Data Assimilation: A Mathematical Introduction* (Cham: Springer)

[32] Anderson B D O and Moore J B 2012 *Optimal Filtering (Dover Books on Electrical Engineering Series)* (New York: Dover)

[33] Freitag M and Potthast R 2013 Synergy of inverse problems and data assimilation techniques *Large Scale Inverse Problems* (*Radon Series on Computational and Applied Mathematics* vol 13) ed M Cullen *et al* (Berlin: Walter de Gruyter)

[34] Kalnay E 2003 *Atmospheric Modeling, Data Assimilation and Predictability* (Cambridge: Cambridge University Press)

[35] Kalnay E, Hunt B, Edward O and Szunyogh I 2006 *Ensemble Forecasting and Data Assimilation: Two Problems with the Same Solution* (Cambridge: Cambridge University Press)

[36] Lewis J M, Lakshmivarahan S and Dhall S 2006 *Dynamic Data Assimilation: A Least Squares Approach* (Cambridge: Cambridge University Press)

[37] Daley R 1993 *Atmospheric Data Analysis (Cambridge Atmospheric and Space Science Series)* (Cambridge: Cambridge University Press)

[38] Cullen M J P 2006 *A Mathematical Theory of Large-Scale Atmosphere/Ocean Flow* (London: Imperial College Press)

[39] Blayo E, Bocquet M, Cosme E and Cugliandolo L F 2014 *Advanced Data Assimilation for Geosciences: Lecture Notes of the Les Houches School of Physics* (Oxford: Oxford University Press)

[40] Enting I G 2005 *Inverse Problems in Atmospheric Constituent Transport* (Cambridge: Cambridge University Press)

[41] Wunsch C 1997 *The Ocean Circulation Inverse Problem* (Cambridge: Cambridge University Press)

[42] Wunsch C 2006 *Discrete Inverse and State Estimation Problems With Geophysical Fluid Applications* (Cambridge: Cambridge University Press)

[43] Gongbing P, Leslie L M and Shao Y 2002 *Environmental Modelling and Prediction* (Berlin: Springer)

[44] Goldberg M D and Weng D F 2006 Advanced technology microwave sounder *Earth Science Satellite Remote Sensing* ed J J Qu *et al* (Berlin: Springer) pp 243–53

[45] John J Q, Gao W, Kafatos M, Murphy R E and Salomonson V V 2007 *Earth Science Satellite Remote Sensing* (*Science and Instruments* vol 1) (Berlin: Springer)

[46] Schäfer K (ed) Remote Sensing of Clouds and the Atmosphere. Society of Photo-optical Instrumentation Engineers, S E de Optica and USNAS Administration 2004 *Proc. SPIE* **5579**

[47] Solimini D 1995 *Microwave Radiometry and Remote Sensing of The Environment* (London: Taylor and Francis)

[48] Coombes S, beim Graben P, Potthast R and Wright J 2014 *Neural Fields* (Berlin: Springer)

[49] Hinze M, Pinnau R, Ulbrich M and Ulbrich S 2009 *Optimization with PDE Constraints Mathematical Modelling: Theory and Applications* vol 23 (Dordrecht: Springer)

[50] Isozaki H and Kurylev Y 2014 *Introduction to Spectral Theory and Inverse Problems on Asymptotically Hyperbolic Manifolds* (*MSJ Memoirs* vol 32) (Tokyo: Mathematical Society of Japan)

[51] Katchalov A, Kurylev Y and Lassas M 2001 *Inverse Boundary Spectral Problems* (*Monographs and Surveys in Pure and Applied Mathematics* vol 123) (London: Chapman Hall)

IOP Publishing

Inverse Modeling

An introduction to the theory and methods of inverse problems and data assimilation

Gen Nakamura and Roland Potthast

Chapter 2

Functional analytic tools

The goal of this chapter is to introduce important tools from mathematical analysis and from functional analysis which are frequently used for the solution of identification and reconstruction problems, and build a basis for the advanced study of data assimilation problems.

The functional analytic language has many strengths and advantages, not only for the mathematical specialist, but also for scientists from various applications. It captures the main features of the problems and brings arguments into a form which applies to many diverse phenomena.

In this chapter, we are making use of parts of lectures by our colleague and teacher Rainer Kress, see [1], but put things into our framework to prepare the upcoming chapters. The following material has been extensively used in lectures at the University of Hokkaido, Japan, at Inha University, Korea, at Reading University, UK, and at the University of Göttingen, Germany.

Today a broad range of functional analysis books are available. But we believe that some classical books are worth reading, for example Reed and Simon [2], Bachman and Narici [3] or the very good book by Harro Heuser [4] (in German).

2.1 Normed spaces, elementary topology and compactness

2.1.1 Norms, convergence and the equivalence of norms

We first collect the main definitions of a distance in linear spaces. The natural idea of a distance between points in a space is usually captured by the mathematical term *metric*. The term *norm* combines these ideas with the *linear structure* of a vector space. Here we move directly into the linear structure.

Definition 2.1.1 (Norm, normed space). *Let X be a real or complex linear space (i.e. vector space). A function $\|\cdot\| : X \to \mathbb{R}$ with the properties*

$$(\text{N1}) \quad \|\varphi\| \geqslant 0 \qquad (\textit{positivity}) \tag{2.1.1}$$

$$(\text{N2}) \quad \|\varphi\| = 0 \ \textit{if and only if} \ \varphi = 0 \qquad (\textit{definiteness}) \tag{2.1.2}$$

$$\text{(N3)} \quad \|\alpha\varphi\| = |\alpha| \, \|\varphi\| \qquad (\textit{homogeneity}) \tag{2.1.3}$$

$$\text{(N4)} \quad \|\varphi + \psi\| \leqslant \|\varphi\| + \|\psi\| \qquad (\textit{triangle inequality}) \tag{2.1.4}$$

for all $\varphi, \psi \in X$ and $\alpha \in \mathbb{R}$ or $\mathbb{C}$ is called a norm on X. A linear space X equipped with a norm is called a normed space. We remark that the dimension of normed space X is defined in terms of the dimension of linear space.

First, consider some examples of normed spaces.

Example 2.1.2 (*n*-dimensional real numbers). *The space $\mathbb{R}^m$ equipped with the* Euclidean norm

$$\|x\|_2 := \left(\sum_{j=1}^{n} |x_j|^2 \right)^{\frac{1}{2}}, \quad x \in \mathbb{R}^m \tag{2.1.5}$$

is a normed space which can be seen by checking the axioms (N1) to (N4). The only difficult part is the triangle inequality, which can be shown based on the Schwarz inequality.

We can also use the norm

$$\|x\|_\infty := \max_{j=1}^{n} |x_j|, \quad x \in \mathbb{R}^m \tag{2.1.6}$$

on $\mathbb{R}^m$. This is called the maximum norm. *Here the establishment of the axioms is straightforward.*

Further, we can define the one norm

$$\|x\|_1 := \sum_{j=1}^{n} |x_j|, \quad x \in \mathbb{R}^m. \tag{2.1.7}$$

It is indeed a norm according to our definition. The one norm is also called the Manhattan metric, *since it fits the travel time with roads as in parts of the city of New York.*

We can now transfer the basic terms from calculus into the setting of normed spaces. Most of the well-known concepts are the same.

Definition 2.1.3 (Convergence, $\epsilon - N$ criterion). *We say that a sequence $(\varphi_n) \subset X$ in a normed space X* converges *to an element $\varphi \in X$, if for every $\epsilon > 0$ there is an integer $N \in \mathbb{N}$ depending on ϵ such that for any $n \geq N$*

$$\|\varphi_n - \varphi\| \leqslant \epsilon. \tag{2.1.8}$$

Then φ is called the limit *of the sequence (φ_n) and we write*

$$\varphi_n \to \varphi, \quad n \to \infty \qquad \text{or} \qquad \lim_{n\to\infty} \varphi_n = \varphi. \tag{2.1.9}$$

Example 2.1.4 (Function spaces). *We consider the space $C([a, b])$ of continuous functions on the interval $[a, b] \subset \mathbb{R}$. The space equipped with the* maximum norm

$$\|\varphi\|_\infty := \max_{x \in [a,b]} |\varphi(x)| \tag{2.1.10}$$

is a normed space.

A different norm on $C([a, b])$ can be defined by the mean square norm

$$\|\varphi\|_2 := \left(\int_a^b |\varphi(x)|^2 \right)^{\frac{1}{2}}. \tag{2.1.11}$$

Convergence in the maximum norm is usually called uniform convergence. *Convergence in the mean square norm is referred to as* mean square convergence. *These two types of convergence are not equivalent. For example, the sequence*

$$\varphi_n(x) := \frac{1}{(1 + x)^n}, \quad x \in [0, 1] \tag{2.1.12}$$

in $C([0, 1])$ is convergent towards $\varphi(x) \equiv 0$ in the mean square norm, but it is not convergent at all in the maximum norm, since at $x = 0$ the functions all take the value $\varphi_n(0) = 1$.

Definition 2.1.5 (Continuity of mappings). *A mapping A from $U \subset X$ with a normed space X into a normed space Y is called continuous at $\varphi \in U$, if*

$$\lim_{n \to \infty} A\varphi_n = A\varphi \tag{2.1.13}$$

for every sequence $(\varphi_n) \subset U$ with $\varphi_n \to \varphi$, $n \to \infty$. The mapping $A : U \to Y$ is called continuous, if it is continuous at all $\varphi \in U$.

Definition 2.1.6 (Equivalence of norms). *We call two norms $\|\cdot\|_1, \|\cdot\|_2$ on a linear space X* equivalent, *if each sequence $(\varphi_n) \subset X$ which converges with respect to $\|\cdot\|_1$ also converges with respect to $\|\cdot\|_2$ and vice versa.*

Theorem 2.1.7 (Estimates for equivalent norms). *Two norms $\|\cdot\|_1, \|\cdot\|_2$ are equivalent if and only if there exist constants $C, c > 0$ such that*

$$c\|\varphi\|_1 \leqslant \|\varphi\|_2 \leqslant C\|\varphi\|_1 \tag{2.1.14}$$

is satisfied for all $\varphi \in X$. Also, the limits with respect to either norm coincide.

Proof. First, if the estimates are satisfied, they immediately imply that convergence of a sequence with respect to $\|\cdot\|_1$ implies convergence of the sequence with respect to $\|\cdot\|_2$ and vice versa.

To show the other direction of the equivalence statement assume that there is no constant $C > 0$ such that the second part of the estimate (2.1.14) is satisfied. Then there is a sequence (φ_n) with $\|\varphi_n\|_1 = 1$ and $\|\varphi_n\|_2 \geqslant n^2$, $n \in \mathbb{N}$. We define

$$\psi_n := \frac{1}{n}\varphi_n, \quad n \in \mathbb{N}. \tag{2.1.15}$$

Then $\psi_n \to 0$, $n \to \infty$ with respect to $\|\cdot\|_1$, but

$$\|\psi_n\|_2 = \frac{1}{n}\|\varphi_n\|_2 \geqslant n \to \infty, \quad n \to \infty, \tag{2.1.16}$$

i.e. the sequence (ψ_n) is not convergent towards 0 with respect to $\|\cdot\|_2$. This shows that for equivalent norms the second part of the estimate must be satisfied. The arguments for the first part with the constant c work in the same way. □

We close this subsection with some observations on finite-dimensional normed spaces. Consider a space

$$X = \mathrm{span}\{u_1, \ldots, u_n\} = \{\alpha_1 u_1 + \cdots + \alpha_n u_n : \alpha_j \in \mathbb{R} \text{ (or } \mathbb{C})\} \tag{2.1.17}$$

with linearly independent elements $u_1, \ldots, u_n \in X$. Then every element $\varphi \in X$ can be expressed as a unique *linear combination*

$$\varphi = \sum_{j=1}^{n} \alpha_j u_j. \tag{2.1.18}$$

In this case we can define a norm in X by

$$\|\varphi\|_\infty := \max_{j=1}^{n} |\alpha_j|. \tag{2.1.19}$$

We obtain the estimate

$$\begin{aligned} \|\varphi\| &= \left\| \sum_{j=1}^{n} \alpha_j u_j \right\| \\ &\leqslant \sum_{j=1}^{n} \|\alpha_j u_j\| \\ &= \sum_{j=1}^{n} |\alpha_j| \, \|u_j\| \\ &\leqslant \max_{j=1}^{n} |\alpha_j| \cdot \sum_{j=1}^{n} \|u_j\| \\ &= C\|\varphi\|_\infty \end{aligned} \tag{2.1.20}$$

with

$$C := \sum_{j=1}^{n} \|u_j\|. \tag{2.1.21}$$

We will also obtain an equivalence in the other direction, stated fully as follows.

Theorem 2.1.8. *On a finite-dimensional linear space all norms are equivalent.*

Proof. For the norm $\|\cdot\|$ of X and the norm $\|\cdot\|_\infty$ defined by (2.1.19) we have already shown one part of the estimate (2.1.14). The other direction is proven as follows. Suppose that there is no $c > 0$ which satisfies $c\|\varphi\|_\infty \leqslant \|\varphi\|$ for all $\varphi \in X$. Then there exists a sequence $(\varphi_n) \subset X$ with $\|\varphi_n\| = 1$ and $\|\varphi\|_\infty \geqslant n$. Define

$$\psi_n := \frac{\varphi_n}{\|\varphi_n\|_\infty}, \quad n \in \mathbb{N} \tag{2.1.22}$$

and consider the representation in terms of the basis elements

$$\psi_n = \sum_{j=1}^{n} \alpha_{n,j} u_j, \quad n \in \mathbb{N}. \tag{2.1.23}$$

The coefficients $\alpha_{n,j}$ are bounded, since we have $\|\psi_n\|_\infty = 1$. Using the theorem of Bolzano–Weierstrass we have convergent subsequences of the sequence of coefficients such that $\alpha_{n(k),j} \to \alpha_{0,j}$, $k \to \infty$ for each $j = 1, \dots, n$. We define

$$\psi_0 := \sum_{j=1}^{n} \alpha_{0,j} u_j \tag{2.1.24}$$

and have the convergence

$$\psi_{n(k)} \to \psi_0, \quad k \to \infty \tag{2.1.25}$$

with respect to $\|\cdot\|_\infty$. By $\|\cdot\| \leqslant C\|\cdot\|_\infty$ we obtain $\psi_{n(k)} \to \psi_0$ with respect to $\|\cdot\|$. But on the other hand we have $\|\psi_n\| = 1/\|\varphi\|_\infty \to 0$, $n \to \infty$. Since the limit is unique this yields $\psi_0 = 0$ and thus $\|\psi_0\|_\infty = 0$. But this contradicts $\|\psi_n\|_\infty = 1$ for all $n \in \mathbb{N}$. Thus also the other part of (2.1.14) must be satisfied for the norms $\|\cdot\|$ and $\|\cdot\|_\infty$.

Finally, given two norms $\|\cdot\|_1$ and $\|\cdot\|_2$ on X we first use the equivalence of $\|\cdot\|_1$ to $\|\cdot\|_\infty$ and then the equivalence of $\|\cdot\|_\infty$ to $\|\cdot\|_2$ to derive the statement of the theorem. □

2.1.2 Open and closed sets, Cauchy sequences and completeness

For working with normed spaces, we need to transfer all the facts about open and closed sets, sequences and completeness in the Euclidean space to this setting. Let X be a normed space and $\varphi \in X$. We call the set

$$B(\varphi, \rho) := \{\psi \in X : \|\psi - \varphi\| < \rho\} \tag{2.1.26}$$

the *open ball* of radius ρ and center φ in X. The set

$$B[\varphi, \rho] := \{\psi \in X : \|\psi - \varphi\| \leqslant \rho\} \tag{2.1.27}$$

is called the *closed ball.*

Definition 2.1.9. *We note the following basic definitions.*

- *A subset U of a normed space X is called open, if for each element $\varphi \in U$ there is $\epsilon > 0$ such that the ball $B(\varphi, \epsilon) \subset U$.*
- *A subset U is called closed, if it contains all limits of convergent subsequences of U.*
- *The closure $\bar{U}$ of a set is the set of all limits of convergent sequences of U.*
- *A set U is called dense in another set V in a normed space X, if $V \subset \bar{U}$.*
- *A set U is called* bounded, *if there is a constant $C > 0$ such that*

$$\|\varphi\| \leqslant C \tag{2.1.28}$$

for all $\varphi \in U$.

We consider some examples.

Example 2.1.10 (Ball in different norms). *Consider the unit ball in different norms on $\mathbb{R}^2$. The ball in the Euclidean norm is what we usually call a* disc. *The ball in the infinity norm is in fact a square. The ball in the one-norm is a diamond. Other norms lead to different shapes.*

Example 2.1.11 (Weierstrass approximation theorem). *Consider the space of continuous functions $C([a, b])$. The* Weierstrass approximation theorem *states that every continuous function φ on $[a, b]$ can be approximated up to any precision by a polynomial*

$$p_n(x) = \sum_{j=0}^{n} a_j x^j \tag{2.1.29}$$

with appropriate coefficients a_j and degree n. In the functional analytic language this means that the space Π of polynomials is dense *in the set $C([a, b])$.*

Note that the space

$$\Pi_n := \left\{ p_n(x) = \sum_{j=0}^{n} a_j x^j \, : \, a_j \in \mathbb{R} \; (or \; \mathbb{C}) \right\} \tag{2.1.30}$$

for fixed $n \in \mathbb{N}$ is not *dense in $C([a, b])$.*

Next, we transfer further elementary definitions from calculus into the environment of normed spaces.

Definition 2.1.12 (Cauchy sequence). *A sequence (φ_n) in a normed space X is called a* Cauchy sequence, *if for every $\epsilon > 0$ there is $N \in \mathbb{N}$ such that*

$$\|\varphi_n - \varphi_m\| \leqslant \epsilon, \quad \forall n, m \geqslant N. \tag{2.1.31}$$

Every convergent sequence is a Cauchy sequence. However, if we have a Cauchy sequence then the limits do not need to be an element of the space. The term Cauchy sequence has been developed to capture convergence without talking about the limit. This is very advantageous, however, now the question is what we can say about the limit.

Example 2.1.13. *Consider the sequence of functions*

$$f_n(x) := \begin{cases} \dfrac{1}{(1+x)^n}, & 0 < x \leqslant 1 \\ 1, & -1 \leqslant x \leqslant 0. \end{cases} \tag{2.1.32}$$

It is a sequence in $C([-1, 1])$ *and also in* $L^2([-1, 1])$. *In the mean square norm the sequence converges towards the function*

$$f(x) := \begin{cases} 0, & 0 < x \leqslant 1 \\ 1, & -1 \leqslant x \leqslant 0. \end{cases} \tag{2.1.33}$$

Clearly, this function is not continuous and, thus, it is not an element of $C([-1, 1])$. *Thus, the sequence does not converge in the normed space* $(C([-1, 1]), \|\cdot\|_2)$.

To describe sets which contain the limits of their Cauchy sequences, we proceed as follows.

Definition 2.1.14. *A subset U of a normed space X is called* complete, *if every Cauchy sequence* (φ_n) *in U converges towards an element* φ *of U. We call a complete normed space X a* Banach space.

2.1.3 Compact and relatively compact sets

We first provide two equivalent definitions of compact sets. One comes from general topology, the other from metric spaces. In our framework both are equivalent. However, the range of the topological definition is far greater, if you want to proceed further into pure mathematics.

Definition 2.1.15. *A subset U of a normed space X is called* compact *if every open covering of U contains a finite subcovering. In more detail, for every family* V_j, $j \in J$ *with some index set J (which is in general infinite, it can be countable or non-countable) of open sets with*

$$U \subset \bigcup_{j \in J} V_j \tag{2.1.34}$$

there is a finite subfamily $V_{j(k)}$, $j(k) \in J$, $k = 1, \ldots, n$ *with*

$$U \subset \bigcup_{k=1}^{n} V_{j(k)}. \tag{2.1.35}$$

The set U is called totally bounded, *if for each* $\epsilon > 0$ *there exists a finite number of elements* $\varphi_1, \ldots, \varphi_n$ *in U such that*

$$U \subset \bigcup_{j=1}^{n} B(\varphi_j, \epsilon), \tag{2.1.36}$$

i.e. each element $\varphi \in U$ *has a distance less than* ϵ *from at least one of the elements* $\varphi_1, \ldots, \varphi_n$.

To obtain some experience with this definition consider examples for non-compact sets.

Example 2.1.16 (Unbounded set). *The set $U := R^+$ in $\mathbb{R}$ is not compact. Consider a covering with bounded open sets*

$$V_j := (j - 2, j + 2), \quad j = 1, 2, 3, \ldots \tag{2.1.37}$$

There is not a finite subcovering which would cover the whole unbounded positive real axis U.

Example 2.1.17 (Not-complete set). *The set $U := (0, 1]$ in $\mathbb{R}$ is not compact. Consider a covering with open sets*

$$V_j := \left(\frac{1}{j+1}, \frac{1}{j - \frac{1}{2}} \right), \quad j = 1, 2, 3, \ldots \tag{2.1.38}$$

There is not a finite subcovering which would cover the whole interval, since any finite subcovering has a maximal index N and then the points smaller than $x = \frac{1}{N+1}$ are not contained in the finite subcovering.

Definition 2.1.18. *A subset U in a normed space X is called* sequentially compact *if every sequence in U has a convergent subsequence in U.*

Remark. It is clear from their definitions that any finite sets are compact and sequentially compact.

Lemma 2.1.19. *A sequentially compact set U is totally bounded.*

Proof. Assume that the set U is not totally bounded. Then there is some number $\epsilon > 0$ for which no finite number N of balls $B(\varphi_j, \epsilon)$ with $\varphi_j \in U$, $j = 1, \ldots, N$ covers U. For every finite number $\varphi_1, \ldots, \varphi_n$ of elements we can find an element $\varphi_{n+1} \in U$ such that $\|\varphi_{n+1} - \varphi_j\| \geqslant \epsilon$ for all $j = 1, \ldots, n$. This leads to a sequence (φ_n) such that

$$\|\varphi_n - \varphi_m\| \geqslant \epsilon, \quad n \neq m. \tag{2.1.39}$$

This sequence cannot contain a convergent subsequence, thus it is not sequentially compact. As a consequence we conclude that sequentially compact sequences are totally bounded. □

Lemma 2.1.20. *For each totally bounded set U there is a dense sequence $(\varphi_n)_{n \in \mathbb{N}}$ of elements.*

Proof. For $n \in \mathbb{N}$ we choose $\epsilon = 1/n$ and collect the finitely many elements $\varphi_1, \ldots, \varphi_N$ for which $B(\varphi_j, \epsilon)$, $j = 1, \ldots, N$ covers U. Putting these together into a sequence for $n = 1, 2, \ldots$ we obtain a dense sequence in U. □

Lemma 2.1.21. *Consider a sequentially compact set U and an open covering V_j, $j \in J$ of U. Then there are $\epsilon > 0$ such that for any $\varphi \in U$ the ball $B(\varphi, \epsilon)$ is contained in one of the domains V_j, $j \in J$.*

Proof. If there is no $\epsilon > 0$, then for $\epsilon = 1/n$, $n \in \mathbb{N}$, there is an element φ_n for which $B(\varphi_n, 1/n)$ is not contained in one of the V_j, $j \in J$. The sequence of these φ_n has a convergent subsequence $(\varphi_{n(k)})_{k\in\mathbb{N}}$ (since U is sequentially compact). Its limit φ is in U and thus in one of the V_j, $j \in J$. Since V_j is open, $\varphi_{n(k)}$ tends to φ and $\epsilon = 1/n$ tends to 0, there is $K \in \mathbb{N}$ such that $B(\varphi_{n(k)}, 1/n(k))$ is in V_j for $k \geqslant K$. But this is a contradiction to our construction above and thus we obtain the statement of the lemma. □

We are now prepared for the following basic equivalence result.

Theorem 2.1.22 (Sequence criterion). *A subset U of a normed space is compact if and only if it is sequentially compact, i.e. if every sequence $(\varphi_n) \subset U$ of elements of U contains a convergent subsequence to an element of U.*

Proof. First, we assume that U is not sequentially compact. In this case there is a sequence (φ_n) in U which does not contain a convergent subsequence in U. Since every $\varphi \in U$ is not a limit point of a subsequence of (φ_n), there is a radius $\rho(\varphi)$ such that the ball $B(\varphi, \rho(\varphi))$ only contains finitely many elements of (φ_n). The set of all these balls clearly is a covering of U. If it contained a finite subcovering, then one of the finitely many balls would necessarily contain an infinite number of elements, thus there is no finite subcovering of this covering and thus U is not compact. This shows that if U is compact it is also sequentially compact.

Next, we show the other direction of the equivalence statement. Assume that U is sequentially compact. Then U is also totally bounded. Consider some open covering V_j, $j \in J$, of U. We need to show that there is a finite subcovering. According to lemma 2.1.21 we know that there is $\epsilon > 0$ such that $B(\varphi, \epsilon)$ is contained in one of the V_j for each $\varphi \in U$. However, since U is totally bounded, there is a finite number of elements $\varphi_1, \ldots, \varphi_n$ such that $B(\varphi_j, \epsilon)$ cover U. But they are all contained in one of the V_j, thus a finite number of the sets V_j will cover U. □

The following theorem of Bolzano and Weierstrass is one of the well-known results of analysis, which we recall here for further use and for illustration.

Theorem 2.1.23 (Bolzano–Weierstrass). *A bounded sequence in $\mathbb{R}^n$ has a convergent subsequence.*

Proof. Let us have a brief look at the proof in $\mathbb{R}$. The n-dimensional version of the proof works analogously.

Let $(z_n)_{n\in\mathbb{N}}$ be a bounded sequence in $\mathbb{R}$. We construct a convergent subsequence via interval partition. Since the sequence is bounded there is $a_1, b_1 \in \mathbb{R}$ such that $a_1 \leqslant z_n \leqslant b_1$ for all $n \in \mathbb{N}$. Then in at least one of the intervals

$$\left[a_1, \frac{a_1 + b_1}{2}\right], \qquad \left(\frac{a_1 + b_1}{2}, b_1\right]$$

there are infinitely many numbers of the sequence. We call this interval $[a_2, b_2]$ with appropriately chosen numbers a_1 and b_2. Then we apply the same argument to this new interval. Again we obtain an interval $[a_3, b_3]$ which contains infinitely many elements of the sequence. We repeat this to obtain two sequences (a_n) and (b_n). The sequence (a_n) is monotonously increasing, the sequence (b_n) is monotonously decreasing. For the distance between the interval boundaries we obtain

$$\|b_n - a_n\| \leqslant \frac{b-a}{2^{n-1}}, \quad n \in \mathbb{N}. \tag{2.1.40}$$

Thus both sequences are convergent towards a point $a = b$. □

Definition 2.1.24. *A subset U of a normed space is called relatively compact if its closure is compact.*

The $\mathbb{R}^m$ is a very important prototype of a normed space, which is studied extensively in calculus modules and courses on linear algebra. Next, we will employ the fact that it is the basic model for any *finite-dimensional* space.

Theorem 2.1.25. *Let U be a bounded and finite-dimensional subset of a normed space, i.e. there is $C > 0$ such that $\|\varphi\| \leqslant C$ for all $\varphi \in U$ and there are linearly independent elements $u_1, \ldots, u_n$ with $n \in \mathbb{N}$*

$$V := \text{span}\{u_1, \ldots, u_n\} = \left\{\alpha_1 u_1 + \cdots + \alpha_n u_n : \alpha_j \in \mathbb{C} \text{ or } \mathbb{R}\right\} \tag{2.1.41}$$

such that $U \subset V$. Then the set U is relatively compact.

Proof. We map $\mathbb{R}^m$ (or $\mathbb{C}^n$) into V via

$$\Psi : \alpha \mapsto \alpha_1 u_1 + \cdots + \alpha_n u_n, \quad \text{for } \alpha = \begin{pmatrix} \alpha_1 \\ \vdots \\ \alpha_n \end{pmatrix} \in \mathbb{R}^m. \tag{2.1.42}$$

This mapping is bijective (one-to-one) and there are constants $C, c > 0$ such that we have the norm estimates

$$c\|\alpha\| \leqslant \|\Psi(\alpha)\| \leqslant C\|\alpha\|, \tag{2.1.43}$$

where $\|\cdot\|$ denotes the Euclidean norm in $\mathbb{R}^m$. Now, via (2.1.43) convergence in V and convergence in $\mathbb{R}^m$ are equivalent. The bounded set U is mapped into a bounded set $W := \Psi^{-1}(U)$ in $\mathbb{R}^m$. We apply the theorem of Bolzano–Weierstrass to conclude that W is relatively compact and again by (2.1.43) this also yields for U, which completes the proof. □

2.2 Hilbert spaces, orthogonal systems and Fourier expansion

2.2.1 Scalar products and orthonormal systems

So far we have used a distance concept which takes into account the linear structure of a space. However, from $\mathbb{R}^m$ we are used to geometry, in particular angles and orthogonality. Can this be transferred and exploited in a more general setting? The answer is yes. The concept of a Hilbert space is the appropriate setting to use orthogonality in a wider framework. The success of this concept has been overwhelming. The whole of modern physics with its quantum processes has been built onto the concept of the Hilbert space.

Definition 2.2.1 (Scalar product). *Let X be a real or complex linear space (i.e. vector space). A function* $\langle \cdot, \cdot \rangle : X \times X \to \mathbb{C}$ *with the properties*

$$\text{(S1)} \quad \langle \varphi, \varphi \rangle \geqslant 0 \qquad (\textit{positivity}) \tag{2.2.1}$$

$$\text{(S2)} \quad \langle \varphi, \varphi \rangle = 0 \text{ if and only if } \varphi = 0 \qquad (\textit{definiteness}) \tag{2.2.2}$$

$$\text{(S3)} \quad \langle \varphi, \psi \rangle = \overline{\langle \psi, \varphi \rangle} \qquad (\textit{symmetry}) \tag{2.2.3}$$

$$\text{(S4)} \quad \langle \alpha\varphi + \beta\psi, \xi \rangle = \alpha \langle \varphi, \xi \rangle + \beta \langle \varphi, \xi \rangle \qquad (\textit{linearity}) \tag{2.2.4}$$

for all $\varphi, \psi, \xi \in X$ *and* $\alpha \in \mathbb{R}$ *or* $\mathbb{C}$ *is called a* scalar product *on X. A linear space X equipped with a scalar product is called a* pre-Hilbert space. *Combining (S3) and (S4) we have*

$$\langle \xi, \alpha\varphi + \beta\psi \rangle = \bar{\alpha} \langle \xi, \varphi \rangle + \bar{\beta} \langle \xi, \psi \rangle. \tag{2.2.5}$$

We call this kind of property anti-linear *not only for what we have here with respect to the second variable of the scalar product.*

A basic inequality for pre-Hilbert spaces is given as follows.

Theorem 2.2.2. *For a pre-Hilbert space X we have the* Cauchy–Schwarz inequality

$$|\langle \varphi, \psi \rangle|^2 \leqslant \langle \varphi, \varphi \rangle \langle \psi, \psi \rangle \tag{2.2.6}$$

for $\varphi, \psi \in X$. *Here the equality holds if and only if* φ *and* ψ *are linearly dependent.*

Proof. For $\varphi = 0$ the inequality is clearly satisfied. For $\varphi \neq 0$ we define

$$\alpha = -\langle \varphi, \varphi \rangle^{-1/2} \overline{\langle \varphi, \psi \rangle}, \qquad \beta = \langle \varphi, \varphi \rangle^{1/2}. \tag{2.2.7}$$

Then we derive

$$0 \leqslant \langle \alpha\varphi + \beta\psi, \alpha\varphi + \beta\psi \rangle \tag{2.2.8}$$

$$= |\alpha|^2 \langle \varphi, \varphi \rangle + 2\,\mathrm{Re}\left(\alpha\bar{\beta} \langle \varphi, \psi \rangle\right) + |\beta|^2 \langle \psi, \psi \rangle \tag{2.2.9}$$

$$= \langle \varphi, \varphi \rangle \langle \psi, \psi \rangle - |\langle \varphi, \psi \rangle|^2. \tag{2.2.10}$$

This proves the inequality. We have equality if and only if $\alpha\varphi + \beta\psi = 0$, i.e. if φ and ψ are linearly dependent.

A scalar product $\langle \cdot, \cdot \rangle$ defines a norm via

$$\|\varphi\| := \langle \varphi, \varphi \rangle^{1/2}, \quad \varphi \in X. \tag{2.2.11}$$

Please check the norm axioms! Thus, a pre-Hilbert space is a normed space. If a pre-Hilbert space is complete, we call it a *Hilbert space*.

We can now introduce elements of geometry into an abstract space using the scalar product.

Definition 2.2.3. *We call two elements φ, ψ of a pre-Hilbert space X* orthogonal, *if*

$$\langle \varphi, \psi \rangle = 0. \tag{2.2.12}$$

In this case we write $\varphi \perp \psi$. Two subsets V, W of a Hilbert space are called orthogonal *if for each pair of elements $\varphi \in V$ and $\psi \in W$ the equation* (2.2.12) *is satisfied. We write $V \perp W$.*

We know orthogonality in $\mathbb{R}^m$. Here, an appropriate scalar product is given by the *Euclidean scalar product*

$$x \cdot y = \sum_{j=1}^{n} x_j y_j, \quad x, y \in \mathbb{R}^m. \tag{2.2.13}$$

For the special case $n = 2, 3$ we know that two vectors x, y are orthogonal if and only if the angle between these vectors is 90° or $\pi/2$. Moreover, using the series definition of $\cos(\theta)$ and $\sin(\theta)$ we can define the angle θ ($0 \leqslant \theta \leqslant 180°$) between vectors x, y via $\|x\|_2 \|y\|_2 \cos(\theta) = x \cdot y$.

In $\mathbb{R}^m$ we know that $e_1, \ldots, e_n \in \mathbb{R}^m$ where

$$e_1 = \begin{pmatrix} 1 \\ 0 \\ \vdots \\ 0 \end{pmatrix}, \quad e_2 = \begin{pmatrix} 0 \\ 1 \\ 0 \\ \vdots \\ 0 \end{pmatrix}, \ldots, \quad e_n = \begin{pmatrix} 0 \\ \vdots \\ 0 \\ 1 \end{pmatrix} \tag{2.2.14}$$

is a basis of orthogonal vectors with respect to the Euclidean scalar product (2.2.13). We have also learned in linear algebra that there are many sets of vectors which span the whole space and are orthogonal to each other. This leads to the definition.

Definition 2.2.4. *A set U of elements $\varphi \in X$ in a pre-Hilbert space X is called an* orthogonal system, *if*

$$\langle \varphi, \psi \rangle = 0, \qquad \varphi \neq \psi \in U. \tag{2.2.15}$$

If in addition we have $\|\varphi\| = 1$ for all $\varphi \in U$, then the set is called an orthonormal system.

Such orthonormal systems are of importance in many mathematical disciplines, for example in approximation theory, interpolation theory and numerical mathematics, but also in very new mathematical areas such as wavelet theory and inverse problems. We will extensively explore orthonormal systems for the study of inversion problems.

We complete this subsection with some further basic definition.

Definition 2.2.5. *Given some set U in a pre-Hilbert space we call*

$$U^{\perp} := \left\{ \varphi \in X : \varphi \perp U \right\} \tag{2.2.16}$$

the orthogonal complement *of U.*

Example 2.2.6. *For example, the orthogonal complement of the set* $U = \{(0, 0, 1)^T\}$ *in* $\mathbb{R}^3$ *is the x–y plane* $\{(s, t, 0) : s, t \in \mathbb{R}\}$. *Given some vector* $y \in \mathbb{R}^3$, $y \neq 0$, *determine its orthogonal complement using the vector product*

$$x \times y = \begin{pmatrix} x_2 y_3 - x_3 y_2 \\ -(x_1 y_3 - x_3 y_1) \\ x_1 y_2 - x_2 y_3 \end{pmatrix} \tag{2.2.17}$$

defined for arbitrary $x, y \in \mathbb{R}^3$. *Note that we have*

$$x \cdot (x \times y) = 0, \qquad y \cdot (x \times y) = 0 \tag{2.2.18}$$

for vectors $x, y \in \mathbb{R}^3$ *with* $x \neq y$, $x \neq 0$, $y \neq 0$. *Thus we choose any vector* $x \in \mathbb{R}^3$ *with* $x \neq y$ *and* $x \neq 0$ *and define vectors*

$$z_1 := x \times y, \qquad z_2 := z_1 \times y. \tag{2.2.19}$$

Then $y \cdot z_1 = 0$ *and* $y \cdot z_2 = y \cdot ((x \times y) \times y) = 0$ *and thus*

$$V := \operatorname{span}\{z_1, z_2\} = \{\alpha z_1 + \beta z_2 : \alpha, \beta \in \mathbb{R}\} \tag{2.2.20}$$

satisfies

$$V = \{y\}^{\perp}. \tag{2.2.21}$$

2.2.2 Best approximations and Fourier expansion

Here we collect some facts about best approximations in pre-Hilbert spaces. These are of interest by themselves, but can also be seen as a basis for the Fourier expansion. We would like to remark that the Fourier theory in Hilbert spaces is much easier than the classical Fourier theory.

Further, note that most solutions of inverse problems can be understood as some type of *best approximation*, for example three-dimensional or four-dimensional variational data assimilation in chapter 5, but also Tikhonov regularization which we introduce in section 3.1.4. Fourier theory is a core ingredient to understand and analyze regularization techniques.

Definition 2.2.7 (Best approximation). *Consider a set* U *in a normed space* X *and* $\varphi \in X$. *Then* $\psi \in U$ *is called the* best approximation *to the element* φ *with respect to* U *if*

$$\|\varphi - \psi\| = \inf_{\xi \in U} \|\varphi - \xi\| \tag{2.2.22}$$

is satisfied. The best approximation is an element with the smallest distance

$$d(\varphi, U) := \inf_{\xi \in U} \|\varphi - \xi\| \tag{2.2.23}$$

of the set U *to the element* φ.

As basic questions of approximation we need to investigate existence and uniqueness of such 'best' approximations. We will provide some results for particular settings for further use.

Theorem 2.2.8. *If U is a finite-dimensional subspace of a normed space X, then for every element $\varphi \in X$ there exists a best approximation with respect to U.*

Proof. We remark that the finite-dimensional subspace U is complete (this is a consequence of the theorem of Bolzano–Weierstrass) and every bounded sequence in U has a convergent subsequence in U. We now construct a minimizing sequence (φ_n) in U, i.e. a sequence with

$$\|\varphi_n - \varphi\| \to d(U, \varphi), \quad n \to \infty. \tag{2.2.24}$$

It is clearly bounded, since $\|\varphi_n\| \leqslant \|\varphi_n - \varphi\| + \|\varphi\|$. Now we choose a convergent subsequence, which tends to an element $\psi \in U$. Passing to the limit for the subsequence $\varphi_{n(k)}$ of φ_n in (2.2.24) we obtain $\|\psi - \varphi\| = d(\varphi, U)$, i.e. ψ is best approximation to φ. □

In normed spaces best approximations do not need to be unique.

Example 2.2.9. *Consider the space $\mathbb{R}^2$ equipped with the maximum norm $\|\cdot\|_\infty$. In fact take $U = \{\lambda e_2 : \lambda \in \mathbb{R}\}$ and $\varphi = (1, 0)$. Then, any point in $\Lambda = \{(0, \mu) : \mu \in \mathbb{R}, |\mu| \leqslant 1\}$ is a best approximation to φ with respect to the maximum norm.*

Next, we employ the scalar product to formulate criteria for best approximations with respect to linear subspaces of pre-Hilbert spaces. In this case we can also answer the uniqueness question.

Theorem 2.2.10. *Let X be a pre-Hilbert space with inner product $\langle\cdot,\cdot\rangle$ and $\varphi \in X$.*

(i) *Let U be a convex subset of X. That is, U satisfies the condition*

$$\alpha\varphi_1 + (1 - \alpha)\varphi_2 \in U, \quad \alpha \in [0, 1] \tag{2.2.25}$$

for any $\varphi_1, \varphi_2 \in U$. Then, $\psi \in U$ is the best approximation to φ if and only if

$$\mathrm{Re}\langle\varphi - \psi, u - \psi\rangle \leqslant 0, \quad u \in U. \tag{2.2.26}$$

(ii) *If U is a linear subspace of X. Then, $\psi \in U$ is the best approximation to φ if and only if*

$$\langle\varphi - \psi, \xi\rangle = 0, \quad \xi \in U, \tag{2.2.27}$$

i.e. if $\varphi - \psi \perp U$.

For both (i) and (ii) there is at most one best approximation.

Proof. Since (ii) easily follows from (i), we only prove (i). Take any $u \in U$ and fix it. Consider the function

$$\begin{aligned} f(\alpha) &= \|\varphi - ((1 - \alpha)\psi + \alpha u)\|^2 \\ &= \|\varphi - \psi\|^2 - 2\alpha\mathrm{Re}\langle\varphi - \psi, u - \psi\rangle + \alpha^2\|u - \psi\|^2 \end{aligned} \tag{2.2.28}$$

for $\alpha \in [0, 1]$. First, assume (2.2.26). Then we have

$$f(1) \geqslant \|\varphi - \psi\|^2 + \|u - \psi\|^2 \geqslant \|\varphi - \psi\|^2. \tag{2.2.29}$$

Hence ψ is the best approximation to φ.

Next we prove the converse. We prove this by a contradictory argument. Let ψ is the best approximation to φ and suppose (2.2.26) does not hold. Then, there exists $u \in U$ such that

$$\mathrm{Re}\langle \varphi - \psi, u - \psi \rangle > 0. \tag{2.2.30}$$

Then by (2.2.28),

$$f(0) > f(\alpha), \qquad 0 < \alpha \ll 1. \tag{2.2.31}$$

Hence, there exist $\psi' = (1 - \alpha)\psi + \alpha u = \psi + \alpha(u - \psi) \in U$ with some $\alpha \in (0, 1)$ such that $\|\varphi - \psi\| > \|\varphi - \psi'\|$, which contradicts the fact that ψ is the best approximation to φ.

Finally, assume that there are two best approximations $\psi_1, \psi_2 \in U$ to φ. Then with $\xi := \psi_2 - \psi_1 \in U$ we obtain

$$0 \leqslant \|\psi_2 - \psi_1\|^2 = \langle \varphi - \psi_1, \psi_2 - \psi_1 \rangle + \langle \varphi - \psi_2, \psi_1 - \psi_2 \rangle \leqslant 0 \tag{2.2.32}$$

thus $\psi_1 = \psi_2$. □

Next, we note some further results about the existence of a unique best approximation preparing our main theorem about the Fourier expansion.

Theorem 2.2.11. *Consider a complete linear subspace U of a pre-Hilbert space X. Then for every element $\varphi \in X$ there exists a unique best approximation to φ with respect to U. Hence, X has the decomposition $X = U \oplus U^{\perp}$, that is, each $\varphi \in X$ can be uniquely written in the form $\varphi = \psi + \chi$ with $\psi \in U$, $\chi \in U^{\perp}$. The mapping which maps each φ to ψ in the above decomposition of φ is called the* orthogonal projection *in X onto U.*

Proof. The idea is to take a minimizing sequence such that

$$\|\varphi - \varphi_n\|^2 \leqslant d(\varphi, U)^2 + \frac{1}{n}, \qquad n \in \mathbb{N}. \tag{2.2.33}$$

For $n, m \in \mathbb{N}$ we estimate

$$\begin{aligned}
&\left\|(\varphi - \varphi_n) + (\varphi - \varphi_m)\right\|^2 + \left\|\varphi_n - \varphi_m\right\|^2 \\
&= \left\|(\varphi - \varphi_n) + (\varphi - \varphi_m)\right\|^2 + \left\|\varphi - \varphi_m - (\varphi - \varphi_n)\right\|^2 \\
&= \left(\text{calculations via } \|v\|^2 = \langle v, v\rangle\right) \\
&= 2\|\varphi - \varphi_n\|^2 + 2\|\varphi - \varphi_m\|^2 \\
&\leqslant 4d(\varphi, U)^2 + \frac{2}{n} + \frac{2}{m}, \qquad n, m \in \mathbb{N}.
\end{aligned} \tag{2.2.34}$$

Since $\|\varphi - 1/2(\varphi_n + \varphi_m)\| \geqslant d(\varphi, U)$ we now obtain

$$\|\varphi_n - \varphi_m\|^2 \leqslant 4d(\varphi, U)^2 + \frac{2}{n} + \frac{2}{m} - 4\left\|\varphi - \frac{1}{2}(\varphi_n + \varphi_m)\right\|^2 \leqslant \frac{2}{n} + \frac{2}{m}$$

for $n, m \in \mathbb{N}$, thus (φ_n) is a Cauchy sequence. Since U is complete, there is a limit ψ of this Cauchy sequence for which from (2.2.33) we obtain (2.2.22), i.e. ψ is the best approximation to φ with respect to U.

The uniqueness is a consequence of the previous theorem 2.2.10. □

The usual term series from analysis can be easily transferred to a normed space setting. Consider a sequence (φ_n) in a normed space X. Then we can define the *partial sums*

$$S_N := \sum_{n=1}^{N} \varphi_n, \quad N \in \mathbb{N}, \tag{2.2.35}$$

which is just a finite sum of elements in X. For each $N \in \mathbb{N}$ the partial sum S_N is an element of X. Thus, we obtain a sequence $(S_N)_{N\in\mathbb{N}}$ in X. If this sequence is convergent towards an element $S \in X$ for $N \to \infty$, then we say that the *infinite series* (in short 'series')

$$\sum_{n=1}^{\infty} \varphi_n \tag{2.2.36}$$

is convergent and has value S. We write $S = \sum_{n=1}^{\infty} \varphi_n$.

We now arrive at the culmination point of this chapter, deriving some quite general theorem about the representation of elements with respect to an orthogonal system.

Theorem 2.2.12 (Fourier representation). *We consider a pre-Hilbert space X with an orthonormal system $\{\varphi_n : n \in \mathbb{N}\}$. Then the following properties are equivalent:*

(1) *The set* span $\{\varphi_n : n \in \mathbb{N}\}$ *is dense in X. Recall that* span{} *is the set of all linear combinations of a* finite *subset of* $\{\varphi_n : n \in \mathbb{N}\}$.

(2) *Each element $\varphi \in X$ can be expanded in a* Fourier series

$$\varphi = \sum_{n=1}^{\infty} \langle \varphi, \varphi_n \rangle \varphi_n. \tag{2.2.37}$$

(3) *For each $\varphi \in X$ there is the* Parseval identity

$$\|\varphi\|^2 = \sum_{n=1}^{\infty} |\langle \varphi, \varphi_n \rangle|^2. \tag{2.2.38}$$

Usually the polarization of this is called the Parseval identity and (2.2.38) *itself is called the Plancherel theorem. If an orthonormal system satisfies these properties it is called* complete.

Proof. First we show that (1) implies (2). According to theorems 2.2.10 and 2.2.11 the partial sum

$$\psi_N = \sum_{n=1}^{N} \langle \varphi, \varphi_n \rangle \varphi_n \tag{2.2.39}$$

is the best approximation to φ with respect to the finite-dimensional linear space $\operatorname{span}\{\varphi_1, \ldots, \varphi_N\}$. Since according to (1) the span is dense in X, the best approximation ψ_N will converge towards φ for $N \to \infty$, which establishes the convergence of (2.2.37).

To show the implication from (2) to (3) we consider

$$\langle \varphi, \psi_N \rangle = \sum_{n=1}^{N} \overline{\langle \varphi, \varphi_n \rangle} \langle \varphi, \varphi_n \rangle = \sum_{n=1}^{N} |\langle \varphi, \varphi_n \rangle|^2 \tag{2.2.40}$$

for the previous ψ_N. Passing to the limit $N \to \infty$ we obtain (2.2.38).

To pass from (3) to (1) we calculate

$$\left\| \varphi - \sum_{n=1}^{N} \langle \varphi, \varphi_n \rangle \varphi_n \right\|^2 = \|\varphi\|^2 - \sum_{n=1}^{N} |\langle \varphi, \varphi_n \rangle|^2. \tag{2.2.41}$$

The right-hand side tends to zero and the left-hand side is the approximation of φ by elements of $\operatorname{span}\{\varphi_n : n \in \mathbb{N}\}$. This yields (1).

With the three implications (1) ⇒ (2), (2) ⇒ (3) and (3) ⇒ (1) we have proven the equivalence of all three statements. □

Remark 2.2.13. *The proof of the previous theorem provides the* Bessel inequality

$$\sum_{n=1}^{\infty} |\langle \varphi, \varphi_n \rangle|^2 \leqslant \|\varphi\|^2 \tag{2.2.42}$$

for any $\varphi \in X$ *when* $\{\varphi_n : n \in \mathbb{N}\}$ *is an orthonormal system in* X.

Before dealing with convergence of the Fourier series of a function $\varphi \in L^2((0, 2\pi))$ in the mean square sense, we prepare the following lemma which is necessary for its necessity.

Lemma 2.2.14. $C^\infty([a, b])$ *is dense in* $L^2((a, b))$. *In particular* $C([a, b])$ *is dense in* $L^2((a, b))$.

Proof. We use the method of mollification for the proof. The details are as follows. Let $f \in L^2((a, b))$. Then for $\delta (0 < \delta < 1)$ we define $f_\delta \in L^2((\alpha, \beta))$ with $\alpha = (a + b)/2 - (b - a)/(2\delta)$, $\beta = (a + b)/2 + (b - a)/(2\delta)$ by

$$f_\delta(x) = f\left(\delta x + \frac{a + b}{2}(1 - \delta) \right), \qquad x \in (\alpha, \beta). \tag{2.2.43}$$

Then it is easy to see by the Lebesgue dominated convergence theorem that $f_\delta \to f (\delta \to 1)$ in $L^2((a, b))$. We extend f_δ to the whole $\mathbb{R}$ by extending it 0 outside (α, β) and denote its extension by $\tilde{f}_\delta$, i.e.

$$\tilde{f}_\delta(x) = \begin{cases} f_\delta(x) & \text{if } x \in (\alpha, \beta) \\ 0 & \text{if otherwise.} \end{cases} \tag{2.2.44}$$

Now we mollify $\tilde{f}_\delta$ by a mollifier $\chi_\varepsilon \in C_0^\infty(\mathbb{R})$ given by $\chi_\varepsilon(x) = \varepsilon^{-1}\chi(\varepsilon^{-1}x)$ with a $\chi \in C_0^\infty(\mathbb{R})$ with the properties $0 \leqslant \chi(x) \leqslant 1$ for any $x \in \mathbb{R}$, $\chi(x) = 0$ if $|x| \geqslant 1$ and $\int_{\mathbb{R}} \chi(x)\mathrm{d}x = 1$. The mollification $\tilde{f}_{\delta,\varepsilon}$ of $\tilde{f}_\delta$ by this mollifier is defined by their convolution, i.e.

$$\tilde{f}_{\delta,\varepsilon}(x) = \left(\tilde{f}_\delta * \chi_\varepsilon\right)(x) = \int_{\mathbb{R}} \tilde{f}_\delta(x-y)\chi(y)\,\mathrm{d}y. \tag{2.2.45}$$

By using $\sup_{|y|\leqslant\varepsilon} \int_{\mathbb{R}} |\tilde{f}_\delta(x-y) - \tilde{f}_\delta(x)|^2\,\mathrm{d}x \to 0(\varepsilon \to)$ because $\tilde{f}_\delta \in L^2(\mathbb{R})$, it can be easily seen that

$$\tilde{f}_{\delta,\varepsilon} \to \tilde{f}_\delta\,(\varepsilon \to 0) \text{ in } L^2(\mathbb{R}). \tag{2.2.46}$$

Then the proof will be finished by just noting that $\tilde{f}_\delta = f_\delta$ in (a, b) and $\tilde{f}_{\delta,\varepsilon} \in C_0^\infty(\mathbb{R})$. □

Example 2.2.15. *As an example consider the set* $L^2([0, 2\pi])$ *of square integrable functions on the interval* $[0, 2\pi]$. *Equipped with the scalar product*

$$\langle\varphi, \psi\rangle := \int_0^{2\pi} \varphi(y)\overline{\psi(y)}\,\mathrm{d}y \tag{2.2.47}$$

this space is a pre-Hilbert space. The norm induced by this scalar product is the mean square norm which we have discussed above.

For $n = 1, 2, 3, \ldots$ *consider the functions*

$$\varphi_{2n}(x) := \frac{1}{\sqrt{\pi}}\sin(nx), \qquad x \in [0, 2\pi], \quad n = 1, 2, 3, \ldots \tag{2.2.48}$$

$$\varphi_{2n+1}(x) := \frac{1}{\sqrt{\pi}}\cos(nx), \qquad x \in [0, 2\pi], \quad n = 0, 1, 2, \ldots. \tag{2.2.49}$$

We note the orthogonality relations

$$\int_0^{2\pi} \cos(nx)\cos(mx)\,\mathrm{d}x = \begin{cases} 0, & n \neq m \\ \pi, & n = m \end{cases} \tag{2.2.50}$$

$$\int_0^{2\pi} \cos(nx)\sin(mx)\,\mathrm{d}x = 0 \tag{2.2.51}$$

$$\int_0^{2\pi} \sin(nx)\sin(mx)\,\mathrm{d}x = \begin{cases} 0, & n \neq m \\ \pi, & n = m \end{cases} \tag{2.2.52}$$

for all $n, m \in \mathbb{N}_0 = \mathbb{N} \cup \{0\}$. *Thus, the set*

$$\mathcal{B} := \left\{\varphi_n : n \in \mathbb{N}\right\} \tag{2.2.53}$$

with φ_n *defined by (2.2.48), (2.2.49) is an orthonormal system in* $L^2([0, 2\pi])$.

From the approximation theorem for trigonometric polynomials we obtain that the set of trigonometric polynomials

$$t_N(x) := a_0 + \sum_{n=1}^{N}(a_n \cos(nx) + b_n \sin(nx)), \quad x \in [0, 2\pi] \tag{2.2.54}$$

where $N \in \mathbb{N}$ *and* $a_j, b_j \in \mathbb{R}$ *(or* $\mathbb{C}$*) is dense in the space* $C([0, 2\pi])$ *by the Weierstrass approximation theorem. By lemma 2.2.14, the set of trigonometric polynomials is also dense in* $X = L^2([0, 2\pi])$ *equipped with the scalar product (2.2.47).*

This shows that property (1) of theorem 2.2.12 is satisfied. As a consequence, all equivalent properties hold and we can represent any function in $L^2([0, 2\pi])$ *in terms of its* Fourier series

$$\varphi(x) = a_0 + \sum_{n=1}^{\infty}(a_n \cos(nx) + b_n \sin(nx)), \quad x \in [0, 2\pi], \tag{2.2.55}$$

where the convergence is valid in the mean square sense.

2.3 Bounded operators, Neumann series and compactness

2.3.1 Bounded and linear operators

Linear mappings in $\mathbb{R}^m$ are well known from elementary courses on linear algebra. Analogously, a mapping $A : X \to Y$ from a linear space X into some linear space Y is called *linear*, if

$$A(\alpha\varphi + \beta\psi) = \alpha A\varphi + \beta A\psi \tag{2.3.1}$$

for all $\varphi, \psi \in X$ and $\alpha, \beta \in \mathbb{R}$ (or $\mathbb{C}$). Linear operators have quite nice properties. First, we note:

Theorem 2.3.1. *A linear operator is continuous in the whole space if it is continuous at one element (for example in* $\varphi = 0$*).*

Proof. If A is continuous everywhere it is clearly continuous in a particular element $\varphi_0 \in X$. If now A is continuous in $\varphi_0 \in X$, i.e. for every sequence $(\varphi_n) \subset X$ with $\varphi_n \to \varphi_0$ for $n \to \infty$, we have $A\varphi_n \to A\varphi_0$, $n \to \infty$. Consider an arbitrary element $\varphi \in X$ and a sequence $(\tilde{\varphi}_n)$ with $\tilde{\varphi}_n \to \varphi$ for $n \to \infty$. We define $\varphi_n := \tilde{\varphi}_n + \varphi_0 - \varphi$, which tends towards φ_0 for $n \to \infty$. Now we conclude

$$A\tilde{\varphi}_n = A\varphi_n + A(\varphi - \varphi_0) \to A\varphi_0 + A(\varphi - \varphi_0) = A\varphi \tag{2.3.2}$$

for $n \to \infty$, which completes the proof.

Definition 2.3.2. *A linear operator* $A : X \to Y$ *with normed spaces* X, Y *is called* bounded, *if there is a constant* $C > 0$ *such that*

$$\|A\varphi\| \leqslant C\|\varphi\|, \quad \varphi \in X. \tag{2.3.3}$$

We call such a constant C *a* bound *for the operator* A.

We can restrict ourselves in the definition of the bound to a set of non-zero elements $\varphi \in X$ with $\|\varphi\| \leqslant 1$ or even with $\|\varphi\| = 1$. This is due to the fact that

$$\|A\varphi\| = \left\| A\left(\frac{\varphi}{\|\varphi\|}\right) \right\| \cdot \|\varphi\| \tag{2.3.4}$$

for all non-zero $\varphi \in X$. Thus, if (2.3.3) is satisfied for all φ with $\|\varphi\| \leqslant 1$, then via (2.3.4) we estimate

$$\|A\varphi\| \leqslant C\left\| \frac{\varphi}{\|\varphi\|} \right\| \cdot \|\varphi\| = C\|\varphi\| \tag{2.3.5}$$

for arbitrary non-zero $\varphi \in X$.

Definition 2.3.3. *An operator is bounded if and only if*

$$\|A\| := \sup_{\|\varphi\|=1} \|A\varphi\| = \sup_{\|\varphi\|\leqslant 1} \|A\varphi\| < \infty. \tag{2.3.6}$$

In this case we call $\|A\|$ *the* operator norm *of* A.

Example 2.3.4. *Examples for bounded operators on* $\mathbb{R}^m$ *are given by matrices. Note that in* $\mathbb{R}^m$ *every linear operator is bounded.*

Example 2.3.5. *Consider the integral operator*

$$(A\varphi)(x) := \int_a^b k(x, y)\varphi(y)\,\mathrm{d}y, \quad x \in [a, b] \tag{2.3.7}$$

with some continuous kernel $k : [a, b] \times [a, b] \to \mathbb{C}$. *Then we can estimate*

$$\begin{aligned} |A\varphi(x)| &= \left| \int_a^b k(x, y)\varphi(y)\mathrm{d}y \right| \\ &\leqslant \int_a^b |k(x, y)| \cdot |\varphi(y)|\,\mathrm{d}y \\ &\leqslant \int_a^b |k(x, y)| \cdot \sup_{y\in[a,b]} |\varphi(y)|\,\mathrm{d}y \\ &= C\|\varphi\|_\infty \end{aligned} \tag{2.3.8}$$

with

$$C := \sup_{x\in[a,b]} \int_a^b |k(x, y)|\,\mathrm{d}y. \tag{2.3.9}$$

This proves that the operator $A : C([a, b]) \to C([a, b])$ *is bounded where* $C([a, b])$ *is equipped with the maximum norm.*

We can add operators $A_1, A_2 : X \to Y$ by pointwise summation

$$(A_1 + A_2)\varphi := A_1\varphi + A_2\varphi, \quad \varphi \in X \tag{2.3.10}$$

and multiply an operator by a real or complex number α via

$$(\alpha A)\varphi := \alpha A\varphi. \tag{2.3.11}$$

Thus, the set of all linear operators is itself a *linear space*. Also, each linear combination of *bounded* linear operators is again a bounded operator. We call this space $BL(X, Y)$, the *space of bounded linear operators* from the normed space X into the normed space Y.

Theorem 2.3.6. *The linear space $BL(X, Y)$ of bounded linear operators for normed spaces X, Y is a normed space with the norm (2.3.6). If Y is a Banach space (i.e. complete), then $BL(X, Y)$ is also a Banach space.*

Proof. It is easy to see that $BL(X, Y)$ is a normed space. Assume that (A_n) is a Cauchy sequence of operators $X \to Y$. Since

$$\|A_n\| \leqslant \|A_n - A_1\| + \|A_1\| \leqslant C, \quad n \in \mathbb{N}, \tag{2.3.12}$$

the sequence A_n is bounded. Then also $(A_n\varphi)$ is a Cauchy sequence in Y for each point $\varphi \in X$. Since Y is a Banach space, $(A_n\varphi)$ converges towards an element $\psi \in Y$. We define the operator A via

$$A\varphi := \psi = \lim_{m\to\infty} A_m\varphi. \tag{2.3.13}$$

Then it is easy to see that A is linear. From

$$\|\psi\| = \lim_{m\to\infty} \|A_m\varphi\| \leqslant C\|\varphi\| \tag{2.3.14}$$

we obtain that A is a bounded operator. We have

$$\begin{aligned}\|A - A_n\| &= \sup_{\|\varphi\|\leqslant 1} \|A_n\varphi - A\varphi\| \\ &\leqslant \sup_{\|\varphi\|\leqslant 1} \limsup_{m\to\infty} \|A_n\varphi - A_m\varphi\| \\ &\leqslant \limsup_{m\to\infty} \|A_n - A_m\| \\ &\leqslant \epsilon \end{aligned} \tag{2.3.15}$$

for all n sufficiently large. Thus A is the operator limit of the sequence (A_n) in $BL(X, Y)$. □

Take note of the two different concepts of convergence for operators. First, there is *pointwise convergence*, when

$$A_n\varphi \to A\varphi, \quad n \to \infty \tag{2.3.16}$$

for each fixed point $\varphi \in X$. Second, there is *norm convergence* if

$$\|A_n - A\| \to 0, \quad n \to \infty. \tag{2.3.17}$$

Example 2.3.7. *A standard example for operators which are pointwise convergent, but not norm convergent, are* interpolation operators. *For $a < b \in \mathbb{R}$ consider the grid points*

$$x_k^{(N)} := a + \frac{b-a}{N}k, \quad k = 0, \ldots, N \tag{2.3.18}$$

and piecewise linear functions $p_k^{(N)}$ in $C([a, b])$ with

$$p_k^{(N)}\left(x_j^{(N)}\right) = \begin{cases} 1, & k = j \\ 0, & \text{otherwise.} \end{cases} \tag{2.3.19}$$

Such a set of functions is called a Lagrange basis *for the interpolation problem to find a piecewise linear function on $C([a, b])$ which is equal to a given function φ in $x_j^{(N)}$, $j = 1, \ldots, N$. A solution to this interpolation problem is given by the interpolation operator*

$$(P_N\varphi)(x) := \sum_{k=1}^{N} p_k^{(N)}(x)\varphi(x_k), \quad x \in [a, b]. \tag{2.3.20}$$

The infinity norm of P_N can be estimated by

$$\begin{aligned} |(P_N\varphi)(x)| &\leqslant \|\varphi\|_\infty \left| \sum_{k=0}^{N} p_k^{(N)}(x) \right| \\ &= \|\varphi\|_\infty \cdot 1. \end{aligned} \tag{2.3.21}$$

For every fixed element $\varphi \in C([a, b])$ we have the convergence

$$\|P_N\varphi - \varphi\|_\infty = \sup_{x\in[a,b]} \left| \sum_{k=1}^{N} p_k^{(N)}(x)\varphi(x_k) - \varphi(x) \right| \to 0 \tag{2.3.22}$$

for $N \to \infty$ due to the continuity of the function φ on the closed interval $[a, b]$. Thus the interpolation operator converges pointwise towards the identity operator. However, the norm difference between P_N and I satisfies

$$\|P_N - I\|_\infty = \sup_{x\in[a,b],\, \|\varphi\|_\infty \leqslant 1} \left| \sum_{k=1}^{N} p_k^{(N)}(x)\varphi(x_k) - \varphi(x) \right| = 2 \tag{2.3.23}$$

for all $N \in \mathbb{N}$. Thus P_N does not converge towards I.

Example 2.3.8. *As a second example consider integral operators on the real axis*

$$(A_n\varphi)(x) := \int_{\mathbb{R}} \underbrace{\frac{1}{1+|x|^2} \cdot \frac{1}{1+|n-y|^3}}_{=:k_n(x,y)} \varphi(y)\,\mathrm{d}y, \quad x \in \mathbb{R} \tag{2.3.24}$$

which we study as an operator from the set $L^2(\mathbb{R})$ into itself. The kernel is visualized via MATLAB *or Scilab in figure 2.1.*

Figure 2.1. The figure shows the kernel of (2.3.24) in dependence on y for $x = 0$ for $n = 2, 5, 8$. The kernel moves along the $\mathbb{R}$-axis.

Let us investigate the operator sequence for a fixed L^2-function φ in $\mathbb{R}$. We remark that given $\epsilon > 0$ we can find $A > 0$ such that

$$\int_{|x|\geqslant a} |\varphi(y)|^2 \, \mathrm{d}y \leqslant \epsilon \tag{2.3.25}$$

for all $a > A$. Further, for fixed b and given ϵ we can find N such that

$$|k_n(x, \ y)| \leqslant \epsilon, \qquad |x|, |y| \leqslant b, \quad n \geqslant N. \tag{2.3.26}$$

Clearly the kernel $k(x, y)$ is bounded by 1 for all $x, y \in \mathbb{R}$.

We estimate using the Cauchy–Schwarz inequality in $L^2(\mathbb{R})$

$$\begin{aligned} \int_{\mathbb{R}} \left| \int_{\mathbb{R}} k_n(x, \ y)\varphi(y)\, \mathrm{d}y \right|^2 \mathrm{d}x &\leqslant \int_{\mathbb{R}} \left(\int_{\mathbb{R}} |k_n(x, \ y)|^2 \, \mathrm{d}y \int_{\mathbb{R}} |\varphi(y)|^2 \, \mathrm{d}y \right) \mathrm{d}x \\ &= \|\varphi\|_2^2 \cdot \int_{\mathbb{R}} \int_{\mathbb{R}} |k_n(x, \ y)|^2 \, \mathrm{d}y \, \mathrm{d}x. \end{aligned} \tag{2.3.27}$$

Since the kernel $k_n(x, y)$ is square integrable over $\mathbb{R} \times \mathbb{R}$, this shows the boundedness of the operators A_n from $L^2(\mathbb{R})$ into $L^2(\mathbb{R})$. Further, it can be seen that the sequence is not norm convergent, but it converges pointwise towards zero for functions in $L^2(\mathbb{R})$.

Bounded linear operators have the following extremely useful property.

Theorem 2.3.9. *A linear operator is continuous if and only if it is bounded.*

Proof. First, assume that an operator $A : X \to Y$ is bounded and consider a sequence (φ_n) in X with $\varphi_n \to 0$, $n \to \infty$. Then from $\|A\varphi_n\| \leqslant C\|\varphi_n\|$ it follows that $A\varphi_n \to 0$, $n \to \infty$. Thus A is continuous in 0 and therefore continuous everywhere according to theorem 2.3.1.

Now let us assume that A is continuous, but not bounded. This means there is not a constant C with $\|A\varphi\| \leqslant C\|\varphi\|$ for all $\varphi \in X$. Thus for every constant $C = n$ we can find φ_n such that the estimate is violated, i.e. such that $\|A\varphi_n\| > n\|\varphi_n\|$. We define $\psi_n := \varphi_n / \|A\varphi_n\|$. Then by construction we have $\|\psi_n\| \to 0$, $n \to \infty$. Since A is continuous we have $A\psi_n \to 0$. But this is a contradiction to $\|A\psi_n\| = 1$ for all $n \in \mathbb{N}$. Thus A must be bounded. □

We close the subsection with some remarks about the *space of square summable sequences*

$$\ell^2 := \left\{ a = (a_j)_{j\in\mathbb{N}} : \text{each } a_j \in \mathbb{R}, \ \sum_{j=1}^{\infty} |a_j|^2 < \infty \right\}. \tag{2.3.28}$$

Equipped with the scalar product

$$\langle a, b\rangle := \sum_{j=1}^{\infty} a_j \overline{b_j} \tag{2.3.29}$$

it is a Hilbert space. The canonical Fourier basis of ℓ^2 is given by the elements $e^{(j)} \in \ell^2$ defined by

$$e_k^{(j)} := \begin{cases} 1 & j = k \\ 0 & \text{otherwise}. \end{cases} \tag{2.3.30}$$

In ℓ^2 the Cauchy–Schwarz inequality

$$\left| \sum_{j=1}^{\infty} a_j \overline{b_j} \right| \leqslant \left(\sum_{j=1}^{\infty} |a_j|^2 \right)^{\frac{1}{2}} \left(\sum_{j=1}^{\infty} |b_j|^2 \right)^{\frac{1}{2}} \tag{2.3.31}$$

is proven analogously to theorem 2.2.2.

Consider a linear operator $L : \ell^2 \to \mathbb{C}$. We define $l = (l_j)_{j \in \mathbb{N}}$ with $l_j := L(e^{(j)}) \in \mathbb{C}$. Then we obtain

$$L(a) = L\left(\sum_{j=1}^{\infty} a_j e^{(j)} \right) = \sum_{j=1}^{\infty} a_j L(e^{(j)}) = \sum_{j=1}^{\infty} l_j a_j = \langle a, \bar{l}\rangle. \tag{2.3.32}$$

On the other hand, every sequence $l = (l_j)_{j \in \mathbb{N}}$ with $l_j \in \mathbb{R}$ defines a linear operator L on ℓ^2 by

$$L(a) := \sum_{j=1}^{\infty} l_j a_j = \langle a, \bar{l}\rangle. \tag{2.3.33}$$

Theorem 2.3.10. *A linear operator L on ℓ^2 with corresponding sequence $l = (l_j)_{j \in \mathbb{N}}$ is bounded if and only if $l \in \ell^2$.*

Proof. If $l = (l_j)_j \in \mathbb{N}$ is bounded in ℓ^2, then by (2.3.31) applied to (2.3.32) the linear operator L is bounded. On the other side, if L is bounded, then there is a constant C such that $|L(a)| < C$ for all $a \in \ell^2$ with $\|a\|_{\ell^2} < 1$. Assume that $l \notin \ell^2$, i.e.

$$\sum_{j=1}^{\infty} |l_j|^2 = \infty. \tag{2.3.34}$$

We define

$$\rho_n := \sum_{j=1}^{n} |l_j|^2 \tag{2.3.35}$$

and set

$$a^{(n)} = \left(a_j^{(n)} \right)_{j \in \mathbb{N}} \text{ with } a_j^{(n)} := \frac{1}{\sqrt{\rho_n}} \cdot \begin{cases} l_j, & j \leqslant n \\ 0, & \text{otherwise}. \end{cases} \tag{2.3.36}$$

Then,

$$\|a^{(n)}\|^2 = \frac{1}{\rho_n}\sum_{j=1}^{n}\left|l_j\right|^2 = 1 \tag{2.3.37}$$

and further

$$L(a^{(n)}) = \sum_{j=1}^{\infty}\overline{l}_j\, a_j^{(n)} = \sum_{j=1}^{n}\overline{l}_j l_j = \rho_n \to \infty \tag{2.3.38}$$

for $n \to \infty$, in contradiction to $|L(a^{(n)})| \leqslant C$. This means that for bounded L the corresponding sequence l needs to be in ℓ^2 and the proof is complete. □

2.3.2 The solution of equations of the second kind and the Neumann series

In the introduction we discussed integral equations of the form

$$(I - A)\varphi = f.$$

Integral equations of this form are called *integral equations of the second kind.* We will develop two different solution approaches to such integral equations, the first for a bounded linear operator A for which the norm is sufficiently small. The second approach consists of Riesz theory for the case when A is compact, which will be defined in the next subsection.

First, recall the meaning of convergence of a series in a normed space as described in (2.2.36). Clearly, this also applies to operator series, i.e. to sums

$$\sum_{k=1}^{\infty} A_k \tag{2.3.39}$$

where $A_k : X \to Y$, $k \in \mathbb{N}$, are bounded linear operators. Here, we will restrict our attention to the case where $A : X \to X$ with a Banach space X, i.e. the operator maps a Banach into itself. In this case we can define *powers* of the operator A recursively via

$$A^k := A(A^{k-1}), \quad k = 1, 2, 3, \ldots \quad A^0 = I. \tag{2.3.40}$$

A simple example is given by the multiple application of a square matrix in $\mathbb{R}^{n\times n}$.

Example 2.3.11. *Consider an integral operator*

$$(A\varphi)(x) := \int_a^b k(x, y)\varphi(y)\,\mathrm{d}y, \quad x \in [a, b] \tag{2.3.41}$$

with continuous kernel $k : [a, b] \times [a, b] \to \mathbb{R}$. *Such an integral operator is a bounded linear operator in* $C([a, b])$.

In analogy to the geometric series we proceed as follows. First we observe that if $q := \|A\| < 1$, then

$$\|A^k\| \leqslant \|A\|^k = q^k \to 0, \quad k \to \infty. \tag{2.3.42}$$

We define the partial sum

$$S_n := \sum_{k=0}^{n} A^k. \tag{2.3.43}$$

For $n > m$, $m, n \in \mathbb{N}$ we estimate

$$\|S_n - S_m\| = \left\| \sum_{k=m+1}^{n} A^k \right\| \leqslant \sum_{k=m+1}^{n} q^k. \tag{2.3.44}$$

We use

$$(1 - q)\sum_{k=0}^{n} q^k = 1 - q^{n+1} \tag{2.3.45}$$

to calculate

$$\sum_{k=0}^{n} q^k = \frac{1 - q^{n+1}}{1 - q} \tag{2.3.46}$$

and

$$\sum_{k=m+1}^{n} q^k = \frac{q^{m+2} - q^{n+1}}{1 - q}. \tag{2.3.47}$$

For $q < 1$ we have $q^l \to 0$, for $l \to \infty$, thus S_n is a Cauchy sequence in $L(X, X)$. Since $L(X, X)$ is a Banach space, we obtain convergence of (S_n) towards some element $S \in L(X, X)$, i.e. a mapping from X into X. Imitating (2.3.45) for the operators A we obtain

$$(I - A)S_n = (I - A)\sum_{k=0}^{n} A^k = 1 - A^{n+1}. \tag{2.3.48}$$

By norm estimates we have $A^{n+1} \to 0$, $n \to \infty$, thus S_n converges towards the inverse $(I - A)^{-1}$ of $I - A$. We collect all results and estimates in the following theorem.

Theorem 2.3.12. *Consider a bounded linear operator $A : X \to X$ on a Banach space X. If the norm of A satisfies $\|A\| < 1$, then the operator $I - A$ has a bounded inverse on X which is given by the* Neumann series

$$(I - A)^{-1} = \sum_{k=0}^{\infty} A^k. \tag{2.3.49}$$

It satisfies the estimates

$$\|(I - A)^{-1}\| = \frac{1}{1 - \|A\|}. \tag{2.3.50}$$

In this case for each $f \in X$ the integral equation of the second kind

$$(I - A)\varphi = f \tag{2.3.51}$$

does have a unique solution $\varphi \in X$ which in the norm depends continuously on the right-hand side f.

The Neumann series can be rewritten in a nice form which is well known in numerical mathematics. We transform the series solution

$$\varphi_n := \sum_{k=0}^{n} A^k f \tag{2.3.52}$$

into the form

$$\varphi_n = \sum_{k=1}^{n} A^k f + f = A \sum_{k=1}^{n} A^{k-1} f + f = A\varphi_{n-1} + f. \tag{2.3.53}$$

This leads to:

Theorem 2.3.13. *Consider a bounded linear operator $A : X \to X$ on a Banach space X. If the norm of A satisfies $\|A\| < 1$, then the sequence φ_n of* successive approximations

$$\varphi_{n+1} := A\varphi_n + f, \quad n = 0, 1, 2, \ldots \tag{2.3.54}$$

with starting value $\varphi_0 = 0$ converges to the unique solution φ of $(I - A)\varphi = f$.

The Neumann series and successive approximations are very beautiful tools which are used often in all areas of mathematics. However, they have drawbacks. The theory applied only for $\|A\| < 1$. This condition, however, is violated often for the integral equations which arise from applications. This was a basic and important problem for mathematics until around 1900, when Fredholm, Riesz and Hilbert made significant progress in the treatment of such integral equations. We will study these and for this purpose next investigate *compact* operators.

2.3.3 Compact operators and integral operators

We first start with a general definition of compact operators.

Definition 2.3.14. *Let $A : X \to Y$ be a linear operator between normed spaces X and Y. Then A is called* compact *if it maps each bounded set in X into a relatively compact set in Y.*

This is a topological definition. Next, we provide an equivalent definition which uses sequences and which will be used again and again in the following arguments. Note that the topological definition has more potential to pass into weak topologies and deeper investigations, but for the beginning and in the framework of normed spaces using sequences is often easier.

Theorem 2.3.15. *A linear operator $A : X \to Y$ between normed spaces X and Y is compact if and only if for each bounded sequence $(\varphi_n) \subset X$ the image sequence $(A\varphi_n)$ contains a convergent subsequence in Y.*

Remark. Note that the image sequence $(A\varphi_n)$ is converging in Y, i.e. the limit element or limit point is an element of Y. If the limit is in a larger space, the completion of Y for example, then the operator is not compact.

Proof. One direction of the equivalence is very quick. If A is compact and (φ_n) is a bounded sequence, then $V := \{A\varphi_n : n \in \mathbb{N}\}$ is relatively compact, i.e. by the series arguments for compact sets its closure $\bar{V}$ is compact. Thus there is a convergent subsequence which converges towards an element of $\bar{V} \subset Y$.

The other direction is slightly more involved. Assume that there is a bounded set $U \subset X$ such that $V = \{A\varphi : \varphi \in U\}$ is not relatively compact, i.e. $\bar{V}$ is not compact. Then there is a sequence in $\bar{V}$ which does not have a convergent subsequence and by approximation arguments there is a sequence (ψ_n) in V which does not have a convergent subsequence. Thus there is a sequence (φ_n) with $\psi_n = A\varphi_n$ in the bounded set U for which $A\varphi_n$ does not have a convergent subsequence. This means that A is not sequentially compact. As a consequence sequentially compact operators must be compact and the proof is complete. □

We next collect basic properties of compact operators.

Theorem 2.3.16. *For compact linear operators we have that:*

(a) *Compact linear operators are bounded.*

(b) *If A, B are compact linear operators and $\alpha, \beta \in \mathbb{R}$ or $\mathbb{C}$, then $\alpha A + \beta B$ is a compact linear operator.*

(c) *Let X, Y, Z be normed spaces, $A : X \to Y$ and $B : Y \to Z$ bounded linear operators. If either A or B is compact, then the product $B \circ A : X \to Z$ is compact.*

Proof.

(a) Relatively compact sets are bounded, thus a compact operator maps the bounded set $B[0, 1]$ into a set bounded by a constant C, which yields $\|A\varphi\| \leqslant C$ for all $\|\varphi\| \leqslant 1$, therefore A is bounded.

(b) This can be easily seen by the definition of compact operator.

(c) Let us consider the case where A is bounded and B is compact. Consider a bounded sequence (φ_n) in X. Then the sequence (ψ_n) with $\psi_n := A\varphi_n$ is bounded in Y. Compactness of B yields the existence of a convergent subsequence of $(B\psi_n)$ in Z. Thus, for every bounded sequence (φ_n) in X the sequence $\chi_n := B(A(\varphi))$ in Z has a convergent subsequence and thus $B \circ A$ is a compact operator.

The case where A is compact and B is bounded works analogously. □

So far we have studied compact operators with series arguments. We now come to a quite important and far reaching result, which will help us to better understand the role of compact operators. They are basically the closure of the set of finite-dimensional operators.

Theorem 2.3.17. *Let X be a normed space and Y a Banach space. If a sequence $A_n : X \to Y$ of compact linear operators is norm convergent towards an operator $A : X \to Y$, then A is compact.*

Remark. Recall that norm convergence means $\|A_n - A\| \to 0$ for $n \to \infty$. Pointwise convergence is not sufficient to obtain this result!

Proof. The proof uses some important standard tools. Let (φ_k) be a bounded sequence in X with $\|\varphi_k\| \leqslant C$ for $k \in \mathbb{N}$. We want to show that $(A\varphi_k)$ has a convergent subsequence. To this end we use the standard *diagonalization procedure*. First, we choose a subsequence $(\varphi_{k(l)})$ of (φ_k) such that $(A_1\varphi_{k(1,l)})$ converges. From this subsequence we choose a subsequence $(\varphi_{k(2,l)})$ again such that $(A_2\varphi_{k(l)})$ converges. We repeat this procedure for all $j = 3, 4, 5, \ldots$. We obtain a grid of sequences

$$\begin{pmatrix} \varphi_{k(1,1)} & \varphi_{k(1,2)} & \varphi_{k(1,3)} & \varphi_{k(1,4)} & \cdots \\ \varphi_{k(2,1)} & \varphi_{k(2,2)} & \varphi_{k(2,3)} & \varphi_{k(2,4)} & \cdots \\ \varphi_{k(3,1)} & \varphi_{k(3,2)} & \varphi_{k(3,3)} & \varphi_{k(3,4)} & \cdots \\ \vdots & \vdots & \vdots & \vdots & \vdots \end{pmatrix} \tag{2.3.55}$$

where each row is a subsequence of the upper rows and for every $n \geqslant j$ the sequence $(A_j\varphi_{k(n,l)})$ arising from an application of A_j to the sequence in row n converges. We now choose the diagonal sequence

$$\psi_j := \varphi_{k(j,j)}, \quad j = 1, 2, 3, \ldots \tag{2.3.56}$$

Since $(\psi_j)_{j \geqslant n}$ is a subsequence of row n we know that for arbitrary fixed $n \in \mathbb{N}$ the sequence $(A_n\psi_j)_{j\in\mathbb{N}}$ converges for $j \to \infty$.

Next, we show that $(A\psi_j)$ is a Cauchy sequence. To this end choose $\epsilon > 0$. First, we find N_1 such that $\|A - A_n\| \leqslant \epsilon/3C$ for $n \geqslant N_1$. Second, since $(A_{N_1}\psi_j)$ converges there is $N_2 \in \mathbb{N}$ such that

$$\left\|A_{N_1}\psi_j - A_{N_1}\psi_l\right\| \leqslant \frac{\epsilon}{3}, \quad j, l \geqslant N_2. \tag{2.3.57}$$

Now, we estimate via the triangle inequality

$$\begin{aligned} \left\|A\psi_j - A\psi_l\right\| &\leqslant \left\|A\psi_j - A_{N_1}\psi_j\right\| + \left\|A_{N_1}\psi_j - A_{N_1}\psi_l\right\| + \left\|A_{N_1}\psi_l - A\psi_l\right\| \\ &\leqslant \frac{\epsilon}{3C} \cdot C + \frac{\epsilon}{3} + \frac{\epsilon}{3C} \cdot C \\ &= \epsilon \end{aligned} \tag{2.3.58}$$

for all $j, l \geqslant N_2 = N_2(\epsilon)$. Hence $(A\psi_j)$ is a Cauchy sequence and thus convergent in the Banach space Y. This shows that A is compact.

To complete our claim from above we need to study finite-dimensional operators.

Theorem 2.3.18. *A bounded operator $A : X \to Y$ with finite-dimensional range $A(X)$ is compact.*

Proof. Consider a bounded set $U \in X$. It is mapped into a bounded set $A(U)$ in the finite-dimensional range $A(X)$, which is norm isomorphic to $\mathbb{R}^m$. But according to the Bolzano–Weierstrass theorem 2.1.23 in the form of theorem 2.1.25 this set $A(U)$ is relatively compact and thus A is compact. □

We want to complete this subsection with a result about the identity operator, which is important for many linear inverse problems. However, we need a well-known lemma as preparation.

Lemma 2.3.19 (Riesz). *Let X be a normed space and $U \underset{\neq}{\subset} X$ a closed subspace. Then for any $\alpha \in (0, 1)$ there exists an element $\psi \in X$ such that*

$$\|\psi\| = 1, \quad \|\psi - \varphi\| \geqslant \alpha, \quad \varphi \in U. \tag{2.3.59}$$

Proof. There is an element $f \in X$ with $f \notin U$. Since U is closed we know that

$$\beta := \inf_{\varphi \in U} \|f - \varphi\| > 0. \tag{2.3.60}$$

Choose an element $g \in U$ which is close to the minimal distance, in particular with

$$\beta \leqslant \|f - g\| \leqslant \frac{\beta}{\alpha}. \tag{2.3.61}$$

Then define a unit vector ψ by

$$\psi := \frac{f - g}{\|f - g\|}. \tag{2.3.62}$$

For all $\varphi \in U$ we have the estimate

$$\begin{aligned} \|\psi - \varphi\| &= \frac{1}{\|f - g\|} \left\| f - \underbrace{g - \|f - g\|\varphi}_{\in U} \right\| \\ &\geqslant \frac{\beta}{\|f - g\|} \geqslant \alpha \end{aligned} \tag{2.3.63}$$

which completes the proof. □

Theorem 2.3.20. *The identity operator $I : X \to X$ is compact if and only if X has finite dimension.*

Proof. If X has finite dimension, then clearly I is compact, since every bounded set is mapped onto itself and it is relatively compact according to theorem 2.1.25. Next, consider the case where X is not finite-dimensional. We will construct a bounded sequence which does not have a convergent subsequence, which then shows that the identity operator cannot be compact. We start with an arbitrary $\varphi_1 \in X$ with $\|\varphi_1\| = 1$. Then $U_1 := \text{span}\{\varphi_1\}$ is a closed linear subspace. According to the Riesz lemma 2.3.19 there is an element $\varphi_2 \in X$ with $\|\varphi_2\| = 1$ and

$$\|\varphi_2 - \varphi_1\| \geqslant \frac{1}{2}. \tag{2.3.64}$$

Now, we define $U_2 := \text{span}\{\varphi_1, \varphi_2\}$. Again, we find $\varphi_3 \in X$ with $\|\varphi_3\| = 1$ and

$$\|\varphi_3 - \varphi_k\| \geqslant \frac{1}{2}, \quad k = 1, 2. \tag{2.3.65}$$

We proceed in the same way to obtain a sequence (φ_n) for which

$$\|\varphi_n\| = 1 \ (n \in \mathbb{N}), \quad \|\varphi_n - \varphi_m\| \geqslant \frac{1}{2}, \quad n \neq m. \tag{2.3.66}$$

This sequence cannot have a convergent subsequence, and the proof is complete. □

We finish this subsection with examples.

Example 2.3.21. *The integral operator (2.3.41) with a continuous kernel is a compact operator on $C([a, b])$. This can be obtained via theorems 2.3.17 and 2.3.18 as follows. First, we construct a polynomial approximation*

$$p_n(x, y) := \sum_{j,k=0}^{n} a_{j,k} x^j y^k, \quad x, y \in [a, b] \tag{2.3.67}$$

to the kernel function $k(x, y)$. According to the Weierstrass approximation theorem given ϵ there is n such that

$$\sup_{x,y \in [a,b]} \left| k(x, y) - p_n(x, y) \right| \leqslant \epsilon \tag{2.3.68}$$

for n sufficiently large. We define the approximation operator

$$(A_n \varphi)(x) := \int_a^b p_n(x, y) \varphi(y) \, \mathrm{d}y, \quad x \in [a, b]. \tag{2.3.69}$$

The range of the operator A_n is the finite-dimensional polynomial space Π_n, since integration with respect to y leads to a polynomial in x of degree n. We have the norm estimate

$$\|A_n - A\|_\infty \leqslant \sup_{x \in [a,b]} \int_a^b \left| k(x, y) - p_n(x, y) \right| \mathrm{d}y, \tag{2.3.70}$$

which can be made arbitrarily small for sufficiently large $n \in \mathbb{N}$. Thus we have approximated the operator A in the norm by a finite-dimensional and thus compact operator A_n. Hence A is compact.

A compact operator $A : X \to Y$ on an infinite-dimensional space X cannot have a bounded inverse, since otherwise $A^{-1} \circ A = I$ would be compact, which is not the case for X of infinite dimension. This is an important conclusion. Together with the previous example it shows that the integral equation of the first kind

$$\int_a^b k(x, y)\varphi(y)\,\mathrm{d}y = f(x), \quad x \in [a, b] \tag{2.3.71}$$

with continuous kernel $k(x, y)$ cannot be boundedly invertible! Here we discover the phenomenon of *ill-posedness*, which we will study in detail later, see section 3.1.1.

2.3.4 The solution of equations of the second kind and Riesz theory

We now study the solution of the integral equation

$$(I - A)\varphi = f \tag{2.3.72}$$

of the second kind with a compact operator A on a normed space. For more than 150 years there was no solution theory for this type of equation, which naturally appear in many applications arising from fluid dynamics, acoustic or electromagnetic waves, potential theory and, since 1920, in quantum mechanics. For most equations from nature the Neumann series results were not applicable. The important breakthrough was made by Fredholm in 1900. Here, we will use the related results of Riesz, which are valid in general normed spaces.

We will survey the three famous theorems of Riesz and an important conclusion. We will not present all proofs here but instead refer to the literature [1]. We define

$$L := I - A \tag{2.3.73}$$

where as usual I denotes the identity operator. Recall the *nullspace*

$$N(L) = \{\varphi \in X : L\varphi = 0\} \tag{2.3.74}$$

and its range

$$L(X) = \{L\varphi : \varphi \in X\}. \tag{2.3.75}$$

Theorem 2.3.22 (First Riesz theorem). *The nullspace of the operator L is a finite-dimensional subspace of X.*

Theorem 2.3.23 (Second Riesz theorem). *The range L(X) of the operator L is a closed linear subspace.*

Theorem 2.3.24 (Third Riesz theorem). *There exists a uniquely determined integer $r \in \mathbb{N}_0$, such that*

$$\begin{aligned} &\text{(i) } \{0\} = N(L^0) \not\subset N(L^1) \not\subset \dots \not\subset N(L^r) = N(L^{r+1}) = \dots \\ &\text{(ii) } X = L^0(X) \not\subset L^1(X) \not\subset \dots \not\subset L^r(X) = L^{r+1}(X) = \dots\dots \end{aligned} \tag{2.3.76}$$

The integer r is called the Riesz number *of the operator A and it is important to know that either (i) or (ii) can determine r. Furthermore, we have*

$$X = N(L^r) \oplus L^r(X). \tag{2.3.77}$$

The Riesz theory has an important application to the solution of (2.3.72).

Theorem 2.3.25 (Riesz). *Let X be a normed space and $A : X \to X$ be a compact linear operator. If $I - A$ is injective, then the inverse $(I - A)^{-1} : X \to X$ exists and is bounded.*

Proof. Let $L = I - A$. By the assumption the operator satisfies $N(L) = \{0\}$. Thus the Riesz number is $r = 0$. From the third Riesz theorem we conclude that $L(X) = X$, i.e. the operator is surjective. The inverse operator $(I - A)^{-1}$ exists. To show that L^{-1} is bounded we assume that this is not the case. Then there is a sequence (f_n) with $\|f_n\| = 1$ such that $(L^{-1}f_n)$ is unbounded. We define

$$\varphi_n := L^{-1}f_n, \qquad g_n := \frac{f_n}{\|\varphi_n\|}, \qquad \psi_n := \frac{\varphi_n}{\|\varphi_n\|} \tag{2.3.78}$$

for $n \in \mathbb{N}$. We conclude that $g_n \to 0$, $n \to \infty$ and $\|\psi_n\| = 1$. A is compact, hence we can choose a subsequence $(\psi_{n(k)})$ such that $(A\psi_{n(k)})$ converges towards $\psi \in X$. By construction we have

$$\psi_{n(k)} - A\psi_{n(k)} = g_{n(k)}, \quad k \in \mathbb{N}, \tag{2.3.79}$$

thus we obtain $\psi_{n(k)} \to \psi$, $k \to \infty$. However, then $\psi \in N(L)$, i.e. $\psi = 0$. But this contradicts $\|\psi_n\| = 1$. Thus L^{-1} must be bounded. □

With the Riesz theory we are able to solve integral equations which arise in many practical problems in the theory of waves. Acoustics, electromagnetics and elasticity lead to a variety of integral equations of the form (2.3.72).

2.4 Adjoint operators, eigenvalues and singular values

The concept of adjoint operators and adjoint problems is of far reaching importance for direct and inverse problems. Studying adjoint relationships between inversion algorithms will be an important part of chapters 12 and 15. The *tangent linear adjoint* solution is the key ingredient for state-of-the-art data assimilation algorithms, see section 5.3.1.

2.4.1 Riesz representation theorem and adjoint operators

In this first subsection we consider a linear bounded operator $A : X \to Y$ from a Hilbert space X into a Hilbert space Y. Recall that bounded linear functionals F are linear mappings $X \to \mathbb{R}$ or $X \to \mathbb{C}$

$$F(\alpha\varphi + \beta\eta) = \alpha F(\varphi) + \beta F(\eta) \tag{2.4.1}$$

for $\varphi, \eta \in X, \alpha, \beta \in \mathbb{R}$ or $\mathbb{C}$ with

$$\sup_{\|\varphi\| \leqslant 1} |F(\varphi)| < \infty. \tag{2.4.2}$$

Consider a bounded linear mapping $G : \mathbb{R} \to \mathbb{R}$. What do these mappings look like? We have $G(\alpha) = \alpha G(1) = G(1) \cdot \alpha$ for all $\alpha \in \mathbb{R}$. This means that we can write the mapping as a multiplication with the real number $G(1)$. In two dimensions $G : \mathbb{R}^2 \to \mathbb{R}$ we obtain

$$\begin{aligned} G(x) &= G(x_1 e_1 + x_2 e_2) = G(e_1)x_1 + G(e_2)x_2 \\ &= \begin{pmatrix} G(e_1) \\ G(e_2) \end{pmatrix} \cdot \begin{pmatrix} x_1 \\ x_2 \end{pmatrix} = g \cdot x \end{aligned} \tag{2.4.3}$$

with a vector $g \in \mathbb{R}^2$ for every $x = (x_1, x_2) \in \mathbb{R}^2$. This result can be obtained in a very general form in a Hilbert space.

Theorem 2.4.1 (Riesz representation). *Consider a Hilbert space X. Then for each bounded linear function $F : X \to \mathbb{C}$ there is a uniquely determined element $f \in X$ such that*

$$F(\varphi) = \langle \varphi, f \rangle, \quad \varphi \in X, \quad \|F\| = \|f\|. \tag{2.4.4}$$

Proof. To show the uniqueness assume that there are two elements $f_1, f_2 \in X$ with $\langle \varphi, f_1 \rangle = \langle \varphi, f_2 \rangle$ for all $\varphi \in X$. This yields $\langle \varphi, f_1 - f_2 \rangle = 0$ for all $\varphi \in X$, in particular for $\varphi = f_1 - f_2$, and thus $\|f_1 - f_2\| = \langle f_1 - f_2, f_1 - f_2 \rangle = 0$, from which we obtain $f_1 = f_2$.

To show the existence of such a function we apply the best approximation theorems from section 2.2.2. If $F = 0$, i.e. $F(\varphi) = 0$ for all $\varphi \in X$, then $f = 0$ is the right element. If $F \neq 0$ there is at least one $w \in X$ for which $F(w) \neq 0$. Further, we remark that the nullspace

$$N(F) = \{\varphi \in X : F(\varphi) = 0\} \tag{2.4.5}$$

is a closed subspace of X by the continuity of F. Thus according to theorems 2.2.10 and 2.2.11 there is a best approximation v to w in $N(F)$, which satisfies

$$w - v \perp N(F) \Leftrightarrow \langle w - v, \psi \rangle = 0, \quad \psi \in N(F). \tag{2.4.6}$$

Now with $g := w - v$ we observe that $\psi := F(g)\varphi - F(\varphi)g \in N(F)$ for arbitrary elements $\varphi \in X$, thus

$$\langle F(g)\varphi - F(\varphi)g, g \rangle = \overline{\langle g, F(g)\varphi - F(\varphi)g \rangle} = \overline{\langle w - v, \psi \rangle} = 0. \tag{2.4.7}$$

Now using the linearity of the scalar product in the first and the anti-linearity in the second component this yields

$$F(\varphi) = \left\langle \varphi, \frac{\overline{F(g)}g}{\|g\|^2} \right\rangle, \quad \varphi \in X, \tag{2.4.8}$$

such that $f := \overline{F(g)}g/\|g\|^2$ is the function with the desired properties. □

Combining this with theorem 2.2.11 we immediately have the following.

Corollary 2.4.2. *Let M be a subspace in a Hilbert space X. Then M is dense in X if and only if we have the following condition for any bounded linear function $F : X \to \mathbb{C}$:*

$$F(\varphi) = 0, \quad \varphi \in M \Longrightarrow F = 0. \tag{2.4.9}$$

We are now able to define adjoint operators. For matrices in $\mathbb{R}^m$ you know the transpose A^T of a matrix A already, defined as the matrix where the first row becomes the first column, the second row the second column and so forth. For vectors $x, y \in \mathbb{R}^m$ and a matrix $A \in \mathbb{R}^{n \times n}$ we have

$$x \cdot Ay = \sum_{j=1}^{n} x_j \left(\sum_{k=1}^{n} a_{jk} y_k \right) = \sum_{k=1}^{n} \left(\sum_{j=1}^{n} a_{jk} x_j \right) y_k = (A^T x) \cdot y. \tag{2.4.10}$$

This can also be extended into a Hilbert space setting. To make the distinction between the scalar products in the Hilbert spaces X and Y we sometimes use the notation $\langle \cdot, \cdot \rangle_X$ and $\langle \cdot, \cdot \rangle_Y$. Let $A \in BL(X, Y)$. Then, for each fixed element $\psi \in Y$ the mapping

$$\varphi \mapsto \langle A\varphi, \psi \rangle_Y \in \mathbb{C} \tag{2.4.11}$$

defines a bounded linear functional on X. Thus by the Riesz theorem 2.4.1 there is $f \in X$ such that $\langle A\varphi, \psi \rangle = \langle \varphi, f \rangle$ for all $\varphi \in X$. This defines a mapping $A^*\psi := f$ from Y into X. This operator is uniquely determined. We have prepared the following definition.

Definition 2.4.3. *For an operator $A \in BL(X, Y)$ between two Hilbert spaces X and Y the* adjoint *operator is uniquely defined via*

$$\langle A\varphi, \psi \rangle_Y = \langle \varphi, A^*\psi \rangle_X, \quad \varphi \in X, \quad \psi \in Y. \tag{2.4.12}$$

Adjoint operators are very important for the understanding of a variety of phenomena in the natural sciences. First, we remark that

$$(A^*)^* = A, \tag{2.4.13}$$

which is a consequence of

$$\langle \psi, (A^*)^* \varphi \rangle_Y = \langle A^*\psi, \varphi \rangle_X = \overline{\langle \varphi, A^*\psi \rangle_X} = \overline{\langle A\varphi, \psi \rangle_Y} = \langle \psi, A\varphi \rangle_Y$$

for all $\varphi \in X$, $\psi \in Y$. Further, we calculate

$$\|A^*\psi\|^2 = \langle A^*\psi, A^*\psi \rangle = \langle AA^*\psi, \psi \rangle \leqslant \|A\| \|A^*\psi\| \|\psi\| \tag{2.4.14}$$

for all $\psi \in Y$ via the Cauchy–Schwarz inequality theorem 2.2.2. Division by $\|A^*\psi\|$ yields the boundedness of the adjoint operator A^* with $\|A^*\| \leqslant \|A\|$. Via (2.4.13) we can exchange the roles of A and A^*, i.e. we have $\|A\| \leqslant \|A^*\|$. We summarize the results in the following theorem.

Theorem 2.4.4. *The norm of the adjoint operator A^* is equal to the norm of its dual A, i.e. we have*

$$\|A^*\| = \|A\|. \tag{2.4.15}$$

Example 2.4.5. *As an example we calculate the adjoint of an integral operator*

$$(A\varphi)(x) := \int_a^b k(x, y)\varphi(y)\,\mathrm{d}s(y), \quad x \in [c, d], \tag{2.4.16}$$

with continuous kernel k in the Hilbert spaces $X = L^2([a, b])$ *and* $Y = L^2([c, d])$ *with* $a < b$ *and* $c < d$ *in* $\mathbb{R}$. *Using Fubini's theorem with functions* $\varphi \in L^2([a, b])$ *and* $\psi \in L^2([c, d])$ *we derive*

$$\begin{aligned}
&\langle A\varphi, \psi\rangle \\
&= \int_c^d \left(\int_a^b k(x, y)\varphi(y)\,\mathrm{d}s(y)\right)\overline{\psi(x)}\,\mathrm{d}s(x) \\
&= \int_a^b \varphi(y)\left(\int_c^d k(x, y)\overline{\psi(x)}\,\mathrm{d}s(x)\right)\mathrm{d}s(y) \\
&= \int_a^b \varphi(y)\overline{\left(\int_c^d \overline{k(x, y)}\psi(x)\,\mathrm{d}s(x)\right)}\mathrm{d}s(y) \\
&= \langle \varphi, A^*\psi\rangle
\end{aligned} \tag{2.4.17}$$

with

$$(A^*\psi)(y) := \int_c^d \overline{k(x, y)}\psi(x)\,\mathrm{d}s(x), \quad y \in [a, b]. \tag{2.4.18}$$

This operator satisfies the equation (2.4.12) and thus is the uniquely determined adjoint operator. Thus for calculation of the adjoint we need to exchange the role of the kernel variables x and y and need to take the complex conjugate of the kernel.

Finally, we state the following without proof.

Theorem 2.4.6. *If* $A : X \to Y$ *is a compact linear operator between Hilbert spaces X and Y, then the adjoint operator* $A^* : Y \to X$ *is also compact.*

For the special case of the integral operator (2.4.16) this is an immediate consequence of example 2.3.21. For the general case, the theorem can be easily shown by using $AA^* : Y \to Y$ is compact and $\|A^*y\|_X^2 = \langle AA^*y, y\rangle_Y$ for $y \in Y$.

2.4.2 Weak compactness of Hilbert spaces

We have seen in the previous subsection that the Riesz representation theorem enables us to define the adjoint A^* of an operator $A \in BL(X, Y)$ for Hilbert spaces X and Y. Another interesting by-product of this theorem is the weak compactness of Hilbert space which means that every bounded sequence in a Hilbert space X has a subsequence which converges weakly in X. We first give the definition of weak convergence in X.

Definition 2.4.7. *Let X be a Hilbert space with inner product* $\langle \cdot,\cdot \rangle$.

(i) *A sequence* (x_n) *in* X *is weakly convergent to* $x \in X$ *if for any* $y \in X$, $\langle x_n, y\rangle \to \langle x, y\rangle$ *as* $n \to \infty$. *This will be denoted by* $x_n \rightharpoonup x$, $n \to \infty$.

(ii) *A sequence* (x_n) *in* X *is called weakly bounded if for any* $y \in X$, *the sequence* $(\langle x_n, y\rangle)$ *is bounded in* $\mathbb{C}$.

Theorem 2.4.8. *A sequence* (x_n) *in a Hilbert space* X *is weakly bounded if and only if* (x_n) *is bounded in* X.

Proof. Since it is clear that the boundedness of a sequence in X implies the weak boundedness of the sequence in X, we will prove the converse of this. Let a sequence (x_n) be weakly bounded in X. Then, we claim that there exist $x \in X$, $r > 0$ and $K > 0$ such that

$$|\langle x_n, y\rangle| \leqslant K, \quad n \in N, \quad y \in B(x, r). \tag{2.4.19}$$

In fact if we assume that this is not true, then using the continuity of the inner product, there exist sequences (n_j) in $\mathbb{N}$, (y_j) in X and (r_j) in $\mathbb{R}$ such that

$$(n_j) \text{ is monotone increasing and for any } 2 \leqslant j \in \mathbb{N},$$

$$0 < r_j < 1/j, \quad B\left[y_{j+1}, r_{j+1}\right] \subset B\left(y_j, r_j\right), \quad \left|\langle x_{n_j}, y\rangle\right| > j \quad \text{for } y \in B\left(y_j, r_j\right).$$

Clearly $\|y_n - y_m\| < 1/m$ for $n > m$. Hence (y_n) is a Cauchy sequence in X. By the completeness of Hilbert space, there exists a unique limit $y \in X$ to this sequence. Since for any fixed $j \in \mathbb{N}$, $y_k \in B[y_j, r_j]$ for all $k > j$. By letting $k \to \infty$, y belongs to all $B[y_j, r_j]$. Hence, $|\langle x_{n_k}, y\rangle| > j$ for any $k > j$. This cannot happen because (x_n) is weakly bounded. Thus we have the claim.

By this claim, we have for any z ($\|z\| < 1$),

$$|\langle x_n, z\rangle| = \frac{1}{r}|\langle x_n, x + rz\rangle - \langle x_n, x\rangle| \leqslant \frac{2K}{r}. \tag{2.4.20}$$

Here we can even have this for any z ($\|z\| \leqslant 1$) by the continuity of the inner product. Hence, $\|x_n\| = \sup_{\|z\|\leqslant 1}|\langle x_n, z\rangle| \leqslant (2K)/r$. □

In order to prove the weak compactness of Hilbert spaces, we prepare a lemma which is very easy to prove.

Lemma 2.4.9. *Let X be a Hilbert space with inner product* $\langle \cdot,\cdot \rangle$ *and it has a dense subset D. Then, if* $x \in X$ *and a sequence* (x_n) *in* X *satisfies*

$$\langle x_n, y\rangle \to \langle x, y\rangle, \quad n \to \infty \text{ for all } y \in D, \tag{2.4.21}$$

then $x_n \rightharpoonup x$, $n \to \infty$, *i.e. it converges weakly in X.*

Now we state and prove the weak compactness of Hilbert spaces as follows.

Theorem 2.4.10 (Weak compactness of Hilbert spaces). *Let X be a Hilbert space. Then for any bounded sequence* (x_n) *in* X, *there exists a weakly convergent subsequence* $(x_{n(k)})$ *of* (x_n).

Proof. Let $D = \{y_n : n \in N\}$, $\langle \cdot,\cdot \rangle$ be the inner product of X and $\|x_n\| \leqslant C$ with some constant $C > 0$ for any $n \in \mathbb{N}$. Then, using the boundedness of (x_n) and the diagonal argument, there exists a subsequence $(x_{n(k)})$ of (x_n) such that for any $y \in \text{span } D$, the limit $\lim_{k\to\infty}\langle x_{n(k)}, y\rangle =: f(y)$ exists. It is easy to see that f is anti-linear and satisfies $\|f\| \leqslant C$. Further, f can be extended to $\overline{\text{span } D}$ preserving all its properties.

Now let $P : X \to \overline{\text{span } D}$ be a orthogonal projection. Then, we can further extend f to the whole X without destroying its properties by considering the mapping $X \ni x \to f(Px)$. We will still use the same notation f for the extension of f. By the Riesz representation theorem, there exists a unique $x \in X$ such that $f(y) = \langle x, y\rangle$ for any $x \in X$. This gives

$$\left\langle x_{n(k)}, y\right\rangle \to \langle x, y\rangle, \quad k \to \infty, \quad y \in X. \tag{2.4.22}$$

□

2.4.3 Eigenvalues, spectrum and the spectral radius of an operator

In linear algebra we study linear mappings in $\mathbb{R}^m$, which can be expressed as matrices $A \in \mathbb{R}^{n\times n}$. A very simple subclass of matrices are diagonal matrices

$$\boldsymbol{D} = \text{diag}\{d_1, \dots, d_n\} = \begin{pmatrix} d_1 & 0 & \dots & 0 \\ 0 & d_2 & 0 & \vdots \\ 0 & 0 & \ddots & 0 \\ 0 & \dots & \dots & d_n. \end{pmatrix} \tag{2.4.23}$$

These matrices are very easy for studying mappings, since on the unit basis vectors $e_1, \dots, e_n$ the matrix $\boldsymbol{D}$ is just a multiplication operator

$$\boldsymbol{D} \circ (\alpha_j e_j) = \left(d_j \alpha_j\right) e_j, \quad j = 1, \dots, n. \tag{2.4.24}$$

If we could find a basis $B = \{\varphi_1, \varphi_2, \varphi_3, \dots\}$ in a space X such that for a matrix A we have

$$A\varphi_j = \lambda_j \varphi_j, \tag{2.4.25}$$

with real or complex numbers λ_j, $j = 1, \dots, n$, then the application of the operator A would be reduced to simple multiplications on the basis functions.

The resolution of a matrix equation

$$A\varphi = f \tag{2.4.26}$$

with an invertible matrix A in such a basis is very simple. Assume that

$$f = \sum_{j=1}^{n} \beta_j \varphi_j. \tag{2.4.27}$$

Then the solution of (2.4.26) is given by

$$\varphi = \sum_{j=1}^{n} \frac{\beta_j}{\lambda_j} \varphi_j, \tag{2.4.28}$$

since

$$A\varphi = \sum_{j=1}^{n} \frac{\beta_j}{\lambda_j} A\varphi_j = \sum_{j=1}^{n} \frac{\beta_j}{\lambda_j} \lambda \varphi_j = \sum_{j=1}^{n} \beta_j \varphi_j = f. \tag{2.4.29}$$

In inverse problems we often find such matrices where λ_j becomes small for large j. In this case inversion is equivalent to the multiplication with large numbers $1/\lambda_j$. Small errors in the coefficients β_j will lead to large errors in the coefficients of the reconstructed quantity φ. This phenomenon is called *instability* of the inverse problem. In numerical mathematics we speak of *ill-conditioned* matrices.

We now collect adequate terms and definitions to treat the general situation in a Banach and Hilbert space setting.

Definition 2.4.11. *Consider a bounded linear operator A on a normed space X. We call $\lambda \in \mathbb{C}$ an* eigenvalue *of A, if there is an element $\varphi \in X$, $\varphi \neq 0$, such that*

$$A\varphi = \lambda\varphi. \tag{2.4.30}$$

In this case φ is called the eigenvector.

If λ is an eigenvalue of A, then the operator $\lambda I - A$ is not injective and thus cannot be invertible. In general, the operator might still not be invertible even if λ is not an eigenvalue of A. This motivates the following definition.

Definition 2.4.12. *If $\lambda I - A$ is boundedly invertible, then we call λ a* regular value *of A. The set $\rho(A) \subset \mathbb{C}$ of all regular values of A is called the* resolvent set, *its complement $\sigma(A) := \mathbb{C}\backslash\rho(A)$ is called the* spectrum *of A. The operator*

$$R(\lambda, A) := (\lambda I - A)^{-1} \tag{2.4.31}$$

is called the resolvent. *The* spectral radius $r(A)$ is the supremum

$$r(A) := \sup_{\lambda \in \sigma(A)} |\lambda|. \tag{2.4.32}$$

Note that since the set of eigenvalues is a subset of the spectrum, all eigenvalues lie in the ball with radius $r(A)$ around the origin in the complex plane $\mathbb{C}$.

For the spectrum of a compact operator we have the following result, which will be an important part of the basis of our study of inverse problems.

Theorem 2.4.13 (Spectrum of a compact operator). *Let $A : X \to X$ be a compact linear operator in an infinite-dimensional space X. Then $\lambda = 0$ is an element of the spectrum $\sigma(A)$ and $\sigma(A)\backslash\{0\}$ consists of at most a countable set of eigenvalues with no point of accumulation except, possibly, $\lambda = 0$.*

Proof. First, consider the case $\lambda = 0$. Then $\lambda = 0$ being regular is by definition equivalent to A being boundedly invertible and thus $I = A^{-1}A$ being compact. But since X is infinite-dimensional, the identity operator cannot be compact and thus λ cannot be regular.

For $\lambda \neq 0$ we apply the Riesz theory. Then in the case where the null space $N(\lambda I - A) = \{0\}$ the operator is injective and thus boundedly invertible. As a consequence λ is regular. Otherwise, we have $N(\lambda I - A) \neq \{0\}$, i.e. λ is an eigenvalue. Thus every point λ in $\mathbb{C}\backslash\{0\}$ is either a regular point or an eigenvalue.

Finally, we need to show that the set of eigenvalues $\sigma(A)\backslash\{0\}$ has no accumulation point other than zero. We do this by constructing sequences of subspaces spanned by the eigenvalues. Assume that we have a sequence (λ_n) of distinct eigenvalues and $\lambda \in \mathbb{C}\backslash\{0\}$ such that $\lambda_n \to \lambda$, $n \to \infty$ which implies $|\lambda_n| \geqslant R > 0$ $(n \in \mathbb{N})$. For each λ_n choose an associated eigenvector φ_n and define the subspaces

$$U_n := \text{span}\{\varphi_1, \ldots, \varphi_n\}, \qquad n \in \mathbb{N}. \tag{2.4.33}$$

It is not hard to see that eigenvectors of distinct eigenvalues are linearly independent. Thus $U_n \subset U_{n+1}$, $U_n \neq U_{n+1}$, $n = 1, 2, 3, \ldots$. Let $\psi_1 = \|\varphi_1\|^{-1}\varphi_1$. Then, by the Riesz lemma 2.3.19 we can choose a sequence $\psi_n \in U_n$ with $\|\psi_n\| = 1$ and

$$d(\psi_n, U_{n-1}) := \inf_{\psi \in U_{n-1}} \|\psi_n - \psi\| \geqslant \frac{1}{2}, \qquad n \geqslant 2. \tag{2.4.34}$$

We can represent ψ_n as a linear combination of the basis elements, i.e.

$$\psi_n = \sum_{j=1}^{n} \alpha_{nj}\varphi_j \quad \text{with} \quad \alpha_{nj} \in \mathbb{C}\ (1 \leqslant j \leqslant n). \tag{2.4.35}$$

For $m < n$ we derive

$$A\psi_n - A\psi_m = \sum_{j=1}^{n} \lambda_j \alpha_{nj}\varphi_j - \sum_{j=1}^{m} \lambda_j \alpha_{mj}\varphi_j = \lambda_n(\psi_n - \psi) \tag{2.4.36}$$

with $\psi \in \text{span}\{\varphi_1, \ldots, \varphi_{n-1}\} = U_{n-1}$. This yields

$$\|A\psi_n - A\psi_m\| \geqslant \frac{|\lambda_n|}{2} \geqslant \frac{R}{2}. \tag{2.4.37}$$

Thus the sequence $(A\psi_n)$ does not contain a convergent subsequence. However, (ψ_n) is a bounded sequence and A was assumed to be compact, such that we obtain a contradiction and our assumption was wrong. This completes the proof of the theorem. □

2.4.4 Spectral theorem for compact self-adjoint operators

We are step-by-step approaching more general results about the representation of operators as multiplication operators in special basis systems. As a preparation we start with the following theorem, which will be an important tool both for the further theory as well as for inversion of linear operator equations in the framework of inverse problems.

Theorem 2.4.14. *For a bounded linear operator $A : X \to Y$ between Hilbert spaces X, Y we have*

$$A(X)^{\perp} = N(A^{*}), \quad N(A^{*})^{\perp} = \overline{A(X)}, \tag{2.4.38}$$

where $N(A^{*})$ and $A(X)$ are the null-space of A^{*} and range of A, respectively.

Proof. We first establish a sequence of equivalences as follows. We have

$$\begin{aligned} g \in A(X)^{\perp} &\Leftrightarrow \langle A\varphi, g\rangle = 0, \quad \varphi \in X \\ &\Leftrightarrow \langle \varphi, A^{*}g\rangle = 0, \quad \varphi \in X \\ &\Leftrightarrow A^{*}g = 0 \\ &\Leftrightarrow g \in N(A^{*}), \end{aligned} \tag{2.4.39}$$

which proves the first equality of spaces. To prove the second part denote by P the orthogonal projection in Y onto the closed subspace $\overline{A(X)}$. For function $\varphi \in (A(X)^{\perp})^{\perp}$, by using $A(X)^{\perp} = \overline{A(X)}^{\perp}$, we have $P\varphi - \varphi \in \overline{A(X)}^{\perp}$. Here, observe that for any $\psi \in \overline{A(X)}^{\perp}$,

$$\langle P\varphi - \varphi, \psi\rangle = \langle P\varphi, \psi\rangle - \langle \varphi, \psi\rangle = \langle \varphi, P\psi\rangle = 0.$$

which implies $\varphi \in \overline{A(X)}$. Hence, $(A(X)^{\perp})^{\perp} \subset \overline{A(X)}$. On the other hand, we have $(A(X)^{\perp})^{\perp} \subset \overline{A(X)}$ due to the fact that $U^{\perp} = \bar{U}^{\perp}$ for any subspace U of Y. Hence, we have obtained

$$(A(X)^{\perp})^{\perp} = \overline{A(X)}, \tag{2.4.40}$$

which together with the first result yields the second equality.

If we know the adjoint operator A^{*} and can evaluate its nullspace $N(A^{*})$, via the previous theorem we also obtain the image space $\overline{A(X)}$. Exchanging the role of A and A^{*}, we can also use it to analyze the null-space of some operator A.

Definition 2.4.15. *Consider an operator $A : X \to X$ mapping a Hilbert space X into itself. It is called* self-adjoint *if $A = A^{*}$, i.e. if*

$$\langle A\varphi, \psi\rangle = \langle \varphi, A\psi\rangle, \quad \varphi, \psi \in X. \tag{2.4.41}$$

We first establish some basic equality between the norm and the quadratic form $\langle A\varphi, \varphi\rangle$ on X. In its proof we make use of the *parallelogram equality*

$$\|\varphi + \psi\|^2 + \|\varphi - \psi\|^2 = 2\big(\|\varphi\|^2 + \|\psi\|^2\big) \tag{2.4.42}$$

for all $\varphi, \psi \in X$, which is derived by elementary application of the linearity of the scalar product.

Theorem 2.4.16. *For a bounded linear self-adjoint operator A we have*

$$q := \sup_{\|\varphi\|=1} |\langle A\varphi, \varphi\rangle| = \|A\|. \tag{2.4.43}$$

Proof. Using the Cauchy–Schwarz inequality for $\varphi \in X$ with $\|\varphi\| = 1$ we have for any $\varphi, \psi \in X$

$$|\langle A\varphi, \varphi\rangle| \leqslant \|A\varphi\| \, \|\varphi\| \leqslant \|A\|, \tag{2.4.44}$$

from which we obtain $q \leqslant \|A\|$. The other direction is slightly tricky. First convince yourself by elementary calculation that we have

$$\begin{aligned} &\langle A(\varphi + \psi), \varphi + \psi\rangle - \langle A(\varphi - \psi, \varphi - \psi\rangle \\ &= 2\{\langle A\varphi, \psi\rangle + \langle A\psi, \varphi\rangle\} \\ &= 4\,\mathrm{Re}\langle A\varphi, \psi\rangle. \end{aligned} \tag{2.4.45}$$

This yields the estimate

$$4\mathrm{Re}\langle A\varphi, \psi\rangle \leqslant q \cdot \left\{\|\varphi + \psi\|^2 + \|\varphi - \psi\|^2\right\} \leqslant 2q \cdot \left\{\|\varphi\|^2 + \|\psi\|^2\right\}, \tag{2.4.46}$$

where we have used (2.4.42). Finally, for φ with $\|\varphi\| = 1$ and $A\varphi \neq 0$, we define $\psi := A\varphi/\|A\varphi\|$ and calculate

$$\|A\varphi\| = \mathrm{Re}\,\frac{\langle A\varphi, A\varphi\rangle}{\|A\varphi\|} = \mathrm{Re}\langle A\varphi, \psi\rangle \leqslant q\frac{\|\varphi\|^2 + \|\psi\|^2}{2} = q. \tag{2.4.47}$$

Hence, $\|A\| \leqslant q$, and the proof is complete.

Theorem 2.4.17. *The eigenvalues of self-adjoint operators are real and eigenvectors to different eigenvalues are orthogonal to each other.*

Proof. First we remark that for an eigenvalue λ with eigenvector φ we have

$$\lambda\langle\varphi, \varphi\rangle = \langle A\varphi, \varphi\rangle = \langle\varphi, A\varphi\rangle = \bar{\lambda}\langle\varphi, \varphi\rangle, \tag{2.4.48}$$

hence $\lambda \in \mathbb{R}$. Let $\lambda \neq \mu$ be two eigenvalues of A with eigenvectors φ and ψ, respectively. Then we have

$$(\lambda - \mu)\langle\varphi, \psi\rangle = \langle A\varphi, \psi\rangle - \langle\varphi, A\psi\rangle = 0, \tag{2.4.49}$$

which yields $\varphi \perp \psi$ and the proof is complete.

Theorem 2.4.18. *Consider a bounded linear self-adjoint operator $A : X \to X$. Then its spectral radius is equal to its norm, i.e.*

$$r(A) = \|A\|. \tag{2.4.50}$$

If A is compact, then there is at least one eigenvalue λ with $|\lambda| = \|A\|$.

Proof. Since the theorem clearly holds for the case $A = 0$, we can assume $A \neq 0$. First, consider a complex number $\lambda \in \mathbb{C}$ with $|\lambda| > \|A\|$. Then we can study the operator $I - B$ with $B := \frac{1}{\lambda}A$. The norm of B satisfies

$$\|B\| = \frac{1}{|\lambda|}\|A\| < 1. \tag{2.4.51}$$

Now, by the Neumann series theorem 2.3.12 the operator $I - B$ is boundedly invertible, hence the same is true for $\lambda I - A = \lambda(I - B)$. Thus λ is a regular value, which yields $r(A) \leqslant \|A\|$.

By theorem 2.4.16 there is a sequence (φ_n) in X with $\|\varphi_n\| = 1$ such that

$$|\langle A\varphi_n, \varphi_n\rangle| \to \|A\|, \quad n \to \infty. \tag{2.4.52}$$

Since the ball with radius $\|A\|$ in $\mathbb{C}$ is compact, there is a λ with $|\lambda| = \|A\|$ such that

$$\langle A\varphi_n, \varphi_n\rangle \to \lambda, \quad n \to \infty. \tag{2.4.53}$$

By multiplication with a real number of modulus 1 to A we may achieve the situation where λ is positive. We now calculate

$$\begin{aligned} 0 \leqslant \|A\varphi_n - \lambda\varphi_n\|^2 &= \|A\varphi_n\|^2 - 2\lambda\langle A\varphi_n, \varphi_n\rangle + \lambda^2\|\varphi_n\|^2 \\ &\leqslant \|A\|^2 - 2\lambda\langle A\varphi_n, \varphi_n\rangle + \lambda^2 \\ &= 2\lambda(\lambda - \langle A\varphi_n, \varphi_n\rangle) \to 0, \quad n \to \infty. \end{aligned} \tag{2.4.54}$$

Therefore

$$(\lambda I - A)\varphi_n \to 0, \quad n \to \infty. \tag{2.4.55}$$

But in this case λ can not be a regular value of A, because

$$1 = \|\varphi_n\| = \|(\lambda I - A)^{-1}(\lambda I - A)\varphi_n\| \to 0, \quad n \to \infty \tag{2.4.56}$$

would generate a contradiction. This shows that $r(A) \geqslant |\lambda| = \|A\|$ and thus (2.4.50).

If A is compact, there exists a subsequence $(A\varphi_{n(k)})$ converging to some $\psi \in X$. Hence, by (2.4.55),

$$\lambda\varphi_{n(k)} \to \psi\ (k \to \infty), \text{ i.e. } \varphi_{n(k)} \to \lambda^{-1}\psi =: \varphi\ (k \to \infty). \tag{2.4.57}$$

Here, due to $\|\varphi_{n(k)}\| = 1\, (k \in \mathbb{N})$ we have $\varphi \neq 0$. Then, by using (2.4.55) once again, we have $A\varphi = \lambda\varphi$. This ends the proof. □

Example 2.4.19. *Consider a two point boundary value problem given by*

$$u''(x) + \kappa^2 u(x) = \varphi(x), \quad u(0) = u(1) = 0, \tag{2.4.58}$$

with $\varphi \in C([0, 1])$ *on* $\Omega := (0, 1)$, *where* $\kappa/(2\pi) > 0$ *is the frequency of vibration with small displacement which can be a vibration of a string with unit mass and cramped at its both ends* $x = 0$, $x = 1$ *under some small external force* $\varphi(x)$. *If* $\kappa/(2\pi)$ *is not a resonant frequency that is* $\kappa/(2\pi) \notin \mathbb{Z}_+ = \mathbb{N} \cup \{0\}$, *then this boundary value problem has a unique solution* $u \in C^2((0, 1) \cap C([0, 1])$ *and it is given by*

$$u(x) = (A\varphi)(x) = \int_\Omega G(x, \xi)\varphi(\xi)\,\mathrm{d}\xi, \ x \in [0, 1], \tag{2.4.59}$$

where $G(x, \xi)$ *is the so-called* Green's function *of* (2.4.58) *and it is given by*

$$G(x, \xi) = \begin{cases} \dfrac{-\sin\kappa\cos(\kappa x) + \cos\kappa\sin(\kappa x)}{\kappa\sin\kappa}\sin(\kappa\xi) & \text{if } 0 \leqslant \xi < x \leqslant 1 \\ \dfrac{\cos\kappa\sin(\kappa\xi) - \sin\kappa\cos(\kappa\xi)}{\kappa\sin\kappa}\sin(\kappa x) & \text{if } 0 \leqslant x < \xi \leqslant 0 \end{cases} \tag{2.4.60}$$

It is easy to see that $G(x, \xi) \in C^0([0, 1] \times [0, 1])$ *and* $G(x, \xi) = G(\xi, x)$ *for all* $x, \xi \in [0, 1]$, *which implies that the operator* A *defined by* (2.4.59) *is a compact self-adjoint operator on* $L^2((a, b))$.

In theorems 2.4.13 and 2.4.17 some spectral properties of compact self-adjoint operators have been given. Based on these properties, we are now prepared for the *spectral theorem for self-adjoint compact operators.* This is a big mile-stone in the study of operators and operator equations.

Theorem 2.4.20 (Spectral theorem). *Consider a non-zero compact self-adjoint operator* A *on a Hilbert space* X. *We order the sequence of non-zero eigenvalues* (λ_n) *according to its size such that*

$$|\lambda_1| \geqslant |\lambda_2| \geqslant |\lambda_3| \geqslant \dots \tag{2.4.61}$$

with no point of accumulation except, possibly, $\lambda = 0$. *Denote the orthogonal projection operator onto the eigenspace* $N(\lambda_n I - A)$ *by* P_n *and let* $Q : X \to N(A)$ *be the orthogonal projection onto the null-space of* A. *Then we have the orthogonal decomposition*

$$X = \bigoplus_{n=1}^{\infty} N(\lambda_n I - A) \oplus N(A) \tag{2.4.62}$$

in the sense that each element $\varphi \in X$ *can be written as*

$$\varphi = \sum_{n=1}^{\infty} P_n\varphi + Q\varphi. \tag{2.4.63}$$

The operator A *can be represented as a multiplication operator by* λ_n *on the eigenspace* $N(\lambda_n I - A)$, *i.e. we have*

$$A = \sum_{n=1}^{\infty} \lambda_n P_n. \tag{2.4.64}$$

If there are only finitely many eigenvalues, then the series degenerate into finite sums.

Proof. We only consider the case where there are infinitely many eigenvalues, because the proof is easier for the other case. We first remark that the orthogonal projections P_n are self-adjoint and bounded with $\|P_n\| = 1$. Further, we have that $P_n P_k = 0$ for $n \neq k$ as a direct consequence of the orthogonality of the eigenspaces. Since their image space is the space $N(\lambda_n I - A)$ and by Riesz theory these spaces have finite dimension, the operators P_n are compact.

Consider the difference between A and the partial sum of (2.4.64), i.e. the operators

$$A_m := A - \sum_{n=1}^{m} \lambda_n P_n, \quad m \in \mathbb{N}. \tag{2.4.65}$$

We want to show that $A_m \to 0$, $m \to \infty$ and we carry this out by studying the eigenvalues of A_m. We show that the non-zero eigenvalues of A_m are given by $\lambda_{m+1}, \lambda_{m+2}, \ldots,$ Let $\lambda \neq 0$ be an eigenvalue of A_m with eigenvector φ. Then for $1 \leqslant n \leqslant m$ we have

$$\lambda P_n\varphi = P_n A_m \varphi = P_n(A\varphi - \lambda_n\varphi).$$

We use this to derive

$$\lambda^2 \|P_n\varphi\|^2 = \langle A\varphi - \lambda_n\varphi, P_n(A\varphi - \lambda_n\varphi)\rangle = \langle \varphi, (A - \lambda_n)P_n(A\varphi - \lambda_n\varphi)\rangle = 0$$

since $P_n = P_n^2$ is self-adjoint and it maps onto the nullspace $N(\lambda_n I - A)$. This yields $P_n\varphi = 0$ and hence $A\varphi = \lambda\varphi$. Next, assume that $\varphi \in N(\lambda_n I - A)$. Then for $n > m$ we obtain $A_m\varphi = \lambda_n\varphi$ and for $n \leqslant m$ we have $A_m\varphi = 0$. Thus, the eigenvalues of A_m which are different from zero are λ_{m+1}, λ_{m+2}, etc, i.e. the construction of A_m removes step-by-step the eigenvalues λ_1, λ_2, ... from A. But now theorems 2.4.13 and 2.4.18 show that

$$\|A_m\| = |\lambda_{m+1}| \to 0, \quad m \to \infty. \tag{2.4.66}$$

This shows the norm convergence of the series (2.4.64). To prove the orthogonal representation (2.4.62), (2.4.63) we remark that

$$\left\langle \varphi, \sum_{n=1}^{m} P_n\varphi \right\rangle = \underbrace{\left\langle \varphi - \sum_{n=1}^{m} P_n\varphi, \sum_{n=1}^{m} P_n\varphi \right\rangle}_{=0} + \left\langle \sum_{n=1}^{m} P_n\varphi, \sum_{n=1}^{m} P_n\varphi \right\rangle = \sum_{n=1}^{m} \|P_n\varphi\|^2. \tag{2.4.67}$$

We now calculate

$$\left\| \varphi - \sum_{n=1}^{m} P_n\varphi \right\|^2 = \|\varphi\|^2 - \sum_{n=1}^{m} \|P_n\varphi\|^2, \tag{2.4.68}$$

which shows that the sum

$$\sum_{n=1}^{\infty} P_n\varphi \tag{2.4.69}$$

is norm convergent in X. An application of the bounded operator A now yields

$$A\left(\varphi - \sum_{n=1}^{\infty} P_n\varphi\right) = A\varphi - \sum_{n=1}^{\infty} \lambda_n P_n\varphi = 0, \tag{2.4.70}$$

hence $\varphi - \sum_{n=1}^{\infty} P_n\varphi \in N(A)$ and we have shown (2.4.63). This completes our proof. □

2.4.5 Singular value decomposition

The spectral theorem as presented above is satisfied only for a very particular class of operators. Integral operators belong to this class only if their kernel is symmetric and real. However, we need to study more general equations and the spectral decomposition for general compact linear operators in Hilbert spaces is known as *singular value decomposition.*

Definition 2.4.21. *Consider a compact linear operator $A : X \to Y$ with Hilbert spaces X, Y and its adjoint A^*. Then the square roots of the non-zero eigenvalues of the self-adjoint compact operator $A^*A : X \to X$ are called the* singular values *of the operator A.*

We are now prepared for the following important result.

Theorem 2.4.22. *We order the sequence (μ_n) of the non-zero singular values of the non-zero compact operator $A : X \to Y$ such that $\mu_1 \geqslant \mu_2 \geqslant \mu_3 \geqslant \ldots$, where the values are repeated according to their multiplicity, i.e. according to the dimension $\dim N(\mu_n^2 I - A^*A)$ of the nullspace of $\mu_n^2 I - A^*A$. By $Q : X \to N(A)$ we denote the orthogonal projection onto the null-space $N(A)$ of A. Then there exist orthonormal sequences $(\varphi_n) \subset X$ and $(g_n) \subset Y$ such that*

$$A\varphi_n = \mu_n g_n, \qquad A^* g_n = \mu_n \varphi_n \tag{2.4.71}$$

for $n \in \mathbb{N}$. For each $\varphi \in X$ we have the singular value decomposition

$$\varphi = \sum_{n=1}^{\infty} \langle \varphi, \varphi_n \rangle \varphi_n + Q\varphi. \tag{2.4.72}$$

The operator A is represented by

$$A\varphi = \sum_{n=1}^{\infty} \mu_n \langle \varphi, \varphi_n \rangle g_n. \tag{2.4.73}$$

Each system (μ_n, φ_n, g_n) with these properties is called a singular system *of A. If the operator A is injective, then the orthonormal system $\{\varphi_n : n \in \mathbb{N}\}$ is complete in X.*

Proof. Let (φ_n) be the orthonormal sequence of eigenvectors of the compact self-adjoint operator A^*A with eigenvalues μ_n^2. We define

$$g_n := \frac{1}{\mu_n} A\varphi_n. \tag{2.4.74}$$

Then (2.4.71) is an immediate consequence of $A^*A\varphi = \mu_n^2 \varphi$. The representation (2.4.63) can be written as

$$\varphi = \sum_{n=1}^{\infty} \langle \varphi, \varphi_n \rangle \varphi_n + Q\varphi, \tag{2.4.75}$$

where Q is the orthogonal projection onto $N(A^*A)$. Let $\psi \in N(A^*A)$. Then $\langle A\psi, A\psi \rangle = \langle \psi, A^*A\psi \rangle = 0$, therefore $A\psi = 0$ and $\psi \in N(A)$. This implies $N(A^*A) = N(A)$ and thus Q is the orthogonal projection onto $N(A)$. We have shown (2.4.72). Finally, the representation of A given in (2.4.73) is obtained by an application of A to (2.4.72). □

We now write the solution of the equation

$$A\varphi = f \tag{2.4.76}$$

in spectral form and to understand the ill-posedness of the problem in terms of multiplication operators.

Theorem 2.4.23 (Picard). *Let $A : X \to Y$ be a compact linear operator between Hilbert spaces X, Y with a singular system (μ_n, φ_n, g_n). The equation (2.4.76) has a solution if and only if $f \in N(A^*)^\perp$ and*

$$\sum_{n=1}^{\infty} \frac{1}{\mu_n^2} |\langle f, g_n\rangle|^2 < \infty. \tag{2.4.77}$$

In this case the solution φ of (2.4.76) is given by

$$\varphi = \sum_{n=1}^{\infty} \frac{1}{\mu_n} \langle f, g_n\rangle \varphi_n. \tag{2.4.78}$$

Proof. The condition $f \in N(A^*)^\perp$ is a consequence of theorem 2.4.14. For a solution φ of (2.4.76) we calculate

$$\mu_n\langle \varphi, \varphi_n\rangle = \langle \varphi, A^* g_n\rangle = \langle A\varphi, g_n\rangle = \langle f, g_n\rangle, \tag{2.4.79}$$

which shows (2.4.78). Further, we obtain the estimate

$$\sum_{n=1}^{\infty} \frac{1}{\mu_n^2} |\langle f, g_n\rangle|^2 = \sum_{n=1}^{\infty} |\langle \varphi, \varphi_n\rangle|^2 \leqslant \|\varphi\|^2, \tag{2.4.80}$$

thus if the equation has a solution the sum (2.4.77) is bounded and the first direction of the theorem is proven.

To show the other direction we assume that $f \in N(A^*)^\perp$ and (2.4.77) is satisfied. Then the boundedness of (2.4.77) shows the convergence of the sum (2.4.78). We apply A to (2.4.78) to calculate

$$\begin{aligned} A\varphi &= \sum_{n=1}^{\infty} \frac{1}{\mu_n} \langle f, g_n\rangle A\varphi_n \\ &= \sum_{n=1}^{\infty} \langle f, g_n\rangle g_n \\ &= f, \end{aligned} \tag{2.4.81}$$

where we used $f \in N(A^*)^\perp$ and (2.4.72) for the singular system (μ_n, g_n, φ_n) of A^* for the last step. □

2.5 Lax–Milgram and weak solutions to boundary value problems

In this subsection we will introduce a theorem called the *Lax–Milgram theorem*. It is very important for solving linear elliptic partial differential equations by variational methods. We first provide the definitions of a *sesquilinear form* and define *X-coercivity*.

Definition 2.5.1. *Let X be a linear space over $\mathbb{C}$ and $a(,) : X \times X \to \mathbb{C}\,(u, v) \to \mathbb{C}$ be a mapping.*

(i) *The map $a(,)$ is called a* sesquilinear form *on V if it is linear with respect to the first variable and anti-linear with respect to the second variable. That is for any $u, u_1, u_2, v, v_1, v_2 \in X$ and $\alpha_1, \alpha_2, \beta_1, \beta_2 \in \mathbb{C}$, we have:*

$$\begin{aligned} a(\alpha_1 u_1 + \alpha_2 u_2, v) &= \alpha_1 a(u_1, v) + \alpha_2 a(u_2, v), \\ a\big(u, \beta_1 v_1 + \beta_2 v_2\big) &= \overline{\beta}_1 a(u, v_1) + \overline{\beta}_2 a(u, v_2). \end{aligned} \tag{2.5.1}$$

(ii) *If X is a normed space, the sesquilinear form $a(,)$ on X is called continuous and X-coercive if it satisfies*

$$|a(u, v)| \leqslant C\|u\|\ \|v\|, \quad u, v \in X \tag{2.5.2}$$

for some constant $C > 0$ independent of u, v and

$$|a(v, v)| \geqslant c\|v\|^2, \quad v \in X \tag{2.5.3}$$

for some another constant $c > 0$ independent of v, respectively.

The Riesz representation theorem 2.4.1 implies the following lemma as a direct consequence.

Lemma 2.5.2. *Let X be a Hilbert space with inner product $\langle\cdot,\cdot\rangle_X$.*

(i) *If $a(,)$ is a continuous sesquilinear form on X, then there exist unique elements A and $B \in BL(X)$ such that*

$$a(u, v) = \langle Au, v\rangle_X = \langle u, Bv\rangle_X, \quad u, v \in X. \tag{2.5.4}$$

(ii) *Denote by $\tilde{X}'$ the set of all anti-linear continuous functions on X. Then, there exists an isometry $J \in BL(\tilde{X}', X)$ such that*

$$f(v) = \langle J(f), v\rangle_X, \quad f \in \tilde{X}', \quad v \in X. \tag{2.5.5}$$

Based on this lemma, we have

Theorem 2.5.3. *Let X and $a(,)$ be a Hilbert space over $\mathbb{C}$ and X-coercive continuous sesquilinear form on X, respectively. Then for any $f \in \tilde{X}'$, there exists a unique $u = u(f) \in X$ such that*

$$a(u, v) = f(v), \quad v \in X, \quad \|u(f)\|_X \leqslant c^{-1}\|f\|_{\tilde{X}'},$$

where $c > 0$ is the constant in (2.5.3).

Proof. By the X-coercivity of $a(,)$, A and B in lemma 2.5.2 are injective and

$$\|Au\|_X \geqslant c\|u\|_X, \quad u \in X, \tag{2.5.6}$$

where $c > 0$ is the constant in (2.5.3). Then, this estimate implies that $A(X)$ is closed subspace in X. Hence, by $A^* = B$,

$$A(X)^\perp = N(B) = \{0\},$$

which implies $A(X) = X$. Therefore, A is an isomorphism on X and from (2.5.6) the operator norm $\|A^{-1}\|$ satisfies

$$\|A^{-1}\| \leqslant c^{-1}. \tag{2.5.7}$$

Now

$$a(u, v) = f(v), \quad v \in X$$

is equivalent to

$$\langle Au, v\rangle_X = \langle J(f), v\rangle_X, \quad v \in X.$$

Therefore, u can be uniquely given by $u = A^{-1}J(f)$ and it satisfies the estimate $\|u\| \leqslant c^{-1}\|f\|_{\tilde{X}'}$. □

Remark 2.5.4. *Let X and $a(,)$ be a Hilbert space over $\mathbb{C}$ and a X-coercive continuous sesquilinear form on X, respectively. Then, it is easy to see that there exists a unique $\mathcal{A} \in BL(X, X')$ with $X' := BL(X, \mathbb{C})$ such that*

$$a(u, v) = \mathcal{A}u(v), \quad v \in X. \tag{2.5.8}$$

Hence for given $f \in \tilde{X}'$, $a(u, v) = f(v)$ for any $v \in X$ is equivalent to $\mathcal{A}u = f$. For solving a boundary value problem for an elliptic equation with some homogeneous boundary condition by the variational method, $\mathcal{A}$ expresses the elliptic operator with the homogeneous boundary condition.

We finish this section with a simple application of the Lax–Milgram result.

Corollary 2.5.5. *Let X be a Hilbert space and $A : X \to X$ a bounded linear operator with*

$$\langle A\varphi, \varphi\rangle \geqslant c\|\varphi\|^2, \quad \varphi \in X. \tag{2.5.9}$$

Then A is invertible with

$$\|A^{-1}\| \leqslant c^{-1}. \tag{2.5.10}$$

Proof. Clearly, by (2.2.1) a sesquilinear form is defined by

$$a(\varphi, \psi) := \langle A\varphi, \psi\rangle, \quad \varphi, \psi \in X.$$

By (2.5.9) it is coercive. For $b \in X$ an element of X' is given by $f(\psi) := \langle b, \psi\rangle$ and we have $\|f\|_{X'} = \|b\|_X$. Then, by theorem 2.5.3 there is $\varphi \in X$ such that

$$\langle A\varphi, \psi\rangle = \langle b, \psi\rangle, \quad \psi \in X,$$

which is equivalent to

$$A\varphi = b,$$

i.e. the operator A is invertible on X, and we have the norm estimate (2.5.10) by (2.5.7). □

2.6 The Fréchet derivative and calculus in normed spaces

The goal here is to collect basic definitions and notations around the Fréchet derivative in $\mathbb{R}^m$ and in normed spaces.

Let X, Y and Z denote arbitrary normed spaces and U and V are open subsets of X and Y, respectively. Recall that by $BL(X, Y)$ we denote the normed space of bounded linear operators from X to Y. For a mapping $A \in BL(X, Y)$ we write Ah for $A(h)$ indicating the linear dependence on $h \in X$. Also, recall the definition of $o(\|h\|)$ denoting a function or operator A which satisfies

$$\frac{\|A(h)\|}{\|h\|} \to 0, \quad \|h\| \to 0. \tag{2.6.1}$$

Definition 2.6.1 (Fréchet differential). *Consider a mapping $F : U \to Y$ and an element $u \in U$. We say that F is (Fréchet) differentiable at $u \in U$, if there exists $F' \in BL(X, Y)$ and a mapping $R : U \to Y$ with $R(h) = o(\|h\|)$ such that*

$$F(u + h) = F(u) + F'h + R(h). \tag{2.6.2}$$

In this case the linear mapping F' is determined uniquely and is called the Fréchet differential of F at u and is sometimes denoted by $F'(u)$. If F is differentiable at all $u \in U$, we say that F is (Fréchet) differentiable in U.

Examples. We next study several examples to gain experience with derivatives. The following examples also prepare the differentiability proofs for shape monitoring and shape reconstruction problems.

(a) For functions $f : \mathbb{R} \to \mathbb{R}$, clearly the Fréchet derivative coincides with the classical derivative. This is also true for functions $F : \mathbb{R}^m \to \mathbb{R}^m$. In this case, $F = (F_1, F_2, \dots, F_m)^T$ a column vector and T denotes the transpose. Also its Fréchet derivative is given by the matrix

$$F = \begin{pmatrix} \frac{\partial F_1}{\partial x_1} & \cdots & \frac{\partial F_1}{\partial x_n} \\ \vdots & & \vdots \\ \frac{\partial F_m}{\partial x_1} & \cdots & \frac{\partial F_m}{\partial x_n} \end{pmatrix}. \tag{2.6.3}$$

We also use the notation $\nabla F := F$.

(b) If F is linear, then $F' = F$. This can be seen immediately from the definition of linearity

$$F(u + h) = F(u) + F(h), \tag{2.6.4}$$

which leads to $F'(h) = F(h)$ and $R \equiv 0$ in (2.6.2).

(d) Consider $F = (F_1, F_2, \dots, F_m)^T$ again with $X = \mathbb{R}^m$ and $Y = \mathbb{R}^m$. Assume that each $F_j\,(1 \leqslant j \leqslant m)$ has a bounded continuous derivative in a neighborhood U of a point $u \in X$ and that F' depends continuously on x in a neighborhood U of a point u, i.e. $F \in BC^1(U, Y)$. Then, with $u = (u_1, \dots, u_n)^T$ and $h = (h_1, \dots, h_n)^T$ the *fundamental theorem of calculus* yields

$$\begin{aligned} F(u+h) &= F(u) + \nabla F h + o\big(\|h\|\big) \\ &= F(u) + \sum_{j=1}^{n} \frac{\partial F}{\partial u_j} h_j + o\big(\|h\|\big). \end{aligned} \tag{2.6.5}$$

(e) Let Z_1 be the space of Hölder-continuously differentiable functions $C^{1,\alpha}(\partial D)$ on ∂D, where D is a domain in $\mathbb{R}^m$ with boundary of class $C^{1,\alpha}$, $\alpha > 0$. We define

$$U := \left\{ r \in Z_1 : \|r\|_{C^{1,\alpha}(\partial D)} < c \right\} \tag{2.6.6}$$

with some constant c. The derivative of the mapping

$$F_1 : U \to \mathbb{R}, \quad F_1(r) := r(x), \quad r \in U \tag{2.6.7}$$

for $x \in \partial D$ fixed is given by

$$F'_1 h = h(x), \quad h \in Z. \tag{2.6.8}$$

For the Fréchet derivative we have the *product rule* and the *chain rule* as follows. Consider $A : U \to Y$ and $B : U \to Y$. Then, we have

$$\begin{aligned} A(u+h) \cdot B(u+h) &= (A(u) + A'h + R(h)) \cdot (B(u) + B'h + S(h)) \\ &= A(u) \cdot B(u) + (A(u) \cdot B'h + A'h \cdot B(u)) \\ &\quad + (A(u) + A'h + R(h)) \cdot S(h) + R(h) \cdot (B'h + S(h)), \end{aligned} \tag{2.6.9}$$

where $\cdot$ can be any multiplicative operation in Y for example multiplication in Y by assuming Y is a Banach algebra. This leads to the *product rule*

$$(A \cdot B)' = A' \cdot B + A \cdot B'. \tag{2.6.10}$$

Assume that $A : U \to Y$ and $B : A(U) \to Z$. Then, for the composition $B \circ A$ of the two mapping we obtain

$$\begin{aligned} (B \circ A)(u+h) &= B(A(u+h)) = B(A(u) + A'h + o(\|h\|)) \\ &= B(A(u)) + B'(A'h + o(\|h\|)) + o(A'h + o(\|h\|) \\ &= B(A(u)) + B' \circ A'h + o(\|h\|), \end{aligned} \tag{2.6.11}$$

where we used $B'o(\|h\|) = o(\|h\|)$, $o(A'h + o(\|h\|) = o(\|h\|)$ and $o(\|h\|) + o(\|h\|) = o(\|h\|)$ in the sense of the definition of the *little o*, leading to the *chain rule*

$$(B \circ A)' = B' \circ A'. \tag{2.6.12}$$

We can build higher Fréchet derivatives, where here the conceptional work is a little bit more difficult than for the first order case. Note that the Fréchet derivative depends on both the point u where it is calculated as well as the argument h, i.e. F' maps $U \times X$ into Y, where the mapping is linear in the second argument.

We denote the set of all mappings

$$A : \underbrace{X \times \cdots \times X}_{n \text{ times}} \to Y, \quad (x_1, \ldots, x_n) \mapsto A(x_1, \ldots, x_n), \tag{2.6.13}$$

which are linear in every argument $x_1, \ldots, x_n$, as multi-linear mappings of order n, which are bounded in the sense that

$$\sup_{0 \neq x_j \, \|x_j\| \leqslant 1, j=1,\ldots,n} \frac{|A(x_1, \ldots, x_n)|}{\|x_1\| \cdots \|x_n\|} < \infty. \tag{2.6.14}$$

We denote this set by $BL^{(n)}(X, Y)$. Note that $BL^{(1)}(\mathbb{R}^m, \mathbb{R}^m)$ coincides with the set of $m \times n$-matrices, the set $BL^{(2)}(X, Y)$ with the set of *quadratic forms* on X with values in Y.

The first Fréchet derivative is a mapping

$$F' : U \to BL^{(1)}(X, Y), \quad u \mapsto F'(u) \tag{2.6.15}$$

via

$$F'(u, x) = F'(u)x \in Y, \quad x \in X. \tag{2.6.16}$$

The Fréchet derivative F'' as a derivative of F' is a mapping

$$F'' : U \to BL^{(2)}(X, Y), \quad u \mapsto F''(u) \tag{2.6.17}$$

via

$$F''(u, x, y) = F''(u)(x, y) \in Y, \quad x, y \in X. \tag{2.6.18}$$

The nth Fréchet derivative $F^{(n)}$ is the derivative of the $(n - 1)$th derivative and thus

$$F^{(n)} : U \to BL^{(n)}(X, Y). \tag{2.6.19}$$

Often, the notation

$$F^{(n)} = \frac{\mathrm{d}^n F}{\mathrm{d}u^n}$$

is used for the nth Fréchet derivative with respect to u. We also employ the abbreviation

$$F^{(n)}(u, x) := F^{(n)}(u)(x, \ldots, x) \tag{2.6.20}$$

when the same argument $x_j = x$ for $j = 1, \ldots, n$ is used.

Now consider the situation where we have a mapping $A : U \to Y$ where $U \subset X$ and Y is the space of operators A on Z. Then, under reasonable conditions, the differentiability of A implies the differentiability of the inverse operator A^{-1} in dependence on $u \in U$.

Theorem 2.6.2. *Let X be a normed space, $U \subset X$ be an open set and Y be a Banach algebra with neutral element e. Here the Banach algebra Y is a Banach space with multiplication y_1y_2 defined for any $y_1, y_2 \in Y$ such that $\|y_1y_2\| \leqslant \|y_1\|\|y_2\|$ and e plays role of the unit of this multiplication. Let $A = A(u) \in Y$ be Fréchet differentiable in $u_0 \in U$. We further assume that $A^{-1} = A^{-1}(u)$ exists in Y for all $u \in U$ near u_0 and the inverse depends continuously on u near u_0. Then $A^{-1}(u)$ is Fréchet differentiable at u_0 with Fréchet derivative*

$$(A^{-1})' = -A^{-1}A'A^{-1}. \tag{2.6.21}$$

$$\text{i.e. } \left((A^{-1})'(u_0)\right)x = -A'(u_0)\left(A'(u_0)x\right)A^{-1}(u_0), \quad x \in X. \tag{2.6.22}$$

Proof. We define

$$z(h) := A^{-1}(u_0 + h) - A^{-1}(u_0) + A^{-1}(u_0)A'(u_0, h)A^{-1}(u_0).$$

We need to show that $z(h) = o(\|h\|)$ to prove the statement of the theorem. We multiply by $A(u_0)$ from the right and from the left and use the continuity of the inverse $A^{-1}(u)$ to derive

$$\begin{aligned} A(u_0)z(h)A(u_0) &= A(u_0)A^{-1}(u_0 + h)A(u_0) - A(u_0) + A'(u_0, h) \\ &= (A(u_0 + h) - A(u_0))A^{-1}(u_0 + h)(A(u_0 + h) - A(u_0)) \\ &\quad - (A(u_0 + h) - A(u_0) - A'(u_0, h)) \\ &= o(\|h\|), \end{aligned} \tag{2.6.23}$$

which completes the proof. □

Usually, differentiability is used to derive high-order estimates for a function in a neighborhood of some point.

Lemma 2.6.3. *Let $f : U \to \mathbb{C}$ be two times continuously Fréchet differentiable. Then, for $h \in X$ sufficiently small and $u \in U$, we have*

$$f(u + h) = f(u) + f'(u)h + f_1(u, h) \tag{2.6.24}$$

with

$$f_1(u, h) = \int_0^1 (1 - t)f''(u + th, h)\,\mathrm{d}t. \tag{2.6.25}$$

Proof. We define a function $g : [0, 1] \to \mathbb{C}$ by $g(t) := f(u + th)$, $t \in [0, 1]$. By the chain rule the derivative of g with respect to t is given by $g'(t) = f'(u + th, h)$ and the second derivative is given by $g''(t) = f''(u + th, h, h)$, for which we use the short notation $f''(u + th, h)$. Now, applying the fundamental theorem of calculus two times we have

$$
\begin{aligned}
g(1) - g(0) &= \int_0^1 g'(t)\,\mathrm{d}t \\
&= \int_0^1 \left(g'(0) + \int_0^t g''(s)\,\mathrm{d}s \right) \mathrm{d}t \\
&= g'(0) + \int_0^1 \int_0^t g''(s)\,\mathrm{d}s\,\mathrm{d}t \\
&= g'(0) + \int_0^1 (1 - \rho) g''(\rho)\,\mathrm{d}\rho,
\end{aligned}
$$

which yields (2.6.24) with (2.6.25).

Remark. The same result as in (2.6.25) holds if we replace $\mathbb{C}$ by some Banach space Y, where one needs to define integrals with values in a Banach space.

The Picard–Lindelöf method to solve the Cauchy problem for first order ordinary differential equations is well known. It transforms the problem to an integral equation and looks for a fixed point. This method can be considered as an application of the Banach fixed-point theorem which we will introduce next. To begin with we define some terminologies.

Definition 2.6.4. *Let X be a Banach space with norm $\|\cdot\|$ and $F : X \to X$ be a mapping. F is called a contraction mapping on X if it satisfies*

$$\|F(x) - F(y)\| \leqslant c\|x - y\|, \quad x, y \in X \tag{2.6.26}$$

for some constant $0 < c < 1$.

Then the Banach fixed-point theorem is given as follows.

Theorem 2.6.5 (Banach fixed-point theorem). *Let F be a contraction mapping on a Banach space X. Then, F has a unique fixed point $x_0 \in X$ that is $F(x_0) = x_0$. Further for any $x \in X$, the sequence (x_n) defined by*

$$x_1 = x, \quad x_{n+1} = F(x_{n-1}), \quad n \in \mathbb{N} \tag{2.6.27}$$

converges to x_0.

Proof. To see the uniqueness of a fixed point, let x, x' be fixed points of F. Then by (2.6.26)

$$\|x - x'\| = \|F(x) - F(x')\| \leqslant c\|x - x'\|. \tag{2.6.28}$$

Since $0 < c < 1$, we must have $x = x'$.

Next we show that the sequence (x_n) converges in X which implies (x_n) converges in X to the unique fixed point $x_0 \in X$ by (2.6.27) and the continuity of F on X due to (2.6.26). To see the convergence of sequence (x_n) it is enough to show for any $n, m \in \mathbb{N}$ with $n > m$

$$\|x_n - x_m\| \leqslant \frac{c^n}{1 - c}\|x_1 - x_0\|. \tag{2.6.29}$$

Observe that for any $n \in \mathbb{N}$

$$\|x_{n+1} - x_n\| = \|F(x_n) - F(x_{n-1})\| \leqslant c\|x_n - x_{n-1}\|. \tag{2.6.30}$$

Hence $\|x_{n+1} - x_n\| \leqslant c^{n-1}\|x_2 - x_1\|$ for any $n \in \mathbb{N}$. This implies

$$\|x_n - x_m\| \leqslant \sum_{j=m}^{n-1} \left\|x_{j+1} - x_j\right\| \leqslant \|x_2 - x_1\| \sum_{j=m}^{n-1} c^j < \frac{c^m}{1 - c}\|x_2 - x_1\|. \tag{2.6.31}$$

Thus (x_n) is a Cauchy sequence which converges towards some $x_* \in X$. Clearly, by choosing $n = m + 1$ we have $\|F(x_m) - x_m\| \to 0$ for $m \to \infty$, such that by continuity $F(x_*) = x_*$, i.e. x_* is a fixed point and by uniqueness we have $x_* = x_0$.

Bibliography

[1] Kress R 1999 *Linear Integral Equations* (*Applied Mathematical Sciences* vol 82) 2nd edn (New York: Springer)

[2] Reed M and Simon B 1980 *Functional Analysis* (New York: Academic)

[3] Bachman G and Narici L 2000 *Functional Analysis* (Mineola, NY: Dover)

[4] Heuser H 1986 *Funktionalanalysis* (Teubner: Leipzig)

Chapter 3

Approaches to regularization

Many classical problems of mathematical modeling behave well in terms of their dependence on boundary conditions and input data: they show a continuous dependence on the input and if you increase the number of grid points for numerical approximations, you decrease the approximation error.

However, as we have already seen for solving integral equations of the first kind in figure 1.3, this type of *regular* behavior does not necessarily take place when you go to *inverse* or *reconstruction* problems. Often, the solutions show discontinuous dependence on the measurements. They amplify measurement noise. And they show some type of *non-uniqueness*.

For numerical algorithms for solving inverse problems, the rule that more effort will lead to better results no longer holds. Taking more grid points for approximations often leads to higher *instability*, such that the solutions inherit more and more noise which is carried on from the measured data. The term *ill-posed* has become popular to describe such phenomena closely linked to inversion.

Various methods have been developed to make ill-posed problems more *regular*, i.e. to approximate them by bounded, continuous and stable algorithms and operators. We call such approaches *regularization* methods, they are the topic of the following sections.

For further reading we suggest Groetsch [1], Engl *et al* [2], Hansen [3], Kirsch [4], Schuster, Kaltenbacher, Hofmann and Kazimierski [5], Kaltenbacher, Neubauer and Scherzer [6], Wang *et al* [7], Potthast [8] Kress [9] and Colton and Kress [10].

3.1 Classical regularization methods

3.1.1 Ill-posed problems

Applications in the natural sciences often involve initial value problems or boundary value problems. For a long time scientists have explored methods for their solution and for the numerical approximation of such solutions. In 1923 Hadamard postulated three properties which needed to be satisfied for what he called a properly posed or *well-posed* problem.

doi:10.1088/978-0-7503-1218-9ch3

Definition 3.1.1. *A problem is called* well-posed, *if we have:*

1. *Existence of a solution.*
2. *Uniqueness of the solution.*
3. *Continuous dependence of the solution on the data.*

If a problem is not well-posed, we call it ill-posed.

The above definition of a well-posed problem in the setting of operator equations is given next.

Definition 3.1.2. *Consider a bounded linear operator $A : X \to Y$ from a normed space X into a normed space Y. Then the equation*

$$A\varphi = f \tag{3.1.1}$$

is called well-posed, *if A is bijective and the inverse $A^{-1} : Y \to X$ is continuous. Otherwise the equation is called* ill-posed.

There are different kinds of ill-posedness, due to the three conditions which might be violated. Note that usually these conditions are not independent. For example, bijective operators on Banach spaces are boundedly invertible by the closed graph theorem.

We have already studied a number of ill-posed operator equations which we can note from the following theorem.

Theorem 3.1.3. *Let $A : X \to Y$ be a compact linear operator from a normed space X into a normed space Y, where X is infinite-dimensional. Then the equation $A\varphi = f$ of the first kind is an ill-posed equation.*

Proof. We have already shown that on the infinite-dimensional space X the operator $I = A^{-1}A$ cannot be compact, thus A^{-1} cannot be bounded. This proves the theorem. □

3.1.2 Regularization schemes

Often, the right-hand side f in ill-posed equations of the type

$$A\varphi = f \tag{3.1.2}$$

is given by measurements. In this case we are in fact given a function $f^{(\delta)} \in Y$ with

$$\| f - f^{(\delta)} \| \leqslant \delta, \tag{3.1.3}$$

where δ is the measurement error. Since the inverse of the operator A is unbounded, the solution $\varphi^{(\delta)}$ of the equation

$$A\varphi^{(\delta)} = f^{(\delta)} \tag{3.1.4}$$

can have an arbitrary distance to the true solution φ, depending on the particular choice of the right-hand side $f^{(\delta)}$. This is due to the fact that for unbounded A^{-1}

there is a sequence $\psi_n \in Y$ with $\|\psi_n\| = 1$ such that $\varphi_n := A^{-1}\psi_n$ satisfies $\|\varphi_n\| \to \infty$, $n \to \infty$. For arbitrary fixed $\delta > 0$ we define

$$f_n^{(\delta)} := f + \delta\psi_n, \qquad \varphi_n^{(\delta)} := \varphi + \delta\varphi_n \tag{3.1.5}$$

to obtain solutions of (3.1.3), (3.1.4) with

$$\left\|\varphi_n^{(\delta)}\right\| \to \infty, \qquad n \to \infty. \tag{3.1.6}$$

We note that 'measurement error' can be given just by numerical cut-off after 8 or 16 digits when computing solutions of an equation by some numerical scheme. In this case $\delta = 10^{-8}$ or $\delta = 10^{-16}$, but still there are functions $\varphi_n^{(\delta)}$ with (3.1.3) which satisfy (3.1.6).

The basic idea for solving ill-posed equations works as follows. We try to find a bounded approximation R_α to the unbounded operator A^{-1} depending on some parameter α, such that for perfect data and $\alpha \to 0$ we have convergence of the corresponding approximate solution φ_α to the true solution of the ill-posed equation.

Definition 3.1.4. *Let $A : X \to Y$ be an injective bounded linear operator between normed spaces X, Y. A* regularization scheme *is a family of bounded linear operators $R_\alpha : Y \to X$, $\alpha > 0$, such that*

$$R_\alpha A\varphi \to \varphi, \qquad \alpha \to 0 \tag{3.1.7}$$

for all $\varphi \in X$. The limit (3.1.7) means that R_α tends pointwise *or* element-wise *to A^{-1}. We also write $R_\alpha \xrightarrow{p} A^{-1}$, $\alpha \to 0$.*

Please note that the convergence $R_\alpha \xrightarrow{p} A^{-1}$ cannot be norm convergence for ill-posed injective equations.

Theorem 3.1.5. *Consider normed spaces X, Y, $\dim X = \infty$, a compact linear operator $A : X \to Y$ and a regularization scheme R_α for A^{-1}. Then the family R_α, $\alpha > 0$ of bounded operators cannot be uniformly bounded with respect to α and the operators R_α cannot be norm convergent.*

Proof. Assume that there is a constant $C > 0$ such that $\|R_\alpha\| \leqslant C$ for all $\alpha > 0$. Then from $R_\alpha f \to A^{-1}f$ for all $f \in A(X)$ from which we derive $\|A^{-1}f\| \leqslant C\|f\|$, thus $A^{-1} : A(X) \to X$ is bounded by C. But then $I = A^{-1}A$ is compact and X finite-dimensional, in contradiction to our assumption. We have shown that R_α cannot be uniformly bounded.

For the second statement assume that norm convergence is given, i.e.

$$\|R_\alpha - A^{-1}\| \to 0, \qquad \alpha \to 0.$$

Then there is $\alpha > 0$ such that

$$\|A^{-1}f\| = \|(A^{-1} - R_\alpha)f + R_\alpha f\| \leqslant \frac{1}{2}\|A^{-1}f\| + \|R_\alpha\|\|f\| \tag{3.1.8}$$

uniformly for all $f \in A(X)$. This yields

$$\|A^{-1}f\| \leqslant 2\|R_\alpha\|\|f\|, \tag{3.1.9}$$

thus A^{-1} is bounded on $A(X)$, and as above we obtain a contradiction. Thus R_α cannot be norm convergent. □

We estimate the error for the reconstruction of the function φ from

$$\varphi_\alpha^{(\delta)} = R_\alpha f^{(\delta)} \quad \text{with } \|f^{(\delta)} - f\| \leqslant \delta \tag{3.1.10}$$

by

$$\begin{aligned} \left\|\varphi_\alpha^{(\delta)} - \varphi\right\| &= \|R_\alpha f^{(\delta)} - R_\alpha f + R_\alpha f - \varphi\| \\ &\leqslant \|R_\alpha f^{(\delta)} - R_\alpha f\| + \|R_\alpha A\varphi - \varphi\| \\ &\leqslant \|R_\alpha\|\delta + \|R_\alpha A\varphi - \varphi\|. \end{aligned} \tag{3.1.11}$$

The *second* term in this estimate purely depends on the approximation of A by R_α and does not contain any influence from measurement errors. We call it the *regularization error*, which is the error arising from the approximation of the operator A. The *first* term comes from the data error of size δ. It is magnified by the application of R_α, however, due to the boundedness of the operator for every fixed $\alpha > 0$ we can control its influence. The first term is called *data error* within the reconstruction.

In general, there is a characteristic behavior of these error types. The regularization error tends to zero for $\alpha \to 0$. We have shown that it is a pointwise convergence. The data error tends to infinity for $\alpha \to 0$, depending on the particular measurements $f^{(\delta)}$. Usually, for large α the regularization error is large and the measurement error is small. This leads to curves as shown in figure 3.1.

We close this subsection with an important further type of convergence. So far we have addressed the question of convergence of reconstructions for the limit $\alpha \to 0$ when perfect data f is given. But it is a practical requirement on the reconstruction method to investigate the *measurement error* convergence $\delta \to 0$, i.e. when we have better and better measurements reducing the error of the right-hand side. In this case we want to choose $\alpha = \alpha(\delta)$ appropriately such that the approximate solution $\varphi^{(\delta)}$ tends to φ.

Definition 3.1.6. *We call a function $\alpha = \alpha(\delta)$ a* strategy *for a regularization scheme R_α if $\alpha \to 0$ for $\delta \to 0$. Such a strategy is called* regular, *if*

$$R_{\alpha(\delta)} f^{(\delta)} \to A^{-1}f, \quad \delta \to 0 \tag{3.1.12}$$

for each $f^{(\delta)}$ with $\|f^{(\delta)} - f\| \leqslant \delta$.

We will give a regular strategy for the Tikhonov regularization as theorem 3.1.16 in the forthcoming subsection 3.1.6.

3.1.3 Spectral damping

In this subsection we will learn how to build regularization schemes on the basis of the singular value decomposition. This is called *spectral damping*, since it employs

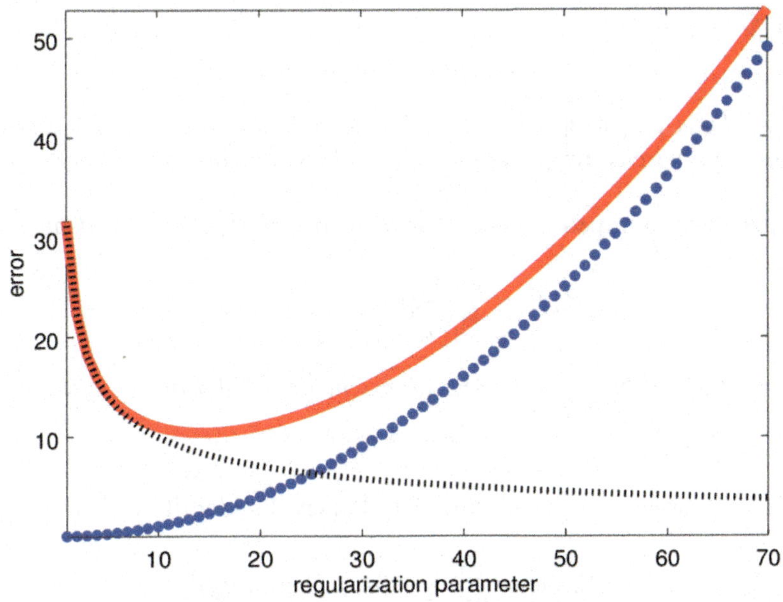

Figure 3.1. The data error contribution (black dotted line), regularization error (blue dotted line) and total error (red line) as modeled by (3.1.11). The *minimum error* for the reconstruction is obtained with some regularization parameter $\alpha > 0$ where the sum of the data error contribution and regularization error has its minimum. Within the area of inverse problems there is a whole industry of methods for the proper choice of the regularization parameter $\alpha > 0$ to come close to this minimum. Also note that Bayesian assumptions within the stochastic perspective as in (4.2.7) and (5.6.3) automatically provide the regularization term via its prior distribution $p(x)$.

the spectral theorem and the eigenvalue decomposition of the self-adjoint operator A^*A. From Picard's theorem 2.4.23 we see that the inverse A^{-1} is given by

$$A^{-1}f = \sum_{n=1}^{\infty} \frac{1}{\mu_n} \langle f, g_n \rangle \varphi_n. \tag{3.1.13}$$

Here the singular values μ_n tend to zero for $n \to \infty$, thus the factors $1/\mu_n$ become large. The idea of spectral damping is to multiply these factors by a damping function q. Clearly, to obtain pointwise convergence we need to construct this function with particular properties. Some adequate conditions on the damping function are established in the following central result.

Theorem 3.1.7. *Consider an injective compact linear operator $A : X \to Y$ with singular system (μ_n, φ_n, g_n). Let $q : (0, \infty) \times (0, \|A\|] \to \mathbb{R}$ be a bounded function satisfying the conditions:*

(a) *For each $\alpha > 0$ there is a constant $c = c(\alpha)$ such that*

$$|q(\alpha, \mu)| \leqslant c(\alpha)\mu, \quad 0 < \mu \leqslant \|A\|, \tag{3.1.14}$$

i.e. $q(\alpha, \mu)$ becomes small for μ small. This controls the regularizing effect.

(b) *We have the limit*

$$q(\alpha, \mu) \to 1, \quad \alpha \to 0 \tag{3.1.15}$$

for all $0 < \mu \leqslant \|A\|$, *i.e. for* $\alpha \to 0$ *the damping is less and less strong, such that pointwise convergence can occur. This controls the convergence features.*

Then the spectrally damped *bounded linear operators* $R_\alpha : Y \to X$ *defined by*

$$R_\alpha f := \sum_{n=1}^{\infty} \frac{1}{\mu_n} q(\alpha, \mu_n)\langle f, g_n\rangle \varphi_n \tag{3.1.16}$$

for $f \in Y$ and $\alpha > 0$ *build a regularization scheme for* A *with the estimate*

$$\|R_\alpha\| \leqslant c(\alpha). \tag{3.1.17}$$

Proof. We first show a norm estimate. Via Parseval's identity (2.2.38) we calculate

$$\begin{aligned}\|R_\alpha f\|^2 &= \sum_{n=1}^{\infty} \frac{1}{\mu_n^2} |q(\alpha, \mu_n)|^2 |\langle f, g_n\rangle|^2 \\ &\leqslant c(\alpha)^2 \sum_{n=1}^{\infty} |\langle f, g_n\rangle|^2 \\ &\leqslant c(\alpha)^2 \|f\|^2\end{aligned} \tag{3.1.18}$$

for all $f \in Y$, which yields the bound (3.1.17). To show the pointwise convergence we first calculate

$$\langle R_\alpha A\varphi, \varphi_n\rangle = \frac{1}{\mu_n} q(\alpha, \mu_n)\langle A\varphi, g_n\rangle = q(\alpha, \mu_n)\langle \varphi, \varphi_n\rangle. \tag{3.1.19}$$

Thus, we derive

$$\begin{aligned}\|R_\alpha A\varphi - \varphi\|^2 &= \sum_{n=1}^{\infty} |\langle R_\alpha A\varphi - \varphi, \varphi_n\rangle|^2 \\ &= \sum_{n=1}^{\infty} \left(q(\alpha, \mu_n) - 1\right)^2 |\langle \varphi, \varphi_n\rangle|^2,\end{aligned} \tag{3.1.20}$$

because (φ_n) is a complete orthonormal system in X due to the injectivity of the operator A.

Now, we use the boundedness of $q(\alpha, \mu)$ and the condition (b) to estimate the above sum. Since $q(\alpha, \mu_n)$ is bounded the above infinite sum is convergent. Given $\epsilon > 0$ we can choose N such that

$$\sum_{n=N+1}^{\infty} \left(q(\alpha, \mu_n) - 1\right)^2 |\langle \varphi, \varphi_n\rangle|^2 \leqslant \frac{\epsilon}{2} \tag{3.1.21}$$

for all $\alpha > 0$. Second, from (b) we observe convergence of $q(\alpha, \mu_n)$, $n = 1, \ldots, N$, towards 1 for $\alpha \to 0$. Thus, there is $\alpha_0 > 0$ such that

$$\left(q(\alpha, \mu_n) - 1\right)^2 \leqslant \frac{\epsilon}{2\|\varphi\|^2}, \quad n = 1, \ldots, N \tag{3.1.22}$$

for $\alpha \leqslant \alpha_0$. Collecting the two parts of the sum together we estimate

$$\|R_\alpha A\varphi - \varphi\|^2 \leqslant \epsilon \tag{3.1.23}$$

for $\alpha \leqslant \alpha_0$. This shows $R_\alpha A\varphi \to \varphi$ for $\alpha \to 0$ and completes the proof.

We will study two well-known damping schemes in the following section.

3.1.4 Tikhonov regularization and spectral cut-off

In this subsection we will establish the theory and convergence of the well-known and widely used schemes called Tikhonov regularization and spectral cut-off.

Theorem 3.1.8. *Let $A : X \to Y$ be an injective compact linear operator between Hilbert spaces X, Y. Then for each $\alpha > 0$ the operator $\alpha I + A^*A$ is boundedly invertible and the operator*

$$R_\alpha := (\alpha I + A^*A)^{-1}A^* \tag{3.1.24}$$

defines a regularization scheme for A with

$$\|R_\alpha\| \leqslant \frac{1}{2\sqrt{\alpha}}. \tag{3.1.25}$$

This is known as Tikhonov regularization.

Remark. The Tikhonov regularization replaces the solution of

$$A\varphi = f \tag{3.1.26}$$

by the approximation equation

$$\alpha\varphi_\alpha + A^*A\varphi_\alpha = A^*f, \tag{3.1.27}$$

which can be obtained from (3.1.26) by multiplication with A^* and addition of the term $\alpha\varphi$. In numerical mathematics the equation $A^*A\varphi = A^*f$ is known as *Euler's equation* or the *normal equation* for the minimization of

$$\mu(\varphi) := \|A\varphi - f\|^2 = \langle A\varphi - f, A\varphi - f\rangle. \tag{3.1.28}$$

The Tikhonov regularization adds $\alpha\|\varphi\|^2$ to stabilize the solution, i.e. it minimizes

$$\mu_{\mathrm{Tik}}(\varphi) := \alpha\|\varphi\|^2 + \|A\varphi - f\|^2. \tag{3.1.29}$$

Proof. To show that $\alpha I + A^*A$ is boundedly invertible we note that A is compact, thus A^*A is compact and we may apply the Riesz theory. To establish injectivity of the operator consider a function $\varphi \in X$ with $\alpha\varphi + A^*A\varphi = 0$. Then we have

$$\alpha\langle\varphi, \varphi\rangle + \langle A\varphi, A\varphi\rangle = \langle\varphi, \alpha\varphi + A^*A\varphi\rangle = 0. \tag{3.1.30}$$

Since the left-hand side of this equation is positive, we obtain $\|\varphi\|^2 = \langle\varphi, \varphi\rangle = 0$, i.e. $\varphi = 0$. Thus the operator is injective and by the Riesz theorem 2.3.25 it is boundedly invertible.

A practical example solving an integral equation of the first kind will be worked out in detail including its brief OCTAVE or MATLAB code in section 6.4.

Next, we study the equation (3.1.27) in terms of its spectral representation. Since A is injective, we have the expression

$$\varphi_\alpha = R_\alpha f = \sum_{n=1}^{\infty} \langle \varphi_\alpha, \varphi_n \rangle \varphi_n. \tag{3.1.31}$$

The application of A^*A is reduced to a multiplication by μ_n^2. Hence

$$(\alpha I + A^*A) \sum_{n=1}^{\infty} \langle \varphi_\alpha, \varphi_n \rangle \varphi_n = \sum_{n=1}^{\infty} \left(\alpha + \mu_n^2\right) \langle \varphi_\alpha, \varphi_n \rangle \varphi_n. \tag{3.1.32}$$

However, since (μ_n, g_n, φ_n) is a singular system for A^*, we have

$$A^* f = \sum_{n=1}^{\infty} \mu_n \langle f, g_n \rangle \varphi_n. \tag{3.1.33}$$

Hence, to have $(\alpha I + A^*A)\varphi_\alpha = A^* f$, each $\langle \varphi_\alpha, \varphi_n \rangle$ has to be given by

$$\langle \varphi_\alpha, \varphi_n \rangle = \frac{\mu_n}{\alpha + \mu_n^2} \langle f, g_n \rangle. \tag{3.1.34}$$

Thus, Tikhonov regularization is identical to a damping scheme with the damping function

$$q(\alpha, \mu) := \frac{\mu^2}{\alpha + \mu^2}, \quad \mu \geqslant 0. \tag{3.1.35}$$

The function q is bounded and it satisfies the conditions (a) and (b) of theorem 3.1.7. Here, the function $c(\alpha)$ can be estimated via the arithmetic–geometric mean inequality

$$\sqrt{\alpha}\,\mu \leqslant \frac{\alpha + \mu^2}{2}, \tag{3.1.36}$$

which leads to

$$c(\alpha) = \frac{1}{2\sqrt{\alpha}}. \tag{3.1.37}$$

Now, according to theorem 3.1.7 the Tikhonov regularization in fact establishes a regularization scheme with the bound (3.1.25). □

For the other regularization method, known as *spectral cut-off*, the inverse of A is approximated by

$$R_\alpha f := \sum_{\mu_n^2 \geqslant \alpha} \frac{1}{\mu_n} \langle f, g_n \rangle \varphi_n, \tag{3.1.38}$$

for $\alpha > 0$.

Theorem 3.1.9. *Let $A : X \to Y$ be an injective compact linear operator with singular system (μ_n, φ_n, g_n), $n \in \mathbb{N}$. Then the spectral cut-off establishes a regularization scheme for the operator A with*

$$\|R_\alpha\| \leqslant \frac{1}{\sqrt{\alpha}}. \tag{3.1.39}$$

Proof. We observe that the spectral cut-off is identical to a spectral damping-scheme with the damping function

$$q(\alpha, \mu) := \begin{cases} 0, & \mu^2 < \alpha \\ 1, & \text{otherwise.} \end{cases} \tag{3.1.40}$$

Here theorem 3.1.7 does not directly apply. There are sharper versions to show analogous results, but here we will use elementary reasoning. Clearly, R_α is bounded by $1/\sqrt{\alpha}$ which is obtained directly from the factor $1/\mu_n$ for $\mu_n^2 \geqslant \alpha \Leftrightarrow \mu_n \geqslant \sqrt{\alpha}$. For $\alpha \to 0$ and $f \in A(X)$ the sum converges towards the inverse

$$A^{-1} f := \sum_{n=1}^{\infty} \frac{1}{\mu_n} \langle f, g_n \rangle \varphi_n, \tag{3.1.41}$$

given by Picard's theorem 2.4.23. This completes the proof. □

We end this section with a result which turns out to be very useful for various sampling and probing methods in chapters 13–15.

Theorem 3.1.10. *Consider a compact linear injective operator $A : X \to Y$ from a Hilbert space X into a Hilbert space Y and the Tikhonov regularization scheme for solving $A\varphi = f$. Then, we have*

$$\|R_\alpha f\| \leqslant C, \quad \alpha > 0, \tag{3.1.42}$$

with some constant $C = C_f$ if $f \in R(A)$, i.e. f is in the range of A, and we have

$$\|R_\alpha f\| \to \infty, \quad \alpha \to 0 \tag{3.1.43}$$

in the case where $f \notin R(A)$. In other words, the regularized solution φ_α of the equation $A\varphi = f$ tends to ∞ if *$f \notin R(A)$.*

Proof. Since R_α is a regularization scheme, for $f = A\varphi$ we have $R_\alpha f \to \varphi$ for $\alpha \to 0$ by (3.1.7). This shows that $R_\alpha f$ is bounded for sufficiently small α and it is clearly bounded for $\alpha \geqslant \alpha_0 > 0$, which proves (3.1.42).

Now assume that $\|R_\alpha f\|$ is bounded. Then there is a weakly convergent subsequence of it, such that $R_{\alpha_n} f \rightharpoonup \varphi_* \in X, n \to \infty$. Since A is compact this implies that $AR_{\alpha_n} f \to A\varphi_*$. We note that for Tikhonov regularization AR_α is bounded uniformly for all $\alpha > 0$. Let $f_n = f_n$ be a sequence in Y with $f_n \to f$ for $n \to \infty$. Then we have

$$\|AR_\alpha f - f\| \leqslant \left\|AR_\alpha(f - f_n)\right\| + \left\|AR_\alpha f_n - f_n\right\| + \left\|f_n - f\right\|. \tag{3.1.44}$$

Given $\epsilon >$ we choose n sufficiently large to make $\|f_n - f\|$ and $\|AR_\alpha(f - f_n)\|$ smaller than $\epsilon/3$ each. Then, we fix n and choose $\alpha > 0$ sufficiently small to make $\|AR_\alpha f_n - f_n\|$ smaller than $\epsilon/3$. This shows that $AR_\alpha f \to f$ for $\alpha \to 0$. Thus we have $A\varphi_* = f$ and $f \in R(A)$. Finally, we conclude that if $f \notin R(A)$ the term $R_\alpha f$ cannot be bounded by Picard's theorem (2.4.23). □

3.1.5 The minimum norm solution and its properties

Having $A \in BL(X, Y)$, $f \in Y$ with Hilbert spaces X, Y, a solution called the *minimum norm solution* for the equation $A\varphi = f$ with respect to φ is very important in studying inverse problems when f can have some error such that $f \notin A(X)$. This is usually the case if real measurements are employed. In this subsection we will give the definition of the *minimum norm solution*, its existence and properties.

Definition 3.1.11. *Let X, Y, A, f be as above. Then for any given $\delta > 0$, we call φ_0 the minimum norm solution of $A\varphi = f$ with discrepancy δ if φ satisfies*

$$\|\varphi_0\| = \inf\{\|\varphi\| : \|A\varphi - f\| \leqslant \delta\}. \tag{3.1.45}$$

Theorem 3.1.12. *For Hilbert spaces X, Y consider $A \in BL(X, Y)$ with a dense range. Then for any given $\delta > 0$, there exists a unique minimum norm solution φ_0 with discrepancy δ.*

Proof. Let $U = \{\varphi \in X : \|A\varphi - f\| \leqslant \delta\}$. Clearly U is a convex set. If $\|f\| \leqslant \delta$, then $\varphi_0 = 0$ is the minimum norm solution with discrepancy δ. Hence let $\|f\| > \delta$. Then $0 \notin U$ and by A having a dense range, $U \neq \emptyset$.

Now by theorem 2.2.10, $\varphi_0 \in U$ is the best approximation to 0 in U if and only if

$$\mathrm{Re}\langle \varphi_0, \varphi_0 - \varphi\rangle \leqslant 0, \quad \varphi \in U \tag{3.1.46}$$

and such a $\varphi_0 \in U$ is unique.

The rest of proof will be given by providing several claims which clarify how the minimum norm solution is given. First of all we claim:

Claim 1. If $\varphi_0 \in X$ satisfies

$$\alpha\varphi_0 + A^*A\varphi_0 = A^*f \tag{3.1.47}$$

for some $\alpha > 0$ and $\|A\varphi_0 - f\| = \delta$, then φ_0 satisfies (3.1.46) and also such a φ_0 is unique.

The uniqueness of claim 1 follows from the uniqueness given in theorem 2.2.10. The rest of this claim can be seen as follows.

$$\begin{aligned}\alpha\mathrm{Re}\langle \varphi_0, \varphi_0 - \varphi\rangle &= \mathrm{Re}\langle A^*(f - A\varphi_0), \varphi_0 - \varphi\rangle \\ &= +\mathrm{Re}\langle A\varphi_0 - f, A\varphi - f\rangle - \|A\varphi_0 - f\|^2 \leqslant \delta(\|A\varphi - f\| - \delta) \leqslant 0\end{aligned} \tag{3.1.48}$$

for any $\varphi \in U$. Next we claim the following which holds even without assuming that A has a dense range.

Claim 2. For any $\alpha > 0$ there exists $(\alpha I + A^*A)^{-1} \in BL(X)$ which depends continuously on α and satisfies the estimate $\|(\alpha I + A^*A)^{-1}\| \leqslant \alpha^{-1}$ for any $\alpha > 0$.

This basically follows from the Lax–Milgram theorem. The details are as follows. Let $T = \alpha I + A^*A$. Then the sesquilinear form $a(,)$ defined by

$$a(\varphi, \psi) = \langle T\varphi, \psi \rangle, \quad \varphi, \psi \in X \tag{3.1.49}$$

is continuous on X. By

$$\mathrm{Re}\langle T\varphi, \varphi \rangle = \alpha\|\varphi\|^2 + \|A\varphi\|^2 \geqslant \alpha\|\varphi\|^2, \quad \varphi \in X, \tag{3.1.50}$$

the sesquilinear form $a(,)$ is coercive and T is injective. By the Lax–Milgram theorem, for any $f \in X$, there exists a unique $\varphi \in X$ such that

$$\langle f, \psi \rangle = a(\varphi, \psi) = \langle T\varphi, \psi \rangle, \quad \psi \in X, \tag{3.1.51}$$

i.e. $T\varphi = f$, and also from (3.1.50) we have $\|T^{-1}\| \leqslant \alpha^{-1}$. The continuity of T on $\alpha > 0$ follows from the Neumann series expansion.

By the Tikhonov regularization scheme, we know that $\varphi_\alpha = (\alpha I + A^*A)^{-1}A^*f$ is the minimizer of the functional $\|A\varphi - f\|^2 + \alpha\|\varphi\|^2$ for $\varphi \in X$. That is

$$\|A\varphi_\alpha - f\|^2 + \alpha\|\varphi_\alpha\|^2 = \inf_{\varphi \in X}(\|A\varphi - f\|^2 + \alpha\|\varphi\|^2). \tag{3.1.52}$$

Next we want to show that there exist $\alpha > 0$ such that $\|\varphi_\alpha - f\| = \delta$. For this consider a continuous function $G(\alpha)$ for $\alpha > 0$ defined by

$$G(\alpha) = \|A\varphi_\alpha - f\|^2 - \delta^2. \tag{3.1.53}$$

By (3.1.50), we have $\alpha\|\varphi_\alpha\| \leqslant \|A^*f\|$ for $\alpha > 0$. Hence $\varphi_\alpha \to 0$, $\alpha \to \infty$ which implies $G(\alpha) \to \|f\|^2 - \delta^2 > 0$, $\alpha \to \infty$.

To see the behavior of $G(\alpha)$ as $\alpha \to 0$, we prepare the final claim.

Claim 3. $A\varphi_\alpha \to f$, $\alpha \to 0$.

This follows from (3.1.52) and A having a dense range. The details are as follows. Since $A(X)$ is dense in Y, we have for any $\varepsilon > 0$ that there exists $\varphi_\varepsilon \in X$ such that $\|A\varphi_\varepsilon - f\|^2 < \varepsilon/2$. By taking $\alpha > 0$ such that $\alpha\|\varphi_\varepsilon\|^2 \leqslant \varepsilon/2$, we have

$$\|A\varphi_\alpha - f\|^2 \leqslant \|A\varphi_\alpha - f\|^2 + \alpha\|\varphi_\alpha\|^2 \leqslant \|A\varphi_\varepsilon - f\|^2 + \alpha\|\varphi_\varepsilon\|^2 < \varepsilon.$$

By claim 3, we have $G(\alpha) \to -\delta^2$ as $\alpha \to 0$. Therefore, by the intermediate value theorem for continuous functions, there exists $\alpha > 0$ such that $G(\alpha) = 0$ and the associated φ_α gives the minimum norm solution with discrepancy δ. Finally we remark that the uniqueness of φ_0 is guaranteed by claim 1. □

Remark 3.1.13. *For minimum norm solutions we observe the following characterization and property.*

(i) *Let φ_0 be the minimum norm solution of $A\varphi = f$ with discrepancy δ. Then the proof of theorem 3.1.12 shows that φ_0 is given by*

$$\varphi_0 = \begin{cases} 0 & \text{if } \|f\| \leqslant \delta \\ \varphi_\alpha := (\alpha I + A^*A)^{-1}A^*f & \\ \text{with } \alpha \text{ satisfying } \|A\varphi_\alpha - f\| = \delta & \text{if } \|f\| > \delta. \end{cases} \tag{3.1.54}$$

(ii) *Let φ^δ be the minimum norm solution of $A\varphi = f^\delta$ with discrepancy $\delta > 0$ and $\|f^\delta - f\| \leqslant \delta$. If $f \notin A(X)$, then $\|\varphi^\delta\| \to \infty$, $\delta \to 0$.*

Proof. (ii) can be proved by a contradictory argument. Its details are as follows. Suppose (φ^δ) is bounded and let $\delta_n = 1/n$, $\varphi_n = \varphi^{\delta_n}$ for $n \in \mathbb{N}$. Then by the weak compactness of Hilbert spaces and $A \in BL(X, Y)$, there exist a subsequence $(\varphi_{n(k)})$ of (φ_n), $\varphi^* \in X$ and $\psi \in Y$ such that

$$\varphi_{n(k)} \rightharpoonup \varphi^* \text{ in } X. \tag{3.1.55}$$

Then it is easy to see that $A\varphi^* = \psi$. Since $\|A\varphi_{n(k)} - f\| \leqslant 1/n(k)$ for $k \in \mathbb{N}$ and $A\varphi_{n(k)} - f \to A\varphi^* - f$, $k \to \infty$, we have

$$\|A\varphi^* - f\| \leqslant \liminf_{k\to\infty} \left\|A\varphi_{n(k)} - f\right\| = 0. \tag{3.1.56}$$

Hence $A\varphi^* = f$ which is a contradiction. □

Next we will give several properties of the minimum norm solutions.

Theorem 3.1.14. *Let X, Y be Hilbert spaces and $A \in BL(X, Y)$ have a dense range. Further, let φ_0 and φ_n be the minimum norm solutions to $A\varphi_0 = f_0$ and $A\varphi_n = f_n$ with discrepancy δ for each $n \in \mathbb{N}$, respectively. Then, if $f_n \to f_0$, $n \to \infty$, (φ_n) weakly converges to φ_0 as $n \to \infty$.*

Proof. From the definition of the minimum norm solution,

$$\begin{aligned} &\left\|A\varphi_n - f_n\right\| \leqslant \delta, \quad \|\varphi_n\| = \inf\left\{\|\varphi\| : \left\|A\varphi - f_n\right\| \leqslant \delta\right\}, \quad n \in \mathbb{N} \\ &\left\|A\varphi_0 - f_0\right\| \leqslant \delta, \quad \|\varphi_0\| = \inf\left\{\|\varphi\| : \left\|A\varphi - f_0\right\| \leqslant \delta\right\}. \end{aligned} \tag{3.1.57}$$

Set $U_n = \{\varphi \in X : \|A\varphi - f_n\| \leqslant \delta\}$ for each $n \in \mathbb{N}$. Since $f_n \to f_0$, $n \to \infty$, there exists a sufficiently large $n \in \mathbb{N}$ such that $\|f_n - f_0\| \leqslant \delta/2$. Further, by $\overline{A(X)} = Y$, there exists $\varphi^* \in X$ such that $\|A\varphi^* - f_0\| \leqslant \delta/2$. Hence, for sufficiently large $n \in N$, we have

$$\|A\varphi^* - f_n\| \leqslant \|A\varphi^* - f_0\| + \|f_n - f_0\| \leqslant \delta, \tag{3.1.58}$$

which means that $\varphi^* \in U_n$ and hence $\|\varphi_n\| = \inf_{\varphi\in U_n}\|\varphi\| \leqslant \|\varphi^*\|$ for sufficiently large $n \in \mathbb{N}$. Therefore the sequences (φ_n) in X and $(A\varphi_n)$ in Y are bounded in X and by the weak compactness of Hilbert spaces, there exists subsequence $(\varphi_{n(k)})$ of (φ_n) such that $\varphi_{n(k)} \to \bar{\varphi}$, $k \to \infty$ weakly in X and $A\varphi_{n(k)} \to A\bar{\varphi}$, $k \to \infty$ weakly in Y, respectively. Hence, for any $\psi \in X$, we have

$$\left\langle \varphi_{n(k)}, \psi \right\rangle \to \langle \bar{\varphi}, \psi \rangle, \quad k \to \infty. \tag{3.1.59}$$

In particular, taking $\psi = \bar{\varphi}$, we have

$$\|\bar{\varphi}\|^2 = \lim_{k\to\infty} \left\langle \varphi_{n(k)}, \bar{\varphi} \right\rangle \leqslant \|\bar{\varphi}\| \lim_{k\to} \left\|\varphi_{n(k)}\right\|. \tag{3.1.60}$$

Hence we have from $f_n \to f_0$, $n \to \infty$ that

$$\begin{aligned} \|\bar{\varphi}\| &\leqslant \lim_{k\to\infty} \left\|\varphi_{n(k)}\right\| = \lim_{k\to\infty} \inf\left\{ \|\varphi\| : \left\|A\varphi - f_{n(k)}\right\| \leqslant \delta \right\} \\ &\leqslant \lim_{k\to\infty} \inf\left\{ \|\varphi\| : \left\|A\varphi - f_0\right\| + \left\|f_0 - f_{n(k)}\right\| \leqslant \delta \right\} \\ &\leqslant \inf\{\|\varphi\| : \|A\varphi - f_0\| \leqslant \delta\} = \|\varphi_0\|. \end{aligned} \tag{3.1.61}$$

On the other hand, we have $\|A\bar{\varphi} - f_0\|^2 = \lim_{k\to\infty} \langle A\varphi_{n(k)} - f_0, A\bar{\varphi} - f_0 \rangle$, and then

$$\begin{aligned} \left\|A\bar{\varphi} - f_0\right\| &\leqslant \lim_{k\to\infty} \left\|A\varphi_{n(k)} - f_0\right\| = \lim_{k\to\infty} \left\{ \left\|A\varphi_{n(k)} - f_0\right\| + \left\|f_{n(k)} - f_0\right\| \right\} \\ &\leqslant \lim_{k\to\infty} \left\{ \delta + \left\|f_{n(k)} - f_0\right\| \right\} = \delta. \end{aligned} \tag{3.1.62}$$

These two estimates (3.1.61) and (3.1.62) imply that $\bar{\varphi}$ is a minimum norm solution of $A\varphi = f$ with discrepancy δ. Due to the uniqueness of the minimum norm solution to $A\varphi = f$ with discrepancy, we have $\bar{\varphi} = \varphi_0$, that is, $\varphi_{n(k)}$ weakly converges to φ_0 as $k \to \infty$.

At last let $(\varphi_{n(\ell)})$ be an arbitrary subsequence of (φ_n). By using the same argument as that for $(\varphi_{n(k)})$ above, there exists a subsequence $(\varphi'_{n(\ell)})$ of $(\varphi_{n(\ell)})$ such that $(\varphi'_{n(\ell)})$ weakly converges to φ_0. Therefore, the sequence (φ_n) weakly converges to φ_0 as $n \to \infty$. □

Theorem 3.1.15. *Let X, Y be Hilbert spaces. Assume that $A \in BL(X, Y)$ is injective and has a dense range. Let φ^δ and φ_0 be the respective minimum norm solutions of $A\varphi = f^\delta$ and $A\varphi = f$ with discrepancy δ, where f^δ satisfies $\|f^\delta - f\| \leqslant \delta$ for some $f \in Y$. Then we have the following statements.*

(i) *If $f \in A(X)$, then $\varphi^\delta \to \varphi_0$, $\delta \to 0$.*

(ii) If $f \in AA^*(Y)$, *then* $\|\varphi^\delta - \varphi_0\| = O(\delta^{1/2})$, $\delta \to 0$.

(iii) If $\|f^\delta\| > \delta$, *then φ_0 in (i) and (ii) can be replaced by $A^{-1}f$.*

Proof. We first prove (i). By remark 3.1.13, (i), we have

$$\begin{aligned}\delta^2 + \alpha\|\varphi^\delta\|^2 &= \|A\varphi^\delta - f^\delta\|^2 + \alpha\|\varphi^\delta\|^2 \\ &= \inf\left\{ \|A\varphi - f^\delta\|^2 + \alpha\|\varphi\|^2 : \varphi \in X \right\} \\ &\leqslant \|A(A^{-1}f) - f^\delta\|^2 + \alpha\|A^{-1}f\|^2 \\ &\leqslant \delta^2 + \alpha\|A^{-1}f\|^2,\end{aligned}$$

if $\|f^\delta\| > \delta$. This implies

$$\|\varphi^\delta\| \leqslant \|A^{-1}f\|, \tag{3.1.63}$$

and we also have

$$\|A\varphi^\delta - f\| \leqslant \|A\varphi^\delta - f^\delta\| + \|f^\delta - f\| \leqslant 2\delta. \tag{3.1.64}$$

Even in the case $\|f^\delta\| \leqslant \delta$, this and (3.1.63) are true due to $\varphi^\delta = 0$ by remark 3.1.13. Hence, for any $g \in Y$,

$$|\langle \varphi^\delta - A^{-1}f, A^*g\rangle| = |\langle A\varphi^\delta - f, g\rangle| \leqslant 2\delta\|g\| \to 0, \quad \delta \to 0.$$

By taking account of the denseness of $A^*(Y)$ in X due to the injectivity of A, this implies

$$\varphi^\delta \to A^{-1}f, \quad \delta \to 0 \text{ weakly in } X. \tag{3.1.65}$$

Therefore by (3.1.63), we have

$$\begin{aligned}\|\varphi^\delta - A^{-1}f\|^2 &= \|\varphi^\delta\|^2 - 2\mathrm{Re}\langle \varphi^\delta, A^{-1}f\rangle + \|A^{-1}f\|^2 \\ &\leqslant 2\mathrm{Re}\langle A^{-1}f - \varphi^\delta, A^{-1}f\rangle \to 0, \quad \delta \to 0\end{aligned} \tag{3.1.66}$$

which completes the proof of (i). Next we prove (ii). By $f \in AA^*(Y)$ and the injectivity of A, there exists $g \in Y$ such that $A^{-1}f = A^*g$. Hence by (3.1.63), we have

$$\begin{aligned}\|\varphi^\delta - A^{-1}f\|^2 &\leqslant 2\mathrm{Re}\langle A^{-1}f - \varphi^\delta, A^*g\rangle = 2\mathrm{Re}\langle f - A\varphi^\delta, g\rangle \\ &\leqslant 2(\|f - f^\delta\| + \|f^\delta - A\varphi^\delta\|)\|g\| \leqslant 4\delta\|g\|,\end{aligned} \tag{3.1.67}$$

which completes the proof of (ii). Finally we note that (iii) is clear from the proofs of (i) and (ii). □

3.1.6 Methods for choosing the regularization parameter

This section gives an introduction to some methods for choosing the regularization parameter. First of all we will give a regular strategy for choosing the regularization parameter $\alpha = \alpha(\delta)$ depending on the discrepancy $\delta > 0$.

Theorem 3.1.16. *Let X, Y and $A \in BL(X, Y)$ be an operator which is injective and compact. For $\varphi \in X, f \in Y$, consider the Tikhonov regularized solution $\varphi_\alpha^\delta = R_\alpha f^\delta$ for $A\varphi = f$ if f is replaced by f^δ such that $\|f^\delta - f\| \leqslant \delta$. If $\alpha = \alpha(\delta)$ satisfies*

$$\frac{\delta^2}{\alpha(\delta)} \to 0 \qquad \text{for } \delta \to 0, \tag{3.1.68}$$

then $\varphi_\alpha^\delta \to \varphi(\delta \to 0)$ in X.

Proof. Observe that

$$\left\|\varphi_{\alpha(\delta)}^{\delta} - \varphi\right\| \leqslant \left\|R_{\alpha(\delta)}\left(f^{\delta} - f\right)\right\| + \left\|R_{\alpha(\delta)}A\varphi - \varphi\right\|$$
$$\leqslant \delta\left\|R_{\alpha(\delta)}\right\| + \left\|R_{\alpha(\delta)}A\varphi - \varphi\right\|. \tag{3.1.69}$$

Here by theorem 3.1.8 and the assumption (3.1.68) of the theorem, the right-hand side of this inequality tends to zero as $\delta \to 0$. □

Note that this theorem gives a regular strategy for Tikhonov regularization, but it does not say that we have

$$\left\|A\varphi_{\alpha}^{\delta} - f\right\| \leqslant \gamma\delta \text{ with a fix } \gamma \geqslant 1. \tag{3.1.70}$$

In practice we cannot let the discrepancy δ tend to zero. Hence, it is natural to ask how to choose the regularization parameter to satisfy (3.1.70) for given fix δ. Concerning this problem we will describe two methods, the first called *Morozov's discrepancy principle* and the second the *L-curve approach*. Finally, we will discuss the *stochastic perspective*, which also leads to a choice of the regularization term.

Discrepancy principle

The argument in (3.1.53) leading to the existence of a minimum norm solution of $A\varphi^{\delta} = f^{\delta}$ with discrepancy δ defines a method to choose the regularization parameter α to satisfy (3.1.70) with $\gamma = 1$ based on the Tikhonov regularization. We choose $\alpha(\delta)$ such that

$$\|AR_{\alpha}f^{\delta} - f^{\delta}\| = \delta. \tag{3.1.71}$$

This is one example of the discrepancy principle.

The spectral cut-off provides another example of a discrepancy principle. More precisely we have the following.

Theorem 3.1.17. *Let X, Y, A, f, f^{δ} be as in theorem 3.1.16 and assume that A has a dense range.*

(i) *For the spectral cut-off R_{α} for A and a fix $\gamma > 1$, there exists $\alpha = \alpha(\delta)$ depending on f^{δ},*

$$\left\|AR_{\alpha(\delta)}f^{\delta} - f^{\delta}\right\| \leqslant \gamma\delta. \tag{3.1.72}$$

(ii) *We obtain the convergence*

$$R_{\alpha(\delta)}f^{\delta} \to A^{-1}f, \quad \delta \to 0. \tag{3.1.73}$$

Proof. Let (μ_n, φ, g_n) be the singular system for A. Then (μ_n, g_n, φ_n) is the singular system for A^* and by the assumption that A has a dense range, A^* is injective and hence $\{g_n\}$ is a complete orthonormal system of Y.

We first prove (i). Write $f^\delta = \sum_{n=1}^{\infty}\langle f^\delta, g_n\rangle g_n$. Then, by the definition of R_α, we have

$$
\begin{aligned}
AR_\alpha f^\delta - f^\delta &= A\sum_{\mu_n^2\geqslant\alpha}\frac{1}{\mu_n}\langle f^\delta, g_n\rangle g_n - \sum_{n=1}^{\infty}\langle f^\delta, g_n\rangle g_n \\
&= -\sum_{\mu_n^2<\alpha}\langle f^\delta, g_n\rangle g_n
\end{aligned}
\tag{3.1.74}
$$

which implies

$$
\|AR_\alpha f^\delta - f^\delta\|^2 = \sum_{\mu_n^2<\alpha}|\langle f^\delta, g_n\rangle|^2 \to 0, \quad \alpha \to 0. \tag{3.1.75}
$$

Hence we can define $\alpha(\delta)$ by

$$
\alpha(\delta) := \sup\left\{\alpha > 0\colon \sum_{\mu_n^2<\alpha}|\langle f^\delta, g_n\rangle|^2 \leqslant (\gamma\delta)^2\right\} \tag{3.1.76}
$$

to obtain (3.1.72).

Next we prove (ii). First we employ (3.1.74) replacing f^δ by f. Taking account of Bessel's inequality (2.2.42) we obtain

$$
\|AR_\alpha - I\| = 1 \text{ for all } \alpha > 0.
$$

Then from the identity

$$
(AR_\alpha f^\delta - f^\delta) - (AR_\alpha f - f) = (AR_\alpha - I)(f^\delta - f),
$$

we derive

$$
\begin{aligned}
\|AR_\alpha f - f\| &\leqslant \delta + \|AR_\alpha f^\delta - f^\delta\| \\
\|AR_\alpha f^\delta - f^\delta\| &\leqslant \delta + \|AR_\alpha f - f\|.
\end{aligned}
\tag{3.1.77}
$$

By the definition of $\alpha(\delta)$ and (3.1.77), we have

$$
\left\|AR_{\alpha(\delta)}f - f\right\| \leqslant (1+\gamma)\delta \to 0, \quad \delta \to 0.
$$

Since

$$
\left\|AR_{\alpha(\delta)}f - f\right\|^2 = -\sum_{\mu_n^2<\alpha(\delta)}|\langle f, g_n\rangle|^2, \tag{3.1.78}
$$

we have the following two cases:

(a) $\alpha(\delta) \to 0$, $\delta \to 0$

(b) there is $\alpha_0 > 0$ such that $f = \sum_{\mu_n^2\geqslant\alpha_0}\langle f, g_n\rangle g_n$ and $\alpha(\delta) \geqslant \alpha_0$ for any $\delta > 0$.

For the case (a), by (3.1.77) we have

$$
\gamma\delta < \left\|AR_{\alpha(\delta)+\varepsilon}f^\delta - f^\delta\right\| \leqslant \delta + \left\|AR_{\alpha(\delta)+\varepsilon}f - f\right\|
$$

for any $\varepsilon > 0$. Hence we have

$$\delta < (\gamma - 1)^{-1}\left\|AR_{\alpha(\delta)+\varepsilon}f - f\right\|.$$

In the rest of our argument proving (ii) for the case (a), we use some $\varepsilon > 0$ and write $\alpha''(\delta) = \alpha(\delta) + \varepsilon$ for simplicity. Observe that

$$\begin{aligned}\left\|R_{\alpha''(\delta)}f^{\delta} - A^{-1}f\right\| &\leqslant \left\|R_{\alpha''(\delta)}\left(f^{\delta} - f\right)\right\| + \left\|R_{\alpha''(\delta)}A\varphi - A^{-1}f\right\| \\ &\leqslant \frac{1}{\gamma - 1}\left\|R_{\alpha''(\delta)}\right\|\left\|A\left(R_{\alpha''(\delta)}A\varphi - \varphi\right)\right\| + \left\|R_{\alpha''(\delta)}A\varphi - A^{-1}f\right\|.\end{aligned}$$

Since the second term on the right-hand side of this inequality tends to 0 as $\alpha''(\delta) \to 0$, we only need to show

$$\left\|R_{\alpha''(\delta)}\right\|\left\|A\left(R_{\alpha''(\delta)}A\varphi - \varphi\right)\right\| \to 0, \quad \alpha''(\delta) \to 0 \tag{3.1.79}$$

for any $\varphi \in X$. By (3.1.39) and (3.1.74), this follows from

$$\begin{aligned}\left\|R_{\alpha''(\delta)}\right\|^2\left\|A\left(R_{\alpha''(\delta)}A\varphi - \varphi\right)\right\|^2 &\leqslant \alpha''(\delta)^{-1}\sum_{\mu_n^2<\alpha''(\delta)} \mu_n^2\,|\langle\varphi, \varphi_n\rangle|^2 \\ &\leqslant \sum_{\mu_n^2<\alpha''(\delta)} |\langle\varphi, \varphi_n\rangle|^2.\end{aligned} \tag{3.1.80}$$

Next we consider the case (b). In this case we clearly have

$$A^{-1}f = \sum_{\mu_n^2\geqslant\alpha_0} \mu_n^{-1}\langle f, g_n\rangle\varphi_n = R_\alpha f \tag{3.1.81}$$

for any $\alpha \leqslant \alpha_0$. Hence

$$\|AR_\alpha f^\delta - f^\delta\| = \|(AR_\alpha - I)(f^\delta - f)\| \leqslant \|f^\delta - f\| \leqslant \delta < \gamma\delta \tag{3.1.82}$$

for any $\alpha \leqslant \alpha_0$. This implies that $\alpha(\delta) = \alpha_0$ by the definition of $\alpha(\delta)$. Therefore, by (3.1.81), we have

$$\left\|R_{\alpha(\delta)}f^\delta - A^{-1}f\right\| = \left\|R_{\alpha_0}\left(f^\delta - f\right)\right\| \leqslant \delta/\sqrt{\alpha_0} \to 0, \quad \delta \to 0. \tag{3.1.83}$$

This completes the proof. □

L-curve approach

Let X, Y, A, f, f^δ be as in theorem 3.1.17. Consider the minimizer

$$\varphi_\alpha^\delta = \arg\min_{\varphi\in X} \mu_{\text{Tik}}(\varphi), \tag{3.1.84}$$

of the cost function $\mu_{\text{Tik}}(\varphi) = \alpha\|\varphi\|^2 + \|A\varphi - f^\delta\|^2$ of Tikhonov regularization for $A\varphi = f^\delta$. As visualized in figure 3.1,

- if $\alpha > 0$ becomes large, then $A\varphi_\alpha^\delta$ does not fit f^δ and the residual $\|A\varphi_\alpha^\delta - f^\delta\|$ becomes large.
- On the other hand, if $\alpha > 0$ becomes small, then the data error interacting with the unboundedness of A^{-1} dominates as we have seen in subsection 3.1.2.

Based on this observation the *L-curve approach* is a heuristic approach which suggests taking the regularization parameter $\alpha = \alpha(\delta)$ as

$$\alpha(\delta) = \text{argmax}_\alpha\, \kappa(\alpha), \tag{3.1.85}$$

where $\kappa(\alpha)$ is the curvature of the *L-curve* defined by

$$\begin{aligned} (p(\alpha), q(\alpha)) &= \{(\log R(\alpha), \log S(\alpha)) :\ \alpha > 0\} \\ &\text{with } R(\alpha) = \left\|A\varphi_\alpha^\delta - f^\delta\right\|^2,\ S(\alpha) = \left\|\varphi_\alpha^\delta\right\|^2. \end{aligned} \tag{3.1.86}$$

Let (μ_n, φ_n, g_n) be the singular system of A. Then, by

$$\varphi_\alpha^\delta = (\alpha I + A^*A)^{-1}A^*f^\delta, \qquad A\varphi_\alpha^\delta - f^\delta = -\alpha(\alpha I + A^*A)^{-1}f^\delta$$

we have

$$R(\alpha) = \sum_{n=1}^{\infty} r_\alpha(\mu_n)|\langle f^\delta, g_n\rangle|^2 \tag{3.1.87}$$

$$S(\alpha) = \sum_{n=1}^{\infty} s_\alpha(\mu_n)|\langle f^\delta, g_n\rangle|^2, \tag{3.1.88}$$

where

$$r_\alpha(\mu) = \frac{\alpha^2}{\left(\mu^2 + \alpha\right)^2}, \qquad s_\alpha(\mu) = \frac{\mu^2}{\left(\mu^2 + \alpha\right)^2}. \tag{3.1.89}$$

Here we have used (μ_n, g_n, φ_n) as the singular system of A^* and the assumption that A has a dense range which implies $f^\delta = \sum_{n=1}^{\infty}\langle f^\delta, g_n\rangle g_n$. Using the relation $\mathrm{d}/(\mathrm{d}\alpha)r_\alpha = -\alpha\ \mathrm{d}/(\mathrm{d}\alpha)s_\alpha$, we have

$$\begin{aligned} \kappa(\alpha) &= \left(p'q'' - p''q'\right)\left(p'^2 + q'^2\right)^{-3/2}(\alpha) \\ &= RS\left(RS/|S'| + \alpha R + \alpha^2 S\right)\left(R^2 + \alpha^2 S^2\right)^{-3/2}(\alpha), \end{aligned} \tag{3.1.90}$$

where the prime and double prime denote the first order derivative and second order derivative with respect to α, respectively. We note that $|S'|$ can be given by

$$|S'(\alpha)| = \sum_{n=1}^{\infty}\left|\frac{\mathrm{d}}{\mathrm{d}\alpha}s_\alpha(\mu_n)\right|\left|\langle f^\delta, g_n\rangle\right|^2. \tag{3.1.91}$$

Regularization via prior densities

We will introduce the stochastic perspective on inverse problems, data assimilation and regularization in chapter 4 and section 5.6. Here, we briefly need to comment on the choice of the regularization parameter from a *stochastic* or more specific *Bayesian* perspective.

A Bayesian approach for Gaussian densities starts with a prior distribution of the form

$$p^{(\text{prior})}(x) = c e^{-\frac{1}{2}\left\{(x-\mu)^T B^{-1}(x-\mu)\right\}}, \quad x \in \mathbb{R}^m \tag{3.1.92}$$

with mean μ, covariance matrix B and normalization constant c. It is a model for our knowledge about our system before we incorporate the measurements y, telling us that we have a most likely state $\mu \in \mathbb{R}^m$ and the probability distribution around μ is Gaussian and has variance and covariance, given by B.

The distribution of the measurement error for a measurement $y \in \mathbb{R}^m$, $m \in \mathbb{N}$, and the simulated measurement $H(x)$ with observation operator H is modeled by

$$p(y|x) = c e^{-\frac{1}{2}\left\{(y-H(x))^T R^{-1}(y-H(x))\right\}}, \quad x \in \mathbb{R}^m, \tag{3.1.93}$$

with data covariance matrix R and normalization c, here written as a generic constant. The posterior distribution after a Bayes step, as we will describe in detail in (4.2.7), will be given by the product of prior and data distributions, i.e. by

$$p^{(\text{post})}(x) = c e^{-\frac{1}{2}\left\{(y-H(x))^T R^{-1}(y-H(x))+(x-\mu)^T B^{-1}(x-\mu)\right\}}, \quad x \in \mathbb{R}^m \tag{3.1.94}$$

with normalization c. For $B = \alpha^{-1}I$ and $R = I$ or by using weighted norms as in (5.2.3) this obtains the form

$$p^{(\text{post})}(x) = c e^{-\frac{1}{2}\left\{\alpha\|x-\mu\|^2+\|y-H(x)\|^2\right\}}, \quad x \in \mathbb{R}^m, \tag{3.1.95}$$

i.e. the posterior is a Gaussian which has the Tikhonov functional $\mu_{\text{Tik}}(x)$ as a weight function. If α is large, $B = \alpha^{-1}I$ means that the variance of the prior distribution is small, the posterior mainly clusters around the prior mean and measurements lead to small corrections only. A small α corresponds to a large variance of the prior distribution and thus a large uncertainty of your prior knowledge; measurements can then cause very large corrections and instability can occur.

From a Bayesian perspective, the prior completely determines the regularization. When you have accepted the prior, the problem becomes automatically well-posed, since as shown in theorem 3.1.8 Tikhonov regularization with fixed $\alpha > 0$ describes a bounded operator. The task which remains is to evaluate the posterior (3.1.94) or (3.1.95), since by your set-up you solve a well-posed problem. We say that *the prior regularizes your problem* and provides a choice of the regularization term or regularization parameter.

Sometimes the argument is used that (3.1.94) or (3.1.95) is a mere product of the prior and the data distribution, thus the Bayesian calculation is completely well-posed and you do not need to solve any operator equation. This is of course true in the sense that the evaluation of the posterior for some state x is well-posed and well-conditioned for any $\alpha > 0$. But it is nothing more than saying that the Tikhonov functional $\mu_{\text{Tik}}(x)$ depends continuously on x and evaluation is stable. Still, for small α the calculation of the minimum becomes very unstable and ill-posedness arises for its dependence on the measurement y. Classical regularization theory provides proper insight into the ill-posedness which is inherent in the Bayesian approach for priors with large variance.

3.2 The Moore–Penrose pseudo-inverse and Tikhonov regularization

We conclude this section with some tools on inversion which are very useful when solving inverse problems. Consider an equation

$$Ax = f \tag{3.2.1}$$

where $A : X \to Y$ is a bounded linear operator, but not invertible as mapping $Y \to X$. This could be the case where X and Y are finite-dimensional with $n = \dim(X)$ and $m = \dim(Y)$ and where $m < n$ such that A is a rank deficient matrix. We also want to treat the more general case where X and Y are Hilbert spaces with some scalar product.

When we try to solve (3.2.1) as well as possible, we might just try to find the minimizer of (3.1.28) and seek an approximate solution by

$$(A^*A)x = A^*f. \tag{3.2.2}$$

Lemma 3.2.1. *Let X, Y be Hilbert spaces. Then the operator A^*A is injective on the range $R(A^*)$ of the operator A^* in X. If X is finite-dimensional, then it is surjective and thus bijective $R(A^*) \to R(A^*)$.*

Proof. We first show the injectivity. Assume that $z \in R(A^*)$ such that $A^*Az = 0$. Then we conclude that $Az \in N(A^*) = R(A)^\perp$ by theorem 2.4.14. Since trivially $Az \in R(A)$ we conclude $Az = 0$. Now $Az = 0$ means $z \in N(A) = R(A^*)^\perp$. By assumption we have $z \in R(A^*)$, such that $z = 0$ follows. This shows that A^*A is injective on $R(A^*)$.

Second, we show the surjectivity of A^*A on $R(A^*)$. Consider $v \in R(A^*)$. Then, there is $\psi \in Y$ such that $v = A^*\psi$. We decompose Y into $N(A^*) \oplus N(A^*)^\perp$, such that $\psi = \psi_1 + \psi_2$ with $\psi_1 \in N(A^*)$ and $\psi_2 \in N(A^*)^\perp$. By theorem 2.4.14 we have $N(A^*)^\perp = \overline{R(A)} = R(A)$ in the case where X is finite-dimensional. Then there is $\varphi \in X$ such that $\psi_2 = A\varphi$, and hence

$$v = A^*\psi = A^*\psi_2 = A^*A\varphi. \tag{3.2.3}$$

A decomposition of φ into $\varphi_1 \in N(A)$ and $\varphi_2 \in N(A)^\perp = \overline{R(A^*)} = R(A^*)$ for finite-dimensional X shows $v = A^*A\varphi_2$ and thus the surjectivity of A^*A on $R(A^*) \subset X$. □

We note that if X is not finite-dimensional, in general A^*A is not surjective into $R(A^*)$, since A is not surjective into $\overline{R(A)} = N(A^*)^\perp$, and then we can show by a contradictory argument that any element v which is given as $v = A^*\psi$ with $\psi \in \overline{R(A)} \setminus R(A)$ is not in $R(A^*)$.

In the finite-dimensional case, we have shown the invertibility of A^*A on $R(A^*)$, such that the *Moore–Penrose pseudo-inverse*

$$A^\dagger := (A^*A)^{-1}A^* \tag{3.2.4}$$

is well defined. It is a mapping $Y \to N(A)^\perp = R(A^*)$. In the infinite-dimensional case, we note that by corollary 2.5.5 we have the invertibility of the coercive operator $Q := \alpha I + A^*A$ for any $\alpha > 0$. Then, the Tikhonov regularization operator

$$R_\alpha = (\alpha I + A^*A)^{-1}A^*$$

defined in (3.1.24) can be understood as a *regularized* version of the Moore–Penrose pseudo-inverse $A^\dagger$ defined in (3.2.4).

Lemma 3.2.2. *Let X be finite-dimensional. Then*

$$\|R_\alpha - A^\dagger\| \leqslant C\sqrt{\alpha}, \quad \alpha > 0, \tag{3.2.5}$$

i.e. we have $R_\alpha \to A^\dagger$ for $\alpha \to 0$ in the sense of operator convergence in $BL(Y, X)$ with a Hölder-type convergence estimate (3.2.5).

Proof. We derive

$$\begin{aligned}\left(R_\alpha - A^\dagger\right)f &= \left((\alpha I + A^*A)^{-1}A^* - (A^*A)^{-1}A^*\right)f \\ &= (A^*A)^{-1}\left\{(A^*A)(\alpha I + A^*A)^{-1} - I\right\}A^*f \\ &= (A^*A)^{-1}\left\{(\alpha I + A^*A - \alpha I)(\alpha I + A^*A)^{-1} - I\right\}A^*f \\ &= -(A^*A)^{-1}\left\{\alpha(\alpha I + A^*A)^{-1}\right\}A^*f. \end{aligned} \tag{3.2.6}$$

Thus, using the norm estimate (3.1.25) for $(\alpha I + A^*A)^{-1}$, we obtain (3.2.5) with some constant C. □

Lemma 3.2.3. *Let X be finite-dimensional. Then, the operator $AA^\dagger$ is an orthogonal projection onto $\overline{R(A)} = N(A^*)^\perp$. Also, the operator $A^\dagger A$ is an orthogonal projection onto $R(A^*) = N(A)^\perp$.*

Proof. First, consider $AA^\dagger = A(A^*A)^{-1}A^*$. Clearly, A^* maps Y into $R(A^*)$, which is mapped into itself by $(A^*A)^{-1}$ and mapped into $R(A) = N(A^*)^\perp$ by A. On $R(A)$ we can see by

$$AA^\dagger(A\varphi) = A(A^*A)^{-1}A^*A\varphi = A\varphi$$

the operator $AA^\dagger$ is the identity, and by continuity of the operators this is the case also on $\overline{R(A)} \subset Y$. The orthogonality of the projection is as follows. Choose $z \in Y$, $z = z_1 + z_2$ with $z_1 \in \overline{R(A)}$ and $z_2 \in R(A)^\perp = N(A^*)$. Then for $\psi \in R(A)$ we have

$$\langle z - AA^\dagger z, \psi\rangle = \langle z_2, \psi\rangle = 0, \tag{3.2.7}$$

which completes the first statement. $A^\dagger A$ is the identity on $N(A)^\perp$ and it is mapping $N(A)$ into 0, so that the second statement is straightforward. □

From this lemma we obtain the following theorem.

Theorem 3.2.4. *Let X, Y, A be as in lemma 3.2.3 and $y = A^\dagger f$* with $f \in Y$. Then,

(i) $\|Ay - f\| \leqslant \|Ax - f\|$ *for any $x \in X$.*

(ii) *If $f \in R(A)$, then $AA^\dagger f = f$ and $\|y\| \leqslant \|x\|$ for any $x \in X$ with $Ax = f$.*

Proof. We first prove (i). Put $P = AA^\dagger$. Then by lemma 3.2.3,

$$A^*(Ay - f) = A^*(Pf - f) = (PA)^*f - A^*f = A^*f - A^*f = 0.$$

Hence we have for any $x \in X$

$$\begin{aligned} \|Ax - y\|^2 &= \|Ay - f\|^2 + 2\mathrm{Re}\langle x - y, A^*(Ay - f)\rangle + \|A(x - y)\|^2 \\ &= \|Ay - f\|^2 + \|A(x - y)\|^2 \geqslant \|Ay - f\|^2. \end{aligned}$$

Next we prove (ii). The first statement is clear from (i). To see the second statement, let $x \in X$, $Ax = f$ and $Q = A^\dagger A$. Then, by lemma 3.2.3 and its proof,

$$\langle y, x - y\rangle = \langle y, Q(x - y)\rangle = \langle y, A^\dagger A(x - y)\rangle = \langle y, A^\dagger Ax - y\rangle = 0.$$

This immediately implies

$$\|x\|^2 = \|y\|^2 + \|x - y\|^2 \geqslant \|y\|^2. \tag{3.2.8}$$

□

Finally, we note that we have

$$A(A^*A) = (AA^*)A, \tag{3.2.9}$$

where we restrict our attention to the finite-dimensional case. By multiplication with $(A^*A)^{-1}$ from the right, which is well defined on $R(A^*)$ and by $(AA^*)^{-1}$ from the left, which according to lemma 3.2.1 applied to A^* instead of A is well-defined on $R(A)$, we obtain

$$(AA^*)^{-1}A = A(A^*A)^{-1}, \tag{3.2.10}$$

on $R(A^*)$. This leads to

$$\begin{aligned} (A^\dagger)^* &= \left((A^*A)^{-1}A^*\right)^* \\ &= A(A^*A)^{-1} \\ &= (AA^*)^{-1}A \\ &= (A^*)^\dagger \end{aligned} \tag{3.2.11}$$

on $R(A^*)$. On $R(A^*)^\perp = N(A)$ the operator $(A^*)^\dagger$ is identical to zero. If we extend $(A^*A)^{-1}$ from $R(A^*)$ to X by being zero on $R(A^*)^\perp$, the above equality (3.2.11) holds on the whole space X.

3.3 Iterative approaches to inverse problems

Iterative approaches to solve inverse problems have one common feature. In the easiest setting they consider the inverse problem as a general operator equation of the kind

$$H(\varphi) = 0 \tag{3.3.1}$$

where φ is the unknown quantity in some Banach space X and H is a nonlinear mapping $X \to Y$ into some Banach space Y. Here (1.2.2) could be of the form

$$H(\varphi) = f. \tag{3.3.2}$$

But if we consider $H(\varphi) - f$ as $H(\varphi)$ itself, then we have (3.3.1). There are several general methods for solving nonlinear equations. From finite-dimensional analysis Newton's method and the gradient method are well known. These methods are built on differentiability properties of the forward problem with respect to the unknown quantity. This might be a refractive index in scattering or the shape of the unknown objects.

From nonlinear analysis fixed point iterations are familiar. Also, scientific computing has developed various iterative, hierarchical and domain-decomposition schemes for large-scale matrix inversion, which can also be applied to the solution of inverse problems.

In the framework of inverse problems for partial differential equations several smart methods for iterative solution of such problems have been suggested and tested, which decompose the problem into different parts which are solved in turn by potential approaches via integral equations.

3.3.1 Newton and quasi-Newton methods

Newton's method is one of the oldest ways to solve nonlinear problems. Here, we will study the Newton method for inverse problems, which involves a number of severe complications compared to its well-posed or finite-dimensional setting.

The basic idea and convergence properties

Newton's method can be applied to an inverse problem in the general form (3.3.1). We may replace the mapping H by its linearized version at some point $\varphi \in X$, which leads to

$$H(\varphi) + H'(\varphi)h \overset{!}{=} 0, \quad \varphi, h \in X, \tag{3.3.3}$$

where the exclamation mark indicates that instead of the original equation (3.3.1) we solve the Newton equation (3.3.3). Usually, $H'(\varphi)$ is the Frechét derivative of $H(\varphi)$ at φ if it is Frechét differentiable, but here we just need to have some linearized version of (3.3.1). From (3.3.3) we obtain an element $h \in X$ by

$$h = -(H'(\varphi))^{-1}H(\varphi). \tag{3.3.4}$$

In fact, usually it is non-trivial to show that $H'(\varphi)$ is in fact injective. Further, if (3.3.3) is ill-posed, then $H'(\varphi)$ does not have a bounded inverse. Usually, it is not surjective onto Y and the inverse is not bounded. This means that solving (3.3.3) can only be carried out approximately by regularization with all the effects which have been described in the sections on linear ill-posed problems, see section 3.1. We can now iterate (3.3.4). Starting with some *initial approximation* $\varphi_0 \in X$ we define

$$\varphi_{k+1} := \varphi_k - \left(H'(\varphi_k)\right)^{-1} H(\varphi_k) \tag{3.3.5}$$

for $k = 0, 1, 2, 3, \ldots$. We call $(\varphi_k)_{k\in\mathbb{N}} \subset X$ the Newton sequence with initial value φ_0.

In general, every iterative method needs to be stopped to be applicable. For well-posed problems we usually stop when the desired accuracy is reached, i.e. when for some accuracy $\epsilon > 0$ we have

$$e_k := \|\varphi_k - \varphi^{(\text{true})}\| \leqslant \epsilon. \tag{3.3.6}$$

If $\varphi^{(\text{true})}$ is not available, an estimate for the error is calculated by

$$\tilde{e}_k := \|\varphi_k - \varphi_{k-1}\|, \quad n \in \mathbb{N}. \tag{3.3.7}$$

One basic approach to solve (3.3.3) is to employ one of the standard regularization schemes R_α with regularization parameter α for linear ill-posed equations in every Newton step. We then obtain

$$\varphi_{k+1} = \varphi_k + R_{\alpha_k} H(\varphi_k), \quad k = 1, 2, 3, \ldots \tag{3.3.8}$$

where α_k is chosen in dependence of $k \in \mathbb{N}$. We call this the *iteratively regularized Gauss–Newton method.*

The method (3.3.8) is computationally expensive since the regularized inverse R_α needs to be calculated in every step of the iteration. *Quasi–Newton methods* try to increase the effectivity of the calculations by keeping R_α fixed for the whole iteration process or for a part of it.

The Newton method for ill-posed problems

There are two different approaches to proving convergence of Newton's method (3.3.5) for ill-posed problems. The first approach is due to Potthast [8] and has been developed in the framework of severely ill-posed shape reconstruction problems. The second approach due to Scherzer *et al* [6] and is valid more generally, but it is limited to problems which are mildly nonlinear.

Let us understand why the classical convergence theory for Newton's method does not go through for ill-posed problems. We start from (3.3.5) and assume that φ^* is the true solution with $H(\varphi^*) = 0$. Further, we use

$$\begin{aligned} H(\varphi_k) - H(\varphi^*) &= \int_0^1 H'\big(\varphi^* + \tau(\varphi_k - \varphi^*)\big)(\varphi_k - \varphi^*)\,\mathrm{d}\tau \\ &= H'(\varphi^*)(\varphi_k - \varphi^*) + O\Big(\big\|\varphi_k - \varphi^*\big\|^2\Big) \end{aligned} \tag{3.3.9}$$

where we assume that H' is continuously depending on φ, such that the derivative depends Lipschitz continuously on the argument φ

$$\|H'(\varphi_1) - H'(\varphi_2)\| \leqslant c\|\varphi_1 - \varphi_2\| \tag{3.3.10}$$

with some constant c for all $\varphi_1, \varphi_2 \in X$. Then subtracting φ^* from (3.3.5) we obtain

$$\begin{aligned} \varphi_{k+1} - \varphi^* &= \varphi_k - \varphi^* - H'(\varphi_k)^{-1}\big(H(\varphi_k) - H(\varphi^*)\big) \\ &= H'(\varphi_k)^{-1}\Big\{\big(H'(\varphi_k) - H'(\varphi^*)\big)(\varphi_k - \varphi^*) + O\Big(\big\|\varphi_k - \varphi^*\big\|^2\Big)\Big\}. \end{aligned} \tag{3.3.11}$$

Now, if H' is boundedly invertible, from (3.3.10) we obtain *local second order convergence*

$$\big\|\varphi_{k+1} - \varphi^*\big\| \leqslant c\,\big\|\varphi_k - \varphi^*\big\|^2 \tag{3.3.12}$$

with some constant c of the Newton sequence $(\varphi_k)_{k\in\mathbb{N}}$ towards the true solution φ^*. **Clearly, if the inverse of H' is not bounded, the estimate (3.3.12) is not possible.** Scherzer *et al* formulated a particular non-linearlity condition to overcome this difficulty, Potthast used extensibility conditions to obtain boundedness of $(H')^{-1}$ on particular subsets which appear when Newton's method is applied. For further reading we refer to the above literature.

3.3.2 The gradient or Landweber method

The *gradient method* is a well-known basic scheme in applied mathematics. It relies on a simple observation that for a mapping $\mu : \mathbb{R}^m \to \mathbb{R}$ the vector of partial derivatives

$$\nabla\mu := \begin{pmatrix} \dfrac{\partial\mu}{\partial x_1} \\ \vdots \\ \dfrac{\partial\mu}{\partial x_k} \end{pmatrix} \tag{3.3.13}$$

provides the direction of steepest increase of μ and, thus, $-\nabla\mu$ is the direction of steepest decrease. To minimize μ we can go in the direction of steepest decrease, which leads to an iteration

$$\varphi_{k+1} := \varphi_k - \eta\nabla\mu(\varphi_k), \quad n = 1, 2, 3, \ldots, \tag{3.3.14}$$

where η is called the *step size* and where we need to supply some starting solution $\varphi_0 \in \mathbb{R}^m$.

In principle, we can apply this gradient method to the general inverse problem (3.3.1) if H is differentiable with respect to φ in its *discretized setting*. We first use a finite-dimensional subspace X_k of X with basis ψ_n, $n = 1, \ldots, N$ and the ansatz

$$\varphi(\alpha) = \sum_{n=1}^{N} \alpha_n \psi_n \tag{3.3.15}$$

with $\alpha := (\alpha_n)_{n=1,\ldots,N} \in \mathbb{R}^N$. Then, we replace the solution of the inverse problem by the minimization of

$$\mu(\alpha) := \|H(\varphi(\alpha))\|^2, \quad \alpha \in \mathbb{R}^m, \tag{3.3.16}$$

where the square of the norm is taken to achieve full differentiability of the functional μ with respect to α. This approach is a kind of *projection method,* since we replace the full inverse problem by a projection of its argument α into the finite-dimensional space X_k. Then, we can apply the classical gradient method (3.3.14).

Often, inverse problems seek full parameter functions or densities defined on surfaces or on open sets. In this case the problem lives in a Banach or Hilbert space X and we seek an element φ in an infinite-dimensional space. It is more natural to reserve the structure of the original problem to understand its full implications, instead of quickly passing into a finite-dimensional setting (3.3.15), which will depend on a particular discretization.

The gradient in Banach and Hilbert spaces

Let X, Y be Banach spaces and $H : X \to Y$ be a nonlinear differentiable mapping from X into Y. The idea of the *gradient method* for the solution of an inverse problem is to search for a minimum of the error functional

$$\mu(\varphi) := \|H(\varphi) - f\|^2, \quad \varphi \in X \tag{3.3.17}$$

by using the direction of *steepest descent*.

We start with some general results on the gradient of a function defined on a Banach or Hilbert space. One can calculate the direction $\eta_{\max} \in X$ of the strongest increase of a function $\mu : X \to \mathbb{R}$ at a point $\varphi \in X$ using the definition of differentiability

$$\mu(\varphi + h) = \mu(\varphi) + \mu'(\varphi)h + \mu_1(\varphi, h) \tag{3.3.18}$$

by

$$\frac{\|\mu(\varphi + h) - \mu(\varphi)\|}{\|h\|} \stackrel{!}{=} \max \tag{3.3.19}$$

for h arbitrarily small. Using $\mu_1(\varphi, h) = o(\|h\|)$ this leads to

$$h^{(\max)} = \arg \max_{h \in X, \|h\|=1} \|H'h\| = \arg \max_{\hat{h} \in \mathbb{S}} \left\|H'\hat{h}\right\|, \tag{3.3.20}$$

where $H'\varphi = \mu'(\varphi)$, we recall that $\mathbb{S} := \{\varphi \in X : \|\varphi\| = 1\}$ denotes the unit sphere in X.

In a Hilbert space setting we can evolve this further and develop most of the notation used in $\mathbb{R}^m$. Let X be a Hilbert space with scalar product $\langle\cdot,\cdot\rangle$ and consider some orthonormal basis ψ_j, $j \in \mathbb{N}$. We recall that in this case every element $h \in X$ has a representation

$$h = \sum_{n=1}^{\infty} \alpha_n \psi_n \tag{3.3.21}$$

with coefficients $\alpha = (\alpha_n)_{n\in\mathbb{N}} \in \ell^2$. As shown in theorem 2.2.12, equation (2.2.38), this establishes an isometric mapping of $\varphi : X \to \ell^2$ where $\varphi \mapsto \alpha$, and we can search for the direction $h_{\max}$ in its ℓ^2-coordinates $\alpha^{(\max)} := (\alpha_n^{(\max)})_{n\in\mathbb{N}}$ defined by (3.3.21). Via the ℓ^2-representation for h, the linearity and continuity of H' we obtain

$$H'h = \sum_{n=1}^{\infty} \alpha_n H'\psi_n, \tag{3.3.22}$$

which is in $\mathbb{R}$ for every $h \in X$. Using the basis $(\psi_n)_{n\in\mathbb{N}}$ we define $d_n := H'\psi_n \in \mathbb{R}$. Since H' is linear and bounded, from theorem 2.3.10 we obtain $d := (d_n)_{n\in\mathbb{N}} \in \ell^2$. In the space X this corresponds to the element

$$\nabla\mu := \sum_{n=1}^{\infty} \left(H'\psi_n\right) \psi_n \tag{3.3.23}$$

in X, which we call the *gradient* of μ in the space X. We also note that

$$\nabla_{\ell^2}\mu := \left(H'\psi_n\right)_{n\in\mathbb{N}} \tag{3.3.24}$$

is the ℓ^2-vector of the coordinates of $\nabla\mu$. We rewrite (3.3.22) into

$$H'h = \langle h, \nabla\mu\rangle = \langle \alpha, \nabla_{\ell^2}\mu\rangle \tag{3.3.25}$$

using the scalar products in X and ℓ^2, respectively. Then, the maximization of problem (3.3.20) can be written as

$$\text{maximize } \langle \alpha, \nabla_{\ell^2}\mu\rangle_{\ell^2} \text{ under the condition } \|\alpha\|_{\ell^2} = 1. \tag{3.3.26}$$

According to the Cauchy–Schwarz equality the maximum is achieved when α and $\nabla_{\ell^2}\mu$ are linearly dependent, i.e. when $\alpha = \nabla_{\ell^2}\mu/\|\nabla_{\ell^2}\mu\|$. We summarize these results in the following lemma.

Lemma 3.3.1. *Let $\{\psi_n : n \in \mathbb{N}\}$ be an orthonormal basis in a Hilbert space X and consider some boundedly differentiable mapping $\mu : X \to \mathbb{R}$. Then the direction of steepest descent in a point $x \in X$ is given by the gradient $\nabla\mu$ defined by (3.3.23).*

A special case is given if $H : X \to Y$ is a *linear* mapping with a real-valued Hilbert space X. Then, we calculate

$$\begin{aligned}\mu'(\varphi)h &= \frac{\partial}{\partial\varphi}(\langle H\varphi - f, H\varphi - f\rangle)h \\ &= 2\langle Hh, H\varphi - f\rangle \\ &= 2\langle h, H^*(H\varphi - f)\rangle\end{aligned} \tag{3.3.27}$$

with the adjoint operator H^* with respect to the scalar product in X, such that the gradient $\nabla\mu \in X$ of μ defined by (3.3.17) is given by

$$\nabla\mu = -2H^*(f - H\varphi). \tag{3.3.28}$$

Then, the gradient method is given by

$$\varphi_{k+1} = \varphi_k + \eta_k H^*\Big(f - H(\varphi_k)\Big), \quad k \in \mathbb{N} \tag{3.3.29}$$

with starting vector φ_0 and step size $\eta_k \in \mathbb{R}$. This is also known as *Landweber iteration*, see for example [2].

If a mapping H is Fréchet differentiable, we can obtain a gradient method by local linearization, i.e. we have

$$\begin{aligned}\mu'(\varphi)h &= \frac{\partial}{\partial\varphi}(\langle H(\varphi) - f, H(\varphi) - f\rangle)h \\ &= 2\mathrm{Re}\Big\langle H'(\varphi)h, H(\varphi) - f\Big\rangle \\ &= 2\mathrm{Re}\Big\langle h, (H'(\varphi))^*(H(\varphi) - f)\Big\rangle,\end{aligned} \tag{3.3.30}$$

which for real-valued Hilbert spaces yields

$$\nabla\mu = -2(H')^*(f - H(\varphi)). \tag{3.3.31}$$

In this case, the gradient or *Landweber* method is given by

$$\varphi_{k+1} = \varphi_k + \eta_k\Big(H'(\varphi_k)\Big)^*\Big(f - H(\varphi_k)\Big), \quad k \in \mathbb{N} \tag{3.3.32}$$

with starting vector φ_0 and step size $\eta_k \in \mathbb{R}$.

Convergence and divergence of the gradient method

If we march in the gradient direction with a sufficiently small step size, in general we can expect to obtain a new vector for which the target functional (3.3.17) is smaller than for the previous iteration. What happens if we keep the step size fixed? And do we obtain convergence of the gradient method when we apply the scheme to an operator H for which the equation is ill-posed?

Let us study the linear case first. For true data $f = H\varphi^{(\text{true})}$ with the true minimum $\varphi^{(\text{true})}$ we obtain from (3.3.29)

$$\begin{aligned}\varphi_{k+1} - \varphi^{(\text{true})} &= \varphi_k - \varphi^{(\text{true})} + \eta_k H^*H\Big(\varphi^{(\text{true})} - \varphi_k\Big) \\ &= \Big(I - \eta_k H^*H\Big)\Big(\varphi_k - \varphi^{(\text{true})}\Big)\end{aligned} \tag{3.3.33}$$

for $k \in \mathbb{N}_0$. If we use a fixed step size $\eta_k = \eta$, this leads to

$$\varphi_k - \varphi^{(\text{true})} = (I - \eta H^*H)^k\Big(\varphi_0 - \varphi^{(\text{true})}\Big), \quad k \in \mathbb{N}. \tag{3.3.34}$$

If H is a compact linear operator, then H^*H is a compact self-adjoint operator in the Hilbert space X. The singular value decomposition (μ_k, ψ_k, g_k) of H is chosen such that μ_k is ordered according to its size with maximal value μ_1. With respect to the orthonormal system $(\psi_k)_{k\in\mathbb{N}}$ the operator $I - \eta H^*H$ corresponds to a multiplication of an element by

$$d_{k,n} = \Big(1 - \eta\mu_n^2\Big)^k, \quad k, n \in \mathbb{N}. \tag{3.3.35}$$

Then, we can always choose $0 < \eta < \mu_1$ such that

$$|d_{1,n}| \leqslant 1, \quad n \in \mathbb{N}. \tag{3.3.36}$$

This yields

$$\|I - \eta H^*H\| \leqslant 1 \tag{3.3.37}$$

and also

$$d_{k,n} \to 0, \quad k \to \infty \tag{3.3.38}$$

for each n for which $\mu_n > 0$. In the theorem given below, we will choose η to satisfy $\|H^*H\| < 1/\eta$. But for compact operators we know from theorem 2.4.13 that $\mu_n \to 0$ for $n \to \infty$, such that we also have

$$\|(I - \eta H^*H)^k\| = 1, \quad k \in \mathbb{N}, \tag{3.3.39}$$

i.e. we do not have norm convergence of the operator $(I - \eta H^*H)^k$. If H is injective, we have $\mu_k > 0$ and in this case (3.3.38) is satisfied for all modes. But keep in mind that it is not uniform for $n \in \mathbb{N}$. Let $(\beta_{k,n})_{n\in\mathbb{N}}$ denote the Fourier coefficients of $\varphi_k - \varphi^{(\text{true})}$ with respect to the basis $(\psi_n)_{n\in\mathbb{N}}$. Then we have

$$\beta_{k,n} = d_{k,n}\,\beta_{0,n} \to 0, \quad k \to \infty \tag{3.3.40}$$

for each $n \in \mathbb{N}$. Since $\beta_{0,n}$ is square summable and $|\beta_{k,n}| < |\beta_{0,n}|$ for all $n \in \mathbb{N}$, we decompose

$$\sum_{n=1}^{\infty} |\beta_{k,n}|^2 = \sum_{n=1}^{N} |\beta_{k,n}|^2 + \sum_{n=N+1}^{\infty} |\beta_{k,n}|^2. \tag{3.3.41}$$

Given $\epsilon > 0$ we can choose N such that the second sum is smaller than $\epsilon/2$ and then $k \in \mathbb{N}$ such that the sum of the finitely many terms of the first sum are smaller than $\epsilon/2$. This yields $\varphi_k - \varphi^{(\text{true})} \to 0$, $k \to \infty$. We summarize the result in the following theorem.

Theorem 3.3.2. *Let H be an operator from a Hilbert space X into a Hilbert space Y bounded and linear. Assume that we have true data $f \in H(X)$. Then, the gradient or Landweber method (3.3.29) with fixed step size η chosen such that $\|H^*H\| < 1/\eta$, converges towards the true solution $\varphi^{(\text{true})} \in X$.*

Of course, in general we cannot expect that true data $f \in Y$ are given, but we will have some data error of size $\delta > 0$. In this case, we need to modify our above analysis. Let $f = f^{(\text{true})} + f^{(\delta)} \in Y$ be given, where $\|f^{(\delta)}\| \leqslant \delta$. Then, from (3.3.33) with $\eta_k = \eta$ we obtain

$$\varphi_{k+1} - \varphi^{(\text{true})} = \big(I - \eta H^*H\big)\big(\varphi_k - \varphi^{(\text{true})}\big) + \eta_k H^* f^{(\delta)}, \tag{3.3.42}$$

leading to

$$e_{k+1} = N e_k + r \tag{3.3.43}$$

with $e_k := \varphi_k - \varphi^{(\text{true})}$, $N := I - \eta H^*H$ and $r = \eta H^* f^{(\delta)}$. By a simple induction the iteration formula (3.3.43) yields

$$\begin{aligned} e_k &= N^k e_0 + \sum_{j=0}^{k-1} N^j r \\ &= N^k e_0 + R^{(k)} f^{\delta}, \quad k \in \mathbb{N}, \end{aligned} \tag{3.3.44}$$

with

$$R^{(k)} := \sum_{j=0}^{k-1} (1 - \eta H^*H)^j \eta H^*, \quad k \in \mathbb{N}. \tag{3.3.45}$$

Clearly, as shown in (3.3.41), we know that $\|N^k e_0\| \to 0$ for $k \to \infty$. For the second term from $\|N\| \leqslant 1$ and $\|\eta H^*\| < 1$ we obtain the estimate

$$\|R^{(k)} f^\delta\| = \left\| \sum_{j=0}^{k-1} N^j r \right\| \leqslant k\delta, \quad k \in \mathbb{N}. \tag{3.3.46}$$

But (as shown in [2]) we can obtain a better estimate using the telescopic sum $\sum_{j=0}^{k-1} N^j (1 - N) = 1 - N^k$, which leads to

$$\left(\sum_{j=0}^{k-1} N^j \right) \eta H^* H = 1 - (1 - \eta H^* H)^k,$$

and has norm smaller or equal to one. Finally, we estimate

$$\|R^{(k)}\|^2 = \|R^k (R^k)^*\| = \left\| \left(\sum_{j=0}^{k-1} N^j \right) \eta^2 H^* H \left(\sum_{j=0}^{k-1} N^j \right) \right\| \leqslant \left\| \sum_{j=0}^{k-1} N^j \right\| \leqslant k,$$

which yields

$$\|R^{(k)} f^\delta\| \leqslant \sqrt{k}\,\delta, \quad k \in \mathbb{N}. \tag{3.3.47}$$

Next, we demonstrate that for the gradient method or Landweber iteration in general we obtain errors growing arbitrarily large for $k \to \infty$. To this end we employ the spectral representation of the operators and show by construction that we have infinite growth if $f^{(\delta)} \not\in H(X)$.

The eigenvalues of N are given by $d_{1,n}$ as defined in (3.3.35). This means that with $f_n^\delta := \langle g_n, f^{(\delta)} \rangle$ for $n \in \mathbb{N}$ the error e_k has the spectral coefficients

$$\begin{aligned} e_{k,n} &= d_{1,n}^k e_{0,n} + \frac{1 - d_{1,n}^k}{1 - d_{1,n}} \eta \mu_n f_n^{(\delta)} \\ &= d_{1,n}^k e_{0,n} + \frac{1 - \left(1 - \eta \mu_n^2\right)^k}{\mu_n} f_n^{(\delta)}, \quad k \in \mathbb{N}. \end{aligned} \tag{3.3.48}$$

If $f^{(\delta)} \not\in H(X)$, we know from theorem 2.4.23 that

$$S_N := \sum_{j=1}^{N} \left| \frac{f_n^{(\delta)}}{\mu_n} \right|^2 \to \infty, \quad N \to \infty. \tag{3.3.49}$$

Further, we know that for $k \to \infty$ we have

$$\left(1 - \eta \mu_n^2\right)^k \to 0$$

for every fixed $n \in \mathbb{N}$. Given a constant $c > 0$ we first choose N such that $S_N \geqslant 2c^2$ and then we choose k such that

$$1 - \left(1 - \eta \mu_n^2\right)^k \geqslant \frac{1}{\sqrt{2}}, \quad n = 1, \ldots, N. \tag{3.3.50}$$

This leads to

$$\begin{aligned}\|R^{(k)}f^{(\delta)}\|^2 &= \sum_{j=1}^{\infty}\left|\left(1-\left(1-\eta\mu_n\right)^k\right)\right|^2\left|\frac{f_n^{(\delta)}}{\mu_n}\right|^2 \\ &\geqslant \frac{1}{2}\sum_{j=1}^{N}\left|\frac{f_n^{(\delta)}}{\mu_n}\right|^2 = \frac{1}{2}S_N \geqslant c^2. \end{aligned} \tag{3.3.51}$$

We have shown that for given constant $c > 0$ we can find k such that $\|R^{(k)}f^{(\delta)}\| \geqslant c$, which yields $\|R^{(k)}f^{(\delta)}\| \to \infty$ for $k \to \infty$. We summarize these results in the following theorem.

Theorem 3.3.3. *Let H be a linear operator from a Hilbert space X into a Hilbert space Y. Assume that we are given data $f^{(\delta)}$ with error of size $\delta > 0$. Then we have the error estimate*

$$\left\|\varphi_k - \varphi^{(\text{true})}\right\| \leqslant \left\|(I - \eta H^* H)^k\left(\varphi_0 - \varphi^{(\text{true})}\right)\right\| + \sqrt{k}\,\delta. \tag{3.3.52}$$

If $f^{(\delta)}$ is not an element of $H(X)$, then the gradient method will diverge for $k \to \infty$ in the sense that

$$\|R^{(k)}f^{(\delta)}\| \to \infty, \qquad \|e_k\| \to \infty \qquad \textit{and} \qquad \|\varphi_k\| \to \infty \tag{3.3.53}$$

for $k \to \infty$.

We have shown that the operator $I - \eta H^*H$ is spectrally diagonal with diagonal elements in (0, 1), with an accumulation point at 1. This means that $(I - \eta H^*H)^k$ will damp spectral values by q^k with some $q \in (0, 1)$. But the speed will be arbitrarily slow since the set of eigenvalues accumulates at 1.

The second term in (3.3.42) with $f^{(\delta)} \notin H(X)$ is growing. The speed of growth is bounded by $\sqrt{k}$, but the growth can also be arbitrarily slow depending on the spectral coefficients of the particular error vector $f^{(\delta)}$. It is clear that we need to stop the iteration at some index k_* depending on δ, if we want to achieve a reasonable reconstruction. This leads us to the following section.

3.3.3 Stopping rules and convergence order

Let us study an iterative method for solving an operator equation $H(\varphi) = f$ in the form

$$\varphi_{k+1} = \varphi_k + R_k\left(f - H(\varphi_k)\right), \quad k = 0, 1, 2, \ldots \tag{3.3.54}$$

with starting state φ_0 and some operator R_k, for example with $R_k = H^*$. There are different types of convergence statements which are of interest.

If we feed in true data $f = f^{(\text{true})} = H(\varphi^{(\text{true})})$, a good method should converge towards the true solution, i.e. we expect that we can prove

$$\|\varphi_k - \varphi^{(\text{true})}\| \to 0, \quad k \to \infty. \tag{3.3.55}$$

We have also already shown that iterative methods with data $f^{(\delta)}$ which are not in the range of the observation operator H in general will diverge, i.e. we have

$$\|\varphi_k\| \to \infty, \quad k \to \infty \tag{3.3.56}$$

in this case. This shows that we need to stop the iteration at some index $k_* \in \mathbb{N}$. The stopping index will depend on the size of the data error. Usually, the iteration first reduces the error, before the divergence effect (3.3.56) takes over.

There is a minimum reconstruction error at some index $k_{\min}$. Usually a *stopping rule* is formulated which tries to find the index $k_{\min}$ in dependence of the data error δ. *A priori* stopping rules do not use the calculated iterations, *a posteriori* stopping rules use both δ and the iteration sequence φ_k.

Assume that we have a stopping rule which is given by a function $k_* = k_*(\delta)$ for $\delta > 0$. Then, we can study the behavior of $\|\varphi_{k_*(\delta)} - \varphi^{(\mathrm{true})}\|$ in dependence on the data error δ for $\delta \to 0$. This is an important type of convergence, the convergence of the regularized solution to the true solution when the data error becomes small. Often, *convergence rates* are studied, i.e. the speed of convergence

$$\left\|\varphi_{k_*(\delta)} - \varphi^{(\mathrm{true})}\right\| \to 0, \quad \delta \to 0. \tag{3.3.57}$$

In general, depending on the data $f \in Y$, the convergence can be arbitrarily slow.

A popular tool to obtain stronger convergence rates for ill-posed problems are so-called *source conditions*. Here, we will demonstrate the set-up and role of source conditions by working out a generic case of an iterative method showing how convergence statements can be achieved and how source conditions speed up the convergence order.

Consider the Landweber iteration as in (3.3.42), leading to

$$\varphi_k - \varphi^{(\mathrm{true})} = (1 - \eta H^* H)^k\left(\varphi_0 - \varphi^{(\mathrm{true})}\right) + O\left(\sqrt{k}\,\delta\right) \tag{3.3.58}$$

according to theorem 3.3.3. With respect to the singular system (μ_j, ψ_j, g_j) of H the first term corresponds to a multiplication by $d_{k,n} = (1 - \eta\mu_n^2)^k$ as defined in (3.3.35).

We remark that for any $n_* \in \mathbb{N}$ fixed and $\varphi^{(\mathrm{true})} := \psi_{n_*}$ we have $\|\varphi^{(\mathrm{true})}\| = 1$. The error in the Landweber iteration is larger than the n_*th coefficient of the reconstruction error vector $\varphi_k - \varphi^{(\mathrm{true})}$, i.e. larger than

$$E(n_*, k_*) := \left(1 - \eta\mu_{n_*}^2\right)^k \beta_{0,n_*}. \tag{3.3.59}$$

Since $\varphi_0 \in X$, its Fourier coefficients with respect to ψ_j will tend to zero for $j \to \infty$, thus $\beta_{0,n_*} \to -1$, $n_* \to \infty$. Since $\mu_n \to 0$ for $n \to \infty$ we have

$$E(n_*, k_*) \to 1, \quad n_* \to \infty \tag{3.3.60}$$

for any k_* fixed. This shows that there can be no uniform convergence for this reconstruction method on $B_1 = \{\varphi^{(\mathrm{true})} \in X : \|\varphi^{(\mathrm{true})}\| \leqslant 1\}$. Whatever stopping rule we choose, we can always pick a true solution which yields an error arbitrarily close to one.

The above situation changes drastically when we introduce *source conditions*. We define the space

$$X^{(s)} := \left\{ \varphi \in X : \|\varphi\|_s^2 := \sum_{n=1}^{\infty} |\langle \varphi, \psi_n \rangle|^2 \mu_n^{-2s} < \infty \right\}. \tag{3.3.61}$$

The space $X^{(s)}$ demands a decay of the Fourier coefficients with respect to the singular system of H with weights given by powers $2s$ of the singular values μ_n. It is equal to the range $((H^*H)^s)X$ of the operator $(H^*H)^s$ in X.

Assume that $\varphi_0 - \varphi^{(\text{true})}$ is in $X^{(s)}$. This means that its Fourier coefficients $\beta_{0,n}$ satisfy

$$\left\| \varphi_0 - \varphi^{(\text{true})} \right\|_s^2 = \sum_{n=1}^{\infty} \beta_{0,n}^2 \mu_n^{-2s} < \infty. \tag{3.3.62}$$

We then estimate $\varphi_k - \varphi^{(\text{true})}$ by

$$\begin{aligned} \left\| \varphi_k - \varphi^{(\text{true})} \right\|_X &\leqslant \left(\sum_{n=1}^{\infty} \left(1 - \eta\mu_n^2\right)^k \beta_{0,n}^2 \right)^{\frac{1}{2}} + O\left(\sqrt{k}\delta\right) \\ &= \left(\sum_{n=1}^{\infty} \left(1 - \eta\mu_n^2\right)^k \mu_n^{2s} \left(\beta_{0,n}^2 \mu_n^{-2s}\right) \right)^{\frac{1}{2}} + O\left(\sqrt{k}\delta\right). \end{aligned} \tag{3.3.63}$$

The supremum of the term $(1 - \eta\mu^2)^k \mu^{2s}$ with respect to $\mu \in (0, \sqrt{1/\eta})$ is obtained by setting its gradient to zero. A little calculation leads to

$$\begin{aligned} \sup_{\mu \in \left(0, \sqrt{1/\eta}\right)} \left| \left(1 - \eta\mu^2\right)^k \mu^{2s} \right| &= \eta^{-s} \left(1 - \frac{s}{s+k}\right)^k \left(1 + \frac{k}{s}\right)^{-s} \\ &\sim \left(\frac{s}{\eta e}\right)^s k^{-s} =: c_0 k^{-s} \end{aligned} \tag{3.3.64}$$

for $k \to \infty$. Now, from (3.3.63) we obtain

$$\left\| \varphi_k - \varphi^{(\text{true})} \right\|_X \sim \sqrt{c_0} k^{-s/2} \left\| \varphi_0 - \varphi^{(\text{true})} \right\|_s + \sqrt{k}\delta, \quad k \to \infty. \tag{3.3.65}$$

Choosing an *a priori* stopping rule

$$k := \delta^{\frac{-2}{s+1}}, \quad \delta > 0 \tag{3.3.66}$$

we calculate

$$k^{-s/2} = \delta^{\frac{s}{s+1}}, \tag{3.3.67}$$

and

$$\sqrt{k}\,\delta = \delta^{\frac{-1}{s+1}+1} = \delta^{\frac{s}{s+1}}. \tag{3.3.68}$$

We collect the estimates into the following result.

Theorem 3.3.4. *Assume that we have the source condition $\varphi_0 - \varphi^{(\mathrm{true})} \in X^s$ and the stopping rule $k_* = \delta^{\frac{-2}{s+1}}$. Then, Landweber iteration (3.3.34) satisfies the convergence estimate*

$$\left\| \varphi_{k_*(\delta)} - \varphi^{(\mathrm{true})} \right\|_X \leqslant c\delta^{\frac{s}{s+1}}, \quad \delta > 0 \tag{3.3.69}$$

with some constant $c > 0$.

Bibliography

[1] Groetsch C W 1993 *Inverse Problems in the Mathematical Sciences* (Leipzig: Vieweg)

[2] Engl H W, Hanke M and Neubauer A 1996 *Regularization of Inverse Problems* (*Mathematics and its Applications* vol 375) (Dordrecht: Kluwer Academic)

[3] Hansen P C 2010 *Discrete Inverse Problems: Insight and Algorithms* (Philadelphia, PA: Society for Industrial and Applied Mathematics)

[4] Kirsch A 1996 *An Introduction to the Mathematical Theory of Inverse Problems* (*Applied Mathematical Sciences* vol 120) (New York: Springer)

[5] Schuster T, Kaltenbacher B, Hofmann B and Kazimierski K S 2012 *Regularization Methods in Banach Spaces* (Berlin: De Gruyter)

[6] Kaltenbacher B, Neubauer A and Scherzer O 2008 *Iterative Regularization Methods for Nonlinear Ill-Posed Problems* (Berlin: De Gruyter)

[7] Wang Y, Yagola A G and Yang C (ed) 2010 *Optimization and Regularization for Computational Inverse Problems and Applications* (Heidelberg: Springer)

[8] Potthast R 2001 On the convergence of a new Newton-type method in inverse scattering *Inverse Problems* **17** 1419–34

[9] Kress R 1999 *Linear Integral Equations* (*Applied Mathematical Sciences* vol 82) 2nd edn (New York: Springer)

[10] Colton D and Kress R 1983 *Integral Equation Methods in Scattering Theory* (New York: Wiley)

Chapter 4

A stochastic view of inverse problems

Stochastic methods have been extremely popular and powerful for the solution of inverse problems. They provide additional insight and tools which go beyond the simple calculation of some solution or expectation value.

Here, we will introduce the basic concepts in a concise way, including the role of estimators for inversion, Gaussian and non-Gaussian densities and Bayesian methods. A particular focal point here is the relationship between *stochastic* and *deterministic* tools and viewpoints.

We have collected all basic stochastic tools and notations on which our arguments are based into section 4.5. For further reading we refer the reader to Kaipio and Somersalo [1], Biegler *et al* [2] and Aster *et al* [3]. Parameter estimation is presented for example in the books of Tarantola [4], Beck and Arnold [5] and Heijden *et al* [6]. For basic stochastics see, for example, to Georgii [7].

4.1 Stochastic estimators based on ensembles and particles

It is well known that we can use n independent draws x_ℓ, $\ell = 1, \ldots, n$ from a probability distribution to obtain an approximation for the expectation value of some random variable (see (4.5.16)). This leads to *stochastic estimators* and to the *theory of estimation.*

Today, *ensemble methods* are very popular in data assimilation. They use an ensemble of states of a dynamical system to estimate the covariance or *uncertainty* of the background states, see figure 4.1, and then feed this covariance into the state estimation when measurements are available.

The term *particle method* is used when in general non-Gaussian distributions are estimated by an ensemble of states, called *the particles.* Particle methods are Monte Carlo methods, where the particles are drawn from some distribution and are then propagated through time by the underlying dynamical system and are employed to estimate the distributions at some given point in time, see the dashed line in figure 4.1.

doi:10.1088/978-0-7503-1218-9ch4

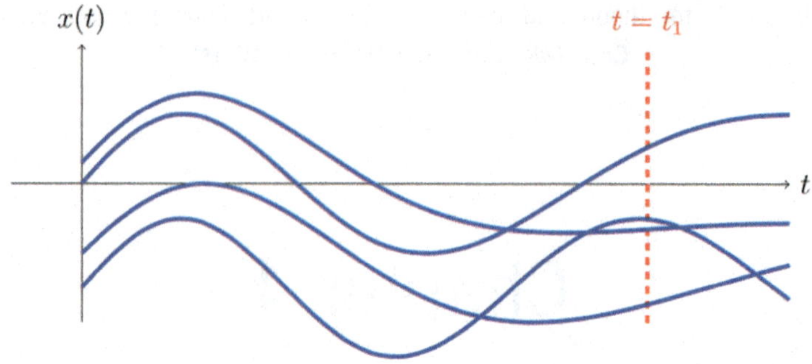

Figure 4.1. An ensemble of states propagated through time can be employed to estimate the mean and variance of some time-dependent distribution, for example at time $t = t_1$ as indicated by the dashed line.

Estimator for some probability measure. Consider a family of probability measures P_ϑ (for a definition see (4.5.2)) depending on some parameter $\vartheta \in \Theta$ and a real-valued function $g(\vartheta)$ which is to be estimated. In the simplest case the function g is the parameter ϑ itself. Let Z be a finite or countable set of draws from the probability distribution P_{ϑ_0} with the unknown parameter ϑ_0. Then, any mapping

$$\hat{g} : Z \to \mathbb{R} \tag{4.1.1}$$

is called an *estimator* of g. Of course, we need estimators with particular properties which allow us to deduce knowledge about the unknown parameter ϑ_0. Before we go into more detail, let us study an example.

Example. We consider a probability distribution P with Gaussian density (4.5.27) on $\mathbb{R}$. The Gaussian density depends on two parameters, its mean μ and the variance σ^2 or standard deviation σ. Assume that σ is known. Then we obtain a family of measures by (4.5.27) with $\vartheta = \mu$. Let A_j, $j \in \mathbb{N}$ be the random variables drawing independently from P, i.e. we obtain x_1 as a measurement from the first draw, x_2 as the measurement from the second draw etc. Then the task is to estimate $g(\vartheta) = \mu$ from $x_1, x_2, \ldots, x_n$ with some $n \in \mathbb{N}$. According to (4.5.26), the probability of measuring $x_1, \ldots, x_n$ is given by the density

$$\begin{aligned} f(x) &= f(x_1) \cdot \cdots \cdot f(x_n) \\ &= \left(2\pi\sigma^2\right)^{-n/2} \exp\left(-\frac{1}{2\sigma^2} \sum_{k=1}^{n} (x_k - \mu)^2 \right), \quad x = (x_1, \ldots, x_n). \end{aligned} \tag{4.1.2}$$

What is a good estimate for μ? If we have only one measurement, what choice of μ is reasonable? Of course for one measurement, everything can happen since for every interval $[a, b] \subset \mathbb{R}$ the probability of measuring $x_1 \in [a, b]$ is positive. But for more measurements we can say more. It is shown in (4.5.16), that the probability of

$$\hat{\mu} = \frac{1}{n} \sum_{k=1}^{n} x_k \tag{4.1.3}$$

to have a distance larger than ϵ from the true expectation value μ tends to zero. Thus (4.1.3) is an *estimator* in the sense of (4.1.1) which for $n \to \infty$ converges in probability (as defined in (4.5.18)) to the true mean of the Gaussian distribution. We immediately see that the expectation for $\hat{\mu}$ is $\mathbb{E}(\hat{\mu}) = \mu$. Since x_k, $k = 1, \ldots, n$ are independent draws, by (4.5.7) and (4.5.9) the variance of (4.1.3) is given by

$$\operatorname{Var}(\hat{\mu}) = \frac{1}{n^2}\sum_{k=1}^{n}\operatorname{Var}(x_k) = \frac{1}{n}\sigma^2 \tag{4.1.4}$$

with the variance σ^2 of x_k.

Maximum likelihood estimators. The estimator (4.1.3) can be obtained by different approaches. One idea is to search for the probability distribution under which the measured values are most likely, i.e. maximizing the probability $P_\vartheta((x_1, \ldots, x_n))$. The corresponding estimator is called the *maximum likelihood estimator*. In the above example we search for the parameter μ for which (4.1.2) is maximal. Searching for the maximum of (4.1.2) is equivalent to minimizing the functional

$$J(\mu) := \sum_{k=1}^{n}(x_k - \mu)^2. \tag{4.1.5}$$

The first order optimality condition $\frac{\mathrm{d}J(\mu)}{\mathrm{d}\mu} = 0$ applied to (4.1.5) leads to

$$\sum_{k=1}^{n}(x_k - \mu) = 0 \Leftrightarrow n\mu = \sum_{k=1}^{n}x_k,$$

which yields (4.1.3), i.e. the estimator (4.1.3) is the maximum likelihood estimator for estimating the mean of a Gaussian density. Using $\frac{\mathrm{d}J(\mu, \sigma)}{\mathrm{d}\sigma} = 0$, by some lines of calculation the maximum likelihood estimator for estimating both the mean and variance of (4.1.2) can be seen to be (4.1.3) and

$$\hat{\sigma}^2 = \frac{1}{n}\sum_{k=1}^{n}(x_k - \hat{\mu})^2.$$

Here, by some careful but elementary calculation using (4.5.7) the expectation $\mathbb{E}(\hat{\sigma}^2)$ of the estimator can be seen to be equal to

$$\mathbb{E}(\hat{\sigma}^2) = \frac{n-1}{n}\sigma^2.$$

To keep the estimator bias free, by multiplication with $\frac{n}{n-1}$ the factor $\frac{1}{n}$ is replaced by $\frac{1}{n-1}$, leading to

$$\hat{\sigma}^2 = \frac{1}{n-1}\sum_{k=1}^{n}(x_k - \hat{\mu})^2. \tag{4.1.6}$$

This estimator is of particular importance for ensemble methods in data assimilation, since it is used for the ensemble Kalman filter (EnKF), which is one of the main starting points for modern large-scale data assimilation techniques.

As a consequence of the weak law of large numbers, lemma 4.5.1, we immediately obtain stochastic convergence of the estimator $\hat{\mu}$ given by (4.1.3), i.e. we have

$$P\left(\left|\frac{1}{n}\sum_{j=1}^{n}A_j - \mu\right| \geqslant \epsilon\right) \leqslant \frac{c}{\epsilon^2 n} \tag{4.1.7}$$

with some constant c. Applying the same result to $\tilde{A}_j := (A_j - \mu)^2$, under the condition that the fourth moments are bounded, we obtain

$$P\left(\left|\frac{1}{n}\sum_{j=1}^{n}(x_j - \mu)^2 - \sigma^2\right| \geqslant \epsilon\right) \leqslant \frac{\tilde{c}}{\epsilon^2 n} \tag{4.1.8}$$

with some constant $\tilde{c}$. Estimators which are stochastically convergent are also called *consistent estimators*. We summarize these results in the following lemma.

Lemma 4.1.1. *The estimators (4.1.3) and (4.1.6) for estimating the mean μ and variance of a distribution with bounded centralized moments up to order four from independent draws x_1, x_2, x_3,... are* consistent.

A multi-dimensional Gaussian density (4.5.28) in $\mathbb{R}^n$ depends on vectorial parameters, the mean $\mu \in \mathbb{R}^m$ and the covariance matrix $B \in \mathbb{R}^{n \times n}$. In this case, there are $n^2 + n$ parameters to determine, where $(n + 1)n/2$ of them are independent due to the symmetry of B. A consistent estimator for the mean and the covariance from measurements $x_\ell \in \mathbb{R}^m$ for $\ell = 1, 2, 3, \ldots$ is given by

$$\hat{\mu} := \frac{1}{n}\sum_{\ell=1}^{n} x_\ell \tag{4.1.9}$$

and

$$\hat{B} := \frac{1}{n-1}\sum_{\ell=1}^{n}(x_\ell - \mu)(x_\ell - \mu)^T, \tag{4.1.10}$$

see (4.5.4) for the definition of the covariance, from which (4.1.10) is obtained as in (4.1.6).

4.2 Bayesian methods

Assume that we consider two quantities $x \in \Omega \subset \mathbb{R}^m$ and $y \in V \subset \mathbb{R}^m$, where we will later assume that y is a measurement of a function of x, i.e. $y = f(x) + e$ with some error e. We denote the joint probability distribution of x and y by its density function $p(x, y)$. This means that

$$p(U, W) := \int_U \int_W p(x, y)\, \mathrm{d}x\, \mathrm{d}y \tag{4.2.1}$$

describes the probability of finding x in $U \subset \Omega$ and y in $W \subset V$. Clearly, for probability densities we have $p(x, y) \geqslant 0$ and we demand the standard normalization condition

$$\int_\Omega \int_V p(x, y)\, \mathrm{d}x\, \mathrm{d}y = 1, \tag{4.2.2}$$

since the probability of finding the variables in any state is equal to one. The conditional probability of x under the condition y is given by

$$p(x|y) = c\, p(x, y), \tag{4.2.3}$$

where c is a normalization constant such that

$$1 = \int_{\Omega} p(x|y)\,\mathrm{d}x = c \int_{\Omega} p(x, y)\,\mathrm{d}x. \tag{4.2.4}$$

Thus, the constant c is the inverse of

$$p(y) = \int_{\mathbb{R}^m} p(x, y)\,\mathrm{d}x, \quad y \in \mathbb{R}^m, \tag{4.2.5}$$

and we have

$$p(x, y) = p(x|y) \cdot p(y), \quad x \in \Omega, \;\; y \in V. \tag{4.2.6}$$

We are now prepared to introduce the well-known *Bayes' formula* for inverse problems.

Theorem 4.2.1 (Bayes' formula for inverse problems). *The posterior probability $p(x|y)$ with measurement $y \in V$ is given by*

$$p(x|y) = \frac{p(x) \cdot p(y|x)}{p(y)}, \quad x \in \Omega, \;\; y \in V. \tag{4.2.7}$$

In the case where $p(y)$ is not available, we write

$$p(x|y) = c\, p(x) \cdot p(y|x), \quad x \in \Omega, \tag{4.2.8}$$

where c is given by

$$c = \left(\int_{\Omega} p(x) p(y|x)\,\mathrm{d}x \right)^{-1}. \tag{4.2.9}$$

Proof. The Bayes' formula (4.2.7) is a direct consequence of (4.2.6), which is written twice with the roles of x and y interchanged, leading to

$$p(x, y) = p(x|y) \cdot p(y) = p(y|x) \cdot p(x), \quad x \in \Omega, \;\; y \in V. \tag{4.2.10}$$

The probability $p(y)$ can be calculated from the normalization condition (4.2.4), leading directly to (4.2.9). □

We need to note that Bayes' formula (4.2.8) is just a generic formula for conditional probabilities, it needs further algorithmic tools to actually *solve* an inverse problem. In the case where we know nothing about $x \in \Omega$, the probability distribution $p(x)$ will be a constant given by

$$p(x) = \frac{1}{\int_{\Omega} \mathrm{d}x}, \quad x \in \Omega. \tag{4.2.11}$$

In this case Bayes' formula just states that the probability of a state x is proportional to the conditional probability of the measurement y given the state x.

When a *prior* density $p(x)$ is given and when we know how measurements y are distributed in the case of a given x, Bayes' formula describes a proper *posterior* probability distribution $p(x|y)$ which coincides with the knowledge given by the prior $p(x)$ and the conditional measurement distribution $p(y|x)$.

We will explore methods to make use of Bayes' formula to actually use these probability distributions by calculating particular quantities based on them (such as its mean or variance), in the following section using Monte Carlo methods, in sections 5.4 and 5.5 using the Kalman filter and the EnKF, and in section 5.6 using Bayesian data assimilation and particle filters.

4.3 Markov chain Monte Carlo methods

Markov chain Monte Carlo (MCMC) methods generate sequences of states $x_1, x_2, x_3, \ldots$ which sample some particular probability distribution p. Based on this sequence particular quantities such as the *mean* or the *variance* can be estimated, e.g. by the estimators (4.1.3) or (4.1.6).

The term *Monte Carlo* refers to a repeated random sampling of some distribution. A *Markov chain* is a sequence (a *chain*) of states such that the subsequent state x_{k+1} depends only on the state x_k at time t_k and not on the previous history of states x_{k-1}, x_{k-2} etc at times $t_{k-1}, t_{k-2}, \ldots$.

For *inverse problems*, MCMC methods are of importance when the probability distribution p under investigation is the distribution $p(x|y)$ of some states $x \in X$ which we would like to reconstruct given some measurements y in a measurement space Y.

Here, x can be either states of some dynamical systems, a set of parameters or a set of unknown distributed quantities, or both states and unknown parameter functions. MCMC methods have the capability to provide both an estimate of the *mean* or *most likely state* of some distribution as well as of its *uncertainty*. With a high sampling density, they are able to provide information about an unknown distribution up to any precision.

Note that the *solution* of a stochastic inverse problems can be either the mean of some distribution, its variance or any property of the full posterior distribution $p_*(x) := p(x|y)$ of states given some measurements $y \in Y$.

Our goal here is to provide a consistent, self-contained approach including its full derivation and proof for MCMC methods in solving inverse problems.

Take an open subset $\Omega \subset \mathbb{R}^m$ for X and let $p(x)$ denote some probability density on Ω. We assume that $p(x)$ is in the space $L^1(\Omega)$ which is the set of all measurable functions whose absolutes are integrable on Ω. Also, by $P(x \to x')$ we denote some probability density which describes the *transition probability* of a Markov chain, i.e. the probability of a time-discrete stochastic process to draw x' at time t_{k+1} when $x \in \Omega$ was drawn in its previous step t_k. For our further derivation we write $P(x \to x')$ as a *kernel* of an integral equation in the form

$$P(x \to x') = k(x', x) + r(x)\delta(x' - x), \quad x, x' \in \Omega. \tag{4.3.1}$$

Here, $\delta(x - x')$ is the Dirac delta function and we assume k to be continuous and bounded on $\Omega \times \Omega$ and the second term includes the case where with some probability we reject draws and keep the previous element $x \in \Omega$. The condition

$$\begin{aligned} 1 &= \int_\Omega P(x \to x')\,\mathrm{d}x' \\ &= \int_\Omega k(x', x)\,\mathrm{d}x' + r(x), \quad x \in \Omega \end{aligned} \tag{4.3.2}$$

yields

$$r(x) = 1 - \int_\Omega k(x', x)\,\mathrm{d}x', \quad x \in \Omega. \tag{4.3.3}$$

Let $p_k(x)$ be the probability to draw x at time t_k and denote the probability to draw x' at t_{k+1} when x was drawn at t_k by $P(x \to x')$. Then, the probability to obtain x' at time t_{k+1} is given by

$$\begin{aligned} p_{k+1}(x') &= \int_\Omega p_k(x) P(x \to x')\,\mathrm{d}x \\ &= \int_\Omega k(x', x) p_k(x)\,\mathrm{d}x + r(x') p_k(x'), \quad x' \in \Omega. \end{aligned} \tag{4.3.4}$$

In general, we denote the transition probability from x at time t_k to $x' \in \Omega$ at time t_{k+m} by $P_{k,k+m}(x \to x')$. We will assume that this transition does not depend on its starting time t_k, such that $P_{k,k+m}$ is independent of $k \in \mathbb{N}$.

Clearly, any path from x at t_k to x' at t_{k+m} will lead through points $z \in \Omega$ at time t_{k+j} for $1 \leqslant j < m$. The sets $G(z)$, $z \in \Omega$, of all the paths through $z \in \Omega$ are disjoint, such that by basic properties of a probability distribution as described in (4.5.2) the probabilities need to be summed up to calculate the probability of all paths from x to x'. This is expressed by the *Chapman–Kolmogorov equation*

$$P_{k,k+m}(x \to x') = \int_\Omega P_{k,k+j}(x \to z) P_{k+j,k+m}(z \to x')\,\mathrm{d}z. \tag{4.3.5}$$

We employ the abbreviations

$$(Kp)(x') := \int_\Omega k(x', x) p(x)\,\mathrm{d}x \tag{4.3.6}$$

$$(Rp)(x') := r(x') p(x') \tag{4.3.7}$$

for $x' \in \Omega$ and

$$A := K + R. \tag{4.3.8}$$

The probability density is *stationary* at density p, if $p_{k+1} = p_k$, i.e. if the density p satisfies

$$p = Ap. \tag{4.3.9}$$

with A defined in (4.3.8). The equation (4.3.9) is the *fixed point equation* of stationary transition probabilities. Using equation (4.3.3), we can transform (4.3.9) into

$$\begin{aligned} p(x') &= \int_\Omega k(x', x)p(x)\,\mathrm{d}x + r(x')p(x') \\ &= \int_\Omega k(x', x)p(x)\,\mathrm{d}x - \int_\Omega k(\xi, x')p(x')\,\mathrm{d}\xi + p(x'). \end{aligned} \tag{4.3.10}$$

This leads to the MCMC *balance equation*

$$\int_\Omega k(x', x)p(x)\,\mathrm{d}x = \int_\Omega k(\xi, x')p(x')\,\mathrm{d}\xi, \quad x' \in \Omega. \tag{4.3.11}$$

Here, our goal is to show the existence of a unique fixed point using the Banach fixed-point theorem. To this end, we first need to transform the equation into a *nonlinear fixed point equation.* We assume that the transition probability $k(x', x)$ satisfies a positivity condition

$$k(x', x) \geqslant \rho > 0, \quad x', x \in \Omega, \tag{4.3.12}$$

for some parameter $\rho > 0$. In this case, we can define

$$k_c(x', x) := k(x', x) - c, \quad x', x \in \Omega, \tag{4.3.13}$$

with $0 < c \leqslant \rho/2$ sufficiently small. Then, $\tilde{k}$ is still positive on Ω. Now, we modify the operator A by defining

$$(A_c p)(x') := \int_\Omega k_c(x', x)p(x)\,\mathrm{d}x + r(x')p(x'), \quad x' \in \Omega. \tag{4.3.14}$$

With the operator A_c given by (4.3.14) the fixed point equation (4.3.9) or (4.3.10), respectively, becomes

$$p(x') = (A_c p)(x') + c, \quad x' \in \Omega, \tag{4.3.15}$$

where we used that p is a probability distribution and has an integral equal to one.

Lemma 4.3.1. *Under the condition (4.3.12) the linear operator A_c defined by (4.3.14) is a contraction mapping in $L^1(\Omega)$ if we take c small enough. That is*

$$\|A_c p\| \leqslant \kappa \|p\| \tag{4.3.16}$$

with some constant $0 < \kappa < 1$ for any $p \in L^1(\Omega)$, where $\|p\| = \|p\|_{L^1(\Omega)} := \int_\Omega |p(x)|\,\mathrm{d}x$.

Proof. We first use $k(x', x) - c > 0$ and (4.3.3) to estimate

$$\begin{aligned} \int_\Omega |k(x', x) - c|\,\mathrm{d}x' &\leqslant \int_\Omega (k(x', x) - c)\,\mathrm{d}x' \\ &= \int_\Omega k(x', x)\,\mathrm{d}x' - |\Omega| c \\ &= 1 - r(x) - |\Omega| c, \end{aligned} \tag{4.3.17}$$

where our assumptions guarantee that the right-hand side is positive. Now, we estimate the $L^1(\Omega)$-norm of $A_c p$ by

$$
\begin{aligned}
||A_c p||_{L^1(\Omega)} &= \int_\Omega \left| \int (k(x', x) - c)p(x)\, dx + r(x')p(x') \right| dx' \\
&\leqslant \int_\Omega \int_\Omega |k(x', x) - c|\, dx' |p(x)|\, dx + \int_\Omega r(x')|p(x')|\, dx' \\
&\leqslant \int_\Omega (1 - r(x) - |\Omega|c)|p(x)|\, dx + \int_\Omega r(x')|p(x')|\, dx' \\
&= (1 - |\Omega|c)||p||_{L^1(\Omega)}.
\end{aligned} \tag{4.3.18}
$$

By our assumptions we have chosen c such that $\kappa := |1 - |\Omega|c| < 1$ by taking c small enough, such that A_c is a contraction in $L^1(\Omega)$. □

By an application of the Banach fixed-point theorem to the nonlinear equation (4.3.15) in $L^1(\Omega)$ we immediately obtain the following important conclusion from the above result.

Theorem 4.3.2. *Under the conditions of lemma 4.3.1 the fixed point equation (4.3.9) has a unique fixed point p_* in $L^1(\Omega)$ and for any starting density p_0 in $L^1(\Omega)$ the sequence of densities p_j defined by*

$$
p_j := Ap_{j-1} = (K + R)p_{j-1}, \quad j = 1, 2, 3, \ldots \tag{4.3.19}
$$

converges towards this fixed point distribution p_*.

Proof. We clearly have

$$
||Aq - Ap|| = ||A_c q - A_c p|| \leqslant \kappa ||q - p|| \tag{4.3.20}
$$

and hence the result follows from the Banach fixed-point theorem. □

Assume that we are starting at some point $x_0 \in \Omega$ and create the sequence x_j, $j \in \mathbb{N}$, of points by drawing from the transition probability $P(x \to x')$. Then, x_1 is generated by drawing from $p_0(x) := P(x_0 \to x)$. By the Chapman–Kolmogorov equation (4.3.5) point x_2 comes from drawing from

$$
\begin{aligned}
p_1(x) &= \int_\Omega p_0(z)P(z \to x)\, dz \\
&= \left(Ap_0\right)(x), \quad x \in \Omega.
\end{aligned} \tag{4.3.21}
$$

In the same way we see that x_j is a draw from p_j defined inductively by (4.3.19). We will denote the probability distribution p_j which was starting with $p_0[x_0](x) := P(x_0 \to x)$, $x \in \Omega$, by

$$
p_j[x_0](x) := \left(A^j p_0[x_0]\right)(x), \quad x \in \Omega,\ j \in \mathbb{N}. \tag{4.3.22}
$$

As a consequence of theorem 4.3.2 we now obtain the following important convergence.

Corollary 4.3.3. *Choose some* $z \in \Omega$ *and assume that the conditions of lemma 4.3.1 are satisfied. Then in* $L^1(\Omega)$ *the sequence of probability densities* $p_j[z]$ *converges towards the unique fixed point* p_* *of equation (4.3.9), i.e.*

$$\left\| p_j[z] - p_* \right\|_{L^1(\Omega)} \to 0, \quad j \to \infty. \tag{4.3.23}$$

If we have convergence of probability distributions in $L^1(\Omega)$, this implies convergence of all main moments, including mean and variance (based on (4.5.6)).

Lemma 4.3.4. *Assume that we have a sequence* p_j *of probability densities such that* $p_j \to p_*$ *in* $L^1(\Omega)$*. Then, for* $\ell \in \mathbb{N}$ *we have*

$$\int_\Omega x^\ell p_j(x)\,\mathrm{d}x \to \int_\Omega x^\ell p_*(x)\,\mathrm{d}x, \tag{4.3.24}$$

i.e. the ℓ*th moment of the distributions converge for* $j \to \infty$.

Proof. We estimate

$$\left| \int_\Omega x^\ell \big(p_j(x) - p_*(x)\big)\,\mathrm{d}x \right| \leqslant C|\Omega|^\ell \left\| p_j - p_* \right\|_{L^1(\Omega)} \tag{4.3.25}$$

for some constant $C > 0$ which tends to 0 for $j \to \infty$, and the proof is complete. □

Based on corollary 4.3.3 we can now formulate strategies to generate a sequence of points $x_0, x_1, x_2, \ldots$ which approximately samples the probability distribution p_*. The simplest approach would be to fix some starting point z and generate $x_0, x_1, x_2, \ldots$ by drawing n times starting from one and the same z again and again. This samples $p_n[z]$. By the weak law of large numbers lemma 4.5.1 we know that the estimator for the mean converges in probability to the mean of $p_n[z]$, and by corollary 4.3.3 and lemma 4.3.4 we have convergence of the mean of this distribution to the mean of p_*. This establishes convergence of the MCMC method for the estimation of the mean of p_* and a similar argument can be applied to higher moments.

In general, MCMC methods do not use a restart from z for every draw, but starting with the current point x_j on average is a much better approximation. This leads to the following standard version of MCMC sampling.

Algorithm 4.3.5 (MCMC sampling). *For realizing the* MCMC method *we start with some point* x_0.

1. *We choose* $n \in \mathbb{N}$ *times according to the transition probability* $P(x \to x')$*, i.e. generating the points* $x_{0,\xi}$, $\xi = 1, 2, 3, \ldots, n$*. For n sufficiently large, according to corollary 4.3.3 this corresponds to choosing from an approximation* $p_j[x_0]$ *to* p_**. We define* $x_1 := x_{0,n}$.
2. *With* x_0 *replaced by* x_1 *we continue to choose points* $x_{1,\xi}$ *for* $\xi = 1, 2, 3, \ldots, n$ *according to the transition probability* $P(x \to x')$*. We define* $x_2 := x_{1,n}$.
3. *Step 2 is repeated, generating a sequence of points* $x_0, x_1, x_2, \ldots, x_k$ *for* $k \in \mathbb{N}_0$.

The sequence x_k, $k \in \mathbb{N}_0$*, consists of all the nth points of the sequence which is obtained by starting from* x_0 *and choosing from* $P(x \to x')$.

The main point and distinguishing feature of different MCMC methods is how to construct transition probability densities $P(x \to x')$ when some probability distribution p_* is given, such that p_* is the unique fixed point of equation (4.3.9) with kernel $K(x', x) := P(x \to x')$. We will introduce two main approaches in the following section 4.4.

4.4 Metropolis–Hastings and Gibbs sampler

In the previous section we have studied an approach for sampling a particular probability density distribution $p(x)$, $x \in \Omega$, when a transition probability $P(x \to x')$ is given such that $p(x)$ is the unique fixed point of the *fixed point equation*

$$p(x') = \int_{\Omega} p(x)P(x \to x')\,\mathrm{d}x, \quad x' \in \Omega. \tag{4.4.1}$$

Here, we will start with the *detailed balance equation*, which is a differential form of the MCMC balance equation (4.3.11), given by

$$k(x', x)p(x) = k(x, x')p(x'), \quad x, x' \in \Omega. \tag{4.4.2}$$

We note that integrating (4.4.2) with respect to x using (4.3.3) leads to (4.3.10) and thus to (4.4.1). In general, the detailed balance is stronger than the stationarity condition given by equation (4.4.1), i.e. there are stationary distributions of Markov chains which do not satisfy the detailed balance equation.

Let us start with some continuous transition probability $q(x', x)$ for the transition of x to x', which we use as a *proposal density* to suggest draws as candidates for our transition probability density $k(x', x)$, $x, x' \in \Omega$. The goal is to construct a correction $\alpha(x', x)$ such that have

$$k(x', x) = \alpha(x', x)q(x', x), \quad x, x' \in \Omega, \tag{4.4.3}$$

is a transition kernel which satisfies the detailed balance equation (4.4.2). The choice

$$\alpha(x', x) := \min\left\{1, \frac{p(x')q(x, x')}{p(x)q(x', x)}\right\} \tag{4.4.4}$$

is known as the *Metropolis–Hastings algorithm*. We obtain (4.4.2) using (4.4.3) with (4.4.4), since for x', x with $p(x')q(x, x') < p(x)q(x', x)$ we have $\alpha(x, x') = 1$ and

$$\begin{aligned} k(x', x)p(x) &= \frac{p(x')q(x, x')}{p(x)q(x', x)} q(x', x)p(x) \\ &= p(x')q(x, x') \\ &= p(x')\alpha(x, x')q(x, x') \\ &= p(x')k(x, x'), \quad x', x \in \Omega, \end{aligned} \tag{4.4.5}$$

and this holds analogously for $p(x')q(x, x') \geqslant p(x)q(x', x)$.

Assume that $p(x)$ can be calculated for any given x, and $q(x', x)$ is a chosen proposal distribution. Then, to draw from $k(\cdot, x)$ can be carried out by first choosing x' from $q(\cdot, x)$. Then, the number $\alpha(x', x)$ is calculated. We randomly choose s from

[0, 1] with uniform distribution and if $s \in [0, \alpha(x', x)]$ we keep x', otherwise we choose $x' = x$. This corresponds to choosing from $k(\cdot, x)$ defined by (4.4.3).

By the arguments of theorem 4.3.2 and lemma 4.3.4 we obtain convergence in probability of the MCMC method with the Metropolis–Hastings kernel for sampling the distribution $p(x)$, $x \in \Omega$.

As an example consider some bimodal probability distribution on a two-dimensional domain $\Omega = [a_1, b_1] \times [a_2, b_2]$ which is chosen as $[0, 10] \times [0, 5]$ in figure 4.2. Here, we choose the prior density as the sum of two Gaussians with centers $x_1 := (3, 2)$ and $x_2 := (8, 3)$ and covariance matrices

$$B_1 := \begin{pmatrix} 2 & 1 \\ 1 & 2 \end{pmatrix}, \quad B_2 := \begin{pmatrix} 2 & -1.8 \\ -1.8 & 2 \end{pmatrix}. \tag{4.4.6}$$

We have used a measurement $y = (5, 4)$ and a measurement error distribution given by a Gaussian with mean at y and variance one. As proposal density $q(x', x) := cN_0^1(|x' - x|)$ we choose a normal distribution N_0^1 with mean zero and variance one. Since we work on a bounded domain, it needs some normalization with the constant $c = c(x)$. Practically, this is achieved by choosing from an unbounded normal distribution and repeating the choice when points outside Ω have been obtained.

We now come to an alternative definition of a transition kernel. With our space dimension n we define

$$K(x', x) := \prod_{\xi=1}^{n} p\left(x'_\xi | x'_1, \ldots, x'_{\xi-1}, x_{\xi+1}, \ldots, x_n\right) \tag{4.4.7}$$

for $x', x \in \Omega$, which is known as a *Gibbs sampler* (or more precisely a *single component Gibbs sampler*). Here, $p(x'_\xi | x'_1, \ldots, x'_{\xi-1}, x_{\xi+1}, \ldots, x_n)$ is the conditional probability of x'_ξ when all other variables are set accordingly.

Let us show that (4.4.1) is satisfied. For simplicity, we restrict our presentation to the case $n = 3$. We embed Ω into a box $Q = [-C, C]^n$ and extend all probability distributions by 0 into Q. Note that

$$\int_{[-C,C]} p(x_1, x_2, x_3)\, \mathrm{d}x_1 = p(x_2, x_3). \tag{4.4.8}$$

We calculate

$$\begin{aligned} &\int_{[-C,C]^3} p(x) K(x', x)\, \mathrm{d}x \\ &= \int_{[-C,C]^3} p(x_1, x_2, x_3) p(x'_1 | x_2, x_3) p(x'_2 | x'_1, x_3) p(x'_3 | x'_1, x'_2)\, \mathrm{d}x \\ &= \int_{[-C,C]^2} (p(x'_1 | x_2, x_3) p(x_2, x_3)) p(x'_2 | x'_1, x_3) p(x'_3 | x'_1, x'_2)\, \mathrm{d}x_2 \mathrm{d}x_3 \\ &= \int_{[-C,C]^2} p(x'_1, x_2, x_3) p(x'_2 | x'_1, x_3) p(x'_3 | x'_1, x'_2)\, \mathrm{d}x_2 \mathrm{d}x_3, \end{aligned} \tag{4.4.9}$$

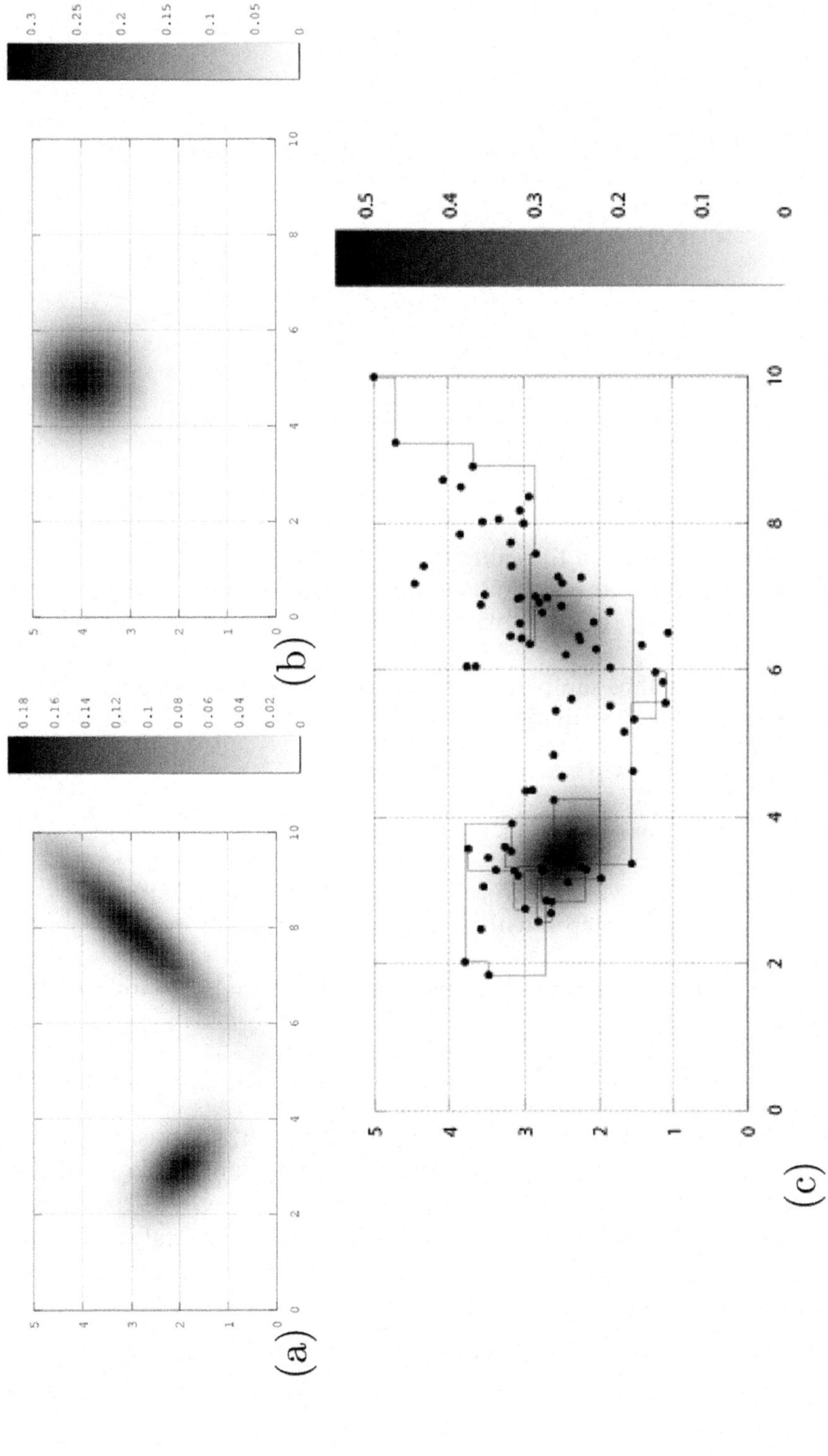

Figure 4.2. Some prior probability distribution $p_{\text{prior}}(x)$, $x \in \Omega$ in (a). A measurement $y = (5, 4)$ provides a conditional measurement distribution $p(y|x)$, $x \in \Omega$ as shown in (b). The posterior distribution is visualized in (c). (c) also shows the result of the MCMC method with the Metropolis–Hastings sampler starting with point (10, 5) in the upper right-hand corner. The first 50 sampling points are connected by lines. Here we show 200 samples altogether.

where we applied equation (4.2.6) to the term in round brackets. In the next step we integrate with respect to x_2 to obtain

$$\begin{aligned}
&\int_{[-C,C]^2} p(x'_1, x_2, x_3)p(x'_2|x'_1, x_3)p(x'_3|x'_1, x'_2)\,\mathrm{d}x_2\mathrm{d}x_3,\\
&= \int_{[-C,C]} (p(x'_2|x'_1, x_3)p(x'_1, x_3))p(x'_3|x'_1, x'_2)\,\mathrm{d}x_3,\\
&= \int_{[-C,C]} p(x'_1, x'_2, x_3)p(x'_3|x'_1, x'_2)\,\mathrm{d}x_3
\end{aligned} \tag{4.4.10}$$

and finally

$$\begin{aligned}
&\int_{[-C,C]} p(x'_1, x'_2, x_3)p(x'_3|x'_1, x'_2)\,\mathrm{d}x_3\\
&= p(x'_3|x'_1, x'_2)p(x'_1, x'_2)\\
&= p(x'_1, x'_2, x'_3)\\
&= p(x').
\end{aligned} \tag{4.4.11}$$

The general proof is obtained in the same way inductively.

By the arguments of theorem 4.3.2 and lemma 4.3.4 we obtain convergence in the probability of the MCMC method with the Gibbs sampler for sampling the distribution $p(x)$, $x \in \Omega$.

The sampling of the Gibbs sampler is now carried out step-by-step according to the conditional probability distributions

$$p\left(x'_\xi \middle| x'_1, \ldots, x'_{\xi-1}, x_{\xi+1}, \ldots, x_n\right) \tag{4.4.12}$$

for $\xi = 1, 2, 3, \ldots$ as follows:

1. We start with $x = x_0 \in \Omega$.
2. Given x, for $\xi = 1, \ldots, n$ we iteratively draw x'_ξ from (4.4.12).
3. Set $x_k = x'$ and go to step 2 with $x = x_k$ and k increased by 1.

The key step here is to draw from (4.4.12), based on the given distribution p. This is carried out using

$$p\left(x'_\xi \middle| x'_1, \ldots, x'_{\xi-1}, x_{\xi+1}, \ldots, x_n\right) = c\ p\left(x'_1, \ldots, x'_{\xi-1}, x'_\xi, x_{\xi+1}, \ldots, x_n\right) \tag{4.4.13}$$

where c is a normalization constant. For the single component Gibbs sampler which we present here, this corresponds to choosing from a one-dimensional probability distribution which is given by $p(s) := c\ p(x'_1, \ldots, x'_{\xi-1}, s, x_{\xi+1}, \ldots, x_n)$ in dependence on $s \in [-C, C]$, where c normalizes this distribution such that its integral is equal to 1.

4.5 Basic stochastic concepts

For the convenience of the reader, here we put together the main stochastic concepts which are employed in the framework of inverse problems based on a *probability space* (X, Σ, P). Here, X is some set, Σ is a *σ-algebra* of subsets of X, i.e. we have

$$\begin{aligned} &(1)\ X \in \Sigma, \\ &(2)\ G \in \Sigma \Rightarrow G^c = X \backslash G \in \Sigma, \\ &(3)\ G_1, G_2, G_3, \ldots \in \Sigma \Rightarrow \bigcup_{j=1}^{\infty} G_j \in \Sigma. \end{aligned} \tag{4.5.1}$$

The mapping $P : \Sigma \to \mathbb{R}$ is a *probability measure* if it satisfies

$$\begin{aligned} &(1)\ P(G) \geqslant 0, \quad G \in \Sigma, \\ &(2)\ P(X) = 1, \\ &(3)\ P\left(\bigcup_{j=1}^{\infty} G_j\right) = \sum_{j=1}^{\infty} P(G_j) \end{aligned} \tag{4.5.2}$$

for disjoint sets G_j, $j \in \mathbb{N}$. Usually the sets $G \in \Sigma$ are called *events*.

Given any set V of subsets of X, we call the smallest σ-algebra which contains V the sigma algebra $\sigma(V)$ generated by V. For $\mathbb{R}^m$, usually a canonical σ-algebra is used, which is generated by the cuboids $U := (a_1, b_1] \times \cdots \times (a_n, b_n]$ for $a_j < b_j \subset \mathbb{R}$. This is called the *Borel algebra* $\mathcal{B}$.

A mapping $A : X \to$ with Y given as $Y = \mathbf{R}^n$ with sigma algebra $\mathcal{B}$ is called *measurable*, if $A^{-1}(U) \in \Sigma$ for all sets $U \in \Sigma'$. A *random variable* is a measurable mapping from X into some other space Y, if for every $y \in Y$ we have $A^{-1}(y) \in \Sigma$. In a probability space, integration of measurable functions is defined by standard tools of measure theory. For random variables A, A' we assume that the *expectation*

$$\mathbb{E}(A) := \int_X A(\omega)\, \mathrm{d}P(\omega), \tag{4.5.3}$$

is well defined. *Variance* and *covariance* are given by

$$\begin{aligned} \mathrm{Var}(A) &:= \mathbb{E}\big((A - \mathbb{E}(A))^2\big), \\ \mathrm{Cov}(A, A') &:= \mathbb{E}((A - \mathbb{E}(A))(A' - \mathbb{E}(A'))), \end{aligned} \tag{4.5.4}$$

where for $A = (A_1, \ldots, A_n)$ having values in $\mathbb{R}^m$ (or $\mathbb{C}^n$), its *covariance matrix* is given by

$$\begin{aligned} B &= \left(\mathbb{E}\left(\left(A_j - \mathbb{E}(A_j)\right)\overline{\left(A_k - \mathbb{E}(A_k)\right)}\right)\right)_{j,k=1,\ldots,n} \\ &= \mathbb{E}\{\mathbf{A}\bar{\mathbf{A}}^T\}, \end{aligned} \tag{4.5.5}$$

where $\mathbf{A} := (A_j - \mathbb{E}(A_j))_{j=1,\ldots,n}$ is the column vector of the centered random variable A. Here, we will not go into the measure theoretic details, but assume that the expectation as well as variance and covariance of random variables are well defined and exist.

Now for real valued random variables, we recall that the *standard deviation* is given by $\sigma := \sqrt{\mathrm{Var}(A)}$ and that $\rho := \mathrm{Cov}(A, A')/(\mathrm{Var}(A)\mathrm{Var}(A'))$ is called *correlation*. If $\mathrm{Cov}(A, A') = 0$, we call two random variables A, A' *uncorrelated*. For expectation values, variances and covariances of random variables $A, A', A_1, \ldots, A_n$ we employ the standard calculation rules

$$\mathrm{Var}(A) = \mathbb{E}(|A|^2) - |\mathbb{E}(A)|^2 \tag{4.5.6}$$

$$\mathrm{Var}(aA + b) = |a|^2 \mathrm{Var}(A) \tag{4.5.7}$$

$$\mathrm{Cov}(A, A') = \mathbb{E}(A\overline{A'}) - \mathbb{E}(A)\overline{\mathbb{E}(A')} \tag{4.5.8}$$

$$\mathrm{Var}(A_1 + \cdots + A_n) = \sum_{j=1}^{n} \mathrm{Var}(A_j) + \sum_{j \neq k, j,k=1}^{n} \mathrm{Cov}(A_j, A_k). \tag{4.5.9}$$

The expectation value $\mathbb{E}(A^k)$ is called the *kth moment* of A, we call $\mathbb{E}((A - \mathbb{E}(A))^k)$ the *kth centralized moment* of A. We also note that if the variances of A and A' exist, then we have the *Cauchy–Schwarz inequality*

$$\left|\mathbb{E}(A\bar{A}')\right|^2 \leqslant \mathbb{E}(|A|^2)\mathbb{E}(|A'|^2). \tag{4.5.10}$$

We call a family $\{G_j : j \in J\}$ of events G_j in (X, Σ) *independent*, if for every finite subset $\{j_1, \ldots, j_n\} \subset J$ we have

$$P(G_{j_1} \cap \cdots \cap G_{j_n}) = P(G_{j_1}) \cdot \cdots \cdot P(G_{j_n}). \tag{4.5.11}$$

A family $\{A_j : j \in J\}$ of random variables $A_j : (X, \Sigma) \to (Y, \Sigma')$ are called *independent*, if for every finite subset $\{j_1, \ldots, j_n\} \subset J$ the equation (4.5.11) is satisfied for $G_{j_\ell} := A_{j_\ell}^{-1} U_{j_\ell} \in \Sigma$, $\ell = 1, \ldots, n$. Independent random variables are uncorrelated.

Basic convergence results for estimators solving inverse or data assimilation problems are based on the following basic results. In a probability space we have *Markov's inequality*

$$P\left(\left\{\omega : |A(\omega)| \geqslant \epsilon\right\}\right) \leqslant \frac{1}{\epsilon}\mathbb{E}(A), \tag{4.5.12}$$

which is a direct consequence of the monotony of the integral (4.5.3) by

$$\mathbb{E}(|A|) \geqslant \epsilon P(|A| \geqslant \epsilon). \tag{4.5.13}$$

A special case of Markov's inequality is known as *Tschebycheff's inequality*. When we replace A by $|A - \mathbb{E}(A)|^2$ we obtain

$$P(|A - \mathbb{E}(A)| \geqslant \epsilon) \leqslant \frac{\mathrm{Var}(A)}{\epsilon^2}. \tag{4.5.14}$$

Consider a sequence $A_1, A_2, A_3, \ldots$ of independent random variables with identical expectation value $\mathbb{E}(A_j) = m$ and bounded variance $\mathrm{Var}(A_j) \leqslant C$. Then we define

$$S_n := \frac{1}{n}\sum_{j=1}^{n} A_j, \quad n \in \mathbb{N}. \tag{4.5.15}$$

An application of the Tschebycheff inequality to S_n yields the following well-known and important result.

Lemma 4.5.1 (Weak law of large numbers). *For a sequence of independent random variables with identical expectation value m and variances bounded by $C > 0$ we have*

$$P\left(\left|\frac{1}{n}\sum_{j=1}^{n} A_j - m\right| \geqslant \epsilon\right) \leqslant \frac{C}{\epsilon^2 n} \to 0, \quad n \to \infty. \tag{4.5.16}$$

The law of large numbers is a basic tool to analyze estimators for inverse problems and to estimate dynamical systems when measurements with random errors are given. The behavior (4.5.16) tells us, for example, that if we independently draw n times from a distribution A, then the probability that the sum

$$\hat{A} := \frac{1}{n}\sum_{j=1}^{n} x_j \tag{4.5.17}$$

of measurement values x_j, $j = 1, \ldots, n$, has a distance larger than ϵ from the expectation value $m = \mathbb{E}(A)$ is going to zero with bound $C/(\epsilon^2 n)$ for $n \to \infty$. Several different concepts of convergence are used for stochastic estimators.

Stochastic convergence. For a probability space (X, Σ, P) assume that a random variable $A : X \to Y$ and a family A_k, $k \in \mathbb{N}$ of random variables $A_k : X \to Y$ is given. We speak of *stochastic convergence* or *convergence in probability* of $(A_k)_{k\in\mathbb{N}}$ towards A if

$$P(\{\omega : |A_k(\omega) - A(\omega)| \geqslant \epsilon\}) \to 0, \quad k \to \infty. \tag{4.5.18}$$

We will write

$$A_k \xrightarrow{p} A, \quad k \to \infty. \tag{4.5.19}$$

Convergence in probability is a basic and important property.

Further convergence concepts. For a sequence A_k, $k \in \mathbb{N}$ of random variables $A_k : X \to Y$ we speak of *almost sure convergence* if

$$P\left(\left\{\omega : \lim_{k\to\infty} A_k(\omega) = A(\omega)\right\}\right) = 0. \tag{4.5.20}$$

This means that we have convergence of the random variable on a set of probability one. For a sequence A_k, $k \in \mathbb{N}$ of random variables $A_k : X \to Y$ we speak of *convergence in the mean* if

$$\mathbb{E}(|A_k - A|) \to 0, \quad k \to \infty. \tag{4.5.21}$$

We will use *densities* to work with probability distributions in $\mathbb{R}^m$. Any integrable function $f : \mathbb{R}^m \to \mathbb{R}$ which satisfies

$$f(x) \geqslant 0, \quad x \in \mathbb{R}^m$$
$$\int_{\mathbb{R}^m} f(x)\, \mathrm{d}x = 1$$

is called a *probability density*. Then, for the cuboid $Q := (a_1, b_1] \times \cdots \times (a_n, b_n]$ the function

$$P(Q) := \int_Q f(x)\, \mathrm{d}x \tag{4.5.22}$$

defines a probability function on cuboids which can be uniquely extended to a probability function on the Borel algebra $\mathcal{B}$. Given a density f the function $F : \mathbb{R}^m \to \mathbb{R}$ defined by

$$F(x) := \int_{-\infty}^{x_1} \ldots \int_{-\infty}^{x_n} f(y)\, \mathrm{d}y, \quad x = (x_1, \ldots, x_n) \in \mathbb{R}^m, \tag{4.5.23}$$

is called *distribution function* on $\mathbb{R}^m$. For a real-valued random variable A a distribution function is given by

$$F_A(x) := P(\{\omega : A(\omega) \leqslant x\}), \quad x \in \mathbb{R}. \tag{4.5.24}$$

If F_A is differentiable, a density is defined by

$$f_A(x) := \frac{\mathrm{d}F_A(x)}{\mathrm{d}x}, \quad x \in \mathbb{R}. \tag{4.5.25}$$

Given independent random variables $A_1, \ldots, A_L$ which have densities $f_1, \ldots, f_L$, then $A := (A_1, \ldots, A_L)$ has a density f which satisfies

$$f(x_1, \ldots, x_n) = f_1(x_1) \cdot \cdots \cdot f_n(x_n), \quad x = (x_1, \ldots, x_n) \in \mathbb{R}^m. \tag{4.5.26}$$

The *Gaussian distribution* or *normal distribution* in one dimension is given by the well-known density

$$f(x) = \frac{1}{\sqrt{2\pi\sigma^2}} \mathrm{e}^{-\frac{(x-\mu)^2}{2\sigma^2}}, \quad x \in \mathbb{R}, \tag{4.5.27}$$

depending on its mean μ and the *standard deviation* σ. In $\mathbb{R}^m$ with a positive symmetric matrix B we obtain the *multivariate Gaussian*

$$f(x) = \frac{1}{\sqrt{(2\pi)^n \det(B)}} \mathrm{e}^{-\frac{1}{2}(x-\mu)^T B^{-1}(x-\mu)}, \quad x \in \mathbb{R}^m, \tag{4.5.28}$$

for which normalization is based on the integral formula

$$\int_{\mathbb{R}^m} \mathrm{e}^{-\frac{1}{2}x^T Bx}\, \mathrm{d}x = \sqrt{\frac{(2\pi)^n}{\det(B)}} \tag{4.5.29}$$

for a positive symmetric matrix $B \in \mathbb{R}^{n \times n}$.

Bibliography

[1] Kaipio J and Somersalo E 2005 *Statistical and Computational Inverse Problems* (*Applied Mathematical Sciences* vol 160) (New York: Springer)

[2] Biegler L, George B, Ghattas O, Heinkenschloss M, Keyes D, Mallick B, Tenorio L, van Bloemen Waanders B, Willcox K and Marzouk Y 2011 *Large-Scale Inverse Problems and Quantification of Uncertainty (Wiley Series in Computational Statistics)* (New York: Wiley)

[3] Aster R C, Borchers B and Thurber C H 2013 *Parameter Estimation and Inverse Problems* (Boston, MA: Academic)

[4] Tarantola A 2005 *Inverse Problem Theory and Methods for Model Parameter Estimation* (Philadelphia, PA: Society for Industrial and Applied Mathematics)

[5] Beck J V and Arnold K J 1977 *Parameter Estimation in Engineering and Science* (New York: Wiley)

[6] van der Heijden F, Duin R, de Ridder D and Tax D M J 2004 *Classification, Parameter Estimation and State Estimation* (New York: Wiley)

[7] Georgii H-O 2013 *Stochastics: Introduction to Probability and Statistics* (Berlin: De Gruyter)

IOP Publishing

Inverse Modeling

An introduction to the theory and methods of inverse problems and data assimilation

Gen Nakamura and Roland Potthast

Chapter 5

Dynamical systems inversion and data assimilation

Inverse problems play a key role in dynamical systems. We need them for the investigation of dynamical systems, to determine the state of dynamical systems from measurements and as a key step towards forecasting techniques.

When you try to simulate some natural or technical dynamical system, you need to make sure that your simulation comes close to nature or reality. This can only be achieved when your system parameters are chosen appropriately and when you continue to use appropriately corrected states for each of your simulation periods.

When the estimation of the initial states is carried out again and again, using measurement data and the first guess from previous estimates, we use the term *data assimilation*. Using data assimilation algorithms measurements are used to control the system simulation and for forecasting.

Here, we will study dynamical systems described by some state vector φ in a space X which might either be a normed space or some probability space (X, Σ, μ) with a sigma algebra Σ and a measure μ. A dynamical system is given by some dynamics, which enables us to calculate the system state $\varphi(t)$ or φ_t at some time t from the knowledge of the system state $\varphi(s)$ or φ_s at time $s \leqslant t$. We can describe such a system by a *model operator*, i.e. a mapping

$$M : U \subset \mathbb{R} \times \mathbb{R} \times X \to X, \quad \varphi \mapsto M(t, s, \varphi). \tag{5.0.1}$$

Here, we have noted the dependence on both times s and t explicitly. We will use the notation

$$M(t, s)\varphi := M(t, s, \varphi)$$

to reflect our view that M operates on the element φ and describes its evolution from time s to time t. An example is visualized in figure 5.1. For consistency we assume that for all $s \in \mathbb{R}$ the operator $M(s, s)$ is the identity and that the development of the system from s to t is given by its development from s to ρ and from ρ to t, i.e.

doi:10.1088/978-0-7503-1218-9ch5

$$M(s, s) = I, \tag{5.0.2}$$

$$M(t, \rho) \circ M(\rho, s) = M(t, s). \tag{5.0.3}$$

This means our states φ_t can be written as

$$\varphi_t = M(t, s)\varphi_s, \quad s \leqslant t. \tag{5.0.4}$$

The restriction to a subset U of $\mathbb{R} \times \mathbb{R} \times X$ takes care of the fact that often nonlinear dynamical systems are only defined for some short period of time $T = [t_1, t_2]$.

Usually, M is called the *evolution operator* or *evolution function* of the dynamical system and X is called the *phase space* or *state space*. The set

$$\gamma(s, \varphi) := \{M(t, s)\varphi \text{ for any } t \text{ with } (t, s, \varphi) \in U\} \tag{5.0.5}$$

is called the *path* or *orbit* through the point φ at time s. As *trajectory* we denote its graph

$$\Gamma(s, \varphi) := \{(t, \varphi_t) = (t, M(t, s)\varphi) \text{ for any } t \text{ with } (t, s, \varphi) \in U\}.$$

Often systems are time independent, i.e. they depend only on the difference $\tau = t - s$, such that one of the time parameters can be dropped and both orbit and trajectory are independent of s. A typical set-up starts with initial states φ_0 at time $s = 0$. Then, we write

$$\varphi_t := M(t, 0)\varphi_0, \quad t \geqslant 0. \tag{5.0.6}$$

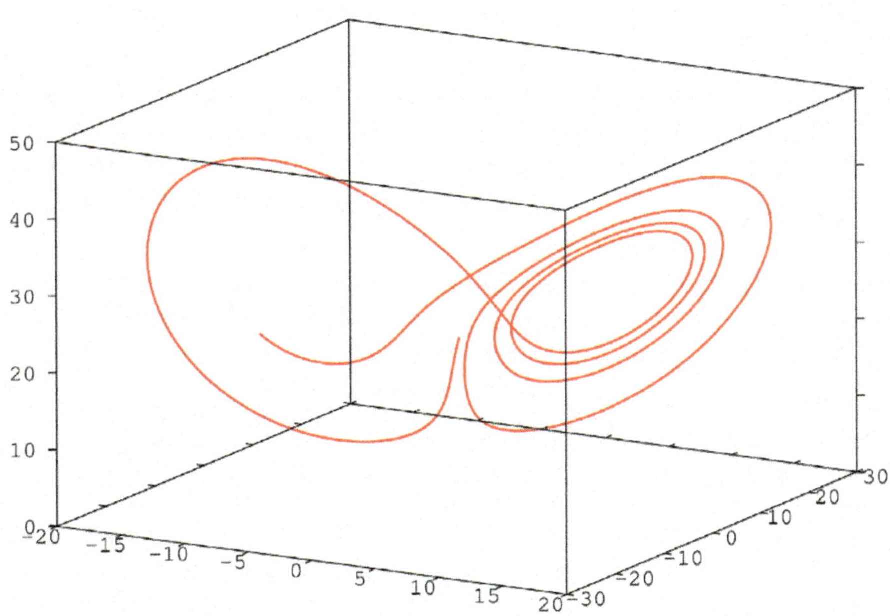

Figure 5.1. A trajectory in a three-dimensional state space. The figure displays the famous butterfly of the Lorenz 1963 model.

5.1 Set-up for data assimilation

For a dynamical system we assume that we carry out measurements of quantities f which depend on the system φ, i.e. we have

$$f = H(t, \varphi) \tag{5.1.1}$$

with some linear or nonlinear mapping

$$H : V \subset \mathbb{R} \times X \to Y,$$

where Y is called the *measurement space* and V is some appropriate subset of $\mathbb{R} \times X$ for which measurements are defined. The quantity f will contain all data which can be obtained, i.e. it will be a vector or a family of objects. We assume that Y is either a normed space or some probability space (Y, Σ, μ). Our presentation here aims to bridge the concepts and viewpoint usually used in estimation theory, functional analysis, inverse problems and data assimilation, in the spirit of Freitag and Potthast [1].

We will assume that we carry out measurements at a sequence of discrete times $t_1, t_2, \ldots, t_L$. The basic data assimilation approach is visualized in figure 5.2. The corresponding measurements are denoted by

$$f_1, f_2, \ldots, f_L \in Y. \tag{5.1.2}$$

Often, in practice you cannot guarantee that all measurements are available at all times t_k. This means that the measurement operator H and the observation space Y in fact depend on the time variable t. However, here we will not go into further detail and assume measurements in some space Y. All methods and results can be transferred to the more general situation of variable measurement spaces in a straightforward way.

Definition 5.1.1 (Set-up for data assimilation). *We assume that we have a dynamical system as in (5.0.1) with properties given by (5.0.2), (5.0.3). We also assume that we have some a priori information about the state of the system at $t_0 = 0$ given by $\varphi_0 \in X$. Further, we are given measurements as in (5.1.2) for a measurement operator (5.1.1) at times $t_1, \ldots, t_L$.*

The *task of data assimilation* is to determine state estimates

$$\varphi_0^{(a)}, \varphi_1^{(a)}, \varphi_2^{(a)}, \ldots, \varphi_L^{(a)} \tag{5.1.3}$$

at times $t_1, \ldots, t_L$ for the state of the dynamical system given by definition 5.1.1. Here, the letter $^{(a)}$ stands for *analysis* which is obtained from assimilating the available data. When we apply the model M without using the data, we call the trajectory through φ_0 given by

$$\varphi_0^{(b)} = \varphi_0, \varphi_1^{(b)} = M(t_1, 0)\varphi_0, \ldots, \varphi_L^{(b)} = M(t_L, 0)\varphi_0 \tag{5.1.4}$$

the *background.* Usually, the initial guess φ_0 is obtained from earlier calculations, incorporating many measurements.

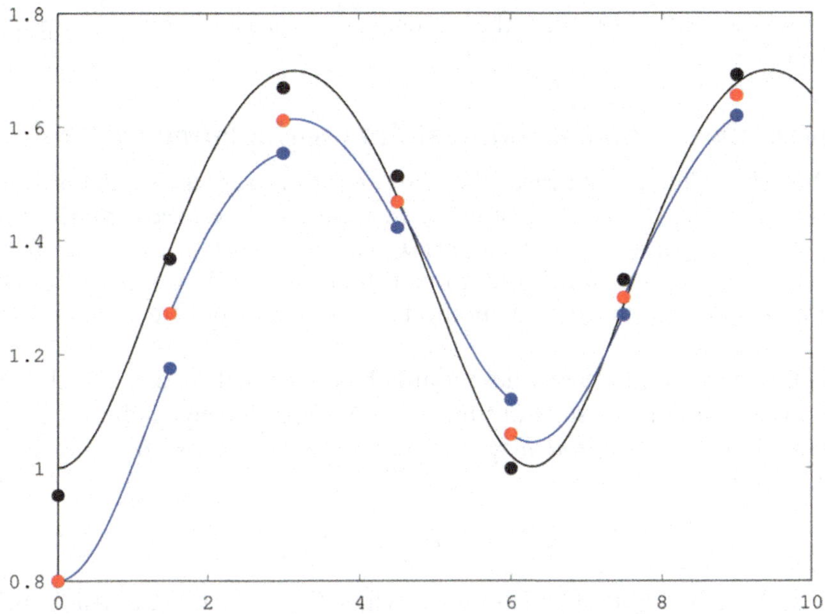

Figure 5.2. Data assimilation uses *measurements* f and knowledge from previous steps in the form of a *first guess* or *background* $x^{(b)}$ to calculate an *analysis* $x^{(a)}$ at each time step. The analysis is then used to calculate the *first guess* for the next assimilation time, using some model M of the underlying dynamical system. This is repeated within the *data assimilation cycle*. In this example, we display measurements f_k as black dots and calculate the analysis (red dots) as the average between the data and background (blue dots). The true curve is shown in black, the model simulations in blue.

The set-up of definition 5.1.1 is still lacking important input to carry out the data assimilation task. In what way can we combine measurements, calculations and *a priori* information?

- Assume, for example, that our *a priori* information and the model M are perfect. Then we do not really need the measurements and we just calculate the system via (5.0.6).
- On the other hand, assume that the *a priori* information φ_0 contains large errors. Then, it should *not* be used strongly when the estimate $\varphi_1^a, \dots, \varphi_L^a$ is calculated, but we should rely more on the current measurements.
- But usually there are not enough measurements to fully determine the state $\varphi \in X$. Collecting measurements over time might help, but for most practically relevant set-ups even if measurements are carried out over long periods of time they do not fully determine $\varphi \in X$.

We need to quantify further knowledge about the validity of the *a priori* information and about the measurements. This can be carried out by an optimization approach with particular *weights*, or by a stochastic approach where an *error distribution* for each of the quantities is specified. In the case of Gaussian distributions we will see that the stochastic approach and the optimization approach coincide.

5.2 Three-dimensional variational data assimilation (3D-VAR)

The 3D-VAR schemes concern the data assimilation as in definition 5.1.1, where measurements $f_k \in Y$ at time $t = t_k$ are assimilated sequentially time step by time step. The name refers to a *variational optimization* in *space*, in contrast to schemes which also treat the time dimension and are then called *four-dimensional*, and in contrast to methods which incorporate further stochastic tools.

Let us first consider the case where X and Y are normed spaces. Usually, we have some knowledge about the state at time t_{k-1}, i.e. we are given a solution $\varphi_{k-1}^{(a)}$ at t_{k-1}. Then, a *first guess* or *background* for the state is calculated by an application of the model, i.e. by

$$\varphi_k^{(b)} := M(t_k, t_{k-1})\varphi_{k-1}^{(a)}. \tag{5.2.1}$$

Now, we seek a new solution $\varphi_k^{(a)}$ at time t_k which fits to the first guess $\varphi_k^{(b)}$ and to the data $f_k = H(t_k, \varphi_k^{(a)})$ as well. A straightforward approach is the reformulation of this problem as an optimization problem to minimize

$$J_{3\mathrm{D},k}(\varphi) := \alpha\left\|\varphi - \varphi_k^{(b)}\right\|_X^2 + \left\|f_k - H(t_k, \varphi)\right\|_Y^2, \quad \varphi \in X, \tag{5.2.2}$$

where $\alpha > 0$ controls the weight between the role of the previous knowledge and the data. The minimizer of (5.2.2) is denoted as $\varphi_k^{(a)}$ and it is known as *the analysis*. A small α means that we put more weight on the measurements, a large α puts more weight on the *a priori* knowledge. Often, α is included into the norm $\|\cdot\|_X$ of X; this is discussed in more detail in the framework of stochastic methods.

For applications, the choice of the norms in X and Y is crucial. For the case $X = \mathbb{R}^m$ for some $n \in \mathbb{N}$ usually the norm in X is chosen as

$$\|\varphi\|_{B^{-1}}^2 := \varphi^T B^{-1}\varphi \tag{5.2.3}$$

with some positive definite (and thus invertible) matrix $B \in \mathbb{R}^{n\times n}$. B is effectively modeling correlation information between the components of $\varphi_k \in \mathbb{R}^m$, compare to chapter 4 on stochastic inverse problems.

In general, when H is a nonlinear operator, the minimization of (5.2.2) is a nonlinear optimization problem. However, if H is linear, the minimization of (5.2.2) corresponds to solving the linear equation

$$H\varphi = f_k \tag{5.2.4}$$

by Tikhonov regularization with regularization parameter $\alpha > 0$ and *a priori* guess $\varphi = \varphi_k^{(b)}$. The solution of the minimization is carried out along the lines of the

Tikhonov regularization scheme as described in section 3.1.4 applied to $A = H$. We incorporate *a priori* knowledge into the minimization of (3.1.29) by

$$\mu_{\mathrm{Tik},2}(\varphi) := \|H\varphi - f\|_Y^2 + \alpha\|\varphi - \varphi_0\|_X^2, \tag{5.2.5}$$

where we use the generic notation φ_0 for the background $\varphi_k^{(b)}$. The normal equations for the minimization of (5.2.5) are obtained by calculating the gradient with respect to φ and setting it to zero. We obtain

$$H^*(H\varphi - f) + \alpha(\varphi - \varphi_0) = 0$$

which by two lines of calculation is transformed into

$$(\alpha I + H^*H)(\varphi - \varphi_0) = H^*(f - H\varphi_0).$$

This leads to the update formula

$$\begin{aligned}\varphi &= \varphi_0 + (\alpha I + H^*H)^{-1}H^*(f - H\varphi_0)\\ &= \varphi_0 + R_\alpha(f - H\varphi_0),\end{aligned} \tag{5.2.6}$$

where R_α as defined by (3.1.24) is usually denoted as *Kalman gain matrix*. The case where the scalar product in X is given by

$$\langle\varphi, \psi\rangle_{B^{-1}} := \langle\varphi, B^{-1}\psi\rangle, \quad \varphi, \psi \in X, \tag{5.2.7}$$

with a positive definite self-adjoint or symmetric matrix B and the scalar product of Y is given by

$$\langle f, g\rangle_{R^{-1}} := \langle f, R^{-1}g\rangle, \quad f, g \in Y, \tag{5.2.8}$$

with some positive definite self-adjoint or symmetric matrix R leads to an update formula given by

$$\varphi = \varphi_0 + (\alpha B^{-1} + H'R^{-1}H)^{-1}H'R^{-1}(f - H\varphi_0), \tag{5.2.9}$$

where H' denotes the adjoint matrix with respect to the standard scalar product on X. Note that we use the notation H^* for the adjoint with respect to the scalar products given by (5.2.7) in X and (5.2.8) in Y. Since we have

$$\begin{aligned}\langle H^*f, \psi\rangle_{B^{-1}} &= \langle f, H\psi\rangle_{R^{-1}} = \langle f, R^{-1}H\psi\rangle\\ &= \langle H'R^{-1}f, \psi\rangle = \langle H'R^{-1}f, BB^{-1}\psi\rangle = \langle BH'R^{-1}f, \psi\rangle_{B^{-1}},\end{aligned} \tag{5.2.10}$$

the relationship of H' and H^* is given by

$$H^* = BH'R^{-1}. \tag{5.2.11}$$

Usually, the dimension of the measurement space Y is much smaller than the dimension of the system space X. In this case H^*H is a much larger matrix than HH^* and we would like to transform our calculations into the measurement space. We remark that we have

$$(\alpha I + H^*H)H^* = \alpha H^* + H^*HH^* = H^*(\alpha I + HH^*).$$

For $\alpha > 0$ the operator $\alpha I + HH^*$ is invertible on Y, and the same is true for $\alpha I + H^*H$ on X, since

$$(\alpha I + H^*H)\varphi = 0$$

implies

$$\alpha\langle\varphi, \varphi\rangle + \underbrace{\langle H\varphi, H\varphi\rangle}_{\geqslant 0} = 0,$$

thus $\varphi = 0$ and $\alpha I + H^*H$ is injective and surjective on X. This leads to

$$(\alpha I + H^*H)^{-1}H^* = H^*(\alpha I + HH^*)^{-1}. \quad (5.2.12)$$

By (5.2.12) we transform (5.2.6) into

$$\varphi_k^{(a)} = \varphi_k^{(b)} + H^*(\alpha I + HH^*)^{-1}\left(f - H\varphi_k^{(b)}\right). \quad (5.2.13)$$

Using (5.2.11), the update equation (5.2.13) can be written as

$$\varphi_k^{(a)} = \varphi_k^{(b)} + BH'(\alpha R + HBH')^{-1}\left(f - H\varphi_k^{(b)}\right). \quad (5.2.14)$$

When H is nonlinear and $\mathbf{H}$ denotes its linearization, then (5.2.13) is replaced by

$$\varphi_k^{(a)} = \varphi_k^{(b)} + \mathbf{H}^*(\alpha I + \mathbf{H}\mathbf{H}^*)^{-1}\left(f - H\left(\varphi_k^{(b)}\right)\right). \quad (5.2.15)$$

In (5.2.13) the calculation of the updates is carried out by first solving the linear system

$$(\alpha I + \mathbf{H}\mathbf{H}^*)\psi = f - H\left(\varphi_k^{(b)}\right) \quad (5.2.16)$$

by an appropriate (usually iterative) scheme and then calculating

$$\varphi_k^{(a)} = \varphi_k^{(b)} + \mathbf{H}^*\psi. \quad (5.2.17)$$

The full data assimilation algorithm can now be formulated as follows.

Definition 5.2.1 (3D-VAR). *Given an initial state φ_0 at time t_0 and measurement data f_k at time t_k, $k = 1,2,3, \ldots,$ 3D-VAR calculates analysis states $\varphi_k^{(a)}$ at time t_k successively by*

1. *starting with $\varphi_0^{(a)} := \varphi_0$,*
2. *propagation of the analysis state $\varphi_{k-1}^{(a)}$ at time t_{k-1} to time t_k by (5.2.1)*
3. *calculation of a new analysis state $\varphi_k^{(a)}$ at time t_k by (5.2.16) and (5.2.17).*

3D-VAR can be viewed as the application of Tikhonov regularization in each assimilation step. Thus, each assimilation step itself is stable. However, errors might accumulate during iteration. The corresponding error analysis is carried out in section 10.5.

In section 6.2 we work out a step-by-step example for 3D-VAR for the nonlinear dynamical system suggested by Lorenz in 1963.

5.3 Four-dimensional variational data assimilation (4D-VAR)

The basic idea of the 4D-VAR scheme is to calculate the analysis $x^{(a)}$ for the data assimilation problem 5.1.1 by a minimization which includes the time steps from t_k to t_{k+K} with some window size $K \in \mathbb{N}$ to calculate the analysis state at t_k, see figure 5.3. The corresponding functional is given by

$$J_{4D,k}(\varphi) := \alpha \left\| \varphi - \varphi_k^{(b)} \right\|_X^2 + \sum_{\xi=1}^{K} \left\| f_{k+\xi} - H\left(t_{k+\xi}, x_{k+\xi}\right) \right\|_Y^2, \tag{5.3.1}$$

where using the notation (5.0.4) we employ

$$x_{k+\xi} := M(t_{k+\xi}, t_k)\varphi, \quad \xi = 1, 2, 3, \ldots \tag{5.3.2}$$

for $\varphi \in X$. The analysis $\varphi_k^{(a)}$ at time t_k is found by minimizing (5.3.1) and the analysis at other points in time can be calculated using the forward model

$$\varphi_{k+\xi}^{(a)} := M(t_{k+\xi}, t_k)\varphi_k^{(a)}, \quad \xi = 1, \ldots, K. \tag{5.3.3}$$

This means that the 4D-VAR scheme (5.3.1)–(5.3.3) forces the solution to fully satisfy the forward model in the interval $[t_k, t_{k+K}]$. 4D-VAR puts the same weight onto all data points to calculate its analysis.

Usually the models which are employed for practical calculations are far from being perfect, how to include model error into the calculations is an important basic question of research. Techniques such as *weak-constraint* 4D-VAR have been suggested to overcome such limitations, see for example [2].

Here, we first describe classical 4D-VAR in section 5.3.1 and then describe a simple ensemble-based approach to 4D-VAR in section 5.3.2.

5.3.1 Classical 4D-VAR

Here, our goal is to understand the basic steps which are needed to realize a 4D-VAR minimization algorithm. If we assume that the norms in X and Y are obtained from scalar products $\langle \cdot, \cdot \rangle_X$ and $\langle \cdot, \cdot \rangle_Y$ and if both H and M are differentiable with respect to its initial conditions, then the Fréchet derivative of (5.3.1) with respect to φ is calculated as

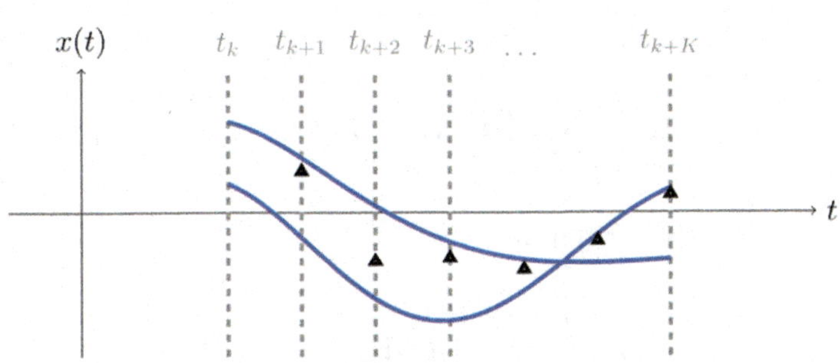

Figure 5.3. The 4D-VAR minimizes a cost functional over a full time window, here between t_k and t_{k+K}. We display measurements as black triangles. The goal is to match the triangles by a blue curve as well as possible.

$$\frac{\partial J_{4\mathrm{D},1}(\varphi)}{\partial \varphi}\delta\varphi = 2\alpha\ \mathrm{Re}\langle \varphi - \varphi_0,\ \delta\varphi\rangle_X$$

$$+ 2\ \mathrm{Re}\ \sum_{\xi=k+1}^{k+K} \left\langle H(t_\xi,\ \varphi_\xi) - f_\xi,\ \left.\frac{\partial H(t_\xi,\ \psi)}{\partial \psi}\right|_{\varphi_\xi} \frac{\partial M(t_\xi,\ t_k,\ \varphi)}{\partial \varphi}\delta\varphi \right\rangle_Y . \tag{5.3.4}$$

We abbreviate

$$\mathbf{H}_\xi := \left.\frac{\partial H(t_\xi,\ \psi)}{\partial \psi}\right|_{M(t_\xi,0,\varphi)}, \qquad \mathbf{M}_{\xi,k} := \frac{\partial M(t_\xi,\ t_k,\ \varphi)}{\partial \varphi} \tag{5.3.5}$$

and note that these are linear mappings

$$\mathbf{H}_\xi : X \to Y, \qquad \mathbf{M}_{\xi,k} : X \to X$$

for $\xi,\ k = 1, \dots, K$. From the property (5.0.3) we know that $M(t_\xi,\ t_k,\ \varphi)$ is the product

$$M(t_\xi,\ t_k)\varphi = M(t_\xi,\ t_{\xi-1}) \circ M(t_{\xi-1},\ t_{\xi-2}) \circ \dots \circ M(t_1,\ t_k)\varphi.$$

Differentiation of (5.3.6) by the chain rule yields

$$\mathbf{M}_{\xi,k} = \mathbf{M}_{\xi,\xi-1}\mathbf{M}_{\xi-1,\xi-2} \cdots \mathbf{M}_{k+1,k} = \prod_{j=\xi-1}^{k} \mathbf{M}_{j+1,j}. \tag{5.3.6}$$

We can now use (5.3.6) to transform (5.3.4) into

$$\frac{\partial J_{4\mathrm{D},k}(\varphi)}{\partial \varphi}\delta\varphi = 2\alpha\ \mathrm{Re}\langle \varphi - \varphi_0,\ \delta\varphi\rangle_X$$

$$+ 2\ \mathrm{Re}\ \sum_{\xi=k+1}^{k+K} \left\langle \mathbf{M}_{\xi,k}^* \mathbf{H}_\xi^* \big(H(t_\xi,\ \varphi_\xi) - f_\xi\big),\ \delta\varphi \right\rangle_X . \tag{5.3.7}$$

$$= 2\alpha\ \mathrm{Re}\langle \varphi - \varphi_0,\ \delta\varphi\rangle_X$$

$$+ 2\ \mathrm{Re} \left\langle \sum_{\xi=k+1}^{k+K} \prod_{j=k}^{\xi-1} \mathbf{M}_{j+1,j}^* \mathbf{H}_\xi^* \big(H(t_\xi,\ \varphi_\xi) - f_\xi\big),\ \delta\varphi \right\rangle_X , \tag{5.3.8}$$

where the last line can be written as

$$+2\ \mathrm{Re} \left\langle \sum_{\xi=k+1}^{k+K} \prod_{j=k}^{\xi-1} \mathbf{M}_{j+1,j}^*\ b_\xi,\ \delta\varphi \right\rangle_X ,$$

with

$$b_0 := 0, \quad b_j := \mathbf{H}_j^*\Big(H\big(t_j, \varphi_j\big) - f_j\Big), \quad j = k, \ldots, k + K. \tag{5.3.9}$$

To avoid unnecessary applications of $\mathbf{M}_{k,k-1}^*$ we can reorder the sum of products using the following iterative definition

$$\psi_{k+K} := b_{k+K}, \tag{5.3.10}$$

$$\psi_{j-1} := \mathbf{M}_{j,j-1}^* \psi_j + b_{j-1} \tag{5.3.11}$$

for $j = k + K, \ldots, k + 1$, see figure 5.4, which for $j = k + 1$ calculates ψ_k.

From (5.3.9)–(5.3.11) we obtain

$$\sum_{\xi=k+1}^{k+K} \prod_{j=k}^{\xi-1} M_{j,j-1}^* b_\xi = \psi_k$$

for the sum over the products in (5.3.8). The scheme (5.3.11) is carried out by a backward integration of the system M by its adjoint linearized model $M_{j,j-1}^*$ with the measurement data increments b_j as forcing terms.

In the framework of *dynamical inverse problems* usually the state $\varphi(t)$ in the state space X is given by a system of differential equations

$$\dot{\varphi} = F(t, \varphi), \quad t \geqslant 0 \tag{5.3.12}$$

with *initial condition*

$$\varphi(0) = \varphi_0 \tag{5.3.13}$$

given some initial state $\varphi_0 \in X$. A main task is to differentiate $\varphi(t)$ with respect to the initial condition φ_0. We define the *model operator* M by

$$M(t, 0)(\varphi_0) := \varphi(t), \quad t \geqslant 0, \tag{5.3.14}$$

where $\varphi(t)$ is the solution to (5.3.12) and (5.3.13). For the Fréchet derivative of M with respect to φ_0, we use the notation

$$\frac{\mathrm{d}M(t, 0)(\varphi_0)}{\mathrm{d}\varphi_0} = \varphi'(t)(\varphi_0). \tag{5.3.15}$$

$$\begin{bmatrix} \mathbf{M}_{k+1,k}^* & & & & b_{k+1} \\ +\mathbf{M}_{k+1,k}^* & \mathbf{M}_{k+2,k+1}^* & & & b_{k+2} \\ +\mathbf{M}_{k+1,k}^* & \mathbf{M}_{k+2,k+1}^* & \mathbf{M}_{k+3,k+2}^* & & b_{k+3} \\ \cdots & \cdots & \cdots & & \\ +\mathbf{M}_{k+1,k}^* & \cdots & \cdots & \mathbf{M}_{k+K,k+K-1}^* & b_{k+K} \end{bmatrix}$$

Figure 5.4. We display the sum by ordering the products and sums into a rectangular scheme. The sum is now carried out iteratively by starting in the lower right-hand corner.

The derivative φ' is a linear mapping from X into X. If $X = \mathbb{R}^m$, then $\varphi(t) \in \mathbb{R}^m$ is a vector and the derivative φ' is a matrix

$$\varphi'(t) = \begin{pmatrix} \frac{d\varphi_1}{d\varphi_{0,1}} & \cdots & \frac{d\varphi_1}{d\varphi_{0,n}} \\ \vdots & & \vdots \\ \frac{d\varphi_n}{d\varphi_{0,1}} & \cdots & \frac{d\varphi_n}{d\varphi_{0,n}} \end{pmatrix}. \tag{5.3.16}$$

If M is twice continuously differentiable with respect to both t and φ_0, then we can exchange the differentiation with respect to t and φ_0, and derive

$$\begin{aligned} \frac{d}{dt}\varphi' &= \frac{d}{dt}\frac{d}{d\varphi_0}\varphi = \frac{d}{d\varphi_0}\frac{d}{dt}\varphi \\ &= \frac{d}{d\varphi_0}F(t, \varphi) = \frac{dF(t, \tilde{\varphi})}{d\tilde{\varphi}}\Big|_{\tilde{\varphi}=\varphi} \circ \frac{d\varphi}{d\varphi_0}, \\ &= F'(t, \varphi(t))\, \varphi'(t) \end{aligned} \tag{5.3.17}$$

where we have employed (5.3.12) in the second step. At time $t = 0$ we have $\varphi(0)(\varphi_0) = \varphi_0$ according to (5.3.13), leading to $\varphi(0)(\varphi_0 + \delta\varphi_0) = \varphi_0 + \delta\varphi_0 = \varphi(0)(\varphi_0) + \delta\varphi_0$, such that we obtain

$$\varphi'(0) = \frac{d\varphi}{d\varphi_0}\Big|_{t=0} = I, \tag{5.3.18}$$

with the identity I in X. We have shown the following result.

Theorem 5.3.1. *The tangent linear model, i.e. the Fréchet derivative of φ with respect to the initial conditions φ_0 of (5.3.12), is given by the solution to the* system of differential equations

$$\frac{d}{dt}\varphi'(t) = F'(t, \varphi)\varphi'(t) \; t \geqslant 0, \tag{5.3.19}$$

with initial condition

$$\varphi'(0) = I, \tag{5.3.20}$$

where F' is the Fréchet derivative of the forcing term $F(t, \varphi)$ of (5.3.12) with respect to its second argument φ.

For state estimation problem (5.3.1) the basic task is to find a state φ_0 at time t_0 which fits given data f_1 at a later point in time t_1. This leads to minimization of expressions of the form

$$J(\varphi_0) = \|H(\varphi(t_1)) - f_1\|^2 = \|H(M(t_1, t_0)(\varphi_0)) - f_1\|^2 \tag{5.3.21}$$

for some fixed $t_1 > 0$ and given or measured $f_1 \in X$. Gradient or Newton methods for minimizing (5.3.21) are based on derivatives with respect to φ_0.

In a Hilbert space environment, where $\|\cdot\|^2 = \langle\cdot,\cdot\rangle$, differentiation with respect to φ_0 leads to

$$\frac{d}{d\varphi_0}(J(\varphi_0))(\delta\varphi_0) = 2\,\mathrm{Re}\big\langle \varphi'(t_1)(\delta\varphi_0),\, (H')^*(H(\varphi(t_1)) - f_1)\big\rangle. \tag{5.3.22}$$

This not yet optimal for the evaluation of the gradient. Let us define a function $\psi(t) \in X$ on the interval $[t_0, t_1]$ by

$$\dot{\psi}(t) = -F'(t, \varphi(t))^*\psi(t), \quad t \in (t_0, t_1) \tag{5.3.23}$$

and *final condition*

$$\psi(t_1) = (H')^*\big(H(\varphi(t_1)) - f_1\big). \tag{5.3.24}$$

Then, we have the following important observation.

Lemma 5.3.2. *For $t \in [t_0, t_1]$, the scalar product*

$$g(t) := \big\langle \varphi'(t)(\delta\varphi_0), \psi(t)\big\rangle \tag{5.3.25}$$

is constant over time for any $\delta\varphi_0 \in X$.

Proof. We differentiate $g(t)$ with respect to t and employ (5.3.17) to calculate

$$\begin{aligned}
\frac{dg(t)}{dt} &= \frac{d}{dt}\langle \varphi'(t)(\delta\varphi_0), \psi(t)\rangle \\
&= \left\langle \frac{d}{dt}\varphi'(t)(\delta\varphi_0), \psi(t)\right\rangle + \left\langle \varphi'(t)(\delta\varphi_0), \frac{d}{dt}\psi(t)\right\rangle \\
&= \big\langle F'(t, \varphi(t))\varphi'(t)(\delta\varphi_0), \psi(t)\big\rangle + \big\langle \varphi'(t)(\delta\varphi_0), -F'(t, \varphi(t))^*\psi(t)\big\rangle \\
&= \big\langle \varphi'(t)(\delta\varphi_0), F'(t, \varphi(t))^*\psi(t)\big\rangle - \big\langle \varphi'(t)(\delta\varphi_0), F'(t, \varphi(t))^*\psi(t)\big\rangle \\
&= 0.
\end{aligned} \tag{5.3.26}$$

Since the derivative of $g(t)$ from (5.3.25) is zero by (5.3.26), we obtain the statement of our lemma. □

We refer to figure 9.6 for a practical demonstration of this result in the case of the Lorenz 1963 system. As a consequence of the previous lemma 5.3.2 we obtain a very efficient way to calculate the gradient of the functional (5.3.21).

Theorem 5.3.3. *Let $\psi(t) \in X$ be the solution of (5.3.23) and (5.3.24) on $[t_0, t_1]$. Then, the gradient (5.3.22) of (5.3.21) is given by*

$$\frac{d}{d\varphi_0}(J(t_1, \varphi_0))(\delta\varphi_0) = 2\,\mathrm{Re}\langle \delta\varphi_0, \psi(0)\rangle. \tag{5.3.27}$$

Proof. We have shown that $g(t)$ from (5.3.25) is constant over time and thus

$$\begin{aligned} \frac{\mathrm{d}}{\mathrm{d}\varphi_0}(J(t_1, \varphi_0))(\delta\varphi_0) &= 2\,\mathrm{Re}\langle \varphi'(t_1)(\delta\varphi_0), \psi(t_1)\rangle \\ &= 2\,\mathrm{Re}\langle \varphi'(0)(\delta\varphi_0), \psi(0)\rangle \\ &= 2\,\mathrm{Re}\langle \delta\varphi_0, \psi(0)\rangle, \end{aligned} \tag{5.3.28}$$

where we used (5.3.18), and the proof is complete. □

We call the function $\psi(t)$ defined by (5.3.23) and (5.3.24) the *tangent linear adjoint* model. It is used to calculate the gradient of functionals involving the dynamical model by

1. first integrating the model over $[t_0, t_1]$ to calculate $\varphi(t)$ for $0 \leqslant t \leqslant t_1$.
2. Then, the data $\varphi(t_1) - f$ are integrated backward in time over $[t_0, t_1]$ according to (5.3.23) to evaluate $\psi(0)$.
3. Finally, the gradient is now given by the expression (5.3.27).

We work out this derivative in detail for the Lorenz 1963 model in section 9.5, where we also provide some simple code examples for the different ingredients and its testing.

5.3.2 Ensemble-based 4D-VAR

Often, the analytic calculation of the derivatives (5.3.5) of H and the model M are a practical challenge[1]. So it is an important problem to perform the minimization when the operator M is provided, but $\mathbf{M}_{\ell,k}$ is not. Also, we need to keep in mind that the application of M is usually quite cost intensive.

In recent years the work with ensembles of solutions has become very popular, and here we want to show how we can employ an ensemble of solutions to obtain an approximate evaluation of $\mathbf{M}_{\ell,k}^*$ on the subspace X_L spanned by the ensemble.

We describe the basic approach in a simple setting for some nonlinear operator $A : X \to X$ where $X = \mathbb{R}^m$. Let $\varphi \in X$ be some initial guess. We consider a subspace

$$U := \mathrm{span}\left\{q^{(1)}, \dots, q^{(L)}\right\} \tag{5.3.29}$$

with linearly independent elements $q^{(\ell)} \in X$, $\ell = 1, \dots, L$. Then we construct an ensemble $\mathcal{E}_Q$ of solutions by

$$\varphi^{(\ell)} := \varphi + q^{(\ell)} \in X, \quad \ell = 1, .., L \tag{5.3.30}$$

for $q^{(\ell)} \in X$, $\ell = 1, \dots, L$, sufficiently small. The ensemble spans the L-dimensional affine subspace $\varphi + U$ of $\mathbb{R}^m$. We define the ensemble matrix $Q \in \mathbb{R}^{n\times L}$ by

$$Q := \left(q^{(1)}, q^{(2)}, \dots, q^{(L)}\right). \tag{5.3.31}$$

[1] Note that this challenge is less on a mathematical level, but on a human resources level to manage very large codes and their derivatives on a stable level for operational use.

Since the columns of Q are linearly independent, from $Q\psi = 0$ we deduce $\psi = 0$ and thus Q is injective. Assume $Q^T Q\psi = 0$. Then $0 = \psi^T Q^T Q\psi = ||Q\psi||^2$, from which we obtain $Q\psi = 0$ and thus $\psi = 0$, i.e. $Q^T Q$ is injective in $\mathbb{R}^{L\times L}$. We obtain the invertibility of $Q^T Q$ in $\mathbb{R}^{L\times L}$. This means that the operator

$$P_Q := Q(Q^T Q)^{-1} Q^T = QQ^\dagger \tag{5.3.32}$$

with the pseudo-inverse $Q^\dagger$ defined in (3.2.4) is well defined. According to lemma 3.2.3 the operator $P_Q \in \mathbb{R}^{n\times n}$ defined by (5.3.32) with Q given by (5.3.31) is an orthogonal projector $X \to U$.

We now use the ensemble to approximate the derivative of the operator A in a neighborhood of φ, which is feasible as long as A is differentiable. We employ the standard *finite difference approximation*

$$\frac{\partial A(\varphi)}{\partial \varphi} q^{(\ell)} \approx \frac{1}{\epsilon_\ell}\Big(A(\varphi + q^{(\ell)}) - A(\varphi)\Big), \quad \ell = 1, \ldots, L, \tag{5.3.33}$$

where

$$\epsilon_\ell := ||q^{(\ell)}||, \quad \ell = 1, \ldots, L.$$

Assume that we apply the nonlinear operator A to all members of the ensemble $\mathcal{E}_Q$ given by (5.3.30). Then by (5.3.33) we obtain an approximation to the matrix

$$S_Q := \frac{\partial A(\varphi)}{\partial \varphi} Q. \tag{5.3.34}$$

The final step is based on the following lemma.

Lemma 5.3.4. *With the projection operator P_Q defined in (5.3.32) and the pseudo-inverse $Q^\dagger = (Q^T Q)^{-1} Q^T$ defined in (3.2.4) we have*

$$P_Q \frac{\partial A(\varphi)^T}{\partial \varphi} = (Q^\dagger)^T S_Q^T. \tag{5.3.35}$$

Proof. From (5.3.34) we obtain

$$S_Q^T = Q^T \frac{\partial A(\varphi)^T}{\partial \varphi}$$

and thus with the help of (3.2.10)

$$\begin{aligned} (Q^\dagger)^T S_Q^T &= Q(Q^T Q)^{-1} Q^T \frac{\partial A(\varphi)^T}{\partial \varphi} \\ &= P_Q \frac{\partial A(\varphi)^T}{\partial \varphi}, \end{aligned} \tag{5.3.36}$$

which is the desired statement. □

The previous lemma tells us that the application of $(Q^{\dagger})^T$ to $S_Q^T \in \mathbb{R}^{L\times n}$ provides the orthogonal projection of

$$\frac{\partial A(\varphi)^T}{\partial \varphi}$$

onto the ensemble subspace U, i.e. it calculates an approximation to the application of the transpose of the linearization of A. The application of $(Q^{\dagger})^T$ needs the inversion of the matrix $Q^T Q \in \mathbb{R}^{L\times L}$, which is a low-dimensional problem.

If the operators $\mathbf{H}_j$ which are used in (5.3.9) are available, we might apply the above procedure to evaluate $\mathbf{M}^*_{j,j-1}$ in (5.3.11). If $\mathbf{H}_j$ is not given either, we might apply the method to the product of nonlinear operators

$$A_\xi = H\big(t_\xi, M(t_\xi, t_k, \varphi)\big), \quad \xi = k + 1, \ldots, k + K \tag{5.3.37}$$

as used in the original full nonlinear functional (5.3.1).

To carry out the ensemble-based 4D-VAR, $L + 1$ runs of the forward code for the full time interval are sufficient, leading to a time-evolution of the subspace U. For the following we will assume that the basis functions $q^{(\ell)}$ are chosen with sufficiently small norm such that we can choose $\epsilon_\ell = 1$. Using the notation

$$\varphi_\xi := M(t_\xi, t_k, \varphi), \; \varphi_\xi^{(\ell)} := M\big(t_\xi, t_k, \varphi + q^{(\ell)}\big), \tag{5.3.38}$$

for $\ell = 1, \ldots, L$ and the time window described by the indices $\xi = k + 1, \ldots, k + K$, the space U is transformed into

$$U_\xi := \text{span}\left\{\varphi_\xi^{(1)} - \varphi_\xi, \ldots, \varphi_\xi^{(L)} - \varphi_\xi\right\}, \quad \ell = 1, \ldots, L, \tag{5.3.39}$$

i.e. the new basis elements of U_ξ are

$$q_\xi^{(\ell)} := \varphi_\xi^{(\ell)} - \varphi_\xi, \quad \ell = 1, \ldots, L. \tag{5.3.40}$$

The L runs of our forward scheme provide the knowledge of the time-evolution of the matrix Q given by

$$Q_\xi := \left(q_\xi^{(1)}, \ldots, q_\xi^{(L)}\right) \tag{5.3.41}$$

for $\xi = k + 1, \ldots, k + K$. The derivative of the model from t_{j-1} to t_j is approximated by

$$\begin{aligned}\frac{\partial M(t_j, t_{j-1}, \varphi)}{\partial \varphi} q_{j-1}^{(\ell)} &\approx \frac{1}{\left\|\varphi_{j-1}^{(\ell)} - \varphi_{j-1}\right\|}\Big(M\big(t_j, t_{j-1}, \varphi_{j-1}^{(\ell)}\big) - M\big(t_j, t_{j-1}, \varphi_{j-1}\big)\Big) \\ &= \frac{\varphi_j^{(\ell)} - \varphi_j}{\left\|\varphi_{j-1}^{(\ell)} - \varphi_{j-1}\right\|}\end{aligned} \tag{5.3.42}$$

for time index $j = k + 1, \ldots, k + K$ and ensemble index $\ell = 1, \ldots, L$, such that our $L + 1$ runs approximate the matrix

$$\mathbf{M}_{Q,j,j-1} := \frac{\partial M(t_j, t_{j-1}, \varphi)}{\partial \varphi} Q_{j-1} \tag{5.3.43}$$

with Q_{j-1} given by (5.3.41), i.e. the matrix S_Q defined in (5.3.34). Now, according to lemma 5.3.4, the orthogonal projection of $\mathbf{M}_{Q,j,j-1}{}^T$, which for real spaces and the ℓ^2-scalar product coincides with $\mathbf{M}^*_{j,j-1}$, onto the ensemble space U_{j-1} is given by

$$\left(Q^\dagger_{(j-1)}\right)^T \left(\mathbf{M}_{Q,j,j-1}\right)^T, \quad j = k + 1, \ldots, k + K.$$

This means that the ensemble allows the calculation of the approximation to the gradient (5.3.4) of the 4D-VAR functional by

$$\begin{aligned}
\frac{1}{2}\nabla_\varphi J_{4\mathrm{D}} &= B^{-1}(\varphi - \varphi_0) + \sum_{\xi=k+1}^{k+K} \prod_{j=k}^{\xi-1} \mathbf{M}^*_{j+1,j}\mathbf{H}^*_\xi R^{-1}\left(H(t_\ell, \varphi_k) - f_k\right). \\
&\approx B^{-1}(\varphi - \varphi_0) \\
&\quad + \sum_{\xi=k+1}^{k+K} \prod_{j=k}^{\xi-1} \left(Q^\dagger_{j-1}\left(\mathbf{M}_{Q,j,j-1}\right)^T\right)\mathbf{H}^*_\xi R^{-1}\left(H\left(t_\xi, \varphi_\xi\right) - f_\xi\right),
\end{aligned} \tag{5.3.44}$$

which for actual calculation we write as

$$\begin{aligned}
&= B^{-1}(\varphi - \varphi_0) \\
&\quad + \sum_{\xi=k+1}^{k+K} \prod_{j=k}^{\xi-1} \left(\left(Q^\dagger_{(j-1)}\right)^T\left(\mathbf{H}_\xi\mathbf{M}_{Q,j,j-1}\right)^T\right) R^{-1}\left(H\left(t_\xi, \varphi_\xi\right) - f_\xi\right).
\end{aligned} \tag{5.3.45}$$

Here, the calculation of $R^{-1}(H(t_\xi, \varphi_\xi) - f_\xi)$ with band-limited matrix R is of order $O(m)$ where m is the dimension of the observation space Y. The calculation of $\mathbf{H}_\xi\mathbf{M}_{Q,j,j-1}$ is of order $O(mL)$ with the ensemble size L. Finally, $Q^\dagger_{j-1}$ has the cost nL^2, where n is the dimension of the state space. This leads to the total cost of order $O(nL^2 + mnL + m)$.

For the Lorenz 1963 system as described in section 6.1 a simple ensemble-based approximation to the gradient is calculated in section 9.5 and compared to its calculation via the tangent linear adjoint equation (5.3.23).

5.4 The Kalman filter and Kalman smoother

Here, we will first present a completely deterministic approach to the Kalman filter, which provides insight into its equivalence to 4D-VAR for linear systems from the very beginning. Let us study the Tikhonov functional

$$J_1(\varphi) := \alpha\|\varphi - \varphi_0\|^2_X + \|f^{(1)} - A\varphi\|^2_Y, \quad \varphi \in X, \tag{5.4.1}$$

with a linear operator $A : X \to Y$, where we assume that X, Y are some Hilbert spaces over $\mathbb{R}$, i.e. the norm $\|\cdot\|$ is given by $\|\varphi\|^2_X = \langle\varphi, B^{-1}\varphi\rangle$ and $\|\psi\|^2_Y = \langle\psi, R^{-1}\psi\rangle$

with positive definite symmetric invertible operators (or matrices) B and R and we have $\langle \varphi, \psi \rangle = \langle \psi, \varphi \rangle$. We interpret the minimization of (5.4.1) as our assimilation step for the data $f^{(1)} \in Y$ with $\varphi^{(1)}$ being the minimizer of $J_1(\varphi)$.

Now assume that we obtain more data $f^{(2)} \in Y$. For simplicity here we first drop the time dynamics and just consider the assimilation of observations for a fixed time. Then with data $f^{(1)}$ and $f^{(2)}$ we would like to proceed in the same way and minimize

$$J_2(\varphi) := \alpha\|\varphi - \varphi_0\|_X^2 + \|f^{(1)} - A\varphi\|_Y^2 + \|f^{(2)} - A\varphi\|_Y^2, \quad \varphi \in X, \tag{5.4.2}$$

analogously to (5.4.1). Further, for our scheme it should not make a difference to first assimilate $f^{(1)}$ and then $f^{(2)}$ or vice versa or both at the same time as in (5.4.2).

A natural solution to keep the assimilation of $f^{(2)}$ consistent with (5.4.2) when $f^{(1)}$ is first assimilated is to change the norm in X such that it includes the information which is obtained in the first step. This means we want to define some norm $\|\varphi\|_{X,1}$ on X such that

$$\alpha\|\varphi - \varphi_0\|_X^2 + \|f^{(1)} - A\varphi\|_Y^2 = \|\varphi - \varphi^{(1)}\|_{X,1}^2 + \tilde{c}, \quad \varphi \in X, \tag{5.4.3}$$

with some constant $\tilde{c}$ depending on $f^{(1)}$ and φ_0. Then we can rewrite J_2 into

$$J_2(\varphi) = \|\varphi - \varphi^{(b),1}\|_{X,1}^2 + \|f^{(2)} - A\varphi\|_Y^2 + \tilde{c}, \quad \varphi \in X. \tag{5.4.4}$$

To find $\|\cdot\|_{X,1}$ we need to study the quadratic functional J_1 which is equal to

$$\begin{aligned} J_1(\varphi) &= \alpha\langle \varphi - \varphi_0, B^{-1}(\varphi - \varphi_0)\rangle_X + \left\langle f^{(1)} - A\varphi, R^{-1}(f^{(1)} - A\varphi)_Y \right\rangle \\ &= \langle \varphi, \alpha B^{-1}\varphi\rangle - 2\langle \varphi, \alpha B^{-1}\varphi_0\rangle + \langle \varphi_0, \alpha B^{-1}\varphi_0\rangle \\ &\quad + \langle f^{(1)}, R^{-1}f^{(1)}\rangle - 2\langle \varphi, A^*R^{-1}f^{(1)}\rangle + \langle \varphi, A^*R^{-1}A\varphi\rangle. \end{aligned} \tag{5.4.5}$$

With

$$G := \alpha B^{-1} + A^*R^{-1}A \tag{5.4.6}$$

we transform (5.4.5) into

$$\begin{aligned} J_1(\varphi) &= \langle \varphi, G\varphi - 2\alpha B^{-1}\varphi_0 - 2A^*R^{-1}f^{(1)}\rangle \\ &\quad + \underbrace{\langle f^{(1)}, R^{-1}f^{(1)}\rangle}_{\text{const}} + \underbrace{\langle \varphi_0, \alpha B^{-1}\varphi_0\rangle}_{\text{const}}. \end{aligned} \tag{5.4.7}$$

The minimum of (5.4.7) is the minimum of the original minimum of the Tikhonov regularization (5.4.1), given by

$$\varphi^{(1)} = \varphi_0 + G^{-1}A^*R^{-1}\left(f^{(1)} - A\varphi_0\right). \tag{5.4.8}$$

We are now prepared for proving the following result.

Lemma 5.4.1. *With G defined by (5.4.6) a norm is defined on X by*

$$\|\varphi\|_{X,1} := \langle \varphi, G\varphi\rangle, \quad \varphi \in X \tag{5.4.9}$$

For the functional J_1 defined in (5.4.1) and $\varphi^{(1)}$ given by the solution of the 3D-VAR (5.4.8) we have

$$J_1(\varphi) = \langle \varphi - \varphi^{(1)}, G(\varphi - \varphi^{(1)}) \rangle + c, \quad \varphi \in X, \tag{5.4.10}$$

with some constant c depending on A, φ_0 and $f^{(1)}$.

Proof. Clearly G is a self-adjoint positive definite operator, thus by (5.4.9) a norm is defined on X. Using (5.4.8) we calculate

$$\begin{aligned} G(\varphi - \varphi^{(1)}) &= G\varphi - G\varphi^{(1)} \\ &= G\varphi - (\alpha B^{-1} + A^* R^{-1} A)\varphi_0 - A^* R^{-1}(f^{(1)} - A\varphi_0) \\ &= G\varphi - \alpha B^{-1}\varphi_0 - A^* R^{-1} f^{(1)}. \end{aligned} \tag{5.4.11}$$

Applying (5.4.11) twice we calculate

$$\begin{aligned} &\left\langle \varphi - \varphi^{(1)}, G\left(\varphi - \varphi^{(1)}\right)\right\rangle \\ &= \langle \varphi - \varphi^{(1)}, G\varphi - \alpha B^{-1}\varphi_0 - A^* R^{-1} f^{(1)} \rangle \\ &= \langle \varphi - \varphi^{(1)}, G\varphi \rangle - \langle \varphi, \alpha B^{-1}\varphi_0 + A^* R^{-1} f^{(1)} \rangle + \text{const} \\ &= \langle G\varphi, \varphi - \varphi^{(1)} \rangle - \langle \varphi, \alpha B^{-1}\varphi_0 + A^* R^{-1} f^{(1)} \rangle + \text{const} \\ &= \left\langle \varphi, G\left(\varphi - \varphi^{(1)}\right)\right\rangle - \langle \varphi, \alpha B^{-1}\varphi_0 + A^* R^{-1} f^{(1)} \rangle + \text{const} \\ &= \langle \varphi, G\varphi - 2\alpha B^{-1}\varphi_0 - 2A^* R^{-1} f^{(1)} \rangle + \text{const}. \end{aligned} \tag{5.4.12}$$

Up to a constant we have equality of (5.4.12) with (5.4.7) and the proof of (5.4.10) is complete. □

As a result of the above lemma 5.4.1 we can now formulate an iterative algorithm which step-by-step incorporates new measurements $f^{(1)}, f^{(2)}, f^{(3)}, \ldots$ which are simulated using observation operators $A_1, A_2, A_3, \ldots$. We define

$$\varphi^{(0)} := \varphi_0, \quad G^{(0)} := \alpha B^{-1}. \tag{5.4.13}$$

and for $k = 1, 2, 3, \ldots$ iteratively update our solution by

$$G^{(k)} := G^{(k-1)} + A_k^* R^{-1} A_k \tag{5.4.14}$$

$$\varphi^{(k)} = \varphi^{(k-1)} + G^{(k)-1} A_k^* R^{-1}\left(f^{(k)} - A_k \varphi^{(k-1)}\right). \tag{5.4.15}$$

The scheme (5.4.13), (5.4.14), (5.4.15) is known as the *Kalman filter* or *Kalman smoother*, we will come to the difference in a moment. Here we first derive a slightly different form of the equations. The matrix

$$K^{(k)} := G^{(k)-1} A_k^* R^{-1} \tag{5.4.16}$$

is called the *Kalman gain matrix*. Using $B^{(k)} := G^{(k)-1}$ as in (5.2.12) we can write

$$K^{(k)} = B^{(k-1)} A_k^* \left(R + A_k B^{(k-1)} A_k^*\right)^{-1}. \tag{5.4.17}$$

The basic point here, in contrast to the (iterated) 3D-VAR scheme, is the update of the matrix B as given by (5.4.14). The update formula is usually written as follows.

Lemma 5.4.2. *The update formula (5.4.14) is equivalent to*

$$B^{(k)} = \left(I - K^{(k)} A_k\right) B^{(k-1)}, \quad k \in \mathbb{N}. \tag{5.4.18}$$

Proof. Multiplying (5.4.16) with $G^{(k)}$ from the left and A_k from the right we obtain

$$G^{(k)} K^{(k)} A_k = A_k^* R^{-1} A_k,$$

which by multiplication with -1, adding $G^{(k)}$ and using (5.4.14) yields

$$\begin{aligned} G^{(k)}\left(I - K^{(k)} A_k\right) &= G^{(k)} - A_k^* R^{-1} A_k \\ &= G^{(k-1)}. \end{aligned} \tag{5.4.19}$$

Multiplication with $B^{(k-1)}$ from the right and $B^{(k)}$ from the left yields (5.4.18), and the proof is complete. □

Algorithm 5.4.3 (Generic Kalman filter). *Assume that for $k = 1, 2, 3, \ldots$ measurements $f^{(k)}$ are given modeled by linear observation operators $A_k\colon X \to Y$, $k = 1, 2, 3, \ldots$ and let $\varphi^{(0)} \in X$ and $B^{(0)}$ be given. Then, we iteratively assimilate the data by*

$$K^{(k)} := B^{(k-1)} A_k \left(R + A_k B^{(k-1)} A_k^*\right)^{-1} \tag{5.4.20}$$

$$\varphi^{(k)} := \varphi^{(k-1)} + K^{(k)}\left(f^{(k)} - A_k \varphi^{(k-1)}\right) \tag{5.4.21}$$

$$B^{(k)} := \left(I - K^{(k-1)} A_k\right) B^{(k-1)}. \tag{5.4.22}$$

We summarize the above derivations into the following basic lemma.

Lemma 5.4.4. *Let $B^{(0)}$ be injective. Then, the function $\varphi^{(k)}$ defined by (5.4.21) is the unique minimizer of the functional*

$$J_k(\varphi) := \|\varphi - \varphi^{(0)}\|^2_{(B^{(0)})^{-1}} + \sum_{\xi=1}^{k} \|f^{(\xi)} - A_\xi \varphi\|^2_{R^{-1}}. \tag{5.4.23}$$

Proof. For the case $k = 1$ the result is just Tikhonov regularization. For $k = 2$ we have carried out the proof in the equations (5.4.1)–(5.4.10). The case for $k > 2$ is now obtained by induction based on the iterative formulation of the generic Kalman filter (5.4.20)–(5.4.22). □

Of course, when the model is not the identity, we need to incorporate the model propagation between different assimilation steps. For a linear model, this can be carried out by using

$$A_k = H_k M_{k,0} \tag{5.4.24}$$

and the above scheme. It will update the model state at time $t = t_0$, from measurements at time t_1, t_2 etc. This is called a *Kalman smoother*, since it updates in a time prior to the measurement time, and the application of the adjoint

$$A_k^* = M_{k,0}^* H_k^* \tag{5.4.25}$$

is usually smoothing the analysis.

For practical applications, we are more interested in the updates at subsequent times t_1, t_2 etc. This can be obtained by a model propagation by $M_{k,k-1}$ of all quantities in each step. Let $\varphi_\xi^{(s)}$ denote the analysis at time t_ξ when data to t_s are assimilated. We propagate the states of step $k-1$ to time t_k by

$$\varphi_k^{(s)} := M_{k,k-1}\varphi_{k-1}^{(s)}, \qquad s, k \in \mathbb{N}, \tag{5.4.26}$$

which can be derived based $\varphi_{k-1} = M_{k,k-1}^{-1}\, \varphi_k$ in the scalar product by $\langle \varphi_{k-1} = G_{k-1}^{(k-1)}\, \varphi_{k-1} \rangle$. Also, the weights need to be propagated. We denote the weight at t_0 after assimilation of data from t_0 to t_{k-1} by $G_0^{(k-1)}$. When we propagate it to t_ξ, we denote it by $G_\xi^{(k-1)}$, $\xi = 1, 2, 3, \ldots$. In each step we first employ a *weight propagation step* via

$$G_k^{(k-1)} := \left(M_{k,k-1}^{-1}\right)^* G_{k-1}^{(k-1)} M_{k,k-1}^{-1}, \tag{5.4.27}$$

which is a pull-back of the state to the previous time step, then the application of the previous weight and the transformation back to the current time step. Then, the *analysis step* is carried out by the *calculation of the minimizer*

$$\varphi_k^{(k)} = \varphi_k^{(k-1)} + G_k^{(k-1)^{-1}} H_k^* R^{-1}\left(f^{(k)} - H_k \varphi_k^{(k-1)}\right), \tag{5.4.28}$$

and the weight update

$$G_k^{(k)} := G_k^{(k-1)} + H_k^* R^{-1} H_k. \tag{5.4.29}$$

In contrast to the Kalman *smoother*, which works at time t_0, the Kalman *filter* follows the data and assimilates data at time t_k into the state φ_k at time t_k. We summarize the steps as follows.

Definition 5.4.5 (Kalman filter). *The Kalman filter carries out data assimilation as defined in definition 5.1.1 based on an initial weight matrix $G_0^{(0)} = G_0$ and an initial state $\varphi_0^{(0)} = \varphi_0$ as follows.*

1. *The analysis state $\varphi_{k-1}^{(k-1)}$ at time t_{k-1} is propagated to time t_k to obtain $\varphi_k^{(k-1)}$ following (5.4.26). The weight matrix $G_{k-1}^{(k-1)}$ is propagated from t_{k-1} to t_k by (5.4.27).*
2. *Using data $f^{(k)}$ at t_k we calculate the analysis $\varphi_k^{(k)}$ at t_k by (5.4.28) and the analysis $G_k^{(k)}$ of the weights at t_k by (5.4.29).*

The Kalman filter is just an *on-line* version of the Kalman smoother, in the sense that the Kalman filter calculates at t_k what the Kalman smoother calculates at t_0.

The schemes are equivalent for linear model dynamics in the sense that they provide the same analysis at time t_k when measurements up to t_k are assimilated and analysis states are propagated to t_k.

Lemma 5.4.6. *For linear systems M and linear observation operators H, the Kalman filter is equivalent to the assimilation of all data up to* t_k *by the Kalman smoother and subsequent propagation of the result to the time* t_k.

Proof. The equivalence is obtained from (5.4.14) and (5.4.15) by a multiplication from the left by $(M^*_{k,k-1})^{-1}$ and on (5.4.14) from the right by $(M_{k,k-1})^{-1}$ using (5.4.24), (5.4.25) and (5.4.27). □

Usually a more common version of the update formula (5.4.29) is used, working with the matrices $B_k^{(s)}$, not their inverses $G_k^{(s)}$. According to lemma 5.4.2 we have

$$B_k^{(k)} = (I - K^{(k)}H_k)B_k^{(k-1)} \tag{5.4.30}$$

where $K^{(k)}$ denotes the Kalman gain matrix (5.4.16), which in terms of $B_k^{(k-1)}$ is given by (5.4.20).

We are now prepared to relate the Kalman filter or Kalman smoother, respectively, to the 4D-VAR algorithm, which for two measurements was given by (5.4.2). The full 4D-VAR for k measurements $f^{(1)}, \ldots, f^{(k)}$ minimizes

$$J_{4\text{D},k}(\varphi) = \alpha\|\varphi - \varphi_0\|_X^2 + \sum_{\xi=1}^{k}\left\|f^{(\xi)} - H_\xi M_{k,0}\varphi\right\|_Y^2, \quad \varphi \in X, \tag{5.4.31}$$

where we assume that we have linear observation operators H_ξ for each measurement $f^{(\xi)}$. We denote the minimizing function by $\varphi^{(4\text{D},k)}$ and include the model propagation into the definition of $A_\xi = H_\xi M_{\xi,0}$ in lemma 5.4.4 to achieve the following result.

Theorem 5.4.7 (Equivalence of Kalman filtering with transported 4D-VAR). *We assume that our model M is linear. For assimilating measurements* $f^{(1)}, f^{(2)}, f^{(3)} \ldots$ *at times* t_ξ, $\xi = 1, 2, \ldots$ *with a linear operator* $H_\xi: X \to Y$ *and with a priori knowledge given by* $\varphi^{(0)} \in X$, *the Kalman smoother given by (5.4.13)–(5.4.15) is equivalent to the 4D-VAR given by the minimization of (5.4.31) in the sense that*

(a) *for* $k = 1, 2, \ldots$ *the functional (5.4.31) is given by*

$$J_{4\text{D},k}(\varphi) = \langle \varphi - \varphi^{(k)}, G^{(k)}(\varphi - \varphi^{(k)})\rangle + c_k, \quad \varphi \in X, \tag{5.4.32}$$

with some constant c_k *(depending on k).*

(b) *for each step of the Kalman smoother we have*

$$\varphi^{(k)} = \varphi^{(4\text{D},k)}, \quad k = 1, 2, \ldots, \tag{5.4.33}$$

i.e. for linear model dynamics 4D-VAR and the Kalman smoother coincide. In particular, the result of the Kalman filter $\varphi_k^{(k)}$ *at time* t_k *coincides with the transported minimizer of the 4D-VAR functional* $\varphi_k^{(4\text{D},k)} = M_{k,0}\varphi^{(4\text{D},k)}$.

Proof. We carry out a proof by induction. First, consider the case $k = 1$. In this case (5.4.15) is just the Tikhonov regularization or 3D-VAR, and clearly the Kalman equation provides the minimizer of the functional (5.4.31).

Let us assume that the statement is proven for $k - 1$. This means that $\varphi^{(k-1)}$ is the minimizer of (5.4.31) for index $k - 1$ and that

$$J_{4D,k-1}(\varphi) = \left\langle \varphi - \varphi^{(k-1)}, G^{(k-1)}\left(\varphi - \varphi^{(k-1)}\right)\right\rangle + c_{k-1} \qquad (5.4.34)$$

with some constant c_{k-1}. We now apply lemma 5.4.1 to the assimilation of $f^{(k)}$ with a background norm B given by $G^{(k-1)-1}$, noting that

$$J_{4D,k}(\varphi) = J_{4D,k-1}(\varphi) + \|f^{(k)} - A\varphi\|_Y^2.$$

We obtain

$$J_{4D,k}(\varphi) = \left\langle \varphi - \varphi^{(k)}, G^{(k)}\left(\varphi - \varphi^{(k)}\right)\right\rangle + c_k \qquad (5.4.35)$$

with some constant c_k, thus we have shown the identity (5.4.32). Clearly, $\varphi^{(k)}$ is the minimizer of the quadratic functional

$$\left\langle \varphi - \varphi^{(k)}, G^{(k)}\left(\varphi - \varphi^{(k)}\right)\right\rangle + c_k$$

and thus it is the minimizer of $J_{4D,k}$. We have shown (5.4.33). The final argument for the Kalman filter comes from transporting the equivalent results of the Kalman smoother and of the 4D-VAR to the time t_k, which completes the proof. □

5.5 Ensemble Kalman filters (EnKFs)

We have introduced several methods for data assimilation in the previous sections, including Tikhonov data assimilation, 3D-VAR, 4D-VAR and the Kalman filter.

Evaluating the different approaches, we note that 3D-VAR or Tikhonov data assimilation work with fixed norms at every time step and do not fully include all the dynamic information which is available from previous assimilations. Since 4D-VAR uses full trajectories over some time window, it implicitly includes such information and we can expect it to be superior to the simple 3D-VAR. However, for linear systems both Bayes data assimilation for Gaussian error distributions and the Kalman filter are equivalent to 4D-VAR and include all available information by updating the weight matrices and propagating them through time. In general, we can expect them to yield results comparable to those of 4D-VAR. For nonlinear systems, it would be better for the general Bayesian approach to incorporate non-Gaussian dynamical distributions.

The need to *propagate* some probability distribution, in particular its covariance information, is a characteristic feature of the Bayes data assimilation and the Kalman filter. It is also their main challenge, since the matrices $B_k^{(a)}$ or $B_k^{(b)}$ have dimension $n \times n$, which for large n is usually not feasible in terms of computation time or storage, even when supercomputers are employed for the calculation as in most operational centers for atmospheric data assimilation. Thus, a key need for these methods is to formulate algorithms which give a reasonable approximation to the weight matrices $B_k^{(b)}$ with smaller computational cost than by the use of (5.4.13) and (5.4.15) or (5.4.28). This leads us to *ensemble methods* for data assimilation.

The basic idea of *ensemble methods* is to use an *ensemble* of states to estimate dynamic information, e.g. the B-matrix, as an input to the cycled analysis steps. This idea is old and has been employed in the framework of *sequential Monte Carlo methods* since the 1960s. In the framework of large-scale geophysical applications, it has obtained much more attention since the 1990s and has been applied within operational forecasting centers since the beginning of the 2010s.

Often, the approach to ensemble methods is carried out via stochastic estimators. Here, we want to stay within the framework of the previous sections and study the ensemble approach from the viewpoint of applied mathematics. The stochastic viewpoint will be discussed in a second step.

Definition 5.5.1 (Ensemble). *An* ensemble *with N members is any finite set of vectors* $\varphi^{(\ell)} \in X$ *for* $\ell = 1, \ldots, N$. *We can propagate the ensemble through time by applying our model dynamics* M *or* M_k. *Starting with an* initial ensemble $\varphi_0^{(\ell)}$, $\ell = 1, \ldots, L$, *this leads to ensemble members*

$$\varphi_k^{(\ell)} = M[t_{k-1}, t_k]\varphi_{k-1}^{(\ell)}, \quad k = 1, 2, 3, \ldots. \tag{5.5.1}$$

For an arbitrary ensemble $\varphi^{(1)}, \ldots, \varphi^{(N)}$ we use the mean estimator $\hat{\mu}$ defined by (4.1.9)

$$\hat{\mu} = \frac{1}{N}\sum_{\ell=1}^{N} \varphi^{(\ell)} \tag{5.5.2}$$

to define the *ensemble matrix*

$$Q := \frac{1}{\sqrt{N-1}}\left(\varphi^{(1)} - \hat{\mu}, \ldots, \varphi^{(N)} - \hat{\mu}\right). \tag{5.5.3}$$

The *ensemble subspace* $U_Q \subset X$ is given by

$$U_Q = \operatorname{span}\left\{\varphi^{(1)} - \hat{\mu}, \ldots, \varphi^{(N)} - \hat{\mu}\right\}. \tag{5.5.4}$$

We call the vectors $\varphi^{(\ell)} - \hat{\mu}$, $\ell = 1, \ldots, N$ the *centered ensemble*. We remark that for a linearly independent ensemble state in X we have $\dim U_Q = N - 1$.

The standard stochastic estimator $\hat{B}$ for the covariance matrix B is given by (4.1.10), which in terms of the ensemble matrix Q is calculated by

$$\hat{B} = QQ^*. \tag{5.5.5}$$

Definition 5.5.2 (EnKF). *The EnKF employs an ensemble of states to estimate dynamic information by cycling the following data assimilation steps.*

1. *First, given an analysis ensemble* $\varphi_{k-1}^{(a,\xi)}$, $\xi = 1, \ldots, N$, *at* t_{k-1}, *as in (5.5.1) the model dynamics* M *is used to propagate the ensemble members to time* t_k *and calculate a first-guess or background ensemble* $\varphi_k^{(b,\xi)}$, $\xi = 1, \ldots, N$.
2. *In an* analysis step *at time* t_k, *the EnKF calculates an estimate for the matrix* $B_k^{(b)}$ *by (5.5.5) based on the background ensemble* $\varphi_k^{(b,1)}, \ldots, \varphi_k^{(b,N)}$.

3. *This is used to calculate an update of the ensemble mean following the Kalman filter (5.4.28). An important step is the generation of an analysis ensemble* $\varphi_k^{(a,1)}, \ldots, \varphi_k^{(a,N)}$*, which fits to the analysis covariance matrix* $B_k^{(a)}$ *as calculated by the Kalman filter.*

Different approaches have been suggested to generate the analysis ensemble. We will come to this shortly, presenting the approach of the *square-root filter* (SRF).

As an important step, we study the construction of a particular family of ensembles generated by the eigenvalue decomposition of some self-adjoint weight matrix $B^{(b)}$. For a self-adjoint matrix, according to theorem 2.4.20 there is a complete set of eigenvectors of B, i.e. we have vectors $\psi^{(1)}, \ldots, \psi^{(n)} \in X$ and eigenvalues $\lambda^{(1)}, \ldots, \lambda^{(n)}$ such that

$$B\psi^{(\ell)} = \lambda^{(\ell)}\psi^{(\ell)}, \quad \ell = 1, \ldots, n. \tag{5.5.6}$$

The eigenvalues are real valued and we will always assume that they are ordered according to their size $\lambda^{(1)} \geqslant \lambda^{(2)} \geqslant \ldots \geqslant \lambda^{(n)}$. With the matrix $\Lambda := \mathrm{diag}\{\sqrt{\lambda^{(1)}}, \ldots, \sqrt{\lambda^{(n)}}\}$ and the orthogonal matrix $U := (\psi^{(1)}, \ldots, \psi^{(n)})$ we obtain

$$B = U\Lambda^2 U^* = (U\Lambda)(U\Lambda)^*, \tag{5.5.7}$$

where we note that $U^* = U^{-1}$. This representation corresponds to the well-known *principle component analysis* of the *quadratic form* defined by

$$F(\varphi, \psi) := \varphi^T B\psi, \quad \varphi, \psi \in X. \tag{5.5.8}$$

Geometrically, B defines a hypersurface of second order with positive eigenvalues, whose level curves form a family of $n - 1$-dimensional ellipses in X. The principal axes of this ellipse are given by the eigenvectors $\psi^{(l)}$, $\ell = 1, \ldots, n$.

The application of B to some vector $\varphi \in X$ according to (5.5.7) is carried out by a projection of φ onto the principle axis $\psi^{(\ell)}$ of B, then the multiplication with $\lambda^{(\ell)}$is applied. This set-up can be a basis for further insight to construct a low-dimensional approximation of B.

Before we continue the ensemble construction we first need to discuss the *metric* in which we want an approximation of the B-matrix. We remark that the role of B in the Kalman filter is mainly in the update formulas (5.4.13)–(5.4.15). Here, to obtain a good approximation of the vector updates in L^2, we need B to be approximated in the operator norm based on L^2 on $X = \mathbb{R}^m$. That is what we will use as our basis for the following arguments.

Lemma 5.5.3. *We construct an ensemble of vectors by choosing the* $N - 1$ *maximal eigenvalues of* B *and its corresponding eigenvectors* $\psi^{(1)}, \ldots, \psi^{(N-1)}$. *We define*

$$Q := \left(\sqrt{\lambda^{(1)}}\psi^{(1)}, \ldots, \sqrt{\lambda^{(N-1)}}\psi^{(N-1)}\right). \tag{5.5.9}$$

Then, we have the error estimate

$$\|B - QQ^*\|^2 = \sup_{j=N,\ldots,n} |\lambda^{(j)}| = |\lambda^{(N)}|. \tag{5.5.10}$$

Proof. The proof is obtained from

$$B - QQ^* = U\tilde{\Lambda}^2 U^* \tag{5.5.11}$$

with $\tilde{\Lambda}^2 = \mathrm{diag}\{0, \ldots, 0, \lambda^{(N)}, \lambda^{(N+1)}, \ldots, \lambda^{(n)}\}$, where there are $N - 1$ zeros on the diagonal of $\tilde{\Lambda}$. Since U is an orthogonal matrix, the norm estimate (5.5.10) is now straightforward. □

From the Courant minimum–maximum principle we know that

$$\lambda^{(\ell)} = \min_{\dim U = \ell - 1} \max_{\varphi \in U^\perp, \|\varphi\|=1} \langle \varphi, B\varphi \rangle. \tag{5.5.12}$$

Then, we have

$$\begin{aligned}
\|B - QQ^*\| &\geqslant \sup_{B\varphi \perp U_Q, \|\varphi\|=1} \|(B - QQ^*)\varphi\| \\
&\geqslant \sup_{B\varphi \perp U_Q, \|\varphi\|=1} \|B\varphi\| \\
&\geqslant \sup_{B\varphi \perp U_Q, \|\varphi\|=1} \langle \varphi, B\varphi \rangle \\
&\geqslant \min_{\dim U = N-1} \sup_{\varphi \perp U, \|\varphi\|=1} \langle \varphi, B\varphi \rangle \\
&= \lambda^{(N)}.
\end{aligned} \tag{5.5.13}$$

The above results are summarized in the following theorem.

Theorem 5.5.4. *Let the eigenvalues $\lambda^{(1)} \geqslant \lambda^{(2)} \geqslant \ldots \geqslant \lambda^{(n)}$ of the self-adjoint weight matrix B be ordered according to its size and let $\varphi^{(1)}, \ldots, \varphi^{(N)}$ with $N \in \mathbb{N}$ be an arbitrary ensemble of states in X. Then, the error for the approximation of the weight matrix B by QQ^* with Q defined in (5.5.9) is estimated from below by*

$$\|B - QQ^*\| \geqslant \lambda^{(N)}. \tag{5.5.14}$$

Remark. The optimal error $\lambda^{(N)}$ can be achieved when the centered scaled ensemble (5.5.9) spans the space of the $N - 1$ eigenvectors $\psi^{(1)}, \ldots, \psi^{(N-1)}$ of B with the largest eigenvalues $\lambda^{(1)}, \ldots, \lambda^{(N-1)}$ with appropriate coefficients as in (5.5.9).

Ensembles can be used to approximate the weight matrix $B_{k+1}^{(\mathrm{b})}$ when the weight matrix $B_k^{(\mathrm{a})}$ is given. If $B_k^{(\mathrm{a})}$ is approximated by the ensemble $\varphi_k^{(1)}, \ldots, \varphi_k^{(N)}$ in the form

$$B_k^{(\mathrm{a})} \approx Q_k^{(\mathrm{a})} \left(Q_k^{(\mathrm{a})}\right)^*, \tag{5.5.15}$$

then by (5.4.27) we derive an approximation for $B_{k+1}^{(\mathrm{b})}$ by

$$\begin{aligned}
B_{k+1}^{(\mathrm{b})} &= M_k B_k^{(\mathrm{a})} M_k^* \\
&\approx M_k Q_k^{(\mathrm{a})} \left(Q_k^{(\mathrm{a})}\right)^* M_k^* \\
&= M_k Q_k^{(\mathrm{a})} \left(M_k Q_k^{(\mathrm{a})}\right)^* \\
&= Q_{k+1}^{(\mathrm{b})} \left(Q_{k+1}^{(\mathrm{b})}\right)^*.
\end{aligned} \tag{5.5.16}$$

Lemma 5.5.5. *If the error of the approximation of $B_k^{(b)}$ by an ensemble $\varphi_k^{(1)}, \ldots, \varphi_k^{(N)}$ with ensemble matrix $Q_k^{(b)}$ is given by*

$$\left\| B_k^{(a)} - Q_k^{(a)} \left(Q_k^{(a)} \right)^* \right\| \leqslant \epsilon, \tag{5.5.17}$$

then the error estimate for the propagated ensemble at time t_{k+1} is given by

$$\left\| B_{k+1}^{(b)} - Q_{k+1}^{(b)} \left(Q_{k+1}^{(b)} \right)^* \right\| \leqslant \|M_k\| \|M_k^*\| \epsilon. \tag{5.5.18}$$

Proof. Based on (5.5.16) the proof is straightforward. □

A key question of ensemble methods is how to update the ensemble in the data assimilation step. Given the data f_k at time t_k, how do we obtain an ensemble which approximates the analysis matrix $B_k^{(a)}$ given an ensemble which approximates the background matrix $B_k^{(b)}$. We know that for the Kalman filter the analysis weight matrix $B_k^{(a)}$ is calculated from $B_k^{(b)}$ by (5.4.30). In terms of the ensemble approximations this means

$$Q_k^{(a)} \left(Q_k^{(a)} \right)^* = (I - K_k H_k) Q_k^{(b)} \left(Q_k^{(b)} \right)^* \tag{5.5.19}$$

with the *ensemble Kalman matrix*

$$K_k := Q_k^{(b)} \left(Q_k^{(b)} \right)^* H_k^* \left(R + H_k Q_k^{(b)} \left(Q_k^{(b)} \right)^* H_k^* \right)^{-1}, \tag{5.5.20}$$

leading to

$$Q_k^{(a)} \left(Q_k^{(a)} \right)^* = Q_k^{(b)} \left\{ \underbrace{I - \left(Q_k^{(b)} \right)^* H_k^* \left(R + H_k Q_k^{(b)} \left(Q_k^{(b)} \right)^* H_k^* \right)^{-1} H_k Q_k^{(b)}}_{=:T} \right\} \left(Q_k^{(b)} \right)^*. \tag{5.5.21}$$

The term T in the curly brackets is self-adjoint and non-negative, which can be seen by using some singular-value decomposition and elementary estimates, i.e. there is a matrix S such that $T = SS^*$. It can be calculated for example by a singular value decomposition. This finally leads to

$$Q_k^{(a)} = Q_k^{(b)} S. \tag{5.5.22}$$

The update formula (5.5.22) for the ensemble is called the *ensemble Kalman square root filter (SRF)*.

In section 6.3 we employ the ensemble Kalman SRF to carry out data assimilation on the Lorenz 1963 system, demonstrating its potential when limited measurement data is available.

We finish with some error estimate for the approximation of the analysis B-matrix $B^{(a)}$ in dependence on the error of the background B-matrix $B^{(b)}$ for the ensemble Kalman SRF in comparison with the full Kalman filter.

Lemma 5.5.6. *Assume that $\varphi_k^{(1)}, \ldots, \varphi_k^{(N)}$ is an ensemble which satisfies*

$$\left\| B_k^{(b)} - Q_k^{(b)}\left(Q_k^{(b)}\right)^* \right\| \leqslant \epsilon. \tag{5.5.23}$$

with some $\epsilon < \|B_k^{(b)}\|$. Then, for the analysis ensemble defined by (5.5.22) and the analysis matrix $B^{(a)}$ defined by the Kalman filter, we have

$$\left\| B_k^{(a)} - Q_k^{(a)}\left(Q_k^{(a)}\right)^* \right\| \leqslant C\epsilon \tag{5.5.24}$$

with some constant C not depending on $Q^{(b)}$.

Proof. Using the notation $K_k^{(\mathrm{true})}$ for the Kalman update matrix with background matrix $B_k^{(b)}$, by (11.2.34) and (5.5.19) we decompose

$$\begin{aligned} B_k^{(a)} - Q_k^{(a)}\left(Q_k^{(a)}\right)^* &= \left(I - K_k^{(\mathrm{true})} H_k\right)\left(B_k^{(b)} - Q_k^{(b)}\left(Q_k^{(b)}\right)^*\right) \\ &\quad + \left(K_k^{(\mathrm{true})} - K_k\right) H_k Q_k^{(b)}\left(Q_k^{(b)}\right)^* \end{aligned} \tag{5.5.25}$$

with K_k defined by (5.5.20). We remark that due to its special structure with regularizer R the norm of the inverse term in (5.5.20) is bounded uniformly not depending on $Q^{(b)}$. Further, the norm

$$\begin{aligned} \left\| Q_k^{(b)}\left(Q_k^{(b)}\right)^* \right\| &= \left\| B_k^{(b)} + \left(Q_k^{(b)}\left(Q_k^{(b)}\right)^* - B_k^{(b)}\right) \right\| \\ &\leqslant \|B_k^{(b)}\| + \epsilon \\ &\leqslant 2\|B_k^{(b)}\| \end{aligned}$$

is bounded uniformly. Using standard identities and estimates of the form

$$P^{-1} - \tilde{P}^{-1} = \tilde{P}^{-1}(\tilde{P} - P)P^{-1} \tag{5.5.26}$$

such that

$$\|P^{-1} - \tilde{P}^{-1}\| \leqslant \|\tilde{P}^{-1}\| \cdot \|P^{-1}\| \cdot \|\tilde{P} - P\| \tag{5.5.27}$$

and

$$\begin{aligned} \|QP^{-1} - \tilde{Q}\tilde{P}^{-1}\| &= \left\| Q\left(P^{-1} - \tilde{P}^{-1}\right) + \left(Q - \tilde{Q}\right)\tilde{P}^{-1} \right\| \\ &\leqslant \|Q\| \cdot \|P - \tilde{P}\| \cdot \|P^{-1}\| \cdot \|\tilde{P}^{-1}\| + \|Q - \tilde{Q}\| \cdot \|\tilde{P}^{-1}\| \end{aligned} \tag{5.5.28}$$

for operators Q, $\tilde{Q}$ and P, $\tilde{P}$, we derive

$$\|K_k^{(\mathrm{true})} - K_k\| \leqslant c\epsilon \tag{5.5.29}$$

with a constant c not depending on $Q^{(b)}$. This estimate applied to (5.5.25) yields the desired result (5.5.24), and the proof is complete.

Localization. We have kept one very important topic for EnKF and other ensemble methods, that is *localization*. If the covariance matrix B is estimated by $B = QQ^*$, where Q is the ensemble matrix defined in (5.5.3), then the *increments* of the Kalman filter analysis mean update

$$\varphi_k^{(a)} = \varphi_k^{(b)} + BH^*(R + HBH^*)^{-1}\left(y - H\left(\varphi_k^{(b)}\right)\right), \quad k \in \mathbb{N}, \tag{5.5.30}$$

are in the subspace spanned by the columns of the matrix Q, i.e. they are in the *ensemble subspace* U given in (5.5.4).

However, for large-scale problems only relatively low numbers of ensemble members can be used, for example for numerical weather prediction (NWP) usually between 40 and 200 ensemble members can be run on modern supercomputers. But the state space has a dimension of 10^8 or 10^9, such that we cannot hope to achieve sufficient approximation quality without further tools. Running the basic EnKF for a large-scale problem does not work, the *spurious correlations* generated by the low-rank approximation of the covariance matrix completely destroy the quality of the analysis $\varphi^{(a)}$.

The key insight comes from the fact that the underlying partial differential equations (PDEs) are mainly local, such that a sufficient approximation to the analysis can be achieved when we solve a local version of the equations. This localization is then carried out in parallel many times, where in the optimal case each local analysis is independent. When the problem is decomposed into L local subproblems, each has 100 degrees of freedom, and we have increased our total number of degrees of freedom from $N = 100$ degrees of freedom to $L \cdot N$ degrees of freedom. We can, for example, localize around a point $x \in \mathbb{R}^m$ by only using observations which are in some ball $B_\rho(x)$ for the assimilation. A local and transformed version of the EnKF has been suggested with the *local ensemble transform Kalman filter* (LETKF) in [3] and related publications, and it has become very successful for large-scale applications.

Localization can be carried out in state space and in observation space, where both approaches have advantages and disadvantages. In this book, localization in state space has been applied to problems of cognitive neuroscience, see section 7.6, where we localize the *inverse neural kernel problem*.

Let us close the section with some historical remarks as well as links to current important developments. In the area of geophysical data assimilation, the basic idea of the EnKF was first proposed by [4] and the idea was applied to global NWP by [5]. [6] developed the theoretical basis of the EnKF methods based on perturbations of the observations. [7] proposed the alternative approach of the SRF. It does not use randomly perturbed observations, but formulates a deterministic calculation of the posterior ensemble.

Further variants of ensemble filters include the *ensemble adjustment Kalman filter* of [8], the ensemble transform Kalman filter of [9] and the serial ensemble SRF filter of [7]. Localization is a key ingredient of the local ensemble Kalman filter by [10] and the LETKF by [3], which assimilate all locally available observations in one step, but only in a limited local domain.

Important current research topics on EnKFs are covariance localization and inflation, see [11–16]. Flow-adaptive localization has been described by [20] and [17–19]; multi-scale localization by [21]; and flow-adaptive inflation by [22, 23] and [24]. *Generalized and transformed localization* is developed in [25].

The investigation of large ensembles has been carried out by [26], see also [27]. This list is, of course, far from being complete, but gives a sense of the strong activity in this very relevant field of research.

5.6 Particle filters and nonlinear Bayesian data assimilation

We now come to *Bayesian* data assimilation. The basic idea here is to base the assimilation step on Bayes' formula (4.2.7), where the prior probability distribution $p(x)$ is calculated from previous time steps. In this section, we employ the stochastic notation $x \in X$ for the states and $y \in Y$ for the observations.

In its most general form, *Bayesian data assimilation* employs the full prior distribution $p(x) = p_k(x)$, not only its mean in the form of a background state $x_k^{(b)}$ at time t_k, and it calculates the full *posterior distribution* $p(x|y) = p_k(x|y_k)$ at time t_k, not only an approximation $x_k^{(a)}$ to its mean.

Algorithm 5.6.1 (Bayesian data assimilation). *We consider the data assimilation problem given by definition 5.1.1, where some initial probability distribution $p_0(x)$ on X is given. We consider the initial distribution as our analysis distribution for time t_0. Then, we solve the data assimilation problem by the following iterative steps.*

1. *Based on the model M we propagate the analysis distribution from t_{k-1} to t_k to obtain the prior or* background *distribution $p_k^{(b)}(x)$.*
2. *Then we employ Bayes' formula to calculate a* posterior *or* analysis *distribution*

$$p_k^{(a)}(x) := p(x|y_k) = cp(y_k|x)p_k^{(b)}(x), \quad x \in X, \tag{5.6.1}$$

where c is a normalization constant such that

$$\int_X p_k^{(a)}(x)\,\mathrm{d}x = 1. \tag{5.6.2}$$

Remark. For some methods to carry out Bayesian data assimilation the normalization is not necessary, for example when a Markov chain Monte Carlo (MCMC) method based on the Metropolis–Hastings sampler is employed.

A key point of Bayesian data assimilation is to describe and work with the probability distributions under consideration. In the case where $p_k^{(b)}$ is a Gaussian distribution (4.5.28) with mean $x_k^{(b)}$ and covariance $B_k^{(b)}$, and where the error distribution $p(y_k|x)$ of the measurements is a Gaussian distribution of the form

$$p(y_k|x) = \tilde{c}e^{-\frac{1}{2}\left\{(y_k-Hx)^T R^{-1}(y_k-H_k x)\right\}}, \quad x \in X,$$

with covariance R and normalization constant $\tilde{c}$, the posterior distribution

$$p_k^{(a)}(x) = ce^{-\frac{1}{2}\left\{(y_k - Hx)^T R^{-1}(y_k - H_k x) + \left(x - x_k^{(b)}\right)^T \left(B_k^{(b)}\right)^{-1}\left(x - x_k^{(b)}\right)\right\}}, \quad x \in X. \tag{5.6.3}$$

To find the mean and covariance of this posterior Gaussian distribution is exactly the calculation step of the *Kalman filter* of section 5.4, i.e. for the case of Gaussian distributions the Kalman filter carries out Bayesian data assimilation exactly. The EnKF of section 5.5 estimates a Gaussian distribution to fit the given ensembles and then carries out a Bayesian assimilation step explicitly.

3D-VAR can be seen as an approximation to Bayesian data assimilation when only the mean of the posterior distribution is updated in each assimilation step, but the distribution itself is kept fixed as a Gaussian distribution with covariance B_0.

We will next introduce several further approaches to carry out or approximate a Bayesian data assimilation step without the limitation of a Gaussian distribution. The first such approach is the classical *particle filter*.

Algorithm 5.6.2 (Classical particle filter). *The* classical particle filter *employs an ensemble $x^{(\ell)}$ of states representing the prior probability distribution $p_k^{(b)}$ at time t_k. The analysis step at time t_k is carried out by a calculation of new weights*

$$w_{k,\ell} := p_k^{(a)}(x^{(\ell)}) = cp\left(y_k \middle| x^{(\ell)}\right)p_k^{(b)}(x^{(\ell)}), \quad \ell = 1, \ldots, L, \tag{5.6.4}$$

for the particles $x^{(\ell)}$ according to the Bayes' formula (5.6.1). Then, a resampling step is carried out where particles are chosen and multiplied according to their relative weight w_ℓ.

The classical particle filter carries out a resampling either by just adapting the multiplicity of the individual particles according to the relative weight $[w_{k,\ell}/L]$, (where the Gaussian bracket $[s]$ denotes the largest integer smaller than s), or by carrying out some real resampling in a neighborhood of the particles with sufficiently large weights $w_{k,\ell}$, for example by randomly choosing $[w_{k,\ell}/L]$ particles from a normal distribution with center $x^{(\ell)}$ and some variance σ, which can either be determined adaptively from the full ensemble or can be used as a tuning parameter.

For large-scale problems, classical particle filters suffer from filter divergence in the sense that mostly only one or very view of the particles obtain a sufficiently large weight. Then, we either end up with L copies of one particle, or we strongly rely on a random resampling in each analysis step, losing a lot of the information which is usually carried by the ensemble.

There are several special cases, where the calculation of the analysis ensemble can be carried out explicitly and in a deterministic way. This is the case for example in the EnKF described in section 5.5, or for the *Gaussian particle filter*, which we will describe next. Here, we use the *Gaussian function* defined in (4.5.28) by

$$g_B(x, z) := \frac{1}{\sqrt{(2\pi)^n \det(B)}} e^{-\frac{1}{2}(x-z)^T B^{-1}(x-z)}, \quad x \in \mathbb{R}^m, \tag{5.6.5}$$

with center $z \in X$ and covariance B.

Algorithm 5.6.3 (Gaussian particle filter). *The Gaussian particle filter treats each particle as a Gaussian probability distribution. The prior or background probability distribution is then given by a Gaussian sum*

$$p_k^{(b)}(x) := \sum_{\ell=1}^{L} g_{B_{k-1}}(x, x^{(b,\ell)}), \quad x \in X. \tag{5.6.6}$$

The analysis distribution for the Gaussian particle filter is given by

$$p_k^{(a)}(x) := \sum_{\ell=1}^{L} g_{B_k}(x, x^{(a,\ell)}), \quad x \in X, \tag{5.6.7}$$

where the analysis ensembles $x^{(a,\ell)}$, $\ell = 1, \dots, L$ are calculated by treating each particle as an individual Gaussian distribution, i.e. we calculate

$$x^{(a,\ell)} := x^{(b,\ell)} + BH_k(R + H_k BH_k^*)^{-1}\big(y_k - H_k x^{(b,\ell)}\big), \quad \ell = 1, \dots, L. \tag{5.6.8}$$

The Gaussian covariance matrices B_k are either kept constant setting $B_k = B_{k-1}$ for all $k = 1, 2, \dots$; or they are updated according to the Kalman update formula (5.4.30), i.e.

$$B_k = \big(I - B_{k-1}H_k(R + H_k B_{k-1} H_k^*)^{-1} H_k\big)B_{k-1}, \quad k = 1, 2, 3, \dots. \tag{5.6.9}$$

For large-scale problems, the update (5.6.9) cannot be carried out efficiently. In this case a fixed B_k in the Gaussian particle filter is usually chosen.

The advantage of this approximation is the fact that the calculation of the new analysis mean $x^{(a,\ell)}$ is a fast and easy formula (5.6.8). It shares this advantage with the EnKF, which also provides an explicit formula for one Gaussian distribution which is estimated on the basis of the full ensemble $x^{(b,\ell)}$.

We need to emphasize that the *Gaussian particle filter* does *not* fully realize Bayes' formula (5.6.1) for the whole distribution, nor for the individual Gaussian particles, but employs only a crude approximation to it by moving the mean of the particles according to an individual application of Bayes' formula to each particle.

The full and correct realization of Bayesian data assimilation is obtained by MCMC methods as described in section 4.3 to sample a distribution and to propagate it from one time t_{k-1} to the next time step t_k. The MCMC method provides a technique to generate an *ensemble* $x^{(\ell)}$, $\ell = 1, \dots, L$ of states which sample some given probability distribution. We have described the *Metropolis–Hastings* and the *Gibbs* sampler in section 4.4.

Algorithm 5.6.4 (Markov chain particle filter (MCPF)). *The* MCPF *employs MCMC methods for the resampling step based on Bayes' formula. Based on an initial ensemble $x_0^{(a,\ell)}$, $\ell = 1, \dots, L$, it carries out cycled assimilation steps as follows.*

1. *The ensembles $x_{k-1}^{(a,\ell)}$ are propagated from t_{k-1} to t_k, leading to the new background ensemble $x_k^{(b,\ell)}$ at time t_k.*

2. *The background distribution is approximated in some basis, for example using* $q \in \mathbb{N}$ Gaussian *basis functions as*

$$p_k^{(b)}(x) := \sum_{\xi=1}^{q} g_B(x, z_\xi), \quad x \in X. \tag{5.6.10}$$

As a standard choice we suggest $q = L$ *and* $z_\ell := x^{(b,\ell)}$ *for* $\ell = 1, \ldots, L$, *i.e. we use Gaussian radial basis functions to approximate the posterior density distribution. The variance B can be estimated using the ensemble, or it can be chosen fixed.*

3. *The posterior or analysis distribution* $p_k^{(a)}(x)$ *is calculated by Bayes' formula (5.6.1), i.e.*

$$p_k^{(a)}(x) = cp(y|x)\sum_{\xi=1}^{q} g_B(x, z_\xi), \quad x \in X. \tag{5.6.11}$$

4. *Finally, a resampling of the ensemble is carried out by the MCMC methods, see algorithm 4.3.5, based on either the Metropolis–Hastings or the Gibbs sampler, leading to the analysis ensemble* $x_k^{(a,\ell)}$, $\ell = 1, \ldots, L$.

The convergence of these methods for the resampling steps has been investigated in theorem 4.3.2, corollary 4.3.3 and lemma 4.3.4.

When we choose $q = 1$ in the second step combined with a standard covariance estimator $\hat{B} = QQ^*$ as chosen in (5.5.5), the posterior distribution $p_k^{(a)}(x)$ is a Gaussian distribution as employed by the EnKF algorithm 5.5.2. In this case, the MCPF samples the same probability distribution as the EnKF, but the method of carrying out the sampling is completely different, and the ensembles which are generated differ significantly.

The MCPF is able to work with fully non-Gaussian probability distributions, for example with multimodal distributions, and has a wide flexibility to adapt its approximations to different applications. Also, it is easy to include further conditions and further knowledge into the probability distribution, for example when particular variables are restricted to some part of the space. The full potential of the MCPF has not yet been exploited in the framework of large-scale data assimilation.

We have discussed the topic of *localization* for the EnKF in the previous section 5.5. It applies to the particle filter as well. The key idea of the LETKF to work in ensemble space locally for a reasonably chosen localization grid can be carried over to particle filters, where we can work in ensemble space and carry out all steps on the ensemble coefficients, where we need to transport our metric into the ensemble space as well. Here, we will not go into further detail about important current research questions which arise from the need of *homogeneity* between different localization areas when solutions of PDEs are re-sampled. Techniques from *optimal transport theory* are being investigated, but also straightforward techniques within the MCPF provide options to achieve sufficient smoothness of solutions.

Let us, again, close this section with some historical remarks and links to current research activities. Several particle filter methods have been formulated and tested for small-dimensional problems, see [28, 29]. More recently, Ades and van Leeuwen [30] have adapted their *equivalent-weights particle filter* to high-dimensional systems using a simple relaxation technique. Further, *Gaussian mixture models* have been developed, based on the estimation of the model error probability distribution by a set of Gaussians, see [31] or [32] for a hybrid method of a Gaussian mixture and the particle filter.

Frei and Künsch [33] develop a hybrid method for an EnKF and a particle filter. A *localized MCPF*, which is basically a particle filter based on a MCMC method in ensemble space, is under development at the German Weather Service [34], combining all the advantages of LETKFs with the fully nonlinear use of the posterior probability distribution for the generation of the posterior ensemble.

We also refer the reader to recent books on nonlinear data assimilation [35, 36].

Bibliography

[1] Freitag M and Potthast R 2013 Synergy of inverse problems and data assimilation techniques *Large Scale Inverse Problems* (*Radon Series on Computational and Applied Mathematics* vol 13) ed M Cullen *et al* (Berlin: de Gruyter)

[2] Navon I M 2009 Data assimilation for numerical weather prediction: a review *Data Assimilation for Atmospheric, Oceanic and Hydrologic Applications* ed S K Park and L Xu (Berlin: Springer) pp 21–65

[3] Hunt B R, Kostelich E J and Szunyogh I 2007 Efficient data assimilation for spatiotemporal chaos: a local ensemble transform Kalman filter *Physica* D **230** 112–26

[4] Evensen G 1994 Sequential data assimilation with a nonlinear quasi-geostrophic model using Monte Carlo methods to forecast error statistics *J. Geophys. Res.* **99** 10143–62

[5] Houtekamer P L and Mitchell H L 1998 Data assimilation using an ensemble Kalman filter technique *Mon. Weather Rev.* **126** 796–811

[6] Burgers G, van Leeuwen P J and Evensen G 1998 Analysis scheme in the ensemble Kalman filter *Mon. Weather Rev.* **126** 1719–24

[7] Whitaker J S and Hamill T M 2002 Ensemble data assimilation without perturbed observations *Mon. Weather Rev.* **130** 1913–24

[8] Anderson J L 2001 An ensemble adjustment Kalman filter for data assimilation *Mon. Weather Rev.* **129** 2884–903

[9] Bishop C H, Etherton B J and Majumdar S J 2001 Adaptive sampling with the ensemble transform Kalman filter. Part I: Theoretical aspects *Mon. Weather Rev.* **129** 420–36

[10] Ott E, Hunt B R, Szunyogh I, Zimin A V, Kostelich E J, Corazza M, Kalnay E, Patil D J and Yorke J A 2004 A local ensemble Kalman filter for atmospheric data assimilation *Tellus* A **56** 415–28

[11] Miyoshi T, Yamane S and Enomoto T 2007 Localizing the error covariance by physical distances within a local ensemble transform Kalman filter (LETKF) *SOLA* **3** 89–92

[12] Miyoshi T and Sato Y 2007 Assimilating satellite radiances with a local ensemble transform Kalman filter *SOLA* **3** 37–40

[13] Campbell W F, Bishop C H and Hodyss D 2010 Vertical covariance localization for satellite radiances in ensemble kalman filter *Mon. Weather Rev.* **138** 282–90

[14] Greybush S J, Kalnay E and Miyoshi T 2011 Balance and ensemble Kalman filter localization techniques *Mon. Weather Rev.* **139** 511–22

[15] Janjić T, Nerger L, Albertella A, Schroeter J and Skachko S 2011 On domain localization in ensemble-based Kalman filter algorithms *Mon. Weather Rev.* **139** 2046–60

[16] Perianez A, Reich H and Potthast R 2015 Optimal localization for ensemble Kalman filter *J. Meteor. Soc. Japan.* at press

[17] Bishop C H and Hodyss D 2007 Flow-adaptive moderation of spurious ensemble correlations and its use in ensemble-based data assimilation *Q. J. R. Meteorol. Soc.* **133** 2029–44

[18] Bishop C H and Hodyss D 2009 Ensemble covariances adaptively localized with ECO-RAP. Part 1: Tests on simple error models *Tellus* A **61** 84–96

[19] Bishop C H and Hodyss D 2009 Ensemble covariances adaptively localized with ECO-RAP. Part 2: A strategy for the atmosphere *Tellus* A **61** 97–111

[20] Anderson J L 2007 Exploring the need for localization in ensemble data assimilation using a hierarchical ensemble filter *Physica* D **230** 99–111

[21] Miyoshi T and Kondo K 2013 A multi-scale localization approach to an ensemble Kalman filter *SOLA* **9** 170–3

[22] Anderson J L 2007 An adaptive covariance inflation error correction algorithm for ensemble filters *Tellus* A **59** 210–24

[23] Anderson J L 2009 Spatially and temporally varying adaptive covariance inflation for ensemble filters *Tellus* A **61** 72–83

[24] Miyoshi T 2011 The Gaussian approach to adaptive covariance inflation and its implementation with the local ensemble transform Kalman filter *Mon. Weather Rev.* **139** 1519–35

[25] Nadeem A and Potthast R 2015 Transformed and generalized localization for ensemble methods in data assimilation *Math. Methods Appl. Sci.* DOI:10.1002/mma.3496 (at press)

[26] Miyoshi T, Kondo K and Imamura T 2014 The 10,240-member ensemble Kalman filtering with an intermediate AGCM *Geophys. Res. Lett.* **41** 5264–71

[27] Hamrud M, Bonavita M and Isaksen L 2014 EnKF and Hybrid Gain Ensemble Data Assimilation *Technical Memorandum* 733

[28] van Leeuwen P J 2009 Particle filtering in geophysical systems *Mon. Weather Rev.* **137** 4089–114

[29] van Leeuwen P J 2010 Nonlinear data assimilation in geosciences: an extremely efficient particle filter *Q. J. R. Meteorol. Soc.* **136** 1991–9

[30] Ades M and van Leeuwen P J 2014 The equivalent-weights particle filter in a high-dimensional system *Q. J. R. Meteorol. Soc.* **141** 484–503

[31] Hoteit I, Pham D-T and Triantafyllou G *et al* 2008 A new approximate solution of the optimal nonlinear filter for data assimilation in meteorology and oceanography *Mon. Weather Rev.* **136** 317–34

[32] Stordal A S, Karlsen H A and Naevdal G 2011 Bridging the ensemble Kalman filter and particle filters: the adaptive Gaussian mixture filter *Comput. Geosci.* **15** 293–305

[33] Frei M and Künsch H R 2013 Bridging the ensemble Kalman and particle filters *Biometrika* **100** 781–800

[34] Potthast R, Reich H, Rhodin A and Schraffm C 2015 A localized Markov chain particle filter (LMCPF) for data assimilation *Deutscher Wetterdiens Report*

[35] Van Leeuwen P J, Cheng Y and Reich S 2015 *Nonlinear Data Assimilation* (Cham: Springer International)

[36] Reich S and Cotter C 2015 *Probabilistic Forecasting and Bayesian Data Assimilation* (Cambridge: Cambridge University Press)

Chapter 6

Programming of numerical algorithms and useful tools

A key part of the area of inverse problems and data assimilation is the development of *algorithms* which solve inverse problems and assimilate data into dynamical models. Algorithms need to be implemented and tested, both on generic situations which are set up to investigate their behavior as well as in real world environments.

Of course, today a wide variety of software *languages* and *tools* is available for implementation. Often, the choice of tools is influenced by the institutional environment of researchers. For example, many operational centers for weather prediction run their code on supercomputers where FORTRAN compilers have been used for decades. Also, programming languages such as C++ and JAVA are very popular among researchers today.

For *developing* new algorithms and for *learning* about inversion and assimilation, *fast implementation and visualization* is needed. To fully understand algorithms, you need to gain some experience of simple examples and more complex problems. To this end, we will provide generic codes for many of the algorithms in an environment which can be easily used by everyone (available for download **here**). The codes can be run with either MATLAB or its free version OCTAVE [1], which is available for download on the Internet.

Here, we assume that the reader is familiar with elementary programming in OCTAVE, such as variables, loops, scripts, functions and elementary plotting. These aspects can be learned within hours from simple online tutorials. We do not assume much further knowledge.

6.1 MATLAB or OCTAVE programming: the butterfly

Let us start with a simple OCTAVE code to generate the butterfly shown in figure 5.1 or figure 6.1, which is one of the standard examples for a dynamical system

doi:10.1088/978-0-7503-1218-9ch6

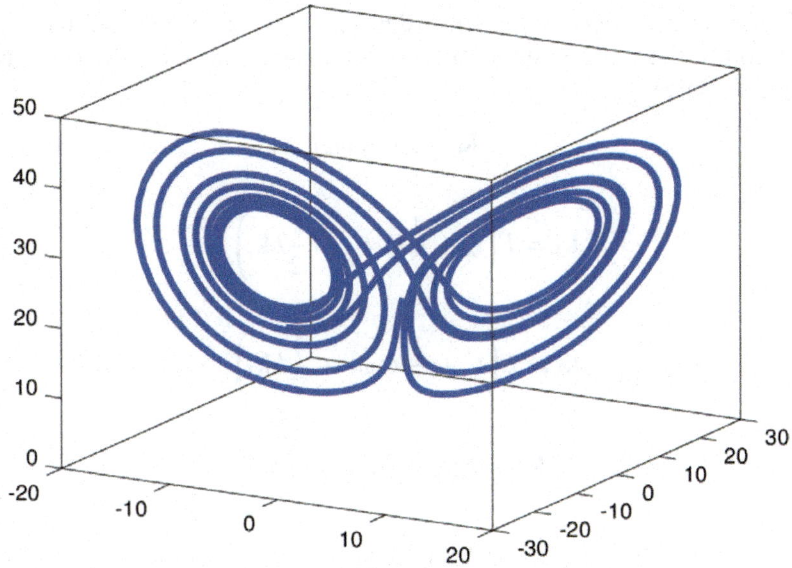

Figure 6.1. The result of code 6.1.3 which displays the solution to the Lorenz 1963 system (6.1.1)–(6.1.3).

introduced by *Lorenz* in 1963. It is very popular as a study case for data assimilation algorithms and consists of three coupled ordinary differential equations

$$\dot{x} = \sigma(y - x) \tag{6.1.1}$$

$$\dot{y} = x(\rho - z) - y \tag{6.1.2}$$

$$\dot{z} = xy - \beta z, \tag{6.1.3}$$

with constants σ, ρ, β, known as the *Prandtl number*, the *Rayleigh number* and a non-dimensional *wave number*, respectively. The classical values $\sigma = 10$, $\beta = 8/3$ and $\rho = 28$ have been suggested by Lorenz.

The state of the system is given by $\varphi = (x, y, z)^T \in \mathbb{R}^3$ and the above system can be summarized into the form

$$\dot{\varphi}(t) = F(t, \varphi), \qquad \varphi(0) = \varphi_0 \tag{6.1.4}$$

with

$$F(t, \varphi) := \begin{pmatrix} \sigma(y - x) \\ x(\rho - z) - y \\ xy - \beta z \end{pmatrix}. \tag{6.1.5}$$

The *Runge–Kutta scheme* of 4th-order (see for example [2]) to calculate a numerical solution to the system (6.1.1)–(6.1.3) with initial values x_0, y_0 and z_0 iteratively calculates

$$\mathbf{k}_1 = F(t_k, \varphi_k) \tag{6.1.6}$$

$$\mathbf{k}_2 = F\left(t_k + \frac{1}{2}h, \varphi_k + \frac{1}{2}h\mathbf{k}_1\right) \tag{6.1.7}$$

$$\mathbf{k}_3 = F\left(t_k + \frac{1}{2}h, \varphi_k + \frac{1}{2}h\mathbf{k}_2\right) \tag{6.1.8}$$

$$\mathbf{k}_4 = F(t_k + h, \varphi_k + h\mathbf{k}_3) \tag{6.1.9}$$

$$\varphi_{k+1} = \varphi_k + \frac{1}{6}h(\mathbf{k}_1 + 2\mathbf{k}_2 + 2\mathbf{k}_3 + \mathbf{k}_4). \tag{6.1.10}$$

Code 6.1.1. *The following* OCTAVE *or* MATLAB *function simulates the solution to the ordinary differential equation (6.1.1)–(6.1.3) by a 4th-order Runge–Kutta scheme.*

```
function [x,y,z]= sim_06_1_1_Lorenz63(N_Time,h,x0,y0,z0,sigma,rho,beta)

% Solving du/dt = F(u), u(0)=u0 for u=[x;y;z] via
% the 4-th order Runge-Kutta scheme.

x    = zeros(N_Time,1);           % vector initialization
y    = zeros(N_Time,1);           % ~
z    = zeros(N_Time,1);           % ~
x(1) = x0;                     % initial value x
y(1) = y0;                     % initial value y
z(1) = z0;                     % initial value z

for i = 1:N_Time
    k1 = sim_06_1_2_F([x(i);y(i);z(i)],sigma,rho,beta);
    k2 = sim_06_1_2_F([x(i);y(i);z(i)]+h*k1/2,sigma,rho,beta);
    k3 = sim_06_1_2_F([x(i);y(i);z(i)]+h*k2/2,sigma,rho,beta);
    k4 = sim_06_1_2_F([x(i);y(i);z(i)]+h*k3,sigma,rho,beta);

    xtmp = [x(i);y(i);z(i)]+(h/6)*(k1+(2*k2)+(2*k3)+k4);
    x(i+1)=xtmp(1);
    y(i+1)=xtmp(2);
    z(i+1)=xtmp(3);
end
```

Here, we use the function `sim_06_1_2_F` as given by the following code.

Code 6.1.2. *Forcing term for the Lorenz 1963 dynamical system.*

```
function xout = sim_06_1_2_F(xin,sigma,rho,beta)

xout(1,1) = sigma*(xin(2)-xin(1));
xout(2,1) = xin(1)*(rho-xin(3))-xin(2);
xout(3,1) = xin(1)*xin(2)-beta*xin(3);
```

MATLAB or OCTAVE allows both functions and scripts. Functions hide the values of variables and allow clean calls of functionality. For code development, often simple scripting is helpful at the beginning. In any case, parameters need to be set and functions need to be called. We recommend carrying this out using well-defined scripts as in the following code 6.1.3, since this allows a clear debugging of how values are set.

Code 6.1.3. *A simple script* `sim_06_1_3_butterfly.m` *to call the Runge–Kutta solution of the Lorenz 1963 system to generate the 'butterfly' shown in figure 5.1.*

```
N_Time = 1000;         % number of time steps
h      = 0.01;         % time spacing
x0     = 0.4;          % initial value x
y0     = -0.7;         % initial value y
z0     = 21;           % initial value z
sigma  = 10;           % choose Lorenz' values
rho    = 28;           % ~
beta   = 8/3;          % ~

% Call Runge Kutta function
[x,y,z] = sim_06_1_1_Lorenz63(N_Time,h,x0,y0,z0,sigma,rho,beta);

% Display the Solution Curve
fo = figure;                          % open figure
po = plot3(x,y,z,'LineWidth',4);      % curve
view(30,20);                          % change viewing angle
ao = get(po,'Parent');                % get axis object identifier
set(ao,'FontSize',14);                % set graphics font size

% Save the Image as png
savefile(fo,'sim_06_1_3_butterfly');
```

In the next section we will show how data assimilation methods can easily be applied to the Lorenz 1963 model.

6.2 Data assimilation made simple

For data assimilation algorithms, we need to be able to apply the model to some initial state and run it some time T into the future. Here, we realize this model by a routine called `Lorenz63`. It is based on the Runge–Kutta solution of the Lorenz 1963 model described in section 6.1. We have set up this model routine such that we can generate an ensemble of $L \in \mathbb{N}$ states by running the model for an ensemble of L initial conditions.

Code 6.2.1. *The dynamics is generated by the* model *routine* `sim_06_2_1_Lorenz63.m`.

```
function xout = sim_06_2_1_Lorenz63(xin,T,sigma)

% xin is the initial state or an ensemble of such states, size 3xL
% T the time to run the system for
% sigma is a variable parameter

rho = 28;          % fixed Parameter for Lorenz63  Dynamics
beta = 8/3;       % fixed Parameter for Lorenz63  Dynamics

L = size(xin,2);    % Size of current input ensemble

% Loop over all ensemble members
for ll=1:L
  x0 = xin(1,ll);    % initial value
  y0 = xin(2,ll);    % initial value
  z0 = xin(3,ll);    % initial value

  h=0.005;           % time grid spacing and the
  N_Time=T/h;        % corresponding number of time steps

  % Model Dynamics
  [x,y,z] = sim_06_1_1_Lorenz63(N_Time,h,x0,y0,z0,sigma,rho,beta);
  % output state(s)
  xout(:,ll) = [x(N_Time,1),y(N_Time,1),z(N_Time,1)];
end
```

The code here calls the routine `sim_06_1_1_Lorenz63.m`, which generates the full trajectory between time $t = 0$ and time $t = T$ with a spacing h of the time steps and N_{Time} such steps to reach T. This has been chosen just for simplicity, of course for other applications one would modify the original routine and not store all intermediate states.

Now, we first run the model to generate some kind of truth. This is usually denoted as a *nature run*. In practical applications the nature run is different from the model—as nature is different from our approximations to it. So it is important to thoroughly choose the parameters and set-up of the nature run to reflect this difference between the truth and the model which is to be used for the data assimilation. Here, we choose to modify the Lorenz 1963 parameter σ for the nature run.

Observations are modeled by the *observation operator H*. Here, we choose a linear operator, which observes either one component of the system, or the sum of two components, or the full state.

Code 6.2.2. *Code* `sim_06_2_2_Generate_Original_and_Data.m` *to generate the* nature run *and the* measurement data.

```
x0 = [0;-10;21];  % Initial state
randn('seed',0) % Use the same random numbers to achieve repeatability

% 1) Setup Observation operator
H = [1 0 0];      % observing the first component of the state only
%H = [1 1 0];     % observing the sum of the first and second component
%H = eye(3,3);    % observing the full state;

% 2) Generate a "nature" run and the "observations"
Nnat    = 60;   % steps for nature run
x       = x0;   % initial state for iteration
dtime   = 0.1;  % time interval between measurements
sigma0  = 10;   % parameter in the Lorentz system
for j=1:Nnat
        x = sim_06_2_1_Lorenz63(x,dtime,sigma0); % Calculate next true state
        xv(:,j) = x;                             % and save it in xv
    y(:,j) = H*x + 0.4*(rand(size(H,1),1)-0.5); % observation with noise
end
```

The next step is running the data assimilation algorithm. Here, we start with three-dimensional variational assimilation (3D-VAR) or Tikhonov regularization. We need to define the covariance matrices, which here we choose by hand as diagonal matrices.

A key point here is to make sure that the model is different from the nature run model. Here, we modify the parameter σ. The initial state for our assimilation is chosen to be identical to the initial state of the nature run.

The *assimilation cycle* takes model runs and assimilation steps in turn. We carry out `Nnat` such cycling steps for all available data y_j at time t_j, $j = 1, \ldots,$ `Nnat`. From the *analysis* `xa` at some time step t_{j-1} we run the model forward and calculate the *background* `xb` at time t_j. Then, the analysis step following equation (5.2.14) with $\alpha = 1$ is carried out and `xa` is calculated at time t_j. We store it in some vector and also calculate the error e_j at time t_j, i.e. the norm difference between the analysis and the truth. Now, the whole analysis cycle can be summarized into a short script.

Code 6.2.3. *Code* `sim_06_2_3_Carry_out_Assimilation.m` *to carry out the data assimilation by a 3D-VAR algorithm.*

```
B = 0.3*eye(3,3);    % background error covariance matrix
m = size(H,1);       % number of observations
R = 0.3*eye(m,m);    % data error covariance matrix

xa = x0;             % start with xa to be set to x0
sigmaA=15;           % a different parameter, representing a slightly
                     % different model when doing the assimilation
for j=1:Nnat
   xb = sim_06_2_1_Lorenz63(xa,dtime,sigmaA);   % calculate the background
   xbv(:,j) = xb;                          % and save it for later display
   xa = xb + B*H'*inv(R+ H*B*H')*(y(:,j)-H*xb); % the analysis step
   xav(:,j)=xa;                            % save it for display
   e(j)=norm(xav(:,j)-xv(:,j));            % calculate the error, we know the truth
end
```

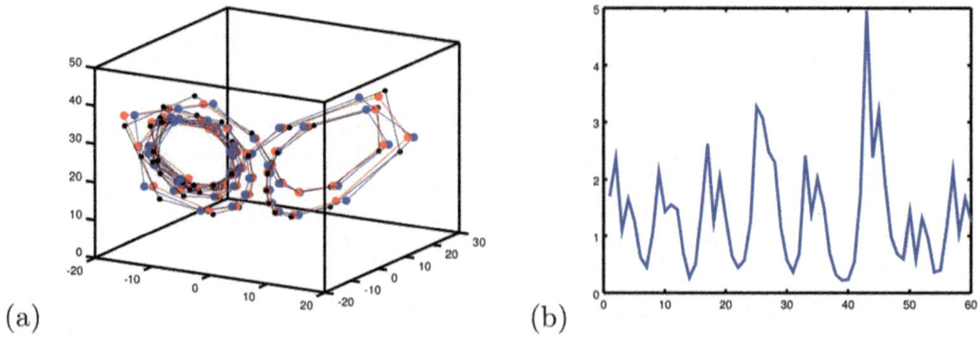

Figure 6.2. The result of code 6.2.3 as generated by code 6.2.4, i.e. the assimilation with observation operator $H = [1\ 1\ 0]$ and the 3D-VAR algorithm. In (a) the original is shown in black, the background in blue and the analysis in red. The error evolution is shown in (b). The assimilation leads to a stable synchronization between the truth and the assimilation cycle.

Finally, we display the nature run, the background states and the analysis states in typical three-dimensional graphics as shown in figure 6.2. The following code also generates a figure with the error curve e_j, $j = 1, \ldots,$ `Nnat`.

Code 6.2.4. *Code* `sim_06_2_4_Plot_Original_and_Analysis.m` *to display the nature run, the background states and the analysis states.*

```
% Generate a figure to display the different curves
fo   = figure;
plot3(xbv(1,:),xbv(2,:),xbv(3,:),'k.-','MarkerSize',15, ...
   'Color',[0 0 1]);        % showing the background curve
hold on;
plot3(xav(1,:),xav(2,:),xav(3,:),'k.-','MarkerSize',15, ...
   'Color',[1 0 0]);        % showing the analysis curve
po = plot3(xv(1,:),xv(2,:),xv(3,:),'k.-','MarkerSize',10, ...
   'Color',[0 0 0]);        % showing the nature run
%legend('background','analysis','original')
ao = get(po,'Parent');   % get handle to axis object
% set font size and line width, set viewpoint and generate legend
set(ao,'FontSize',14, 'LineWidth',3); view(30,20);
% save image as png
savefile(fo,'sim_06_2_assimilation_results');    % save image
% Generate a Figure to display the first guess errors
fobj = figure;
po=plot(e,'LineWidth',3);  % plot error evolution vector
ao = get(po,'Parent');      % get handle to axis object
% set font size and line width, set viewpoint and generate legend
set(ao,'FontSize',14, 'LineWidth',3);
savefile(fo,'sim_06_2_error_evolution');  % save image
```

Finally, we note that these scripts can be called by a control script to run them all one by one, just call `sim_06_2_5_Control.m`.

Code 6.2.5. *Script to control the nature run, the data assimilation cycle and visualization.*

```
clear all; close all;
sim_06_2_2_Generate_Original_and_Data;
sim_06_2_3_Carry_out_Assimilation;
sim_06_2_4_Plot_Original_and_Analysis;
```

6.3 Ensemble data assimilation in a nutshell

The goal of this section is to show how the *ensemble Kalman square root filter* (SRF) can be set up in a simple way for the Lorenz 1963 system to carry out the assimilation. We will also show that for some observation operators this ensemble data assimilation system is better than 3D-VAR.

The generation of the *nature run* and the simulation of measurements is described in code 6.2.2. Then, we need to generate an initial ensemble and carry out the data assimilation step in a cycled way as described by equation (5.2.14) with the covariance matrix B approximated according to equation (5.5.15) and the calculation of the analysis ensemble by the SRF as described in (5.5.22). The control code is given as follows, generating figure 6.3.

Code 6.3.1. *Script* `sim_06_3_1_control_EnKF.m` *to carry out the ensemble Kalman SRF assimilation cycle for the Lorenz 1963 system.*

```
clear all; close all;
sim_06_2_2_Generate_Original_and_Data;
sim_06_3_2_EnKF_square_root_filter;
sim_06_2_4_Plot_Original_and_Analysis;
```

And the ensemble Kalman filter is carried out by code 6.3.2.

Code 6.3.2. *Script* `sim_06_3_2_EnKF_square_root_filter.m` *to carry out the ensemble Kalman SRF assimilation cycle for the Lorenz 1963 system.*

```
L = 5;                   % number of ensemble members
m = size(H,1);           % number of observations
R = 0.3*eye(m,m);        % data error covariance matrix

xa = repmat(x0,1,L) + 0.9*(rand(3,L)-0.5); % generate initial ensemble
sigmaA=15;               % parameter for model in assimilation cycle
for j=1:Nnat
  xb = sim_06_2_1_Lorenz63(xa,dtime,sigmaA);  % calculate the background ensemble
  xbm = (sum(xb'))'/L;                 % and its mean
  xbv(:,j) = xbm;                      % save it in xbv
  Q = (xb-repmat(xbm,1,L))/sqrt(L-1); % setup matrix of differences
  B = Q*Q';                            % ensemble covariance matrix
  K = B*H'*inv(R+H*B*H');              % Kalman Gain Matrix
  xam = xbm + K*(y(:,j)-H*xbm);        % analysis mean
  % Square Root Filter for Analysis Ensemble
  TT = eye(L,L) - Q'*H'*inv(R+H*B*H')*H*Q;
  [U,S,V]=svd(TT);                     % take SVD, i.e. TT = U*S*V'
  T = U*sqrt(S)*V';                    % square root of TT
  Qad = Q*T*1.2;                       % calculate analysis differences
  xa = repmat(xam,1,L)+Qad*sqrt(L-1); % full analysis ensemble
  xav(:,j)=xam;                        % save analysis mean
  e(j)=norm(xav(:,j)-xv(:,j));         % calculate the error of the mean
end
```

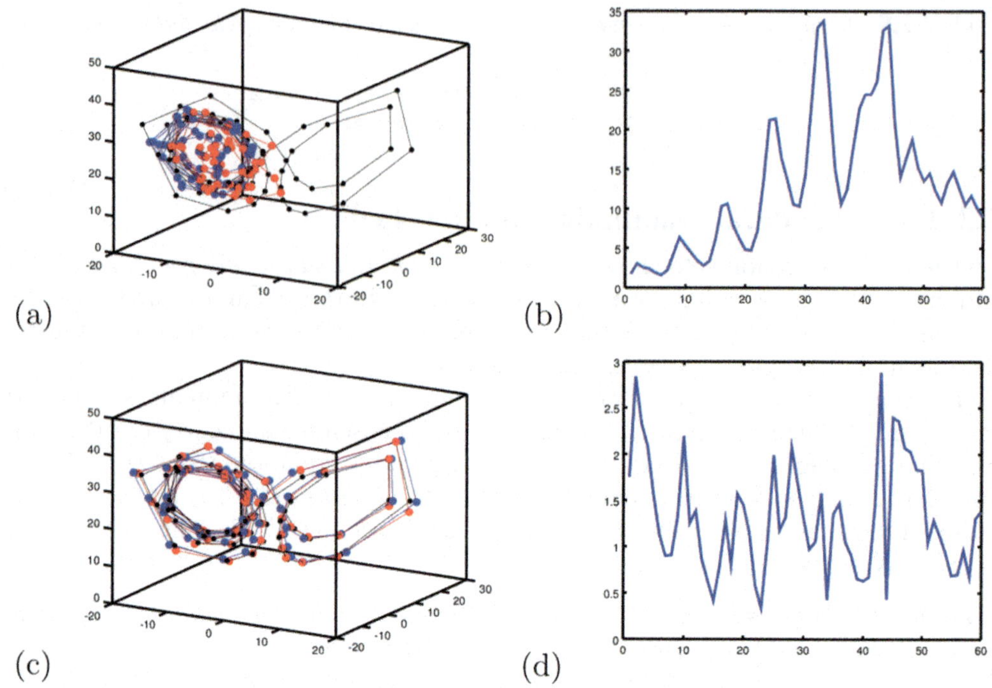

Figure 6.3. The result of code 6.2.3 with the 3D-VAR algorithm; in contrast to figure 6.2 we employed the observation operator $H = [1\ 0\ 0]$. In (a) and (c) the original is shown in black, the background in blue and the analysis in red. The error evolution is shown in (b) and (d). The assimilation does not lead to a stable synchronization between the truth and the assimilation cycle. In (c) and (d) we show the results of code 6.3.2, where the *ensemble Kalman SRF* is used. Here, with the ensemble Kalman filter using the dynamic covariances between the different variables, the assimilation is stable.

We note that for a three-dimensional system such as Lorenz 1963 we do not need techniques such as localization. Also, we need to remark that we employed some *inflation* by a factor $\gamma = 1.2$ in line 19 of code 6.3.2, i.e. we have pushed the spread of the analysis ensemble up by a multiplicative factor.

If we do not use some kind of *inflation* to take care of the model error, the spread of the background ensemble is only increased by the divergence of the model dynamics when different ensemble initial conditions are used. In this case, it will become too small and not properly take into account the difference between the truth (the nature run) and the model ensembles.

6.4 An integral equation of the first kind, regularization and atmospheric radiance retrievals

To learn more about regularization methods let us solve an integral equation of the first kind with some Gaussian kernel. Such equations can be employed as models for atmospheric temperature or humidity reconstruction in an atmospheric column from measured infrared or microwave radiation.

Let $k : [0, H] \times [0, H] \to \mathbb{R}$ for some maximal height $H > 0$ be a continuous kernel function, $f \in C([0, H])$ be some function representing observations and $\varphi \in C([0, H])$ be an unknown profile function. In *atmospheric radiance retrievals*, φ could be the temperature or humidity in an atmospheric column and $f(\nu)$ could be measurements of *brightness temperature* at particular frequencies ν. Here we consider f as a function of the maximum value of its kernel $k(\nu, \cdot)$ in $[0, H]$—such that ν becomes a height variable and is no longer a frequency or some channel number. We search for a solution to

$$\int_0^H k(\nu, s)\varphi(s)\,\mathrm{d}s = f(\nu), \quad \nu \in [0, H] \tag{6.4.1}$$

by a *collocation method* as follows. Let $\mathbf{x} \in \mathbb{R}^m$ be a discretization of the function φ at quadrature points

$$s_\xi = \frac{\xi}{n} \cdot H \in [0, H], \quad \xi = 1, \ldots, n. \tag{6.4.2}$$

We replace the integral by a *quadrature formula*, which in the most simple case is the *rectangular rule*, evaluate the formula for $\nu_\xi := \xi/n \cdot H$ for $\xi = 1, \ldots, n$, employ the definitions

$$\mathbf{y} := \left(f(\nu_1), \ldots, f(\nu_n)\right)^T \in \mathbb{R}^m, \quad \gamma_\xi := \frac{H}{n}, \tag{6.4.3}$$

for the right-hand side at the points y_ξ, $\xi = 1, \ldots, n$ and with quadrature weights γ_ξ, to obtain the *finite-dimensional discrete linear system*

$$\sum_{\xi=1}^{n} k(s_j, s_\xi)\gamma_\xi\, \mathbf{x}_\xi = \mathbf{y}_j, \quad j = 1, \ldots, n. \tag{6.4.4}$$

The integral operator $A : C([0, H]) \to C([0, H])$ defined by

$$(A\varphi)(\nu) := \int_0^H k(\nu, s)\varphi(s)\,\mathrm{d}s, \quad \nu \in [0, H], \tag{6.4.5}$$

is approximated by the matrix

$$\mathbf{A} := \left(k(s_j, s_\xi)\gamma_\xi\right)_{j,\xi=1,\ldots,n}, \tag{6.4.6}$$

such that the operator equation (6.4.1) is approximated by

$$\mathbf{A}\mathbf{x} = \mathbf{y}. \tag{6.4.7}$$

Let k be a Gaussian kernel, i.e.

$$k(\nu, s) = \mathrm{e}^{-\sigma|\nu - s|^2}, \quad \nu, s \in [0, H]. \tag{6.4.8}$$

We define a true solution $\mathbf{x}_{\mathrm{true}} := (\cos(s_\xi))_{\xi=1,\ldots,n}$, data by $\mathbf{y} := \mathbf{A}\mathbf{x}_{\mathrm{true}} + \delta$ where δ is some random error generated by a uniform distribution between ± 0.01. The unregularized solution is calculated by $x = \mathbf{A}^{-1}\mathbf{y}$ and the regularized solution via

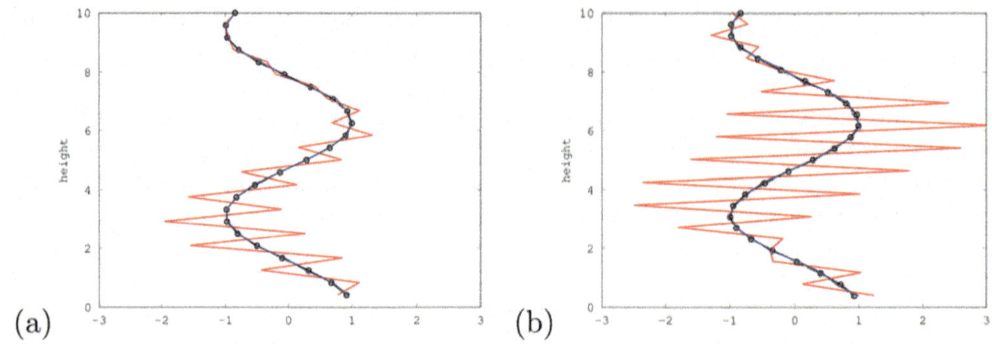

Figure 6.4. The true function x_{true} (black dots), the reconstruction x without regularization (red) and the regularized solution x_α in blue. (a) is carried out with $n = 24$, (b) with $n = 26$, where the ill-posedness strongly increases, as seen by the oscillations which take over in the reconstruction (red curve). The images are generated using code 6.4.1.

Tikhonov regularization (3.1.24) $\mathbf{x}_\alpha := (\alpha I + \mathbf{A}^*\mathbf{A})^{-1}\mathbf{A}^* y$. The OCTAVE code is given by code 6.4.1 and the results are displayed in figure 6.4.

Code 6.4.1. *Script* `sim_06_4_1_radiance_igl.m` *to demonstrate the implicit ill-posedness of radiance data retrieval and radiance assimilation.*

```
H = 10; n = 24;        % Interval, number of quadrature pointss
h = H/n;               % integration grid spacing

s = (h:h:H)';          % vector with integration points
sm = repmat(s,1,n);    % matrix with integration points

sigma = 2;             % parameter for Gaussian kernel
k = exp(-sigma*(sm-sm').^2);  % Gaussian kernel

xtrue = cos(s);        % true solution
A     = k*h;           % discrete version of integral operator
y     = A*xtrue + 0.02*(rand(n,1)-0.5);     % right-hand side
[U,S,V]=svd(A);        % singular value decomposition of A
x = A\y;               % solve equation Ax = y

alpha = 1e-3;          % regularization parameter
xa = (alpha*eye(n,n) + A'*A)\(A'*y); % Tikhonov regularization

% Graphics
fo = figure;                % generate figure
% plot naive solution to IGL in red
po = plot(xtrue,s,'ko-','LineWidth',5); % plot true solution
ao = get(po,'Parent');      % get axis object to control font size
hold on;                    % keep previous curve in figure
plot(x,s,'r','LineWidth',5);       % plot unregularized solution
plot(xa,s,'b.-','LineWidth',3);  % plot regularized solution
set(ao,'FontSize',16);      % set font size
axis([-3 3 0 H]);           % axis control
ylabel('height');  % set x label
filename = ['sim_06_4_1_radiance_igl.png']; % current date and time
savefile(fo,filename);                  % save image
```

Figure 6.4 demonstrates the unregularized and regularized solutions to the integral equation (6.4.1), where we chose regularization parameter $\alpha = 10^{-3}$. The behavior of the solutions depends on the number of quadrature points chosen. The more quadrature points we choose, the higher the ill-posedness of the inverse problem, reflected in figure 6.4(a) with $n = 24$ quadrature points and (b) with $n = 26$ quadrature points.

6.5 Integro-differential equations and neural fields

The following chapter 7 will introduce inverse theory for neural fields. Here, we want to carry out the basic first steps to simulate such fields. The neural field equation in its simplest form is an *integro-differential equation* of the form

$$\tau \dot{u}(x, t) = -u(x, t) + \int_{\Omega} w(x, y) f(u(y, t))\,\mathrm{d}y \tag{6.5.1}$$

for $x \in \Omega$, $t \in [0, T]$ on some domain Ω in $\mathbb{R}^m$ for $m = 1, 2, 3$ and time interval $[0, T]$, where τ is some parameter, w is a neural kernel and f is a *sigmoidal function*

$$f(s, \eta) = \frac{1}{1 + \mathrm{e}^{-(s-\eta)}}. \tag{6.5.2}$$

As is well known, it is possible to numerically solve such an equation by explicit time iteration known as an *Euler scheme* defined by

$$\frac{\tau}{h_t}(\mathbf{u}_{k+1} - \mathbf{u}_k) = -\mathbf{u}_k + \mathbf{W} f(\mathbf{u}_k, \eta), \quad k = 0, 1, 2, \ldots, \tag{6.5.3}$$

for the state vectors $\mathbf{u}_\xi \in \mathbb{R}^m$ at time t_ξ, where we have chosen a discretization

$$(\mathbf{u}_k)_\xi := u(x_\xi, t_k), \quad k = 0, 1, 2, \ldots, \quad \xi = 1, \ldots, n \tag{6.5.4}$$

where x_ξ, $\xi = 1, \ldots, n$ with $n \in \mathbb{N}$ points is a discretization of the domain Ω and $t_k := h_t k$, $k = 0, 1, 2, \ldots$ with the time grid spacing $h_t > 0$. The kernel $\mathbf{W}$ based on $w(x_\xi, x_j)$ for $\xi, j = 1, \ldots, n$ is defined as the matrix

$$\mathbf{W} := \Big(\big(w(x_\xi, x_j) s_j\big)\Big)_{\xi, j=1,\ldots,n} \tag{6.5.5}$$

with integration weight s_j at point x_j, $j = 1, \ldots, n$. From (6.5.3) we obtain

$$\mathbf{u}_{k+1} = \mathbf{u}_k + \frac{h_t}{\tau}(-\mathbf{u}_k + \mathbf{W} f(\mathbf{u}_k, \eta)), \quad k = 0, 1, 2, \ldots. \tag{6.5.6}$$

For our example, we choose a kernel defined by

$$w(x, y) := c\mathrm{e}^{-\sigma_1 \cdot (x_1 - y_1)^2 - \sigma_2 (x_2 - y_2)^2} \cdot (x_1 - y_1), \quad x, y \in \Omega, \tag{6.5.7}$$

which we denote as a *directed Gaussian* in direction x_1. We start some simulation in a *rectangular neural patch* Ω with the initial state given by a Gaussian

$$u(x, 0) := \mathrm{e}^{-\sigma |x - x_0|^2}, \quad x \in \Omega, \tag{6.5.8}$$

with some $x_0 \in \Omega$.

For many examples we need to quickly generate points on a regular grid. This is carried out in an elementary way in lines 1–13 of code 6.5.1, where `p1v` and `p2v` contain the x_1 and x_2 coordinates of the calculation nodes on a regular grid on $\Omega = [0, a_1] \times [0, a_2]$. This is explained in detail in figure 8.2 and section 8.1.

Based on these points, the $n \times n$-matrix representing the neural kernel on the grid with n points is defined in lines 14–18, compare figure 6.5. Initialization of the neural field equation is carried out in lines 19–25. The Euler scheme (6.5.6) is performed by the loop of the lines 26–30.

Code 6.5.1. *Script* `sim_06_5_1_neural_simulation.m` *to calculate the temporal evolution of some neural activity function according to the neural field equation (6.5.1). Further, we show the code for the sigmoidal function* `f.m`.

```
1  % I. Setup grid of calculation points
2  a1 = 6; a2 = 4;             % rectangular area  [0 a1] x [0 a2]
3  n1  = 15;                   % number of points in x1-direction
4  n2  = n1+1;                 % number of points in x2-direction
5  n   = n1*n2;                % total number of grid points
6  h1 = a1/n1;                 % grid spacing in x1 direction
7  h2 = a2/n2;                 % grid spacing in x2 direction
8  p1  = h1:h1:a1;             % coordinates of points in x1 direction
9  p2  = h2:h2:a2;             % coordinates of points in x2 direction
10 p1m = repmat(p1,n2,1);      % matrix of coordinates in x1 direction
11 p2m = repmat(p2',1,n1);     % matrix of coordinates in x2 direction
12 p1v = reshape(p1m,n,1);     % vector of x1-coordinates of all nodes
13 p2v = reshape(p2m,n,1);     % vector of x2-coordinates of all nodes

14 % II. Define the neural kernel w
15 sigma1 = 3;  sigma2 = 5;  % some decay constants
16 Wmat = 30*exp(- sigma1*(repmat(p1v,1,n)-repmat(p1v',n,1)).^2 ...
17              - sigma2*(repmat(p2v,1,n)-repmat(p2v',n,1)).^2 ...
18                  ).*(repmat(p1v,1,n)-repmat(p1v',n,1))*h1*h2;

19 % III. Setup for Neural Field Equation
20 tau = 1; eta = 0.5; % parameter tau and eta for sigmoidal function
21 sigma = 3;           % decay constant for initial state
22 Nt  = 40;            % number of time steps
23 ht  = 4/Nt;          % time stepsize
24 uv = zeros(n,Nt);    % setup vector of states for Nt time steps
25 uv(:,1) = exp(-sigma*((p1v-p1v(40)).^2+(p2v-p2v(40)).^2));  % initial state

26 % IV. Simulate the neural field equation by an Euler scheme
27 for k=1:Nt
28         du = ht/tau*(-uv(:,k) + Wmat*f(uv(:,k),eta) );
29         uv(:,k+1) = uv(:,k)+du;
30 end
```

```
function s_out = f(s_in,eta)
% sigmoidal function
s_out = 1./(1+exp(-8*(s_in-eta*ones(size(s_in)))));
```

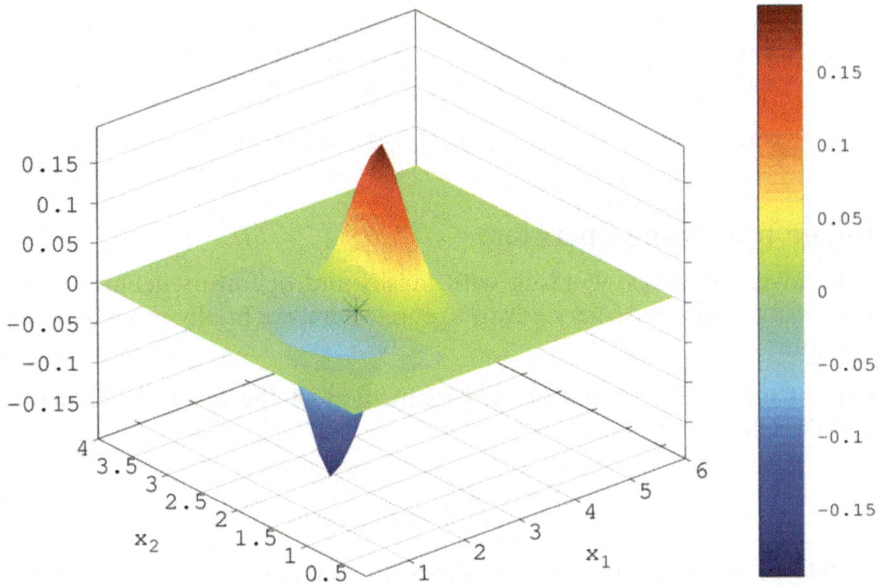

Figure 6.5. One column of the kernel (6.5.7), showing that a point (shown as *) excites points in the x_1-direction and inhibits nodes in the negative x_1-direction, the figure is generated by `sim_06_5_0_directed_Gaussian.m`.

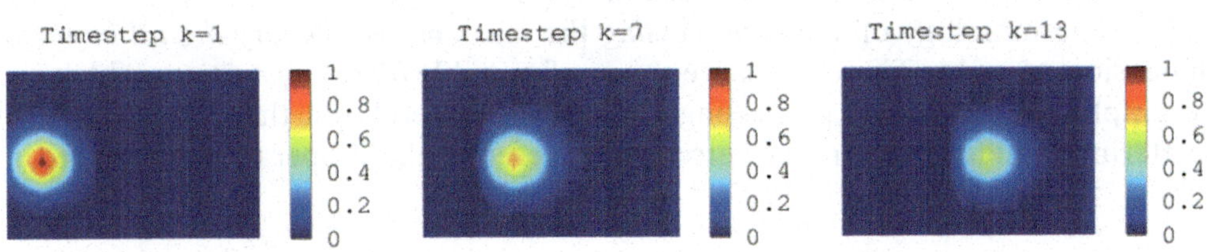

Figure 6.6. We display the activity distribution as described by some neural field $u(x, t)$ at three different time steps, when the kernel **W** is given by (6.5.7) and some Gaussian excitation is used as initial state $u(\cdot, 0)$.

We display the state evolution of the neural activity function $u(x, t)$ in figure 6.6 by showing the activity at three different time steps. The OCTAVE code to calculate these images is shown in code 6.5.2.

Code 6.5.2. *Script* `sim_06_5_1_show_graphics.m` *to display the temporal evolution (figure 6.6) of some neural activity function according to the neural field equation (6.5.1).*

```
fo = figure;                    % generate figure
for kk=1:3                      % show states for 4 different times
  subplot(1,3,kk);              % different subfigures
  k = 1+floor(Nt/6)*(kk-1); % choose time step to show
  umat = reshape(uv(:,k),n2,n1); % reshape vector into matrix
  so = surf(p1,p2,umat);    % generate the surface plot
  ao = get(so,'Parent');    % get axis object handle
  set(ao,'Visible','off'); % set axis labels invisible
```

```
    view(2);  % set view, interpolation, coloring and colorbar
    shading interp; axis equal; axis tight; caxis([0 1]);
    colorbar('FontSize',10); title(['Timestep k=' num2str(k)])
end
savefile(fo,'sim_06_5_1_neural_simulation');
```

6.6 Image processing operators

For many inversion tasks, working with masks and operators defined on two- or three-dimensional images is very advantageous. Here, we briefly provide a survey of such operators.

Definition of masks. A straightforward realization of the masks is carried out as follows. We define

$$\chi_G(x) := \begin{cases} 1 & x \in G \\ 0 & \text{otherwise.} \end{cases} \tag{6.6.1}$$

The discretized version of the mask in the form of a column vector is defined by

$$\chi_G := \left(\chi_G(p_l)\right)_{l=0,\ldots,M}. \tag{6.6.2}$$

Examples are given in figure 6.7. We reorder the column vector according to (8.1.17) to obtain a matrix version of the mask $\mathbf{m}_G$ of dimension $M_1 \times M_2$.

In the terminology of *image processing* the mask $\mathbf{m}_G$ is a *binary digital image* of dimension $M_1 \times M_2$. There is a large range of possible *binary operations* which can be employed for image processing and is very useful for the construction of scattering domains. Before we introduce several of these operations, we need to

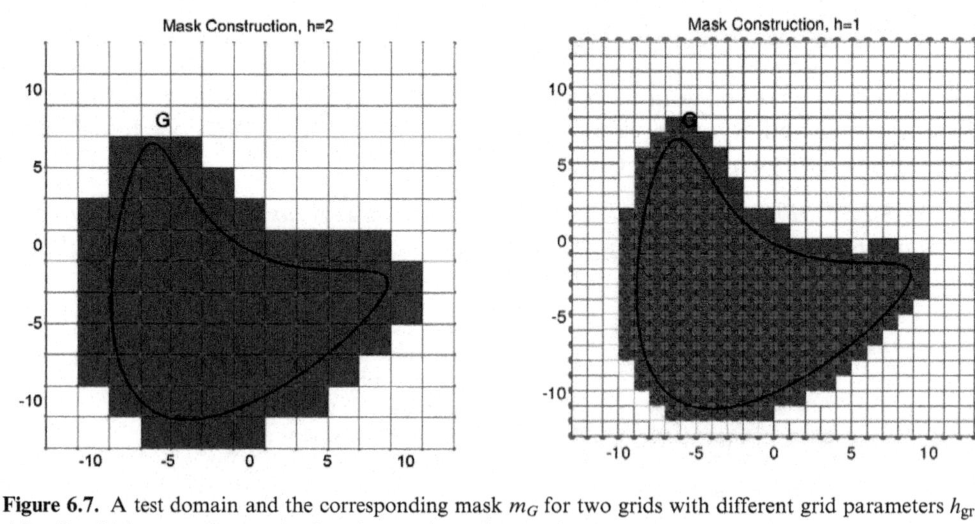

Figure 6.7. A test domain and the corresponding mask m_G for two grids with different grid parameters h_{grid}. All cells which come close to the domain are shown in gray here.

consider the construction of binary masks from boundary representations of domains.

Construction of masks. The construction of the masks can be a non-trivial task, depending on the particular form of the test domain G. If G is a ball with center z and radius R, then the mask can be easily constructed by testing the distance of a point $x \in \mathcal{G}$ to z. The same procedure can be applied to all star-like domains which are given in parametrized form.

For more general domains it is more difficult to decide whether a point $x \in \mathcal{G}$ is inside or outside G. We briefly mention two possibilities.

(a) It is possible to restrict one's attention to a simple class of domains for which every point x outside G has a straight line connecting x with infinity. In this case we can test all lines passing from a point x as to whether they intersect ∂G. If we find one which is not intersecting ∂G we know that x is on the outside, otherwise it is in $\bar{G}$.

(b) In two dimensions, using the *winding number* of a curve, we can calculate a mask for some domain G with given boundary parametrization. In complex coordinates $z = z_1 + iz_2$ with $z_1, z_2 \in \mathbb{R}$, it is carried out by the Cauchy integral

$$f(z) := \frac{1}{2\pi i} \int_{\partial G} \frac{1}{y - z} \mathrm{d}y, \quad z \in \mathbb{R}^2. \tag{6.6.3}$$

For our purposes transform (6.6.3) into the form

$$f(z) = \frac{1}{2\pi} \int_0^{2\pi} \left\{ \frac{(z_1 - y_1) \cdot \Gamma_2'(t)}{|z - y|^2} - \frac{(z_2 - y_2) \cdot \Gamma_1'(t)}{|z - y|^2} \right\} \mathrm{d}t \tag{6.6.4}$$

with the tangential vector $\Gamma(t)'$ of the curve Γ parametrized over $[0, 2\pi]$. Figure 6.8 demonstrates the generation of masks via the Cauchy integral (6.6.4).

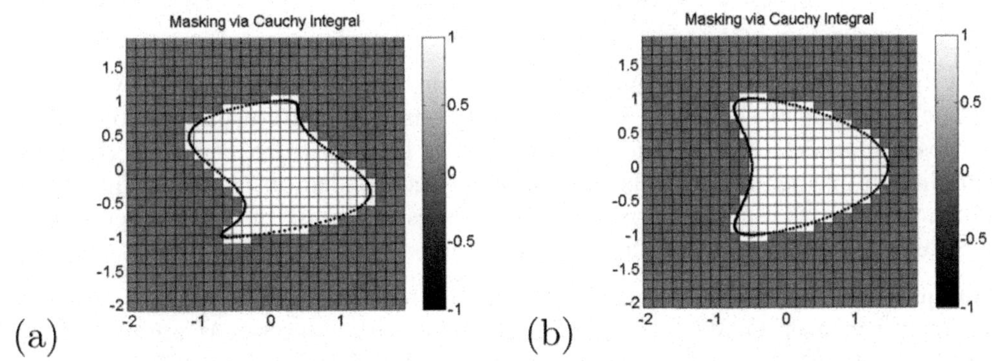

Figure 6.8. The generation of masks via the Cauchy integral for two different domains whose boundary is described by some trigonometric polynomial. For the Cauchy integral we used $n = 200$ quadrature points and evaluated the integral on a regular grid with $M = 50$ points in each direction.

Masking operations. The elementary set theoretic operations such as intersections and unions of sets can easily be implemented via masking operations. Let G and H denote two domains in $\mathbb{R}^2$ with masks $\mathbf{m}_G$ and $\mathbf{m}_H$. Then the intersection is carried out by

$$\mathbf{m}_{G\cap H} = \mathbf{m}_G \odot \mathbf{m}_H, \tag{6.6.5}$$

where the symbol $\odot$ denotes pointwise (element-wise) multiplication of matrices or vectors. Unions of G and H are calculated by

$$\mathbf{m}_{G\cup H} = \operatorname{sign}(\mathbf{m}_G + \mathbf{m}_H), \tag{6.6.6}$$

which means that we add the masks of G and H by regular matrix addition and then pointwise take the signum function. Both operations (6.6.5) and (6.6.6) can be generalized to products or sums with n terms, i.e. we have

$$\mathbf{m}_{G_1\cap\ldots\cap G_m} = \mathbf{m}_{G_1} \odot \ldots \odot \mathbf{m}_{G_m}, \tag{6.6.7}$$

$$\mathbf{m}_{G_1\cup\ldots\cup G_m} = \operatorname{sign}(\mathbf{m}_{G_1} + \cdots + \mathbf{m}_{G_m}). \tag{6.6.8}$$

Another useful operation is the following *shrinking process*. On a binary matrix $\mathbf{m}$ we define

$$S\mathbf{m}_{i,j} := \min\{\mathbf{m}_{i,j}, \mathbf{m}_{i+1,j}, \mathbf{m}_{i-1,j}, \mathbf{m}_{i,j+1}, \mathbf{m}_{i,j-1}\}, \tag{6.6.9}$$

i.e. the matrix element $S\mathbf{m}_{i,j}$ is one if and only if all the corresponding adjacent matrix elements of $\mathbf{m}$ are one. The application of the shrinking operator S removes all boundary elements of a domain G given by the binary matrix $\mathbf{m}$. Analogously, we can define the *thickening operator*

$$T\mathbf{m}_{i,j} := \max\{\mathbf{m}_{i,j}, \mathbf{m}_{i+1,j}, \mathbf{m}_{i-1,j}, \mathbf{m}_{i,j+1}, \mathbf{m}_{i,j-1}\}. \tag{6.6.10}$$

A *smoothing operator* of *level* ℓ can be defined by

$$L_\ell \mathbf{m}_{i,j} := \begin{cases} 1, & \sum\{\mathbf{m}_{i,j}, \mathbf{m}_{i+1,j}, \mathbf{m}_{i-1,j}, \mathbf{m}_{i,j+1}, \mathbf{m}_{i,j-1}\} > \ell \\ 0, & \text{otherwise.} \end{cases} \tag{6.6.11}$$

This operator removes pixels which have less or equal to ℓ neighbors of value 1.

The operators of shrinking, thickening and smoothing are usually employed for purposes of *noise removal* and *smoothing*. We employ them for our probe and sampling implementations. The smoothing effect is demonstrated in figure 6.9, where we apply the shrinking operation two times to a noisy mask and then thicken it by an application of the thickening operator.

Multiplication operators. Multiplication operators will be helpful for several inversion schemes. Given a function $f \in L^2(M)$ with some set M and a function $g \in C(M)$, let M_g be the operator which maps $L^2(M)$ into itself by multiplication of every function f by g, i.e.

$$\left(M_g f\right)(z) := g(z) \cdot f(z), \quad z \in M. \tag{6.6.12}$$

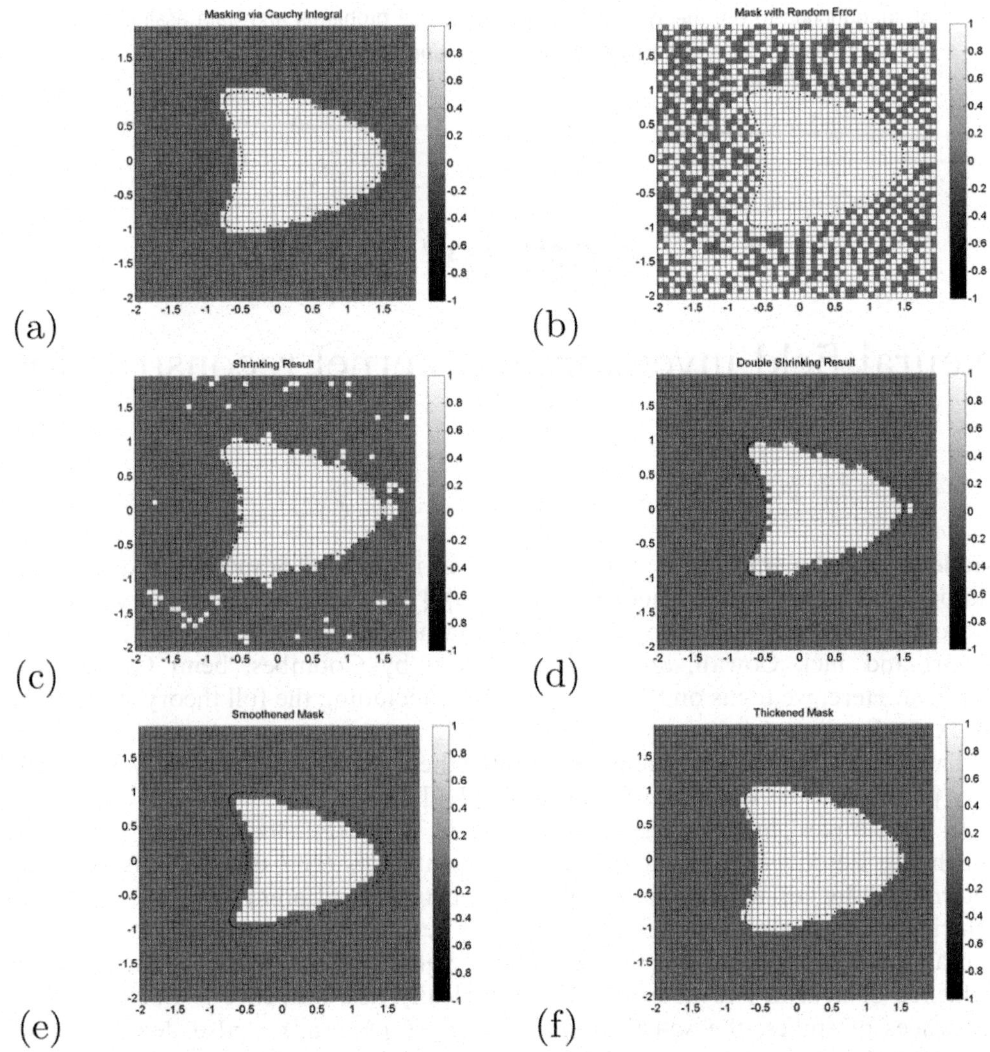

Figure 6.9. A mask for a kite-shaped domain (a) on a 50 × 50 grid. Random error is added to the mask in (b). An application of the shrinking process (6.6.9) generates (c). We apply (6.6.9) a second time to obtain a fully cleaned mask (d). Clearly, the mask in (d) is slightly smaller than the original mask, since two layers of boundary points have been removed. We then apply the smoothing operator L_3 and L_4 to generate (e). Finally, a thickening operation leads to (f).

Bibliography

[1] Eaton J W, Bateman D, Hauberg S and Wehbring R 2015 *GNU Octave Version 4.0.0 Manual: a High-Level Interactive Language for Numerical Computations* (Boston, MA: Free Software Foundation)

[2] Kress R 1998 *Numerical Analysis (Graduate Texts in Mathematics* vol 181) (Berlin: Springer)

IOP Publishing

Inverse Modeling

An introduction to the theory and methods of inverse problems and data assimilation

Gen Nakamura and Roland Potthast

Chapter 7

Neural field inversion and kernel reconstruction

Modeling the dynamics of neural activity is an important branch of mathematical biology and computational neuroscience, see [1] for a recent introductory book on the topic, with contributions by the founding fathers of neural field theory: Shun-ichi Amari and Jack Cowan, and with a tutorial by Coombes, beim Graben and Potthast. Here, we focus on the *inverse problem* including the full theory and hands-on examples.

The study of simple neuron models and the theory of neural networks was already very popular in the 1940s, see for example [2–9].

In the simplest model we study a system of N neurons as shown in figure 7.1, which are described by their activation potential u_i. The dynamics is basically determined by an excitation or inhibition between different neurons, where in principle every neuron is linked to every other neuron. The connectivity between the neural units is modeled by the synaptic weights w_{ij}, given the strength of the influence of neuron j to neuron i. Physiologically, the influence is transmitted by sequences of spikes, the so-called *spike trains* $r_i(t) = f(u_i(t))$, also denoted as an *activation function*, with some nonlinear function f.

A discrete neuron model which has a reasonable physiological significance is known as a *leaky integrator unit* [2, 3, 6, 8–10], which is usually described by the system of ordinary differential equations

$$\tau \frac{\mathrm{d}u_i(t)}{\mathrm{d}t} + u_i(t) = \sum_{j=1}^{N} w_{ij} f\big(u_j(t)\big). \tag{7.0.1}$$

Here $u_i(t)$ is the time-dependent membrane potential of neuron number i.

The left-hand side of (7.0.1) describes the intrinsic dynamics of a leaky integrator unit, i.e. an exponential decay of membrane potential with time constant τ. The right-hand side of (7.0.1) represents the *net-input* to unit i: the weighted sum of activity delivered by all units j that are connected to unit i ($j \to i$). Therefore, the

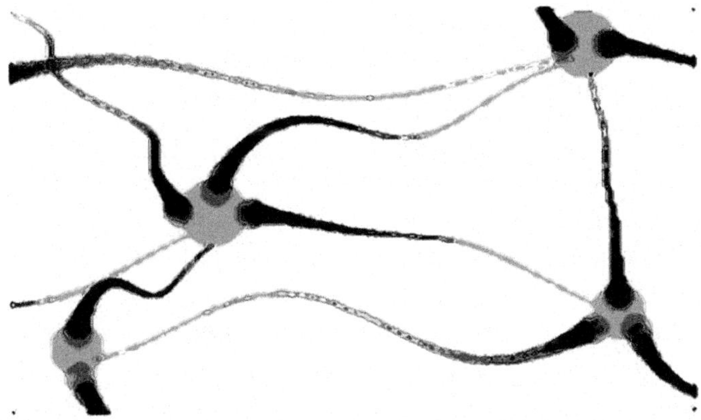

Figure 7.1. A schematic image of a network of neurons. A neural field is a continuous version of such a network, with connectivity strength given by $w(x, y)$ between points $x, y \in \Omega$, $d = 1, 2$ or $d = 3$.

weight matrix $W = (w_{ij})$ includes three different kinds of information: (1) unit j is connected to unit i if $w_{ij} \neq 0$ (connectivity, network topology), (2) the *synapse* $j \to i$ is *excitatory* ($w_{ij} > 0$), or *inhibitory* ($w_{ij} < 0$), (3) the *strength* of the synapse is given by $|w_{ij}|$.

For the activation function f, essentially two different approaches are common. On the one hand, a *deterministic* McCulloch–Pitts neuron [7] is obtained from a Heaviside step function

$$f(s) := \begin{cases} 0, & s < \eta \\ 1, & s \geqslant \eta \end{cases} \tag{7.0.2}$$

for $s \in \mathbb{R}$ with an *activation threshold* η describing the *all-or-nothing-law* of action potential generation. Supplementing (7.0.1) with a resetting mechanism for the membrane potential, the Heaviside activation function provides a *leaky integrate and fire* neuron model [2].

On the other hand, a *stochastic* neuron model leads to a continuous activation function $f(s) = \text{Prob}(s \geqslant \eta)$ describing the probability that a neuron fires if its membrane potential is above threshold [2]. In computational neuroscience this probability is usually approximated by the sigmoidal *logistic function*

$$f(s) = \frac{1}{1 + \mathrm{e}^{-\sigma(s-\eta)}}, \qquad s \in \mathbb{R}, \tag{7.0.3}$$

with constant $\sigma > 0$.

Analyzing and simulating large neural networks with complex topology is a very difficult problem, due to the nonlinearity of f and the large number of synapses (approximately 10^4 per neuron) and neurons (approximately 10^{12}) in the human cortex. Instead of analytically or numerically computing the sum in the right-hand side of equation (7.0.1), substituting it by an integral over a continuous neural tissue often facilitates such examinations. Therefore, *continuum approximations* of neural networks have been proposed since the 1960s [2, 11–27].

Starting with the leaky integrator network equation (7.0.1), the sum over all units is replaced by an integral transformation of a neural field quantity $u(x, t)$, where the continuous parameter $x \in \Omega$ now indicates the position i in the network. Correspondingly, the synaptic weight matrix w_{ij} turns into a kernel function $w(x, y)$. Then, (7.0.1) assumes the form of a *neural field equation* as discussed in [11, 25, 28]

$$\tau \frac{\partial u(x, t)}{\partial t} + u(x, t) = \int_{\Omega} w(x, y) f(u(y, t)) \, \mathrm{d}y, \quad x \in \Omega, \quad t > 0 \tag{7.0.4}$$

for the *neural potential* $u(x, t)$ with *activation function* f and *initial condition*

$$u(x, 0) = u_0(x), \quad x \in \Omega. \tag{7.0.5}$$

For quite some time, neural field equations have been investigated under serious restrictions upon the integral kernel $w(x, y)$, including homogeneity $w(x, y) = w(x - y)$ and isotropy $w(x, y) = w(|x - y|)$. In these cases, the technique of Green's functions allows the derivation of partial differential equations for the neural waves $u(x, t)$ assuming special kernels such as exponential, locally uniform or 'Mexican hat' functions [13, 15, 19, 23, 24]. Solutions for such neural field equations have been obtained for macroscopic, stationary neurodynamics in order to predict spectra of the electroencephalogram [21–23, 26], or bimanual movement coordination patterns [18, 19]. Heterogeneous kernels and thalamo–cortical loops in addition to homogeneous cortico–cortical connections have been discussed in [20] and [12, 21, 26], respectively.

Here, we present a field theoretic approach to neural modeling based on the Amari and Cowan–Wilson model (7.0.4) and (7.0.5). Our analysis serves as a model for various variations and generalizations of neural field equations which are currently being investigated for applications in the field of cognitive neurodynamics.

In section 7.1 we first examine the solvability of the *integro-differential equation* (7.0.4) with tools from functional analysis, the theory of ordinary differential equations and integral equations. We will provide a proof of global existence of solutions and study their properties in dependence on the smoothness of the synaptic kernel function w and the smoothness of the activation function f.

We will introduce *inverse neural field problems* in section 7.2. Our focus is on the reconstruction of the neural kernel w given some dynamics $u(\cdot, t)$, $t \in [0, T]$ on region of interest Ω. Basic results for this inverse problem are derived, in particular, we derive a mathematical version of the *Hebb learning rule*.

A practical solution of the inverse problem is worked out in section 7.3, where we also provide simple code examples to study the inversion. The case of *homogeneous kernels* is investigated in section 7.4, where we can rewrite the inverse problem into a linear integral equation of the first kind. We will look at the kernel reconstruction problem from the viewpoint of *bi-orthogonal basis functions* in section 7.5.

The inverse neural field problem is a demanding problem in terms of computational resources, since it connects a large number of neurons or neural points with each other. To obtain stable and feasible solutions, we will suggest a *localization* of the inverse problem in section 7.6. Here, we solve a high-dimensional inverse problem by reducing it to a family of *lower-dimensional* inverse problems. A full solution is then constructed by *imbedding*.

We close this introduction by defining some notations which will be used in this chapter. First for an open set $E \subset \mathbb{R}^m$ and a non-negative integer k, $BC^k(E)$ denotes the set of bounded continuous functions defined in E whose derivatives up to order k are bounded continuous in E. If $k = 0$ we simply write $BC^k(E)$ by $BC(E)$. We also define for an interval I, $BC^k(I)$ similarly with the convention that the derivatives are one sided derivatives at any end point of I if the interval is not open. Further for a non-negative integer ℓ, $BC^k(E) \boxtimes BC^\ell(I)$ stands for the set of all $BC^k(E)$ valued functions with respect to their first variable whose Fréchet derivatives up to order k are bounded continuous on I up to order ℓ with respect to their second variable.

7.1 Simulating neural fields

The goal of this section is to prove the existence of the Amari or Cowan–Wilson neural field equation (7.0.4) based on the Banach fixed-point theorem. The approach is well known in integro-differential equations, see for example [29]. Here, we follow [30] with an elementary and easily accessible version of the arguments. Using the operator

$$(Fu)(x, t) := \frac{1}{\tau}\left(-u(x, t) + \int_\Omega w(x, y) f(u(y, t))\, dy\right) \tag{7.1.1}$$

for $x \in \Omega$, $t > 0$ the neural field equation (7.0.4) can be reformulated as

$$\dot{u} = F(u), \tag{7.1.2}$$

where $\dot{u}$ is the time-derivative of u. Our analysis will be based on the integral form of the equation. To this end we define the operator

$$(Au)(x, t) := \int_0^t (Fu)(x, s)\, ds, \quad x \in \Omega, \quad t > 0. \tag{7.1.3}$$

By integration with respect to t we equivalently transform the neural field equation (7.0.4) or (7.1.2), into a *Volterra integral equation*

$$u(x, t) = u(x, 0) + \int_0^t (Fu)(x, s)\, ds, \quad x \in \Omega, \quad t > 0, \tag{7.1.4}$$

which, with A defined in (7.1.3), can be written in the form of a *fixed point equation*

$$u(x, t) = u(x, 0) + (Au)(x, t), \quad x \in \Omega, \quad t > 0. \tag{7.1.5}$$

We note that the neural field equation (7.0.4) is equivalent to the fixed point equation (7.1.5) in a lemma.

Lemma 7.1.1 (Neural field fixed-point formulation). *Assume the basic conditions*

$$\begin{aligned} &w(x, \cdot) \in L^1(\Omega), \quad x \in \Omega \\ &f \in BC(\mathbb{R}) \end{aligned} \tag{7.1.6}$$

for f and w. Then, the Volterra equation (7.1.4) or (7.1.5) is solvable on $\Omega \times (0, \rho)$ for $\rho > 0$ if and only if the neural field equation (7.0.4) or (7.1.2), respectively, is *solvable. In particular, solutions to the Volterra equation (7.1.4) are in* $BC^1([0, \rho])$ *with respect to their time variable.*

Proof. If the neural field equation is solvable with some continuous function $u(x, t)$, we obtain the Volterra integral equation (7.1.4) for the solution u by integration with respect to t. To show that a solution $u(x, t)$ to the Volterra integral equation (7.1.4) in $BC(\Omega) \boxtimes BC([0, \rho])$ satisfies the neural field equation (7.0.4) we need to ensure sufficient regularity, since solutions to equation (7.0.4) need to be differentiable with respect to t. This is the case, since a function

$$g_x(t) := \int_0^t (Fu)(x, s)\,ds, \quad t > 0 \tag{7.1.7}$$

is differentiable with respect to t with derivative $(Fu)(x, t)$. Now, the derivation of (7.0.4) from (7.1.4) by differentiation is straightforward. □

Besides the basic conditions we assume quite general properties on the weight kernel w and the activation function f to obtain existence of global solutions to the neural field equations.

Definition 7.1.2 (Neural kernel conditions). *Let the synaptic integral kernel* w *satisfy*

$$\|w(x, \cdot)\|_{L^1(\Omega)} \leqslant C_w, \quad x \in \Omega \tag{7.1.8}$$

$$\|w(x, \cdot) - w(\tilde{x}, \cdot)\|_{L^1(\Omega)} \leqslant c_w|x - \tilde{x}|, \quad x, \tilde{x} \in \Omega. \tag{7.1.9}$$

with constants C_w, $c_w > 0$ *and*

$$|w(x, y)| \leqslant C_\infty, \quad x, y \in \Omega. \tag{7.1.10}$$

We assume that the activation function $f \in BC^1(\mathbb{R})$ *has Lipschitz constant* L *and satisfies*

$$f(\mathbb{R}) \subset [0, 1]. \tag{7.1.11}$$

The neural field equation (7.0.4) allows general and global estimates for the above kernels, which also guarantee existence of solutions.

Lemma 7.1.3 (Global bounds of neural potential). *Let* $u_0 \in BC(\Omega)$ *be an initial field (7.0.5) for the neural field equation (7.0.4) and assume that the kernel* w *satisfies the conditions of definition 7.1.2. Then the solution* $u(x, t)$ *to the neural field equation is bounded by*

$$C_{\text{tot}} := \max(|u_0(x)|, |C_w|), \tag{7.1.12}$$

i.e. we have the general estimate

$$|u(x, t)| \leqslant C_{\text{tot}}, \quad x \in \Omega, \quad t \geqslant 0. \tag{7.1.13}$$

for solutions to the neural field equation.

Proof. It is straightforward that the term

$$(Ju)(x, t) := \int_{\Omega} w(x, y) f(u(y, t))\, dy, \quad x \in \Omega, \quad t > 0. \tag{7.1.14}$$

can be estimated by

$$|(Ju)(x, t)| \leqslant C_w, \quad x \in \Omega, \quad t \geqslant 0. \tag{7.1.15}$$

Next, we observe that the derivative $\dot{u}(t)$ in the neural field equation is bounded by

$$\dot{u}(x, t) \leqslant -bu(x, t) + \frac{C_w}{\tau}, \qquad \dot{u}(x, t) \geqslant -bu(x, t) - \frac{C_w}{\tau} \tag{7.1.16}$$

with $b = 1/\tau$. Thus, the value of $u(t)$ will be bounded by the solution to the ordinary differential equations

$$u' = -bu + c \tag{7.1.17}$$

with $a = u_0(x)$, $b = 1/\tau$ and $c = \pm C_w/\tau$. By elementary integration of the equation (7.1.17) the solution u is given by

$$u(t) = \frac{c}{b}\left(1 - e^{-bt}\right) + ae^{-bt}, \quad t \geqslant 0. \tag{7.1.18}$$

which is bounded by the bound C_{tot} defined in (7.1.12). This proves the estimate (7.1.13). □

In every small time interval of its evolution the neural field equation describes an exponential relaxation of its activity potential $u(x, t)$ towards the excitation $J(u)$ defined by (7.1.14). Thus, the neural field equation can be seen as a dynamic version of an equation of the type (7.1.17) with its solution (7.1.18).

As preparation for *existence* results we define an appropriate Banach space, which for $\rho > 0$ is chosen as $X_\rho := BC(\Omega) \boxtimes BC([0, \rho])$ equipped with the norm

$$\|u\|_\rho := \sup_{x \in \Omega, t \in [0, \rho]} |u(x, t)|. \tag{7.1.19}$$

Recall that an operator A from a normed space X into itself is called a *contraction*, if there is a constant q with $0 < q < 1$ such that

$$\|Au_1 - Au_2\| \leqslant q\|u_1 - u_2\| \tag{7.1.20}$$

is satisfied for all $u_1, u_2 \in X_\rho$. A point $u_* \in X_\rho$ is called a *fixed point* of A if

$$u_* = Au_* \tag{7.1.21}$$

is satisfied. We are now prepared for studying the properties of A on X_ρ.

Lemma 7.1.4 (Contraction lemma). *Under the conditions of definition 7.1.2 and for* $\rho > 0$ *chosen sufficiently small such that*

$$\frac{\rho}{\tau}(1 + LC_w) < 1, \tag{7.1.22}$$

the operator A *is* a contraction *on the space* X_ρ *defined in (7.1.19).*

Proof. We estimate $Au_1 - Au_2$ and abbreviate $u := u_1 - u_2$. We decompose $A = A_1 + A_2$ into two parts with the linear operator

$$(A_1 v)(x, t) := \frac{-1}{\tau} \int_0^t v(x, s)\, \mathrm{d}s, \quad x \in \Omega, \quad t > 0, \tag{7.1.23}$$

and the nonlinear operator

$$(A_2 v)(x, t) := \frac{1}{\tau} \int_0^t \int_\Omega w(x, y) f(v(y, s))\, \mathrm{d}y\, \mathrm{d}s, \quad x \in \Omega, \quad t > 0. \tag{7.1.24}$$

We can estimate the norm of A_1 by

$$\|A_1 u\|_\rho \leqslant \frac{\rho}{\tau} \|u\|_\rho, \tag{7.1.25}$$

which is a contraction if ρ is sufficiently small. Since $f \in BC^1(\mathbb{R})$ there is a constant L such that

$$\left| f(s) - f(\tilde{s}) \right| \leqslant L\, |s - \tilde{s}|, \quad s, \tilde{s} \in \mathbb{R}. \tag{7.1.26}$$

This yields

$$\begin{aligned} \left| (Ju_1)(x, t) - (Ju_2)(x, t) \right| &\leqslant \int_\Omega |w(x, y)| \left| f(u_1(y, t)) - f(u_2(y, t)) \right| \mathrm{d}y \\ &\leqslant L \int_\Omega |w(x, y)| \left| u_1(y, t) - u_2(y, t) \right| \mathrm{d}y \\ &\leqslant LC_w \|u_1 - u_2\|_\rho. \end{aligned} \tag{7.1.27}$$

Finally, by an integration with respect to t we now obtain the estimate

$$\|A_2 u_1 - A_2 u_2\|_\rho \leqslant \frac{\rho}{\tau} LC_w \|u_1 - u_2\|_\rho. \tag{7.1.28}$$

For ρ sufficiently small the operator A_2 is a contraction on the space X_ρ. For ρ chosen such that (7.1.22) is satisfied, the operator $A = A_1 + A_2$ is a contraction on X_ρ. □

Now, local existence is given by the following theorem.

Theorem 7.1.5 (Local existence for neural fields). *Assume that the synaptic weight kernel w and the activation function f satisfy the conditions of definition 7.1.2 and let $\rho > 0$ be chosen such that (7.1.22). Then we obtain existence of solutions to the neural field equations on the interval* $[0, \rho]$.

Remark. The result is a type of *Picard–Lindelöf theorem* well known in the theory for ordinary differential equations for the neural field equation (7.0.4) under the conditions of definition 7.1.2 and $f \in BC^1(\mathbb{R})$.

Proof. We employ the Banach fixed-point theorem to the operator equation (7.1.5). We have shown that the operator A is a contraction on X_ρ defined in (7.1.19).

Then, also the operator $\tilde{A}u := u_0 + Au$ is a contraction on the complete normed space X_ρ. Now, according to the Banach fixed-point theorem the equation

$$u = \tilde{A}u \tag{7.1.29}$$

as a short form of the Volterra equations (7.1.5) or (7.1.4), has one and only one fixed point u_*. This proves the unique solvability of (7.1.4). Finally, by the equivalence lemma 7.1.1 we obtain the unique solvability of the neural field equation (7.0.4) on $t \in [0, \rho]$. □

In the last part of this section we combine the global estimates with local existence to obtain a global existence result.

Theorem 7.1.6 (Global existence for neural fields). *Under the conditions of definition 7.1.2 we obtain existence of global bounded solutions to the neural field equation.*

Proof. We first remark that the neural field equation does not explicitly depend on time. As a result we can apply the local existence result with the same constant ρ to any interval $[t_0, t_0 + \rho] \subset \mathbb{R}$ when initial conditions $u(x, t_0) = u_0$ for $t = t_0$ are given. This means we can use theorem 7.1.5 iteratively.

First, we obtain existence of a solution on an interval $I_0 := [0, \rho]$ for ρ chosen such that (7.1.22) is satisfied. Then, the function $u_1(x) := u(x, \rho)$ serves as new initial condition for the neural field equation on $t > \rho$ with initial conditions u_1 at $t = \rho$. We again apply theorem 7.1.5 to this equation to obtain existence of a solution on the interval $I_1 = [\rho, 2\rho]$.

This process is continued to obtain existence on the intervals

$$I_n := [n\rho, (n+1)\rho], \quad n \in \mathbb{N},$$

which shows existence for all $t \in \mathbb{R}$. A global bound for this solution has been derived in lemma 7.1.3. □

The neural field equation (7.0.4) is only the most basic version of a whole collection of neural field and neural mass models, which bear the potential to simulate neural activity and cognitive functions. We refer to numerical examples given by code 6.5.1 in section 6.5 visualized in figure 6.6.

The above derivation provides a basis to work with stable neural dynamics for a wide range of applications and as a basis for the *inverse problem*, which we describe next.

7.2 Integral kernel reconstruction

The *inverse neural field problem* is of huge importance for applications. It is very difficult to measure *neural connectivity* directly—both for living and dead tissue. But to measure electrical activity as a time series is a standard today, the quality of which has strongly evolved over the past decades. Thus, we search for the *wiring* of the brain, i.e. the *connectivity* between different neurons and collections of neurons. In our field theory it is given by the kernel $w(x, y)$ for $x, y \in \Omega$.

In a first attempt, we can think of w as a continuous function of its variables x and y. We will first investigate the situation searching for bounded continuous kernels, i.e. $w \in BC(\Omega \times \Omega)$.

However, in human or animal brains, connectivity between different regions does not change continuously. This leads to spaces of discontinuous functions. Since we need integrability, we will search for kernels in L^1 or L^2. The Hilbert space structure of $L^2(\Omega \times \Omega)$ will enable us to gain a lot of further insight, it will be the basis of much of our analysis starting with equation (7.2.18) in this section and continued in section 7.5.

As a background for our dynamics we focus on the neural field $u(x, t)$ depending on the space variable $x \in \Omega$ with some bounded domain Ω and the time $t \in [0, T]$ with $T > 0$ governed by the Amari equation (7.0.4) with *initial condition* (7.0.5) and with $f : \mathbb{R} \to [0, 1]$ being some given smooth function.

For the *direct problem* the kernel w and the initial condition u_0 are given. The task is to calculate the neural field $u(x, t)$, $x \in \Omega$, $t \geqslant 0$ in space and time. The *inverse problem* assumes that we are given u and u_0 partially or completely. The inverse problem is then to determine some kernel w which has $u(x, t)$ as its solution for $x \in \Omega$ and $t \in [0, T]$.

To formulate the inverse problem in a mathematical form we define the *synaptic weight operator*

$$(W\varphi)(x) := \int_{\Omega} w(x, y)\varphi(y)\,\mathrm{d}y, \qquad x \in \Omega. \tag{7.2.1}$$

When applied to $f(u(y, t))$ it is acting on the space variable of the functions under consideration. Then, the neural field equation (7.0.4) takes the form

$$\tau\frac{\partial u}{\partial t}(x, t) = -u(x, t) + \{Wf(u(\cdot, t))\}(x), \qquad x \in \Omega, \qquad t > 0. \tag{7.2.2}$$

We can write this in a shorter form by definition of the functions

$$\psi(x, t) := \tau\frac{\partial u}{\partial t}(x, t) + u(x, t), \qquad x \in \Omega, \qquad t \geqslant 0 \tag{7.2.3}$$

and

$$\varphi(x, t) := f(u(x, t)), \qquad x \in \Omega, \qquad t \geqslant 0. \tag{7.2.4}$$

With φ and ψ given by (7.2.3) and (7.2.4) the inverse neural field equation can be equivalently written as

$$\psi(x, t) = \int_{\Omega} w(x, y)\varphi(y, t)\,\mathrm{d}y, \qquad x \in \Omega, \qquad t \in [0, T], \tag{7.2.5}$$

or

$$\psi = W\varphi \quad \text{on } \Omega \times [0, T]. \tag{7.2.6}$$

This is a *generalized inverse operator problem*, where W is an unknown integral operator with kernel w.

Definition 7.2.1 (Full field neural inverse problem). *Let X be either the Banach space of bounded continuous functions $BC(\Omega \times \Omega)$ defined in $\Omega \times \Omega$ or the space of square-integrable functions $L^2(\Omega \times \Omega)$. Given a function*

$$v \in BC(\Omega) \boxtimes BC^1([0, T])$$

for some $T > 0$ and with $u_0(x) := v(x, 0)$, $x \in \Omega$, the full field neural inverse problem *is to construct a kernel w in X such that the solution $u(x, t)$ of the neural field equation (7.0.4) is given by v.*

By the transformation (7.2.3) and (7.2.4) we have shown that the inverse neural field problem is equivalent to solving (7.2.5) or (7.2.6) where the functions (ψ, φ) are elements of the set

$$U = \left\{ \left(\tau \frac{\partial v}{\partial t} + v, f(v) \right) : v \in BC(\Omega) \boxtimes BC^1([0, T]) \right\}. \tag{7.2.7}$$

Note that although the integral equation (7.2.5) is linear with respect to w, the time behavior of the neural fields u under consideration is usually highly *nonlinear*.

The study of the inverse problem given by definition 7.2.1 or (7.2.5) includes an analysis of *uniqueness*, *existence* and the *stability* or *instability* (hence *ill-posedness*) of the kernel reconstruction. For practical applications a thorough understanding, in particular of uniqueness and stability issues, is of great importance. We will provide simple code examples for the inverse problem in section 7.3.

The *instability* of the inverse neural field problem is obtained as follows. In equation (7.2.5) we consider x as a parameter and use

$$w_x(y) := w(x, y), \qquad k(t, y) := \varphi(y, t) \quad \text{and} \quad \psi_x(t) := \psi(x, t)$$

for $x, y \in D$ and $t \in [0, T]$. Then, it is rewritten into the *family of linear integral equations of the first kind*

$$\psi_x(t) = \int_\Omega k(t, y) w_x(y) \, \mathrm{d}y, \quad t \in [0, T], \tag{7.2.8}$$

with given $\psi_x \in BC([0, T])$ and unknown functions $w_x(\cdot) \in L^1(\Omega)$ for $x \in \Omega$. If $\varphi(y, t)$ is continuous in $y \in \Omega$ and $t \in [0, T]$ we have an integral equation of the first kind with continuous kernels.

Lemma 7.2.2. *The full field neural inverse problem as described in definition 7.2.1 is* ill-posed *in the sense of definition 3.1.1.*

Proof. The equations (7.2.8) are ill-posed in the sense of definition 3.1.1, which is a consequence of theorem 3.1.3 of section 3.1.1 in combination with example 2.3.21. As a consequence, the inverse neural field problem described in definition 7.2.1 is ill-posed because the stability implies the stability for each fix x. □

As a first step to study the solvability of the inverse problem we note some basic *regularity properties* of solutions to the neural field equation. They put necessary conditions on the prescribed functions v which can be a solution to some neural field equation (7.0.4), (7.1.2) or (7.2.2).

Lemma 7.2.3. *If $f \in BC^n(\mathbb{R})$, then the solution to the neural field equation (7.0.4) is in $BC^{n+1}([0, T])$ with respect to time, i.e. $t \mapsto u(\cdot, t)$ is $n + 1$ times continuously differentiable as a function on $[0, T]$ and its derivatives are bounded.*

Proof. We employ the Volterra integral equation form (7.1.4) of the neural field equation (7.0.4). If u is in $BC^k([0, T])$ for $k \leqslant n$, then also Fu defined by (7.1.1) is in $BC^k([0, T])$ with respect to the time variable. Then, by the integration with respect to time in (7.1.4) as in (7.1.7) we obtain that u is in $BC^{k+1}([0, T])$. An iterative application of this argument for $k = 0$ to $k = n$ yields the statement of the lemma. □

We conclude that for the sigmoidal function f defined in (7.0.3), which is an entire function and thus in $BC^\infty(\mathbb{R})$, the solution $u(x, t)$ for $x \in \Omega$ and $t \in [0, T]$, to the neural field equation (7.0.4) is infinitely many times boundedly differentiable.

The inverse neural field problem (7.2.5) can be understood as the search for an operator W which has some prescribed mapping properties:

1. It maps a given family of states $\varphi(\cdot, t)$ onto a given family of images $\psi(\cdot, t)$.
2. It is an integral operator with kernel $w(x, y)$, $x, y \in \Omega$.

In the following steps we study mapping properties of the operator W defined in (7.2.1) as an operator on $BC(\Omega)$ as well as the properties of the operator which maps kernels w into the operator space $BL(BC(\Omega), BC(\Omega))$ of bounded linear functions on $BC(\Omega)$. Our goal is to provide conditions under which there is an operator W which satisfies target 1 and when it is an integral operator, i.e. when it satisfies target 2.

Lemma 7.2.4. *Equip $BC(\Omega)$ with canonical norm $\|\varphi\|_\infty := \sup_{x\in\Omega}|\varphi(x)|$ and denote the norm of the operator $W : BC(\Omega) \to BC(\Omega)$ by $\|W\|_\infty$. Then we have*

$$\|W\|_\infty = \sup_{x\in\Omega} \int_\Omega |w(x, y)|\, dy. \tag{7.2.9}$$

Proof. We estimate

$$\begin{aligned} \|W\varphi\|_{BC(\Omega)} &= \sup_{x\in\Omega}\left|\int_\Omega w(x, y)\varphi(y)\, dy\right| \\ &\leqslant \left(\sup_{x\in\Omega}\int_\Omega |w(x, y)|\, dy\right) \cdot \sup_{y\in\Omega}|\varphi(y)|, \end{aligned}$$

which proves

$$\|W\|_\infty \leqslant \sup_{x,y\in\Omega} \int_\Omega |w(x, y)|\, dy. \tag{7.2.10}$$

To show the equality we first consider the case where $w(x, y) \neq 0$ on $\Omega \times \Omega$. We choose a sequence $(x_\xi)_{\xi\in\mathbb{N}} \subset \Omega$ such that

$$\int_\Omega |w(x_\xi, y)|\, dy \to \sup_{x\in\Omega}\int_\Omega |w(x, y)|\, dy, \quad \xi \to \infty.$$

Then, we define $\varphi_\xi(y) := w(x_\xi, y)/|w(x_\xi, y)|$, which is in $BC(\Omega)$ and has norm $\|\varphi_\xi\|_\infty = 1$. We calculate

$$\left|\left(W\varphi_\xi\right)(x_\xi)\right| = \left|\int_\Omega w(x_\xi, y)\frac{w(x_\xi, y)}{\left|w(x_\xi, y)\right|}\,dy\right| \to \sup_{x\in\Omega}\int_\Omega |w(x, y)|\,dy$$

for $\xi \to \infty$, which proves equality in (7.2.10), under the assumptions that w has no zero we define $N_\xi := \{y \in \Omega : w(x_\xi, y) = 0\}$. We could try to define $\tilde{\varphi}_\xi(y) := w(x_\xi, y)/|w(x_\xi, y)|$ for $y \notin N_\xi$ and set $\tilde{\varphi}_\xi = 0$ on N_ξ. However, in general this function $\tilde{\varphi}_\xi$ is not continuous on Ω, i.e. it is not in $BC(\Omega)$. To construct an appropriate sequence of continuous functions we define

$$\chi_{\xi,n}(y) := \begin{cases} 0, & d(y, N_\xi) \geqslant 1/n, \\ 1 - n\,d(y, N_\xi), & 0 \leqslant d(y, N_\xi) \leqslant 1/n, \end{cases} \tag{7.2.11}$$

for $y \in \Omega\backslash N_\xi$, where $d(y, N_\xi) := \inf_{z\in N_\xi}|y - z|$. Since Ω is a bounded set, we obtain

$$0 \leqslant \int_{\Omega\backslash N_\xi} \chi_{\xi,n}(y)\,dy \leqslant \int_{\{y\notin N_\xi : d(y,N_\xi)\leqslant 1/n\}} dy \to 0, \quad n \to \infty. \tag{7.2.12}$$

Then, we define

$$\varphi_{\xi,n}(y) := \begin{cases} \left(1 - \chi_{\xi,n}(y)\right)\dfrac{w(x_\xi, y)}{\left|w(x_\xi, y)\right|}, & y \notin N_\xi, \\ 0, & y \in N_\xi. \end{cases}$$

By construction we have $\varphi_{\xi,n} \in BC(\Omega)$ and we observe $\|\varphi_{\xi,n}\| = 1$. We calculate

$$\begin{aligned}
\left|\left(W\varphi_{\xi,n}\right)(x_\xi)\right| &= \left|\int_\Omega w(x_\xi, y)\varphi_{\xi,n}(y)\,dy\right| \\
&= \left|\int_{\Omega\backslash N_\xi} w(x_\xi, y)\left[\left(1 - \chi_{\xi,n}\right)\frac{w(x_\xi, y)}{\left|w(x_\xi, y)\right|}\right]dy\right| \\
&= \left|\int_{\Omega\backslash N_\xi}\left(1 - \chi_{\xi,n}\right)\left|w(x_\xi, y)\right|dy\right| \\
&\to \int_\Omega \left|w(x_\xi, y)\right|dy, \quad n \to \infty, \\
&\to \sup_{x\in\Omega}\int_\Omega |w(x, y)|dy, \quad \xi \to \infty,
\end{aligned} \tag{7.2.13}$$

where we use (7.2.12) and $w(x_\xi, y) = 0$ for $y \in N_\xi$. The limit (7.2.13) now proves equality in (7.2.10) in the general case. □

To study the mapping $w \mapsto W$ of the kernel w onto the operator W we follow lemma 7.2.4 and equip the space $BC(\Omega \times \Omega)$ with the norm

$$\|w\|_{BC(\Omega\times\Omega)} := \sup_{x\in\Omega} \int_\Omega |w(x, y)|\,dy. \tag{7.2.14}$$

Theorem 7.2.5. *For a bounded set* Ω *the mapping*

$$K : BC(\Omega \times \Omega) \to BL(BC(\Omega), BC(\Omega)), \quad w \mapsto W \tag{7.2.15}$$

with W *defined in (7.2.1) is a linear, bounded and injective mapping, but not surjective.*

Proof. Clearly, the mapping K is linear, since

$$\begin{aligned} K(w_1 + w_2)\varphi &= \int_\Omega (w_1(\cdot, y) + w_2(\cdot, y))\varphi(y)\,dy \\ &= \int_\Omega w_1(\cdot, y)\varphi(y)\,dy + \int_\Omega w_2(\cdot, y)\varphi(y)\,dy \\ &= (Kw_1)\varphi + (Kw_2)\varphi. \end{aligned}$$

The norm of K can be calculated based on

$$\|K(w)\varphi\|_{BC(\Omega)} = \sup_{x\in\Omega}\left|\int_\Omega w(x, y)\varphi(y)\,dy\right|.$$

From lemma 7.2.4 we now obtain $\|K\| = 1$ as a mapping from $BC(\Omega \times \Omega)$ equipped with the norm (7.2.14) into $BL(BC(\Omega), BC(\Omega))$.

To show the injectivity of K assume that w is a kernel such that $W\varphi = 0$ for all $\varphi \in BC(\Omega)$. We study a sequence of bounded continuous functions $\delta_{z,n}(\cdot)$ approximating a delta function δ_z. Then, we have

$$0 = (W\delta_{z,n})(x) = \int_\Omega w(x, y)\delta_{z,n}(y)\,dy \to w(x, z), \quad n \to \infty, \tag{7.2.16}$$

which shows $w(x, z) = 0$ for all $x, z \in D$, and thus the operator K is injective.

The operator K is not surjective. This is shown by constructing examples V of bounded linear operators which cannot be represented by integral operators with continuous kernels. We note for example the *point evaluation operator*

$$(V\varphi)(x) := \varphi(x), \quad x \in D, \tag{7.2.17}$$

which clearly is bounded, linear, but cannot be represented as an integral operator with a continuous kernel. Further examples for bounded linear operators which cannot be written as integral operators with continuous kernels are operators with singular kernels of various types. □

In the second step we study bounded linear operators on the space $L^2(\Omega)$. By the Cauchy–Schwarz inequality (2.2.2) we obtain the norm estimate

$$\| W\varphi\|_{L^2(\Omega)}^2 \leqslant \int_\Omega \left| \int_\Omega w(x, y)\varphi(y)\,\mathrm{d}y \right|^2 \mathrm{d}x$$
$$\leqslant \left(\int_{\Omega\times\Omega} |w(x, y)|^2 \,\mathrm{d}y\,\mathrm{d}x \right) \|\varphi\|_{L^2(\Omega)}^2, \tag{7.2.18}$$

which proves boundedness of the mapping

$$K : L^2(\Omega \times \Omega) \to BL\big(L^2(\Omega), L^2(\Omega)\big), \quad w \mapsto W. \tag{7.2.19}$$

However, here the estimate (7.2.18) is only an upper estimate, but note that the norm of W can be given by the square root of spectral radius $\rho(W^*W)$, see (2.4.32).

When is W an integral operator on $L^2(\Omega)$ with kernel in $L^2(\Omega \times \Omega)$? Consider any orthonormal basis $\{\varphi_n, n \in \mathbb{N}\}$ in X. Then, every element $\varphi \in X$ can be written as

$$\varphi = \sum_{n=1}^{\infty} \beta_n \varphi_n, \quad \beta = \big(\beta_1, \beta_2, \ldots\big) \in \ell^2. \tag{7.2.20}$$

The image sequence $\psi_n := W\varphi_n$ satisfies

$$\left\| \sum_{n=N_1}^{N_2} \beta_n \psi_n \right\|^2 = \left\| \sum_{n=N_1}^{N_2} \beta_n (W\varphi_n) \right\|^2$$
$$= \left\| W \sum_{n=N_1}^{N_2} \beta_n \varphi_n \right\|^2$$
$$\leqslant \| W \|^2 \cdot \sum_{n=N_1}^{N_2} |\beta_n|^2, \tag{7.2.21}$$

which proves that

$$\sum_{n=1}^{\infty} \beta_n \psi_n = W\varphi.$$

Using $\beta_n = \langle \varphi_n, \varphi \rangle_{L^2(\Omega)} = \int_\Omega \varphi_n(y)\varphi(y)\,\mathrm{d}y$ we can write W as

$$(W\varphi)(x) = \lim_{N\to\infty} \sum_{n=1}^{N} \psi_n(x) \int_\Omega \varphi_n(y)\varphi(y)\,\mathrm{d}y,$$
$$= \lim_{N\to\infty} \int_\Omega \left(\sum_{n=1}^{N} \psi_n(x)\varphi_n(y) \right) \varphi(y)\,\mathrm{d}y, \quad x \in D. \tag{7.2.22}$$

In general, we cannot expect the terms in round brackets to converge to a function in the space $L^2(\Omega \times \Omega)$. For example, when $\psi_n = \varphi_n$ for all n we obtain a representation of the delta function $\delta(x - y)$ with respect to the variable y:

$$\int_\Omega \varphi(z)\left(\sum_{n=1}^{N} \varphi_n(x)\varphi_n(z)\right) \mathrm{d}z = \sum_{n=1}^{N} \varphi_n(x)\int_\Omega \varphi_n(z)\varphi(z)\, \mathrm{d}z$$

$$= \sum_{n=1}^{N} \varphi_n(x)\beta_n \to \varphi(x), \quad N \to \infty, \tag{7.2.23}$$

therefore

$$\sum_{n=1}^{\infty} \varphi_n(x)\varphi_n(z) = \delta(x - z) \tag{7.2.24}$$

to be interpreted in $L^2(\Omega \times \Omega)$ in the sense of (7.2.23). Clearly (7.2.24) is not an element of $L^2(\Omega \times \Omega)$.

Now, we formulate the following result, which can be seen as a special case of the theory of *Hilbert–Schmidt integral operators.*

Theorem 7.2.6. *A bounded linear operator W is an integral operator on $L^2(\Omega)$ if and only if the sum*

$$w(x, y) := \sum_{n=1}^{\infty} \psi_n(x)\varphi_n(y) \tag{7.2.25}$$

with $\psi_n := W\varphi_n$, $n \in \mathbb{N}$, is convergent in $L^2(\Omega \times \Omega)$ for any orthonormal basis $\{\varphi_n : n \in \mathbb{N}\}$.

Proof. If the sum is convergent in $L^2(\Omega \times \Omega)$, then we can rewrite (7.2.22) as integral operator

$$(W\varphi)(x) = \int_\Omega \left(\sum_{n=1}^{\infty} \psi_n(x)\varphi_n(y)\right)\varphi(y)\, \mathrm{d}y,$$

$$= \int_\Omega w(x, y)\varphi(y)\, \mathrm{d}y, \quad x \in \Omega. \tag{7.2.26}$$

By (7.2.18) the operator is a linear bounded operator on $L^2(\Omega)$. Now, assume that W is an integral operator with kernel $w \in L^2(\Omega \times \Omega)$. Then, for any orthonormal basis $\{\varphi_n : n \in \mathbb{N}\}$ and $\psi_n = W\varphi_n$ we obtain $\psi_n \in L^2(\Omega)$ and

$$\sum_{n=1}^{N} \psi_n(x)\varphi_n(y) = \sum_{n=1}^{N}\left(\int_\Omega w(x, z)\varphi_n(z)\, \mathrm{d}z\right)\varphi_n(y)$$

$$= \int_\Omega w(x, z)\left(\sum_{n=1}^{N} \varphi_n(y)\varphi_n(z)\right) \mathrm{d}z$$

$$\to w(x, y), \quad x, y \in \Omega, \quad N \to \infty, \tag{7.2.27}$$

which is due to (7.2.23) is satisfied in $L^2(\Omega \times \Omega)$. □

Note that the kernel construction (7.2.25) corresponds to the *Hebb learning rule* for a finite linear perceptron with orthogonal training vectors extended to infinite-dimensional function spaces, telling you that a connection between a neuron at y and a neuron at x should be proportional to the sum of the products of input $\varphi_n(y)$ and corresponding output $\psi_n(x)$ for all available input patterns φ_n. The Hebb learning rule was originally suggested by Hebb [31], as a psychological mechanism based on physiological arguments. Here, it appears as a consequence of the Fourier theorem for orthonormal systems, see also [32].

One of the implications of the above theorem is that in general we cannot assume that we can control the neural field over an infinite time.

Theorem 7.2.7. *In general, the full field neural inverse problem with infinite time $T = \infty$ is not solvable in $L^2(\Omega \times \Omega)$.*

Proof. Consider some orthonormal system $\{\varphi_n, n \in \mathbb{N}\}$ in $L^2(\Omega)$ constructed out of Haar type basis functions, i.e. φ_n has values either zero or one on Ω. For simplicity, here we assume that $f(0) = 0$, the general case can be treated analogously. In this case $f(\varphi_n) = c \cdot \varphi_n$ where $c = f(1)$. Consider $t_n := n$ for $n \in \mathbb{N}$ and let $\{\chi_n : n \in \mathbb{N}\}$ be some C^n-smooth partition of unity

$$1 = \sum_{n=1}^{\infty} \chi_n(t), \quad t \geqslant 0, \tag{7.2.28}$$

such that

$$\chi_n(t) = 0, \quad t \notin B(t_n, 1 - \epsilon), \tag{7.2.29}$$

with some $0 < \epsilon < 1/4$ and

$$\frac{d\chi_n}{dt}(t_n) = 0, \quad n \in \mathbb{N}. \tag{7.2.30}$$

Now, we define

$$v(x, t) := \sum_{n=1}^{\infty} \chi_n(t)\varphi_n(x), \quad x \in D, \quad t > 0. \tag{7.2.31}$$

By the definitions of $\varphi(\cdot, t)$ and $\psi(\cdot, t)$, we obtain

(a) $\varphi(\cdot, t_n) = f(v(\cdot, t_n)) = c \cdot \varphi_n(\cdot)$ for $n \in \mathbb{N}$,

(b) $\frac{dv}{dt}(\cdot, t_n) = 0$ for $n \in \mathbb{N}$, and thus

(c) $\psi(\cdot, t_n) = \varphi_n$ for $n \in \mathbb{N}$.

This means that to solve the full field neural inverse problem we need to construct an operator W mapping φ_n onto $c^{-1} \cdot \varphi_n = W\varphi_n$ for $n \in \mathbb{N}$. However, as shown in (7.2.23) and theorem 7.2.6 the kernel $w(x, y)$ defined in (7.2.25) is not an element of $L^2(\Omega \times \Omega)$. This proves the statement of the theorem for $L^2(\Omega \times \Omega)$. □

Compactness. We complete this section with basic compactness statements which by theorem 3.1.3 imply instability of the kernel construction problem if we seek kernels in spaces of differentiable functions or in Sobolev spaces. We note that by the compactness of the embedding

$$BC^n(\Omega \times \Omega) \to BC(\Omega \times \Omega), \tag{7.2.32}$$

the mapping K defined in (7.2.15) is compact if it is considered in the spaces

$$K : BC^n(\Omega \times \Omega) \to BL(BC(\Omega), BC(\Omega)) \tag{7.2.33}$$

for $n \geqslant 1$ is *compact*. Also, using H^s smooth kernels in L^2 we obtain compactness of the mapping

$$K : H^s(\Omega \times \Omega) \to BL\big(L^2(\Omega), L^2(\Omega)\big) \tag{7.2.34}$$

We summarize the consequences of this compactness in the following theorem.

Theorem 7.2.8. *In the spaces (7.2.33) and (7.2.34) the mapping $K : w \mapsto W$ is compact. The mapping cannot have a bounded inverse and, thus, the kernel w depends unstably on the right-hand side.*

7.3 A collocation method for kernel reconstruction

The goal of this section is to describe the numerical solution of the *inverse neural field* problem by solving the *kernel reconstruction problem* introduced and analyzed in section 7.2. The numerical simulation of the neural field equation has been introduced in section 6.5.

We will employ a *collocation method* for the inverse problem, i.e. we use a quadrature formula for the discretization of the integrals and base our inversion on Tikhonov regularization as introduced in section 3.1.4.

The numerical discretization of equations (7.2.3) and (7.2.4) for some $t \in \mathbb{R}$ by collocation is based on collocation points $x_\xi \in \Omega$, $\xi = 1, \dots, n$, and vectors

$$\boldsymbol{\varphi}(t) := \big(\varphi(x_\xi, t)\big)_{\xi=1,\dots,n}, \qquad \boldsymbol{\psi}(t) := \big(\psi(x_\xi, t)\big)_{\xi=1,\dots,n} \tag{7.3.1}$$

in $\mathbb{R}^m$ as discretized versions of $\varphi(\cdot, t)$ and $\psi(\cdot, t)$. For $K \in \mathbb{N}$ time discretization points

$$t_1 < t_2 < \cdots < t_K \tag{7.3.2}$$

in $[0, T]$ we abbreviate

$$\boldsymbol{\varphi}_k := \boldsymbol{\varphi}(t_k), \quad \boldsymbol{\psi}_k := \boldsymbol{\psi}(t_k), \quad k = 1, \dots, K. \tag{7.3.3}$$

The discretization of the integral operator W based on some quadrature formula with collocation points x_j, $j = 1, \dots, n$ and quadrature weights s_j has been described in (6.5.5), which gives an $n \times n$ matrix, $\mathbf{W}$.

As the next step we define the $\mathbb{R}^{n \times K}$-matrices

$$\mathbf{A} := \big(\boldsymbol{\varphi}^{(1)}, \dots, \boldsymbol{\varphi}^{(K)}\big) \quad \text{and} \quad \mathbf{B} := \big(\boldsymbol{\psi}^{(1)}, \dots, \boldsymbol{\psi}^{(K)}\big). \tag{7.3.4}$$

The collocation method to approximately solve (7.2.5) is now given by the equation

$$\mathbf{B} = \mathbf{W}\mathbf{A} \tag{7.3.5}$$

to find the unknown kernel $\mathbf{W}$. If $K < n$, the equation (7.3.5) cannot have a unique solution, since $\mathbf{A}$ is rank deficient and not invertible.

We can try to calculate a solution using the *Moore–Penrose pseudo-inverse*

$$\mathbf{A}^{\dagger} := (\mathbf{A}^{*}\mathbf{A})^{-1}\mathbf{A}^{*} \tag{7.3.6}$$

defined in (3.2.4). We note that

$$\mathbf{A}\mathbf{A}^{\dagger} = \mathbf{A}(\mathbf{A}^{*}\mathbf{A})^{-1}\mathbf{A}^{*} = P_{R(A)} \tag{7.3.7}$$

according to lemma 3.2.3, which is the projection onto $R(A) \in X$ and leads to

$$\mathbf{B}\mathbf{A}^{\dagger} = \mathbf{W}\mathbf{A}\mathbf{A}^{\dagger} = \mathbf{W}P_{R(A)}. \tag{7.3.8}$$

An alternative is to use the adjoint (or transpose equation, which for a real scalar product is identical to the adjoint)

$$\mathbf{B}^{*} = \mathbf{A}^{*}\mathbf{W}^{*}. \tag{7.3.9}$$

Then, again using lemma 3.2.3, we obtain

$$(\mathbf{A}^{*})^{\dagger}\mathbf{B}^{*} = P_{R(A)}\mathbf{W}^{*}. \tag{7.3.10}$$

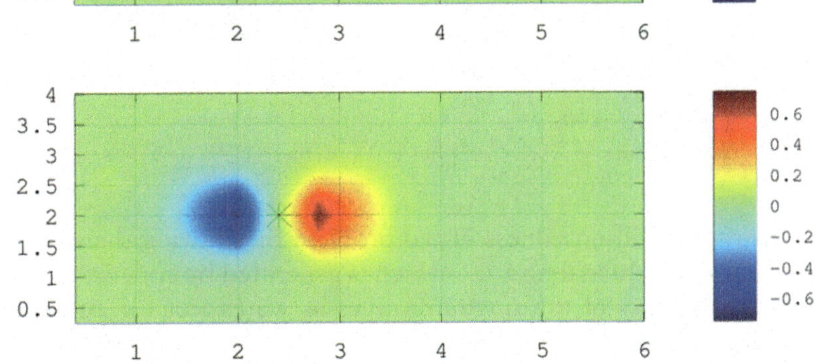

Figure 7.2. The original (top) and the reconstructed kernel (bottom) at $p = (2.4, 2)$, using the dynamics of code 6.5.1 visualized in figure 6.6 generated by code 7.3.1.

Using (3.2.11), equations (7.3.8) and (7.3.10) can be transformed into each other.

In general $\mathbf{A}^{\dagger}$ is strongly *ill-conditioned* and the condition number quickly increases with K. For regularization we employ the Tikhonov inverse

$$\mathbf{R}_{\alpha} := (\alpha I + \mathbf{A}^{*}\,\mathbf{A})^{-1}\mathbf{A}^{*} \tag{7.3.11}$$

with $\alpha > 0$. The regularized kernel is now calculated by

$$\mathbf{W}_{\alpha} = \mathbf{B}(\alpha I + \mathbf{A}^{*}\mathbf{A})^{-1}\mathbf{A}^{*} \tag{7.3.12}$$

or

$$\mathbf{W}_{\alpha} = \left((\alpha I + \mathbf{A}\mathbf{A}^{*})^{-1}\mathbf{A}\mathbf{B}^{*}\right)^{*}, \tag{7.3.13}$$

which is obtained from (7.3.12) by using

$$\begin{aligned}&(\alpha I + \mathbf{A}\mathbf{A}^{*})\mathbf{A} = \mathbf{A}(\alpha I + \mathbf{A}^{*}\mathbf{A})\\ &\quad\Leftrightarrow \mathbf{A}(\alpha I + \mathbf{A}^{*}\mathbf{A})^{-1} = (\alpha I + \mathbf{A}\mathbf{A}^{*})^{-1}\mathbf{A}.\end{aligned} \tag{7.3.14}$$

Next, we carry out the kernel reconstruction problem based on the neural fields introduced in section 6.5. The original and resulting kernel is visualized in figure 7.2.

Code 7.3.1. *Script* `sim_07_3_1_neural_inversion.m` *to reconstruct a neural kernel given some neural field on a domain* Ω *based on the temporal evolution of some neural activity function according to the neural field equation (6.5.1).*

```
for k=1:Nt
  psi(:,k) = tau*(uv(:,k+1)-uv(:,k))/ht + uv(:,k);
  phi(:,k) = f(uv(:,k),eta);
end

% solve the inverse problem B = W A or B' = A' W'
A = phi';

rhs = psi';
alpha = 0.001;
nn = size(A,2);
% solving A*Wmat =
Wmat_alpha = (inv( alpha*eye(nn,nn)+A'*A )*A'*rhs)';
```

We demonstrate the ill-posedness of the problem by showing the singular values of the matrix A for $K = 40$ in figure 7.3, which is generated by code 7.3.2. Clearly, we see an exponential decay of the singular values as we expect for an exponentially ill-posed problem.

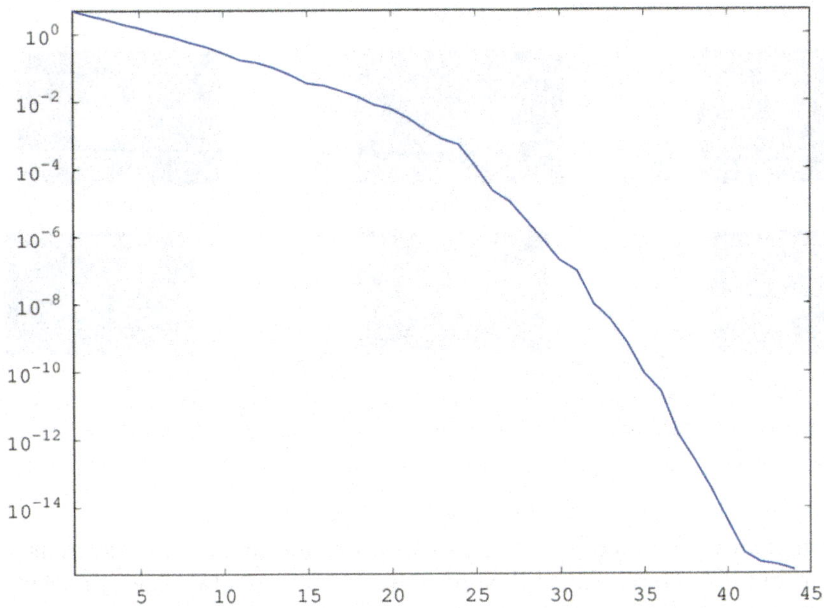

Figure 7.3. The singular values of the matrix $\mathbf{A}$ discretizing $\varphi(x, t_k)$ for $k = 1, \ldots, K$ for $K = 40$ on a logarithmic scale, showing that the singular values behave as $ce^{-\rho n}$, $n = 1, \ldots, Nt$ with some constants $c, \rho > 0$.

Code 7.3.2. *Script* `sim_07_3_3_singular_values.m` *to visualize the ill-posedness of the neural kernel reconstruction problem.*

```
[U,S,V] = svd(A);  % singular value decomposition
s = diag(S);       % diagonal elements of S

fo = figure;
po = semilogy(s,'LineWidth',3);    % plot values
ao = get(po,'Parent');             % get parent handle
set(ao,'FontSize',14);             % set font size
axis tight;                        % axis control
% save figure as png
savefile(fo,'sim_07_3_3_singular_values');
```

Finally, we need to test whether the neural dynamics with the reconstructed kernel coincides with the original neural dynamics. Here, we need to avoid using the same time grid, since otherwise our test would merely test a finite set of dynamical vectors. Figure 7.4 shows the dynamics when we use $K = 45$.

7.4 Traveling neural pulses and homogeneous kernels

In neural field theory it is very popular to study homogeneous kernels

$$w(x, y) = \tilde{w}(x - y).$$

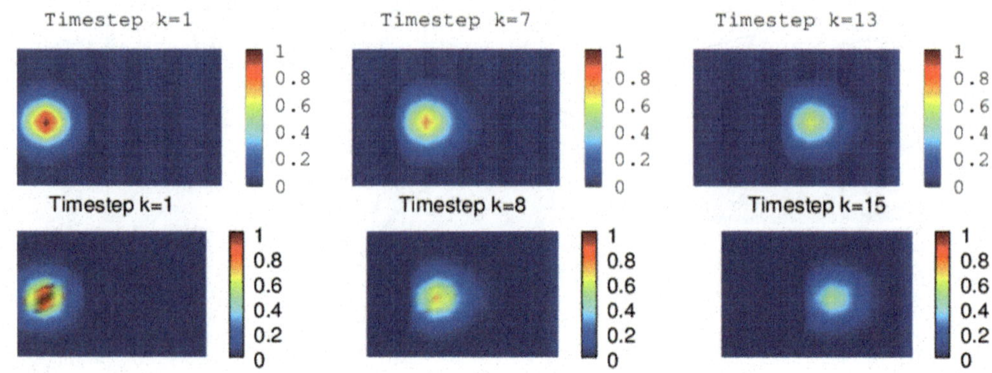

Figure 7.4. A test of the neural dynamics (second row) based on the reconstructed kernel W_α in comparison to the prescribed dynamics (first row with $K = 40$ shown in figure 6.6) where for the second row we chose $K = 45$ time steps.

Homogeneity is also linked to the investigation of special solutions such as traveling pulses or bumps. Here, we will study the inverse neural field problem in the translation invariant case.

Let us study an initial field $u_0(x)$ and a traveling pulse defined by

$$u(x, t) := u_0(x - qt), \quad x \in \Omega, \quad t \geqslant 0 \tag{7.4.1}$$

with $0 \neq q \in \mathbb{R}^3$ [13, 33]. Then, $u(x, t)$ is a traveling wave with direction $\hat{q} = \frac{q}{|q|}$ and wave speed $|q|$. With $\varphi_0(x) := \varphi(x, 0)$ we obtain

$$\varphi(x, t) = f(u(x, t)) = f(u_0(x - qt)) = \varphi_0(x - qt), \tag{7.4.2}$$

and with $\psi_0(x) := \psi(x, 0) = \lim_{t\to+0}(\tau\partial_t u(x, t) + u(x, t))$ we derive

$$\begin{aligned}\psi(x, t) &= \tau\frac{\partial u_0(x - qt)}{\partial t} + u_0(x - qt)\\ &= -\tau q \cdot \nabla u_0(x - qt) + u_0(x - qt)\\ &= \psi_0(x - qt)\end{aligned} \tag{7.4.3}$$

for $x \in D$, $t \geqslant 0$, such that $\psi(x, t) = \psi_0(x - qt)$. Given φ, ψ by (7.4.2) and (7.4.3) we study the inverse neural field problem in integral form with a kernel $w(x, y) = \tilde{w}(x - y)$. Let φ_0 and ψ_0 be a solution to the integral equation

$$\begin{aligned}\psi_0(x) &= \int_\Omega \tilde{w}(x - \tilde{y})\varphi_0(\tilde{y})\,\mathrm{d}\tilde{y},\\ &= \int_{x-\Omega} \varphi_0(x - y)\tilde{w}(y)\,\mathrm{d}y \quad x \in \Omega,\end{aligned} \tag{7.4.4}$$

where we used the substitution $x - \tilde{y} = y$ for $\tilde{y} \in \Omega$. Then, the solution of the equation $\psi = W\varphi$ can be obtained from $\psi_0 = W\varphi_0$ by

$$\begin{aligned}\psi(x, t) &= \psi_0(x - qt)\\ &= \int_{\Omega} w(x - qt, y)\varphi_0(y)\,\mathrm{d}y\\ &= \int_{\Omega} \tilde{w}(x - qt - y)\varphi_0(y)\,\mathrm{d}y\\ &= \int_{\Omega} \tilde{w}(x - (y + qt))\varphi_0(y)\,\mathrm{d}y\\ &= \int_{\Omega+qt} \tilde{w}(x - \tilde{y})\varphi_0(\tilde{y}-qt)\,\mathrm{d}\tilde{y}\\ &= (W\varphi)(x, t),\end{aligned} \tag{7.4.5}$$

where we made use of $\tilde{y} = y + qt$. We have shown the following result.

Theorem 7.4.1 (Integral equation for a homogeneous problem). *Given a field u as a traveling wave (7.4.1), there is a solution of the full field inverse neural problem with translation invariant kernel*

$$w(x, y) = \tilde{w}(x - y)$$

if and only if the function $\tilde{w}$ satisfies the integral equation

$$\psi_0(x) = \int_{\tilde{\Omega}} \rho(x, y)\varphi_0(x - y)\tilde{w}(y)\,\mathrm{d}y, \quad x \in \Omega, \tag{7.4.6}$$

with

$$\rho(x, y) = \begin{cases} 1, & y \in x - \Omega \\ 0, & \textit{otherwise} \end{cases} \tag{7.4.7}$$

and

$$\tilde{\Omega} := \{x - y : x, y \in \Omega\}. \tag{7.4.8}$$

The integral equation (7.4.6) is an integral equation of the first kind with a kernel in $L^2(\Omega)$. The integral operator is compact in $BC(\Omega)$ and in $L^2(\Omega)$, as we will see in the following theorem (see [30]).

Theorem 7.4.2 (Ill-posedness of a homogeneous problem). *In general, the equation (7.4.6) does not have a solution. If it exists, the solutions depend unstably on the right-hand side in both $BC(\Omega)$ and $L^2(\Omega)$. Thus, the full field inverse neural problem with traveling wave fields and homogeneous kernels is ill-posed in the sense of Hadamard (3.1.1).*

Proof. We study the operator

$$(Ag)(x) := \int_{\tilde{\Omega}} k(x, y)g(y)\,\mathrm{d}y, \quad x \in \Omega. \tag{7.4.9}$$

For $k(x, y) := \rho(x, y)\varphi_0(x - y)$ we first remark that in this case $k(x, \cdot) \in L^2(\tilde{\Omega})$ for every fixed $x \in \Omega$ and the norm of this function is smaller than $\|\varphi_0\|_{L^2(\Omega)}$. Then we estimate

$$\|Ag\|^2_{L^2(\Omega)} \leqslant \int_\Omega \left| \int_{\tilde{\Omega}} \rho(x, y)\varphi_0(x - y)g(y)\,\mathrm{d}y \right|^2 \mathrm{d}x$$
$$= |\tilde{\Omega}| \cdot \|\varphi_0\|^2_{L^2(\Omega)} \|g\|^2_{L^2(\tilde{\Omega})}, \tag{7.4.10}$$

such that A is bounded from $L^2(\tilde{\Omega})$ into $L^2(\Omega)$. A typical argumentation to show compactness of an operator of the type (7.4.9) with piecewise continuous kernel we first construct a continuous approximation as in (7.2.11) and then employ a finite-dimensional approximation of the continuous kernel as in the examples 2.1.11 and 2.3.21 to construct a finite-dimensional approximation sequence A_n to A. Then, by theorems 2.3.17 and 2.3.18 the operator A is a compact operator from $L^2(\tilde{\Omega})$ into $L^2(\Omega)$. Hence the solution of (7.4.6) depends unstably on the right-hand side due to theorem 3.1.3 and the problem is ill-posed in the sense of Hadamard. This completes the proof. □

7.5 Bi-orthogonal basis functions and integral operator inversion

We have seen that the inverse neural field problem can be transformed into a family of integral equations of the first kind. Solution methods for such integral equations have been studied for a long time, see for example [34–36], and we have developed a regularized collocation approach for the solution of the inverse problem in section 7.3.

Here, our goal is to look at the inversion from the viewpoint of *bi-orthogonal basis functions* as in [30]. We study the construction of a bi-orthogonal set in particular for a Riesz basis $\{\varphi_n, n = 1, 2, 3, \dots\}$ of X, which is a basis of X with the property: there are constants $c_1, c_2 > 0$ such that

$$c_1 \sum_{j=1}^{\infty} |\alpha_j|^2 \leqslant \left\| \sum_{j=1}^{\infty} \alpha_j \varphi_j \right\|^2 \leqslant c_2 \sum_{j=1}^{\infty} |\alpha_j|^2 \tag{7.5.1}$$

for all $\alpha = (\alpha_j)_{j\in\mathbb{N}} \in \ell^2$. Bi-orthogonal sets can be an important ingredient to understand the behavior of the solutions of our dynamic kernel reconstruction problem formulated in definition 7.2.1.

In a Hilbert space X with scalar product $\langle \cdot, \cdot \rangle$ two linearly independent sets of functions $Q = \{\varphi_1, \varphi_2, \dots\}$ and $R = \{\rho_1, \rho_2, \dots\}$ are called *bi-orthogonal*, if

$$\langle \rho_i, \varphi_k \rangle = 0 \quad \text{for all } k \neq i, \quad \langle \rho_i, \varphi_i \rangle = c_i, \quad i \in \mathbb{N}, \tag{7.5.2}$$

where $c_i > 0$ for $i \in \mathbb{N}$. The construction of a bi-orthonormal set R to Q is usually carried out as follows. We define

$$V_k := \operatorname{span}\{\varphi_1, \dots, \varphi_{k-1}, \varphi_{k+1}, \dots\}, \quad k \in N \tag{7.5.3}$$

denote its orthogonal space by $V_k^\perp$ and remark that $X = \bar{V}_k \oplus V_k^\perp$. We conclude that $\varphi_k \notin \bar{V}_k$, since it is linearly independent of V_k. Thus, its orthogonal projection $\tilde{\rho}_k$ onto $V_k^\perp$ cannot be zero. The bi-orthogonal elements are now given by

$$\rho_k := \frac{\tilde{\rho}_k}{\|\tilde{\rho}_k\|^2}, \qquad k = 1, 2, 3, \ldots \tag{7.5.4}$$

For $i \neq k$ we have $\rho_k \in V_k^\perp$ and thus $\langle \rho_k, \varphi_i \rangle = 0$. Further, we note that $\varphi_k = \tilde{\rho}_k + \tilde{\varphi}_k$ where $\tilde{\varphi}_k \in V_k$ and thus

$$\langle \tilde{\rho}_k, \varphi_k \rangle = \langle \tilde{\rho}_k, \tilde{\rho}_k + \tilde{\varphi}_k \rangle = \langle \tilde{\rho}_k, \tilde{\rho}_k \rangle = \|\tilde{\rho}_k\|^2. \tag{7.5.5}$$

This yields $\langle \rho_k, \varphi_i \rangle = \delta_{ki}, \quad k, i \in \mathbb{N}$, and, thus, by (7.5.4) we define a set $R := \{\rho_1, \rho_2, \ldots\}$ which satisfies (7.5.2) with constants $c_i = 1$, $i \in \mathbb{N}$.

To show that the elements of R are linearly independent assume that

$$\sum_{j=1}^{n} \beta_j \rho_{k_j} = 0$$

with constants $\beta_j \in \mathbb{C}$ and $k_j \in \mathbb{N}$ for $j = 1, \ldots, n$. We multiply the term by φ_{k_i} for $i = 1, \ldots, n$ to obtain

$$0 = \left\langle \sum_{j=1}^{n} \beta_j \rho_{k_j}, \varphi_{k_i} \right\rangle = \sum_{j=1}^{n} \beta_j \langle \rho_{k_j}, \varphi_{k_i} \rangle = \beta_i, \quad i = 1, \ldots, n, \tag{7.5.6}$$

and thus R is linearly independent.

For a Riesz basis $\{\varphi_n, n = 1, 2, 3, \ldots\}$ of X the mapping

$$A : \ell^2 \to X, \quad \alpha \mapsto \sum_{j=1}^{\infty} \alpha_j \varphi_j \tag{7.5.7}$$

is a bounded and boundedly invertible mapping from ℓ^2 onto $A(\ell^2) \subset X$. We have

$$\langle A\alpha, \psi \rangle_X = \sum_{j=1}^{\infty} \alpha_j \langle \varphi_j, \psi \rangle_X, \tag{7.5.8}$$

such that $(\langle \varphi_j, \psi \rangle_X)_{j \in \mathbb{N}} \in \ell^2$. Thus, a dual operator $A' : X \to \ell^2$ with respect to the scalar products $\langle \cdot, \cdot \rangle_{\ell^2}$ and $\langle \cdot, \cdot \rangle_X$ defined by $\langle A\alpha, \psi \rangle_X = \langle \alpha, A'\psi \rangle_{\ell^2}$ is given by

$$A' : X \to \ell^2, \quad \psi \mapsto \left(\langle \varphi_j, \psi \rangle_X \right)_{j \in \mathbb{N}}. \tag{7.5.9}$$

We estimate

$$\langle \alpha, A'A\alpha \rangle_{\ell^2} = \langle A\alpha, A\alpha \rangle_X \leqslant c_2 \|\alpha\|_{\ell^2}^2, \tag{7.5.10}$$

and

$$\langle \alpha, A'A\alpha \rangle_{\ell^2} = \langle A\alpha, A\alpha \rangle_X \geqslant c_1 \|\alpha\|_{\ell^2}^2, \tag{7.5.11}$$

thus according to the Lax–Milgram theorem the operator $A'A$ is boundedly invertible in ℓ^2 with a bound given by $1/c_1$. Note that if c_1 is small, the inverse can have a large norm and the equation is highly ill-conditioned. We calculate

$$A'A\alpha = \left(\langle \varphi_k, A\alpha\rangle_X\right)_{k\in\mathbb{N}} = \left(\sum_{j=1}^{\infty}\langle \varphi_k, \varphi_j\rangle_X \alpha_j\right)_{k\in\mathbb{N}}, \tag{7.5.12}$$

thus the operation of $A'A$ on ℓ^2 can be expressed as a matrix multiplication with the symmetric *metric tensor* M defined by

$$M := \left(\langle \varphi_k, \varphi_j\rangle\right)_{k,j\in\mathbb{N}}. \tag{7.5.13}$$

According to (7.5.11) M is boundedly invertible on ℓ^2 with inverse M^{-1}, for which an infinite matrix is defined by

$$(M^{-1})_{k,j} := \langle e_k, M^{-1}e_j\rangle_{\ell^2}, \quad k, j = 1, 2, 3, \ldots$$

for $e_j \in \ell^2$ defined by 1 in its jth entry and zero otherwise, where by symmetry of M according to

$$\langle \beta, M^{-1}\alpha\rangle = \langle M\beta', M^{-1}M\alpha'\rangle = \langle M\beta', \alpha'\rangle = \langle \beta', M\alpha'\rangle = \langle M^{-1}\beta, \alpha\rangle$$

for $\alpha = M\alpha'$ and $\beta = M\beta'$ also M^{-1} is symmetric and each row or line of M^{-1} is in ℓ^2. We are now prepared to define

$$\rho_k := \sum_{j=1}^{\infty}(M^{-1})_{k,j}\varphi_j, \quad k \in \mathbb{N}, \tag{7.5.14}$$

which is convergent by (7.5.1). It satisfies

$$\langle \rho_k, \varphi_i\rangle_X = \left\langle \sum_{j=1}^{\infty}(M^{-1})_{k,j}\varphi_j, \varphi_i \right\rangle = \sum_{j=1}^{\infty}(M^{-1})_{k,j}\langle \varphi_j, \varphi_i\rangle = (M^{-1}M)_{ki} = \delta_{ki}$$

for $k, i \in \mathbb{N}$. Thus, equation (7.5.14) provides a constructive method to calculate the bi-orthonormal basis functions.

Given two linearly independent sets of elements $Q = \{\varphi_1, \varphi_2, \ldots\} \subset X$ and $S = \{\psi_1, \psi_2, \ldots\} \subset X$ we can now construct linear operators

$$V_n\varphi := \sum_{j=1}^{n}\psi_j\langle \rho_j, \varphi\rangle, \qquad V\varphi := \sum_{j=1}^{\infty}\psi_j\langle \rho_j, \varphi\rangle, \tag{7.5.15}$$

where $R = \{\rho_1, \rho_2, \ldots\}$ is a bi-orthogonal basis for Q.

Lemma 7.5.1. *The operator V_n is linear and bounded on H. If both Q and S are Riesz bases of H, then V is linear and bounded on H as well. Further, we have*

$$V_n\varphi_i = \begin{cases} \psi_i, & i = 1, \dots, n \\ 0, & i > n, \end{cases}$$
$$V\varphi_i = \psi_i, \quad i \in \mathbb{N}. \tag{7.5.16}$$

Proof. We first note that $(\langle \rho_j, \varphi \rangle)_{j\in\mathbb{N}}$ is in ℓ^2 because M is boundedly invertible on ℓ^2. Then, the boundedness of V is a consequence of (7.5.1) for the Riesz basis $(\psi_j)_{j\in\mathbb{N}}$.

Practically, our space will be finite-dimensional, i.e. $X = \mathbb{R}^m$ with $n \in \mathbb{N}$, as in (7.3.3), and we will be given elements $\varphi_k \in \mathbb{R}^m$ for $k = 1, \dots, K$, i.e. the matrices $\mathbf{A}$ and $\mathbf{B}$ in $\mathbb{R}^{n\times K}$ as in (7.3.4). When these elements are linearly independent, we can find a bi-orthonormal set of functions by calculating $\mathbf{A}^\dagger$, since

$$\mathbf{A}^\dagger \mathbf{A} = (\mathbf{A}^* \mathbf{A})^{-1}\mathbf{A}^*\mathbf{A} = I_K$$

according to lemma 3.2.1, i.e. the rows of $\mathbf{A}^\dagger \in \mathbb{R}^{K\times n}$ are bi-orthonormal to the columns of $\mathbf{A}$. The metric tensor M is given by the matrix $\mathbf{A}^*\mathbf{A}$ and M^{-1} by the inverse $(\mathbf{A}^*\mathbf{A})^{-1}$ which is calculated in $\mathbb{R}^{K\times K}$ as a step to calculate $\mathbf{A}^\dagger$. The sum (7.5.14) is then given by $\mathbf{A}(\mathbf{A}^*\mathbf{A})^{-1} \in \mathbb{R}^{n\times K}$, such that the operator V_n defined in (7.5.15) obtains the form

$$\mathbf{V} = \mathbf{B}(\mathbf{A}^* \mathbf{A})^{-1}\mathbf{A}^*,$$

i.e. it is the unregularized version of the reconstruction of the kernel $\mathbf{W}$. The above construction of bi-orthogonal basis functions can be considered as a particular viewpoint to understand the collocation method (7.3.12) or (7.3.13) of section 7.3.

7.6 Dimensional reduction and localization

Neural kernel reconstruction is a strongly ill-posed problem described by the integral equation (7.2.5). We have demonstrated the ill-posedness of its solution in figure 7.3 by the exponential decay of its singular values.

Here, our goal is to introduce techniques of *dimensional reduction* and *localization* to reduce the inverse task to a set of smaller tasks which are less ill-posed.

Our starting point is the family of integral equations (7.2.8), which shows that the full field neural inverse problem of definition 7.2.1 naturally decouples into a family of independent problems

$$\psi_x(t) = \int_\Omega k(t, y) w_x(y)\, \mathrm{d}y, \quad t \in [0, T], \tag{7.6.1}$$

with unknown functions $w_x(\cdot)$, where for each $x \in \Omega$ the equation (7.6.1) can be solved independently. Recall that the kernel $k(t, y) = \varphi(y, t) = f(u(y, t))$ reflects the underlying dynamics.

Let us assume that the connectivity kernel w is *local* in the sense that

$$w(x, y) = 0 \quad \text{for all} \quad |x - y| \geqslant \rho. \tag{7.6.2}$$

Then for each $x \in \Omega$ the integration in (7.6.1) only takes place on $B(x, \rho)$. Further, if we have $f(0) = 0$ we can restrict our attention to times $t \in [0, T]$ where $f(u(\cdot, t)) > \epsilon$ on $B(x, \rho)$ with some parameter $\epsilon > 0$. We call the corresponding set of times $G_x \subset [0, T]$. This leads to a *localized* system

$$\psi_x(t) = \int_{B(x,\rho)} k(t, y) w_x(y)\, \mathrm{d}y, \qquad t \in G_x. \tag{7.6.3}$$

The solutions $w_x(\cdot)$ of (7.6.3) on $B(x, \rho)$ are *imbedded* into w and extended by zero to a full kernel.

Lemma 7.6.1. *Under the condition (7.6.2) and for $\epsilon = 0$ the solution of (7.2.5) is equivalent to the solution of n localized systems (7.6.3) on the localization domains $B(x_\xi, \rho)$ for $\xi = 1, \ldots, n$.*

The discretized form of (7.6.3) decouples (7.3.9) into a family of equations

$$\mathbf{b}^{(\xi)} = \mathbf{K}^{(\xi)} \mathbf{w}^{(\xi)}, \qquad \xi = 1, \ldots, n \tag{7.6.4}$$

where for $\xi = 1, \ldots, n$ we define

$$\mathbf{b}^{(\xi)} := (\psi(x_\xi, t_k))_{t_k \in G_x}, \qquad \mathbf{K}^{(\xi)} := \Big(\big(\varphi(y_\eta, t_k)\big)\Big)_{t_k \in G_x, y_\eta \in B(x,\rho)} \tag{7.6.5}$$

and

$$\mathbf{w}^{(\xi)} := \big(w(x_\xi, y_\eta)\big)_{y_\eta \in B(x_\xi,\rho)} \in \mathbb{R}^{\tilde{n}} \tag{7.6.6}$$

for some $\tilde{n} \ll n$. In the case where ρ is sufficiently large and $G_x = [0, T]$ we have $\mathbf{K} = \mathbf{A}^*$ and $\mathbf{w}^{(x)}$ is a row of $\mathbf{W}$. The vector $\mathbf{b}^{(\xi)} \in \mathbb{R}^{\tilde{K}}$ with $\tilde{K} \leqslant K$ is part of a row of $\mathbf{B}$. We have $\mathbf{K}^{(\xi)} \in \mathbb{R}^{\tilde{K}\times\tilde{n}}$ and $\mathbf{w}^{(\xi)} \in \mathbb{R}^{\tilde{n}}$, such that (7.6.4) is a system with $\tilde{K}$ equations and $\tilde{n}$ unknowns. We can now apply the solution steps (7.3.13) to calculate a regularized solution

$$\mathbf{w}_\alpha^{(\xi)} := \big(\alpha I + \mathbf{K}^{(\xi)}(\mathbf{K}^{(\xi)})^*\big)^{-1}(\mathbf{K}^{(\xi)})^* \mathbf{b}^{(\xi)}, \qquad \xi = 1, \ldots, n. \tag{7.6.7}$$

to (7.6.4). In figure 7.6 we demonstrate that the localized reconstruction of w by (7.6.3) yields comparable results to the full solution of (7.2.8) as visualized in figure 7.2. However, the ill-posedness of the equations is significantly reduced, as shown in figure 7.5, where the smallest singular value of the matrices $\mathbf{K}^{(\xi)}$ is factor 10^7 larger than the smallest singular value of the full matrix $\mathbf{A}$.

Code 7.6.2. *Script* `sim_07_6_1_localized_inversion.m` *to reconstruct a neural kernel by a localized method. Run script* `sim_06_5_1_neural_simulation.m` *first to generate* `uv` *and some further variables. Figure 7.6 is then generated by script* `sim_07_3_2_show_kernels.m`.

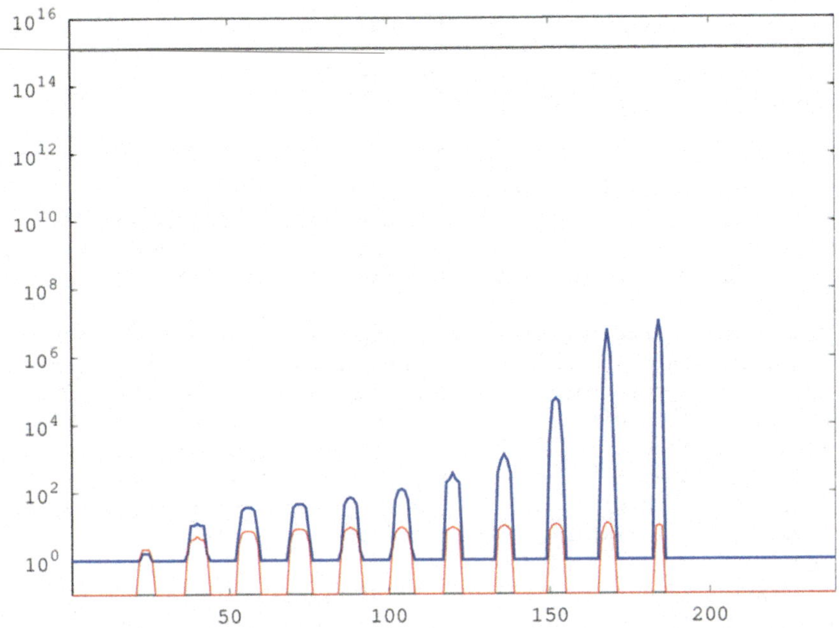

Figure 7.5. The reverse value of the minimal singular value of the matrices $\mathbf{K}^{(\xi)}$, $\xi = 1, \ldots, 240$ (blue line) compared to the reverse minimal singular value of the matrix $\mathbf{A}$ (black line) defined in (7.3.4). For comparison, the red line also displays the size of the temporal index of $\mathbf{K}$.

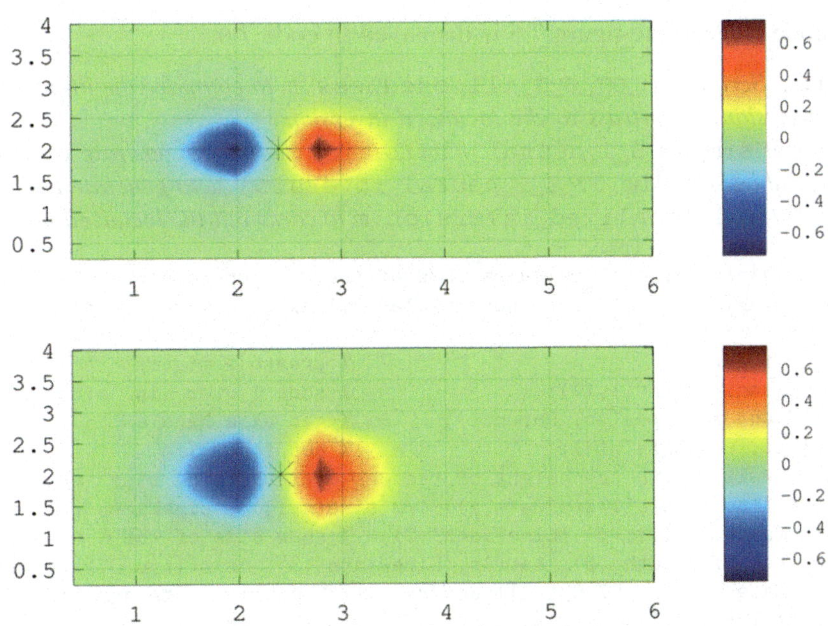

Figure 7.6. The original (top) and the reconstructed kernel (bottom) at $p = (2.4, 2)$, using the dynamics of code 6.5.1 visualized in figure 6.6 generated by the localized inversion procedure as given by code 7.6.2.

```
% first we set up th vectors varphi and psi
for k=1:Nt
  psi(:,k) = tau*(uv(:,k+1)-uv(:,k))/ht + uv(:,k);
  varphi(:,k) = f(uv(:,k),eta);
end

% solve the localized inverse problems b = K w in a loop
rho  = 1.5;                        % set localization radius
Wmat_alpha = zeros(n,n);           % initialization by a default valus 0
nv   = ones(n,1);                  % initialization by a default value 1
for xi=1:n
  ind0 = find(varphi(xi,:)>0.06); % find all time steps where f(u)> eps
  % and calculate the indices for the current local. domain B_rho(x_xi)
  ind1 = find( sqrt((p1v-p1v(xi)).^2+(p2v-p2v(xi)).^2)<rho );
  K = phi(ind1,ind0)';             % the current operator K = K_\xi
  sv(xi,:) = size(K);              % store the size of K in a vector sv
  if( size(K,1)>1 )                % if K is not empty, calc its minimal
    [U,S,V] = svd(K);              % sing. value by carrying out an SVD
    nv(xi,1) = 1/min(diag(S));     % store it in nv
  end
  b = psi(xi,ind0)';               % set right-hand side b
  alpha = 0.01;                    % regularization parameter
  nn = size(K,2);                  % size of current
  % solving b = K*w by Tikhonov regularization
  w_alpha = (inv( alpha*eye(nn,nn)+K'*K )*K'*b);
  Wmat_alpha(xi,ind1) = w_alpha'; % put the vector into Wmat_alpha
end
```

The code to generate figure 7.5 is displayed in code 7.6.3.

Code 7.6.3. *Script* sim_07_6_3_illposedness.m *to compare the ill-posedness of the full inversion in contrast to the localized inversion.*

Run script sim_06_5_1_neural_simulation.m *first to generate* uv *and some further variables, run* sim_07_3_1_neural_inversion.m *to generate the matrix* A and sim_07_6_1_localized_inversion.m *to carry out the localized inversion.*

```
[U,S,V]=svd(A);    % calculate SVD of the full operator A
s = diag(S);       % put the singular values of A into a vector

fo = figure;                              % create a figure
po = semilogy(nv(:,1));                   % show 1 over min. sing. value
hold on; ao = get(po,'Parent');           % get axis handle
set(ao,'FontSize',14);                    % set font size
% draw line with the height of the one over the min. sing. value of A
lo = line([1 n],[1/s(size(s,1)) 1/s(size(s,1))],'LineWidth',4);
plot(sv(:,1)+0.1,'r','LineWidth',2);  % show size of the K time index
set(po,'LineWidth',5); set(lo,'LineWidth',3); axis([1 n 0.1 1e16]); %
filename = ['sim_07_6_1_illposedness']; % current date and time
savefile(fo,filename);                    % save image
```

Bibliography

[1] Coombes S, beim Graben P, Potthast R and Wright J (ed) 2014 *Neural Fields* (Berlin: Springer)

[2] beim Graben P, Drenhaus H, Brehm E, Rhode B, Saddy S and Frisch D 2007 Enhancing dominant modes in nonstationary time series by means of the symbolic resonance analysis *Chaos* **17** 043106

[3] beim Graben P, Liebscher T and Kurths J 2008 Neural and cognitive modeling with networks of leaky integrator units *Lectures in Supercomputational Neuroscience: Dynamics in Complex Brain Networks* (Berlin: Springer)

[4] Freeman W J 1987 Simulation of chaotic EEG patterns with a dynamic model of the olfactory system *Biol. Cybern.* **56** 139–50

[5] beim Graben P and Kurths J 2008 Simulating global properties of electroencephalograms with minimal random neural networks *Neurocomputing* **71** 999–1007

[6] Hertz J, Krogh A and Palmer R G 1991 *Introduction to the Theory of Neural Computation* (*Lecture Notes of the Santa Fe Institute Studies in the Science of Complexity* vol 1) (Cambridge, MA: Perseus)

[7] McCulloch W S and Pitts W 1943 A logical calculus of ideas immanent in nervous activity *Bull. Math. Biophys.* **5** 115–33

[8] Stein R B, Leung K V, Mangeron D and Oğuztöreli M N 1974 Improved neuronal models for studying neural networks *Kybernetik* **15** 1–9

[9] Wilson H R and Cowan J D 1972 Excitatory and inhibitory interactions in localized populations of model neurons *Biophys. J.* **12** 1–24

[10] beim Graben P 2006 Pragmatic information in dynamic semantics *Mind Matter* **4** 169–93

[11] Amari S-I 1977 Neural theory of association and concept-formation *Biol. Cybern.* **26** 175–85

[12] Breakspear M, Roberts J A, Terry J R, Rodrigues S, Mahant N and Robinson P A 2006 A unifying explanation of primary generalized seizures through nonlinear brain modeling and bifurcation analysis *Cerebral Cortex* **16** 1296–313

[13] Coombes S, Lord G J and Owen M R 2003 Waves and bumps in neuronal networks with axo-dendritic synaptic interactions *Physica* D **178** 219–41

[14] Griffith J S 1963 A field theory of neural nets. I: Derivation of field equations *Bull. Math. Biol.* **25** 111–20

[15] Hutt A and Atay F M 2005 Analysis of nonlocal neural fields for both general and gamma-distributed connectivities *Physica* D **203** 30–54

[16] Hutt A and Atay F M 2006 Effects of distributed transmission speeds on propagating activity in neural populations *Phys. Rev.* E **73** 021906

[17] Jirsa V K 2004 Information processing in brain and behavior displayed in large-scale scalp topographies such as EEG and MEG *Int. J. Bifurcation Chaos* **14** 679–92

[18] Jirsa V K and Haken H 1996 Field theory of electromagnetic brain activity *Phys. Rev. Lett.* **77** 960–3

[19] Jirsa V K and Haken H 1997 A derivation of a macroscopic field theory of the brain from the quasi-microscopic neural dynamics *Physica* D **99** 503–26

[20] Jirsa V K and Kelso J A S 2000 Spatiotemporal pattern formation in neural systems with heterogeneous connection topologies *Phys. Rev.* E **62** 8462–5

[21] Rennie C J, Robinson P A and Wright J J 2002 Unified neurophysical model of EEG spectra and evoked potentials *Biol. Cybern.* **86** 457–71

[22] Richardson K A, Schiff S J and Gluckman B J 2005 Control of traveling waves in the mammalian cortex *Phys. Rev. Lett.* **94** 028103

[23] Robinson P A, Rennie C J and Wright J J 1997 Propagation and stability of waves of electrical activity in the cerebral cortex *Phys. Rev.* E **56** 826–40

[24] Venkov N A, Coombes S and Matthews P C 2007 Dynamic instabilities in scalar neural field equations with space-dependent delays *Physica* D **232** 1–5

[25] Wilson H R and Cowan J D 1973 A mathematical theory of the functional dynamics of cortical and thalamic nervous tissue *Kybernetik* **13** 55–80

[26] Wright J J, Rennie C J, Lees G J, Robinson P A, Bourke P D, Chapman C L, Gordon E and Rowe D L 2003 Simulated electrocortical activity at microscopic, mesoscopic, and global scales *Neuropsychopharmacol.* **28** 80–93

[27] Wright J J, Rennie C J, Lees G J, Robinson P A, Bourke P D, Chapman C L, Gordon E and Rowe D L 2004 Simulated electrocortical activity at microscopic, mesoscopic and global scales *Int. J. Bifurcation Chaos* **14** 853–72

[28] Amari S-I 1977 Dynamics of pattern formation in lateral-inhibition type neural fields *Biol. Cybern.* **27** 77–87

[29] Hale J K and Lunel S M V 1993 *Introduction to Functional Differential Equations* (Berlin: Springer)

[30] Potthast R and beim Graben P 2010 Existence and properties of solutions for neural field equations *Math. Methods Appl. Sci.* **33** 935–49

[31] Hebb D O 1949 *The Organization of Behavior* (New York: Wiley)

[32] Peter B G and Potthast R 2009 Inverse problems in dynamic cognitive modelling *Chaos* **19** 015103

[33] Ermentrout G B and McLeod J B 1993 Existence and uniqueness of travelling waves for a neural network *Proc. R. Soc. Edinburgh* A **123** 461–78

[34] Colton D and Kress R 1998 *Inverse Acoustic and Electromagnetic Scattering Theory* (*Applied Mathematical Sciences* vol 93) 2nd edn (Berlin: Springer)

[35] Engl H W, Hanke M and Neubauer A 1996 *Regularization of Inverse Problems* (*Mathematics and its Applications* vol 375) (Dordrecht: Kluwer Academic)

[36] Potthast R 2001 *Point Sources and Multipoles in Inverse Scattering Theory* (*Chapman and Hall/CRC Research Notes in Mathematics* vol 427) (Boca Raton, FL: CRC)

IOP Publishing

Inverse Modeling

An introduction to the theory and methods of inverse problems and data assimilation

Gen Nakamura and Roland Potthast

Chapter 8

Simulation of waves and fields

In this chapter we introduce the reader to the simulation of waves and fields. Our task is easy access to the use of potential operators and integral equations over boundaries using the MATLAB or OCTAVE programming interfaces as a preparation for the inverse problems of subsequent chapters. Also, we want the reader to understand the practical value of the language of *operators* and *jump relations*, which allows us to develop a sincere understanding of important wave phenomena.

Integral equation methods play an important role for many inversion techniques. Here we introduce the basic integral operators and potential operators for use in the subsequent sections and chapters. This part of our presentation can also be used to obtain a quick and easy realization of basic potential operators which can then be the basis for more advanced schemes. We want to enable a student to program the scattering problems rather quickly without being dragged into the need to achieve high convergence orders.

8.1 Potentials and potential operators

We start with potential operators for the time-harmonic wave equation, the *Helmholtz equation*

$$\Delta u + \kappa^2 u = 0. \tag{8.1.1}$$

We consider some domain G in $\mathbb{R}^d$ with dimension $d = 2$ or $d = 3$ with at least a piecewise C^2-smooth boundary. For the rest of the book we will assume that the exterior of our domains D or G is connected with infinity. The *single-layer potential* is defined by

$$(\tilde{S}\varphi)_G(x) := \int_{\partial G} \Phi(x, y)\varphi(y)\,\mathrm{d}s(y), \quad x \in \mathbb{R}^d \tag{8.1.2}$$

doi:10.1088/978-0-7503-1218-9ch8

for $\varphi \in L^2(\partial G)$ with the fundamental solution Φ defined by

$$\Phi(x, y) := \begin{cases} H_0^{(1)}(\kappa|x - y|), & d = 2, \\ \dfrac{e^{i\kappa|x-y|}}{4\pi|x - y|}, & d = 3, \end{cases} \tag{8.1.3}$$

where $H_0^{(1)}$ denotes the Hankel function of the first kind of order zero. For simulation of scattering problems we will also use the *double-layer potential*

$$(\tilde{K}\varphi)_G(x) := \int_{\partial G} \frac{\partial \Phi(x, y)}{\partial \nu(y)} \varphi(y) \, \mathrm{d}s(y), \quad x \in \mathbb{R}^m \backslash \partial G \tag{8.1.4}$$

for $\varphi \in L^2(\partial G)$, where ν denotes the exterior unit normal vector to the domain G and

$$\frac{\partial \Phi(x, y)}{\partial \nu(y)} := \nu(y) \cdot \nabla_y \Phi(x, y).$$

To obtain some feeling for these potential operators, we recommend studying a simple realization of the integrals. In two dimensions we parametrize our boundaries over the interval $[0, 2\pi)$ and employ algorithms based on a uniform mesh defined by

$$t_j := \frac{2\pi\,(j - 1)}{N}, \quad j = 1, \ldots, N. \tag{8.1.5}$$

We describe the boundary Γ_R of circular domains $B_R(z)$ by

$$\Gamma_R(t) := R \cdot (\cos(t), \sin(t))^T + z, \quad t \in [0, 2\pi), \tag{8.1.6}$$

with $z = (z_x, z_y)$. More generally we will work with domains defined by a boundary

$$\Gamma(t) := \begin{pmatrix} p_1(t) \\ p_2(t) \end{pmatrix}, \quad t \in [0, 2\pi) \tag{8.1.7}$$

with functions p_1 and p_2. One standard example often used in the literature is the *kite-shaped* domain

$$p_1(t) = \cos(t) + 0.65\cos(2t) - 0.65, \quad p_2(t) = 1.5\sin(t), \tag{8.1.8}$$

which can be scaled by a factor $r > 0$, shifted by some vector $q \in \mathbb{R}^2$ and rotated by multiplication with the rotation matrix

$$M_\beta := \begin{pmatrix} \cos(\beta) & -\sin(\beta) \\ \sin(\beta) & \cos(\beta) \end{pmatrix} \tag{8.1.9}$$

with rotation angle $\beta \in \mathbb{R}$. Figure 8.1 shows an example where we employed the values $r = 6$, $q = (r, 0)$ and $\beta = 4\pi/3$ to plot the graph of

$$\Gamma(t) := M_\beta(r \cdot \Gamma(t) + q), \quad t \in [0, 2\pi). \tag{8.1.10}$$

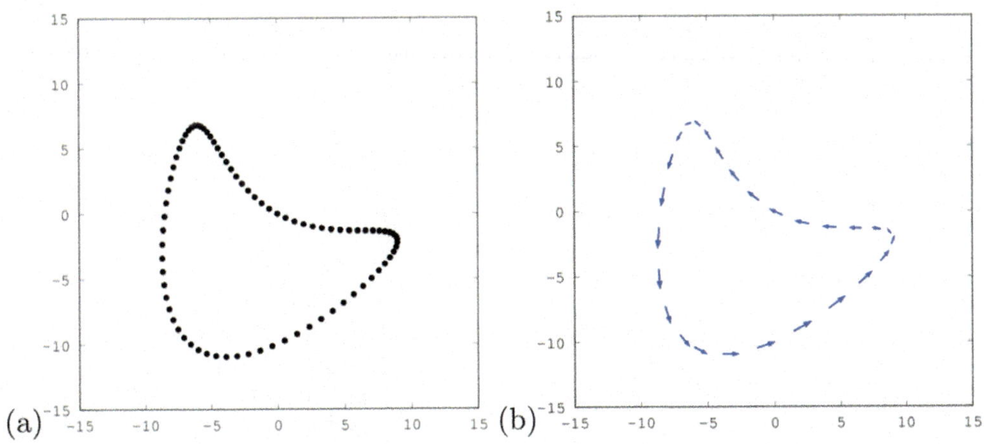

Figure 8.1. The discretization points $x_j = \Gamma(t_j)$ arising from some parametrization Γ for the boundary of a domain G (a). Then, for every third point the tangent vectors are displayed. The figure is generated by `sim_08_1_8_show_domain.m` (code repository).

For numerical calculations we use the boundary points

$$y_j = \Gamma(t_j), \quad j = 1, \ldots, N \tag{8.1.11}$$

and use $x_k := y_k$, $k = 1, \ldots, N$ when we need to discretize $|x - y|$ for $x, y \in \Gamma$.

Code 8.1.1. *Setting up the boundary points (8.1.11) and the tangential vectors (8.1.14) of the scattering domain by the file* `sim_08_1_1_scattering_domain.m`.

```
% I Preparations
N      = 100;              % number of discretization points for curve
ht     = 2*pi/N;           % grid size for curve discretization
t      = 0:ht:2*pi-ht;     % discretization points for parametrization
kappa  = 0.01;             % wave number
eps    = .1;               % cut parameter for singular kernel
domain = 1;                % choose which domain to use

% Definition of domain and tangential vectors:
switch( domain )
case 1  % kite shaped domain

  y1   = cos(t) + 0.65 * cos(2*t) - 0.65;     y2   = 1.5 * sin(t);
  dy1  = -sin(t) - 0.65*2*sin(2*t);           dy2  = 1.5 * cos(t);
case 2  % circle
  y1   = cos(t) ;     y2   = sin(t);
  dy1  = -sin(t);     dy2  = cos(t);
end
```

The integral of a function ψ on ∂G is approximated by some quadrature rule for the parametrized integral. In general, the integration can be written as

Table 8.1. The exponential convergence of the trapezoidal rule for integrating $\psi(x) = \sin(x_1) \cdot \cos(x_2)$ over the boundary Γ shown in figure 8.1. Here, N denotes the number of discretization points and `Int` is given by (8.1.12), see `sim_08_1_7_convergence.m` (code repository).

N	Int
16	16.85153520034177
32	6.455990317606402
64	6.456748650385978
128	6.456750251394578
256	6.456750251384122
512	6.456750251384121
1024	6.456750251384121
2048	6.456750251384121

$$\int_{\partial G} \psi(y)\, \mathrm{d}s(y) \approx \sum_{j=0}^{N-1} \omega_j \psi(\Gamma(t_j)) \tag{8.1.12}$$

with weights $\omega_j \in \mathbb{R}$, $j = 0, \ldots, N-1$. We will employ the *trapezoidal rule* for which

$$\omega_j = \frac{2\pi}{N}|T(t_j)| = \frac{2\pi}{N}\sqrt{T_1(t_j)^2 + T_2(t_j)^2}, \quad j = 1, \ldots, N, \tag{8.1.13}$$

where $T(t)$ is the tangential vector defined as

$$T(t) = \frac{\partial \Gamma(t)}{\partial t} \in \mathbb{R}^2, \quad t \in [0, 2\pi). \tag{8.1.14}$$

For the kite-shaped domain (8.1.8) the tangential vector is given by

$$T_1(t) = -\sin(t) - 1.3\sin(2t), \quad T_2(t) = 1.5\cos(t). \tag{8.1.15}$$

The trapezoidal rule exhibits *exponential convergence* for analytic functions ψ defined on closed analytic surfaces, compare table 8.1. More generally, the convergence order is directly linked to the smoothness of the functions ψ and Γ.

For the evaluation of the single-layer potential $\tilde{S}\varphi$ and other potentials in the simplest case we can use the numerical integration scheme (8.1.12). Here we will describe some generic realization which will be a tool for many inversion algorithms in later chapters. It is a good exercise to implement these potential operators and gain experience with their behavior. The programs might also be useful for colleagues to explore the ideas of various inversion algorithms in a well-defined environment.

We choose some evaluation rectangle $\Omega = [a_1, b_1] \times [a_2, b_2] \subset \mathbb{R}^2$ as shown in figure 8.2 with boundaries at a_1, b_1 in one direction and a_2, b_2 in the other direction, where $a_1 < b_1$, $a_2 < b_2$ are real. A *regular grid* $\mathcal{G}$ on Ω with

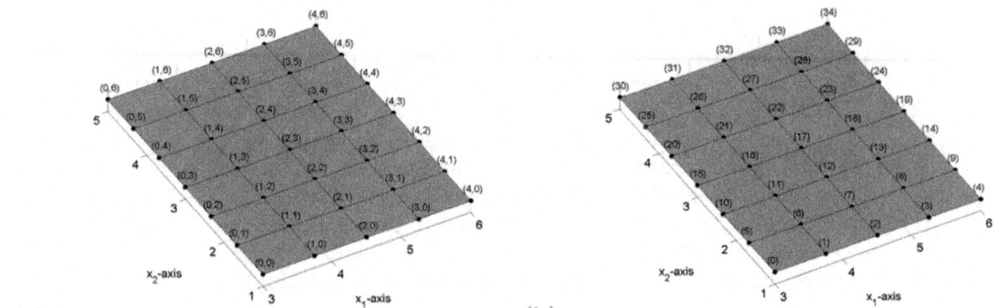

(a) (b)

Figure 8.2. A regular grid on the rectangle $\Omega = [3, 6] \times [1, 5]$ of size (5,7). In (a) the points are ordered into a matrix structure with coordinates (k, j) for $k = 0, \ldots, 4$ and $j = 0, \ldots, 6$. In (b) the points are sequentially labeled by $l = 0, \ldots, 34$. The mapping of two-dimensional numbering onto sequential numbering is described in equation (8.1.17).

M_1 grid points in the x_1-direction and M_2 grid points in the x_2-direction is defined by

$$
\begin{aligned}
x_{1,k} &= a_1 + \frac{b_1 - a_1}{M_1 - 1} \cdot (k - 1), \quad k = 1, \ldots, M_1, \\
x_{2,j} &= a_2 + \frac{b_2 - a_2}{M_2 - 1} \cdot (j - 1), \quad j = 1, \ldots, M_2.
\end{aligned}
\tag{8.1.16}
$$

We use two different ways to label the points p of the grid, either by the *two-dimensional indices* $(x_{1,k}, x_{2,j})$ for $k = 1, \ldots, M_1$ and $j = 1, \ldots, M_2$ or *sequentially* by p_l for $l = 1, \ldots, M$ where $M = M_1 \cdot M_2$ is the total number of evaluation points in the grid. Examples of both ways of labeling are shown in figure 8.2 where $M_1 = 5$ and $M_2 = 7$. Let the *Gauss brackets* $j := [l/M_1]$ denote the largest integer j smaller or equal to l/M_1. The *mod function* calculates the remainder when dividing a natural number l by M_1, i.e. it maps a natural number $l \in \mathbb{N}$ onto $k \in \{0, \ldots, M_1 - 1\}$ such that with some $j \in \mathbb{N}$ we have

$$
l = M_1 \cdot (j - 1) + k. \tag{8.1.17}
$$

With this we obtain the relation between the two-dimensional and the sequential numbering by

$$
p_l = (x_{1,k}, x_{2,j}) \quad \text{where} \quad k = l \bmod M_1, \quad j = [l/M_1] \tag{8.1.18}
$$

when l is given and the two-dimensional index (k, j) needs to be calculated via (8.1.17) if (k, j) is prescribed and the sequential index l needs to be evaluated.

Code 8.1.2. *Setting up the points (8.1.18) of the evaluation domain* Ω *by the file* `sim_08_1_2_evaluation_domain.m`. *Here, we also calculate the differences* $p_\ell - y_j$ *for* $\ell = 1, \ldots, M$ *and* $j = 1, \ldots, N$.

```
% Evaluation domain:
M1       = 50;        % number of points in x1 direction for evaluation
M2       = 51;        % number of points in x2 direction for evaluation
M        = M1*M2;     % total number of evaluation points
a1 = -3; b1 = 3; a2 = -4; b2 = 4;  % definition of domain Q
h1       = (b1-a1)/(M1-1);   % grid size for x1 direction
q1       = a1:h1:b1;         % grid points in x1 direction
h2       = (b2-a2)/(M2-1);   % grid size for x2 direction
q2       = a2:h2:b2;         % grid points in x2 direction

q1mat    = repmat(q1,M2,1);     % preparations for building up grid vector
q2mat    = repmat(q2.',1,M1);   % ~
pvec1    = reshape(q1mat,M,1); % x1 coordinates of grid points
pvec2    = reshape(q2mat,M,1); % x2 coordinates of grid points

% Matrix of the norm differences of the grid points to the curve points:
epsmat   = eps*ones(M,N);       % matrix for cutting singularity
rmat1    = repmat(pvec1,1,N)-repmat(y1,M,1);
rmat2    = repmat(pvec2,1,N)-repmat(y2,M,1);
rmat     = max(sqrt(rmat1.^2 + rmat2.^2),epsmat);
drmat    = repmat(sqrt(dy1.^2 + dy2.^2),M,1);
dytmat1  = repmat(dy1,M,1);
dytmat2  = repmat(dy2,M,1);
```

Next, we define a simple toolbox for the numerical evaluation of the potentials over a curve in $\mathbb{R}^2$ on the grid $\mathcal{G}$. For an integral of the form

$$(Q\varphi)(x) = \int_\Gamma k(x, y)\varphi(y)\, \mathrm{d}s(y), \quad x \in \mathbb{R}^2 \tag{8.1.19}$$

with sufficiently smooth kernel k we employ a quadrature formula (8.1.12) with weights ω_j, $j = 1, \dots, N$ given by (8.1.13) for each p_l, $l = 0, \dots, M - 1$ in $\mathcal{G}$. This leads to

$$(Q_N\varphi)(p_l) = \sum_{j=0,\dots,N-1} k(p_l, y_j)\varphi(y_j)\omega_j, \quad l = 0, \dots, N - 1 \tag{8.1.20}$$

where $y_j := \Gamma(t_j)$, $j = 0, \dots, N - 1$. In *matrix notation* we obtain

$$(Q_N\varphi)(p_l) = \mathbf{Q}\boldsymbol{\varphi} \tag{8.1.21}$$

where the $M \times N$ matrix $\mathbf{Q}$ and the column vector $\boldsymbol{\varphi}$ of length N are given by

$$\mathbf{Q} := \Big(\big(k(p_l, y_j)\omega_j\big)\Big)_{l=0,\dots,M-1,\, j=0,\dots,N-1}, \quad \boldsymbol{\varphi} = \big(\varphi(y_j)\big)_{j=0,\dots,N-1}. \tag{8.1.22}$$

If the kernel $k(x, y)$ has some singularity for $x = y$ we need to be careful in the case where $x \in \Gamma$ or $\mathrm{d}(x, \Gamma) \leqslant \rho \ll 1$. If the singularity is integrable, a simple way to treat this case is to replace the term $k(x, y)$ by

$$k_\rho(x, y) := k(x, y) \cdot \chi_\rho(|x - y|) + c \cdot \big(1 - \chi_\rho(|x - y|)\big) \tag{8.1.23}$$

with some appropriate constant c and the well-known *sigmoidal function* χ defined by

$$\chi_{\sigma,\rho}(t) := \frac{1}{1 - e^{\sigma(t-\rho)}}, \quad t \geqslant 0 \tag{8.1.24}$$

with a constant $L \in \mathbb{N}$. This leads to a scheme which is easy to implement and has linear convergence order. Higher order schemes can be achieved by spectral methods as for example worked out in [1]. As a limit case we use the simple formula

$$r_\epsilon(x, y) := \max(|x - y|, \epsilon), \quad x, y \in \partial G, \tag{8.1.25}$$

to cut the singularity, for example in code 8.1.2, line 18.

We now apply (8.1.21) to the single- and double-layer potential operators (8.1.2) and (8.1.4). We define

$$\tilde{\mathbf{S}} := \Big(\big(\Phi_\rho(p_l, y_j)\omega_j\big)\Big)_{l=0,\dots, M, j=0,\dots, N}, \tag{8.1.26}$$

$$\tilde{\mathbf{K}} := \left(\left(\frac{\partial \Phi_\rho(p_l, y_j)}{\partial \nu(y_j)}\omega_j\right)\right)_{l=0,\dots,M, j=0,\dots,N}. \tag{8.1.27}$$

Then, the numerical approximation to the potential operators is calculated by

$$\mathbf{w}_1 := \tilde{\mathbf{S}} \circ \boldsymbol{\varphi}, \tag{8.1.28}$$

$$\mathbf{w}_2 := \tilde{\mathbf{K}} \circ \boldsymbol{\varphi}. \tag{8.1.29}$$

The potentials $\mathbf{w}_1$ or $\mathbf{w}_2$ are given in the form of a column vector of size M. Via (8.1.17) their values can be reordered into $M_1 \times M_2$-matrices $\mathbf{W}_1$ or $\mathbf{W}_2$, respectively, which is carried out by the `OCTAVE` function `reshape`. A generic realization of the potential evaluation is shown in the codes 8.1.2 and 8.1.3.

The following generic code can be used to realize a single-layer potential as defined by (8.1.2) and (8.1.28) and the double-layer potential (8.1.4) and (8.1.29). It corresponds to singularity dumping as shown in figure 8.3 with $\sigma = \infty$. We first set up the points and coordinate differences in codes 8.1.1 and 8.1.2. The calculation of the operators (8.1.28) and (8.1.29) is now rather short.

Code 8.1.3. *The calculation of the single- and double-layer potentials defined on ∂D evaluated on the domain Q. The calculation is carried out by the file* `sim_08_1_3_tS_tK_ws.m`.

```
tau     = 1; % set to 1 to choose double-layer potential
eta     = 0; % set to 1 to choose single-layer potential

% II Potential Operators tS, tK
tS  = i/4*besselh(0,1,kappa*rmat).*drmat*ht;
tK  = i*kappa/4*(dytmat2.*rmat1-dytmat1.*rmat2).* ...
        besselh(1,1,kappa*rmat)./rmat*ht;
varphi = ones(N,1);                    % set density to 1
%varphi = 3*cos(t');                    % set density to cos(t)
ws  = (tau*tK-i*eta*tS)*varphi; % calculation of potential on Q
```

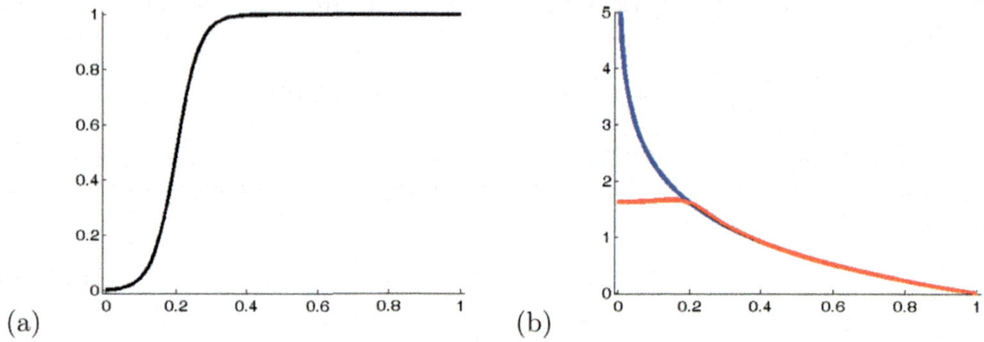

Figure 8.3. (a) A graph of the function $\chi_{\sigma,\rho}(t)$ on $t \in [0, 1]$ for $\rho = 0.2$ and $\sigma = 30$. (b) demonstrates the damped kernel $k_\rho(x, y)$ (red) defined in (8.1.23) with $c = -\log(\rho)$ compared to an original kernel $k(x, y) = -\log(|x - y|)$ (blue) with logarithmic singularity. The singularity is damped out in a neighborhood of radius ρ around $x = 0$.

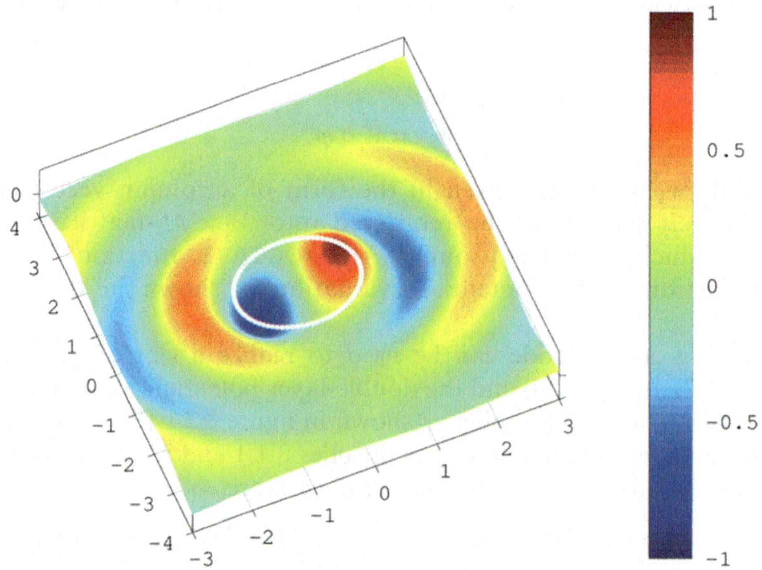

Figure 8.4. Demonstration of the real part of a single-layer potential over the unit circle with density given by $3\cos(t)$, $t \in [0, 2\pi]$. The figure can be generated by code 8.1.4 based on codes 8.1.1–8.1.3.

Finally, we include the graphical representation of the potential as shown in figure 8.4 and save its image in the png format. As a control script we have added `sim_08_1_5_control.m` to the code repository.

Code 8.1.4. *The display is generated as follows by* `sim_08_1_4_ws_surface.m`.

```
% III Graphical representation
fo   = figure; colormap default; % open figure
wmat = reshape(ws,M2,M1); % prepare function in the format M2 x M1 points
sobj = surf(q1,q2,abs(wmat));  % plot surface of potential
view(-23,82); % work on image presentation
axis tight; %axis([a1 b1 a2 b2 -1.1 1.1]);
shading interp; aobj = get(sobj,'Parent'); set(aobj,'FontSize',14);
hold on; plot3(y1,y2,1*ones(size(y1)),'w.','MarkerSize',10);
caxis([-1 1]); colorbar('FontSize',14); set(aobj,'ztick',[0,1]);

savefile(fo,'sim_08_1_4_ws_surface')
```

The properties of the potential operators $\tilde{S}$ and $\tilde{K}$ are crucial ingredients for the simulation of waves and identification of objects. Figure 8.5 shows that the double-layer potential has a jump across the boundary Γ. This jump plays a crucial role for the solution of scattering problems. The single-layer potential is continuous in the whole space. We summarize the theoretical results in the following theorem, for the proofs we refer the reader to [2].

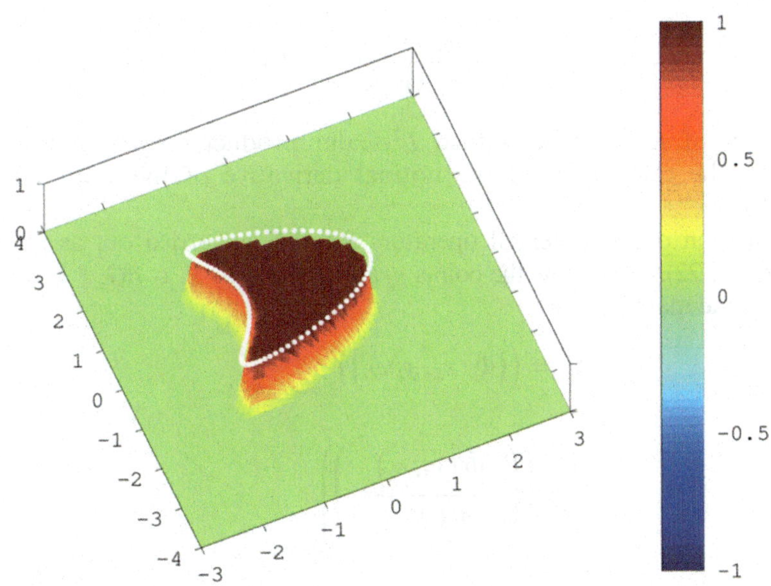

Figure 8.5. Demonstration of the real part of a double-layer potential over the domain (8.1.8) with a constant density given by 1, $t \in [0, 2\pi]$, where we used $\kappa = 0.01$ here. We expect this to be close to zero outside the domain and jump to a constant value of 1 over the boundary, which is known as a *jump relation*. This behavior is clearly observed in the picture. The visualization has been carried out with $N = 100$ points on the boundary Γ (8.1.7) by codes 8.1.1–8.1.4.

Theorem 8.1.5 (Jump relations). *The single-layer potential (8.1.2) with continuous density φ is a continuous function in $\mathbb{R}^m$, $m = 2, 3$. The double-layer potential (8.1.4) is analytic in $\mathbb{R}^m \backslash \bar{D}$ and D. It can be continuously extended from $\mathbb{R}^m \backslash \bar{G}$ into $\mathbb{R}^m \backslash G$ and G into $\bar{G}$ with limiting values*

$$\left(\tilde{K}\varphi\right)_{\pm}(x) = \int_{\partial G} \frac{\partial \Phi(x, y)}{\partial \nu(y)} \varphi(y) \, \mathrm{d}s(y) \pm \frac{1}{2}\varphi(x), \quad x \in \partial G. \tag{8.1.30}$$

On ∂G the double-layer potential has a jump of size $\varphi(x)$.

Layer potentials which are evaluated on the boundary ∂G are usually called *boundary integral operators.* We use the notation

$$(S\varphi)(x) := 2 \int_{\partial G} \Phi(x, y)\varphi(y) \, \mathrm{d}s(y), \quad x \in \partial G \tag{8.1.31}$$

for the *single-layer operator* and

$$(K\varphi)(x) := 2 \int_{\partial G} \frac{\partial \Phi(x, y)}{\partial \nu(y)} \varphi(y) \, \mathrm{d}s(y), \quad x \in \partial G \tag{8.1.32}$$

for a *double-layer potential* where $\varphi \in L^2(\partial G)$ and ν denotes the exterior unit normal vector to the domain G. Here we chose to incorporate the factor 2 in the definition of these operators which naturally arises when one wants to avoid fractions in (8.1.30). Further, the operator

$$(K'\varphi)(x) := \int_{\partial D} \frac{\partial \Phi(x, y)}{\partial \nu(x)} \varphi(y) \, \mathrm{d}s(y), \quad x \in \partial D, \tag{8.1.33}$$

is the adjoint of K in a real-valued L^2-scalar product on ∂D. It is needed for example for the calculation of the normal derivative of the scattered field, see section 9.4.

The evaluation of the potential operators S and K is carried out as introduced in (8.1.20) and (8.1.22), where now the points p_l are replaced by $x_\xi \in \partial G$, $\xi = 0, \ldots, N-1$. This leads to the matrices

$$\mathbf{S} := \Big(\big(\Phi_\rho(x_\xi, y_j)\omega_j \big) \Big)_{\xi, j=0,\ldots,N-1}, \tag{8.1.34}$$

$$\mathbf{K} := \left(\left(\frac{\partial \Phi_\rho(x_\xi, y_j)}{\partial \nu(y_j)} \omega_j \right) \right)_{\xi, j=0,\ldots,N-1}. \tag{8.1.35}$$

For solving boundary integral equations and various inversion techniques later we provide a generic code 8.1.6 for calculation of the operators $\mathbf{S}$ and $\mathbf{K}$. The use of these operators for simulating the propagation of waves will be described in the next section.

Code 8.1.6. *The realization of the single-layer potential operator and the double-layer potential operator as defined by (8.1.31) and (8.1.32) can be carried out by the following code* `sim_08_1_6_S_K.m`. *Here, we describe the implementation for a kite-shaped domain given by (8.1.8).*

```
% I Preparations
N     = 100;             % number of discretization points for curve
ht    = 2*pi/N;          % grid size for curve discretization
t     = 0:ht:2*pi-ht;    % discretization points for curve parametrization
kappa = 2;               % wave number
beta  = 4;               % direction of incidence

% definition of domain and tangential vectors
y1     = cos(t) + 0.65 * cos(2*t) - 0.65;
y2     = 1.5 * sin(t) + 0.3*cos(2*t);
dy1    = -sin(t) - 0.65*2*sin(2*t);
dy2    = 1.5 * cos(t) - 0.3*2*sin(2*t);

ymat1  = repmat(y1,N,1);   % matrix of domain points component 1
ymat2  = repmat(y2,N,1);   % matrix of domain points component 2
dymat1 = repmat(dy1,N,1);  % matrix of domain derivative comp.1
dymat2 = repmat(dy2,N,1);  % matrix of domain derivative comp.2
eps = ht/3.6;              % set cut parameter for singularity
epsmat=eps*ones(N,N);      % matrix of cut parameter

rmat1=ymat1.'-ymat1;  % matrix of point differences compotent 1
rmat2=ymat2.'-ymat2;  % matrix of point differences component 2
rmat=max(sqrt(rmat1.^2 + rmat2.^2),epsmat); % matrix of ||x-y||
drmat=sqrt(dymat1.^2 + dymat2.^2);

% II Potential Operators S and K
S = i/2*besselh(0,1,kappa*rmat).*drmat*ht;
K = i*kappa/2*(dymat2.*rmat1-dymat1.*rmat2).* ...
    besselh(1,1,kappa*rmat)./rmat*ht;
```

8.2 Simulation of wave scattering

The simulation of waves being reflected, diffracted and scattered from objects can be carried out in a very efficient way via integral equations. Here we will introduce this approach for some basic model problems as a basis for the techniques of the subsequent chapters.

Definition 8.2.1 (Scattering problem). *We consider an incident time-harmonic wave* u^i *scattered by a scatterer* D. *Then there is a scattered field* u^s *which satisfies the* Helmholtz equation

$$\Delta u^s + \kappa^2 u^s = 0 \tag{8.2.1}$$

in the exterior $\mathbb{R}^d \backslash \bar{D}$ *of the scatterer* D*. Depending on the physical properties of the scatterer the* total field

$$u = u^{\mathrm{i}} + u^{\mathrm{s}} \tag{8.2.2}$$

satisfies either a Dirichlet boundary condition

$$u|_{\partial D} = 0, \tag{8.2.3}$$

a Neumann boundary condition

$$\left.\frac{\partial u}{\partial \nu}\right|_{\partial D} = 0 \tag{8.2.4}$$

or some impedance boundary condition

$$\left.\frac{\partial u}{\partial \nu}\right|_{\partial D} + \lambda u|_{\partial D} = 0, \tag{8.2.5}$$

where the possibly complex function $\lambda \in C(\partial D)$ *is known as* impedance. *Further, the scattered field* u^{s} *is required to satisfy the* Sommerfeld radiation condition

$$\left(\frac{\partial}{\partial r} - \mathrm{i}\kappa\right) u^{\mathrm{s}} \to 0, \quad r = |x| \to \infty \tag{8.2.6}$$

uniformly for all directions.

For solving the scattering problem introduced in definition 8.2.1 with Dirichlet boundary condition (8.2.3) we use the potential ansatz

$$u^{\mathrm{s}}(x) = \int_{\partial D} \left\{ \frac{\partial \Phi(x, y)}{\partial \nu(y)} - \mathrm{i}\Phi(x, y) \right\} \varphi(y) \, \mathrm{d}s(y), \quad x \in \mathbb{R}^d \backslash \partial D, \tag{8.2.7}$$

also known as the *Brakhage–Werner* approach. For $x \in \mathbb{R}^m \backslash \partial D$ the differentiation with respect to x below the integral is possible and since $\Phi(x, y)$ solves the Helmholtz equation (8.2.1) on $\mathbb{R}^m \backslash \{y\}$ for each y fixed, also the integral (8.2.7) satisfies the Helmholtz equation in $\mathbb{R}^m \backslash \partial D$. Second, since the fundamental solution $\Phi(x, y)$ and its normal derivative satisfy the Sommerfeld radiation condition (8.2.6), the same is true for the integral (8.2.7) due to the compactness of the integration curve ∂D. Finally, to achieve $u^{\mathrm{s}} = -u^{\mathrm{i}}$ on the boundary we employ the jump relations (8.1.30). This leads to the *Brakhage–Werner integral equation*

$$(I + K - \mathrm{i}S)\varphi = -2u^{\mathrm{i}} \quad \text{on } \partial D \tag{8.2.8}$$

for the unknown density $\varphi \in C(\partial D)$. The unique solvability of (8.2.8) can be shown as follows. Since we have the compactness of $K - \mathrm{i}S$ on $C(\partial D)$ by the mapping properties of S and K, theorem 2.3.25 tells us that we only need to show the uniqueness of solution $\varphi \in C(\partial D)$ of (8.2.8). To show this we consider (8.2.8) with the right-hand side equal 0 which implies that u^{s} given by (8.2.7) is equal to

0 on ∂D. Hence by the uniqueness of the Dirichlet boundary value problem in $\mathbb{R}^3 \backslash \bar{D}$, $u^s = 0$ in $\mathbb{R}^3 \backslash D$. Hence by the jump formulas for single- and double-layer potentials, we have

$$-u^s = \varphi, \quad -\partial_\nu u^s = \mathrm{i}\varphi \text{ on } \partial D. \tag{8.2.9}$$

Here the traces of u^s, $\partial_\nu u^s$ are taken from inside D and ν denotes the unit normal normal vector directed outside D. Hence, using Green's formula (17.5) we have

$$\mathrm{i} \int_{\partial D} |\varphi|^2 = \int_{\partial D} u^s \partial_\nu u^s = \int_D \left(|\nabla u^s|^2 - \kappa^2 \, |u^s|^2 \right).$$

Taking the imaginary part for κ real valued this yields $\varphi = 0$ *on* ∂D.

A generic code with a *collocation method* for solving the Brakhage–Werner integral equation in two dimensions is given in code 8.2.2.

Code 8.2.2. *The solution to the Brakhage–Werner integral equation (8.2.8) with* $u^i(x) := e^{i\kappa x \cdot d}$ *for some* $d \in \mathbb{S}$ *can be calculated by the following script* `sim_08_2_2_BW_varphi.m`, *when code 8.1.6 is called first.*

```
tau     = 1; % set to 1 to choose double-layer potential
eta     = 1; % set to 1 to choose single-layer potential

% III Calculate right-hand side
d1  = -cos(beta);  % direction of incidence comp.1
d2  = -sin(beta);   % direction of incidence comp.2
rhs = exp(1i*kappa*(y1*d1+y2*d2)).';

% IIIb Factor or Test for right-hand side
fac  = exp(1i*pi/4)/sqrt(8*pi*kappa);
% rhs  = 1/fac*(i/4)*besselh(0,1,kappa*sqrt(y1.^2+y2.^2)).';

 % IV Solve integral equation
 varphi =(tau*(eye(N,N)+K)-1i*eta*S)\(-2*rhs);
 BWinv = inv(tau*(eye(N,N)+ K)-1i*eta*S); % Brackhage Werner inverse
```

With the density φ the scattered field u^s is now calculated by (8.2.7) and the total field is given by (8.2.2). The corresponding code is introduced in code 8.2.3. A graphical representation, as shown in figure 8.6 where we used an incident field with $d = (-1, 0)$, is generated by code 8.2.4.

Code 8.2.3. *The evaluation of the integral (8.2.7) on some cuboid can be carried out by the following code. This code assumes that code 8.1.6 and code 8.2.2 are run first.*

```
% Evaluation domain:
M1         = 100;          % number of points in x1 direction for evaluation
M2         = 101;          % number of points in x2 direction for evaluation
M          = M1*M2;      % total number of evaluation points
a1         = -8;         % left border of cuboid
b1         = 8;          % right border of cuboid
```

```
a2      = -7;           % bottom border of cuboid
b2      = 7;            % top border of cuboid
h1      = (b1-a1)/(M1-1);   % grid size for x1 direction
q1      = a1:h1:b1;         % grid points in x1 direction
h2      = (b2-a2)/(M2-1);   % grid size for x2 direction
q2      = a2:h2:b2;         % grid points in x2 direction

q1mat   = repmat(q1,M2,1);     % preparations for building up grid vector
q2mat   = repmat(q2.',1,M1);   %
pvec1   = reshape(q1mat,M,1);  % definition of x1 coordinates of grid points
pvec2   = reshape(q2mat,M,1);  % definition of x2 coordinates of grid points

% Matrix of the norm differences of the grid points to the curve points:
epsmat  = eps*ones(M,N);         % matrix for cutting singularity
rmat1   = repmat(pvec1,1,N)-repmat(y1,M,1);
rmat2   = repmat(pvec2,1,N)-repmat(y2,M,1);
rmat    = max(sqrt(rmat1.^2 + rmat2.^2),epsmat);
drmat   = repmat(sqrt(dy1.^2 + dy2.^2),M,1);
dytmat1 = repmat(dy1,M,1);
dytmat2 = repmat(dy2,M,1);

% V Potential Operators S, K
tS  = i/4*besselh(0,1,kappa*rmat).*drmat*ht;
tK  = i*kappa/4*(dytmat2.*rmat1-dytmat1.*rmat2).* ...
      besselh(1,1,kappa*rmat)./rmat*ht;
ws  = (tau*tK-i*eta*tS)*varphi; % calculation of potential on Q
wi  = exp(i*kappa*(d1*pvec1+d2*pvec2)); % incident field on Q
w   = wi + ws; % total field on Q
```

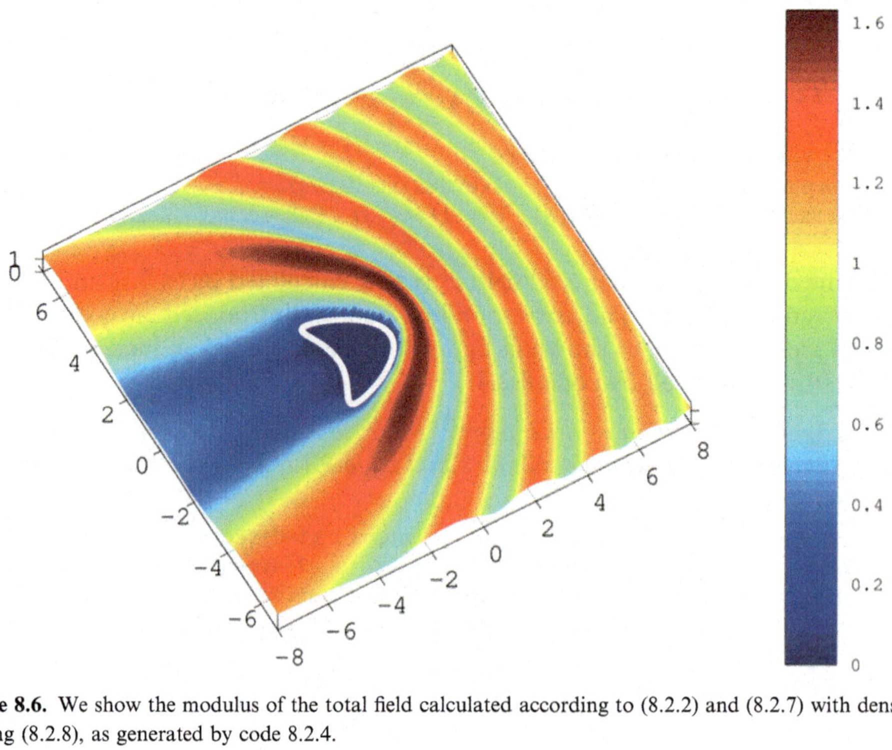

Figure 8.6. We show the modulus of the total field calculated according to (8.2.2) and (8.2.7) with density φ solving (8.2.8), as generated by code 8.2.4.

For the graphical representation of the potential we want to set the unphysical field inside the scatterer to zero. We need a function which is zero inside D and 1 outside. This is generated using the Cauchy mask (6.6.3). The image which the following script generates is shown in figure 8.6.

Code 8.2.4. *The script* `sim_08_2_4_w_Graphics.m` *generates a graphical representation of the total field for scattering by some obstacle as shown in figure 8.6 using a masking operation to damp the layer potentials to zero inside the scatterer. This assumes that the codes 8.1.6, 8.2.2 and 8.2.3 are executed first, as in* `sim_08_2_5_Control.m` *(code repository).*

```
% III Masking via the Cauchy integral
eps2 = 0.1;                          % cut parameter for mask
maskvec  = 1+1/2/pi*(rmat1./max(rmat.^2,eps2).*repmat(dy2,M,1)*ht - ...
    rmat2./max(rmat.^2,eps2).*repmat(dy1,M,1)*ht)*ones(N,1);
masktmp = reshape(maskvec,M2,M1);    % mask in matrix format
mask = masktmp>.8;                   % cleaning mask on the boundary

% IV Graphical representation
fo   = figure; colormap default;            % open figure
wmat = reshape(abs(w),M2,M1);               % prepare for plot
so   = surf(q1,q2,real(wmat.*mask));        % surface plot
%so   = contour(q1,q2,real(wmat.*mask),'LineWidth',2);  % contour plot
view(-30,87); shading interp;               % graphics controls
hold on; po = plot3(y1,y2,0*ones(size(y1)),'k.','MarkerSize',7);
ao = get(po,'Parent');                      % get parent handle
set(ao,'FontSize',14); colorbar;            % colorbar and font size
axis tight;  set(ao,'ztick',[0,1]);         % work on image presentation

savefile(fo,'sim_08_2_4_w_Graphics');       % save image as png
```

8.3 The far field and the far field operator

A solution to the Helmholtz equation (8.2.1), which satisfies the radiation condition (8.2.6), behaves like an outgoing spherical wave

$$u^s(x) = \Phi(x, x_0)\{u^\infty(\hat{x}) + O(|x|^{-\tau})\}, \quad \hat{x} = \frac{x}{|x|}, \quad r = |x| \to \infty, \tag{8.3.1}$$

for $x_0 = 0 \in \mathbb{R}^m$ and some $\tau > 0$ with a factor $u^\infty(\hat{x})$ which depends on the direction $\hat{x}$ only, also see (1.2.7). Henceforth we use the word *outgoing* asking for the Sommerfeld radiation condition to be satisfied. The function u^∞ on $\mathbb{S}$ is called the *far field pattern* of u^s. As shown in [3, 4] the far field pattern of the single-layer potential is given by $u^\infty = S^\infty\varphi$ with

$$(S^\infty\varphi)(\theta) := \gamma_m \int_\Gamma e^{-i\kappa\theta\cdot y}\varphi(y)\, ds(y), \quad \theta \in \mathbb{S} \tag{8.3.2}$$

where for dimension $m = 2$, 3 we use the factor

$$\gamma_m := \begin{cases} \dfrac{e^{i\pi/4}}{\sqrt{8\kappa\pi}}, & m = 2 \\ \dfrac{1}{4\pi}, & m = 3. \end{cases} \tag{8.3.3}$$

The corresponding expression for the double-layer potential is given by

$$(K^\infty\varphi)(\theta) = \gamma_d \int_\Gamma (-i\kappa\nu(y)\cdot\theta)\, e^{-i\kappa\theta\cdot y}\varphi(y)\, ds(y), \quad \theta \in \mathbb{S}. \tag{8.3.4}$$

The numerical evaluation of the far field (8.3.2) of the single-layer potential and (8.3.4) of the double-layer potential is carried out as in (8.1.20). We first choose a discretization of the evaluation set $\mathbb{S}$. Here, we work with L discretization points

$$\theta_k := \begin{pmatrix} \cos(\tau_k) \\ \sin(\tau_k) \end{pmatrix}, \quad k = 1, \ldots, L \tag{8.3.5}$$

where

$$\tau_k := (k-1)\cdot\frac{A}{L} - \frac{A}{2}, \quad k = 1, \ldots, L \tag{8.3.6}$$

with $A = 2\pi$ in the *full-aperture* or $A \in (0, 2\pi)$ in th *limited-aperture* case. Then, we define the discretized operator

$$\mathbf{S}^\infty := \gamma_m\Big(\big(e^{-i\kappa\theta_k\cdot y_j}\omega_j\big)\Big)_{k=1,\ldots,L,\, j=1,\ldots,N} \tag{8.3.7}$$

with y_j given by (8.1.11). The discretized version of the double-layer operator (8.3.4) is given by

$$\mathbf{K}^\infty := \gamma_m\Big(\Big(\big(-i\kappa\theta_k\cdot y_j\big)e^{-i\kappa\theta_k\cdot y_j}\omega_j\Big)\Big)_{k=1,\ldots,L,\, j=1,\ldots,N}. \tag{8.3.8}$$

We are now prepared to calculate the far field pattern of the Brakhage–Werner potential (8.2.7) combining the discretized operators from code 8.2.2, (8.3.7) and (8.3.8) into

$$\mathbf{u}^\infty = (\mathbf{K}^\infty - i\eta\mathbf{S}^\infty)(\mathbf{I} + \mathbf{K} - i\eta\mathbf{S})^{-1}(-2\mathbf{u}^i), \tag{8.3.9}$$

where $\mathbf{u}^i$ denotes the vector of the incident field u^i evaluated at x_ξ for $\xi = 1, \ldots, N$ and $\mathbf{u}^\infty$ is the far field pattern evaluated at θ_k for $k = 1, \ldots, L$. The code for implementing the evaluation of the far field pattern is given in code 8.3.1, the corresponding visualization for the setting is created by code 8.2.2 and is shown in figure 8.7.

Code 8.3.1. *Here we realize the expression (8.3.9) on an aperture of size A as in (8.3.6) as a script* `sim_08_3_1_ffS_ffK_ff.m`*. This code assumes that code 8.1.6 and code 8.2.2 are run first.*

```
% I Preparations for far field evaluations
ffN    = 200; % number of discretization points for curve
hff    = 2*pi/ffN; % grid size for curve discretization
tff    = (0:hff:2*pi-hff)-pi; % discretization points for curve parametrization
% definition of domain and tangential vectors
yff1      = cos(tff);
yff2      = sin(tff);
yffmat1  = repmat(yff1.',1,N);
yffmat2  = repmat(yff2.',1,N);
yfmat1   = repmat(y1,ffN,1);
yfmat2   = repmat(y2,ffN,1);
dyfmat1  = repmat(dy1,ffN,1);
dyfmat2  = repmat(dy2,ffN,1);
drffmat=sqrt(dyfmat1.^2 + dyfmat2.^2);

% II Potential Operators ffS, ffK
ffS  = fac*exp(-i*kappa*(yffmat1.*yfmat1+yffmat2.*yfmat2)).*drffmat*ht;
ffK  = fac*(-i)*kappa*(yffmat1.*dyfmat2-yffmat2.*dyfmat1) ...
       .*exp(-i*kappa*(yffmat1.*yfmat1+yffmat2.*yfmat2))*ht;
ffBW = (tau*ffK-i*eta*ffS); % density to farfield operator
ff   = ffBW*varphi;          % calculation of far field pattern
```

The following visualization code 8.3.2 assumes that codes 8.1.6, 8.2.2 and 8.3.1 are run first. You might also use the control script `sim_08_3_0_Control.m` from the code repository.

Code 8.3.2. *A graphical representation of the far field pattern generated by the script* `sim_08_3_2_ff_Graphics.m`, *see figure 8.7.*

```
% III Graphics for far field pattern
fo    = figure;                              % open figure
po    = plot(tff,real(ff),tff,imag(ff));    % plot Re/Im(ff)
hold on; po2 = plot(tff,abs(ff),'r-.');     % plot abs(ff)
set(po,'LineWidth',6); set(po2,'LineWidth',8); grid on;
mymax=max(abs(ff));                         % maximal values
axis([-pi pi -mymax*1.1 mymax*1.1]);        % control axis
ao = get(po2,'Parent');                     % get parent handle
set(ao,'FontSize',14);                      % font size

savefile(fo,'sim_08_3_2_ff_Graphics');      % save image
```

Superposition of incident fields. For several inversion algorithms superpositions of plane waves play a central role. For a fixed wave number κ the superposition of all plane waves

$$u^{i}(x, d) := e^{i\kappa x\cdot\theta}, \quad x \in \mathbb{R}^m \tag{8.3.10}$$

with directions $\theta \in \mathbb{S}$ distributed over the unit circle or unit ball $\mathbb{S}$ is known as the *Herglotz wave function*

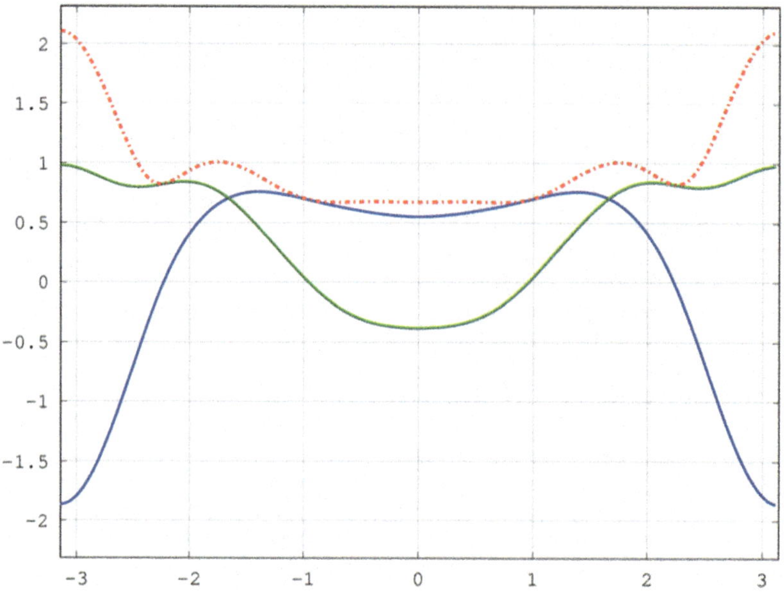

Figure 8.7. The far field pattern of the field u^s defined in (8.2.7) with density φ solving (8.2.8). The real and imaginary part of u^∞ is shown in blue and green, the modulus in red.

$$(v_g)(x) := \int_{\mathbb{S}} e^{i\kappa x\cdot\theta} g(\theta)\, ds(\theta), \quad x \in \mathbb{R}^m \tag{8.3.11}$$

for $g \in L^2(\mathbb{S})$. An alternative notation is $\tilde{H}g := v_g$. The operator H is defined by

$$(Hg)(x) := \int_{\mathbb{S}} e^{i\kappa x\cdot\theta} g(\theta)\, ds(\theta), \quad x \in \partial G, \tag{8.3.12}$$

for $g \in L^2(\mathbb{S})$. Often, the Herglotz wave function is considered as an operator from $L^2(\mathbb{S})$ into $L^2(\partial G)$. The adjoint operator H^* of the Herglotz operator H is given by

$$(H^*\psi)(\theta) := \int_{\partial G} e^{-i\kappa x\cdot\theta} \psi(x)\, ds(x), \quad \theta \in \mathbb{S}. \tag{8.3.13}$$

The operator H^* is also important for direct scattering, up to the factor γ_d defined in (8.3.3) it coincides with the operator S^∞ defined in (8.3.2).

By an application of (8.1.12) and (8.1.22) the integral operator $\tilde{H}$ defined in (8.3.11) can be discretized via the matrix

$$\tilde{\mathbf{H}} := \left(\left(e^{i\kappa p_l\cdot\theta_k} s_k\right)\right)_{l=0,\dots,M-1,k=0,\dots,L-1} \tag{8.3.14}$$

with quadrature weight $s_k = 2\pi/L$ for evaluation on the grid $\mathcal{G}$. Next, consider the Herglotz wave operator H given by (8.3.12), which is an operator from $L^2(\mathbb{S})$ into $L^2(\partial G)$. Its discrete version is

$$\mathbf{H} := \left(\left(e^{i\kappa x_j\cdot\theta_k} s_k\right)\right)_{j=0,\dots,N-1,k=0,\dots,L-1} \tag{8.3.15}$$

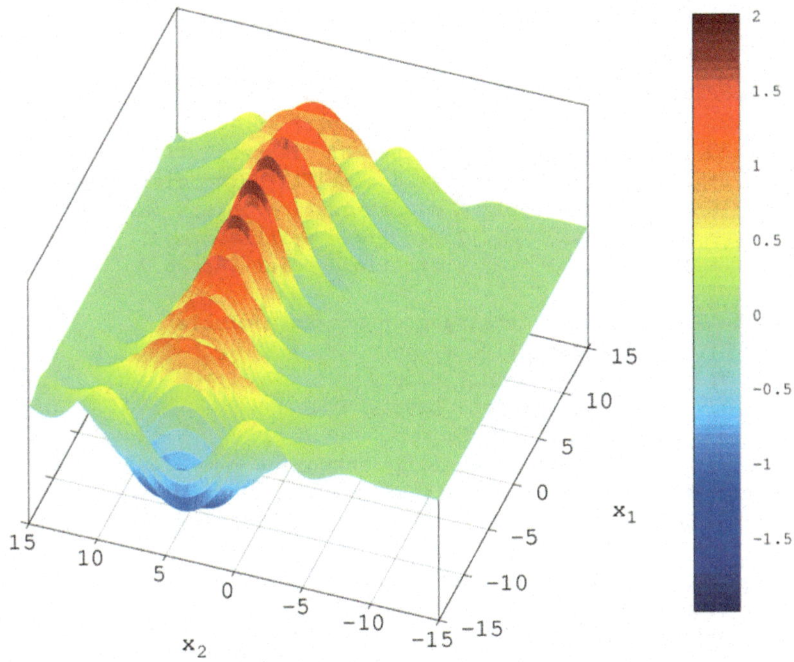

Figure 8.8. We show the *Gaussian beam* generated by a Herglotz wave function (8.3.11) via (8.3.15) with density g defined by (8.3.17). Here, the parameters were $\sigma = 7$and$z = (0, 4)$, and the figure is generated by code 8.3.3 in combination with `sim_08_3_3_c_Graphics.m` from the code repository, with a script very similar to code 8.2.4.

with the points $x_j \in \partial G$ defined in (8.1.11) and $\theta_k \in \mathbb{S}$. The adjoint operator H^* is discretized by

$$\mathbf{H}^* := \left(\left(e^{-i\kappa\theta_k \cdot x_j} \omega_j \right) \right)_{k=0,\ldots,L-1,\, j=0,\ldots,N-1} \tag{8.3.16}$$

with the quadrature weights ω_j defined by (8.1.13). Note that $\mathbf{H}^*$ does *not* coincide with the adjoint matrix $\mathbf{H}' = \bar{\mathbf{H}}^T$ due to the surface measure $|T|$ which is included in ω_j.

Herglotz waves functions can be used to generate particular *Gaussian beams* focusing at a point $z = (z_1, z_2)$ in space as shown in figure 8.8. To this end we employ the density

$$g(\theta) := e^{i\kappa\theta \cdot z} \cdot e^{-\sigma|\theta - \theta_0|^2}, \quad \theta \in \mathbb{S} \tag{8.3.17}$$

with some constant $\sigma > 0$. Here, σ controls the direction of the wave around the direction given by θ_0 and z controls the location of the focus point.

A generic script for calculating the operator $\tilde{\mathbf{H}}$ given by (8.3.14) and the Herglotz wave function with density g defined in (8.3.17) is given in code 8.3.3.

Code 8.3.3. *Here we realize the expression (8.3.11) via (8.3.15) with density g defined by (8.3.17) as a script* `sim_08_3_3_b_herglotz_function.m`*. To set up* *the evaluation grid either call code 8.2.3 or* `sim_08_3_3_a_evaluation_grid.m` *from the code repository.*

```
% I Preparations for far field evaluations
kappa = 2;                         % wave number
ffN   = 80;                        % number of discretization points for curve
hff   = 2*pi/ffN;                  % grid size for curve discretization
tff   = (0:hff:2*pi-hff)-pi;  % discretization points far field
yff1      = cos(tff);              % far field points, comp.1
yff2      = sin(tff);              % far field points, comp.2

% Set the Gaussian beam density g on the unit circle
tff0  = 0;          % around tff0 value
z1    = 0;          % focal point, comp.1
z2    = 4;          % focal point, comp.2
Ampl  = 3;          % amplitude factor
sigma = 7;          % exponential decay factor
g0 = Ampl*(exp(-i*kappa*(yff1*z1+yff2*z2))...
     .*exp(-sigma*(min(abs(tff-tff0),2*pi-abs(tff-tff0))).^2)).';

% II Setup and evaluate Herglotz wave function
hmat1 = repmat(pvec1,1,ffN).*repmat(yff1,M,1);
hmat2 = repmat(pvec2,1,ffN).*repmat(yff2,M,1);
tH    = exp(i*kappa*(hmat1 + hmat2))*hff; % Herglotz wave operator
vg    = tH*g0; % Herglotz wave function with density g0
```

The far field operator. If the far field pattern $u^{\infty}(\cdot, \theta)$ is measured on the unit ball or unit sphere for all directions of incidence $\theta \in \mathbb{S}$, we can define the operator

$$(Fg)(\hat{x}) := \int_{\mathbb{S}} u^{\infty}(\hat{x}, \theta) g(\theta) \, \mathrm{d}s(\theta), \quad \hat{x} \in \mathbb{S} \tag{8.3.18}$$

for $g \in L^2(\mathbb{S})$. It is known as the *far field operator* and will be the basis of several sampling methods. The far field operator basically contains all the knowledge about the unknown scatterer by a complete set of measurements.

The discretized version of the far field operator is now immediately obtained from (8.3.9) and (8.3.15). We define

$$\mathbf{F} := -2(\mathbf{K}^{\infty} - \mathrm{i}\eta \mathbf{S}^{\infty})(\mathbf{I} + \mathbf{K} - \mathrm{i}\eta \mathbf{S})^{-1}\mathbf{H}. \tag{8.3.19}$$

Code 8.3.4. *Here we realize the code to calculate the far field operator F defined in (8.3.18) and (8.3.19) in a script* `sim_08_3_4_F.m`*. We first need to run the codes 8.1.6, 8.2.2 and 8.3.1. We also collected these into the control script* `sim_08_3_5_Control.m` *(code repository).*

```
% I Setup and evaluate Herglotz wave operator H
hmat1    = repmat(y1.',1,ffN).*repmat(yff1,N,1);
hmat2    = repmat(y2.',1,ffN).*repmat(yff2,N,1);
H      = exp(i*kappa*(hmat1 + hmat2))*hff; % Herglotz wave operator

% II Setup farfield operator F
F = -2*ffBW*BWinv*H;
```

8.4 Reciprocity relations

Reciprocity relations formulate fundamental symmetry properties of wave fields arising in many parts of the mathematics of waves. The following two theorems are valid for sound-soft and sound-hard scatterers (see equation (8.2.3) or (8.2.4)) as well as for the Robin boundary condition (8.2.5).

Theorem 8.4.1 (Far field reciprocity). *For the far field pattern $u^\infty(\hat{x}, \theta)$ for scattering of a plane wave with direction of incidence $\theta \in \mathbb{S}$ evaluated at the direction $\hat{x} \in \mathbb{S}$ we have the* far field reciprocity relation

$$u^\infty(\hat{x}, \theta) = u^\infty(-\theta, -\hat{x}), \quad \hat{x}, \theta \in \mathbb{S}. \tag{8.4.1}$$

Proof. By Green's theorem, we know that

$$\int_{\partial D} \left\{ u^{\mathrm{i}}(y, \theta) \frac{\partial u^{\mathrm{i}}}{\partial \nu}(y, -\hat{x}) - u^{\mathrm{i}}(y, -\hat{x}) \frac{\partial u^{\mathrm{i}}}{\partial \nu} u^{\mathrm{i}}(y, \theta) \right\} \mathrm{d}s(y) = 0 \tag{8.4.2}$$

and

$$\int_{\partial D} \left\{ u^{\mathrm{s}}(y, \theta) \frac{\partial u^{\mathrm{s}}}{\partial \nu}(y, -\hat{x}) - u^{\mathrm{s}}(y, -\hat{x}) \frac{\partial u^{\mathrm{s}}}{\partial \nu} u^{\mathrm{s}}(y, \theta) \right\} \mathrm{d}s(y) = 0. \tag{8.4.3}$$

By changing the roles of $\hat{x}$ and θ in the representation formula of the far field $u^\infty(\hat{x}, \theta)$

$$\gamma_d u^\infty(\hat{x}, \theta) = \left\{ u^{\mathrm{s}}(y, \theta) \frac{\partial u^{\mathrm{i}}}{\partial \nu}(y, -\hat{x}) - u^{\mathrm{i}}(y, -\hat{x}) \frac{\partial u^{\mathrm{s}}}{\partial \nu} u^{\mathrm{i}}(y, \theta) \right\} \mathrm{d}s(y) \tag{8.4.4}$$

with $\gamma_m = \frac{\mathrm{e}^{\mathrm{i}\pi/4}}{\sqrt{8\pi\kappa}}$, $(m = 2)$, $\frac{1}{4\pi}$, $(m = 3)$ (see (8.3.3)), we derive

$$\gamma_m u^\infty(-\theta, -\hat{x}) = \int_{\partial D} \left\{ u^{\mathrm{s}}(y, -\hat{x}) \frac{\partial u^{\mathrm{i}}}{\partial \nu}(y, \theta) - u^{\mathrm{i}}(y, \theta) \frac{\partial u^{\mathrm{s}}}{\partial \nu} u^{\mathrm{i}}(y, -\hat{x}) \right\} \mathrm{d}s(y). \tag{8.4.5}$$

From (8.4.4) and (8.4.5) we obtain

$$\begin{aligned} &\gamma_m (u^\infty(\hat{x}, \theta) - u^\infty(-\theta, -\hat{x})) \\ &= \int_{\partial D} \left\{ u^{\mathrm{s}}(y, \theta) \frac{\partial u^{\mathrm{i}}}{\partial \nu}(y, -\hat{x}) - u^{\mathrm{i}}(y, -\hat{x}) \frac{\partial u^{\mathrm{s}}}{\partial \nu}(y, \theta) \right. \\ &\quad \left. - u^{\mathrm{s}}(y, -\hat{x}) \left(\frac{\partial u^{\mathrm{i}}}{\partial \nu}(y, \theta) + u^{\mathrm{i}}(y, \theta) \frac{\partial u^{\mathrm{s}}}{\partial \nu}(y, -\hat{x}) \right) \right\} \mathrm{d}s(y). \end{aligned} \tag{8.4.6}$$

Compare the integrand of (8.4.6) with

$$\begin{aligned} &u(y,\theta)\frac{\partial u}{\partial \nu}(y,-\hat{x}) - u(y,-\hat{x})\frac{\partial u}{\partial \nu}(y,\theta) \\ &= \left(u^{\mathrm{i}}(y,\theta) + u^{\mathrm{s}}(y,\theta)\right)\left(\frac{\partial u^{\mathrm{i}}}{\partial \nu}(y,-\hat{x}) + \frac{\partial u^{\mathrm{s}}}{\partial \nu}(y,-\hat{x})\right) \\ &\quad - \left(u^{\mathrm{i}}(y,-\hat{x}) + u^{\mathrm{s}}(y,-\hat{x})\right)\left(\frac{\partial u^{\mathrm{i}}}{\partial \nu}(y,\theta) + \frac{\partial u^{\mathrm{s}}}{\partial \nu}(y,\theta)\right). \end{aligned} \tag{8.4.7}$$

Then, if we integrate (8.4.7) over ∂D, the terms with the same indices will become zero due to (8.4.2) and (8.4.3). Hence, the integrand of (8.4.6) is nothing but $u(y,\theta)\frac{\partial u}{\partial \nu}(y,-\hat{x}) - u(y,-\hat{x})\frac{\partial u}{\partial \nu}(y,\theta)$. Therefore, we have obtained

$$\begin{aligned} &\gamma_m(u^\infty(\hat{x},\theta) - u^\infty(-\theta,-\hat{x})) \\ &= \int_{\partial D}\left\{u(y,\theta)\frac{\partial u}{\partial \nu}(y,-\hat{x}) - u(y,-\hat{x})\frac{\partial u}{\partial \nu}(y,\theta)\right\}\,\mathrm{d}s(y). \end{aligned} \tag{8.4.8}$$

Now, if we consider the case that ∂D is sound soft, then the integrand of (8.4.8) becomes zero because $u(y,\theta) = u(y,-\hat{x}) = 0$. We also have the same when ∂D is sound hard and we put the Robin boundary condition on ∂D. □

We also obtain the following *mixed reciprocity relation.*

Theorem 8.4.2 (Mixed reciprocity relation). *For the far field pattern $\Phi^\infty(\hat{x}, z)$ of the scattered field for an incident point source $\Phi(\cdot, z)$, $z \in \mathbb{R}^m\backslash\bar{D}$, and the scattered field $u^s(\cdot, d)$ of an incident plane wave with direction $d \in \mathbb{S}$ we have the* mixed reciprocity relation

$$\Phi^\infty(\hat{x}, z) = \gamma_m u^{\mathrm{s}}(z, -\hat{x}), \quad z \in \mathbb{R}^m, \quad \hat{x} \in \mathbb{S}. \tag{8.4.9}$$

Proof. By the representation formula of $\Phi^{\mathrm{s}}(x, z)$

$$\Phi^{\mathrm{s}}(x,z) = \int_{\partial D}\left\{\Phi^{\mathrm{s}}(y,z)\frac{\partial \Phi}{\partial \nu}(y,x) - \Phi(y,x)\frac{\partial \Phi^{\mathrm{s}}}{\partial \nu}(y,z)\right\}\,\mathrm{d}s(y) \tag{8.4.10}$$

for $x \in \mathbb{R}^m\backslash\bar{D}$. Letting $|x| \to \infty$, we have

$$\Phi^\infty(\hat{x},z) = \gamma_m\int_{\partial D}\left\{\Phi^{\mathrm{s}}(y,z)\frac{\partial}{\partial \nu}\mathrm{e}^{-\mathrm{i}\kappa\hat{x}\cdot y} - \mathrm{e}^{-\mathrm{i}\kappa\hat{x}\cdot y}\frac{\partial \Phi^{\mathrm{s}}}{\partial \nu}(y,z)\right\}\,\mathrm{d}s(y). \tag{8.4.11}$$

Here, we note that by Green's formula and the radiation condition,

$$\int_{\partial D}\left\{\Phi^{\mathrm{s}}(y,z)\frac{\partial u^{\mathrm{s}}}{\partial \nu}(y,-\hat{x}) - u^{\mathrm{s}}(y,-\hat{x})\frac{\partial \Phi^{\mathrm{s}}}{\partial \nu}(y,z)\right\}\,\mathrm{d}s(y) = 0. \tag{8.4.12}$$

Thus, we have

$$\Phi^\infty(\hat{x}, z) = \gamma_m \int_{\partial D} \left\{ \Phi^s(y, z) \frac{\partial u}{\partial \nu}(y, -\hat{x}) - u(y, -\hat{x}) \frac{\partial \Phi^s}{\partial \nu}(y, z) \right\} ds(y), \quad z \in \mathbb{R}^m \backslash \bar{D}, \quad \hat{x} \in \mathbb{S}. \tag{8.4.13}$$

For the sound-soft case, this becomes

$$\Phi^\infty(\hat{x}, z) = \gamma_m \int_{\partial D} \left\{ \Phi^s(y, z) \frac{\partial u}{\partial \nu}(y, -\hat{x}) \right\} ds(y), \quad z \in \mathbb{R}^m \backslash \bar{D}, \quad \hat{x} \in \mathbb{S}. \tag{8.4.14}$$

On the other hand, by

$$u^s(z, -\hat{x}) = \int_{\partial D} \left\{ u^s(y, -\hat{x}) \frac{\partial \Phi}{\partial \nu}(y, z) - \Phi(y, z) \frac{\partial u^s}{\partial \nu}(y, -\hat{x}) \right\} ds(y) \tag{8.4.15}$$

for $z \in \mathbb{R}^m \backslash \bar{D}$ and $\hat{x} \in \mathbb{S}$ and

$$\int_{\partial D} \left\{ u^i(y, -\hat{x}) \frac{\partial \Phi}{\partial \nu}(y, z) - \Phi(y, z) \frac{\partial u^i}{\partial \nu}(y, -\hat{x}) \right\} ds(y) = 0 \tag{8.4.16}$$

for $z \in \mathbb{R}^m \backslash \bar{D}$ and $\hat{x} \in \mathbb{S}$ we have

$$u^s(z, -\hat{x}) = \int_{\partial D} \left\{ u(y, -\hat{x}) \frac{\partial \Phi}{\partial \nu}(y, z) - \Phi(y, z) \frac{\partial u}{\partial \nu}(y, -\hat{x}) \right\} ds(y) \tag{8.4.17}$$

for $z \in \mathbb{R}^m \backslash \bar{D}$ and $\hat{x} \in \mathbb{S}$. This becomes

$$u^s(z, -\hat{x}) = \int_{\partial D} \Phi^s(y, z) \frac{\partial u}{\partial \nu}(y, -\hat{x})\, ds(y), \quad z \in \mathbb{R}^m \backslash \bar{D}, \quad \hat{x} \in \mathbb{S} \tag{8.4.18}$$

for the sound-soft case, because we have $\Phi(y, z) + \Phi^s(y, z) = 0$ for $y \in \partial D$ and $z \in \mathbb{R}^m \backslash \bar{D}$. Equations (8.4.14) and (8.4.18) immediately imply the mixed reciprocity relation for the sound-soft case. We can also treat the sound-hard case, and the case when we put the Robin boundary condition on ∂D, with a corresponding result. □

8.5 The Lax–Phillips method to calculate scattered waves

In this section we introduce the *Lax–Phillips method* for solving the direct obstacle scattering problem. Define the operator H by

$$H := \Delta + k^2.$$

Let $D \subset \mathbb{R}^m$ ($m = 2$ or 3) be a bounded open domain with C^2 boundary ∂D. For any function f defined in $E \subset \mathbb{R}^m$, we define its usual support by $\operatorname{supp} f$. That is the closure of the set $\{x \in E : f(x) \neq 0\}$ and the closure is taken with respect to the relative topology of E in $\mathbb{R}^m$.

The direct obstacle scattering problem can be formulated as: given a function $f \in L^2(\mathbb{R}^m \backslash \bar{D})$ with compact supp f, we look for the solution to

$$\begin{cases} Hv = f & \text{in } \mathbb{R}^m \backslash \bar{D}, \\ Bv = 0 & \text{on } \partial D, \\ \lim\limits_{r \to \infty} r^{\frac{m-1}{2}} \left(\dfrac{\partial v}{\partial r} - ikv \right) = 0, & r = |x|, \end{cases} \tag{8.5.1}$$

where B is the boundary operator. Moreover, if B has coefficients, we let them belong to class C^2. For later use, we first give the following assumptions:

(i) The boundary value problem (8.5.1) has at most one solution.

(ii) There exists a unique solution $v \in H^2_{\text{loc}}(\mathbb{R}^m)$ to

$$\begin{cases} Hv = g & \text{in } \mathbb{R}^m, \\ \lim\limits_{r \to \infty} r^{\frac{m-1}{2}} \left(\dfrac{\partial v}{\partial r} - ikv \right) = 0, & r = |x|, \end{cases} \tag{8.5.2}$$

for given function $g \in L^2(\mathbb{R}^m)$ with compact support.

(iii) The boundary value problem (8.5.3) given later is well-posed.

Let $\Omega_j \subset \mathbb{R}^m$ $(j = 1, 2)$ be bounded domains satisfying $\bar{D} \subset\subset \Omega_1 \subset\subset \Omega_2$. We suppose that the boundary $\partial\Omega_2$ is of class C^2 and the function f in (8.5.1) satisfies supp $f \subset \Omega_2$. Define

$$L^2_\diamond(\mathbb{R}^m \backslash \bar{D}) := \left\{ f^* : \ f^* \in L^2(\mathbb{R}^m \backslash \bar{D}), \ \operatorname{supp} f^* \subset \Omega_2 \right\}.$$

For $f^* \in L^2_\diamond(\mathbb{R}^m \backslash \bar{D})$, we set

$$\tilde{f}^*(x) := \begin{cases} f^*(x), & x \in \mathbb{R}^m \backslash \bar{D}, \\ 0, & x \in \bar{D}. \end{cases}$$

By assumption (ii), there exists a function $w \in H^2_{\text{loc}}(\mathbb{R}^m)$ meeting

$$\begin{cases} Hw = \tilde{f}^* & \text{in } \mathbb{R}^m, \\ \lim\limits_{r \to \infty} r^{\frac{m-1}{2}} \left(\dfrac{\partial w}{\partial r} - ikw \right) = 0, & r = |x|. \end{cases}$$

According to the trace theorem, we can find a function $\tilde{w} \in H^2(\mathbb{R}^m \backslash \bar{D})$ such that

$$B\tilde{w} = -Bw \text{ on } \partial D;$$
$$\operatorname{supp} \tilde{w} \subset \Omega_2.$$

Define

$$\hat{f} := f^* + H\tilde{w}.$$

It follows that $\hat{f} \in L^2_\diamond(\mathbb{R}^m \backslash \bar{D})$. Let

$$W := w|_{\mathbb{R}^m \backslash \bar{D}} + \tilde{w} \in H^2_{\text{loc}}(\mathbb{R}^m \backslash \bar{D}).$$

Then we can easily show

$$\begin{cases} HW = \hat{f} & \text{in } \mathbb{R}^m \backslash \bar{D}, \\ BW = 0 & \text{on } \partial D, \\ \lim_{r\to\infty} r^{\frac{m-1}{2}}\left(\frac{\partial W}{\partial r} - ikW\right) = 0 & r = |x|. \end{cases}$$

We now consider the following boundary value problem

$$\begin{cases} HV = \hat{f} & \text{in } \Omega_2 \backslash \bar{D}, \\ V = 0 & \text{on } \partial\Omega_2, \\ BV = 0 & \text{on } \partial D. \end{cases} \tag{8.5.3}$$

By assumption (iii), there exists a unique solution $V \in H^2(\Omega_2 \backslash \bar{D})$ to (8.5.3). Since the functions f^* and $\hat{f}$ are equivalent, $W = W(\hat{f})$ and $V = V(\hat{f})$ are two linear bounded operators.

Next we look for the solution v to (8.5.1) in the form

$$v = W - \chi(W - V) \tag{8.5.4}$$

with $\chi \in C_0^\infty(\Omega_2)$ and $\chi = 1$ on $\bar{\Omega}_1$.

By direct calculations, we have

$$\begin{aligned} f = Hv &= HW - \chi H(W - V) - 2\nabla\chi \cdot \nabla(W - V) - (\Delta\chi)(W - V) \\ &= \hat{f} - \chi(\hat{f} - \hat{f}) + K\hat{f} \\ &= \hat{f} + K\hat{f} \quad \text{in } \Omega_2 \backslash \bar{D}, \end{aligned} \tag{8.5.5}$$

where we let

$$K\hat{f} = -2\nabla\chi \cdot \nabla(W - V) - (\Delta\chi)(W - V).$$

Note that $K : L_\diamond^2(\mathbb{R}^m \backslash \bar{D}) \to H^1(\mathbb{R}^m \backslash \bar{D}) \cap L_\diamond^2(\mathbb{R}^m \backslash \bar{D})$ is bounded and $H^1(\mathbb{R}^m \backslash \bar{D}) \cap L_\diamond^2(\mathbb{R}^m \backslash \bar{D})$ is compactly imbedded into $L_\diamond^2(\mathbb{R}^m \backslash \bar{D})$, we obtain that K is compact from $L_\diamond^2(\mathbb{R}^m \backslash \bar{D})$ into itself. Thus, equation (8.5.5) is of Fredholm type and hence the solvability comes from the uniqueness of its solutions.

Let $f = 0$ in (8.5.5). Then, we have $Hv = 0$ in $\Omega_2 \backslash \bar{D}$. In addition, it follows from (8.5.4) that we have $Hv = HW = \hat{f} = 0$ $\mathbb{R}^m \backslash \Omega_2$, and therefore $Hv = 0$ in $\mathbb{R}^m \backslash \bar{D}$. Noting that $Bv = BV = 0$ on ∂D, we obtain from assumption (i) that

$$v = 0 \quad \text{in } \mathbb{R}^m \backslash \bar{D},$$

which leads to

$$W = \chi(W - V) \quad \text{in } \mathbb{R}^m \backslash \bar{D}. \tag{8.5.6}$$

On the other hand, we have

$$H(W - V) = HW - HV = \hat{f} - \hat{f} = 0 \quad \text{in } \Omega_2 \backslash \bar{D}, \tag{8.5.7}$$

$$B(W - V) = BW - BV = 0 - 0 = 0 \quad \text{on } \partial D. \tag{8.5.8}$$

Since χ is zero on $\partial\Omega_2$, we know from (8.5.6) that

$$W = \chi(W - V) = 0 \quad \text{on } \partial\Omega_2. \tag{8.5.9}$$

From the definition of V, we also have

$$V = 0 \quad \text{on } \partial\Omega_2. \tag{8.5.10}$$

Combining (8.5.9) with (8.5.10) yields

$$W - V = 0 \quad \text{on } \partial\Omega_2. \tag{8.5.11}$$

By assumption (iii), (8.5.7), (8.5.8) and (8.5.11) imply that

$$W - V = 0 \quad \text{in } \Omega_2 \backslash \bar{D}. \tag{8.5.12}$$

Using (8.5.6) again, we have

$$W = 0 \quad \text{in } \Omega_2 \backslash \bar{D},$$

and hence

$$\hat{f} = 0.$$

Thus, the proof of uniqueness for solutions to (8.5.5) is complete.

Bibliography

[1] Kress R 1999 *Linear Integral Equations* (*Applied Mathematical Sciences* vol 82) 2nd edn (New York: Springer)

[2] Colton D and Kress R 1983 *Integral Equation Methods in Scattering Theory* (New York: Wiley)

[3] Colton D and Kress R 1998 *Inverse Acoustic and Electromagnetic Scattering Theory* (*Applied Mathematical Sciences* vol 93) 2nd edn (Berlin: Springer)

[4] Potthast R 2001 *Point Sources and Multipoles in Inverse Scattering Theory* (*Chapman and Hall/CRC Research Notes in Mathematics* vol 427) (Boca Raton, FL: CRC)

IOP Publishing

Inverse Modeling

An introduction to the theory and methods of inverse problems and data assimilation

Gen Nakamura and Roland Potthast

Chapter 9

Nonlinear operators

Most practical inverse or data assimilation problems are nonlinear. Usually, there are nonlinearities in the operator H which maps the system states or unknown parameter functions onto the measurements. In data assimilation, the dynamics of the underlying dynamical system M is also nonlinear.

A key tool both for analysis and for the algorithms is the differentiability of distributions and functions with respect to states or parameter functions. Since states are usually multi-dimensional functions, we need to work with the Fréchet derivative in a Banach space environment.

Derivatives are needed to study the *properties* of the mappings, to evaluate linear or higher-order *approximations*, for *iterative methods* solving inverse problems and in data assimilation when *optimization techniques* or the extended Kalman filter are used. In any of these cases, a thorough study of the underlying mathematics is mandatory.

In this chapter we will prove the differentiability of *boundary integral operators* with respect to variations of the boundary in section 9.1. The results for boundary value operators are used to show the differentiability of *boundary value problems* in section 9.2. We introduce alternative approaches to domain derivatives in section 9.3. Newton's method and the gradient method for an inverse scattering problem are presented in section 9.4. Finally, we introduce the calculation of derivatives of *dynamical models* in section 9.5.

9.1 Domain derivatives for boundary integral operators

The task of this section is to study the dependence of boundary integral operators on variations of the boundary. In particular, we want to show that the standard boundary integral operators of single-layer and double-layer potentials are Fréchet differentiable of arbitrary order.

To study operators which live on some boundary ∂D of a domain $D \subset \mathbb{R}^3$ in dependence on this boundary ∂D, we need to transform the operators onto some reference boundary. A simple set-up is obtained when we study some vector field $r : \partial D \to \mathbb{R}^3$. If ∂D is a boundary of a domain D of class C^ℓ for some $\ell \in \mathrm{N}$ and if we want to stay within this class of boundaries, we need to study vector fields $r \in C^\ell(\partial D, \mathbb{R}^3)$ of the same smoothness level. We will restrict our attention to $C^\ell(\partial D, \mathbb{R}^3)$ and denote its norm by $\|r\|$, but all arguments work similarly in the Hölder space $C^{\ell,\alpha}(\partial D, \mathbb{R}^3)$ for $0 < \alpha \leqslant 1$. For $\|r\|$ sufficiently small we have that

$$\partial D_r := \{x + r(x) : x \in \partial D\} \tag{9.1.1}$$

is again a boundary of a domain D_r of class C^ℓ. Our first goal is to study integral operators of the form

$$(A[r]\varphi)(x) := \int_{\partial D_r} k(x, y)\varphi(y)\,\mathrm{d}s(y), \quad x \in \partial D_r, \tag{9.1.2}$$

where the kernel k is weakly singular, such as the kernels of our single- or double-layer potentials (8.1.31), (8.1.32). More precisely, $k(x, y)$ in $C((\partial D_r \times \partial D_r)\backslash\{x = y\})$ with the estimate $|k(x, y)| \leqslant M|x - y|^{-2+\alpha}$, $x, y \in \partial D_r$, $x \neq y$ for some constants $0 < \alpha \leqslant 2$ and $M > 0$. For $\|r\|$ sufficiently small we choose $\partial D = \partial D_0$ as a reference domain, transform the integrals over ∂D_r onto ∂D and consider $A = A[r]$ as an operator depending on $r \in C^\ell(\partial D, \mathbb{R}^3)$.

Our basic idea is to show that the classical argument to pull the derivative into the integral remains valid under suitable assumptions for weakly singular integral equations, as first suggested in [1]. A generalization to strongly singular integral equations has been worked out in [2]. We work will well-known parametrization techniques. The surface ∂D is parametrized locally by a set of C^ℓ-differentiable mappings

$$\Psi_j : U_j \to \partial D, \quad u \mapsto \Psi_j(u)$$

defined on open sets $U_j \subset \mathbb{R}^2$, $j = 1,\ldots, N$, with $0 \in U_j$ such that

$$\partial D = \bigcup_{j=1}^{N} V_j$$

with $V_j := \Psi_j(U_j)$, $j = 1,\ldots, N$. We employ a *partition of unity* χ_j such that χ_j is compactly supported in the open set V_j and

$$\sum_{j=1}^{N} \chi_j(y) \equiv 1, \quad y \in \partial D.$$

Then, the surface element $\mathrm{d}s$ is given by

$$\mathrm{d}s = \left| \frac{\partial \Psi_j}{\partial u_1} \times \frac{\partial \Psi_j}{\partial u_2} \right| \mathrm{d}u_1\, \mathrm{d}u_2$$

and an integral of function $f \in C(\partial D)$ over ∂D can be written as

$$\int_{\partial D} f(y)\,\mathrm{d}s(y) = \sum_{j=1}^{N} \int_{\partial D} f(y)\chi_j(y)\,\mathrm{d}s(y) \text{ in } V_j$$

$$= \sum_{j=1}^{N} \int_{U_j} \left(f\chi_j\right) \circ \Psi_j(u) \left| \frac{\partial \Psi_j}{\partial u_1} \times \frac{\partial \Psi_j}{\partial u_2} \right| \mathrm{d}u_1\, \mathrm{d}u_2 \qquad (9.1.3)$$

with the notation

$$y = \Psi_j(u), \qquad u = (u_1, u_2),$$

and

$$f(y)\chi_j(y) = \left(f\chi_j\right)(y) = \left(f\chi_j\right)\left(\Psi_j(u)\right) = \left(f\chi_j\right) \circ \Psi_j(u).$$

We assume that the partition of unity is constructed such that each $\chi_j(\Psi_j(u))$ is in $C^\ell(U_j)$, $j = 1, \ldots, N$. A partition of unity and local coordinates (9.1.3) is a convenient tool for studying the differentiability properties of boundary integral operators. We note that the integral thus defined does not depend on the choice of parameterization of ∂D and partition of unity.

The surface element $\mathrm{d}s$ on ∂D_r is transformed by $\mathrm{d}s_r(y) = J_T(r, y)\,\mathrm{d}s(y)$, where J_T is the Jacobi determinant defined by

$$J_T(r, y) := \frac{\left| \dfrac{\partial\left(\Psi_j + r\left(\Psi_j\right)\right)}{\partial u_1} \times \dfrac{\partial\left(\Psi_j + r\left(\Psi_j\right)\right)}{\partial u_2} \right|}{\left| \dfrac{\partial \Psi_j}{\partial u_1} \times \dfrac{\partial \Psi_j}{\partial u_2} \right|} \Bigg|_{u(y)} \qquad (9.1.4)$$

for $y \in \partial D$, where $u = u(y)$ is defined via $y = \Psi_j(u(y))$. Then, the integral operator $A[r]$ can be written in the form

$$(A[r]\varphi)(x) = \int_{\partial D} k[r](x, y)\varphi[r](y)\,\mathrm{d}s(y), \qquad x \in \partial D \cap V_j, \qquad (9.1.5)$$

where we absorb the transformation of the surface element into the kernel by

$$k[r](x, y) := k(x + r(x), y + r(y))J_T(r, y),$$

and employ the notation $\varphi[r](y) = \varphi(y + r(y))$.

Theorem 9.1.1 (Domain derivatives of boundary integral operators). *Let the weakly singular kernel $k[r](x, y)$ be two times continuously Fréchet differentiable with respect to r for all fixed $x \neq y$, $x, y \in \partial D$, such that $k[r](x, y)$ and the derivatives $k'[r](x, y)$,*

$k''[r](x, y)$ are weakly singular uniformly for all $r \in U$ in a small neighborhood $U \subset C^\ell(\partial D, \mathbb{R}^3)$ of $r \equiv 0$, i.e. for $\ell = 0, 1, 2$

$$\left|k^{(\ell)}[r, h](x, y)\right| \leqslant \frac{c\|h\|^\ell}{|x - y|^{2-\alpha}}, \quad x \neq y, \quad r \in U \tag{9.1.6}$$

with some constants $c > 0$, $0 < \alpha \leqslant 2$ and $(k^{(\ell)}[r]h)(x, y) = k^{(\ell)}[r, h](x, y)$ depends continuously on x and y for $x \neq y$. Then, the integral operator $A[r]$ defined by (9.1.5) is Fréchet differentiable with respect to r in the spaces of bounded linear operators on $C(\partial D)$ and $L^2(\partial D)$, with the derivative given by

$$(A'[r, h]\varphi)(x) = \int_{\partial D} k'[r, h](x, y)\varphi(y)\, \mathrm{d}s(y), \quad x \in \partial D. \tag{9.1.7}$$

Proof. We apply lemma 2.6.3 to $k[r](x, y)$ to obtain

$$k[r + h](x, y) = k[r](x, y) + (k'[r]h)(x, y) + k_1[r, h](x, y) \tag{9.1.8}$$

for $x \neq y \in \partial D$ with

$$k_1[r](h)(x, y) = \int_0^1 (1 - t)k''[r + th](h)(x, y)\, \mathrm{d}t, \tag{9.1.9}$$

where we used the notation $k''[r + th](h) = k''[r + th](h, h)$. We now multiply (9.1.8) by $\varphi(y)$ for $\varphi \in C(\partial D)$ or $L^2(\partial D)$ and then integrate the whole equation over ∂D. Then we have

$$A[r + h]\varphi = A[r]\varphi + \tilde{A}[r, h]\varphi + A_1[r, h]\varphi \tag{9.1.10}$$

with

$$(\tilde{A}[r, h]\varphi)(x) = \int_{\partial D} (k'[r]h)(x, y)\varphi(y)\, \mathrm{d}s(y), \quad x \in \partial D, \tag{9.1.11}$$

and

$$(A_1[r, h]\varphi)(x) = \int_{\partial D} k_1[r](h)(x, y)\varphi(y)\, \mathrm{d}s(y), \quad x \in \partial D. \tag{9.1.12}$$

Since the integral operators with weakly singular kernels are well defined as bounded linear operators on $C(\partial D)$ and $L^2(\partial D)$, the theorem follows from

$$\left|k''[r + h](h)(x, y)\right| \leqslant \frac{c}{|x - y|^{2-\alpha}}\|h\|^2, \quad x \neq y, \quad x, y \in \partial D, \tag{9.1.13}$$

for some constant $C > 0$ independent of x, y and $h \in C^\ell(\partial D, \mathbb{R}^3)$ with $\|h\| \ll 1$ which together with (9.1.9) leads to the desired estimate for A_1 and completes the proof. □

To apply the above result to the single- and double-layer potentials we need to prove the Fréchet differentiability of the kernels and establish the bounds (9.1.6). Here the kernel $k_S[r]$ of the single-layer potential is given by

$$k_S[r](x, y) = \Phi(x + r(x), y + r(y))\, J_T(r, y) \quad x, y \in \partial D \tag{9.1.14}$$

with the fundamental solution $\Phi(z, w) = \frac{e^{i\kappa|z-w|}}{4\pi|z - w|}$, for $z \neq w$. We note that for $j = 1, 2$ the mappings

$$g_j : C^1(\partial D, \mathbb{R}^3) \to C(U_k, \mathbb{R}^3), \quad r \mapsto \frac{\partial r \circ \Psi_k}{\partial u_j} \tag{9.1.15}$$

are linear in r and thus Fréchet differentiable, where we have fixed k and suppressed putting into the notation g_j. The Fréchet differentiability of $J_T(r, y)$ with respect to $r \in C^1(\partial D, \mathbb{R}^3)$ infinitely many times is now obtained by an application of the chain rule. We also remark that the mapping

$$g_3 : C^1(\partial D, \mathbb{R}^3) \to \mathbb{R}^3, \quad r \mapsto x + r(x) - y - r(y) \tag{9.1.16}$$

for $x, y \in \partial D$ fixed is a constant plus a linear function in r and hence Fréchet differentiable infinitely many times with first order derivative

$$g_3' : C^1(\partial D, \mathbb{R}^3) \times C^1(\partial D, \mathbb{R}^3) \to \mathbb{R}^3, \quad (r, h) \mapsto h(x) - h(y)$$

for fixed $x, y \in \partial D$. According to the chain rule, the function

$$g_4 : C^1(\partial D, \mathbb{R}^3) \to \mathbb{R}, \quad r \mapsto \frac{1}{\left|x + r(x) - y - r(y)\right|} \tag{9.1.17}$$

is Fréchet differentiable with derivative $g_4' : C^1(\partial D, \mathbb{R}^3) \times C^1(\partial D, \mathbb{R}^3) \to \mathbb{R}$ given by

$$g_4[r](h) = -\frac{\left\langle x + r(x) - y - r(y), h(x) - h(y)\right\rangle}{\left|x + r(x) - y - r(y)\right|^3} \tag{9.1.18}$$

for $x \neq y \in \partial D$ fixed. We have $|x + r(x) - y - r(y)| \leqslant c|x - y|$ with some constant c and $|h(x) - h(y)| \leqslant \|h\| |x - y|$ with the $C^1(\partial D, \mathbb{R}^3)$-norm $\|h\|$ of h. We further remark that $|x + r(x) - y - r(y)| \geqslant \tilde{c}|x - y|$ for $x, y \in \partial D$ with some constant $\tilde{c}$. Hence we obtain an estimate

$$\left|g_4'[r](h)(x, y)\right| \leqslant \frac{c\|h\|}{|x - y|}, \quad x \neq y \in \partial D \tag{9.1.19}$$

with a constant c. A similar formula and analogous estimate can be carried out for all higher derivatives of g_4. The derivative of the exponential term

$$g_5 : C^1(\partial D, \mathbb{R}^3) \to \mathbb{R}, \; r \mapsto e^{i\kappa|x+r(x)-y-r(y)|} \tag{9.1.20}$$

is given by

$$g_5'[r](h)(x, y) = -\mathrm{i}\kappa e^{\mathrm{i}\kappa|x+r(x)-y-r(y)|} \cdot \frac{\left\langle x + r(x) - y - r(y), h(x) - h(y)\right\rangle}{\left|x + r(x) - y - r(y)\right|}, \tag{9.1.21}$$

which gives an estimate $|g_5'[r](h)(x, y)| \leqslant c\|h\|$ with some constant c. Again, all higher derivatives of g_5 are also bounded for $x, y \in \partial D$. This leads to the infinite Fréchet differentiability of the kernel $k_S[r](x, y)$ of the single-layer potential and the estimate

$$\left|\frac{\partial^\xi}{\partial r^\xi}\left(k_S[r](x, y)\right)\right| \leqslant \frac{c_\xi}{|x - y|}, \quad x \neq y \in \partial D, \tag{9.1.22}$$

with a constant $c_\xi > 0$ for $\xi \in \mathbb{N}_0$.

The differentiation of the normal derivative $\partial\Phi(x, y)/\partial\nu(y)$ as it appears in the double-layer potential is slightly more involved, since we need an estimate for the derivative of $\nu[r](y) \cdot (x + r(x) - y - r(y))$ by a constant times at least $|x - y|^{1+\alpha}$ with $\alpha > 0$. The normal vector $\nu[r]$ is given by

$$\nu[r](u) = \frac{N[r](u)}{|N[r](u)|}, \tag{9.1.23}$$

with

$$N[r](u) := \frac{\partial\left(\Psi_j + r \circ \Psi_j\right)}{\partial u_1} \times \frac{\partial\left(\Psi_j + r \circ \Psi_j\right)}{\partial u_2}. \tag{9.1.24}$$

For the further arguments we will drop the index j for reasons of brevity. We remark that

$$\begin{aligned} x + r(x) - y - r(y) &= \Psi(u) + r \circ \Psi(u) - \Psi(v) - r \circ \Psi(v) \\ &= \frac{\partial}{\partial u_1}\left(\Psi(u) + r \circ \Psi(u)\right)\Big|_{\tilde{u}} \cdot (u_1 - v_1) \\ &\quad + \frac{\partial}{\partial u_2}\left(\Psi(u) + r \circ \Psi(u)\right)\Big|_{\tilde{u}} \cdot (u_2 - v_2) \end{aligned} \tag{9.1.25}$$

in $\partial D \cap V_j$ with some point $\tilde{u}$ on the line between u and v. We will use the notation

$$\nabla_u(\Psi + r \circ \Psi)|_{\tilde{u}} \cdot (u - v)$$

for the right-hand side of (9.1.25). We now decompose the term by

$$\begin{aligned} &\Psi(u) + r \circ \Psi(u) - \Psi(v) - r \circ \Psi(v) \\ &= \nabla_u(\Psi + r \circ \Psi)|_u \cdot (u - v) \\ &+ \left(\nabla_u(\Psi + r \circ \Psi)|_{\tilde{u}} - \nabla_u(\Psi + r \circ \Psi)|_u\right) \cdot (u - v). \end{aligned} \tag{9.1.26}$$

With

$$\left(\frac{\partial(\Psi + r\circ\Psi)}{\partial u_1} \times \frac{\partial(\Psi + r\circ\Psi)}{\partial u_2}\right)\cdot \frac{\partial}{\partial u_k}(\Psi + r\circ\Psi)|_u = 0$$

for $k = 1, 2$, we write $\nu[r](y)\cdot(x + r(x) - y - r(y))$ as the sum of

$$T_k := \frac{N[r](u)}{|N[r](u)|}\cdot\left(\frac{\partial}{\partial u_k}(\Psi + r\circ\Psi)|_{\tilde{u}} - \frac{\partial}{\partial u_k}(\Psi + r\circ\Psi)|_u\right)(u_k - v_k)$$

for $k = 1$ and $k = 2$, where $N[r]$ is given by (9.1.24). For T_k we now obtain its infinite Fréchet differentiability with respect to r and, when ∂D and r are Hölder continuously differentiable of order $\alpha \in (0,1)$, we have the estimate

$$|T_k^{(\xi)}(u, v)| \leqslant c\,|u - v|^{1+\alpha} \tag{9.1.27}$$

for all Fréchet derivatives of order $\xi = 0, 1, 2,\ldots$, which leads to

$$\left|\frac{\partial^\xi}{\partial r^\xi}(\nu[r]\cdot(x + r(x) - y - r(y)))\right| \leqslant c_\xi\,|x - y|^{1+\alpha},\quad x \neq y \in \partial D \tag{9.1.28}$$

with some constant c_ξ. With this estimate we then obtain the Fréchet differentiability of the kernels

$$k_K[r](x, y) := \frac{\partial\Phi(x + r(x), y + r(y))}{\partial\nu[r](y)}, \tag{9.1.29}$$

and

$$k_{K'}[r](x, y) := \frac{\partial\Phi(x + r(x), y + r(y))}{\partial\nu[r](x)} \tag{9.1.30}$$

and estimates analogous to (9.1.22). The Fréchet differentiability of the potential operators S, K and K' is now a consequence of theorem 9.1.1.

We collect these results into the following theorem.

Theorem 9.1.2 (Domain derivatives of single- and double-layer potential operators). *The single- and double-layer potential operators S, K and K' are infinitely Fréchet differentiable with respect to variations of the boundary ∂D in the sense that the mappings $r \mapsto S[r]$, $r \mapsto K[r]$ and $r \mapsto K'[r]$ are infinitely Fréchet differentiable from $C^1(\partial D, \mathbb{R}^3)$ into $BL(C(\partial D))$ and $BL(L^2(\partial D))$ for the single-layer potential and from $C^{1,\alpha}(\partial D, \mathbb{R}^3)$ into $BL(C(\partial D))$ and into $BL(L^2(\partial D))$ for the double-layer potential assuming that ∂D is of C^1 class and $C^{1,\alpha}$ class, respectively. The derivatives of the operators are obtained by differentiating their kernels with respect to r.*

In the same way, we can show the differentiability of the operators S^∞ and K^∞, which map densities on the boundary of a domain ∂D into the far field pattern of their single- and double-layer potential in $L^2(\mathbb{S})$.

Theorem 9.1.3 (Domain derivatives of far field operators). *The far field operators S^∞ and K^∞ defined by (8.3.2) and (8.3.4) are infinitely Fréchet differentiable with respect to variations of the boundary ∂D from $C^1(\partial D, \mathbb{R}^3)$ into $L^2(\mathbb{S})$. The derivatives of the operators are obtained by differentiating their kernels with respect to r.*

Proof. A proof is obtained based on the decomposition (9.1.8) and (9.1.10) with the derivative (9.1.11) and the rest (9.1.12). Differentiability of the kernels is obtained as in (9.1.20) and (9.1.23). Here, all kernels will be continuous on compact sets ∂D and $\mathbb{S}$, such that the mapping properties are readily obtained and the differentiability statement follows from (9.1.10). □

9.2 Domain derivatives for boundary value problems

To solve an *inverse obstacle scattering problem* by *Newton's method* or by the *gradient method*, we need to establish the Fréchet differentiability of the mapping of the scatterer onto the measurements. Of course, there are many reasons to study the differentiability of the problem. For sensitivity studies, local uniqueness questions and the quest for a full understanding of the inversion problem, differentiability studies are a basic and important first step.

Here, we assume that we measure either the scattered field u^s or the far field pattern $u^\infty(\cdot, d)$ for scattering of one time-harmonic plane wave $u^i(\cdot, d)$ with direction $d \in \mathbb{S}$. For an impenetrable scatterer $D \subset \mathbb{R}^m$, $m = 2, 3$ this is a mapping

$$F : \partial D \mapsto u^s[\partial D](\cdot, d), \qquad F^\infty : \partial D \mapsto u^\infty[\partial D](\cdot, d).$$

We assume the same regularity condition as in the previous section and will focus on the scattering problem described in definition 8.2.1 with the Dirichlet boundary condition (8.2.3) and show how the Fréchet differentiability can be established and how to calculate the derivative. The layer potentials and boundary operators are treated in the space of continuous functions, because the densities arising from an incident field $u^i(\cdot, d)$ are continuous on ∂D_r.

We start with the potential approach (8.2.7) with the integral equation (8.2.8). Then, the mapping F is given by

$$F(\partial D) = -2(\tilde{K} - i\eta\tilde{S})(I + K - i\eta S)^{-1}u^i(\cdot, d)\Big|_{\partial D}, \tag{9.2.1}$$

and F^∞ is defined by

$$F^\infty(\partial D) = -2(K^\infty - i\eta S^\infty)(I + K - i\eta S)^{-1}\, u^i(\cdot, d)|_{\partial D} \tag{9.2.2}$$

with $\tilde{S}$, $\tilde{K}$, S^∞, K^∞ defined by (8.1.2), (8.1.4), (8.3.2), (8.3.4) and S, K given in (8.1.31), (8.1.32). All operators K^∞, S^∞, K and S depend on the boundary ∂D. Fréchet differentiability of the mappings F and F^∞ with respect to ∂D can now be achieved using the set-up and results of section 9.1. We study domains

$$\partial D_r = \{x + r(x) : x \in \partial D\}$$

with two times differentiable vector fields $r \in C^2(\partial D, \mathbb{R}^3)$. Then, the differentiability of

$$F^\infty : r \mapsto u^\infty[r](\cdot, d)$$

is obtained by collecting the results of theorems 9.1.2 and 9.1.3 using the product rule.

Theorem 9.2.1 (Shape differentiability of scattering problems).

(a) *The mapping of $r \in C^2(\partial D, \mathbb{R}^3)$ onto the scattered field $u^s[r](\cdot, d)$ defined by (9.2.1) with $\partial D = \partial D_r$ is infinitely Fréchet differentiable. Its derivative is obtained by using the product and chain rules by differentiating the boundary integral operators under consideration.*

(b) *The mapping of $r \in C^2(\partial D, \mathbb{R}^3)$ onto the far field pattern $u^\infty[r](\cdot, d)$ defined by (9.2.2) with $\partial D = \partial D_r$ is infinitely Fréchet differentiable. Its derivative is obtained by using the product and chain rules by differentiating the boundary integral operators under consideration.*

We are now able to calculate the Fréchet derivative of the far field mapping $F^\infty[r] : \partial D_r \mapsto u^\infty[r]$ and also those of $F : \partial D_r \mapsto u^s[r]$, where $u^s[r]$ denotes u^s when ∂D is ∂D_r. But for instance the calculation

$$\begin{aligned}
F'[r](h) = &\Big(\tilde{K}'[r](h) - \mathrm{i}\eta\tilde{S}'[r](h)\Big)\Big(I + K[r] - \mathrm{i}\eta S[r]\Big)^{-1}\Big(-2u^{\mathrm{i}}(\cdot, d)|_{\partial D_r}\Big) \\
&- \Big(\tilde{K}[r] - \mathrm{i}\eta\tilde{S}[r]\Big)\Big(I + K[r] - \mathrm{i}\eta S[r]\Big)^{-1}\big(K'[r](h) - \mathrm{i}\eta S'[r](h)\big) \\
&\Big(I + K[r] - \mathrm{i}\eta S[r]\Big)^{-1}\Big(-2u^{\mathrm{i}}(\cdot, d)|_{\partial D_r}\Big) \\
&+ \Big(\tilde{K}[r] - \mathrm{i}\eta\tilde{S}[r]\Big)\Big(I + K[r] - \mathrm{i}\eta S[r]\Big)^{-1}\left(-2\frac{\mathrm{d}}{\mathrm{d}r}\left\{u^{\mathrm{i}}(\cdot, d)|_{\partial D_r}\right\}h\right)
\end{aligned} \tag{9.2.3}$$

involves many different products and sums of boundary integral operators. This approach is not efficient enough to use it for calculations. The solution is given by the following *characterization* of the Fréchet derivative as a solution to a boundary value problem with particular boundary values.

Theorem 9.2.2 (Derivative characterization). *For every $h \in C^2(\partial D, \mathbb{R}^3)$, the Fréchet derivative $(u^s)'(h)$ of the scattered field u^s acting to the increment h for the scattering of incident field u^{i} by sound-soft obstacle D is given by*

$$(u^s)'(h) = -\frac{\partial u}{\partial \nu}\, \nu \cdot h, \quad \text{on } \partial D \tag{9.2.4}$$

where $u = u^s + u^{\mathrm{i}}$ denotes the total field and ν is the outer unit normal of ∂D directed to the exterior of D.

Proof. The boundary condition on the domain ∂D_r can be written as

$$u^s[r](x_r) = -u^{\mathrm{i}}(x_r), \quad x \in \partial D,$$

for any $r \in C^2(\partial D, \mathbb{R}^3)$. Since the right-hand side is Fréchet differentiable with respect to r, this also holds for the left-hand side of the equation and we obtain

$$\left((u^s)'[r](h)\right)(x_r) + \nabla u^s[r](x_r) \cdot h(x) = -\nabla u^i(x_r) \cdot h(x), \quad x \in \partial D. \quad (9.2.5)$$

This yields on the reference boundary ∂D, the boundary condition

$$\begin{aligned}\left((u^s)'[r](h)\right)(x_r) &= -\left(\nabla u^s[r](x_r) + \nabla u^i(x_r)\right) \cdot h(x) = -\nabla u[r](x_r) \cdot h(x)\\ &= -\frac{\partial u[r]}{\partial \nu_r}(x_r)\left(\nu_r(x_r) \cdot h(x)\right),\end{aligned} \quad (9.2.6)$$

where we used that the tangential derivative of a constant function $u[r]$ on ∂D_r is zero. This already shows that the boundary values must be given by (9.2.4).

It remains to show that in fact the Fréchet derivative $(u^s)'[r](h)$ of the scattered field $u^s[r]$ has its trace to the boundary ∂D_r. The derivative is given by the formula (9.2.3). The terms two and three of this formula are given by single-layer and double-layer potentials with continuous densities, for which it is well known by the mapping property of that they can be continuously extended from $\mathbb{R}^3 \backslash \bar{D}$ into $\mathbb{R}^3 \backslash D$.

The first term is slightly more difficult, since here the Fréchet derivatives are involved. It consists of a term with the Fréchet derivatives of the single-layer potential and that of the double-layer potential. We will only show that the Fréchet derivative of the double-layer potential $\tilde{K}'[r](h)$ can have the trace to ∂D_r, because it is easier to show for the former one.

First of all, we note that for showing this, the singularity of the outgoing fundamental solution $\Phi(x, y)$ of the Helmholtz equation with wave number $k > 0$ is the key object to be handled. Hence, we can replace $\Phi(x, y)$ by the fundamental solution $\Phi_0(x, y) = (4\pi|x - y|)^{-1}$, because the structure of singularities of Φ and Φ_0 are the same. Also for a density $\psi \in C(\partial D_r)$, the dominant in the decomposition

$$\begin{aligned}\partial_{\nu_r(y_r)}\Phi_0\left(z_r^\tau, y_r\right)\psi(y_r) &= \partial_{\nu_{r(y_r)}}\Phi_0\left(z_r^\tau, y_r\right)\psi(z_r)\\ &\quad + \partial_{\nu_{r(y_r)}}\Phi_0\left(z_r^\tau, y_r\right)\left(\psi(y_r) - \psi(z_r)\right)\end{aligned} \quad (9.2.7)$$

is the first term, where $\tau > 0$, $y_r = y + r(y)$, $z_r = z + r(z)$, $z_r^\tau = z_r + \tau\nu(z_r)$ with $y, z \in \partial D$. Further observe that for this first term

$$\begin{aligned}4\pi\partial_{\nu_{r(y_r)}}\Phi_0\left(z_r^\tau, y_r\right) &= \left\langle \nu_{r(y_r)}, z_r^\tau - y_r\right\rangle / \left|z_r^\tau - y_r\right|^3\\ &\quad + \left\langle \nu_{r(y_r)}, z_r^\tau - z_r\right\rangle / \left|z_r^\tau, -y_r\right|^3.\end{aligned} \quad (9.2.8)$$

Since

$$\left|\left\langle \nu_{r(y_r)}, z_r - y_r\right\rangle\right| \lesssim \left|z_r - y_r\right|^2, \quad \left|z_r^\tau - y_r\right| \sim \left|z_r - y_r\right| + \tau, \quad (9.2.9)$$

the first term is the dominant term in (9.2.8) and by the decomposition $\nu_{r(y_r)} = \nu_{r(z_r)} + (\nu_{r(y_r)} - \nu_{r(z_r)})$, the dominant part of the first term is $\langle \nu_{r(z_r)}, z_r - y_r\rangle$, where the notations

'$\lesssim$' and '$\sim$' stand for '$\leqslant$' modulo positive constant multiplicity and equivalence, respectively. Based on these it is enough to focus on estimating the Fréchet derivative of $\langle \nu_{r(z_r)}, z_r^\tau - y_r \rangle / |z_r^\tau - y_r|^3$ acting to an incremental h which is given by

$$\begin{aligned} &-3\tau \left| z_r^\tau - y_r \right|^{-5} \left\langle z_r^\tau - y_r, h(z_r) + \tau(\nu'[r](h))(z) - h(y) \right\rangle \\ &= -3\left| z_r^\tau - y_r \right|^{-5} (I_1 + I_2 + I_3 + I_4)(\tau, y, z), \end{aligned} \tag{9.2.10}$$

where

$$\begin{aligned} I_1(\tau, y, z) &:= \tau \left\langle z_r - y_r, h(z) - h(y) \right\rangle, \\ I_2(\tau, y, z) &:= \tau^2 \langle \nu_{z_r}, h(z) - h(y) \rangle, \\ I_3(\tau, y, z) &:= \tau^2 \left\langle z_r - y_r, \left(\nu'[r]h\right)(z) \right\rangle, \\ I_4(\tau, y, z) &:= \tau^3 \left\langle \nu_{z_r}, \left(\nu'[r]h\right)(z) \right\rangle. \end{aligned}$$

Here note that $I_4(\tau, y, z) = 0$ due to

$$\left\langle \nu_{z_r}, \left(\nu'[r]h\right)(z) \right\rangle = \frac{1}{2} \frac{\mathrm{d}}{\mathrm{d}t} \bigg|_{t=0} \langle \nu(z_r + th(z)), \nu(z_r + th(z)) \rangle = 0. \tag{9.2.11}$$

Now using the second estimate given in (9.2.9) and the formulas

$$\int_0^\infty \frac{\tau^{1+\gamma}\rho}{\left(\rho^2 + \tau^2\right)^{(3+\gamma)/2}} \mathrm{d}\rho = \frac{1}{1+\gamma}, \quad \gamma = 0, 1, 2, \tag{9.2.12}$$

we can easily show that the integrals $J_j(\tau, z) = \int_{\partial D_r \cap B(z,\delta)} |I_j(\tau, y, z)| \mathrm{d}\sigma(y)$ for $j = 1, 2, 3$ with a small fix $\delta > 0$ are uniformly bounded for $0 < \tau \ll 1$, $z \in \partial D$ by introducing polar coordinates around z in $\partial D_r \cap B(z, \delta)$ based on the tangent plane of ∂D at z. With these estimates for $J_j(\tau, z)$, we can immediately conclude that $\tilde{K}'[r](h)$ has its trace to ∂D_r by using the Lebesgue dominated convergence theorem. This completes the proof. □

9.3 Alternative approaches to domain derivatives

There are at least two alternative approaches to show the differentiability of a boundary value problem. The first one uses a *variational approach* and is based on showing that the variational form which defines a weak solution to the problem is Fréchet differentiable with respect to variations of the boundary. The other approach is based on the *implicit function theorem*.

9.3.1 The variational approach

Let ∂D_r be a surface given as in (9.1.1) with $r = r(x) \in C^2(\partial D, \mathbb{R}^3)$ under the assumption $\|r\|_{C^2(\partial D, \mathbb{R}^3)} \ll 1$, where we are assuming the same condition as before for the domain D and its boundary ∂D. Further, consider the far field operator

$$F^\infty(\partial D_r) = u^\infty[r],$$

where $u^{\infty}[r]$ is the far field pattern of scattered field $u^{s}[r]$ which yields from the scattering of an incident wave $u^{i} = e^{ikx\cdot d}$ with direction $d \in \mathbb{S}$ by the bounded domain D_r enclosed by the surface ∂D_r.

With this notation we have the following result on the variational domain approach given by Kirsch in [3].

Theorem 9.3.1. *$F^{\infty}[r]$ is Fréchet differentiable and its Fréchet derivative $F^{\infty\prime}$ is given by*

$$F^{\infty\prime}h = u^{\infty\prime}(h), \qquad h \in C^2(\partial D_r, \mathbb{R}^3), \tag{9.3.1}$$

where $u^{\infty\prime}$ is the far field pattern of $u' \in C^2(\mathbb{R}^3\backslash\overline{D_r}) \cap C(\mathbb{R}^3\backslash D_r)$ which solves

$$\begin{cases} (\Delta + k^2)u' = 0 \text{ in } \mathbb{R}^3\backslash\overline{D_r}, \\ u' = -(h\cdot\nu)\partial_\nu u_0 \text{ on } \partial D_r, \\ \text{Sommerfeld radiation condition (8.2.6)} \end{cases} \tag{9.3.2}$$

with the outer unit normal vector ν of ∂D_r directed into the outside D_r and the solution $u_0 \in H^2_{\mathrm{loc}}(\mathbb{R}^3\backslash\overline{D_r})$ of

$$\begin{cases} (\Delta + k^2)u_0 = 0 \text{ in } \mathbb{R}^3\backslash\overline{D_r}, \\ u_0 = 0 \text{ on } \partial D_r, \\ u_0 - u^{i}(\cdot, d) \text{ satisfies Sommerfeld radiation condition (8.2.6).} \end{cases} \tag{9.3.3}$$

Proof. To simplify the notation we will only give the proof for the case $r = 0$. Assume that $R > 0$ is chosen to satisfy $\bar{D} \subset B_{R/2}$ and define Ω_R by $\Omega_R := B_R \cap (\mathbb{R}^3\backslash\bar{D})$. By Green's formula we have

$$\int_{\Omega_R} (\nabla u_0 \cdot \nabla\bar{v}) = \int_{\partial B_R} \partial_\nu u_0 \bar{v} \tag{9.3.4}$$

for any $v \in \widetilde{H_0^1}(\Omega_R) := \{v \in H^1(\Omega_R) : v|_{\partial D} = 0\}$. Let

$$L : H^{1/2}(\partial B_R) \to H^{-1/2}(\partial B_R), \quad g \mapsto Lg := \partial_\nu w$$

be the Dirichlet-to-Neumann map where w solves

$$\begin{cases} (\Delta + k^2)w = 0 \text{ in } \mathbb{R}^3\backslash\overline{B_R}, \\ w|_{\partial B_R} = g, \\ \text{Sommerfeld radiation condition (8.2.6).} \end{cases} \tag{9.3.5}$$

Also, let $L_0 := L$ when $k = 0$ and replace the Sommerfeld radiation condition by the condition that $|x|w$, $|x|^2\nabla w$ are bounded.

Now we state two claims whose brief proofs will be given later.

Claim 1. The operator $-L_0$ is strictly coercive, i.e. there exists a constant $C > 0$ such that

$$-\langle L_0 v, v\rangle \geqslant C\|g\|^2_{H^{1/2}(\partial B_R)} \quad \text{for } g \in H^{1/2}(\partial B_R), \tag{9.3.6}$$

where $\langle \ell, s\rangle$ is the pairing for $\ell \in H^{-1/2}(\partial B_R)$ and $s \in H^{1/2}(\partial B_R)$.

Claim 2. The operator

$$L - L_0 : H^{1/2}(\partial B_R) \to H^{-1/2}(\partial B_R)$$

is compact.

Consider a variational equation for $u \in \widetilde{H_0^1}(\Omega_R)$ which solves

$$\begin{aligned} S(u, v) &:= \int_{\Omega_R} (\nabla u \cdot \nabla \bar{v} - k^2 u\bar{v}) - \langle Lu, v\rangle \\ &= \int_{\partial B_R} (\partial_\nu u^{\mathrm{i}} - Lu^{\mathrm{i}})\bar{v} \end{aligned} \tag{9.3.7}$$

for all $v \in \widetilde{H_0^1}(\Omega_R)$. In terms of the scattered field u^{s} defined by definition 8.2.1, this is nothing but

$$S(u^{\mathrm{s}}, v) = 0, \quad v \in \widetilde{H_0^1}(\Omega_R), \tag{9.3.8}$$

and a way to transform the scattering problem for acoustic waves in the infinite domain $\mathbb{R}^3 \backslash \bar{D}$ with a sound-soft obstacle D to a variational problem in the finite domain Ω_R. Decompose S into $S = S_0 + S_1$ with

$$S_0(u, v) := \int_{\Omega_R} (\nabla u \cdot \nabla \bar{v} + u\bar{v}) - \langle L_0 u, v\rangle, \tag{9.3.9}$$

and

$$S_1(u, v) := -(k^2 + 1)\int_{\Omega_R} u\bar{v} + \int_{\partial B_R} (L_0 - L)u\bar{v}. \tag{9.3.10}$$

Clearly $S_0 > 0$ is strictly coercive. By the Lax–Milgram theorem 2.5.3, there exists a unique isomorphism $T_0 : \widetilde{H_0^1}(\Omega_R) \to \widetilde{H_0^1}(\Omega_R)$ such that

$$S_0(u, v) = (T_0 u, v)_{H^1(\Omega_R)} \tag{9.3.11}$$

for $u, v \in \widetilde{H_0^1}(\Omega_R)$. Also, by the Riesz representation theorem 2.4.1, there exist a unique bounded linear operator $T_1 : \widetilde{H_0^1}(\Omega_R) \to \widetilde{H_0^1}(\Omega_R)$ and $\zeta \in \widetilde{H_0^1}(\Omega_R)$ such that

$$S_1(u, v) = (T_1 u, v)_{H^1}, \quad \int_{\partial B_R} \left(\partial_\nu u^{\mathrm{i}} - Lu^{\mathrm{i}}\right)\bar{v} = (\zeta, v)_{H^1} \tag{9.3.12}$$

for $u,\ v \in \widetilde{H_0^1}(\Omega_R)$, where $(,)_{H^1}$ denotes the inner produce to the Hilbert space $\widetilde{H_0^1}(\Omega_R)$. Hence we can write (9.3.7) as

$$u + T_0^{-1}T_1 u = T_0^{-1}\zeta \text{ in } \widetilde{H_0^1}(\Omega_R). \tag{9.3.13}$$

Here we note that:

Claim 3. $T_1 : \widetilde{H_0^1}(\Omega_R) \to \widetilde{H_0^1}(\Omega_R)$ is compact.

A brief proof of this claim will be given below. By the uniqueness of the boundary value problem

$$\begin{cases} (\Delta + k^2)u = 0 \text{ in } \Omega_R \\ u|_{\partial D} = 0, \quad \partial_\nu u|_{\partial B_R} = \partial_\nu u_0 \end{cases} \tag{9.3.14}$$

in $\widetilde{H_0^1}(\Omega_R)$ and the Fredholm alternative, u_0 is the unique solution of (9.3.13) which means that $u = u_0$ is the solution to (9.3.7). Hence we have obtained the representation of u_0 as the solution to the variational equation (9.3.7).

Next, for given $h \in C^2(\partial D, \mathbb{R}^3)$, extend h to $h \in C^2(\mathbb{R}^3\backslash D, \mathbb{R}^3)$ with $h(\mathrm{y}) = 0\ (|y| \geqslant R/2)$. Let $\varepsilon > 0$ be small and $\phi^\varepsilon(y) := y + \varepsilon h(y)$. Then, $\phi^\varepsilon : \overline{B_R}\backslash D \to \overline{B_R}\backslash D_{\varepsilon h}$ is a diffeomorphism. We further denote $\psi^\varepsilon := (\phi^\varepsilon)^{-1} : \overline{B_R}\backslash D_{\varepsilon h} \to \overline{B_R}\backslash D$ and make the change of variables $x = \phi^\varepsilon(y)$. The total field $u_\varepsilon \in H_{\mathrm{loc}}(\mathbb{R}^3\backslash \overline{D_{\varepsilon h}})$ of the scattering of incident wave u^{i} by sound-soft obstacle $D_{\varepsilon h}$ can be represented in terms of the solution of the variational equation as follows. Observe that for $u_\varepsilon := u^{\mathrm{s}}[\varepsilon h] + u^{\mathrm{i}}$, we have

$$\int_{\Omega_R[\varepsilon r]} (\nabla u_\varepsilon \cdot \nabla \bar{v} - k^2 u_\varepsilon \bar{v}) = \int_{\Omega_R} \left(\sum_{i,j=1}^{3} b_{ij}^\varepsilon \partial_{y_i} \widetilde{u_\varepsilon} \partial_{y_j} (\bar{v} \circ \phi^\varepsilon) - k^2 \widetilde{u_\varepsilon} (\bar{v} \circ \phi^\varepsilon) \right) \det J_{\phi^\varepsilon} \tag{9.3.15}$$

for any $v \in H_0^1(\Omega_R[\varepsilon h])$, where $\tilde{u}_\varepsilon = u_\varepsilon \circ \phi^\varepsilon$, J_{ϕ^ε} is the Jacobian of ϕ^ε and $b_{ij}^\varepsilon = \sum_{\ell=1}^3 \partial_{x_\ell}\psi_i \partial_{x_\ell}\psi_j^\varepsilon$ with $\psi^\varepsilon = (\psi_1^\varepsilon, \psi_2^\varepsilon, \psi_3^\varepsilon)$. Then, since $r = 0$ near ∂B_R, L remains invariant under the change of variable ϕ^ε and hence the variational equation for $\widetilde{u_\varepsilon}$ is given as

$$\begin{aligned} S^\varepsilon(u_\varepsilon, v) &:= S[\varepsilon h](u\varepsilon, v) \\ &= \int_{\Omega_R} \left(\sum_{i,j=1}^{\infty} b_{ij}^\varepsilon \partial_{y_i} \widetilde{u_\varepsilon} \partial_{y_j} \bar{v} - k^2 \widetilde{u_\varepsilon} \bar{v} \right) \det J_{\phi^\varepsilon} - \left\langle L\widetilde{u_\varepsilon}, v \right\rangle \\ &= \int_{\partial B_R} (\partial_\nu u^{\mathrm{i}} - Lu^{\mathrm{i}})\bar{v} \ \left(v \in \widetilde{H_0^1}(\Omega_R) \right). \end{aligned} \tag{9.3.16}$$

Based on (9.3.15) and (9.3.16), we will compute the strong limit

$$\lim_{\varepsilon \to 0} \varepsilon^{-1}(\widetilde{u_\varepsilon} - u_0) \quad \text{in } \widetilde{H_0^1}(\Omega_R).$$

Observe that for any $v \in \widetilde{H_0^1}(\Omega_R)$, we have

$$\begin{aligned} S(\varepsilon^{-1}(\widetilde{u}_\varepsilon - u_0), v) &= \varepsilon^{-1}S(\widetilde{u}_\varepsilon, v) - \varepsilon^{-1}S(u_0, v) \\ &= -\varepsilon^{-1}(S^\varepsilon(\widetilde{u}_\varepsilon, v) - S(\widetilde{u}_\varepsilon, v)) \\ &= -\varepsilon^{-1}\int_{\Omega_R}\left(\sum_{i,j=1}^{3}\left(b_{ij}^\varepsilon \det J_{\phi^\varepsilon} - \delta_{ij}\right)\partial_{y_i}\widetilde{u}_\varepsilon\partial_{y_j}\bar{v}\right. \\ &\quad \left. -k^2\left(\det J_{\phi^\varepsilon} - 1\right)\widetilde{u}_\varepsilon\bar{v}\right), \end{aligned}$$

where δ_{ij} is the Kronecker delta and we have used

$$S(u_0, v) = \int_{\partial B_R}(\partial_\nu u^{\mathrm{i}} - Lu^{\mathrm{i}})\bar{v} = S^\varepsilon(\widetilde{u}_\varepsilon, v).$$

Here by direct computations, we have

$$\begin{aligned} \varepsilon^{-1}\left(\det J_{\phi^\varepsilon} - 1\right) &= \operatorname{div} h + O(\varepsilon), \\ \varepsilon^{-1}\left(b_{ij}^\varepsilon \det J_{\phi^\varepsilon} - \delta_{ij}\right) &= (\operatorname{div} h)\delta_{ij} - \left(\partial_{y_j}h_i + \partial_{y_i}h_j\right) + O(\varepsilon) \end{aligned}$$

uniformly on $\overline{\Omega_R}$ for any i, j $(1 \leqslant i, j \leqslant 3)$ as $\varepsilon \to 0$, where $h = (h_1, h_2, h_3)$. Hence we have

$$\begin{aligned} &S(\varepsilon^{-1}(\widetilde{u}_\varepsilon - u_0), v) \\ &= \int_{\Omega_R}\left\{\sum_{i,j=1}^{3}\left(\partial_{y_j}h_i + \partial_{y_i}h_j - (\operatorname{div} h)\delta_{ij}\right)\partial_{y_i}u_0\partial_{y_j}\bar{v} + k^2u_0\bar{v}\operatorname{div} h\right\} + O(\varepsilon)\|v\|_{H^1} \\ &= \int_{\Omega_R}\left\{\left(J_h + J_h^\top - (\operatorname{div} h)I\right)\nabla u_0 \cdot \nabla\bar{v} + k^2u_0\bar{v}\operatorname{div} h\right\} + O(\varepsilon)\|v\|_{H^1}\left(v \in \widetilde{H_0^1}(\Omega_R)\right) \end{aligned}$$

as $\varepsilon \to 0$, where $\|v\|_{H^1} = \sqrt{(v, v)_{H^1}}$. Thus there exists a unique $\tilde{u} \in \widetilde{H_0^1}(\Omega_R)$ such that $\tilde{u}$ is the strong limit of $\varepsilon^{-1}(\widetilde{u}_\varepsilon - u_0)$ as $\varepsilon \to 0$ and it satisfies

$$S(\tilde{u}, v) = \int_{\Omega_R}\left\{\left(J_h + J_h^\top - (\operatorname{div} h)I\right)\nabla u_0 \cdot \nabla\bar{v} + k^2u_0\bar{v}\operatorname{div} h\right\} \qquad (9.3.17)$$

for any $v \in \widetilde{H_0^1}(\Omega_R)$. Since $r = 0$ near ∂B_R, this implies $\partial_\nu\tilde{u} = L\tilde{u}$ on ∂B_R. Therefore we can extend $\tilde{u}$ outside Ω_R by solving the exterior Dirichlet problem for $\Delta + k^2$ in $\mathbb{R}^3\backslash\bar{B}_R$ with Dirichlet data $\tilde{u}|_{\partial B_R}$. By the well-known regularity result we have $u_0 \in \widetilde{H_0^1}(\Omega_R) \cap H^2(\Omega_R)$. Also, by a direct computation, we have for any $v \in \widetilde{H_0^1}(\Omega_R) \cap H^2(\Omega_R)$

$$\begin{aligned} \left(J_h + J_h^\top - (\operatorname{div})I\right)\nabla u_0 \cdot \nabla\bar{v} &= \operatorname{div}\left[(h \cdot \nabla u_0)\nabla\bar{v} + (h \cdot \nabla\bar{v})\nabla u_0 - \left(\nabla u_0 \cdot \nabla\bar{h}\right)h\right] \\ &\quad -(h \cdot \nabla u_0)\Delta\bar{v} - (h \cdot \nabla\bar{v})\Delta u_0. \end{aligned}$$

Hence, by the Gauss theorem and $r = 0$ near ∂B_R, we have for any $v \in \widetilde{H_0^1}(\Omega_R)$

$$S(\tilde{u},v) = \int_{\Omega_R} \{k^2 u_0 \bar{v} \operatorname{div} h - (h \cdot \nabla u_0)\Delta \bar{v} - (h \cdot \nabla \bar{v})\Delta u_0\}$$
$$- \int_{\partial D} \{(h \cdot \nabla u_0)\nabla \bar{v} + (h \cdot \nabla \bar{v})\nabla u_0 - (\nabla u_0 \cdot \nabla \bar{v})h\} \cdot \nu.$$

It is a bit tedious but not difficult to see that integrating the right-hand side of this by parts yields

$$S(\tilde{u}, v) = k^2 \int_{\Omega_R} \nabla \cdot (u_0 \bar{v} h) + \int_{\Omega_R} \{\nabla(h \cdot \nabla u_0) \cdot \nabla \bar{v} - k^2(h \cdot \nabla u_0)\bar{v}\}$$
$$= \int_{\Omega_R} \{\nabla(h \cdot \nabla u_0) \cdot \nabla \bar{v} - k^2(h \cdot \nabla u_0)\bar{v}\}$$

for any $v \in \widetilde{H_0^1}(\Omega_R)$. Hence by the definition of $S(\tilde{u}, v)$, we have obtained for any $v \in \widetilde{H_0^1}(\Omega_R)$

$$\int_{\Omega_R} (\nabla \tilde{u} \cdot \nabla \bar{v} - k^2 \tilde{u}\bar{v}) - \langle L\tilde{u}, v\rangle = \int_{\Omega_R} \{\nabla(h \cdot \nabla u_0) \cdot \nabla \bar{v} - k^2(h \cdot \nabla u_0)\bar{v}\}.$$

This implies that

$$(\Delta + k^2)(\tilde{u} - h \cdot \nabla u_0) = 0 \text{ in } \Omega_R$$
$$(\tilde{u} - h \cdot \nabla u_0)|_{\partial D} = -(h \cdot \nu)\partial_\nu u_0, \quad \partial_\nu \tilde{u}|_+ = L\tilde{u} = \partial_\nu \tilde{u}|_-,$$

where '$|_+$' and '$|_-$' denote the traces to $\partial\Omega_R$ from outside and inside B_R, respectively. Hence $\tilde{u} \in C^2(\mathbb{R}^3)$ and by the definition of u', we have $u' = \tilde{u} - h \cdot \nabla u_0$ which yields $u' = -(h \cdot \nu)\partial_\nu u_0$ on ∂D.

Since $\tilde{u} = \lim_{\varepsilon \to 0} \varepsilon^{-1}(\widetilde{u_\varepsilon} - u_0) = \tilde{u}$ strongly in $\widetilde{H_0^1}(\Omega_R)$, $\widetilde{u_\varepsilon} = u_\varepsilon \circ \phi^\varepsilon$ and $r = 0$ near ∂B_R, we have the strong limit

$$\lim_{\varepsilon \to 0} \varepsilon^{-1}(u_\varepsilon - u_0) = u' \text{ in } H^{1/2}(\partial B_R). \tag{9.3.18}$$

To finish the proof we recall the well-known representation formula of the far field pattern in terms of the scattered field

$$u_0^\infty(\hat{x}) = \frac{1}{4\pi} \int_{\partial B_R} \left(u_0^s(y)\partial_{\nu_y} e^{-ik\hat{x}\cdot y} - \partial_\nu u_0^s(y) e^{-ik\hat{x}\cdot y}\right) ds(y)$$
$$= \frac{1}{4\pi}\left\{\int_{\partial B_R} (u_0 - u^i)(y)\partial_{\nu_y} e^{-ik\hat{x}\cdot y}\, ds(y) - \left\langle L\left(u_0 - u^i\right), e^{ik\hat{x}\cdot(\cdot)}\right\rangle\right\} \tag{9.3.19}$$

for $\hat{x} \in \mathcal{S}$ where $u_0^s = u_0 - u^i$. There is a similar formula for u_ε^∞ that is the formula obtained by replacing u_0^∞ and u_0^s in (9.3.19) by u_ε^∞ and $\tilde{u}_\varepsilon$, respectively. Hence we have

$$\varepsilon^{-1}\Big(u_\varepsilon^\infty(\hat{x}) - u_0^\varepsilon(\hat{x})\Big) = \frac{1}{4\pi}\int_{\partial B_R} \varepsilon^{-1}(\widetilde{u_\varepsilon} - u_0)\partial_\nu e^{-ik\hat{x}\cdot(\cdot)} - \Big\langle \varepsilon^{-1}L(\widetilde{u_\varepsilon} - u_0), e^{ik\hat{x}\cdot(\cdot)}\Big\rangle \tag{9.3.20}$$

for any $\hat{x} \in \mathcal{S}$, and taking the limit as $\varepsilon \to 0$, we have from (9.3.18)

$$\begin{aligned}\lim_{\varepsilon\to 0}(u_\varepsilon^\infty(\hat{x}) - u_0^\infty(\hat{x})) &= \frac{1}{4\pi}\Big\{\int_{\partial B_R} u'\partial_\nu e^{-ik\hat{x}\cdot(\cdot)} - \Big\langle Lu', e^{ik\hat{x}\cdot(\cdot)}\Big\rangle\Big\}\\ &= \frac{1}{4\pi}\int_{\partial B_R}\Big(u'\partial_\nu e^{-ik\hat{x}\cdot(\cdot)} - \partial_\nu u' e^{-ik\hat{x}\cdot(\cdot)}\Big)\\ &= u'^\infty(\hat{x})\end{aligned}$$

uniformly for $\hat{x} \in \mathcal{S}$.

Finally, we will give brief proofs of claim 1–claim 3. The proofs of claims 1 and 2 are similar and will be given at the same time as follows. It is well known that w_0 and w have the following representations in terms of spherical harmonics $\{Y_n^m\}$

$$\begin{aligned} w_0(x) &= \sum_{n=0}^{\infty}\sum_{|m|\leqslant n} g_n^m \frac{R^{n+1}}{\rho^{n+1}} Y_n^m(\hat{x})\\ w(x) &= \sum_{n=0}^{\infty}\sum_{|m|\leqslant n} g_n^m \frac{H_n^{(1)}(k\rho)}{h_n^{(1)}(kR)} Y_n^m(\hat{x}),\end{aligned} \tag{9.3.21}$$

where $\rho = |x|$, the spherical Bessel function of the first kind $h_n^{(1)}$ and

$$g = \sum_{n=0}^{\infty}\sum_{|m|\leqslant n} g_n^m Y_n^m \quad \text{with} \sum_{n=0}^{\infty}\sum_{|m|\leqslant n}(n+1)|g_n^m|^2 < \infty.$$

Then, using $L_0 = \partial_\rho$ and the orthonormality of $\{Y_n^m\}$, we have for any $g \in H^{1/2}(\partial B_R)$

$$-\langle L_0 g, g\rangle = R^{-1}\sum_{n=0}^{\infty}\sum_{|m|\leqslant n}(n+1)|g_n^m|^2 \geqslant C\|g\|^2_{H^{1/2}(\partial B_R)} \tag{9.3.22}$$

for some constant $C > 0$ independent of g. By $L = \partial_\rho$ and the asymptotics $kRh_n^{(1)\prime}/h_n^{(1)}(kR) = -(n+1)(1 + O(1/n))$ as $n \to \infty$, $L - L_0$ is compact.

To show claim 3, let a sequence (u_n) be bounded in $\widetilde{H_0^1}(\Omega_R)$. Then by the compactness of the embedding $\widetilde{H_0^1}(\Omega_R) \hookrightarrow L^2(\Omega_R)$ and $L - L_0 : H^{1/2}(\partial B_R) \to H^{-1/2}(\partial B_R)$, there is a subsequence (u_n') of (u_n) such that (u_n') and $((L_0 - L)u_n')$ are convergent in $L^2(\Omega_R)$ and $H^{-1/2}(\partial B_R)$, respectively. Hence

$$\begin{aligned}\left\| T_1 u_n' - T_1 u_m' \right\| &= \sup_{\|v\|_{H^1}\leqslant 1} \left| S_1\left(u_n' - u_m', v\right) \right| (k^2 + 1)\|u_n' - u_m'\|_{L^2(\Omega_R)} \\ &\quad + \left\| (L_0 - L)\left(u_n' - u_m'\right) \right\|_{H^{-1/2}(\partial B_R)} \\ &\to 0 \quad \text{for } n, m \to \infty, \end{aligned} \tag{9.3.23}$$

i.e. we have shown that there is a bounded subsequence of $(T_1 u_n)$, thus T_1 is compact by theorem 2.3.15. □

9.3.2 Implicit function theorem approach

Assume that we have a mapping $f : Z \to Y$ from a Banach space Z into a Banach space Y, which depends on a function $r \in X$ in a Banach space X. We note the dependence of f on r by f_r or $f[r]$. Now, let an element $z \in Z$ be defined by

$$f_r(z) = 0. \tag{9.3.24}$$

Then, by (9.3.24) the function or element z depends on r, i.e. $z = z_r$.

To obtain local uniqueness for the nonlinear equation (9.3.24) we assume that $f_r : Z \to Y$ is Fréchet differentiable with Fréchet derivative $\partial_z f_r(z)$ at $z \in Z$. Then, we have

$$f_r(z + \tau) = f_r(z) + \partial_z f_r(z)\tau + \tilde{f}_r(z, \tau) \tag{9.3.25}$$

where $\tilde{f}_r(z, \tau) = o(\|\tau\|)$. Local uniqueness is obtained when $\partial_z f_r(z) : Z \to Y$ is boundedly invertible, since then for all sufficiently small $\tau \in Z$

$$\frac{1}{\|\tau\|}\left(\partial_z f_r(z)\right)^{-1}\left(f_r(z + \tau) - \underbrace{f_r(z)}_{=0} \right) = \frac{\tau}{\|\tau\|} + \left(\partial_z f_r(z)\right)^{-1}\frac{\tilde{f}_r(z, \tau)}{\|\tau\|} \neq 0,$$

which implies that we cannot have $f_r(z + \tau) = 0$. We assume that we have bounded invertibility of the derivative $\partial_z f_r(z)$. Now, let $f[r]$ be Fréchet differentiable with respect to $r \in X$, i.e. we have

$$f[r + h](z) = f[r](z) + \left(\partial_r f[r](z)\right)(h) + f_1[r, h](z), \quad h \in X \tag{9.3.26}$$

with $f_1[r, h](z) = o(\|h\|)$ as $\|h\| \to 0$. This now leads to

$$0 = f[r + h](z_{r+h}) = f[r](z_{r+h}) + \left(\partial_r f[z_{r+h}]\right)(h) + f_1[r, h](z_{r+h}) \tag{9.3.27}$$

for $h \in X$. Following (9.3.25) we evolve

$$f[r](z_{r+h}) = f[r](z_r) + \left(\partial_z f[r](z_r)\right)(z_{r+h} - z_r) + o(\|z_{r+h} - z_r\|). \tag{9.3.28}$$

By combining (9.3.27), (9.3.28) and using $f[r](z_r) = 0$, we have

$$\left(\partial_z f[r](z_r)\right)(z_{r+h} - z_r) = -\left(\partial_r f[r](z_{r+h})\right)(h) - f_1[r, h](z_{r+h}) + o(\|z_{r+h} - z_r\|).$$

We now multiply by the inverse of $\partial_z f[r](z_r)$ to obtain

$$\begin{aligned} z_{r+h} - z_r = &-\left(\partial_z f[r](z_r)\right)^{-1}(\partial_r f[r](z_{r+h}))(h) \\ &+ o(\|h\|) + o(\|z_{r+h} - z_r\|). \end{aligned} \qquad (9.3.29)$$

From the last equation by taking the norm we first obtain the estimate

$$\|z_{r+h} - z_r\| \leqslant c\|h\| + o(\|z_{r+h} - z_r\|)$$

with some constant $c > 0$, leading to

$$\|z_{r+h} - z_r\| \left(1 - \frac{o\left(\|z_{r+h} - z_r\|\right)}{\|z_{r+h} - z_r\|}\right) \leqslant c\|h\|$$

and thus

$$\|z_{r+h} - z_r\| \leqslant \tilde{c}\|h\| \qquad (9.3.30)$$

with some constant $\tilde{c} > 0$ for all $\|h\|$ sufficiently small and for z_{r+h} in a neighborhood of z_r, which proves the local Lipschitz continuity of z_r with respect to r. Now, by

$$\frac{o\left(\|z_{r+h} - z_r\|\right)}{\|h\|} = \frac{o\left(\|z_{r+h} - z_r\|\right)}{\|z_{r+h} - z_r\|} \cdot \frac{\|z_{r+h} - z_r\|}{\|h\|} \to 0, \quad \|h\| \to 0$$

the estimate (9.3.30) implies that $o(\|z_{r+h} - z_r\|) = o(\|h\|)$, and thus (9.3.29) proves the Fréchet differentiability of z_r with respect to r and also provides a formula for the derivative. We summarize this in the following theorem.

Theorem 9.3.2 (Implicit function theorem). *Let X, Y, Z be Banach spaces. Assume that $f_r : Z \to Y$ is Fréchet differentiable and that $f_r = f[r]$ in the space of Y-valued functions on Z is Fréchet differentiable with respect to $r \in X$, then the implicit solution z_r of $f[r](z) = 0$ depends Fréchet differentiably on $r \in X$ with derivative*

$$\frac{\partial z_r}{\partial r}(h) = -\left(\partial_z f[r](z_r)\right)^{-1}\left(\partial_r f[r](z_r)\right)(h), \quad r, h \in X \qquad (9.3.31)$$

provided that $\partial_z f[r](z_r)$ has a bounded inverse.

Remark. We note that (9.3.31) is quite similar to the implicit function theorem in the usual calculus in $\mathbb{R}^m$. □

Next we want to show how the above implicit function theorem can be used to prove the Fréchet differentiability of the scattering problem.

First, we take $R > 0$ large enough so that the obstacle D is compactly embedded in B_R which is an open ball with radius R centered at the origin and extend the vector field $r \in C^2(\partial D, \mathbb{R}^3)$ into a vector field $r \in C^2(B_R)$ which is compactly supported in the open ball B_R. This can be easily obtained by setting

$$r(x + \tau\nu(x)) := \chi(\tau)r(x), \quad x \in \partial D, \quad \tau \in (-\rho, \rho), \tag{9.3.32}$$

where $\chi : (-\rho, \rho) \mapsto \mathbb{R}$ is a compactly supported C^2 function and $\rho > 0$ is sufficiently small. Also, we transform the Laplace operator Δ on $\mathbb{R}^3\backslash\overline{D_r}$ into an operator Δ_r defined on $\mathbb{R}^3\backslash\bar{D}$ by using the transform $R : x \mapsto x + r(x)$ of the space $\mathbb{R}^3\backslash\bar{D}$ into $\mathbb{R}^3\backslash\overline{D_r}$.

Now, we define the function f on the space Z of functions which are C^2 in $\mathbb{R}^3\backslash\bar{D}$, have boundary values in $C(\partial D)$ and satisfy the Sommerfeld radiation condition (8.2.6) for $|x| \to \infty$ uniformly for all directions. We set

$$f[r](u^s) := \begin{pmatrix} \Delta_r u^s + \kappa^2 u^s \\ u^s|_{\partial D} + u^i(\cdot + r(\cdot)) \end{pmatrix} \tag{9.3.33}$$

which has values in $Y := BC(\mathbb{R}^3\backslash\bar{D}) \times C(\partial D)$. Clearly, the mapping f is Fréchet differentiable with respect to its argument u^s, since it is of the form 'linear plus constant'. Its Fréchet derivative is given by

$$\big(\partial_z f[r](z)\big)(\tau) = \begin{pmatrix} \Delta_r\tau + k^2\tau \\ \tau|_{\partial D} \end{pmatrix}, \quad \tau \in Z. \tag{9.3.34}$$

We also note that the solution of the equation

$$\big(\partial_z f[r](z)\big)(\tau) = \begin{pmatrix} q \\ b \end{pmatrix} \tag{9.3.35}$$

with a function q defined on $\mathbb{R}^3\backslash\bar{D}$ and compactly supported in B_R and a continuous function $b \in C(\partial D)$ can be uniquely solved, since it corresponds to an inhomogeneous exterior problem for the Helmholtz equation with inhomogeneous term $q \circ R^{-1}$ and Dirichlet boundary data $b \circ R^{-1}$ in $\mathbb{R}^3\backslash\bar{D}$, and the solution depends continuously on the boundary data and the inhomogeneous term $q \circ R^{-1}$ is supported in B_R. Hence $\partial_z f[r](z)$ has a bounded inverse.

The function $f[r]$ is Fréchet differentiable with respect to r, which is an immediate consequence of the differentiability of the coefficients of the transformed Laplacian Δ_r and the differentiability of the function $u^i(x_r)$ with respect to r. Then, we can apply theorem 9.3.2 to obtain basically part (a) of theorem 9.2.1 as follows.

Theorem 9.3.3. *The mapping of $r \in C^2(\partial D, \mathbb{R}^3)$ onto the scattered field $u^s[r](\cdot, d)$ defined by (8.2.1)–(8.2.3) and the Sommerfeld radiation condition (8.2.6) with $\partial D = \partial D_r$ is Fréchet differentiable. Its derivative is obtained by applying (9.3.31) to the function f defined in (9.3.33).*

Remark. As for the solution u^s to 9.3.34, the regularity of the Fréchet derivative up to the boundary is obtained by standard regularity results for these problems, such that we can now obtain the result of theorem 9.2.2 without the difficulties which we had for justifying (9.2.6).

9.4 Gradient and Newton methods for inverse scattering

The goal of this section is to work out the *gradient method* and the *Newton method* to solve the inverse obstacle scattering problem based on the domain derivative according to theorem 9.2.2 when the far field pattern u^∞ for scattering of some incident wave u^i is given.

Let us start with the Newton iteration (3.3.5), or its regularized form (3.3.8), which we apply to finding the unknown boundary ∂D of the scatterer D. Newtonian iterations are shown in figures 9.1 and 9.2 below. The update formula becomes

$$\partial D_{k+1} = \partial D_k - R_{\alpha_k} H(\partial D_k), \quad k = 1, 2, 3, \ldots, \tag{9.4.1}$$

where the regularization of the inverse $(H')^{-1}$ of H' is carried out by Tikhonov regularization.

$$R_\alpha = (\alpha I + (H')^* H)^{-1} (H')^*, \quad \alpha > 0, \tag{9.4.2}$$

is based on the derivative H' of the far field pattern u^∞ with respect to the boundary shape ∂D. Here, the mapping $H(\partial D)$ is given by

$$H(\partial D) := u^\infty(\partial D) - u^\infty(\partial D_{\text{true}}). \tag{9.4.3}$$

To evaluate the derivative H' according to theorem 9.2.2 for a sound-soft scatterer D, we need to calculate the normal derivative

$$\frac{\partial u}{\partial \nu} = \frac{\partial u^s}{\partial \nu} + \frac{\partial u^i}{\partial \nu} \quad \text{on } \partial D \tag{9.4.4}$$

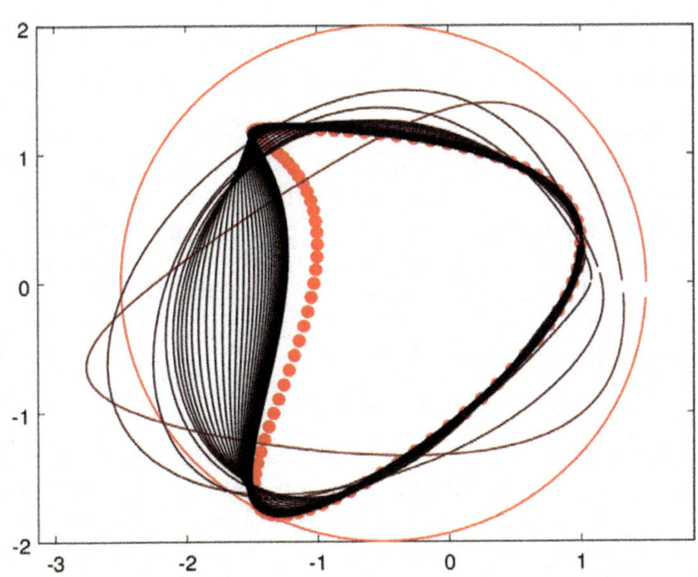

Figure 9.1. Iterations of Newton's method to reconstruct some scatterer (red dots) from its far field pattern u^∞.

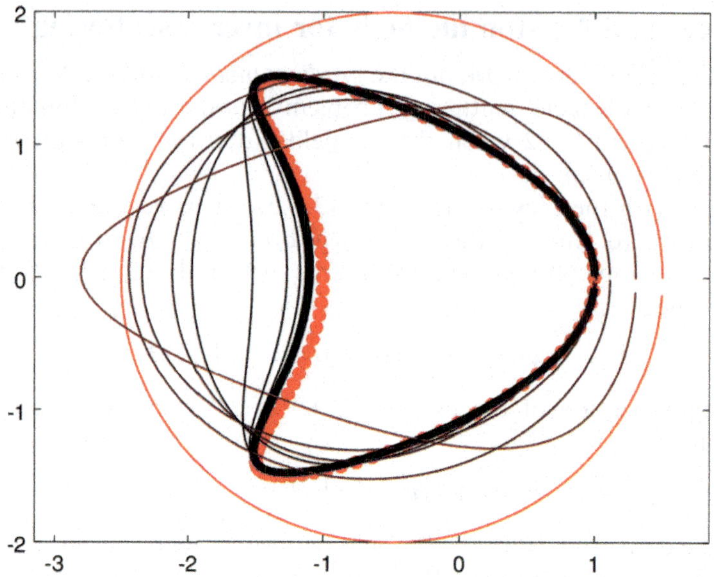

Figure 9.2. The iterations 1, 2, 3, 4, 5, 10,…, 50 for Newton's method applied to the inverse sound-soft scattering problem with 5% random noise on the far field pattern. Here, we apply Newton's method in parameter space, i.e. we used it to reconstruct the Fourier coefficients of the boundary curve following code 9.4.2 with regularization parameter $\alpha = 0.5$. The incident wave is coming from the right, the 50th iteration is shown in bold black. The total field is shown in figure 8.6.

first. Then, to calculate H' in direction $h \in C^2(\partial D, \mathbb{R}^m)$ we need to evaluate $\langle h, \nu \rangle$ and then solve the Dirichlet exterior boundary value problem with boundary data given by

$$g := -\langle \nu, h \rangle \frac{\partial u}{\partial \nu}. \tag{9.4.5}$$

The far field pattern u_g^∞ of its solution is the Fréchet derivative of $u^\infty(\partial D)$ with respect to the variation of the domain by h.

For the calculation of the normal derivative of u^s let us use a single-layer approach for solving the scattering problem, i.e. we search a solution u^s in the form

$$u^s(x) := \int_{\partial D} \Phi(x, y)\varphi(y)\, ds(y), \quad x \in \mathbb{R}^m \backslash D, \tag{9.4.6}$$

with boundary values

$$S\varphi = -u^i \quad \text{on } \partial D. \tag{9.4.7}$$

Clearly, the single-layer potential (9.4.6) satisfies the Helmholtz equation (8.2.1) in $\mathbb{R}^m \backslash \bar{D}$ and it satisfies the Sommerfeld radiation condition (8.2.6). The operator S maps $L^2(\partial D)$ into $H^1(\partial D)$ and if the interior homogeneous Dirichlet problem for D has only the trivial solution, it is boundedly invertible. With a solution φ of (9.4.7) it also satisfies the sound-soft boundary condition (8.2.3), such that u^s defined by

(9.4.6) and (9.4.7) solves the scattering problem. The normal derivative of the field u^s is given by the *jump relation*

$$\frac{\partial u^s}{\partial \nu_\pm}(x) = \int_{\partial D} \frac{\partial \Phi(x, y)}{\partial \nu(x)} \varphi(y)\, \mathrm{d}s(y) \mp \frac{1}{2}\varphi(x), \quad x \in \partial D, \tag{9.4.8}$$

for a proof for continuous densities φ we refer to [4]. We denote the integral part of (9.4.8) by K' as defined in (8.1.33), although it is the adjoint of K only with respect to a real-valued scalar product.

Based on the codes 8.1.6 and 8.2.2 we can calculate the discretized operator K' and the normal derivative of u^s on ∂D in a simple way shown in code 9.4.1.

Code 9.4.1. *The calculation of the adjoint double-layer potential operator K' defined on the boundary ∂D. The calculation is carried out by the file* `sim_09_4_1_c_Kp_dus.m`.

```
nu1    = dy2./sqrt(dy1.^2+dy2.^2);   % normal component 1
nu2    = -dy1./sqrt(dy1.^2+dy2.^2);  % normal component 2
nu1mat = repmat(nu1',1,N);           % normal matrix
nu2mat = repmat(nu2',1,N);           % ~

% Operator K'
Kp = -i*kappa/2*(nu1mat.*rmat1+nu2mat.*rmat2).* ...
     besselh(1,1,kappa*rmat)./rmat.*drmat*ht;

% Normal derivative of scattered field
dus = 0.5*( Kp*varphi - varphi);
```

We have added a script `sim_09_4_1_e_dus_test.m` into the code repository which compares the normal derivative $\partial u^s/\partial \nu$ calculated by (9.4.8) with its approximation by a finite difference quotient. The comparison is shown in figure 9.3.

We can now use the Brakhage–Werner approach (8.2.7) and (8.2.8) to solve the scattering problem and calculate the derivative H' by evaluating (8.3.2) and (8.3.4) as in (8.3.9). With the standard discretization of the boundary ∂D by vectors `y1` and `y2` we write the term (9.4.5) in the form of a matrix operating on the h discretized as

$$\mathbf{h} := \begin{pmatrix} \mathtt{h1} \\ \mathtt{h2} \end{pmatrix} \tag{9.4.9}$$

with the $N \times 1$-vectors `h1` and `h2`. We define

$$\mathbf{N}_j := \mathrm{diag}\{\nu_j\} \in \mathbb{R}^{N\times N}, \quad j = 1, 2, \qquad N := (N_1, N_2) \in \mathbb{R}^{N\times(2N)} \tag{9.4.10}$$

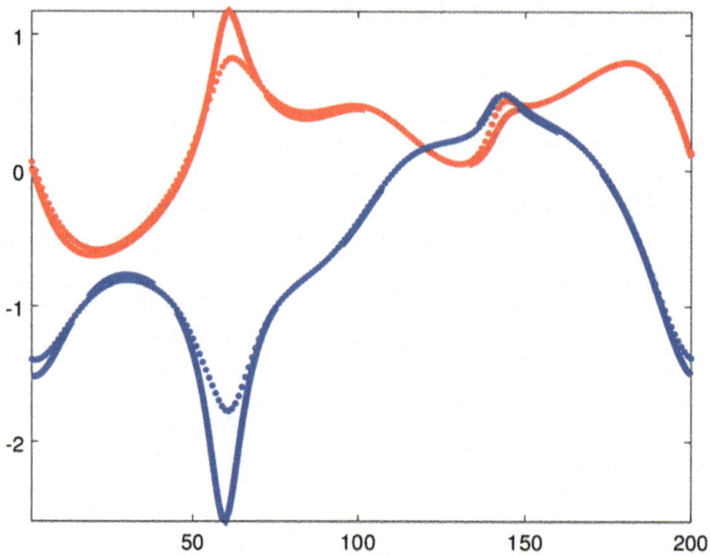

Figure 9.3. A comparison of the field $\partial u^s/\partial \nu$ calculated by (9.4.8) with its finite difference approximation calculated by `sim_09_4_1_e_dus_test.m` with $N = 100$ boundary points (in the case of linear convergence due to cutting the singularity). The approximation is shown by dotted lines, the real part in red and the imaginary part in blue.

such that $\langle \nu, h \rangle$ is carried out by

$$\mathbf{N}\mathbf{h} = \mathbf{N}_1 \mathtt{h1} + \mathbf{N}_2 \mathtt{h2} = \left(\nu_1(x_\xi)\, h_1(x_\xi) + \nu_2(x_\xi)\, h_2(x_\xi)\right)_{\xi=1,\ldots,N}. \tag{9.4.11}$$

We also set

$$\mathbf{u} := \left(\frac{\partial u^s}{\partial \nu}(x_\xi)\right)_{\xi=1,\ldots,N}, \qquad \mathbf{U} = \mathrm{diag}\{\mathbf{u}\}. \tag{9.4.12}$$

We now define the linear operator

$$\mathbf{G} = (\mathbf{K}^\infty - i\eta\mathbf{S}^\infty)(\mathbf{I} + \mathbf{K} - i\eta\mathbf{S})^{-1}(2\mathbf{U}\mathbf{N}) \tag{9.4.13}$$

mapping a vector field $h \in C^2(\partial D)$ onto the Fréchet derivative $H'(h)$ of u^∞ with respect to ∂D. $\mathbf{G}$ can be understood as an $L \times 2N$-matrix when L far field points and N discretization points on ∂D are chosen.

We are now prepared for testing Newton's method as formulated in (9.4.1). Here, we will just prepare a simple version without a *stopping rule* as introduced in section 3.3.3. The following code employs a parametrization of the boundary ∂D by its *Fourier coefficients* and determines an update to the Fourier coefficients in each Newton step. With a low-order approximation this guarantees that the regularity of the boundary stays C^∞. The result is shown in figures 9.1 and 9.2.

Code 9.4.2. *Newton's method for an inverse sound-soft scattering problem. The script is* `sim_09_4_2_Newton.m`. *The output is shown in figure 9.2.*

```
close all; clear all;
% Generate far field data for true domain
N     = 100;           % number of discretization points for curve
kappa = 2;             % wave number
ffN   = 100;           % number of far field evaluation points
beta  = 0*pi;          % direction of wave incidence
rand("seed",1);        % set random seed to make random function repeatable
ht    = 2*pi/N;        % grid size for curve discretization
t     = 0:ht:2*pi-ht;  % discretization points for curve parametrization
PAR = [ones(size(t')) sin(t') cos(t') sin(2*t') cos(2*t') zeros(N,5); ...
       zeros(N,5) ones(size(t')) sin(t') cos(t') sin(2*t') cos(2*t')];

% definition of true domain yo with components y1o and y2o
%p0        = [-0.65 0 1 0 0.65 0 1.5 0 0 0]';
p0         = [-0.65 0 1 0 0.65 0 1.5 0 0 0.3]';
yo         = PAR*p0;    % the points y0 = [yo1; y02]
% calculate far field pattern ffo, add random error
ffotmp     = sim_08_3_7_calc_ff(yo,ffN,kappa,beta);
ffo        = ffotmp.*(1+0.05*1i*(rand(size(ffotmp))-0.5));

% Initialization of dD_0
yp         = [-0.5 0 2 0 0 0 2 0 0 0]';
ypa(:,1) = yp;          % store iteration domains parameters

% Newton Loop
alpha    = 1;                   % regularization parameter for Newton step
Nstep    = 30;                     % number of Newton steps (no stopping rule)
Np       = size(p0,1);             % size of parameter space
for kk =1: Nstep
  ff           = sim_08_3_7_calc_ff(PAR*yp,ffN,kappa,beta);
  eff(kk)      = norm(ff-ffo);   % calculate far field error
  eyp(kk)      = norm(yp-p0);    % calculate parameter error
  ffG          = sim_09_4_7_calc_ffG(PAR*yp,ffN,kappa,beta); % H' on boundary
  A            = ffG*PAR;        % H' on parameter space
  yp           = yp + real((alpha*eye(Np,Np)+A'*A)\(A'*(ff-ffo))); % Newton step
  ypa(:,kk+1) = yp; % store iterate
end
ff             = sim_08_3_7_calc_ff(PAR*yp,ffN,kappa,beta);
eff(Nstep+1)   = norm(ff-ffo);   % calculate last far field error

eyp(Nstep+1)   = norm(yp-p0);    % calculate last parameter error

% Visualization
fo = figure;                        % show some iterations, first true domain
po = plot3(yo(1:N),yo((N+1):2*N),-0.1*ones(1,N),'r.','MarkerSize',13);
hold on; ao = get(po,'Parent'); set(ao,'FontSize',14); % axis controls
for kk=[1,2,3,4,5:ceil(Nstep/Nstep):(Nstep+1)]  % Show some iteration domains
  y = PAR*ypa(:,kk);                % calculate the boudary points
  plot(y(1:N),y((N+1):2*N),'k','LineWidth',1,'Color',[1/kk 0 0]);
end
y = PAR*ypa(:,Nstep+1);             % calculate last iteration boundary
plot(y(1:N),y((N+1):2*N),'k','LineWidth',5,'Color',[0 0 0]); % plot it
axis equal;  view(2);               % axis control

savefile(fo,'sim_09_4_2_Newton');   % save file
```

Gradient method for obstacle scattering. The *gradient method* is the second basic approach to solve a nonlinear inverse problem, see equation (3.3.14), (3.3.29) or (3.3.31) in section 3.3.2. Here, the functional μ is given by

$$\mu(\partial D) := \left\| u^{\infty}[\partial D] - u^{\infty}_{\text{meas}} \right\|^2, \tag{9.4.14}$$

where the norm is $L^S(\mathbb{B}S)$ norm, and the update equation is given by

$$\partial D_{k+1} = \partial D_k + \eta_k (H')^* \left(u^{\infty}_{\text{meas}} - H(\partial D) \right), \quad k = 1, 2, 3, \ldots. \tag{9.4.15}$$

The derivative H' of u^{∞} with respect to ∂D is given in theorem 9.2.2. We have calculated its finite-dimensional approximation by a boundary integral equation in (9.4.13).

It is well known that the *gradient method* exhibits a low convergence, since when you approach the minimum the derivative tends to zero and thus does the gradient. This is demonstrated by figure 9.4. A directional search can speed up the minimization significantly.

Code 9.4.3. *The* gradient method *for an inverse sound-soft scattering problem. This is the script* `sim_09_4_3_b_Gradient.m`, *where the set-up given in script* `sim_09_4_3_a_Setup.m` *is the same as for the Newton method and so is the visualization* `sim_09_4_3_c_Visualization.m`. *The output is shown in figure 9.5.*

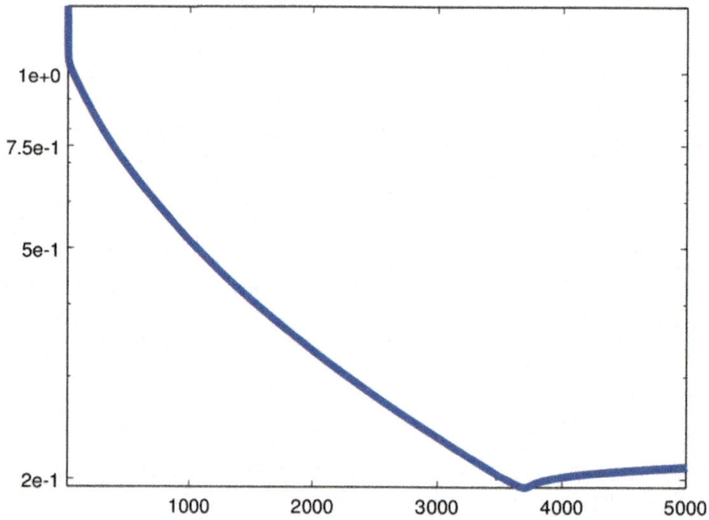

Figure 9.4. The convergence of the error in parameter space for the *gradient method* (9.4.15). The gradient method needs many iterations to find the minimum.

```
% Gradient Loop
eta         = 0.0025;              % step size for gradient step
Nstep       = 500;                 % number of gradient steps
Np          = size(p0,1);          % size of parameter space
for kk =1: Nstep
  ff             = sim_08_3_7_calc_ff(PAR*yp,ffN,kappa,beta);
  eff(kk)        = norm(ff-ffo);   % calculate far field error
  eyp(kk)        = norm(yp-p0);    % calculate parameter error
  ffG            = sim_09_4_7_calc_ffG(PAR*yp,ffN,kappa,beta); % H'
  A              = ffG*PAR;        % H' on parameter space
  yp             = yp - eta*real(A'*(ffo-ff)); % Gradient step
  ypa(:,kk+1)    = yp; % store iterate
end
ff               = sim_08_3_7_calc_ff(PAR*yp,ffN,kappa,beta);
eff(Nstep+1)     = norm(ff-ffo);   % calculate last far field error
eyp(Nstep+1)     = norm(yp-p0);    % calculate last parameter error
```

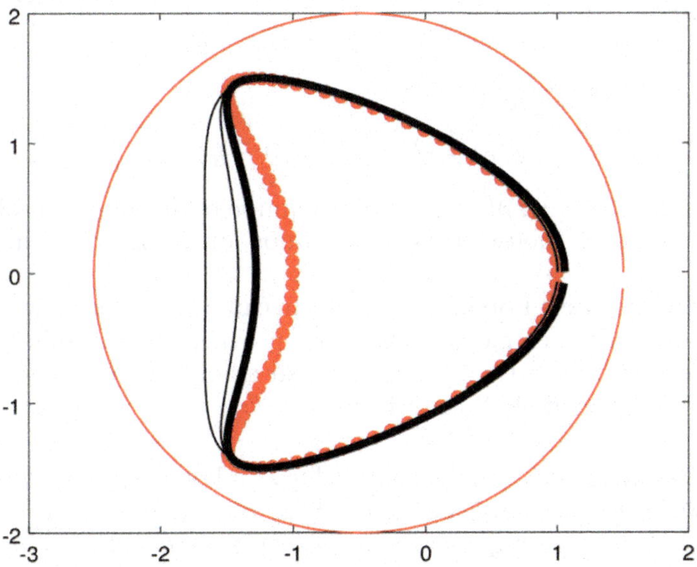

Figure 9.5. The iterations 1, 1000,..., 5000 for the *gradient method* applied to the inverse sound-soft scattering problem with 5% random noise on the far field pattern. Here, we apply the gradient method in parameter space, i.e. we used it to reconstruct the Fourier coefficients of the boundary curve following code 9.4.3. The incident wave is coming from the right. The total field is shown in figure 8.6.

9.5 Differentiating dynamical systems: tangent linear models

As an example of differentiating a dynamical model we choose the system introduced by Lorenz in 1963 as an approximation to thermal convection. The *Lorenz 1963 equations* written in the form of three independent variables x, y, z have been given by (6.1.1)–(6.1.3). Recall that the function F as given in (6.1.5) is defined by

$$F(t, \varphi) = \begin{pmatrix} -\sigma(x(t) - y(t)) \\ \rho x(t) - y(t) - x(t)z(t) \\ x(t)y(t) - \beta z(t) \end{pmatrix}$$

for $t \geqslant 0$ and for the state

$$\varphi(t) = \begin{pmatrix} x(t) \\ y(t) \\ z(t) \end{pmatrix} \in \mathbb{R}^3.$$

A trajectory of the dynamic system is shown in figure 6.1. According to theorems 5.3.1 and 5.3.3 the *tangent linear adjoint* ψ satisfies the equation

$$\frac{\mathrm{d}}{\mathrm{d}t}\psi(t) = -(F'(\varphi))^*\psi(t) = -\begin{pmatrix} -\sigma & +\sigma & 0 \\ \rho - z(t) & -1 & -x(t) \\ y(t) & x(t) & -\beta \end{pmatrix}^T \psi(t) \tag{9.5.1}$$

with *final condition* (5.3.24), i.e.

$$\psi(t_1) = (H')^*\big(H(\varphi(t_1)) - f_1\big), \tag{9.5.2}$$

where $\varphi(t) = (x(t), y(t), z(t))^T$ is given by a solution of the original problem (6.1.1)–(6.1.3). Here, we will choose linear observation operators H as in sections 6.2 and 6.3.

A solution to the direct Lorenz 1963 problem can be found in code 6.1.1, where a Runge–Kutta scheme is employed. The script provides the full solution $\varphi(t)$ at all time steps. Here, we now solve the tangent linear adjoint equations (9.5.1) with final condition (9.5.2) by a simple Euler scheme.

Code 9.5.1. *The calculation of the* tangent linear adjoint $\psi(t)$ *for the Lorenz 1963 dynamical system. This is the script* `sim_09_5_1_c_TLA.m`, *where the set-up given in script* `sim_09_5_1_a_Setup.m` *and the forward problem is taken out of section 6.1 based on the script* `sim_06_1_1_Lorenz63.m`.

```
% initialize tangent linear adjoint, terminal values
psi(:,N_Time+1)   = -2*H'*(y-H*x);
% march back in time propagating the tangent linaear adjoint function
for jj = N_Time:(-1):1
   Fp = sim_09_5_1_b_Fp(xo(jj+1),yo(jj+1),zo(jj+1),sigma,rho,beta);
   psi(:,jj) = psi(:,jj+1) + h*Fp'*psi(:,jj+1);
end
% evaluate the gradient at the end
df2 = psi(:,1);
```

The script `sim_09_5_1_b_Fp.m` *provides the forcing* F' *given in* (9.5.1).

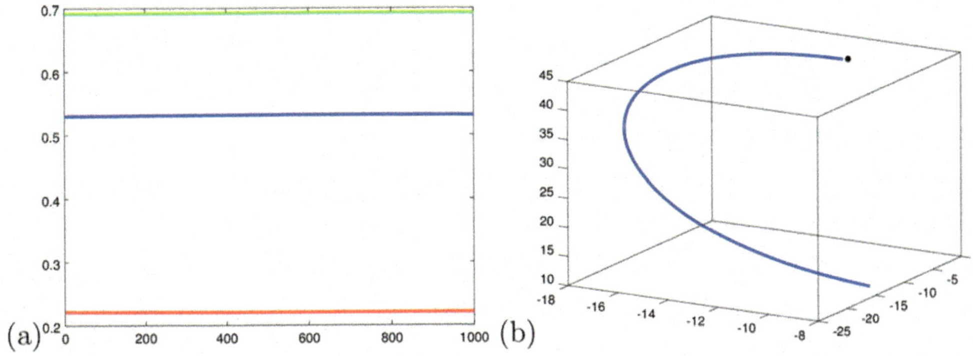

Figure 9.6. (a) The term $g(t)$ defined in (5.3.25) for three different directions $\delta\varphi_0 = e_i$, $i = 1, 2, 3$ in red, green and blue, respectively. (b) Integrating $d\psi/dt$ backwards along the curve $\varphi(t)$ starting with the data y shown as a black dot. This is generated by script `sim_09_5_1_e_test_TLA.m`.

```
function Fp = sim_09_5_1_b_Fp(x,y,z,sigma,rho,beta)
% forcing term for tangent linear adjoint calculation depending
% on the original curve, here given by x,y,z in vectorial form.
for jj=1:size(x,1)      % (optional) loop over the time steps
  Fp(:,:,jj) = [-sigma, sigma, 0; ...
                rho-z(jj), -1, -x(jj); ...
                y(jj), x(jj), -beta];
end
```

We also provide some script to test the statement of lemma 5.3.2 which claims that the scalar product

$$g(t) := \left\langle \varphi'(t)(\delta\varphi_0),\, \psi(t)\right\rangle$$

remains constant over time. This is demonstrated in figure 9.6(a), where we display $g(t)$ for three different directions $\delta\varphi_0 := e_i$, $i = 1, 2, 3$, while we integrate backward from the measurement y shown as a black dot in figure 9.6(b) along the blue curve.

Further, the gradient (5.3.27) given by theorem 5.3.3 is compared when calculated via a finite difference approximation as suggested in ensemble-based four-dimensional variational assimilation in section 5.3.2 in contrast to the tangent linear adjoint integration. For the set-up given by script `sim_09_5_1_e_test_TLA.m`we obtain

$$\nabla J_{\text{FD}} = \begin{pmatrix} 0.5335 \\ 0.69391 \\ 0.22052 \end{pmatrix}, \qquad \nabla J_{\text{TLA}} = \begin{pmatrix} 0.52939 \\ 0.69058 \\ 0.22083 \end{pmatrix}$$

for the gradient evaluated by ensemble finite differences (FD) and based on the integration of the tangent linear adjoint equation (TLA) with $N_{\text{Time}} = 1000$ time steps. Gradient or Newton methods can now be formulated as described in sections 3.3 and 9.4.

Bibliography

[1] Potthast R 1994 Frechet differentiability of boundary integral operators in inverse acoustic scattering *Inverse Problems* **10** 431

[2] Potthast R 1996 Domain derivatives in electromagnetic scattering *Math. Methods Appl. Sci.* **19** 1157–75

[3] Kirsch A 1993 The domain derivative and two applications in inverse scattering theory *Inverse Problems* **9** 81–96

[4] Colton D and Kress R 1983 *Integral Equation Methods in Scattering Theory* (New York: Wiley)

Chapter 10

Analysis: uniqueness, stability and convergence questions

Algorithms in *inverse problems* reconstruct some quantity x from measurements y. In *data assimilation* this is usually a state $x \in X$ of some dynamical system, and it is reconstructed again and again within the *analysis cycle*. With the reconstruction task, basic questions arise:

- Is it possible to reconstruct x at all, in the sense that there is some x which will lead to measurements y? This is the question of *existence*.
- But there could be many such states x leading to y, i.e. we need to get an idea of *uniqueness*.
- Also, the reconstruction could be *unstable*, i.e. it is possible that there are many different states x which lead to basically very similar data y.

We have already discussed these issues as the criteria of Hadamard for a *well-posed* problem, see definition 3.1.1. We have also observed that very many *inverse* problems are *ill-posed* in the sense that they are either *non-unique* or *unstable* and need *stabilization* or *regularization*, respectively. We have introduced several such regularization methods in chapter 3, either for *linear* problems by spectral damping or for *iterative* methods by stopping rules.

So far, we have not discussed the *uniqueness* question for inverse problems in more detail. Here, as an example, we will present some selected uniqueness results on the *inverse obstacle scattering* problem in section 10.2.

Before we go into these specific questions, we will realize in section 10.1 that the classical concepts of uniqueness and stability are not sufficient to completely describe the situation for inverse wave problems. We suggest complementing the concepts with what has been called *ϵ-uniqueness*, i.e. some type of *approximate* uniqueness when particular measurements coincide.

When you regularize an ill-posed inverse problem, you approximate the unbounded inverse mapping by a bounded approximate inverse mapping. Thus,

doi:10.1088/978-0-7503-1218-9ch10

the algorithm is stable. For data assimilation, we carry out such regularized inversion in each step of a *cycled* data assimilation scheme. But what happens to the bounded error over time, since it is propagated to the next time step and then fed into the next reconstruction? We will investigate the *instability of cycled data assimilation* in section 10.5.

Clearly we try to formulate numerical algorithms to find x for given y. For such algorithms, we would like to have

- *Stability* in the sense that they calculate an approximation x_α to the solution of the inverse or data assimilation problem which depends continuously on the measurements y.
- *Convergence* in the sense that for *true data* $y = y^{(\text{true})} = H(x^{(\text{true})})$ they have algorithmic parameters α such that for $\alpha \to 0$ we have $x_\alpha \to x^{(\text{true})}$ of the approximate solution x_α to the true solution $x^{(\text{true})}$.

But measurements for inverse problems almost never provide *true* data. We are lucky if we know reasonable error bounds $\delta > 0$ for the data in the sense that $\|y^{(\text{true})} - y\| \leqslant \delta$ for the measured data y and true data $y^{(\text{true})} = H(x^{(\text{true})})$. Thus, mathematical algorithms try to find a choice $\alpha = \alpha(\delta)$ such that

$$x_{\alpha(\delta)} \to x^{(\text{true})}, \quad \delta \to 0, \tag{10.0.1}$$

i.e. the approximate solution tends to the true solution if the data error δ tends to zero. We have analyzed such algorithms already in sections 3.1.6 and 3.3.3. This type of convergence is called the *regularity* of regularization parameter choice strategy for a regularization schemes.

For data assimilation we will discuss *stability*, *synchrony* and *convergence* in section 10.6.1. This includes the discussion of stochastic convergence and what can be achieved within real-world cycled data assimilation methods.

When we study particular inverse problems, further basic questions of convergence arise. For example, for inverse scattering problems we might be able to stably reconstruct some scattered field u^s or some *indicator function* μ defined in $\mathbb{R}^m$, $m = 2, 3$ such that the scatterers are reconstructed by a condition such as $\mu \geqslant c$. What does the convergence for the reconstruction of such indicator functions tell us about the domains. We will discuss the questions in section 10.6.2. It is relevant for field reconstructions, sampling methods, probe methods and analytic continuation tests, which are the topics of chapters 12–15, respectively. We will discuss the relationship between discretized problems and their continuous version in section 10.3.

There are many *basic mathematical questions* on the stability of an inverse problem, which can give some estimate on its ill-posedness. Methods for estimating the *degree of ill-posedness* have been investigated, see [1].

Here, we we will focus on the *reconstruction* of a solution, with an emphasis on *non-iterative reconstruction schemes*. Non-iterative reconstruction methods can be seen as reconstruction methods in their own right, but they are also viewed as an easy tool to obtain a good initial guess to start *iterative* reconstruction methods, see section 9.4.

10.1 Uniqueness of inverse problems

Here, we will provide generic arguments and concepts on uniqueness and present specific arguments for uniqueness and stability of inverse obstacle scattering problems in section 10.2.

Uniqueness is of crucial importance for inverse problems. In broad terms uniqueness asks: do given measurements y, together with the information that they can be modeled by a particular mapping F such that $y = F(x)$ for $x \in X$, determine the unknown quantity x? Uniqueness results are often profound and important results in different areas of inverse modeling.

Definition 10.1.1 (Uniqueness). *Given the general set-up of definition 1.2.1 we say that the inverse problem is* uniquely solvable *for $V \subset Y$ if for every $y \in V$ there is one and only one element $x \in U$ such that $F(x) = y$. If there is $y \in Y$ such that the solution set*

$$F^{-1}(y) = \{\rho \in U : F(x) = y\} \tag{10.1.1}$$

has more than one element, the inverse problem is called non-unique. Here we used F instead of H to denote the mapping in definition 1.2.1.

Definition 10.1.2 (Local uniqueness). *If the set $F^{-1}(y)$ is discrete for every $y \in V$, i.e. if for $x \in F^{-1}(y)$ there is an open set V with $x \in V$ such that there is no other element $x' \in F^{-1}(y)$ with $x' \neq x$ and $x' \in V$, then we speak of* local uniqueness.

We have discussed local uniqueness in the framework of the *implicit function theorem* in section 9.3.2, which is based on *local uniqueness* for the inverse obstacle scattering problem.

Usually it is both important and challenging to show *uniqueness* for a given inverse problem. When you do not have uniqueness, a regularized reconstruction algorithm usually provides some best approximation in the sense of definition 2.2.7.

But for many inverse problems we cannot show full *uniqueness*, but still obtain *information* about the unknown object which, if accumulated for many different measurements, leads to a better and better approximation to the true unknown quantity $x \in X$. Thus, a slightly more general concept to *uniqueness* or *stability*, called *ϵ-uniqueness*, was introduced in [2, 3].

Definition 10.1.3 (ϵ-uniqueness). *Let $y_1, y_2, \ldots, y_k, \ldots$ be a sequence of measurements carried out with parameters p_j, $j = 1, 2, 3, \ldots$ which can be taken for an inverse problem with an unknown quantity $x \in U \subset X$ in some normed space X. We denote the measurement map by*

$$F_j : U \to Y, \quad x \mapsto F_j(x) = y_j, \quad j \in \mathbb{N}. \tag{10.1.2}$$

Then we have ϵ-uniqueness for the reconstruction of x from measurements y_j in Y, if for any given $\epsilon > 0$ there is a $k = k(\epsilon) \in \mathbb{N}$ such that

$$F_j(x_1) = F_j(x_2), \quad j = 1, \ldots, k \tag{10.1.3}$$

yields

$$\|x_1 - x_2\| \leqslant \epsilon. \tag{10.1.4}$$

Please note that ϵ-uniqueness is not *stability*, since we have full coincidence of the measurements $F_j(x_\xi)$, $\xi = 1, 2$ and $j = 1, \dots, k$. Also, we do not have *uniqueness*, since the measurements $y_j = F_j(x_1)$, $j = 1, \dots, k$ do not uniquely determine x.

Lemma 10.1.4. *Assume we have ϵ-uniqueness for an inverse problems*

$$F_j(x) = y_j \tag{10.1.5}$$

for $j = 1, 2, 3, \dots$, $y_j \in Y$ and $x \in U \subset X$. If we have local uniqueness *for the determination of x from y_j by F_j for $j = 1, \dots, k_0$, there is $k_* \in \mathbb{N}$ such that we have uniqueness for the reconstruction of $x \in U$ from $y_j = F_j(x)$ for $j = 1, \dots, k_*$.*

Proof. Consider some $x_1 \in U$. Local uniqueness implies that for $x_1 \in U$ there is an open set $V(x_1)$ in U such that

$$\left(F_j(x_2)\right)_{j=1,\dots,k_0} \neq \left(F_j(x_1)\right)_{j=1,\dots,k_0}, \quad x_2 \in V(x_1). \tag{10.1.6}$$

There is some $\epsilon > 0$ such that the ball $B(x_1, \epsilon) \subset V(x_1)$. Now, (10.1.5) implies that there is $k(\epsilon)$ such that (10.1.3) for $j = 1, \dots, k(\epsilon)$ implies (10.1.4). Take $k_* = \max\{k(\epsilon), k_0\}$. Then (10.1.3) implies $x_2 \in B(x_1, \epsilon)$ and local uniqueness insures (10.1.6) for k_0 replaced by k_*. Together we conclude that for any $x_2 \in U$ with (10.1.5) for $j = 1, \dots, k_*$ we have (10.1.6) for $j = 1, \dots, k_*$, i.e. the data $y_1, \dots, y_{k_*}$ uniquely determine $x \in U$. □

10.2 Uniqueness and stability for inverse obstacle scattering

This problem has a long history, see [4, 5] and [6]. Although some of the content in these references can be generalized to other equations and systems other than the Helmholtz equation, let us restrict ourselves to the Helmholtz equation to review some of the well-known and recent uniqueness results of inverse scattering by obstacles.

For sound-soft obstacles (see definition 8.2.1) with Lipschitz boundaries, the first original idea to prove such uniqueness results goes back to Schiffer, see [4, 7], who reduces this question to the estimates of the eigenvalues of the Laplacian with a Dirichlet boundary condition. This approach has since been used by many authors. The first complete uniqueness result was given by Colton–Sleemann [8]. They gave an estimate of the number of incident plane waves in order to have a unique identification of the obstacle. In particular, they showed that one incident wave is enough to uniquely identify any sound-soft obstacle if its size is small enough. Then, Stefanov–Uhlmann [9] showed that one incident wave is enough to distinguish two sound-soft obstacles if they are sufficiently close. Further, Gintides [10] generalized these two types of results by weakening the smallness and the closeness conditions.

The related stability results to these uniqueness results were given by Isakov in [11, 12] who showed a *log–log* type estimate for small obstacles and Sincich–Sini who showed a *log* type estimate for small or close obstacles in [13]. Potthast [14] showed a *log* estimate for the convex hull of arbitrary obstacles.

Compared with the sound-soft obstacles, related results for sound-hard and other obstacles with a Robin boundary condition are not known so far.

On the other hand if we assume the obstacles have particular shapes or their boundaries are not analytic, then several results are known. For polygonal or polyhedral shaped obstacles, several results were given by Cheng–Yamamoto [15], Elschner–Yamamoto [16], Alessandrini–Rondi [17], Rondi [6] and Liu–Zou [18, 19] in the recent past. In particular, they showed that one incident wave and m incident waves are enough to uniquely identify sound-soft and sound-hard obstacles in $\mathbb{R}^m$ with $m = 2, 3$, respectively. For obstacles with non-analytic boundaries, Honda–Nakamura–Sini [20] showed that one incident wave and $m - 1$ incident waves are enough to uniquely identify sound-soft obstacles with Lipschitz smooth boundaries and sound-hard obstacles with C^1 smooth boundaries. There is a similar result by Ramm [21] for a convex sound-soft obstacle with a non-analytic boundary. These results for obstacles with non-analytic boundaries together with the other results mentioned above suggest that the remaining basic open problem for the uniqueness of inverse scattering problem is to show the unique identifiability of obstacles with piece-wise analytic boundaries by observing the far field patterns of the scattered waves of finitely many given incident waves.

As an example we now show how to prove the uniqueness for the inverse scattering problem to determine the domain D of the scattering problem of definition 8.2.1 from the far field pattern $u^\infty(\cdot, d)$ for scattering of incident plane waves $u^i(\cdot, d)$ with all directions $d \in \mathbb{S}$ defined in (8.3.10). Let $u_j^\infty(\hat{x}, d)\,(j = 1, 2)$ be the corresponding far field patterns. For simplicity we will only give it for the scattering by sound-soft obstacles.

Let $D_j \subset \mathbb{R}^m\,(j = 1, 2)$ be bounded domains which are considered as sound-soft obstacles. As the basis for the following uniqueness result we remark that for $z_0 \in \partial D$ the Dirichlet boundary condition yields

$$\Phi^s(z, z) \to \infty, \quad z \to z_0. \tag{10.2.1}$$

This can be seen from the boundedness of

$$g(y) := \Phi(y, z_0 + h\nu(z_0)) - \Phi(y, z_0 - h\nu(z_0)), \quad y \in \partial D,$$

for $h \geqslant 0$ sufficiently small, writing the solution to the scattering problem with incident field $\Phi(\cdot, z)$ as a sum $u^s + \Phi(\cdot, z_0 - h\nu(z_0))$ for $z = z_0 + h\nu(z_0)$, where u^s has boundary values given by g on ∂D. Then, we have the following uniqueness theorem.

Theorem 10.2.1. *If for all* $\hat{x}, d \in \mathbb{S}$ *we have*

$$u_1^\infty(\hat{x}, d) = u_2^\infty(\hat{x}, d), \tag{10.2.2}$$

then $D_1 = D_2$.

Proof. Suppose that $D_1 \neq D_2$. Let D_e be the unbounded connected component of $\mathbb{R}^m \backslash (\overline{D_1 \cup D_2})$. By the assumption and the Rellich lemma, we have $u_1^s(z, d) = u_2^s(z, d)\,(z \in D_e,\, d \in \mathbb{S})$. Then, the mixed reciprocity relation (8.4.9) implies $\Phi_1^\infty(d, z) = \Phi_2^\infty(z, d)\,(z \in D_e,\, d \in \mathbb{S})$. Once again using the Rellich lemma, we have $\Phi_1^s(x, d) = \Phi_2^s(x, d)\,(x, z \in D_e)$.

Now, based on what we have shown, we will derive a contradiction. By $D_1 \neq D_2$, we have either $\partial D_1 \backslash \overline{D_2} \neq \emptyset$ or $\partial D_2 \backslash \overline{D_1} \neq \emptyset$. Since the proof is the same for both cases, we only show the proof for the former case. Since $\mathbb{R}^m \backslash \overline{D_1}$ is connected, there is a point $z_0 \in \partial D_1 \backslash \overline{D_2}$ which can be joined with a point in D_e by a continuous curve which does not touch $\overline{D_2}$. Hence we have

$$\infty > \Phi_2^s(z_0, z_0) = \lim_{z \in D_e, z \to z_0} \Phi_2^s(z, z) = \lim_{z \in D_e, z \to z_0} \Phi_1^s(z, z) = \infty,$$

which is a contradiction. As a consequence, we conclude $D_1 = D_2$. □

The basic uniqueness result of the inverse scattering problem for sound-soft obstacles by Colton and Sleeman (see [8]) is as follows.

Theorem 10.2.2. *Let $D_j \subset \mathbb{R}^3$ $(j = 1, 2)$ be two sound-soft obstacles in an a priori known ball of radius R denoted by B_R. Let N be the number given in terms of the multiplicities of zeros of the spherical Bessel functions j_n $(n = 0, 1, \ldots)$ which is defined by*

$$N = \sum_{t_{n\ell} < \kappa R} (2n + 1), \tag{10.2.3}$$

where $t_{n\ell}$ are the zeros of the spherical Bessel functions j_n of order n. Then, we have $D_1 = D_2$, if we assume

$$u_1^\infty(\hat{x}; d) = u_2^\infty(\hat{x}; d), \quad \hat{x} \in \mathbb{S} \tag{10.2.4}$$

for $N + 1$ independent directions $d = d_j \in \mathbb{S}$ $(j = 1, 2, \ldots, N + 1)$, where each $u_j^\infty(\cdot; d)$ denotes the far field patterns of the scattered fields for the obstacle D_j associated to the incident plane wave with incident direction $d \in \mathbb{S}$.

Proof. Suppose $D_1 \neq D_2$. Then either $D_2 \backslash \overline{D_1} \neq \emptyset$ or $D_1 \backslash \overline{D_2} \neq \emptyset$. Hence assume for instance $D_2 \backslash \overline{D_1} \neq \emptyset$. Let E be the unbounded component of $\mathbb{R}^3 \backslash (\overline{D_1 \cup D_2})$ and assume without loss of generality that $D^* = (\mathbb{R}^3 \backslash \overline{E}) \backslash \overline{D_2} \neq \emptyset$. By Rellich's lemma, $u_j^\infty(\cdot; d)$ determines the scattered wave $u_j^s(x; d)$ and hence

$$u_1^s(x; d) = u_2^s(x; d)\,(x \in E) \quad \text{for} \quad d = d_j\ (j = 1, 2, \ldots, N + 1). \tag{10.2.5}$$

This holds for x in $D^* \subset E$. Hence, if we denote the total field $u = u_2^s(x; d) + e^{i\kappa x \cdot d}$ of the scattered wave $u_2^s(x; d)$, $u(x; d)$ is the eigenfunction of Dirichlet eigenvalue $-\kappa^2$ in D^*. It is well known that by the minimum–maximum principle for the eigenvalues of $-\Delta$ in a domain with Dirichlet boundary condition, the eigenvalues are strictly decreasing as the domain increases, which we call the *monotonicity of eigenvalues*.

Let κ^2 be the mth Dirichlet eigenvalue of $-\Delta$ in D^* and M be the multiplicity of the eigenvalue κ^2, where we count the multiplicity by ordering the eigenvalues. Then, due to the monotonicity of eigenvalues, M should be less than or equal to the sum of the multiplicities of the eigenvalues for $-\Delta$ in B_R.

Now the eigenfunctions of $-\Delta$ in B_R satisfying the Dirichlet boundary condition are $j_n(\kappa|x|)Y_n^\ell(\hat{x})$ with the eigenvalues given in terms of the zeros $\mu_{n\ell} = t_{n\ell}^2/R^2$ $(|\ell| \leqslant n)$ of the spherical Bessel function j_n and each of their multiplicities is $2n+1$, where $Y_n^\ell(\hat{x})$ $(|\ell| \leqslant n)$ are the spherical harmonics of degree n. Hence, we have $M \leqslant N$. However, this contradicts to the fact that there are $N+1$ linearly independent eigenfunctions of $-\Delta$ in the non-empty set D^* with Dirichlet eigenvalue κ^2, because the $N+1$ incident waves with linearly independent directions create $N+1$ linearly independent eigenfunctions of $-\Delta$ in D^* with Dirichlet eigenvalue κ^2 which can be continued outside D^* without touching D_1. □

10.3 Discrete versus continuous problems

An important topic of inverse modeling is the discussion of discretization. Often, when you study particular discrete subproblems of inversion, the uniqueness and stability of some equation

$$H(x) = y \tag{10.3.1}$$

can be obtained for such subclasses, but the full inverse problem is not unique or it is not stable. An example will be shown in chapter 11 with the inverse magnetic source problem.

Here, we provide some generic insight into the *stability* of such discretized problems when they are used as an approximation to the continuous full problem.

Definition 10.3.1 (Projection method). *Assume that X, Y are Hilbert spaces and let $\{\varphi_j : j \in \mathbb{N}\}$ be a Riesz basis of X (see (7.5.1)). We can discretize the general inverse problem (1.2.1) by using the finite-dimensional subspace X_N of X with basis φ_j, $j = 1, \ldots, N$ and the* ansatz

$$x_\beta = \sum_{j=1}^{N} \beta_j \varphi_j \tag{10.3.2}$$

with $\alpha := (\alpha_j)_{j=1,\ldots,N} \in \mathbb{R}^N$. Then, we replace the solution of the inverse problem $H(x) = y$, $y \in Y$, as given in definition 1.2.1, by the minimization of

$$g(\beta) := \left\| H(x_\beta) - y \right\|^2, \quad \alpha \in \mathbb{R}^m. \tag{10.3.3}$$

Let us assume that $H : X \to Y$ is linear, injective and compact with dense range in Y, i.e. the full problem (10.3.1) is ill-posed in the sense that its inverse cannot be bounded. In this case the inverse of the mapping $H|_{X_N}$ (i.e. H restricted to X_N) cannot be uniformly bounded for $N \to \infty$, since otherwise the sequence

$$x_N := H_N^{-1} y = H_N^{-1} H x$$

is bounded by $C\|y\|$ in X with some constant C and thus weakly convergent to $x_* \in X$, which leads to

$$Hx_N \to Hx_* = y,$$

such that $x_* = x$, i.e.

$$\|x\| = \sup_{\|z\| \leqslant 1} |\langle z, x\rangle| = \sup_{\|z\| \leqslant 1} \lim_{N\to\infty} |\langle z, x_N| \rangle \leqslant C\|y\|$$

with some constants C. But that contradicts the ill-posedness of (10.3.1).

Usually, if you use some discretization of an ill-posed problem, it has a natural interpretation as a projection method, for example by using some splines such as piecewise constant or piecewise linear functions. We conclude that discretization inherits severe instabilities if you increase the number of discretization points or parameters.

Further, we remark that with the orthogonal projector P_N of X onto X_N for $x \in X_N$ we have $H_N = HP_N$ and further using this we have

$$\begin{aligned} \langle x, H_N^* y\rangle_X &= \langle H_N x, y\rangle_Y = \langle Hx, y\rangle_Y \\ &= \langle x, H^* y\rangle_X = \langle x, P_N H^* y\rangle_X + \underbrace{\langle x, (1 - P_N)H^* y\rangle_X}_{=0}, \end{aligned} \qquad (10.3.4)$$

thus

$$H_N^* y = P_N H^* y, \quad y \in Y, \qquad (10.3.5)$$

and since for $z \in X_N$ the identity $H_N^* H_N z = 0$ implies $\langle H_N x, H_N z\rangle = 0$ for all $x \in X_N$ and thus $H_N z = 0$ which implies $z = 0$, the adjoint H_N^* is injective on $H_N X_N$.

In the case where the spaces X_N are defined by a singular system (μ_j, φ_j, g_j) of the operator H, the projection method coincides with a spectral cut-off, since in this case the minimization of (10.3.3) is equivalent to solving

$$\begin{aligned} \sum_{j=1}^{N} \mu_j^2 \beta_j^{(N)} \varphi_j &= P_N H^* H x_N \\ &= H_N^* y = H_N^* H x = P_N H^* H x \\ &= P_N H^* H P_N x + P_N H^* H (1 - P_N) x \\ &= \sum_{j=1}^{N} \mu_j^2 \beta_j \varphi_j + P_N \left(\sum_{j=N+1}^{\infty} \mu_j^2 \beta_j \varphi_j \right) \\ &= \sum_{j=1}^{N} \mu_j^2 \beta_j \varphi_j \end{aligned}$$

where β_j, $j \in \mathbb{N}$, are the coefficients of x with respect to $(\varphi_j)_{j\in\mathbb{N}}$ and $\beta_j^{(N)}$ are the corresponding coefficients of $x_N \in X_N$. Then, by theorem 3.1.9 the method establishes a *regularization method* as defined in (3.1.7) for the solution to equation 10.3.1.

In general, the second term $P_N H^* H(1 - P_N)x$ can be non-zero and by $x_N - x = (H_N^* H_N)^{-1} H_N^* Hx - x = (H_N^* H_N)^{-1}(P_N H^* Hx - P_N H^* H P_N x) = (H_N^* H_N)^{-1} P_N H^* H\,(1 - P_N)x$ the term

$$e(N) := \left\|(H_N^* H_N)^{-1} P_N H^* H(1 - P_N)x\right\| \tag{10.3.6}$$

is the error of the *projection method* defined in definition 10.3.1.

10.4 Relation between inverse scattering and inverse boundary value problems

When you study inverse wave or field problems, you can either measure on the boundary of some bounded domain Ω, or you can work with the asymptotic behavior of fields, leading to the far field pattern u^∞, see section 8.3. One approach usually leads to measurements given by the *Dirichlet-to-Neumann map*, i.e. by the mapping which maps function values of all the solutions of a partial differential equation in Ω on its boundary $\partial\Omega$ to their normal derivatives on $\partial\Omega$. The goal of this section is to show that both approaches are equivalent in terms of the information they provide for solving the inverse problem.

To begin with, we first formulate the forward problem for the inverse boundary value problem. For that let $\Omega \subset \mathbb{R}^m$ be a bounded domain and $D \subset \Omega$ be an open set compactly embedded in Ω with boundaries $\partial\Omega$ and ∂D which are both of C^3 classes, respectively. Then, the forward problem is the following boundary value problem given as

$$\begin{cases} (\Delta + \kappa^2)u = 0 \text{ in } \Omega\backslash\bar{D} \\ u = 0 \text{ on } \partial D,\; u = f \in H^{1/2}(\partial\Omega) \text{ on } \partial\Omega. \end{cases} \tag{10.4.1}$$

For the well-posedness of (10.4.1), we have to assume:

Assumption 1. That is if $f = 0$, then (10.4.1) only admits a trivial solution $u = 0$ in $\Omega\backslash\bar{D}$.

Under this assumption, we have the following well-known result on the well-posedness for (10.4.1). For the case of continuous boundary values, we have worked out the proof in section 8.2.

Theorem 10.4.1. *For any $f \in H^{1/2}(\partial\Omega)$ there exist a unique solution $u = u^f$ in $H^1(\Omega\backslash\bar{D})$ to (10.4.1). Further, the mapping*

$$f \mapsto u^f, \quad H^{1/2}(\partial\Omega) \to H^1(\Omega\backslash\bar{D})$$

is continuous.

Based on this theorem, we define the Dirichlet-to-Neumann map

$$\Lambda_D : H^{1/2}(\partial\Omega) \to H^{-1/2}(\partial\Omega)$$

as follows.

Definition 10.4.2. *For any* $f \in H^{1/2}(\partial\Omega)$,

$$\Lambda_D(f) = \frac{\partial u^f}{\partial \nu}\Big|_{\partial\Omega}, \tag{10.4.2}$$

where u^f *is the solution to (10.4.1) and* ν *is the outer unit normal vector field of* $\partial\Omega$.

Now the inverse boundary value problem is to identify the unknown sound-soft obstacle D when we know Ω and $\Lambda_D(f)$, but we do not know D. The inverse scattering problem is to find D when u^∞ is given for all incident plane waves $u^i(\cdot, d)$ with direction of incidence $d \in \mathbb{S}$. Our next goal is to show that these two scenarios are equivalent. For this we assume:

Assumption 2. In Ω a solution to the Helmholtz equation with zero boundary values is identical to zero.

Now for our convenience we introduce the terminology *non-vibrating* by the following definition.

Definition 10.4.3. *Let* Ω *be a bounded domain with a* C^2 *boundary such that* $\mathbb{R}^m \backslash \bar{\Omega}$ *is connected and satisfies assumption 2. Then such a* Ω *is called a non-vibrating domain or simply non-vibrating.*

Then we have the following result.

Theorem 10.4.4. *Assume that* Ω *is non-vibrating. Then, knowing the Dirichlet-to-Neumann map* Λ_D *and far field pattern* u^∞ *is equivalent.*

Proof. Without losing any generality in the proof but in order to simplify specifying some constants, we restrict ourselves to the case $m = 3$. For any given $f \in H^{1/2}(\partial\Omega)$ we will show that we can compute $\partial_\nu u^f$ on $\partial\Omega$ from u^∞. By the point source method in section 12.4, for each $d \in \mathbb{S}$, we can compute $u^s(y; d)\,(y \notin \bar{D})$ from $u^\infty(\cdot; d)$. Then, by the mixed reciprocity relation (8.4.9), we can compute

$$\Phi^\infty(\hat{x}, y) = \frac{1}{4\pi} u^s(y; -\hat{x}) \ \ (y \notin \bar{D},\ \hat{x}\,(= d) \in \mathbb{S}). \tag{10.4.3}$$

Now let G be a non-vibrating domain such that $\bar{D} \subset G$, $y \notin \bar{G}$, and $g_y \in L^2(\mathbb{S})$ be a density function of Herglotz wave function

$$\left(Hg_y\right)(\cdot) := \int_{\partial\Omega} e^{i\kappa\hat{x}\cdot(\cdot)} g_y(\hat{x})\, ds(\hat{x}) \tag{10.4.4}$$

such that $Hg_y \approx \Phi(\cdot, y)$ in $L^2(\partial G)$. Then, by the argument we will give later to show the principle of the *singular sources method* (14.1.10),

$$\Phi^s(y, z) \approx 4\pi \int_{\mathbb{S}} \Phi^\infty(-\hat{x}, z) g_y(\hat{x})\, ds(\hat{x}) \ \ (z \in \mathbb{R}^3 \backslash \bar{D}). \tag{10.4.5}$$

Here, if we consider a slightly larger G, then due to the interior regularity of solutions to the Helmholtz equation, the approximation $(Hg_y)(w) \approx \Phi(w, y) = \Phi(y, w)$ with fixed y can be considered in $C^1(\bar{G})$ for the original G. Therefore, we have

$$\frac{\partial \Phi^s}{\partial \nu(z)}(y, z) \approx 4\pi \int_{\mathbb{S}} \frac{\partial \Phi^\infty}{\partial \nu(z)}(-\hat{x}, z) g_y(\hat{x})\, \mathrm{d}x(\hat{x}) \quad (z \in \partial\Omega). \tag{10.4.6}$$

In these formulas (10.4.5) and (10.4.6), we take $z = x$. Then, by the reciprocity relation $\Phi^s(x, y) = \Phi^s(y, x)$ which follows from $G_D(x, y) = G_D(y, x)$ for the outgoing Green function $G_D(x, y) = \Phi(x, y) + \Phi^s(x, y)$ for the scattering problem by sound-soft obstacle D, we have shown that we can compute

$$\Phi^s(x, y), \frac{\partial \Phi^s}{\partial \nu(x)}(x, y), \frac{\partial \Phi^s}{\partial \nu(y)}(x, y) \quad (x, y \in \partial\Omega)$$

from u^∞. By Green's formula, we have

$$u^f(y) = \int_{\partial\Omega} \left(G_D(x, y) \frac{\partial u^f}{\partial \nu}(x) - f(x) \frac{\partial G_D}{\partial \nu(x)}(x, y) \right) \mathrm{d}s(x) \tag{10.4.7}$$

for $y \in \Omega \backslash \bar{D}$. Since $\Phi^s(\cdot, y)$ is a function of C^∞ class for $y \in \partial\Omega$, $G_D(x, y)$ has the jump formula similar to $\Phi(x, y)$ at $\partial\Omega$. Hence, for each $y \in \partial\Omega$, we have

$$\begin{aligned} \frac{\partial u^f}{\partial \nu}(y) = &\int_{\partial\Omega} \frac{\partial G_D}{\partial \nu(y)}(x, y) \frac{\partial u^f}{\partial \nu}(x)\, \mathrm{d}s(x) + \frac{1}{2} \frac{\partial u^f}{\partial \nu}(x) \\ &- \frac{\partial}{\partial \nu(y)} \int_{\partial\Omega} f(x) \frac{\partial G_D}{\partial \nu(x)}(x, y)\, \mathrm{d}x(x). \end{aligned} \tag{10.4.8}$$

By using $\frac{\partial G_D}{\partial \nu(y)}(x, y) = \frac{\partial G_D}{\partial \nu(y)}(y, x)$, (10.4.8) becomes

$$\left(I - K_D^*\right) \frac{\partial u^f}{\partial \nu} = -2 \frac{\partial}{\partial \nu(y)} \int_{\partial\Omega} f(x) \frac{\partial G_D}{\partial \nu(x)}(x, y)\, \mathrm{d}s(x) \text{ on } \partial\Omega, \tag{10.4.9}$$

where K_D is the double-layer potential on $\partial\Omega$ defined by using G_D instead of Φ. We note that the left-hand side belongs to $H^{-1/2}(\partial\Omega)$ (see McLean's book [22] section 7.3, p 218). Since $K_D^* : H^{-1/2}(\partial\Omega) \to H^{1/2}(\partial\Omega)$ is compact, the solvability of (10.4.9) follows from its uniqueness. To see the uniqueness, let $\phi \in H^{-1/2}(\partial\Omega)$ satisfy $(I - K_D^*)\phi = 0$. Define w by

$$w(y) = \int_{\partial\Omega} G_D(y, x) \phi(x)\, \mathrm{d}s(x). \tag{10.4.10}$$

Then, by the jump formula of the single-layer potential and assumption on ϕ, we have

$$\left(\frac{\partial w}{\partial \nu}\right)_+ = \frac{1}{2}(K_D^* - I)\phi = 0$$

on $\partial\Omega$. It is clear that w satisfies the Helmholtz equation in $\mathbb{R}^3\backslash\bar{\Omega}$ and the Sommerfeld radiation condition. Hence, we have by the uniqueness of the exterior Neumann boundary value problem for Helmholtz equation in $\mathbb{R}^3\backslash\bar{\Omega}$, we have $w = 0$ in $\mathbb{R}^3\backslash\bar{\Omega}$. We further have $w_- = w_+ = 0$ on $\partial\Omega$ by the continuity of the single layer potential and w satisfies the Helmholtz equation in $\Omega\backslash\bar{D}$. Hence, by the uniqueness assumption, $w = 0$ in $\Omega\backslash\bar{D}$. Then, by the usual argument using the jump formula for the normal derivative of single-layer potential, we have $\phi = 0$. To sum up, based on $u^\infty(\cdot, d)$ for $d \in \mathbb{S}$ we have constructed an operator (10.4.9) for Λ_D.

Next, we will show how to obtain u^∞ from Λ_D. Since u^∞ can be obtained by the mapping $u^{\mathrm{i}} \mapsto u^{\mathrm{s}}$, we will show how to obtain this mapping from Λ_D. So, suppose we are given Λ_D and u^{i}, we will show how to compute either u^{s} or $u = u^{\mathrm{i}} + u^{\mathrm{s}}$. The key point here is to calculate the total field u on $\partial\Omega$ given u^{i} and Λ_D, but not knowing D. By the representation formula of the solution, we have

$$u^{\mathrm{i}}(x) = \int_{\partial\Omega}\left(\Phi(y, x)\frac{\partial u^{\mathrm{i}}}{\partial\nu}(y) - \frac{\partial\Phi}{\partial\nu}(y, x)u^{\mathrm{i}}(y)\right)\mathrm{d}s(y) \quad (x \in \Omega). \tag{10.4.11}$$

Further, using the Sommerfeld radiation condition, we can derive

$$\int_{\partial\Omega}\left(\Phi(y, x)\frac{\partial u^{\mathrm{s}}}{\partial\nu}(y) - \frac{\partial\Phi}{\partial\nu}(y, x)u^{\mathrm{s}}(y)\right)\mathrm{d}s(y) = 0 \quad (x \in \Omega). \tag{10.4.12}$$

From (10.4.11) and (10.4.12), for $x \in \Omega$ we obtain

$$u^{\mathrm{i}}(x) = \int_{\partial\Omega}\left(\Phi(y, x)\frac{\partial u}{\partial\nu}(y) - \frac{\partial\Phi}{\partial\nu}(y, x)u(y)\right)\mathrm{d}s(y). \tag{10.4.13}$$

Hence, by the jump formula of double-layer potential (8.1.30), we obtain an operator equation for u given by

$$u^{\mathrm{i}} = \frac{1}{2}SN(u|_{\partial\Omega}) - \frac{1}{2}K(u|_{\partial\Omega}) + \frac{1}{2}u|_{\partial\Omega}, \tag{10.4.14}$$

where N is defined by

$$N(u|_{\partial\Omega}) = \Lambda_D(u|_{\partial\Omega}). \tag{10.4.15}$$

What we need to show is the unique solvability of (10.4.14), that is the existence of $(SN - K + I)^{-1}$. Since SN is not compact, we need to work a little bit to show that $SN - K + I$ can be written in the form identity plus compact, such that the Riesz–Fredholm theory applies.

In the rest of the argument, we use the indices 1 and 2 to denote potential operators defined over $\partial\Omega$ and ∂D, respectively. Also in order to distinguish any potential operators from their associated boundary integral operators, we put $\sim$ to the potential operators. For example,

$$\left(\tilde{K}_1\phi\right)(x) = 2\int_{\partial\Omega}\frac{\partial\Phi}{\partial\nu(y)}(x, y)\phi(y)\,\mathrm{d}s(y) \quad (x \notin \partial\Omega)$$

and

$$\left(K_1\hat{\phi}\right)(x) = 2\int_{\partial\Omega} \frac{\partial\Phi}{\partial\nu(y)}(x, y)\hat{\phi}(y)\, \mathrm{d}s(y) \;\; (x \in \partial\Omega).$$

By looking for $u = u^{\mathrm{i}} + u^{\mathrm{s}}$ in the form

$$u = \frac{1}{2}\tilde{K}_1\psi + \frac{1}{2}\tilde{K}_2\phi, \quad \phi \in H^{1/2}(\partial D), \quad \psi \in H^{1/2}(\partial\Omega), \tag{10.4.16}$$

direct computation and jump formulas for $\tilde{K}_1\psi$, $\tilde{K}_2\phi$ lead us to give the following representations of ϕ, ψ in terms of $u|_{\partial\Omega}$.

$$\psi = A(u|_{\partial\Omega}), \qquad \phi = B(u|_{\partial\Omega}) \tag{10.4.17}$$

with

$$\begin{aligned} &A = 2(I + R), \;\; R = -(I - K_1)^{-1}K_1 + (I - K_1)^{-1}\tilde{K}_2 Q^{-1}\tilde{K}_1(I - K_1)^{-1}, \\ &B = -2Q^{-1}\tilde{K}_1(I - K_1)^{-1}, \end{aligned} \tag{10.4.18}$$

where

$$Q = I + \tilde{Q}, \quad \tilde{Q} = K_2 + \tilde{K}_1(I - K_1)^{-1}K_2. \tag{10.4.19}$$

Here we note that $\tilde{Q} \in BL(H^{1/2}(\partial D), H^{1/2}(\partial D))$, $R \in BL(H^{1/2}(\partial\Omega), H^{1/2}(\partial\Omega))$ are compact due to ∂D, $\partial\Omega$ are of C^3 class (see [4], p 44, theorem 3.6) and the invertibility of Q follows from assumption 1. Further, since Ω is non-vibrating, $(I - K_1)^{-1} \in BL(L^2(\partial\Omega), L^2(\partial\Omega))$ exists.

Now we will show that $u^{\mathrm{i}} = 1/2(SN - K + I)u\,|_{\partial\Omega} = 1/2(S_1N - K_1 + I)u|_{\partial\Omega}$ can be rewritten in the form

$$u^{\mathrm{i}} = (I + C)u|_{\partial\Omega} \tag{10.4.20}$$

with the compact operator C: $H^{1/2}(\partial\Omega) \mapsto H^{1/2}(\partial\Omega)$ given by

$$C = -\frac{1}{2}K_1^2(I + R) + \frac{1}{4}\partial_\nu\left(\tilde{K}_2 B.\right) - \frac{1}{2}K_1. \tag{10.4.21}$$

To see this we have again from $u = 1/2\tilde{K}_1\psi + 1/2\tilde{K}_2\phi$,

$$N(u|_{\partial\Omega}) = \frac{\partial u}{\partial\nu}|_{\partial\Omega} = \frac{1}{2}T_1\psi + \frac{1}{2}\frac{\partial(\tilde{K}_2\phi)}{\partial\nu}, \tag{10.4.22}$$

where $(T_1\hat{\phi})(x) = 2\int_{\partial\Omega} \frac{\partial\Phi}{\partial\nu(y)}(x, y)\hat{\phi}(y)\mathrm{d}s(y)$ $(x \in \partial\Omega)$. Then, using $S_1T_1 = K_1^2 - I$, a direct computation gives us (10.4.20).

Finally, by the Fredholm alternative theorem, the existence of

$$(SN - K + I)^{-1} = (S_1N - K_1 + I)^{-1}$$

follows if we show the uniqueness for the equation $(SN - K + I)f = 0$. Let $f \in H^{1/2}(\partial\Omega)$ satisfy $(SN - K + I)f = 0$ and

$$\tilde{u}(x) = \int_{\partial\Omega} \left(\frac{\partial\Phi}{\partial\nu(y)}(x, y)f(y) - \Phi(x, y)(Nf)(y) \right) \mathrm{d}s(y) \qquad (10.4.23)$$

for $x \notin \partial\Omega$. By the jump formula, we have

$$\tilde{u}_- = \frac{1}{2}Kf - \frac{1}{2}f - \frac{1}{2}SNf = 0. \qquad (10.4.24)$$

Also, $\tilde{u}$ clearly satisfies the Helmholtz equation in Ω. Hence, by the assumption that Ω is non-vibrating, we have $\tilde{u} = 0$ in Ω. Now we observe that on $\partial\Omega$ we can derive

$$\begin{aligned} \tilde{u}_+ &= \tilde{u}_+ - \tilde{u}_- \\ &= \left(\frac{1}{2}Kf + \frac{1}{2}f - SNf\right) - \left(\frac{1}{2}Kf - \frac{1}{2}f - SNf\right) = f = u^f, \end{aligned} \qquad (10.4.25)$$

and

$$\begin{aligned} \frac{\partial\tilde{u}_+}{\partial\nu} &= \frac{\partial\tilde{u}_+}{\partial\nu} - \frac{\partial\tilde{u}_-}{\partial\nu} \\ &= \left(\frac{1}{2}Tf - \frac{1}{2}K^*Nf + \frac{1}{2}Nf\right) - \left(\frac{1}{2}Tf - \frac{1}{2}K^*Nf - \frac{1}{2}Nf\right) \\ &= Nf = \frac{\partial u^f}{\partial\nu}. \end{aligned}$$

Hence, let

$$\hat{u} = \begin{cases} \tilde{u} \text{ in } \mathbb{R}^3\backslash\Omega \\ u^f \text{ in } \Omega\backslash\bar{D}, \end{cases} \qquad (10.4.26)$$

then, $\hat{u}$ satisfies $(\Delta + \kappa^2)\hat{u} = 0$ in $\mathbb{R}^3\backslash\bar{D}$, $\hat{u} = 0$ on ∂D and the Sommerfeld radiation condition. Thus, by the uniqueness of the exterior Dirichlet boundary value problem, $\hat{u} = 0$ in $\mathbb{R}^3\backslash D$. In particular, $f = u^f = \hat{u} = 0$ on $\partial\Omega$. □

As a consequence of the above theorem we conclude that if we know u^∞ for all incident plane waves, this is equivalent to knowledge of the Dirichlet-to-Neumann map Λ_D and vice versa.

10.5 Stability of cycled data assimilation

We have discussed the stability of the individual estimation steps of data assimilation methods in section 5.2. Clearly, the state estimation by three-dimensional (3D-VAR) or four-dimensional variational assimilation, or the Kalman filter is stable in each assimilation step. But what happens when you cycle such schemes? Here, we provide a full error analysis for basic *cycled* data assimilation schemes. As a generic case we will focus on iterated 3D-VAR with a constant model $M = I$.

To study the behavior of the iterated 3D-VAR we use the spectral representation as worked out in section 3.1.4. Let X, Y be Hilbert spaces and $A : X \to Y$ be a compact, injective operator with a dense range. Consider a singular system (μ_n, φ_n, g_n) of the operator $A : X \to Y$ as introduced in theorem 2.4.22. With respect to the orthonormal basis sets $\{\varphi_n : n \in \mathbb{N}\}$ and $\{g_n : n \in \mathbb{N}\}$ the application of A corresponds to a multiplication by the singular value μ_n, $n \in \mathbb{N}$. Picard's theorem 2.4.23 shows that the inverse A^{-1} corresponds to a spectral multiplication by $1/\mu_n$, $n \in \mathbb{N}$. That is the term-wise multiplication of $f = \sum_{n=1}^{\infty} \langle f, g_n \rangle g_n$ by $1/\mu_n$. The operator R_α is carried out by a spectral multiplication by

$$\frac{\mu_n^2}{\mu_n^2 + \alpha}.$$

We now consider measurements $f^{(k)} \in Y$ for $k = 0, 1, 2, \ldots$ and we use the update formula (5.2.6), where φ_0 is replaced successively by φ^k to calculate φ^{k+1}, $k = 0, 1, 2, \ldots$. This means we define

$$\varphi^{(k+1)} := \varphi^{(k)} + R_\alpha\left(f^{(k)} - A\varphi^{(k)}\right), \quad k = 0, 1, 2, \ldots. \tag{10.5.1}$$

We assume that the data $f^{(k)} = f^{(\text{true})} + f^{(\delta,k)}$ are the sum of the true data $f^{(\text{true})} = A\varphi^{(\text{true})}$ and some additive error $f^{(\delta,k)}$. Let $\alpha_n^{(\text{true})}$ be the spectral coefficients of the true solution, i.e.

$$\varphi^{(\text{true})} = \sum_{n=0}^{\infty} \alpha_n^{(\text{true})} \varphi_n \tag{10.5.2}$$

and $\delta_n^{(k)}$ be the spectral coefficients of $f^{(\delta,k)}$, i.e.

$$f^{(\delta,k)} = \sum_{n=0}^{\infty} \delta_n^{(k)} g_n, \quad k = 0, 1, 2, \ldots. \tag{10.5.3}$$

The spectral coefficients of $\varphi^{(k)}$ are denoted by $\gamma_n^{(k)}$, i.e. we have

$$\varphi^{(k)} = \sum_{n=0}^{\infty} \gamma_n^{(k)} \varphi_n, \quad k = 0, 1, 2, \ldots. \tag{10.5.4}$$

We can now derive a formula for the behavior of the coefficients $\gamma_n^{(k)}$.

Lemma 10.5.1. *The spectral coefficients $\gamma_n^{(k)}$ of the solutions φ_k for data assimilation by iterated 3D-VAR (10.5.1) with $M = I$ are given by*

$$\gamma_n^{(k+1)} = q_n^{k+1} \gamma_n^{(0)} + \left(1 - q_n^{k+1}\right) \alpha_n^{(\text{true})} + \frac{1 - q_n}{\mu_n} \sum_{\xi=0}^{k} q_n^{k-\xi} \delta_n^{(\xi)}, \tag{10.5.5}$$

where we used $q_n := \alpha/(\mu_n^2 + \alpha)$.

Proof. We use induction on k starting with $k = 0$. Note that the spectral coefficients of $R_\alpha f^{(k)}$ are given by

$$\frac{\mu_n^2}{\mu_n^2 + \alpha} \alpha_n^{(\text{true})} + \frac{\mu_n}{\mu_n^2 + \alpha} \delta_n^{(k)}, \quad n \in \mathbb{N}_0.$$

The term $R_\alpha A\varphi^{(k)}$ has spectral coefficients

$$\frac{\mu_n^2}{\mu_n^2 + \alpha}\gamma_n^{(k)}, \qquad n \in \mathbb{N}_0.$$

The update formula $\varphi^{(k+1)} := \varphi^{(k)} + R_\alpha(f^{(k)} - A\varphi^{(k)})$ thus leads to its spectral representation

$$\begin{aligned}
\gamma_n^{(k+1)} &= \gamma_n^{(k)} + \frac{\mu_n^2}{\mu_n^2 + \alpha}\alpha_n^{(\text{true})} + \frac{\mu_n}{\mu_n^2 + \alpha}\delta_n^{(k)} - \frac{\mu_n^2}{\mu_n^2 + \alpha}\gamma_n^{(k)} \\
&= \frac{\alpha}{\mu_n^2 + \alpha}\gamma_n^{(k)} + \frac{\mu_n^2}{\mu_n^2 + \alpha}\alpha_n^{(\text{true})} + \frac{\mu_n}{\mu_n^2 + \alpha}\delta_n^{(k)} \\
&= q_n\gamma_n^{(k)} + \left(1 - q_n\right)\alpha_n^{(\text{true})} + \frac{1 - q_n}{\mu_n}\delta_n^{(k)} \qquad (10.5.6)
\end{aligned}$$

for $n \in \mathbb{N}_0$. Now, for $k = 0$ this coincides with (10.5.5). As induction step we insert (10.5.5) into (10.5.6) with k replaced by $k + 1$. This leads to

$$\begin{aligned}
\gamma_n^{(k+2)} &= q_n\left(q_n^{k+1}\gamma_n^{(0)} + \left(1 - q_n^{k+1}\right)\alpha_n^{(\text{true})} + \frac{1 - q_n}{\mu_n}\sum_{\xi=0}^{k} q_n^{k-\xi}\delta_n^{(\xi)}\right) \\
&\quad + \left(1 - q_n\right)\alpha_n^{(\text{true})} + \frac{1 - q_n}{\mu_n}\delta_n^{(k+1)} \\
&= q_n^{k+2}\gamma_n^{(0)} + \left(q_n - q_n^{k+2}\right)\alpha_n^{(\text{true})} + \frac{1 - q_n}{\mu_n}\sum_{\xi=0}^{k} q_n^{k+1-\xi}\delta_n^{(\xi)} \\
&\quad + (1 - q_n)\alpha_n^{(\text{true})} + \frac{1 - q_n}{\mu_n}\delta_n^{(k+1)}, \qquad (10.5.7)
\end{aligned}$$

which is equal to (10.5.5) with k replaced by $k + 1$.

If the limit of the sum of $q_n^{k-\xi}\delta_n^{(\xi)}$ for $\xi = 0, \ldots, k$ exists, in the limit $k \to \infty$ from (10.5.5) we obtain

$$\lim_{k\to\infty} \gamma_n^{(k)} = \alpha_n^{(\text{true})} + \frac{1 - q_n}{\mu_n}\lim_{k\to\infty}\left(\sum_{\xi=0}^{k} q_n^{k-\xi}\delta_n^{(\xi)}\right). \qquad (10.5.8)$$

Lemma 10.5.2. *If $f^{(\delta,k)} = 0$ for all $k \in \mathbb{N}_0$, then the iterated 3D-VAR (10.5.1) converges in norm to the true solution $\varphi^{(\text{true})}$, i.e.*

$$\varphi^{(k)} \to \varphi^{(\text{true})}, \quad k \to \infty. \qquad (10.5.9)$$

If we have $f^{(\delta)} \in \text{Range}(A)$ and apply (10.5.1) with the same data $f^{(k)} = f^{(\text{true})} + f^{(\delta)}$, then we have convergence

$$\varphi^{(k)} \to \varphi^{(\text{true})} + A^{-1}(f^{(\delta)}), \quad k \to \infty. \tag{10.5.10}$$

If $f^{(\delta)} \not\in$ Range(A) but $f^{(\delta)} \in \overline{R(A)}$, then we have divergence of the iterated 3D-VAR (10.5.1), i.e.

$$\|\varphi^{(k)}\| \to \infty, \quad k \to \infty. \tag{10.5.11}$$

Proof. For the first case we have $\delta_n^{(\xi)} = 0$ for all $\xi \in \mathbb{N}_0$, $n \in \mathbb{N}_0$ in (10.5.8) and we obtain the convergence $\gamma_n^{(k)} \to \alpha_n^{(\text{true})}$ for every $n \in \mathbb{N}$ of the spectral coefficients. Further, we remark that

$$\left|q_n^k\right| \leqslant 1, \ \left|1 - q_n^{k+1}\right| \leqslant 1, \quad \forall\, n \in \mathbb{N}_0, \quad k \in \mathbb{N}_0.$$

This means that

$$\left|\gamma_n^{(k)}\right| \leqslant \left|\gamma_n^{(0)}\right| + |\alpha_n^{(\text{true})}|, \quad n \in \mathbb{N}_0, \tag{10.5.12}$$

thus

$$\sum_{n=L}^{\infty} \left|\gamma_n^{(k)}\right|^2 \to 0, \quad L \to \infty \tag{10.5.13}$$

uniformly for all $k \in \mathbb{N}_0$. From (10.5.13) and the pointwise convergence of the spectral coefficients we obtain norm convergence (10.5.9).

Next, we consider the case where for all $\xi \in \mathbb{N}_0$ we have $f^{(\delta,\xi)} = f^{(\delta)} \in A(X)$. In this case we set $\delta_n^{(\xi)} = \delta_n$ for any $\xi \in N_0$ and using $|q_n| < 1$ we derive

$$\sum_{\xi=0}^{k} \left|q_n^{k-\xi}\delta_n^{(\xi)}\right| = |\delta_n| \sum_{\xi=0}^{k} \left|q_n^{\xi}\right| < \infty, \quad k \in \mathbb{N}_0$$

for each $n \in \mathbb{N}_0$ and thus the limit (10.5.8) exists. With $\sum_{\xi=0}^{\infty} q_n^{\xi} = \frac{1}{1-q_n}$ from (10.5.8) we obtain

$$\lim_{k\to\infty} \gamma_n^{(k)} = \alpha_n^{(\text{true})} + \frac{1}{\mu_n}\delta_n \tag{10.5.14}$$

with an error $\gamma_n^{(k+1)} - \alpha_n^{(\text{true})}$ proportional to $q_n^{k+1}\delta_n/\mu_n$ for $n \in \mathbb{N}_0$. Since $f^{(\delta)} \in A(X)$ we know that there is a square integrable sequence $(\beta_n)_{n\in\mathbb{N}_0}$ such that $\delta_n = \mu_n\beta_n$. Inserting this into (10.5.14) we can employ the same argument as in (10.5.9) to obtain the norm convergence (10.5.10).

Finally, we study the case where $f^{(\delta)} \not\in A(X)$. In this case according to theorem 2.4.23 we have

$$\sum_{n=0}^{L} \left|\frac{\delta_n}{\mu_n}\right|^2 \to \infty, \quad L \to \infty. \tag{10.5.15}$$

We use the formula (10.5.5) and the general estimate $|a - b|^2 \geqslant |a|^2/2 - |b|^2$ to derive

$$\begin{aligned} \|\varphi^{(k)}\|^2 &\geqslant \sum_{n=0}^{L} \left|\gamma_n^{(k)}\right|^2 \\ &= \sum_{n=0}^{L} \left| q_n^k \gamma_n^{(0)} + \left(1 - q_n^{k+1}\right)\alpha_n^{(\text{true})} + \frac{\delta_n}{\mu_n}\left(1 - q_n^{k+1}\right)\right|^2 \\ &\geqslant \frac{1}{2} \sum_{n=0}^{L} \left|\frac{\delta_n}{\mu_n}\right|^2 \left(1 - q_n^{k+1}\right)^2 - c \end{aligned} \tag{10.5.16}$$

with some constant c. We know that $q_n^{k+1} \to 0$ for $k \to \infty$ for each fixed n. Given some constant C we can now first choose L such that the sum in (10.5.15) is larger than $C + 2c$. Then for all k large enough the sum in (10.5.16) is larger than C. This proves the divergence statement (10.5.11) and our proof is complete. □

As a consequence of lemma 10.5.2 we need to formulate an important remark: if we apply 3D-VAR iteratively and if we have an ill-posed observation operator, in the case of measurement error in general the scheme will be instable.

Usually, the model M has smoothing and damping properties. This can have a significant influence on the asymptotic behavior of errors. If a model is sufficiently damping higher modes, the divergent parts of the iteration will be damped out and the cycled dynamical system becomes stable. For more details we refer the reader to [23].

10.6 Review of convergence concepts for inverse problems

The goal of this section is to provide a brief review of all the different convergence concepts which are useful for studying inversion and data assimilation and for the understanding and monitoring of realistic systems.

Our basic set-up is to work with a normed state space X and a normed observation space Y. We have already seen the *convergence of a regularized reconstruction from true data* in (3.1.7), where the convergence

$$x_\alpha = R_\alpha y = R_\alpha Hx \to x, \quad \alpha \to 0, \tag{10.6.1}$$

of the reconstruction x_α to the true solution x defines a *regularization method.* The convergence (10.6.1) can only be a pointwise convergence in the sense that it holds for individual $x \in X$, but cannot hold in the operator norm.

We have also discussed the *convergence of a regularized solution from data with error*

$$x_{\alpha(\delta)} := R_{\alpha(\delta)} y^\delta \to x, \quad \delta \to 0, \tag{10.6.2}$$

where $y^\delta = Hx + r_\delta$ with $\|r_\delta\| \leqslant \delta$, i.e. the convergence of the approximate solution to the true solution when the error δ in the measurements tends to zero. The convergence of type (10.6.2) defines a *regular* strategy to choose $\alpha = \alpha(\delta)$, see (3.1.12). We have also studied the *speed* or *rate* of the convergence (10.6.2) when *source conditions* are given, see (3.3.69).

The above convergence concepts find their counterpart in stochastics, in particular in Bayesian methods. Convergence (10.6.1) corresponds to the *limit of a flat prior*, i.e. the case where *a priori* all states are equally likely. In the case of a flat prior and true data the maximum likelihood estimator will calculate the most likely state with respect to the data only.

Many inverse problems are linked to waves and fields, in particular in acoustics and electromagnetics, where unknown regions of space are probed by pulses. Then, the inversion task naturally splits into a field reconstruction task and the reconstruction of unknown parameter distributions or scatterers. Both tasks are linked and we will present recent convergence concepts in section 10.6.2. Here, we first study the role of convergence in the framework of dynamical systems.

10.6.1 Convergence concepts in stochastics and in data assimilation

Data assimilation is interested in the cycled reconstruction of some state $x_k \in X$, where usually $k \in \mathbb{N}$ is a discrete time index of time t_k and where states x_k are coupled by the model propagation (5.2.1). Of course, for each single time-slice t_k the convergence concepts (10.6.1) and (10.6.2) can be applied. But the cycled coupled problem carries further questions of convergence and stability over time. Usually, stochastic concepts are used to study convergence.

To explain the basic approach, let us consider a *constant* model dynamics M in the sense that $\dot{x} = 0$, and consider measurements y_k which are directly sampling the state space X. Then, we can consider the measurement as a random draw with some probability distribution with expectation value $m = x^{(\text{true})}$ and bounded variances. Measurements y_k correspond to random draws by independent random variables.

The assimilation by Tikhonov/3D-VAR (5.2.6) with $B = I$, $H = I$ and initial state x_0 at t_0 then corresponds to $x_{k+1}^{(b)} = x_k^{(a)}$ and

$$
\begin{aligned}
x_k^{(a)} &= x_k^{(b)} + \left(\alpha I + H^*H\right)^{-1} H\left(y_k - Hx_k^{(b)}\right) \\
&= x_k^{(b)} + \frac{1}{1+\alpha}(y_k - x_k^{(b)}) \\
&= \frac{\alpha}{1+\alpha} x_k^{(b)} + \frac{1}{1+\alpha} y_k, \quad k = 1, 2, 3, \ldots
\end{aligned}
\tag{10.6.3}
$$

which inductively yields

$$
x_k^{(a)} = \sum_{\xi=1}^{k} q^{k-\xi} p y_\xi + q^k x_0, \quad k = 1, 2, 3, \ldots \tag{10.6.4}
$$

where $q = \frac{\alpha}{1+\alpha}$ and $p = 1 - q = \frac{1}{1+\alpha}$. Note that we have

$$
\begin{aligned}
\mathbb{E}\left\{x_k^{(a)}\right\} &= \sum_{\xi=1}^{k} q^{\xi-1} p \,\mathbb{E}\{y_\xi\} + q^k x_0 = \frac{1-q^k}{1-q} p \,\mathbb{E}\{y_\xi\} + q^k x_0 \\
&= \left(1 - q^k\right) x^{(\text{true})} + q^k x_0.
\end{aligned}
\tag{10.6.5}
$$

By the *Tschebycheff inequality* (4.5.14) we obtain *stochastic convergence*

$$P\left(\left|x_k^{(a)} - x^{(true)} + q^k(x^{(true)} - x_0)\right| \geqslant \epsilon\right) \leqslant \frac{C}{\epsilon^2 k}, \quad k \to \infty, \tag{10.6.6}$$

such that up to an exponentially decaying term $x_k^{(a)}$ is a *consistent estimator*, see (4.1.7). If we continue to measure and to assimilate, we obtain a better and better estimate of the state.

Of course, in general we do not have a constant model and we do not observe the full state, but have some observation operator H with a null-space $N(H) \subset X$. The model M will be different from the true model $M^{(true)}$, such that the propagation will *spread* the probability distribution in each propagation step, while the assimilation *contracts* the spread in each assimilation step. For Gaussian densities, we explicitly obtain

$$\left(B_k^{(a)}\right)^{-1} = \left(B_k^{(b)}\right)^{-1} + H^* R_k^{-1} H, \quad k = 1, 2, 3, \dots$$

or in another form

$$\frac{1}{\left(\sigma_k^{(a)}\right)^2} = \frac{1}{\left(\sigma_k^{(b)}\right)^2} + \frac{1}{\left(\sigma_k^{(obs)}\right)^2}, \quad k = 1, 2, 3, \dots \tag{10.6.7}$$

as derived in detail in (5.4.14), leading to $\sigma_k^{(a)} < \sigma_k^{(b)}$.

We speak of *synchrony* if the assimilation step is able to counteract the spread and keep the state x_k in some bounded distance to the truth $x_k^{(true)}$. However, what distance corresponds to synchrony and what distance is *divergence* will depend on the particular dynamical system under consideration.

Even if $N(H) = \{0\} \subset X$, ill-posedness of the observation operator can lead to severe *instabilities*. We have shown in lemma 10.5.2 that for $f \notin H(X)$ even for *constant* model dynamics the state estimate $x_k^{(a)}$ by a classical data assimilation method will diverge.

Here, we close with remarks on the influence of a *model bias* on the behavior of a data assimilation system. In the easiest possible *constant* case, the true model is keeping the state fixed and the model M is adding some bias $b \in X$. Then, instead of (10.6.3) and (10.6.4) we obtain the iteration

$$x_{k+1}^{(b)} = q x_k^{(b)} + p y_k + b, \quad k = 1, 2, 3, \dots \tag{10.6.8}$$

and

$$\begin{aligned} x_k^{(a)} &= \sum_{\xi=1}^{k} q^{k-\xi} p y_\xi + q^k x_0 + \sum_{\xi=1}^{k-1} q^\xi b, \\ &= \sum_{\xi=1}^{k} q^{k-\xi} p y_\xi + q^k x_0 + \left(1 - q^{k-1}\right)\frac{q}{1-q} b, \quad k = 1, 2, 3, \dots . \end{aligned} \tag{10.6.9}$$

This leads to the estimator

$$\mathbb{E}\left\{x_k^{(a)}\right\} = \left(1 - q^k\right)x^{(\text{true})} + q^k x_0 + \left(1 - q^{k-1}\right)\frac{q}{1-q}b,$$

which by $\frac{q}{1-q} = \alpha$ tends to $x^{(\text{true})} + \alpha b$. This means that:

- the total bias will remain bounded, even if the model tries to run away from the data over time, but
- asymptotically the cycled data assimilation system can inherit a significant accumulated bias of size αb if by choosing q to 1 we put a lot of weight to the model calculations, reflected by a large parameter α.

The thorough treatment of model biases is one of the important tasks and interaction points of model developers and researchers working on data assimilation algorithms.

10.6.2 Convergence concepts for reconstruction methods in inverse scattering

Convergence of reconstructions for either wave fields or unknown scatterers such as inclusions and cavities is one of the key questions in inverse problems and data assimilation. Here, we will discuss what is usually meant when speaking about convergence and how convergence results interact.

Definition 10.6.1 (Convergence of field reconstruction). *Convergence for the field reconstruction problem given in definition 1.2.5 is measured locally in points $x \in \mathbb{R}^m$ or on some set $M \subset \mathbb{R}^m$. We say that the field reconstruction of u^s by some reconstructed field u^s_α depending on the parameter α is convergent in x if*

$$|u^s_\alpha(x) - u^s(x)| \to 0, \quad \alpha \to 0, \tag{10.6.10}$$

and it is uniformly convergent on M if

$$\sup_{x\in M} |u^s_\alpha(x) - u^s| \to 0, \quad \alpha \to 0. \tag{10.6.11}$$

For the reconstruction of the location and shape of objects $D \subset \mathbb{R}^m$ we need some understanding of the convergence of subsets of $\mathbb{R}^m$. We define the *Hausdorff distance* of a point x to a set U by

$$d(x, U) := \min_{y\in U} \|x - y\|, \quad x \in \mathbb{R}^m. \tag{10.6.12}$$

The *Hausdorff metric*

$$d(U, V) := \max\left\{\sup_{x\in U} d(x, V),\ \sup_{x\in V} d(x, U)\right\} \tag{10.6.13}$$

defines a distance between two sets U and V.

Definition 10.6.2 (Convergence of domain reconstructions). *We speak of convergence of domain reconstructions D_α by some reconstruction scheme to the true solution D if we have*

$$d(D_\alpha, D) \to 0, \quad \alpha \to 0. \tag{10.6.14}$$

Indicator functions. We usually use either the description of some domain D by its boundary $\Gamma := \partial D$, by the set of its interior points D, by some *level set* function L or by an *indicator* function I. Here, we focus our attention on indicator functions, since they are at the core of *sampling and probe methods* for *inverse source* and *inverse scattering* problems which will be the main topic of the chapters 13–15. For the rest of this chapter we assume that the exterior $D^e := \mathbb{R}^m \setminus \bar{D}$ of D is connected with infinity, i.e. there is a curve γ continuous on $[0, \infty)$ such that

$$\gamma : (0, \infty) \to D^e, \quad \gamma(0) \in \partial D, \quad \gamma(s) \to \infty \quad \text{for} \quad s \to \infty. \tag{10.6.15}$$

We will assume that $\gamma(s)$ is parametrized according to the length of the curve γ such that $|\gamma(s') - \gamma(s)| \leqslant |s' - s|$ for $s, s' > 0$.

Definition 10.6.3 (Indicator functions). *An* indicator function (I_α, I) *for a domain D is a family of non-negative continuous functions I_α defined on $\mathbb{R}^m$ for $\alpha > 0$ which for $\alpha \to 0$ and on $\mathbb{R}^m \setminus \bar{D}$ converge towards a non-negative continuous function I in $\mathbb{R}^m \setminus \bar{D}$, i.e.*

$$I_\alpha(x) \to I(x), \quad x \in \mathbb{R}^m \setminus \bar{D}, \tag{10.6.16}$$

uniformly on compact subsets of $\mathbb{R}^m \setminus \bar{D}$, such that

$$I(x) \to \infty, \quad x \to y \in \partial D \tag{10.6.17}$$

uniformly for $y \in \partial D$. We further assume that I is bounded by some constant C on $\mathbb{R}^m \setminus B_R$ for some $R > 0$ such that $B_R \supset \bar{D}$.

The behavior (10.6.17) can be used to locate some unknown inclusion or unknown scatterer, and reconstruct its shape and further properties. Clearly, by continuity the indicator function I is bounded on compact subsets of $\mathbb{R}^m \setminus \bar{D}$. Let γ be a curve as defined in (10.6.15). Then from (10.6.17) and the continuity and boundedness of I on $\mathbb{R}^m \setminus B_R$ we know that for any $\epsilon > 0$ there is a constant $c > 0$ such that

$$I(\gamma(s)) < c \quad \text{for } s \in (\epsilon, \infty). \tag{10.6.18}$$

Further, we know that

$$I(\gamma(s)) \to \infty, \quad s \to 0. \tag{10.6.19}$$

The consequence of (10.6.18) and (10.6.19) is that given $c > 0$ and defining $z(c) \in B_R$ to be the first point on $\gamma(s)$ as $s \to 0$ (i.e. coming from the outside) for which $I(\gamma(s)) = c$ and $s(c)$ to be the first s for which $I(\gamma(s)) = c$, we know that

$$z(c) \to \partial D \quad \text{and} \quad s(c) \to 0 \quad \text{as } c \to \infty. \tag{10.6.20}$$

Suppose there is a family of continuous curves γ_x for each $x \in \partial D$ connecting x with infinity such that $s_x(c)$ depends continuously on x for each large enough c and

$s_x(c) \to 0$ as $c \to \infty$ uniformly for $x \in \partial D$, where $s_x(c)$ is as in (10.6.20). We define the *level curves* of the indicator function by

$$\Gamma_c := \{z_x(c) = \gamma_x(s_x(c)) \colon \ x \in \partial D\}. \tag{10.6.21}$$

Then, by (10.6.20) we have $s_x(c) \to 0$ and $\gamma_x(s_x(c)) \to x$ as $c \to \infty$ uniformly for $x \in \partial D$. We summarize this result into the following basic convergence theorem.

Theorem 10.6.4 (Indicator functions determine domains). *Assume that an indicator function* (I_α, I) *is given according to definition* 10.6.3, *and let the set* Γ_c *be defined as in (10.6.21) using family of curves* γ_x *and* I *with the properties that* $s_x(c)$ *is continuous on* ∂D *for each large enough* c *and* $s_x(c) \to 0$ *for* $c \to \infty$ *uniformly for* $x \in \partial D$. *Then,*

$$\Gamma_c \to \partial D, \quad c \to \infty \tag{10.6.22}$$

i.e. we have convergence of the set Γ_c *to the unknown boundary* ∂D *with respect to the Hausdorff distance, i.e. in the sense of set convergence (10.6.14).*

A key algorithmic challenge of probing methods is to choose curves γ and to define the indicator functions (I_α, I) such that the convergence (10.6.16) and the behavior (10.6.17) is obtained in $\mathbb{R}^m \backslash \bar{D}$. We will discuss different strategies in the upcoming chapters. Here, we finish with some basic stability result.

Theorem 10.6.5 (Conditional stability of domain reconstructions). *Assume that the reconstruction of the indicator functions* I_α *is stable with respect to the measurements* u^∞ *on each compact subset* M *of* D^e, *i.e. given* $\rho > 0$ *there is* $\delta > 0$ *such that*

$$|I[u_1^\infty](x) - I[u_2^\infty](x)| \leqslant \rho \quad \text{for } \|u_1^\infty - u_2^\infty\| \leqslant \delta \tag{10.6.23}$$

for all $x \in M$. *Further, assume that the divergence (10.6.17) is uniform on a set of domains U. Then, also the domain reconstruction is stable, i.e. we have*

$$d(D_1, D_2) \to 0 \quad \text{for } \|u_1^\infty - u_2^\infty\| \to 0. \tag{10.6.24}$$

Proof. Given $\epsilon > 0$ we need to show that there is $\delta > 0$ such that

$$\|u_1^\infty - u_2^\infty\| \leqslant \delta \Rightarrow d(D_1, D_2) \leqslant \epsilon.$$

We choose $\partial D = \partial D_1$, curves γ_x for $x \in \partial D$ and Γ_c defined by (10.6.21) such that for $s \in [0, s_0]$, $\gamma_x(s) = x + s\nu(x)$ with the outer unit normal vector ν to ∂D. Then given $\epsilon > 0$ there is $c > 0$ such that $I(\gamma_x(s)) < c$ for $s \geqslant \epsilon$. Further, there is $\sigma_x > 0$ such that $|\gamma_x(\sigma_x)| = R$. Then, given $\rho > 0$ there is $\delta > 0$ such that (10.6.23) is satisfied with $M_x := \{\gamma_x(s) : s \in [\epsilon, \sigma_x]\}$. Then, on M_x we have $I[u_2^\infty](x) < c + \rho$ so that ∂D_2 does not intersect M_x, which yields $D_2 \subset \{x \in \mathbb{R}^m \colon d(x, D) \leqslant \epsilon\}$. The same is satisfied with the role of D_1 and D_2 exchanged. As a consequence we obtain $d(D_1, D_2) \leqslant \epsilon$, and thus the stability (10.6.24). □

Domain sampling. Some inversion schemes employ the principle of *domain sampling*. For a family $\mathcal{G}$ of test domains testing to find set D or its approximation domain they provide an *indicator function* (I_α, I) defined on $\mathcal{G}$ such that for $D \subset G \in \mathcal{G}$ we have

$$I_\alpha(G) \to I(G), \quad \alpha \to 0 \tag{10.6.25}$$

and for $D \not\subset G \in \mathcal{G}$

$$I_\alpha(G) \to \infty, \quad \alpha \to 0. \tag{10.6.26}$$

Then, the domain D is approximated by taking the intersection

$$D_{\alpha,\mathcal{G}} := \bigcap\nolimits_{G \in \mathcal{G}, I_\alpha(G) \leqslant c} G \tag{10.6.27}$$

of any test domain G which passes the test with threshold $c > 0$. Clearly, if the set of test domains is sufficiently rich, by the conditions (10.6.25) and (10.6.26) we obtain convergence of domain sampling in the Hausdorff norm.

Bibliography

[1] Hofmann B 1986 *Regularization for Applied Inverse and Ill-Posed Problems: A Numerical Approach* (*Teubner-Texte zur Mathematik*) (Leipzig: Teubner)

[2] Potthast R 1998 On a concept of uniqueness in inverse scattering for a finite number of incident waves *SIAM J. Appl. Math.* **58** 666–82

[3] Potthast R 2001 *Point Sources and Multipoles in Inverse Scattering Theory* (*Chapman and Hall/CRC Research Notes in Mathematics* vol 427) (Boca Raton, FL: CRC)

[4] Colton D and Kress R 1998 *Inverse Acoustic and Electromagnetic Scattering Theory* (*Applied Mathematical Sciences* vol 93) 2nd edn (Berlin: Springer)

[5] Isakov V 1998 *Inverse Problems for Partial Differential Equations* (*Springer Series in Applied Mathematical Science* vol 127) (Berlin: Springer)

[6] Rondi L 2003 Unique determination of non-smooth sound-soft scatterers by finitely many far-field measurements *Indiana Univ. Math. J.* **52** 1631–62

[7] Lax P D and Phillips R S 1967 *Scattering Theory* (*Pure and Applied Mathematics* vol 26) (New York: Academic)

[8] Colton D and Sleeman B D 1983 Uniqueness theorems for the inverse problem of acoustic scattering *IMA J. Appl. Math.* **31** 253–9

[9] Stefanov P and Uhlmann G 2004 Local uniqueness for the fixed energy angle inverse problem in obstacle scattering *Proc. Am. Math. Soc.* **132** 1351–4

[10] Gintides D 2005 Local uniqueness for the inverse scattering problem in acoustics via the Faber–Krahn inequality *Inverse Problems* **21** 1195–205

[11] Isakov V 1991 Stability estimates for obstacles in inverse scattering *J. Comput. Appl. Math.* **42** 79–89

[12] Isakov V 1993 New stability results for soft obstacles in inverse scattering *Inverse Problems* **9** 535–43

[13] Sincich E and Sini M 2008 Local stability for soft obstacles by a single measurement *Inv. Prob. Imaging* **2** 301–15

[14] Potthast R 2000 Stability estimates and reconstructions in inverse acoustic scattering using singular sources *J. Comput. Appl. Math.* **114** 247–74

[15] Cheng J and Yamamoto M 2003 Uniqueness in an inverse scattering problem within non-trapping polygonal obstacles with at most two incoming waves *Inverse Problems* **19** 1361–84

[16] Elschner J and Yamamoto M 2006 Uniqueness in determining polygonal sound-hard obstacles with a single incoming wave *Inverse Problems* **22** 355–64

[17] Alessandrini G and Rondi L 2005 Determining a sound-soft polyhedral scatterer by a single far-field measurement *Proc. Am. Math. Soc.* **133** 1658–91

[18] Liu H and Zou J 2006 Uniqueness in an inverse acoustic obstacle scattering problem for both sound-hard and sound-soft polyhedral scatterers *Inverse Problems* **22** 515–24

[19] Liu H and Zou J 2007 On unique determination of partially coated polyhedral scatterers with far field measurements *Inverse Problems* **23** 308

[20] Honda N, Nakamura G and Sini M 2013 Analytic extension and reconstruction of obstacles from few measurements for elliptic second order operators *Math. Ann.* **355** 401–27

[21] Ramm A G 2005 Uniqueness of the solution to inverse obstacle scattering problem *Phys. Lett.* A **347** 157–9

[22] McLean W 2000 *Strongly Elliptic Systems and Boundary Integral Equations* (Cambridge: Cambridge University Press)

[23] Moodey A J F, Lawless A S, Potthast R W E and van Leeuwen P J 2013 Nonlinear error dynamics for cycled data assimilation methods *Inverse Problems* **29** 025002

IOP Publishing

Inverse Modeling

An introduction to the theory and methods of inverse problems and data assimilation

Gen Nakamura and Roland Potthast

Chapter 11

Source reconstruction and magnetic tomography

Magnetic tomography is concerned with the reconstruction of current densities and conductivity distributions from measured magnetic fields. It is a classical example of an *inverse source problem*. It is of great importance for medical applications such as *magnetoencephalography* as well as emerging in industrial applications such as *magnetic tomography* for *fuel cells*.

In the following section we will first briefly describe the set-up and numerical solution of the *forward model* which leads to a modeling of the current density j and then introduce the *inverse problem* based on the Biot–Savart operator W which is mapping the current distribution j onto its magnetic field H. The inverse problem consists of the reconstruction of j from the given $H = Wj$, i.e. the solution of a linear integral equation of the first kind.

First, in section 11.1 we describe the mathematical framework for current simulation and study its numerical solution by the *finite integration technique*. Here, we develop this approach with a discretization based on a wire grid and classical *know rules* and *mesh rules*.

In section 11.2 we will provide basic insight into the *uniqueness question*, showing that in general the source reconstruction problem is highly non-unique. A characterization of the null-space of the Biot–Savart operator will be given. Then, we employ and investigate Tikhonov regularization for solving such problems. In particular, we will investigate the use of further conditions from the direct problem throughout the inversion process. We will show that additional constraints improve the stability of the inverse problem and can lead to improved reconstructions.

Section 11.3 deals with *dynamic source problems*. For many applications, the currents j will depend on time, and there will be repeated measurements $H_1, H_2, H_3, \ldots$ at times $t_1, t_2, t_3, \ldots$. This leads to what is known as *data assimilation*, the repeated solution of an inverse problem. Usually, the dynamics which maps currents j_k at time t_k into currents j_{k+1} at time t_{k+1} is given by some model $M_k : X \to X$, where X is the state space for the currents j under consideration.

doi:10.1088/978-0-7503-1218-9ch11

Data assimilation methods are introduced and studied in chapter 5. Here, we will focus on the situation when the model M is not completely known, but depends on some parameter function $p \in Z$. Then, the task is not only to reconstruct the currents j, but also the parameter function p. This is known as the *parameter estimation problem.*

The final section 11.4 of this chapter investigates *classification methods* in the framework of inverse problems. Often, in applications it is not necessary to fully solve the inverse problem when measurements H are given, but one needs to classify the states based on the measurements. We will see that when the *full inverse problem* is ill-posed, this property carries over to the *classification problem* in image space. We will develop the underlying theory; it applies to any ill-posed linear source problem.

11.1 Current simulation

Current distributions appear in many natural or industrial phenomena. In the brain currents are generated by microscopic processes in neural cells, leading to particular areas of high current activity which are often linked to illnesses such as epilepsy. In fuel cells they arise from a chemical process, where they are linked to potentials through the membrane of the cells.

If a conductivity distribution $\sigma = \sigma(x) \in L^\infty(\Omega)$ is given in some simply connected domain $\Omega \subset \mathbb{R}^3$ and when current is fed into the area at its surface $\partial\Omega$, then we obtain a current distribution $j = j(x)$ for $x \in \Omega$ based on *Ohm's law*. We remarked that we have assumed that Ω is simply connected to simplify the proof of theorem 11.2.4 given in subsection 11.2.1 below. In the following section 11.1.1 we will study the simulation of current distributions as a basis for current reconstruction. This conductivity-based current distribution has also been used as a basis for magnetic tomography for fuel cells (see [1–3]), where an *effective conductivity* was used as a macroscopic approximation to the complex microscopic processes taking place in a fuel cell.

11.1.1 Currents based on the conductivity problem

To derive the equations modeling a current distribution we need to start with the time-independent Maxwell equations in a simply connected bounded domain $\Omega \subset \mathbb{R}^3$ with piece-wise C^2 boundary $\partial\Omega$ with well-behaved edges and corners. Following [2] we use standard notation, which employs E for the electric field, D for the electric flux density, H for the magnetic field strength and B for the magnetic flux density.

We denote the electric *permittivity* and the magnetic *permeability* of Ω by ϵ and μ, respectively. The constants ϵ_0 and μ_0 are the well-known permittivity and permeability for the vacuum. The *static Maxwell equations* for the magnetic field H and electric field E for given electric current $j \in (L^2(\Omega))^3$ and electric charge density ρ in the dual space $H^1(\Omega)^*$ of $H^1(\Omega)$ are

$$\nabla \times H = j, \quad \nabla \times E = 0 \tag{11.1.1}$$

$$\nabla \cdot D = \rho, \quad \nabla \cdot B = 0. \tag{11.1.2}$$

They are complemented by the *material equations*

$$D = \epsilon\epsilon_0 E, \quad B = \mu\mu_0 H \tag{11.1.3}$$

and by *Ohm's law* with conductivity σ

$$j = \sigma E. \tag{11.1.4}$$

We assume here that ϵ, $\mu \in L^\infty(\Omega)$ and the matrix $\sigma \in (L^\infty(\Omega))^9$ satisfies

$$\epsilon, \mu, \sigma \geqslant c_0 > 0 \text{ a.e. in } \Omega \tag{11.1.5}$$

and the coercivity given as

$$a^\top \sigma(x) a \geqslant c_0 |a|^2 \text{ for all } a \in R^3 \text{ and for a.e. } x \in \Omega \tag{11.1.6}$$

for some constant c_0.

The above equations can be transformed into an elliptic boundary value problem as follows. Because of Ω being simply connected and $\nabla \times E = 0$, there is an *electric potential* $\varphi_E \in H^1(\Omega)$ such that $E = \nabla\varphi_E$. Then, by (11.1.4) for the current density j we have the equation $j = \sigma\nabla\varphi_E$. We use the identity $\nabla \cdot \nabla \times A = 0$, which is valid for any arbitrary sufficiently smooth vector field A to derive the equation

$$\nabla \cdot j = \nabla \cdot \nabla \times H = 0 \tag{11.1.7}$$

from the Maxwell equations (11.1.1), i.e. the current distribution is *divergence free*. By Ohm's law (11.1.4) from (11.1.7) we now obtain the second order elliptic partial differential equation

$$\nabla \cdot \sigma\nabla\varphi_E = 0 \quad \text{in } \Omega \tag{11.1.8}$$

for the electric potential φ_E. Additionally, from our modeling set-up we have the *boundary condition*

$$\nu \cdot j = g \in H^{-1/2}(\partial\Omega) \quad \text{on } \partial\Omega \tag{11.1.9}$$

with some given function g on $\partial\Omega$ modeling the flow of currents through the boundary $\partial\Omega$. We employ standard Sobolev function spaces, where $H^{-\frac{1}{2}}(\partial\Omega)$ is the space of weak derivatives of functions in $H^{\frac{1}{2}}(\partial\Omega)$, and by the *trace theorem* the space $H^{\frac{1}{2}}(\partial\Omega)$ is the space of all boundary traces of functions in $H^1(\Omega)$, i.e. the space of functions which have the first order derivatives in $L^2(\Omega)$.

Note that the equations (11.1.8), (11.1.9) for φ_E cannot be unique, since adding any constant c to a solution φ_E will also provide a solution. However, by the *normalization condition*

$$\int_\Omega \varphi_E \mathrm{d}y = 0 \tag{11.1.10}$$

uniqueness can be enforced.

In general, the conductivity $\sigma(x)$ at a point $x \in \Omega$ is a *matrix function* and the conductivity problem (11.1.8)–(11.1.10) is an *anisotropic* problem to calculate the current distribution. Often, a *weak form* of the conductivity problem is obtained by multiplication of (11.1.8) by some sufficiently smooth *test function* ψ and partial integration, leading to

$$\int_{\Omega} \nabla\psi^{\mathrm{T}}(x)\sigma(x)\nabla\varphi(x)\,\mathrm{d}x = \int_{\partial\Omega} \psi(x)(\nu(x)\cdot\sigma\nabla\varphi(x))\,\mathrm{d}s(x) \qquad (11.1.11)$$

for $\varphi - \varphi_E$. Searching for $\varphi \in H^1(\Omega)$, equation (11.1.11) is valid for all $\psi \in H^1(\Omega)$. In this case, the function $\nu \cdot \sigma\nabla\varphi$ is an element of the Sobolev space $H^{-1/2}(\partial D)$ and the right-hand side of (11.1.11) is understood in the sense of a dual space scalar product between $H^{-1/2}(\partial\Omega)$ and $H^{1/2}(\partial\Omega)$. Then, the bounded sesquilinear form

$$a(\psi, \varphi) := \int_{\Omega} \nabla\psi^{\mathrm{T}}(x)\sigma(x)\nabla\varphi(x)\,\mathrm{d}x \qquad (11.1.12)$$

for $\varphi, \psi \in H^1(\Omega)$ is *coercive* as defined in (2.5.3). Further, given a current $j \in H^{-1/2}(\partial\Omega)$ we define the linear functional

$$F(\psi) := \int_{\partial\Omega} \psi(x)(\nu(x)\cdot j(x))\,\mathrm{d}s(x) = \int_{\partial\Omega} \psi(x)g(x)\,\mathrm{d}s(x). \qquad (11.1.13)$$

Then (11.1.8) and (11.1.9) in the weak form (11.1.11) can be written as

$$a(\psi, \varphi) = F(\psi) \quad \text{for all } \psi \in H^1(\Omega). \qquad (11.1.14)$$

Now, according to the Lax–Milgram theorem 2.5.3 there is a unique element $\varphi \in H^1(\Omega)$ which is a weak solution of the equation ((11.1.8)) with boundary condition ((11.1.9)).

11.1.2 Simulation via the finite integration technique

The basic idea of the finite integration technique for current simulation as used by Kühn and Potthast [2] is to use a resistor network as approximation to a continuous current density distribution.

Consider a wire network with K nodes and N wires connecting the nodes as displayed in figure 11.1. Every wire connects exactly two nodes and has a direction,

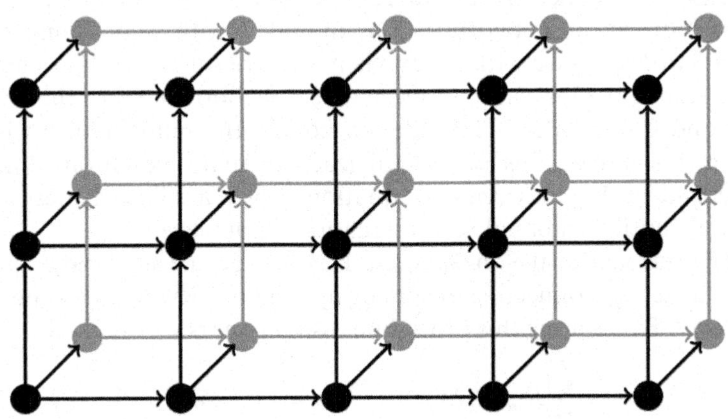

Figure 11.1. A grid of nodes or knots with connecting wires, as employed for current reconstructions. The current flowing through a wire with index ξ is denoted by J_ξ, its resistance by R_ξ.

it is outgoing from one and inbound for the other node. Here, we use the *adjacency matrix*

$$T = (T_{kl})_{k=1,\dots K, l=1,\dots N}, \tag{11.1.15}$$

where T_{kl} is 1 if wire l starts at knot k, -1 if wire l ends in knot k and 0 otherwise. Further, define $\boldsymbol{b} = (b_1, \dots, b_K)^T$ to be the input and outflow into the network, for example

$$b_k = \begin{cases} I_k, & \text{if } k \text{ is an injection knot of the current} \\ -I_k, & \text{if } k \text{ is an outflow knot of the current} \\ 0, & \text{otherwise} \end{cases} \tag{11.1.16}$$

with the pointwise current I_k. We denote the total current by I. For the discrete model we assume that the resistance of wire l is given by R_l. The current J_l flowing through wire l, the resistance R_l of wire l and the voltage U_l between the two endpoints of wire l are connected via *Ohm's law*

$$U_l = R_l J_l. \tag{11.1.17}$$

As is known in standard circuit theory, a current flow in a resistor network can be calculated via *mesh* and *knot rules*, also known as *Kirchhoff's circuit laws*.

- The knot rules are obtained from integrating the divergence equation (11.1.7), such that the sum of all currents flowing into and out of some node is zero.
- The mesh rules are obtained from integrating $\nabla \times E = 0$ together with Ohm's law (11.1.4) $E = j/\sigma$, such that the weighted integral of the currents j with weight σ^{-1} over any closed loop is zero by Stokes' theorem, where $R = \sigma^{-1}$ is the resistance in the corresponding wire.

On a wire grid, the knot rules are given by

$$\sum_{l=1}^{N} T_{kl} J_l = b_k, \quad k = 1, \dots, K. \tag{11.1.18}$$

The construction of the mesh rules can been realized as follows. For every knot k we collect closed loops in our network starting in node k with an outgoing wire. Setting the matrix S_{ml} to 1 if a wire with index l is in a current loop with orientation along the flow direction of the loop and -1 if the wire direction is against the flow direction of the loop and 0 otherwise. The index m counts the number of all loops under consideration. Of course, this in general leads to many equations which are not linearly independent. We may remove equations which are linearly dependent on the other equations until a minimal set of equations is achieved.

For special grids such as the one in figure 11.1, this can be achieved by just using the loops which arise by connecting neighboring points, which are sometimes called *elementary loops*. By iterating this procedure over all knots we obtain the linear system

$$\sum_{l=1}^{N} S_{ml}(R_l J_l) = 0, \quad m = 1, \dots, M, \tag{11.1.19}$$

where M is the maximal number of independent loops in the mesh.

Here, we clearly want to obtain a matrix S which has maximal rank, i.e. there are no linearly dependent rows or columns. For many grids this can be achieved by choosing elementary loops, although for general graphs this might be a challenging task. Here, we focus on grids which consist of regularly arranged nodes with wires which only connect next-neighbor points in three dimensions. Then, elementary loops can be easily constructed by going one step into one direction, one step into another and then closing the loop by going back the first direction and going back the second.

The above procedure results in $K + M$ equations for the $K + M - 1$ currents, with one redundant equation due to the fact that the current density is divergence free. Here, we drop one appropriate equation to obtain a uniquely solvable system.

A convergence theorem for the *finite integration technique* towards the solution of the continuous problem (11.1.8) and (11.1.9) has been shown in [2], theorems 5 and 7, where the discretized problem is extended into the three-dimensional space by interpolation chosen to be either constant or linear in the directions x, y and z.

Theorem 11.1.1. *Let σ be a coercive matrix. Then, the equation system arising from mesh and knot rules has a unique solution for each discretization. As all the numbers of grid points of the discretization in each direction tends to infinity,* it converges *towards the true solution of the boundary problem (11.1.8) and (11.1.9) with respect to $L^2(\Omega)$.*

Finally, let us provide some simple code to realize the equations (11.1.18) and (11.1.19) and simulate some current distribution.

Code 11.1.2. *For a simple uniform grid filling some cuboid the set-up of the nodes and wires is carried out in* `sim_11_1_2_a_wires.m`. *Here we first generate the grid. Each wire is given by its starting node and its direction perpendicular to the x_1, x_2 or x_3 axis. We use tensorial form to obtain fast access to neighboring points and the full indices of the wires in the variables* `wi1`, … , `wi3`.

```
clear all; close all;
a1 = 4; a2 = 1; a3 = 3;     % size of cuboid [0 a1] x [0 a2] x [0 a3]
N1 = 5; N2 = 2; N3 = 4;     % number of discretization points
h1 = a1/(N1-1); h2 = a2/(N2-1); h3 = a3/(N3-1); % grid spacing in 3 directions
N = N1*N2*N3;               % total number of points in the grid

x1 = 0:h1:a1; x2 = 0:h2:a2; x3=0:h3:a3;   % vectors of grid coordinates
x1m = repmat(x1',1,N2,N3);  % fill tensor with x1 coordinates
x2m = repmat(x2,N1,1,N3);   % ~
x3m = repmat(reshape(x3,1,1,N3),N1,N2,1); % ~
p1v = reshape(x1m,N,1);     % vector with x1 coordinates of grid nodes
p2v = reshape(x2m,N,1);     % ~       ~      x2 ~
p3v = reshape(x3m,N,1);     % ~       ~      x3 ~

w11m = x1m(1:(N1-1),:,:);     % x1 coord of nodes where wires in x1 dir start
w12m = x2m(1:(N1-1),:,:);     % x2 coord ...
w13m = x3m(1:(N1-1),:,:);     % x3 coord ... etc
w21m = x1m(:,1:(N2-1),:); w22m = x2m(:,1:(N2-1),:); w23m = x3m(:,1:(N2-1),:);
w31m = x1m(:,:,1:(N3-1)); w32m = x2m(:,:,1:(N3-1)); w33m = x3m(:,:,1:(N3-1));
```

```
Nw1 = (N1-1)*N2*N3;          % number of wires in x1 direction
Nw2 = N1*(N2-1)*N3;          % ~                      x2 ~
Nw3 = N1*N2*(N3-1);          % ~                      x3 ~
Nw  = Nw1+Nw2+Nw3;           % total number of wires
wi1 = reshape(1:Nw1,(N1-1),N2,N3);              % index of wires in x1 direction
wi2 = reshape((1:Nw2)+Nw1,N1,(N2-1),N3);        % ~                      x2
wi3 = reshape((1:Nw3)+Nw1+Nw2,N1,N2,(N3-1)); % ~                         x3
w11v = reshape(w11m,Nw1,1); w12v = reshape(w12m,Nw1,1);  % vectors of
w13v = reshape(w13m,Nw1,1); w21v = reshape(w21m,Nw2,1);  % wire coordinates
w22v = reshape(w22m,Nw2,1); w23v = reshape(w23m,Nw2,1);  % ...
w31v = reshape(w31m,Nw3,1); w32v = reshape(w32m,Nw3,1);
w33v = reshape(w33m,Nw3,1);
```

A simple script to simulate the currents in a regular grid is shown next, it is the script `sim_11_1_2_b_currents.m`. *We have one loop over all nodes collecting all in- and outgoing wires into one equation each. Then we have a loop over all nodes which is setting up the mesh equations for elementary loops and after each loop checks it for linear independence (and removes it if it is linearly dependent).*

```
% Define the anisotropic resistence on the grid points
R1 = 1*ones(N1-1,N2,N3);
R2 = 10*ones(N1,N2-1,N3);
R3 = 1*ones(N1,N2,N3-1);

% The knot rules for each node
en = 1; % knot rule equation counter
for j3=1:N3
  for j2=1:N2
     for j1=1:N1  % add equation for in and outgoing wires at node
        if( j1<N1 ) nrules(en, wi1(j1,j2,j3))=-1; end
        if( j1>1  ) nrules(en, wi1(j1-1,j2,j3))=1;  end
        if( j2<N2 ) nrules(en, wi2(j1,j2,j3))=-1; end
        if( j2>1  ) nrules(en, wi2(j1,j2-1,j3))=1;  end
        if( j3<N3 ) nrules(en, wi3(j1,j2,j3))=-1; end
        if( j3>1  ) nrules(en, wi3(j1,j2,j3-1))=1;  end
        en = en+1;
     end
  end
end
nrules=nrules(1:(en-2),:); % remove last linearly dependent equation

% The mesh rules for up to three elementary loops for each node
em = 1; % mesh rule equation counter
mrules=zeros(1,Nw);
for j1=1:N1
  for j2=1:N2
     for j3=1:N3
        if( j1<N1 && j2<N2) % add equation for elementary loop in x1-x2
           mrules(em, wi1(j1,j2,j3))=R1(j1,j2,j3);
           mrules(em, wi1(j1,j2+1,j3))=-R1(j1,j2+1,j3);
           mrules(em, wi2(j1+1,j2,j3))=R2(j1+1,j2,j3);
           mrules(em, wi2(j1,j2,j3))=-R2(j1,j2,j3);
           if( rank(mrules)==em ) em = em+1;       % remove equation if
           else  mrules=mrules(1:(em-1),:); end  % linearly dependent
        end
```

```
        if( j2<N2 && j3<N3) % add equation for elementary loop in x2-x3
           mrules(em, wi2(j1,j2,j3))=R2(j1,j2,j3);
           mrules(em, wi2(j1,j2,j3+1))=-R2(j1,j2,j3+1);
           mrules(em, wi3(j1,j2+1,j3))=R3(j1,j2+1,j3);
           mrules(em, wi3(j1,j2,j3))=-R3(j1,j2,j3);
           if( rank(mrules)==em ) em = em+1;      % remove equation if
           else  mrules=mrules(1:(em-1),:); end  % linearly dependent
       end
        if( j1<N1 && j3<N3) % add equation for elementary loop in x1-x3
           mrules(em, wi1(j1,j2,j3))=R1(j1,j2,j3);
           mrules(em, wi1(j1,j2,j3+1))=-R1(j1,j2,j3+1);
           mrules(em, wi3(j1+1,j2,j3))=R3(j1+1,j2,j3);
           mrules(em, wi3(j1,j2,j3))=-R3(j1,j2,j3);
           if( rank(mrules)==em ) em = em+1;      % remove equation if
           else  mrules=mrules(1:(em-1),:); end  % linearly dependent
        end
     end
  end
end

rules = [nrules;mrules];  % compose full matrix with knot and mesh rules

% setup right-hand side, in and outflow node
in_ind = 3;                % index of inflow node
out_ind = N1*N2-N1+in_ind;  % index of outflow node
I = 1;                     % total inflowing and outflowing current
b = zeros(Nw,1); b(in_ind,1)=-I; b(out_ind,1)=I;    % right-hand side

j = rules\b;               % solve know and mesh equations
```

The result of this code is displayed in figure 11.2, the script to display these graphics by some quiver plots is in the code repository, it is very similar to code 11.2.2.

11.2 The Biot–Savart operator and magnetic tomography

Let us now assume that we know the current distribution j in the domain $\Omega \subset \mathbb{R}^3$. Magnetic fields H of currents j are calculated via the *Biot–Savart integral operator*, defined by

$$(Wj)(x) := \frac{1}{4\pi} \int_{\Omega} \frac{j(y) \times (x - y)}{|x - y|^3} \mathrm{d}y, \quad x \in \mathbb{R}^3 \tag{11.2.1}$$

for a current density distribution $j \in L^2(\Omega)^3$. The task of *magnetic tomography* in its general form reduces to solving the equation

$$Wj = H \text{ on } \partial G, \tag{11.2.2}$$

where G is some domain with sufficiently smooth boundary such that $\bar{\Omega} \subset G$ and H denotes some measured magnetic field on ∂G. We will call ∂G the *measurement surface*. By $\mathrm{curl}_x(\Phi(x, y)j(y)) = \nabla_x \Phi(x, y) \times j(y)$ for

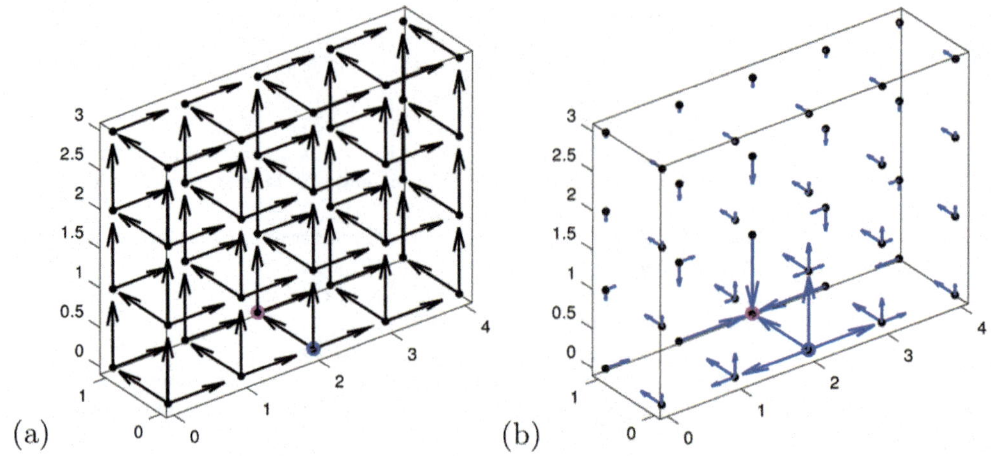

Figure 11.2. (a) The nodes and wires of a regular wire grid as set up by the script `sim_11_1_1_a_wires.m`. The blue node is the inflowing node and the magenta node the outflowing node. The orientation of the wires is given by the arrows. (b) The result of the current simulation as calculated by code 11.1.2 with inflowing and outflowing nodes. The length of the arrows reflects the strength of the current in the wire, the direction is shown by the arrow direction.

$$\Phi(x, y) := \frac{1}{4\pi}\frac{1}{|x - y|}, \quad x \neq y \in \mathbb{R}^3, \tag{11.2.3}$$

the Biot–Savart operator W can be readily seen as the curl of the volume potential

$$(Vj)(x) := \int_\Omega \Phi(x, y)j(y)\,\mathrm{d}y, \quad x \in \mathbb{R}^3, \tag{11.2.4}$$

i.e. we have

$$Wj = \mathrm{curl}\; Vj. \tag{11.2.5}$$

The discretization of the integral operator can be based on the wire grid and measurement points $x_k \in \partial G$, $k = 1, \ldots, m$. Having calculated the current vector

$$J = (J_1, J_2, \ldots, J_{K+M})^T \tag{11.2.6}$$

by (11.1.18) and (11.1.19) we are able to calculate the magnetic field at x_k via the discrete Biot–Savart operator for this wire grid. This reduces to calculating

$$(\mathbf{WJ})_k = \frac{1}{4\pi}\sum_{\ell=1}^{K+M}\int_{\gamma_\ell}\frac{\tilde{J}_\ell \times (x_k - p)}{|x_k - p|^3}\,\mathrm{d}s(p), \tag{11.2.7}$$

with wires γ_ℓ and evaluation points $x_k \in \partial G$ for $k = 1, \ldots, m$, current $\tilde{J}_\ell$ given by

$$\tilde{J}_\ell := \begin{cases} (J_\ell, 0, 0)^T, & \text{if } \gamma_\ell \| e_x, \\ (0, J_l, 0)^T, & \text{if } \gamma_\ell \| e_y \\ (0, 0, J_l)^T, & \text{if } \gamma_\ell \| e_z. \end{cases} \tag{11.2.8}$$

We may use a numerical method to calculate the integrals in (11.2.7) or employ an exact integration of the magnetic field for a straight wire as explicitly given by equation (4.9) in [4]. The measurements are taken at x_k, $k = 1, \ldots, m$, such that we have a measurement vector

$$H = (H_1, H_2, \ldots, H_m)^T. \tag{11.2.9}$$

With finite element and finite volume methods, there are of course higher order methods available for solving the partial differential equation (11.1.8) and higher-order quadrature formulas for calculating the integrals (11.2.1). But the above scheme is simple to implement and it can also be fully realized physically by real wire grids, such that the numerics fully fit to an experimental test setting.

Code 11.2.1. *In its easiest form the set-up of the discrete Biot–Savart operator can be performed by some simple loops, as carried out by the script* `sim_11_2_1_Biot_Savart_Operator.m`. *Here, we first define measurement points on some ellipse containing the wire grid. Then, we set up the operator W based on (11.2.7).*

```
tic;    % to evaluate run time
% setup elliptic evaluation surface
Ng1 = 20; Ng2 = 11; Ng = Ng1*Ng2; hp=.3; % number of evaluation points
hb = 2*pi/Ng1; hg = (pi-2*hp)/Ng2; % angular spacing in polar coordinates
beta  = 0:hb:2*pi-hb;                % vector of angles
gamma = -pi/2+hp:hg:pi/2-hg-hp;      % ~
betam = repmat(beta',1,Ng2); gammam = repmat(gamma,Ng1,1);
betav = reshape(betam,Ng,1); gammav = reshape(gammam,Ng,1);
A1  = 1.2*a1; A2 = 1.2*a2; A3 = 1.2*a3; cv = [a1/2 a2/2 a3/2];
xg1 = cv(1) + A1*cos(betav).*cos(gammav);    % components 1,2 and 3 of the
xg2 = cv(2) + A2*sin(betav).*cos(gammav);    % measurement points
xg3 = cv(3) + A3*sin(gammav);                % ~

Nint = 3;     % number of integration points for integration over wire
W1 = zeros(Ng,Nw); W2 = zeros(Ng,Nw); W3 = zeros(Ng,Nw); % initialization
for k = 1:Ng % loop over all measurement points
  x = [xg1(k); xg2(k); xg3(k)];  % current evaluation point
  for jj=1:Nw1                   % loop over currents in x1 direction
    vtmp = 0;                    % variable for summation
    for xi=1:Nint                % integration loop
      p_xi = [w11v(jj,1);w12v(jj,1);w13v(jj,1)] + [1;0;0]*(xi-1)*h1/Nint;
      vtmp = vtmp + cross([1;0;0],x-p_xi)/norm(x-p_xi)^3;
    end
    W1(k,jj) = 1/4/pi*vtmp(1);  % Biot-Savart Matrix for H1 component
    W2(k,jj) = 1/4/pi*vtmp(2);  % H2 component and
    W3(k,jj) = 1/4/pi*vtmp(3);  % H3 component, for currents in x1 dir.
  end
  for jj=1:Nw2                    % loop over currents in x2 direction
    vtmp = 0;
    for xi=1:Nint                 % integration loop
      p_xi = [w21v(jj,1);w22v(jj,1);w23v(jj,1)] + [0;1;0]*(xi-1)*h2/Nint;
      vtmp = vtmp + cross([0;1;0],x-p_xi)/norm(x-p_xi)^3;
    end
```

```
    W1(k,Nw1+jj) = 1/4/pi*vtmp(1);  % Biot-Savart matrix for currents
    W2(k,Nw1+jj) = 1/4/pi*vtmp(2);  % in x2 direction
    W3(k,Nw1+jj) = 1/4/pi*vtmp(3);
  end
  for jj=1:Nw3                     % loop over currents in x3 direction
    vtmp = 0;
    for xi=1:Nint                  % integration loop
      p_xi = [w31v(jj,1);w32v(jj,1);w33v(jj,1)] + [0;0;1]*(xi-1)*h3/Nint;
      vtmp = vtmp + cross([0;0;1],x-p_xi)/norm(x-p_xi)^3;
    end
    W1(k,Nw1+Nw2+jj) = 1/4/pi*vtmp(1);  % Biot-Savart matrix for
    W2(k,Nw1+Nw2+jj) = 1/4/pi*vtmp(2);  % currents in x3 direction
    W3(k,Nw1+Nw2+jj) = 1/4/pi*vtmp(3);
  end
end
W = [W1;W2;W3];   % Biot-Savar Operator mapping j onto [H1;H2;H3]
t = toc; disp(['time neede t=' num2str(t)]);
```

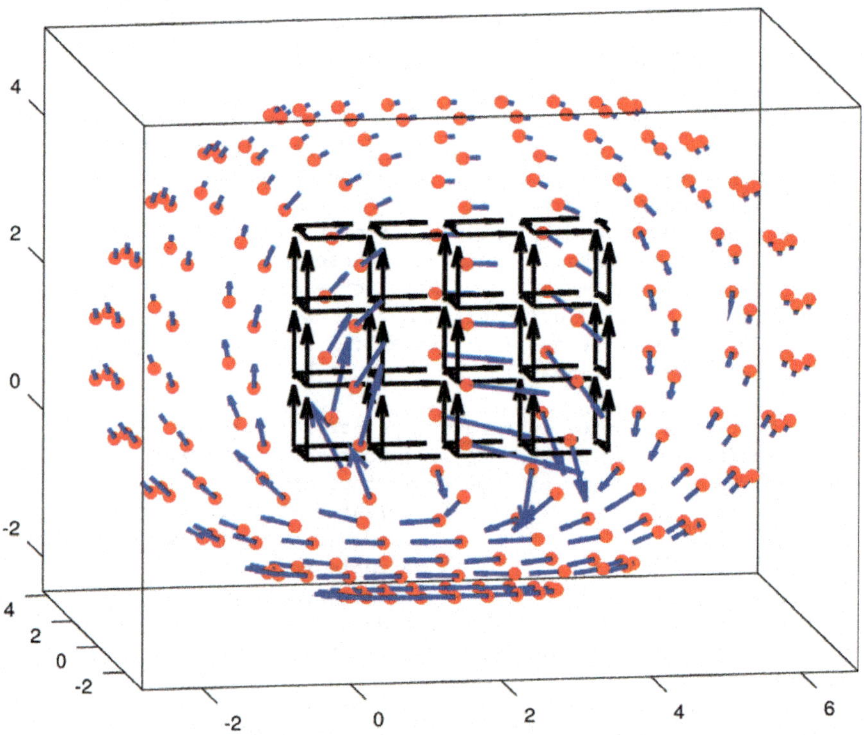

Figure 11.3. The magnetic field of the currents displayed in figure 11.2. The overall set-up can be tested using the right-hand rule and the knowledge that the current is mainly flowing from the front to the back side of the wire grid. This is generated by code 11.2.1 and 11.2.2.

We finish by showing some lines to generate figure 11.3.

Code 11.2.2. *The graphical display of the magnetic field of some current distribution is generated by the script* `sim_11_2_2_H_view.m` *shown here.*

```
% calculate magnetic field of current density
H1 = W1*j; H2=W2*j; H3 = W3*j;  % three vectorial components of H

fo = figure; hold on;            % generate figure
% first show the wire grid
quiver3(w11v,w12v,w13v,ones(Nw1,1),zeros(Nw1,1),zeros(Nw1,1),1, ...
   'k','LineWidth',2);
quiver3(w21v,w22v,w23v,zeros(Nw2,1),ones(Nw2,1),zeros(Nw2,1),.8, ...
   'k','LineWidth',2);
quiver3(w31v,w32v,w33v,zeros(Nw3,1),0.01*ones(Nw3,1),ones(Nw3,1), ...
   1,'k','LineWidth',2);

plot3(xg1,xg2,xg3,'r.','MarkerSize',10);          % measurement points
quiver3(xg1,xg2,xg3,H1,H2,H3,2,'Linewidth',2); % magnetic field arrows
view(-10,10); axis equal;                          % axis controls

savefile(fo,'sim_11_2_2_H_view');                 % save file
```

11.2.1 Uniqueness and non-uniqueness results

We first remark that, as shown in [1], it is sufficient to know the normal component $\nu \cdot H$ of H on the measurement surface ∂G. However, for reconstructions one usually works with the full field H, i.e. with redundant data, which is used in applications to obtain better control of measurement error. Sensors which measure all three components of the magnetic field H are available today with a variety of characteristics for diverse applications.

The *null-space* of the Biot–Savart integral operator is studied in detail in [4], where a characterization of $N(W)$ and its orthogonal space $N(W)^{\perp}$ has been derived by elementary arguments. Here, we will base it on the *Helmholtz decomposition*, which for a sufficiently smooth vector field v in some domain $G \subset \mathbb{R}^3$ establishes the decomposition

$$v = -\nabla\varphi + \nabla \times A \tag{11.2.10}$$

with

$$\varphi(x) = \frac{1}{4\pi}\int_G \frac{\nabla \cdot v(y)}{|x-y|}\mathrm{d}y - \frac{1}{4\pi}\int_{\partial G} \frac{v(y)\cdot\nu(y)}{|x-y|}\mathrm{d}s(y), \quad x \in G, \tag{11.2.11}$$

and

$$A(x) = \frac{1}{4\pi}\int_G \frac{\nabla \times v(y)}{|x-y|}\mathrm{d}y - \frac{1}{4\pi}\int_{\partial G} \frac{v(y)\times\nu(y)}{|x-y|}\mathrm{d}s(y), \quad x \in G, \tag{11.2.12}$$

where $\nu(y)$ denotes the normal vector to ∂G directed outside G at $y \in \partial G$. As a preparation we note that for j with div $j = 0$, we obtain

$$
\begin{aligned}
\operatorname{div}(Vj)(x) &= \int_\Omega \operatorname{div}_x(\Phi(x, y)j(y))\, dy \\
&= \int_\Omega \operatorname{grad}_x \Phi(x, y) \cdot j(y)\, dy \\
&= -\int_\Omega \operatorname{grad}_y \Phi(x, y) \cdot j(y)\, dy \\
&= -\int_\Omega \operatorname{div}_y(\Phi(x, y)j(y))\, dy \\
&= -\int_{\partial\Omega} \Phi(x, y)\nu(y) \cdot j(y)\, ds(y) \\
&= -S(\nu \cdot j)(x), \quad x \in \mathbb{R}^3.
\end{aligned} \tag{11.2.13}
$$

And in $\mathbb{R}^3\backslash\bar{\Omega}$ by

$$
\begin{aligned}
\operatorname{curl} Wj &= \operatorname{curl}\operatorname{curl} Vj \\
&= -\Delta Vj + \operatorname{grad}\ \operatorname{div} Vj \\
&= -\Delta Vj - \operatorname{grad}\ S(\nu \cdot j)
\end{aligned} \tag{11.2.14}
$$

we have

$$
\operatorname{curl} Wj = -\operatorname{grad} S(\nu \cdot j). \tag{11.2.15}
$$

Further, by div curl = 0 we have

$$
\operatorname{div} Wj = 0 \quad \text{in } \mathbb{R}^3. \tag{11.2.16}
$$

Theorem 11.2.3. *The null-space $N(W)$ of the Biot–Savart operator W from $H_{\mathrm{div}=0}(\Omega)$ into $L^2(\partial G)$ is given by*

$$
N(W) = \left\{\operatorname{curl} v : v \in \left(H_0^1(\Omega)\right)^3,\ \operatorname{div} v = 0\right\} \tag{11.2.17}
$$

with $H_0^1(\Omega) := \{v \in (H^1(\Omega))^3 : v|_{\partial\Omega} = 0\}$.

Proof. For the proof consider $v \in H^1(\Omega)$ with $v|_{\partial\Omega} = 0$ and extend it by zero into G, such that all integrals and derivatives of (11.2.10)–(11.2.12) are well defined. We first assume that we have a current density $j = \operatorname{curl} v$ with $v|_{\partial G} = 0$ and $\operatorname{div} v = 0$. Then, for (11.2.10)–(11.2.12) we conclude $\varphi \equiv 0$ in G and thus $v = \operatorname{curl} A$ with

$$
A(x) = \frac{1}{4\pi}\int_G \frac{\nabla \times v(y)}{|x - y|}\, dy = V(\operatorname{curl} v)(x) \tag{11.2.18}
$$

for $x \in \bar{G}$. By assumption we have $v|_{\partial G} = 0$, and thus

$$
0 = v|_{\partial G} = \operatorname{curl} A|_{\partial G} = \operatorname{curl} V(\operatorname{curl} v) = W(\operatorname{curl} v) = Wj, \tag{11.2.19}
$$

which proves that all elements of the form curl v with $v|_{\partial G} = 0$ and $\operatorname{div} v = 0$ in Ω are in the null-space of W.

To show the other direction, consider $j \in N(W)$ with $\operatorname{div} j = 0$. We need to find a field $v \in (H_0^1(\Omega))^3$ such that $j = \operatorname{curl} v$ and $\operatorname{div} v = 0$. We note that in this case Wj has boundary values $Wj = 0$ on ∂G. Each component of Wj satisfies the Laplace equation in $\mathbb{R}^3 \backslash \bar{\Omega}$ and is decaying fast enough together with its first order derivatives, which yields $Wj = 0$ in the exterior of Ω and by (11.2.15) also $S(\nu \cdot j) = 0$ in $\mathbb{R}^3 \backslash \bar{\Omega}$. By the continuity of the single-layer potential S and the uniqueness of the homogeneous interior solution to the Laplace equation in Ω and jump relations we conclude that $j \cdot \nu = 0$ on $\partial\Omega$. Then, by the Helmholtz decomposition (11.2.10)–(11.2.12) and

$$
\begin{aligned}
(Wj)(x) &= \operatorname{curl}(Vj)(x) \\
&= \operatorname{curl}_x \int_G \Phi(x, y) j(y) \, \mathrm{d}y \\
&= \int_G \nabla_x \Phi(x, y) \times j(y) \, \mathrm{d}y \\
&= -\int_G \left(\nabla_y \Phi(x, y)\right) \times j(y) \, \mathrm{d}y \\
&= -\int_G \nabla_y \times (\Phi(x, y) j(y)) \, \mathrm{d}y + \int_G \Phi(x, y) \nabla \times j(y) \, \mathrm{d}y \\
&= -\frac{1}{4\pi} \int_{\partial G} \frac{j(y) \times \nu(y)}{|x - y|} \mathrm{d}s(y) + V(\operatorname{curl} j)(x),
\end{aligned}
\qquad (11.2.20)
$$

we have $j = \operatorname{curl} Wj$ in Ω. Clearly, $Wj = \operatorname{curl} Vj$ is divergence free due to div curl =0 and we have $Wj|_{\partial G} = 0$. Since Wj satisfies the Laplace equation in $\mathbb{R}^3 \backslash \bar{\Omega}$ and is bounded, which first yields $Wj = 0$ in $\mathbb{R}^m \backslash G$, by analyticity of the field Wj in $\mathbb{R}^m \backslash \bar{\Omega}$ this is also the case in $\mathbb{R}^m \backslash \bar{\Omega}$, such that $Wj|_{\partial\Omega} = 0$ and by continuity of Wj on $\partial\Omega$ the proof is complete. □

By theorem 2.4.14 standard Tikhonov regularization for a non-injective operator calculates a solution in $N(W)^\perp$. Thus, a main question of the inverse theory is to study and investigate this space. This study is carried out in detail in [4], here we will present the main results with a simplified proof.

Theorem 11.2.4. *The orthogonal complement $N(W)^\perp$ of $N(W)$ with respect to the L^2-* scalar *product on Ω is given by $N(W)^\perp = Y$, where*

$$
Y = \{j \in (H_{\operatorname{div}=0}(\Omega))^3 : \operatorname{curl} j = \operatorname{grad} q \text{ with some } q \in L^2(\Omega)\} \qquad (11.2.21)
$$

with $(H_{\operatorname{div}=0}(\Omega))^3 := \{v \in L^2(\Omega) : \operatorname{div} v = 0\}$.

Proof. The adjoint W^* is given by

$$
(W^*\psi)(x) = \frac{1}{4\pi} \int_{\partial G} \frac{\psi(y) \times (x - y)}{|x - y|^3} \mathrm{d}s(y), \quad x \in \Omega, \qquad (11.2.22)
$$

for $\psi \in L^2(\partial G)$. We note that we have

$$\begin{aligned}(W^*\psi)(x) &= \frac{1}{4\pi}\int_{\partial G} \frac{\psi(y) \times (x-y)}{|x-y|^3}\,\mathrm{d}s(y)\\ &= \int_{\partial G} \psi(y) \times \nabla_x \Phi(x, y)\,\mathrm{d}s(y)\\ &= \operatorname{curl}\int_{\partial G} \psi(y)\Phi(x, y)\,\mathrm{d}s(y)\\ &= (\operatorname{curl} S\psi)(x), \quad x \in \Omega, \end{aligned} \tag{11.2.23}$$

with the single-layer potential S over ∂G. Hence for j defined by $j := W^*\psi$ in $(H_{\mathrm{div}=0}(\Omega))^3$ on Ω we calculate

$$\begin{aligned}\operatorname{curl} j &= \operatorname{curl}\operatorname{curl} S\psi\\ &= \nabla(\nabla \cdot S\psi) - \underbrace{\Delta S\psi}_{=0} = \nabla q\end{aligned} \tag{11.2.24}$$

with $q := \nabla \cdot S\psi$ in $L^2(\Omega)$. Thus, $j \in Y$ and hence $R(W^*) \subset Y$.

To proceed further we need the following two facts (i) and (ii) from the literature.

(i) For $f \in (H^1(\Omega)^*)^3$, there exists $q \in L^2(\Omega)$ such that $f = \operatorname{grad} q$ if and only if

$$\int_\Omega f \cdot v\,\mathrm{d}x = 0 \quad \text{for all } v \in \left(H_0^1(\Omega)\right)^3 \cap (H_{\mathrm{div}=0}(\Omega))^3. \tag{11.2.25}$$

The if part is clear using the Gauss theorem and we refer to lemma 2.1 in [5] for the only if part.

(ii) As we have assumed that $\Omega \subset \mathbb{R}^3$ is simply connected, we have the following characterization of the range of curl:

$$\begin{aligned}&\left\{\operatorname{curl} v : v \in \left(H^1(\Omega)\right)^3, v \times \nu|_{\partial\Omega} = 0\right\}\\ &\quad = \left\{u \in \left(L^2(\Omega)\right)^3 : \operatorname{div} u = 0,\, u \cdot \nu|_{\partial\Omega} = 0\right\}.\end{aligned} \tag{11.2.26}$$

For a proof we refer to p 226, proposition 4 of [6]. By using this, we prove that Y is a closed subset of $(L^2(\Omega))^3$. To see that Y is closed, let a sequence (j_n) in Y converge to j in $(L^2(\Omega))^3$ as $n \to \infty$. Then, it is easy to see $j \in (H_{\mathrm{div}=0}(\Omega))^3$ and $\operatorname{curl}\operatorname{curl} j = 0$ in Ω. By the fact (ii), for any $v \in (H_0^1(\Omega))^3 \cap (H_{\mathrm{div}=0}(\Omega))^3$, $v = \operatorname{curl} w$ for some $w \in (H^1(\Omega))^3$ such that $w \times \nu|_{\partial\Omega} = 0$. Note that such w can be approximated by $C_0^\infty(\Omega)^3$. Hence we have

$$\int_\Omega \operatorname{curl} j \cdot v\,\mathrm{d}x = \int_\Omega \operatorname{curl} j \cdot \operatorname{curl} w\,\mathrm{d}x = \int_\Omega \operatorname{curl}\operatorname{curl} j \cdot w\,\mathrm{d}x = 0. \tag{11.2.27}$$

By the fact (i), there exists $q \in (L^2(\Omega))^3$ such that $\operatorname{curl} j = \operatorname{grad} q$. Thus we have proved $j \in Y$ and hence Y is closed.

Finally we show $Y \subset N(W)^{\perp}$. Let $j \in Y$ and v in the space $(H_0^1(\Omega))^3 \cap (H_{\mathrm{div}=0}(\Omega))^3$. By the Gauss divergence theorem applied to

$$\mathrm{div}(j \times v) = \mathrm{curl}\, j \cdot v - j \cdot \mathrm{curl}\, v \tag{11.2.28}$$

using (i) we have

$$0 = \int_{\Omega} \mathrm{curl}\, j \cdot v \, \mathrm{d}x = \int_{\Omega} j \cdot \mathrm{curl}\, v \, \mathrm{d}x. \tag{11.2.29}$$

Here we note that $\int_{\Omega} \mathrm{div}(j \times v) = 0$ follows from the denseness of $C^{\infty}(\bar{\Omega})$ in $H(\mathrm{div};\Omega) := \{j \in (L^2(\Omega))^3 : \mathrm{div}\, j \in L^2(\Omega)\}$ and the trace of $j \in H(\mathrm{div};\Omega)$ is in $H^{-1/2}(\partial\Omega)$. Since $\mathrm{curl}\, v \in N(W)$ by lemma 11.2.3 we have $j \in N(W)^{\perp}$ (see [5]). This completes the proof because $N(W)^{\perp} = \overline{R(W^*)} \subset \bar{Y} = Y \subset N(W)^{\perp}$. □

The null-space of the observation operator cannot be reconstructed or seen by a static inversion method. We do not see what is in $N(W)$, but we can reconstruct the projection of j onto $N(W)^{\perp}$. If some current density is in $N(W)^{\perp}$, it can be fully reconstructed.

Let j satisfy (11.1.4) with a homogeneous conductivity distribution σ. For a homogeneous conductivity σ we have $\nabla \times j = \sigma \nabla \times \nabla E = 0$ from (11.1.4). Thus, for $\tilde{j} \in N(W)$, i.e. $\tilde{j} = \mathrm{curl}\, v$ with $v \in H_0^1(\Omega)$, with the help of (11.2.28) we derive

$$\begin{aligned} \langle j, \tilde{j} \rangle_{L^2(\Omega)} &= \langle j, \mathrm{curl}\, v \rangle_{L^2(\Omega)} \\ &= -\int_{\Omega} \nabla \cdot (j \times v) \, \mathrm{d}y + \Big\langle \underbrace{\mathrm{curl}\, j}_{=0}, v \Big\rangle_{L^2(\Omega)} \\ &= -\int_{\partial\Omega} v \cdot \Big(\underbrace{j \times v}_{=0 \text{ on } \partial\Omega} \Big) \mathrm{d}s(y) \\ &= 0, \end{aligned} \tag{11.2.30}$$

i.e. $j \in N(W)^{\perp}$ and, thus, in this special case we obtain *full reconstructability* of j from its magnetic field $H = Wj$ on ∂G.

11.2.2 Reducing the ill-posedness of the reconstruction by using appropriate subspaces

A linear ill-posed equation $Wj = H$ can be solved by any of the regularization methods described in section 3.1, for example by Tikhonov regularization (3.1.24). Our key goal in this section is to investigate appropriate conditions on the current density distributions j to reduce the ill-posedness of the inversion and to improve the reconstructions which can be achieved by Tikhonov regularization. In particular, we study these three algorithms:

1. standard Tikhonov regularization,
2. the use of the condition (11.1.7) in addition to equation (11.2.2), and
3. the incorporation of the full boundary value problem (11.1.8) and (11.1.9) into equation (11.2.2).

We will show how the ill-posedness of the inversion is reduced via the conditions 2 and 3. In particular, (a) we derive *qualitative* estimates of the singular values of the operators under consideration and (b) we provide numerical results about the *quantitative* improvements which can be gained.

These additional conditions establish *improvement to the reconstruction quality* which adds to improvements which can be achieved by changing central inversion parameters such as the distance of the measurement points to the area Ω of the current density j.

1. Tikhonov regularization for source reconstruction. The discretized form of $Wj = H$ (cf equation (11.2.2)) is given by

$$\mathbf{W}\mathbf{J} = \mathbf{H}, \tag{11.2.31}$$

where $\mathbf{W}$ is given by (11.2.7), $\mathbf{J}$ by (11.2.6) and $\mathbf{H}$ by (11.2.9). We have seen in theorem 11.2.3 that the continuous operator W is *not* injective, but there is a large null-space. It has been shown in [4], theorem 4.1, that the semi-discrete operator of $\mathbf{W}$ is *injective*, when the measurements are taken on the full set ∂G. The basic observation here is that the magnetic field has a singularity at a discrete wire, which is proportional to the current flowing in it. If the magnetic field is zero, then there cannot be any current.

This situation needs some further attention. We first show full injectivity of $\mathbf{W}$ when sufficiently many measurements x_k, $k = 1, \dots, m$ are taken on ∂G. Then, we investigate the relationship between the discrete and the continuous operator.

Lemma 11.2.5. *Let* $\{x_k : k = 1, 2, 3, \dots\}$ *be a dense sequence in* ∂G. *If the number of discrete measurement points* x_k, $k = 1, \dots, m$ *on* ∂G *is sufficiently large, then* $\mathbf{W}$ *defined in (11.2.7) is injective.*

Proof. We start our proof with the injectivity of the semi-discrete operator

$$(\tilde{\mathbf{W}}\mathbf{J})(x) = -\frac{1}{4\pi}\sum_{\ell=1}^{L}\int_{\gamma_\ell}\frac{\tilde{J}_\ell \times (x-p)}{|x-p|^3}\,\mathrm{d}s(p), \quad x \in \partial G. \tag{11.2.32}$$

Then, its adjoint $\tilde{\mathbf{W}}^*\psi = ((\tilde{W}^*\psi)_\ell : \ell = 1, \dots, L)$ is given by

$$(\tilde{\mathbf{W}}^*\psi)_\ell = \frac{1}{4\pi}\int_{\partial G}\int_{\gamma_\ell}\frac{\psi(x)\times(x-p)}{|x-p|^3}\,\mathrm{d}s(p)\,\mathrm{d}s(x), \quad \ell = 1, \dots, L \tag{11.2.33}$$

has a dense range. We now employ an approximation of $\tilde{\mathbf{W}}^*\psi$ by the fully discrete operator

$$(\mathbf{W}^*\psi)_\ell = \sum_{k=1}^{m}\int_{\gamma_\ell}\frac{\psi_k \times (x_k - p)}{|x_k - p|^3}\,\mathrm{d}s(p)s_k \tag{11.2.34}$$

where $\boldsymbol{\psi}_k$ is given by $\psi(y_k)$ and s_k is a quadrature weight. By an appropriate ordering of the vectors $\boldsymbol{\psi}_k$ and $(\mathbf{W}^*\boldsymbol{\psi})_\ell$, the operator $\mathbf{W}^*$ can be identified with an

$3L \times 3\,m$-matrix. For sufficiently large m the approximation of a standard orthonormal basis in $\mathbb{R}^{3L}$ leads to rank($\mathbf{W}^*$) = $3L$, such that also $\mathbf{W} \in \mathbb{R}^{3\,m \times 3L}$ has rank $3L$ and, thus, is injective. This ends the proof. □

As described in detail in section 3.1.4, the basic principle of *regularization methods* for operators is to approximate the unbounded operator W^{-1} by a bounded operator R_α with regularization parameter $\alpha > 0$ such that

- for W injective and true data f we have the pointwise convergence $R_\alpha H \to j := W^{-1}f$, $\alpha \to 0$, for $H = Wj$.
- If W is not injective, we cannot expect $R_\alpha H \to j$, but we usually have $R_\alpha H \to Pj$, $\alpha \to 0$, with some orthogonal projection operator P, which for Tikhonov regularization is the orthogonal projection onto $N(W)^\perp$.
- For the discrete case, we employ Tikhonov regularization as introduced in equation (3.1.24)

$$\mathbf{R}_\alpha := (\alpha \mathbf{I} + \mathbf{W}^*\mathbf{W})^{-1}\mathbf{W}^* \tag{11.2.35}$$

  with regularization parameter $\alpha > 0$, where $\mathbf{W}^*$ denotes the complex conjugate transpose matrix of $\mathbf{W}$. Since in this case $\mathbf{W}$ is injective, we have convergence $\mathbf{J}_\alpha R_\alpha \mathbf{H} \to J$ for $\alpha \to 0$ towards $\mathbf{J} = \mathbf{W}^{-1}\mathbf{H}$.

Clearly, the non-uniqueness of the continuous problem must show itself in the discrete problem, when the discretization parameter $n \in \mathbb{N}$ becomes sufficiently large. The interpolated field $J_\alpha^{(n)} \in L^2(\Omega)$ of the vector $\mathbf{J}_\alpha = \mathbf{J}_\alpha^{(n)}$ consists of two components $PJ_\alpha^{(n)} \in N(W)^\perp$ and $(I - P)J_\alpha^{(n)} \in N(W)$, where

$$(I - P)J_\alpha^{(n)} \to 0, \quad n \to \infty, \tag{11.2.36}$$

and

$$PJ_\alpha^{(n)} \to Pj_\alpha, \quad n \to \infty. \tag{11.2.37}$$

For the choice of the regularization parameter we refer to section 3.1.6 or to standard results in [7].

We now describe the use of Tikhonov regularization with additional constraints, which can be understood as a *regularized projection method.* We show how the constraints move the singular values and lead to improved stability. Theoretical estimates are derived and numerical simulations are carried out.

2. Divergence free Tikhonov regularization. We recall that the current distribution j is *divergence free* and it is natural to incorporate condition (11.1.7) into the inversion. We expect this condition to decrease the solution space to provide a more accurate solution and will provide qualitative estimates and quantitative results in the next sections.

For the discrete realization of the inversion we use the condition introduced in equation (11.1.18) for choosing the finite subset X_n of the solution space $(L^2(\Omega))^3$.

This condition corresponds to *Kirchhoff's knot rule*, so we set up the matrix **T** as described in section 11.1.2. Since the discrete solution **J** solves

$$\mathbf{TJ} = \mathbf{b}, \tag{11.2.38}$$

with a *particular solution* j_0 the *general solution* $j_{gen}$of this equation is given by

$$j_{gen} = j_0 + \mathbf{N}z, \tag{11.2.39}$$

with a basis **N** of the null-space of **T** and arbitrary vector z. Inserting this representation into equation (11.2.31) we derive

$$(\mathbf{WN})z = \mathbf{H} - \mathbf{W}j_0 \tag{11.2.40}$$

which can be solved via Tikhonov regularization by setting

$$\mathbf{R}_\alpha := (\alpha \mathbf{I} + (\mathbf{WN})^*(\mathbf{WN}))^{-1}(\mathbf{WN})^*, \quad \alpha > 0. \tag{11.2.41}$$

Now a solution **J** of (11.2.31) can be obtained by setting

$$\mathbf{J} := j_0 + \mathbf{NR}_\alpha\left(\mathbf{H} - \mathbf{W}j_0\right). \tag{11.2.42}$$

Numerical results for the algorithm are shown below. We estimate the singular values of **WN** in (11.2.65).

3. A projection method with special basis functions. In this section we will take into account the information that our currents solve the boundary value problem (11.1.8) and (11.1.9). This leads to a projection method with a special basis $\{j_k : k = 1, \dots N\}$. Here, we construct this basis such that its z-components approximately build a Haar basis, i.e. they are approximately a multiple of 1 in some wire and close to zero in all others.

As background we remind the reader that for the fuel cell application the conductivities in the directions perpendicular to z-axis $x - y$ layers are usually large and uniform due to metallic end plates and carbon layers between the different single fuel-cells. Thus, we use a uniform high conductivity σ_0 in all wires in these directions. We have varying conductivity only in the wires in z-direction, which we label from $k = 1$ to $k = N$. Now, we choose numbers σ_l, $l = 0, 1, 2$ with

$$\sigma_2 \ll \sigma_1 \ll \sigma_0. \tag{11.2.43}$$

Then, the idea is to use the ansatz

$$j = \sum_{k=1}^{N} \xi_k j_k, \tag{11.2.44}$$

where j_k is a current distribution with a conductivity which is set to σ_1 in wire k in z-direction and to σ_2 in all other wires in the z-direction. We use the notation $X_N := \text{span}\{j_k: \ k = 1, \dots, N\}$. The current distributions j_k, $k = 1, \dots, N$ are solutions to the forward problem, i.e. they are calculated via *mesh* and *knot rules* as described in section 11.1.2. The magnetic field of an element $j \in X_N$ at a point $x \in \mathbb{R}^3$ can be calculated via

$$(\mathbf{W}j)(x) = \sum_{k=1}^{N} \xi_k (\mathbf{W}j_k)(x). \tag{11.2.45}$$

Defining $H_k := \mathbf{W}j_k$ by evaluating the Biot–Savart operator, we set up the matrix $\mathbf{H}_s := (H_1, \ldots, H_N)$. Then, we solve the ill-posed linear system

$$\mathbf{H}_s \xi = \mathbf{H} \tag{11.2.46}$$

for the coefficients $\xi := (\xi_1, \ldots, \xi_N)^T$, where we employ Tikhonov regularization for regularization of this system. The solution J of the original problem is obtained by calculating

$$\mathbf{J} := \sum_{k=1}^{N} \xi_k j_k. \tag{11.2.47}$$

In view of our upcoming analysis in theorem 11.2.10, we note that in general the j_k are not an orthonormal basis of the discrete space X_N. However, we can orthonormalize j_k to obtain a basis $\{j_{k,\mathrm{o}} : k = 1, \ldots, N\}$ of X_N and define $\mathbf{H}_{s,\mathrm{o}} := (\mathbf{W}j_{1,\mathrm{o}}, \ldots, \mathbf{W}j_{N,\mathrm{o}})$. Numerical results for the algorithm are shown below; we also estimate the singular values of the orthonormalized version $\mathbf{H}_{s,\mathrm{o}}$ of $\mathbf{H}_s$.

After the set-up of the Biot–Savart matrix in code 11.2.1 the inversion of the Biot–Savart operator by a regularization method is merely two lines of OCTAVE code. The result is shown in figures 11.4 and 11.5.

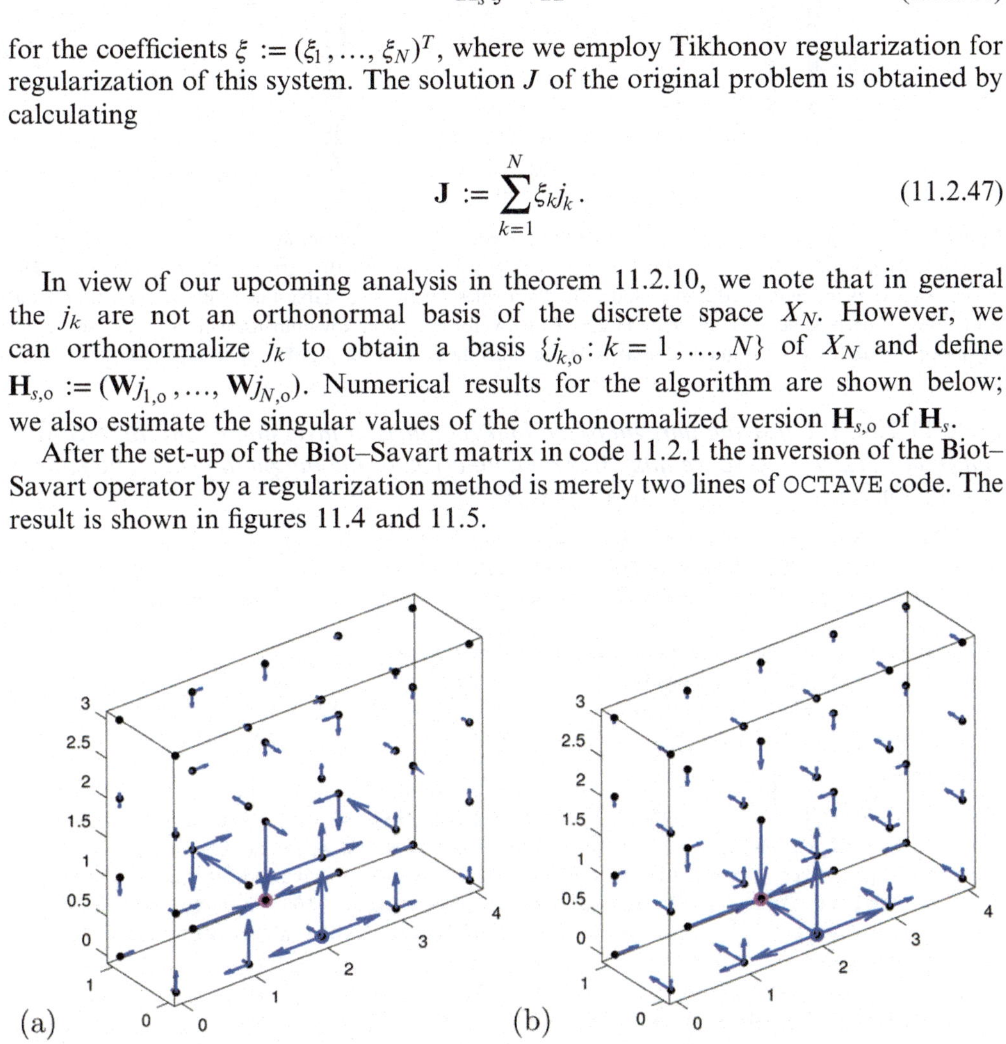

Figure 11.4. The reconstructed currents from figure 11.2 by an unregularized (a) and a regularized (b) reconstruction, where we add 10% error to the magnetic field. This is generated by code 11.2.6. Already for this very low-dimensional example we obtain significant instability for the full-rank matrix W, see also figure 11.5.

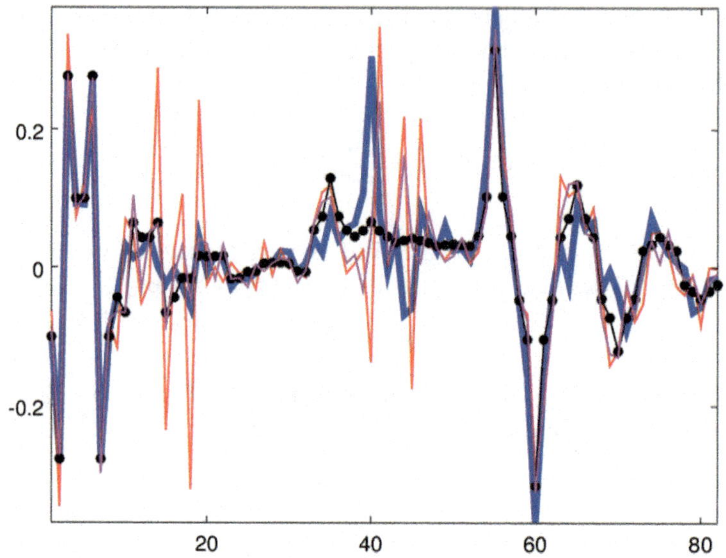

Figure 11.5. The original (blue) and reconstructed current components from figure 11.2 by an unregularized (red) and a regularized (black with dots) reconstruction, where we add 10% error to the magnetic field. The magenta curve is unregularized, but includes the knot rules—it is more stable than without these additional conditions. This is generated by code 11.2.6. Already for this very low-dimensional example we obtain significant instability for the full-rank matrix **W**, see also figure 11.4 for a different visualization.

Code 11.2.6. *We test the unregularized and regularized inversion of the Biot–Savart equation (11.2.1) where we add* 10% *random error to the magnetic field. The code is provided in* `sim_11_2_6_inversion.m`.

```
delta = 0.1;      % random pointwise error in percent
H = [H1;H2;H3]; % setup full right-hand side
Hdelta = H+delta*max(abs(H)).*(rand(size(H))-0.5)*2; % add error

jrec = W\Hdelta; % unregularized reconstruction;
alpha = 1e-8;     % regularization parameter
jalpha = (alpha*eye(Nw,Nw)+W'*W)\W'*H; % regularized reconstruction
jrec2 = [W;nrules]\[Hdelta;b(1:size(nrules,1),1)]; % rec with knot rules

fo = figure; hold on;                 % generate figure
po = plot(j,'b-','LineWidth',4);     % plot original current components
ao = get(po,'Parent'); set(ao,'FontSize',14);  % axis controls
po = plot(jrec,'r-','LineWidth',1); % show unregularized reconstruction
po = plot(jalpha,'k.-','LineWidth',1,'MarkerSize',10); % regularized rec
po = plot(jrec2,'m','LineWidth',1,'MarkerSize',10); % rec with knot rules
axis tight;

savefile(fo,'sim_11_2_6_inversion'); % save image
```

Next, we study the stability of the above algorithms via their singular values and derive estimates for the singular values of the methods.

Usually, for estimating the ill-posedness of inverse problems the size of the singular values is taken as central measure. For discrete inverse problems the condition number defined below is the key quantity. Here, we are mostly interested in the singular values as becomes clear from the following reasons. Consider a linear system of

$$Ax = b \tag{11.2.48}$$

with invertible matrix A which we solve for x with some data error $e^{(\delta)}$, i.e. we calculate $x^{(\delta)}$ by

$$A(x^{(\delta)}) = b + e^{(\delta)}. \tag{11.2.49}$$

In general one estimates the error by

$$\|x - x^{(\delta)}\| \leqslant \|A^{-1}\| \cdot \|e^{(\delta)}\| \tag{11.2.50}$$

and calculates the relative error

$$\frac{\|x - x^{(\delta)}\|}{\|x\|} \leqslant \|A^{-1}\| \cdot \|b\| \cdot \frac{\|e^{(\delta)}\|}{\|b\|}. \tag{11.2.51}$$

Thus, the condition number which is the division of the largest singular value by the smallest singular value of A provides an upper bound for the relative numerical error. Now, we consider two matrices A_1, A_2 with $A_1x = b$ and $A_2x = b$, where the singular values of A_2 are larger than the singular values of A_1, thus the norm $\|A_2^{-1}\|$ is smaller than the norm $\|A_1^{-1}\|$. Still, the condition of A_2 might be larger than the condition of A_1, which is partly the case for our setting of magnetic tomography. If we solve the same problem with given data we need to consider the case where we keep $e^{(\delta)}$ fixed. Then, we solve the systems

$$A_1x_1^{(\delta)} = b + e^{(\delta)}, \qquad A_2x_2^{(\delta)} = b + e^{(\delta)}. \tag{11.2.52}$$

Estimating the reconstruction error $\|x - x^{(\delta)}\|$ as in (11.2.50) we have

$$\|x - x^{\delta}\| \leqslant \|A_2^{-1}\| \cdot \|e^{(\delta)}\| \leqslant \|A_1^{-1}\| \cdot \|e^{(\delta)}\|, \tag{11.2.53}$$

i.e. we have a better estimate for the data error from the second system with A_2 than for the system with A_1. Here, the true solution x is fixed. In this case from (11.2.53) we get a better estimate for the relative error via the second system

$$\frac{\|x - x_2^{(\delta)}\|}{\|x\|} \leqslant \frac{\|A_2^{-1}\| \cdot \|b\|}{\|x\|}\frac{\|e^{(\delta)}\|}{\|b\|}, \tag{11.2.54}$$

compared to the estimate for the first system

$$\frac{\|x - x_1^{(\delta)}\|}{\|x\|} \leqslant \frac{\|A_1^{-1}\| \cdot \|b\|}{\|x\|}\frac{\|e^{(\delta)}\|}{\|b\|}. \tag{11.2.55}$$

If the error is a multiple of the eigenvector with smallest singular value of A_1, then the estimate (11.2.55) will be sharp and the improvement in the error is fully given by the improvement in the estimate (11.2.53) of the norm of the inverse via the singular values. The estimate (11.2.54) fully carries over to (11.2.40): one calculates

$$
\begin{aligned}
&(\mathbf{WN})z = \mathbf{H} - \mathbf{W}j_0 \Rightarrow (\mathbf{WN})(z^{(\delta)} - z) = e^{(\delta)} \\
&\Rightarrow j_{\text{gen}}^{(\delta)} - j_{\text{gen}} = \mathbf{N}(z^{(\delta)} - z) = \mathbf{N}(\mathbf{WN})^{-1}e^{(\delta)}.
\end{aligned} \tag{11.2.56}
$$

Since N has orthonormal columns we obtain (11.2.54) also for a setting of the form (11.2.40).

For general estimates of the singular values we will use the Courant minimum–maximum principle as a key tool. First, we order the eigenvalues of a self-adjoint matrix operator $A : \mathbb{C}^n \to \mathbb{C}^n$ according to their size and multiplicity $\lambda_1 \geqslant \lambda_2 \geqslant \cdots \geqslant \lambda_n$. Then, the Courant minimum–maximum principle states that

$$
\lambda_{n+1-k} = \min_{\dim U = k} \max_{x \in U} \frac{\langle Ax, x \rangle}{\langle x, x \rangle}, \quad k = 1, \ldots, n. \tag{11.2.57}
$$

We use this to prove some useful properties. Roughly speaking, the aim of this section is to show that incorporating some *a priori* knowledge leads to larger singular values of the corresponding Tikhonov matrix. We set up a general framework in terms of subspaces and apply this to our setting of magnetic tomography.

Definition 11.2.7 (*A priori* knowledge via subspace setting). *We denote* $X = \mathbb{C}^n$ *and* $Y = \mathbb{C}^m$ *and we consider an operator* $W : X \to Y$*. For a subspace* $V \subset X = \mathbb{C}^n$ *we define* $W_V : V \to Y$ *by* $W_V = W|_V$ *and* W_V^* *to be its adjoint operator* $Y \to V$ *determined by*

$$
\langle Wx, y \rangle_Y = \langle x, W_V^* y \rangle_V, \quad x \in V, \quad y \in Y. \tag{11.2.58}
$$

It is well known that

$$
N\left(W_V^*\right) = W(V)^\perp. \tag{11.2.59}
$$

Clearly, the operator $A_V := W_V^* W$ *is a self-adjoint operator on* V*, since for* $x, z \in V$ *we have*

$$
\langle z, W_V^* W x \rangle_V = \langle W_V z, Wx \rangle_Y = \langle W_V^* W z, x \rangle_V. \tag{11.2.60}
$$

First, we collect some properties of the adjoint operators arising from the subspaces $\tilde{V} \subset V \subset X$.

Lemma 11.2.8. *Let* $\tilde{V} \subset V \subset X$ *be subspaces and consider the adjoint operators* W_V^* *of* $W_V : V \to Y$ *and* $W_{\tilde{V}}^*$ *of* $W_{\tilde{V}} : \tilde{V} \to Y$ *arising from the restriction of* W *to* V *or* $\tilde{V}$*, respectively. Further, let* $P : V \to \tilde{V}$ *denote the orthogonal projection operator from* V *onto* $\tilde{V} \subset V$*. Then we obtain*

$$
W_{\tilde{V}}^* = P W_V^*. \tag{11.2.61}
$$

Proof. We denote $Q := I - P$ and decompose $x = Qx + Px$ for $x \in V$. For $x \in \tilde{V}, y \in Y$ we obtain $Qx = 0$ and in this case we calculate

$$\begin{aligned}\langle x, PW_{\tilde{V}}^{*}y\rangle &\overset{P=I-Q}{=} \left\langle x, W_{\tilde{V}}^{*}y - QW_{\tilde{V}}^{*}y\right\rangle \overset{Qx=0}{=} \left\langle Px, W_{\tilde{V}}^{*}y - QW_{\tilde{V}}^{*}y\right\rangle \\ &\overset{P(V)\perp Q(V)}{=} \langle x, W_{\tilde{V}}^{*}y\rangle = \langle W_{V}x, y\rangle \overset{x\in\tilde{V}\subset V}{=} \langle W_{\tilde{V}}x, y\rangle = \langle x, W_{\tilde{V}}^{*}y\rangle. \qquad (11.2.62)\end{aligned}$$

This yields $\langle x, (P_{V}^{*} - W_{\tilde{V}}^{*})y\rangle = 0$ for all $x \in \tilde{V}$, from which (11.2.61) follows. □

We now prove a monotonicity property for singular values which directly applies to the setting of magnetic tomography.

Theorem 11.2.9. *Let V, $\tilde{V}\subset\mathbb{C}^n$ be subspaces of $\mathbb{C}^n$ with $\tilde{V}\subset V$ and denote the singular values of W_V or $W_{\tilde{V}}$, respectively, by μ_j or $\tilde{\mu}_j$, $j = 1, 2, 3, \ldots$. We denote the dimensions of V, $\tilde{V}$ by n, $\tilde{n}$ and note that $\tilde{n} \leqslant n$ by $\tilde{V}\subset V$. Then we obtain the estimates*

$$\mu_{n+1-j} \leqslant \tilde{\mu}_{\tilde{n}+1-j}, \quad j = 1, 2, 3, \ldots, \tilde{n}. \qquad (11.2.63)$$

Moreover, in this estimate we will obtain equality for $k \in \mathbb{N}$ if and only if the eigenspace E_k of $W_V^{*}W_V$ with eigenvalue λ_{n+1-k} is a subset of $\tilde{V}$.

Proof. We employ the Courant minimum–maximum principle applied to the eigenvalues λ_k of $W_V^{*}W_V$ and the eigenvalues $\tilde{\lambda}_k$ of $W_{\tilde{V}}^{*}W_{\tilde{V}}$ to derive

$$\begin{aligned}\lambda_{n+1-k} &= \min_{U\subset V,\dim U=k}\left(\max_{x\in U}\frac{\langle W_V^{*}W_Vx, x\rangle}{\langle x, x\rangle}\right) \\ &= \min_{U\subset V,\dim U=k}\left(\max_{x\in U}\frac{\langle W_Vx, W_Vx\rangle}{\langle x, x\rangle}\right) \\ &\leqslant \min_{U\subset \tilde{V},\dim U=k}\left(\max_{x\in U}\frac{\langle W_Vx, W_Vx\rangle}{\langle x, x\rangle}\right) \\ &= \min_{U\subset \tilde{V},\dim U=k}\left(\max_{x\in U}\frac{\langle W_{\tilde{V}}x, W_{\tilde{V}}x\rangle}{\langle x, x\rangle}\right) \\ &= \min_{U\subset \tilde{V},\dim U=k}\left(\max_{x\in U}\frac{\langle W_{\tilde{V}}^{*}W_{\tilde{V}}x, x\rangle}{\langle x, x\rangle}\right) \\ &= \tilde{\lambda}_{\tilde{n}+1-k} \qquad (11.2.64)\end{aligned}$$

for $k = 1, 2, \ldots, \tilde{n}$. In this estimate we will obtain equality for $k \in \mathbb{N}$ if and only if all subspaces U with a dimension $dim(U) = k$ of the eigenspace E_k corresponding to the eigenvalue λ_{n+1-k} (which are spaces U where the Courant minimum–maximum principle attains its minimum) are subsets of $\tilde{V}$. This is equivalent to the eigenspace E_k being a subset of $\tilde{V}$. □

We apply the result to our algorithms for magnetic tomography as follows.

Theorem 11.2.10. *Consider the three solution methods (1)–(3) from section 11.2.2:*

(A) *the Biot–Savart equation given by (11.2.31),*

(B) *the divergence free Biot–Savart equation as in (11.2.40) and*

(C) *the special basis Biot–Savart equation (11.2.46).*

Here, for (11.2.46) *we assume that the basis currents* $j_1, \ldots, j_N$ *build an orthonormal set in the space of all currents. Then for the singular values* $\mu_k^{(1)}$, $\mu_k^{(2)}$ *and* $\mu_k^{(3)}$ *of the matrices* $\mathbf{W}^{(1)} := \mathbf{W}$, $\mathbf{W}^{(2)} := \mathbf{WN}$ *and* $\mathbf{W}^{(3)} := \mathbf{H}_{s,o}$ *we obtain the estimates*

$$\mu^{(1)}_{n_A+1-k} \leqslant \mu^{(2)}_{n_B+1-k} \leqslant \mu^{(3)}_{n_C+1-k} \tag{11.2.65}$$

for $k = 1, 2, \ldots, n_C$.

Proof. The generic case is provided by a matrix $\mathbf{W}$ and a matrix $\mathbf{N}$ with orthonormal columns. In this case we can interpret the mapping $z \mapsto \mathbf{N}z$ as a restriction of the mapping $\mathbf{W}$ to the image space $V := \{v_1, \ldots, v_m\}$ where v_j are the columns of $\mathbf{N}$. Let $U \subset \mathbb{C}^m$ be a subset. Then

$$\min_{U \subset \mathbb{C}^m, \dim U = k} \left(\max_{z \in U} \langle \mathbf{WN}z, \mathbf{WN}z \rangle \right) = \min_{\tilde{U} \subset V, \dim \tilde{U} = k} \left(\max_{x \in \tilde{U}} \langle \mathbf{WN}x, \mathbf{WN}x \rangle \right) \tag{11.2.66}$$

with $V := N(\mathbb{C}^m)$, because $\mathbf{N}$ maps the set of subspaces of $\mathbb{C}^m$ bijectively onto the set of subspaces of V. From this follows that $\mathbf{WN}$ and $\mathbf{W}|_V$ have the same singular values.

The mapping $z \mapsto \mathbf{N}z$ is norm preserving. Now, an application of theorem 11.2.9 proves an estimate of the form (11.2.65) for the comparison of (A) and (B).

For the comparison of (B) and (C) we remark that the matrix arising from $\mathbf{W}j_k$ can be written as $\mathbf{WJ}$ with the orthonormal matrix $\mathbf{J} = (j_1, \ldots, j_N)$. We remark that since the j_k are calculated from equations (11.1.18) and (11.1.19) and are divergence free, the estimate is then obtained as above from theorem 11.2.9. □

We have shown that the use of *a priori* knowledge via knot equations and special basis functions which incorporate the background knowledge leads to better estimates for inversion than the general Biot–Savart equation. Moreover, the estimate of the singular values provides a strong spectral analysis of the situation, which can be used in more detail. Code 11.2.6 provides a numerical study of the situation which as shown in figure 11.5 confirms the above estimates and can be used to evaluate the actual size of the constants for some important sample settings frequently used for the practical application of magnetic tomography.

11.3 Parameter estimation in dynamic magnetic tomography

Dynamic inverse problems study the situation where the object which is to be reconstructed changes over time. For *magnetic tomography*, we obtain a dynamic problem when the current densities $j \in X = (L^2(\Omega))^3$ are time-dependent, i.e. when

$$j = j_k := j(t_k), \quad k = 0, 1, 2, \ldots \tag{11.3.1}$$

and when we have repeated measurements of the magnetic fields

$$H_k := H(t_k) \in Y = (L^2(\partial G))^3, \quad k = 1, 2, 3, \ldots$$

of currents j_k. This leads to a sequence of equations

$$H_k = Wj_k, \quad k = 0, 1, 2, \ldots \tag{11.3.2}$$

with the compact linear operator $W : X \to Y$, where each equation needs to be solved using the knowledge of previous reconstructions transported from time t_{k-1} to time t_k.

An introduction into *inverse methods for dynamical systems* or *data assimilation* is given in chapter 5. Let some dynamics in a Hilbert space X be given. For magnetic tomography a model M_k is a map mapping of a current distribution $j_k \in X$ at time t_k into a current distribution

$$j_{k+1} = M_k(j_k) \in X, \quad k = 0, 1, 2, \ldots \tag{11.3.3}$$

at time t_{k+1}. To bring real data into such a simulation, we can apply the methods of chapter 5, i.e. *three-dimensional or four-dimensional variational assimilation*, or the *Kalman filter*. For magnetic tomography this has been worked out by Marx [8]. The basic idea here is to calculate a regularized reconstruction $j_k^{(a)}$ at time t_k, which is calculated using the background

$$j_k^{(b)} := M_{k-1}\left(j_{k-1}^{(a)}\right), \quad k \in \mathbb{N}, \tag{11.3.4}$$

by

$$j_k^{(a)} = j_k^{(b)} + R_\alpha\left(H_k - Wj_k^{(b)}\right), \tag{11.3.5}$$

where the regularization operator $R_\alpha : Y \to X$ in the case of Tikhonov regularization is given canonically by

$$R_\alpha := (\alpha I + W^* W)^{-1} W^*. \tag{11.3.6}$$

Clearly, we can *cycle* the above approach, which means that

1. given some initial state $j_0^{(a)}$ we calculate $j_1^{(b)}$ first according to (11.3.4),
2. then, we use data H_1 to calculate a reconstruction $j_1^{(a)}$ following (11.3.5) and (11.3.6), also called the *analysis*,
3. then, we repeat these two steps for $k = 2, 3, 5, \ldots$.

If the model M is well known, the cycled Tikhonov approach (11.3.4)–(11.3.6) or other classical or emerging data assimilation algorithms can be employed to calculate an approximation for the *trajectory* $\{j(t) | t \geqslant 0\}$ from (11.3.2) and (11.3.3).

But often, in applications the model M is known only partially, such that we are given some model operator $M[p]$ depending on unknown parameters or parameter functions $p \in U$ with some subset U of a Hilbert space Z. Note that the following steps can be carried out for any compact linear operator $W : X \to Y$ between Hilbert spaces X, Y and a dynamics given by model M on X.

Definition 11.3.1 (Dynamic parameter estimation). *Given measurements H_k for $k = 1, 2, 3, \ldots$ and a family of models $M[p]$ depending on $p \in U \subset Z$ with some subset U of a Hilbert space Z, the* dynamic parameter estimation problem *is to calculate a best estimate for the states j_k at time t_k and the parameter function or vector $p \in U$ from the measurements H_k.*

By augmenting the state vector and the dynamics, the dynamic parameter estimation problem can be considered as a normal *dynamic inverse problem* or *data assimilation problem*. We define

$$\varphi := \begin{pmatrix} j \\ p \end{pmatrix} \tag{11.3.7}$$

and we augment the dynamics of the model by defining

$$\tilde{M}_k := \begin{pmatrix} M_k & O \\ O & I|_Z \end{pmatrix}, \quad k \in \mathbb{N}. \tag{11.3.8}$$

Here, the model $\tilde{M}_k$ maps φ_k into

$$\varphi_{k+1} = \begin{pmatrix} j_{k+1} \\ p \end{pmatrix} = \begin{pmatrix} M_k(j_k) \\ p \end{pmatrix} = \tilde{M}_k \begin{pmatrix} j \\ p \end{pmatrix} = \tilde{M}_k(\varphi_k). \tag{11.3.9}$$

The *dynamic inverse problem* coincides with the *dynamic parameter estimation problem*, i.e. we try to reconstruct φ_k from

$$H_k = \tilde{W}\varphi_k, \quad k = 1, 2, 3, \ldots, \tag{11.3.10}$$

where

$$\tilde{W} : X \times Z \mapsto Y, \quad \tilde{W} := (W, 0) \tag{11.3.11}$$

We remark that due to the special structure of $\tilde{W}$, the value of $p \in Z$ as a part of $\varphi \in X \times Z$ is not changing the data $H = \tilde{W}\varphi \in Y$. This also means that if we just try to solve (11.3.10), without any further background knowledge, it is impossible to reconstruct $p \in U$. We summarize this insight as a lemma.

Lemma 11.3.2. *The space $O \times Z$ is a subset of the null-space $N(\tilde{W})$, such that a reconstruction of $p \in Z$ from given data $H \in Y$ at just one time t_k is not possible.*

Proof. Since the mapping of p onto $H = \tilde{W}(\varphi)$ satisfies

$$(j, p)^T \mapsto \tilde{W}(j, p)^T = Wj$$

we conclude that $\tilde{W}(j, p_1) = \tilde{W}(j, p_2)$ for any $p_1, p_2 \in Z$. This means that the reconstruction of $p \in Z$ from $H = \tilde{W}(j, p)^T$ is not possible. □

However, we have assumed that M depends on p. This means that p influences the development of j and thus also H over time. Hence taking into account the development of *j* with several time steps, there is a possibility that reconstructability can be achieved. Here, we will present some viewpoints, arising from *control theory* and from *statistical methods* of *data assimilation.*

First, the above reconstructability question coincides with the *observability* question in *control theory*. We call a system state φ *observable*, if measurements f uniquely determine φ. Assume that we have a linear model M and observe $y_\xi = HM^\xi\varphi_0$ for $\xi = 1, \ldots, L$, where M^ξ is the composition of M defined by $M^0 = I, M^\xi = M \circ M^{\xi-1}$ $(\xi = 1, 2, \ldots, L)$. Then, we can write this as a linear problem

$$\begin{pmatrix} y_1 \\ y_2 \\ \vdots \\ y_L \end{pmatrix} = H \begin{pmatrix} M^1 \\ M^2 \\ \vdots \\ M^L \end{pmatrix} \varphi_0 =: A\varphi_0. \tag{11.3.12}$$

A simple consequence of the definition is noted in the following lemma.

Lemma 11.3.3. *If for some* $L > 0$ *the matrix or operator* A *is injective, we have observability of* φ_0 *given for L measurements* $y_1, \ldots, y_L$.

Proof. Clearly, if A is injective, the observations determine the state φ_0 uniquely and by definition this implies observability. □

A very simple example for observability of some object over time is if M is some rotation and the observation H is a projection of a three-dimensional object onto the observation plane by some photographic image. Then, after some time the rotation can lead to a full observability of the object.

A standard approach to find the parameters p from (11.3.10) is to use the Kalman filter or the ensemble Kalman filter. Since the Kalman filter and its ensemble version employ the dynamical statistical relationship between different variables of the state space, they can lead to an update of the parameters p, if there are statistical relationships between p and j implicitly given by the dynamics M.

11.4 Classification methods for inverse problems

For many engineering applications, the *states* $\varphi \in X$ or *effects* $f = A\varphi$ of some *underlying processes* given by an operator $A : X \to Y$ between Hilbert spaces X, Y are measured, which need to be judged and classified. For example, it is a long standing tradition in acoustics to use sound recording to monitor the quality of machines and production.

For many of these problems, the solution of the *inverse* problem would be far too costly and is not needed, if the only initial action which is desired needs to report 'something is wrong here!' For example in acoustics, this is achieved by recording some sound signal $U(t)$ and employing a filter which detects changes in the windowed frequency spectrum $\mathcal{F}(U|_{[T,T+\Delta T]})$ of U with time $T \in \mathbb{R}$ and window size $\Delta T > 0$.

In the easiest case, classification leads to a real-valued function $\rho(f)$, where $\rho(f) < \rho_0$ is identified with one class, $\rho(f) \geqslant \rho_0$ with another. *Classification* seems to be an easy alternative to the full solution of the *inverse problem*. But it is important to take into account the nature of the underlying inverse problem.

Here, we will study the *ill-posedness* of classification problems when applied to the data of an ill-posed inverse problem. We will show that the ill-posedness of the inverse problem carries over to the classification task, such that *regularization* is needed for classification as it is for the inversion process itself. The original work was carried out by Lowery *et al* [9].

In general, *classification algorithms* determine appropriate classes $U_1 \subset X$ and $U_2 \subset X$, such that a sample φ is classified as $\varphi \in U_1$ or $\varphi \in U_2$.

Definition 11.4.1 (Supervised classification). *If we are given a set of samples* $f_1^{(1)}, \ldots f_{N_1}^{(1)}$ *and a set of samples* $f_1^{(2)}, \ldots, f_{N_2}^{(2)}$, *we call the problem to determine classes* $U_1, U_2 \subset X$ *such that* $f_1^{(1)}, \ldots, f_{N_1}^{(1)} \in U_1$ *and* $f_1^{(2)}, \ldots, f_{N_2}^{(2)} \in U_2$ *the* supervised classification problem. *The sets of samples are usually called* training sets.

If the classes U_1 and U_2 are linear half-spaces of X, we call the classification problem *linear*. Note that by linear half-space we also mean *affine* half-spaces, i.e. we allow the classification to *not* contain the origin. In general, classification problems will be nonlinear in the sense that the boundaries of U_1 and U_2 are manifolds in X. For example, the class U_1 could be given by some ellipse

$$E := \left\{ \varphi \in X : \sum_{\xi=1}^{\infty} \frac{\langle \varphi, \varphi_\xi \rangle^2}{a_\xi^2} < 1 \right\}, \tag{11.4.1}$$

in *canonical form*, where $\{\varphi_\xi : \xi \in \mathbb{N}\}$ is some complete orthonormal system in X and $\{a_\xi : \xi \in \mathbb{N}\}$ is the set of *principle axis* of the ellipse. The class U_2 might be the exterior of the ellipse $U_2 := X \backslash \bar{U}_1$. Alternatively, U_2 could be another ellipse not intersecting U_1.

Definition 11.4.2 (Stability of classes). *We call a classification into a class* U_1 *and* U_2 *stable, if there is some* $\epsilon > 0$ *such that*

$$d(U_1, U_2) \geqslant \epsilon. \tag{11.4.2}$$

Otherwise, we call the classification unstable.

A linear class U is *parametrized* by a normal vector $\nu \in X$ with $\|\nu\| = 1$ and a threshold parameter ρ, such that

$$U = \{\varphi \in X : \langle \varphi, \nu \rangle \geqslant \rho\}, \tag{11.4.3}$$

or the analogous definition with $\geqslant$ replaced by $\leqslant$. For our further study, we consider two classes U_1 and U_2 in X, as well as a compact linear operator A: $X \to Y$. We use the notation

$$\tilde{U}_\xi := A(U_\xi), \quad \xi = 1, 2, \tag{11.4.4}$$

for the *image classes* $\tilde{U}_1$ and $\tilde{U}_2$ in Y of U_1 and U_2 in X.

Our task here is to investigate the stability of classification methods. Usually, classification methods choose a finite number of samples as described by definition 11.4.1 and constructs the classifications $\tilde{U}_1$ and $\tilde{U}_2$ in Y.

Here, we obtain stable classification, if we can find classes $\tilde{U}_1$ and $\tilde{U}_2$, such that (11.4.2) is satisfied. If this is not the case, the classification is called unstable. The particular choice of the training set is usually random, such that any of the elements $\varphi \in U_1$ could be chosen to generate $f_\xi^{(1)}$ for $\xi = 1, \ldots, N_1$, and any element $\varphi \in U_2$ could lead to $f_\xi^{(2)}$, $\xi = 1, \ldots, N_2$. Depending on the particular choices of elements, the classification could be more or less stable, in the sense that the largest possible distance $\tilde{\epsilon}>0$ between the classes is larger or very small. This is of course a crucial observation. If the stability of our classification depends on a random choice of samples, it is not reliable. This leads to the following terminology.

Definition 11.4.3 (Stability of image classification). *We call the classification of the samples of definition 11.4.1 by classes $\tilde{U}_1$ and $\tilde{U}_2$ stable, if there is $\epsilon > 0$ such that for any choices of original elements $\varphi_\xi^{(1)} \in U_1, \xi = 1, \ldots, N_1$ and $\varphi_\xi^{(2)} \in U_2, \xi = 1, \ldots, N_2$, with*

$$\tilde{U}_\ell = A(U_\ell), \quad f_\xi^{(\ell)} = A\left(\varphi_\xi^{(\ell)}\right), \quad \ell = 1, 2, \quad \xi = 1, \ldots N_\ell, \tag{11.4.5}$$

the estimate (11.4.2) is satisfied for $\tilde{U}_1$ and $\tilde{U}_2$.

Here, we study the linear problem, i.e. the case where the classes U_1 and U_2 are half-spaces in X. In this case, clearly the boundaries of U_1 and U_2 need to be parallel affine hyperplanes in X. To study the distance of the corresponding image classes in Y, we derive the following basic result.

Theorem 11.4.4. *Consider a linear half-space U in X defined by its normal vector v and some distance ρ to the origin by*

$$U = \{x \in X : \langle x, v\rangle \geqslant \rho\} \tag{11.4.6}$$

and let A be a compact injective linear operator $A : X \to Y$. If $v \notin A^*Y$, then the distance of $\tilde{U} = AU$ to the origin is zero. If A is not injective, then the same conclusion holds under the condition that $Pv \notin A^*Y$ where P is the orthonormal projection in X onto $N(A)^\perp$.

Proof. From (11.4.6) we obtain the image class

$$\tilde{U}=\{y \in AX : \langle A^{-1}y, v\rangle \geqslant \rho\}. \tag{11.4.7}$$

We consider the singular system, (μ_n, φ_n, g_n) $n \in \mathbb{N}$ of A, such that

$$A\varphi_n = \mu_n g_n, \quad A^* g_n = \mu_n \varphi_n, \tag{11.4.8}$$

where μ_n are the singular values of A and $\varphi_n \in X$ and $g_n \in Y$ are orthonormal bases. Then for each $y \in Y$ we have

$$A^*y = \sum_{n=1}^{\infty} \mu_n \langle y, g_n \rangle \varphi_n. \tag{11.4.9}$$

By Picard's theorem, $v \notin A^* Y$ yields

$$\sum_{n=N}^{M} \left| \frac{\langle v, \varphi_n \rangle}{\mu_n} \right|^2 \to \infty \text{ as } M \to \infty \tag{11.4.10}$$

for any fixed N. Let

$$\psi^{(N,M)} := \beta^{(N,M)} \sum_{n=N}^{M} \frac{|\langle v, \varphi_n \rangle|}{\mu_n^2} \varphi_n \tag{11.4.11}$$

with some $\beta^{(N,M)} \in \mathbb{R}$. Then, $\psi^{(N,M)}$ lies on the boundary of U if and only if $\langle \psi^{(N,M)}, v \rangle = \rho$, i.e.

$$\beta^{(N,M)} = \frac{\rho}{\sum_{n=N}^{M} \frac{1}{\mu_n^2} |\langle v, \varphi_n \rangle|^2}. \tag{11.4.12}$$

Now, taking such $\beta^{(N,M)}$,

$$A\psi^{(N,M)} = \beta^{(N,M)} \sum_{n=N}^{M} \frac{|\langle v, \varphi_n \rangle|}{\mu_n} g_n. \tag{11.4.13}$$

Therefore,

$$\begin{aligned} \|A\psi^{(N,M)}\|^2 &= |\beta^{(N,M)}|^2 \sum_{n=N}^{M} \frac{|\langle v, \varphi_n \rangle|^2}{\mu_n^2} \\ &= \frac{\rho^2}{\sum_{n=N}^{M} \frac{|\langle v, \varphi_n \rangle|^2}{\mu_n^2}}. \end{aligned} \tag{11.4.14}$$

Thus, the distance between $A\psi^{(N,M)}$ and the origin in Y tends to zero for $M \to \infty$. This shows that $d(\tilde{U}, 0) = 0$, and for injective A the proof is complete.

If A is not injective, we note that $N(A)^\perp = \overline{A^* Y}$. The vector v can be decomposed into

$$v = Pv + (I - P)v \in N(A)^\perp \oplus N(A). \tag{11.4.15}$$

We now consider $A|_{N(A)^\perp}$, on which it is a compact injective linear operator and the above arguments apply. This is because by definition of P we have $\langle v, \varphi_n \rangle = \langle Pv, \varphi_n \rangle$ for all $n \in \mathbb{N}$, and by $Pv \notin A^* Y$ but $Pv \in \overline{A^* Y}$ the divergence (11.4.10) holds also for the original vector v. This completes the proof.

Corollary 11.4.5 (Ill-posedness of classification). *Under the conditions of theorem 11.4.4 the classification of two classes along the vector $v \notin A^*Y$ is instable in the sense that the image classes in Y cannot be separated by a positive distance $\tilde{\rho}>0$. An analogous statement holds for A not injective.*

Proof. Assume we are given two classes U_1 and U_2 in a space X with some positive distance ρ and assume we have $v \notin A^*Y$ with injective operator $A: X \to Y$. Then, we have shown that the image classes $\tilde{U}_1$ and $\tilde{U}_2$ touch the origin in Y, i.e. they cannot have a positive distance $\tilde{\rho}>0$. An analogous statement applies when A is not injective. □

As a consequence, it is impossible to overcome the ill-posedness of a problem by restricting one's attention to classification only. Linear classification methods will inherit the instability of the inverse problem and need *regularization*. This has been worked out in detail in [9].

Bibliography

[1] Kress R, Kühn L and Potthast R 2002 Reconstruction of a current distribution from its magnetic field *Inverse Problems* **18** 1127–46

[2] Potthast R and Kühn L 2003 On the convergence of the finite integration technique for the anisotropic boundary value problem of magnetic tomography *Math. Methods Appl. Sci.* **26** 739–57

[3] Hauer K-H, Potthast R and Wannert M 2008 Algorithms for magnetic tomography—on the role of *a priori* knowledge and constraints *Inverse Problems* **24** 045008

[4] Hauer K-H, Kühn L and Potthast R 2005 On uniqueness and non-uniqueness for current reconstruction from magnetic fields *Inverse Problems* **21** 955

[5] Girault V and Raviart P A 1986 *Finite Element Approximation of the Navier–Stokes Equations* (New York: Springer)

[6] Dautray R and Lions J-L 2000 *Mathematical Analysis and Numerical Methods for Science and Technology* vol 3 (New York: Springer)

[7] Engl H W, Hanke M and Neubauer A 1996 *Regularization of Inverse Problems* (*Mathematics and its Applications* vol 375) (Dordrecht: Kluwer Academic)

[8] Marx B A 2011 *Dynamic Magnetic Tomography* (Der Andere: Uelvesbüll)

[9] Lowery N, Potthast R, Vahdati M and Holderbaum W 2012 On discrimination algorithms for ill-posed problems with an application to magnetic tomography *Inverse Problems* **28** 065010

Chapter 12

Field reconstruction techniques

In this chapter we study methods for the reconstruction of a field u^s from measurements of its far field patterns u^∞, its trace $u^s|_\Lambda$ on some part Λ of a closed surface or its Cauchy values $u^s|_\Lambda$, $\partial u^s/\partial\nu|_\Lambda$ on some open surface Λ.

First, field reconstruction can be seen as a first step towards the inverse shape reconstruction problem. Methods which first calculate the fields and then exploit them further are also called *decomposition methods*, since the *ill-posed* and *nonlinear* shape reconstruction problem from the far field pattern is decomposed into an ill-posed but *linear* field reconstruction problem and a *nonlinear* but well-posed local optimization problem.

A *second* important point is that the field reconstruction techniques are used as tools for several probe and sampling schemes in the following chapters.

Here, we will first study some classical approaches to field reconstructions by particular field representations by a series expansion in section 12.1 and by plane wave or Fourier representations in section 12.2. We will also discuss the limitations of these methods when used for inverse problems.

In section 12.3 we will base the field reconstruction on a *potential approach* following Kirsch and Kress, but employ recent new ideas by Sylvester, Kusiak, Potthast and Schulz. In particular, we will overcome some difficulties with the convergence proof of the original Kirsch–Kress scheme. To this end we will employ a *masking technique* which identifies areas of convergence of the field reconstructions in $\mathbb{R}^m$ and composes them to full reconstructions in the exterior of the unknown domain.

Section 12.4 presents the *point source method* for field reconstruction, which is based on the Green formula and an approximation of the fundamental solution by a superposition of plane waves. The point source method is the basis for the probe and singular sources method used later and is strongly related to the no-response test. We will explore these relations in later chapters. Here we focus on the point source method itself.

doi:10.1088/978-0-7503-1218-9ch12

As an important result in section 12.5 we present a recent proof of *duality* of the point source method and the potential method by Liu and Potthast [1]. Its consequence is the equivalence of the point source method and the potential method if the geometrical settings are identical. We provide explicit results on the choice of the regularization parameters of both methods to obtain identical field reconstructions.

12.1 Series expansion methods

The general idea of series expansion methods starts with some representation of a field u on a set M of the form

$$u = \sum_{n=1}^{\infty} \alpha_n \varphi_n \qquad (12.1.1)$$

with basis functions φ_n for $n \in \mathbb{N}$. If u is known on a subset $V \subset M$, then we can try to find u on M from the knowledge of u on V by determination of α_n via solving the system

$$u|_V = \sum_{n=1}^{\infty} \alpha_n \varphi_n \Big|_V \qquad (12.1.2)$$

and then using the general representation (12.1.1).

If the far field pattern of some scattered field u^s is measured, the setting is slightly more complicated, since instead of (12.1.2) one needs to use some far field representation of the elements φ_n of the form

$$u^\infty = \sum_{n=1}^{\infty} \alpha_n \varphi_n^\infty \qquad (12.1.3)$$

with φ_n^∞ being the far field pattern for φ_n. In this case we determine α_n from (12.1.3) and then represent u by (12.1.1).

The use of series expansions for field extension is quite popular, but it also can have serious limitations which are caused by the ill-posedness of the extension and the missing convergence of the series on the complete domain M where we search for a reconstruction. We will discuss this in detail in the next subsections.

12.1.1 Fourier–Hankel series for field representation

Let (r, φ) be the polar coordinates around the origin in $\mathbb{R}^2$. We call the series of the form

$$\sum_{n=-\infty}^{\infty} a_n H_{|n|}^{(1)}(\kappa r) e^{in\varphi} \qquad (12.1.4)$$

with coefficients $a_n \in \mathbb{C}$, $n \in \mathbb{Z}$, the *Fourier–Hankel series*, where $H_{|n|}^{(1)}$ is the Hankel function of order $|n|$ of the first kind. We can represent any two-dimensional radiating field u outside a circle B_R with radius $R > 0$ and center 0 by the Fourier–Hankel series. More precisely, we have the following result.

Theorem 12.1.1. *Let u be a radiating field in $|x| > R$. Then, u admits a Fourier–Hankel series representation*

$$u(x) = \sum_{n=-\infty}^{\infty} a_n H_{|n|}^{(1)}(\kappa r) e^{in\varphi} \tag{12.1.5}$$

which converges absolutely and uniformly on any compact subset of $|x| > R$. Conversely, if this series (12.1.5) is $L^2(\{|x| = R\})$ convergent, then it converges absolutely and uniformly on any compact set of $|x| > R$ and represents a radiating field.

Proof. Let y with $|y| = \tilde{R}$, $R < \tilde{R} < |x|$. Also, let φ, ψ be the angles of x, y in terms of the polar coordinates, respectively. Then, by the well-known addition theorem for Bessel functions (see [2] that the fundamental solution $\Phi(x, y) = \frac{i}{4} H_0^{(1)}(\kappa|x - y|)$ admits an expansion

$$\Phi(x, y) = \frac{i}{4} \sum_{n=-\infty}^{\infty} H_{|n|}^{(1)}(\kappa|x|) J_{|n|}(\kappa|y|) e^{in(\varphi-\psi)} \tag{12.1.6}$$

which converges together with its term-wise first derivatives absolutely and uniformly with respect to x, y on any compact subset of $|x| > |y|$, where $J_{|n|}$ is the Bessel function of order $|n|$. By inserting the expansion (12.1.6) into the representation formula of u

$$u(x) = \int_{|y|=\tilde{R}} \left\{ u(y) \frac{\partial \Phi(x, y)}{\partial \nu(y)} - \frac{\partial u(y)}{\partial \nu} \Phi(x, y) \right\} ds(y),$$

we obtain the series representation (12.1.5), which converges absolutely and uniformly on any compact set of $|x| > \tilde{R}$, where

$$a_n = \frac{i}{4} \int_{|y|=\tilde{R}} \left\{ \kappa u(y) J_{|n|}'(\kappa|y|) - \frac{\partial u(y)}{\partial |y|} J_{|n|}(\kappa|y|) \right\} e^{-in\varphi} \, ds(y). \tag{12.1.7}$$

By Green's second theorem the coefficient a_n defined by (12.1.7) does not depend on the particular choice of $\tilde{R} > R$. This proves the first half of the theorem.

Now, let the series (12.1.5) be L^2 convergent on $|x| = R$. Then, by the Parseval equality (2.2.38), we have

$$\sum_{n=-\infty}^{\infty} |a_n|^2 \left| H_{|n|}^{(1)}(\kappa R) \right|^2 < \infty. \tag{12.1.8}$$

Recall the asymptotic behavior of $H_n^{(1)}(t) = (-1)^n H_{-n}^{(n)}(t)$ for $n \in \mathbb{N}$

$$H_n^{(1)}(t) = \frac{2^n(n-1)!}{\pi i t^n}(1 + O(1/n)), \quad n \to \infty, \tag{12.1.9}$$

uniformly on a compact subset of $(0, \infty)$ (see equation (3.58) of [3]). Then, with the help of (12.1.8) by using the Cauchy–Schwarz inequality (2.2.2) and the asymptotic behavior of the Hankel function, we derive for $R < R_1 \leqslant |x| \leqslant R_2$ and positive integers M, N with $M < N$ the estimate

$$\begin{aligned} \left| \sum_{M \leqslant |n| \leqslant N} a_n H_{|n|}^{(1)}(\kappa|x|)e^{in\varphi} \right|^2 &= \left| \sum_{M \leqslant |n| \leqslant N} a_n \frac{H_{|n|}^{(1)}(\kappa|x|)}{H_{|n|}^{(1)}(\kappa R)} H_{|n|}^{(1)}(\kappa R)e^{in\varphi} \right|^2 \\ &\leqslant \left\{ \sum_{M \leqslant |n| \leqslant N} |a_n|^2 |H_{|n|}^{(1)}(\kappa R)|^2 \right\} \cdot \left\{ C \sum_{M \leqslant |n| \leqslant N} \left(\frac{R}{|x|} \right)^{2|n|} \right\} \end{aligned} \tag{12.1.10}$$

for some constant $C > 0$ depending only on R, R_1, R_2. The last term of (12.1.10) is bounded by a constant times a geometric series in $R/|x|$. Hence, the series (12.1.5) converges absolutely and uniformly on any compact set of $|x| > R$. By an analogous argument, the term-wise derivative of the series (12.1.5) also converges absolutely and uniformly on any compact subset of $|x| > R$.

In order to show that u satisfies the Helmholtz equation and Sommerfeld radiation condition, it is enough to show that

$$u(x) = \int_{|y|=\tilde{R}} \left\{ u(y)\frac{\partial \Phi(x, y)}{\partial \nu(y)} - \frac{\partial u(y)}{\partial \nu}\Phi(x, y) \right\} ds(y) \tag{12.1.11}$$

for $|x| > \tilde{R} > R$. We denote by $W(f, g) := f \cdot g' - g' \cdot f$ the Wronskian of real-valued functions f and g. Then, by the asymptotic behavior of the Bessel function and Hankel function at infinity, we have

$$W\left(H_{|n|}^{(1)}\left(\kappa\tilde{R}\right), J_{|n|}\left(\kappa\tilde{R}\right)\right) = -\frac{2i}{\pi\kappa\tilde{R}}. \tag{12.1.12}$$

Hence,

$$\begin{aligned} &\int_{|y|=\tilde{R}} \left\{ u(y)\frac{\partial \Phi(x, y)}{\partial \nu(y)} - \frac{\partial u(y)}{\partial \nu}\Phi(x, y) \right\} ds(y) \\ &= \frac{i\kappa}{4} \int_{|y|=\tilde{R}} \sum_{n=-\infty}^{\infty} a_n W\left(H_{|n|}^{(1)}\left(\kappa\tilde{R}\right), J_{|n|}\left(\kappa\tilde{R}\right)\right) H_{|n|}^{(1)}(\kappa|x|)e^{in\varphi}\, ds(y) \\ &= \sum_{n=-\infty}^{\infty} a_n H_{|n|}^{(1)}(\kappa|x|)e^{in\varphi} = u(x) \end{aligned}$$

for $x(|x| > \tilde{R} > R)$. This proves the latter half of the theorem.

In the proof of the above theorem 12.1.1 with (12.1.7) we are given an explicit integral formula for the coefficients a_n, $n \in \mathbb{Z}$, of the Fourier–Hankel expansion (12.1.5) for field u defined outside some ball B_R with radius R around zero, which can be used for the stable calculation of these coefficients analogously to the calculation of the Fourier coefficients $\langle \varphi, \varphi_n \rangle$ in (2.2.37).

Alternatively, we can calculate the coefficients numerically by a matrix inversion. Assume that we are given regularly distributed points x_ℓ with polar coordinates (R, φ_ℓ), $\ell = 1, \ldots, N_m$ on the boundary $\mathbb{S}_R$ of the circle B_R. Then, we need to solve the equation

$$u(x_\ell) = \sum_{n=-N_n}^{N_n} a_n H_{|n|}^{(1)}(\kappa R) e^{in\varphi_\ell}, \quad \ell = 1, \ldots, N_m \tag{12.1.13}$$

with $2N_{n+1}$ unknown coefficients a_n for $n = -N_n, \ldots, N_n$. Equation (12.1.13) can be written in matrix form

$$\mathbf{u} = \mathbf{A}_{FH}\, \mathbf{a} \tag{12.1.14}$$

with vectors

$$\mathbf{u} = \begin{pmatrix} u(x_1) \\ \vdots \\ u(x_{N_m}) \end{pmatrix}, \quad \mathbf{a} = \begin{pmatrix} a_{-N_n} \\ \vdots \\ a_{N_n} \end{pmatrix} \tag{12.1.15}$$

and the matrix

$$A_{FH} := \left(\left(H_{|n|}^{(1)}(\kappa R) e^{in\frac{2\pi\ell}{N_m}} \right) \right)_{\substack{\ell=1,\ldots,N_m, \\ n=-N_n,\ldots,N_n}} . \tag{12.1.16}$$

Since for $N_m \geqslant 2N_n + 1$ the exponential basis vectors

$$\left(e^{in\varphi_\ell} \right)_{\ell=1,\ldots,N_m} = \left(e^{in\frac{2\pi\ell}{N_m}} \right)_{\ell=1,\ldots,N_m} \tag{12.1.17}$$

for $n = -N_n, \ldots, N_n$ are linearly independent in $\mathbb{R}^{N_m}$, the matrix is boundedly invertible.

We conclude this section with some hands-on experiments working with Fourier–Hankel series in code 12.1.2.

Code 12.1.2. *We evaluate the representation of some test field chosen as a point source with source point z by its Fourier–Hankel series. Here we match the series representation and the original on a circle of radius R according to (12.1.13) and (12.1.14). We need to first run some preparations setting up the vectors of evaluation points* `pvec1` *and* `pvec2`. *The scripts are analogous to code 8.2.3, see the file* `sim_12_1_2_a_preparations.m`. *The calculations are carried out by the script* `sim_12_1_2_b_Fourier_Hankel_representation.m`, *which we show next.*

```
Nn          = 30;                  % 2Nn+1: number of terms of FH-series
kappa       = 2;                   % wave number
nvec     = (1:Nn) - ceil(Nn/2);    % indices for coefficients n
Ampl        = 20;                  % amplitude for display
Nm          = 100;                 % number of points on matching boundary
hm          = 2*pi/Nm;             % grid spacing on matching boundary
R           = 5;                   % radius of matching boundary
rR          = R*ones(Nm,1);        % radius vector
tR          = (0:hm:(2*pi-hm)).'; % angle vector
[x1m,x2m] = pol2cart(tR,rR);       % cartesian coordinates of boundary
z1   = -3; z2  = 0;                % centre point coord of original field
[thz,rz] = cart2pol(z1,z2);        % and its polar coordinates
rmvec     = sqrt( (x1m-z1).^2 + (x2m-z2).^2); % vector of differences
rhs = Ampl*i/4*besselh(0,1,kappa*rmvec);      % matching function

% set ups Fourier-Hankel series on matching circle
rvec  = sqrt( (x1m).^2 + (x2m).^2 );
FHn   = besselh(repmat(abs(nvec),Nm,1),1,repmat(kappa*rvec,1,Nn));
enphi = exp(i*repmat(nvec,Nm,1).*repmat(tR,1,Nn)); % exponential basis
FH    = FHn.*enphi;  % Fourier-Hankel matrix A_FH

% evaluate coefficients
alpha = 1e-9;                        % regularization parameter
an    = inv(alpha*eye(Nn,Nn) + FH'*FH)*FH'*rhs; % FH coefficients
```

The evaluation of the original wave field and its Fourier–Hankel representation is carried out as follows, this is the script `sim_12_1_2_c_FH_evaluation.m`.

```
% polar coordinates for evaluation points
[tphimat,rvec] = cart2pol(pvec1,pvec2);          % polar coordinates

% setup matrix tFH
tFHn    = besselh(abs(nvec),1,kappa*rvec);      % radial evaluation part
tenphi  = exp(i*repmat(nvec,M,1).*repmat(tphimat,1,Nn)); % eval angle
tFH     = tFHn.*tenphi;   % evaluation Fourier-Hankel matrix

% evaluation
rivec = sqrt( (pvec1-z1).^2 + (pvec2-z2).^2); % vector of differences
wi    = Ampl*i/4*besselh(0,1,kappa*rivec);    % original field
ws    = tFH*an;                               % FH representation
```

A demo output is shown in figure 12.1, where the graphics is generated using `sim_12_1_2_d_graphics.m` (not shown here, see code repository).

12.1.2 Field reconstruction via exponential functions with an imaginary argument

Since the far field pattern u^{∞} is analytic on the unit circle $\mathbb{S}$, we have an expansion

$$u^{\infty}(\hat{x}) = \sum_{n=-\infty}^{\infty} b_n e^{in\varphi}. \tag{12.1.18}$$

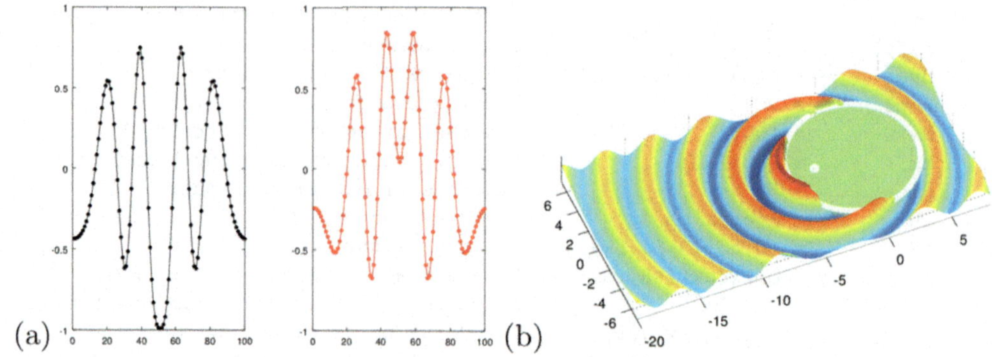

Figure 12.1. (a) The values of an original point source Φ with center $z = (-3, 0)$ (the real and imaginary parts of the field) and its approximation by a Fourier–Hankel series on a circle with radius $R = 5$ around the origin, where we used $N_n = 20$. (b) The field of a Fourier–Hankel series as generated by code 12.1.2. Here, the original and the series representation images are identical, the approximation error is decaying exponentially for increasing number of terms in the Fourier–Hankel expansion in the exterior of the circle B_R and we have set the function to zero inside B_R.

The next two results theorem 12.1.3 and theorem 12.1.4 clarify the relation between the two series (12.1.5) and (12.1.18).

Theorem 12.1.3. *Let u be a radiating field in* $|x| > R > 0$ *with the expansion (12.1.5). Then, the far field pattern* u^∞ *of u is given by the absolutely and uniformly convergent series*

$$u^\infty(\hat{x}) = \frac{1 - \mathrm{i}}{\sqrt{\pi\kappa}} \sum_{n=-\infty}^{\infty} \frac{1}{\mathrm{i}^n} a_n \mathrm{e}^{\mathrm{i}n\varphi}. \tag{12.1.19}$$

The coefficients a_n *in (12.1.19) satisfy the estimate*

$$\sum_{n=-\infty}^{\infty} |a_n|^2 \frac{((2|n|-1)!!)^2}{(\kappa r)^2|n|} < \infty \quad (r > R). \tag{12.1.20}$$

Proof. We start with the expansion (12.1.19) with coefficients

$$b_n = \frac{1}{2\pi} \int_0^{2\pi} u^\infty(\varphi)\mathrm{e}^{-\mathrm{i}n\varphi}\, \mathrm{d}\varphi, \quad n \in \mathbb{Z}, \tag{12.1.21}$$

which is valid in the $L^2([0, 2\pi))$ sense, see example 2.2.15. On the other hand, we have

$$u(x) = \sum_{n=-\infty}^{\infty} a_n H_{|n|}^{(1)}(\kappa|x|)\mathrm{e}^{\mathrm{i}n\varphi} \tag{12.1.22}$$

with

$$a_n H_{|n|}(\kappa|x|) = \frac{1}{2\pi} \int_0^{2\pi} u(|x|, \varphi)\, \mathrm{e}^{-\mathrm{i}n\varphi}\, \mathrm{d}\varphi \quad (n \in \mathbb{Z}).$$

By the well-known asymptotic behavior of the radiating field

$$u(x) = \frac{e^{i\kappa|x|}}{\sqrt{|x|}}\left(u^{\infty}(\varphi) + O\left(\frac{1}{\sqrt{|x|}}\right)\right), \quad |x| \to \infty, \tag{12.1.23}$$

which is satisfied uniformly for all directions, we have by using the asymptotic behavior of the Hankel function of the first kind for large order

$$\begin{aligned} b_n &= \frac{1}{2\pi}\int_0^{2\pi}\left(\lim_{r\to\infty}\sqrt{r}e^{-i\kappa r}u(r, \varphi)\right)e^{-in\varphi}\, d\varphi \\ &= \lim_{r\to\infty}\sqrt{r}e^{-i\kappa r}\frac{1}{2\pi}\left(\int_0^{2\pi} u(r, \varphi)e^{-in\varphi}\, d\varphi\right) \\ &= \lim_{r\to\infty}\sqrt{r}e^{-i\kappa r}a_n H_{|n|}^{(1)}(\kappa r) \\ &= \frac{1 - i}{\sqrt{\pi\kappa}}\frac{a_n}{i^n} \end{aligned} \tag{12.1.24}$$

for $n \in \mathbb{Z}$. Using the Parseval equality we obtain

$$r\sum_{n=-\infty}^{\infty}|a_n|^2\left|H_{|n|}^{(1)}(\kappa r)\right|^2 = \int_{|x|=R}|u(x)|^2\, ds(x). \tag{12.1.25}$$

Combining this with the asymptotic behavior of the Hankel function for large order, we have (12.1.20). We can easily see that by using the Schwarz inequality and (12.1.20) the series (12.1.19) converges absolutely and uniformly. □

Theorem 12.1.4. *Let the coefficients b_n in (12.1.18) of $u^{\infty} \in L^2(\mathbb{S})$ satisfy the estimate*

$$\sum_{n=-\infty}^{\infty}|b_n|^2\frac{((2|n| - 1)!!)^2}{(\kappa R)^{2n}} < \infty \tag{12.1.26}$$

with some $R > 0$. Then

$$u(x) = (1 + i)\frac{\sqrt{\pi\kappa}}{2}\sum_{n=-\infty}^{\infty} i^{|n|}b_n H_{|n|}^{(1)}(\kappa|x|)e^{in\varphi}(|x| > R) \tag{12.1.27}$$

is a radiating field with far field pattern u^{∞}.

Proof. By (12.1.26) and the asymptotic behavior of the Hankel function for large order, the series (12.1.27) converges in the $L^2(\{|x| = R\})$ sense. Then, by theorem 12.1.1, u is the radiating field. Further, by theorem 12.1.3, the far field pattern of u coincides with u^{∞}. □

As shown in figure 12.2 there are strong limitations to the reconstruction of scattered fields using the Fourier–Hankel series. If the field u^s is defined in the exterior of the circle B_R, but is not analytically extensible into a smaller circle $B_{R'}$ with $R' < R$, then the series (12.1.27) will converge outside B_R, but will diverge inside B_R. So we cannot reconstruct the field inside this circle.

Usually, a field u^s which is defined outside some scatterer D will have singular points inside the domain D. Let R_0 be the smallest radius such that $D \subset B_{R_0}$.

Then by the above Fourier series approach we obtain a reconstruction of u^s outside B_{R_0}, but in general it does not converge inside B_{R_0}. If the domain is not located around the origin, usually R_0 is large.

A standard option to obtain reconstructions inside B_{R_0} is to consider Fourier–Hankel expansions with shifted center z. This corresponds to multiplication of the far field pattern with the factor

$$M_z(\hat{x}) := e^{-i\kappa\hat{x}\cdot z}, \quad \hat{x} \in \mathbb{S}. \tag{12.1.28}$$

Instead of shifting the Hankel functions we can first shift the far field pattern by the vector $-z$, then carry out the reconstruction and finally evaluate the terms on the points shifted by z. This is carried out in code 12.1.5, with the result shown in figure 12.2. A much better coverage of the field reconstructions is achieved when the center of the series is moved into the domain D, shown in figure 12.2(d) with error visualized in panel (e).

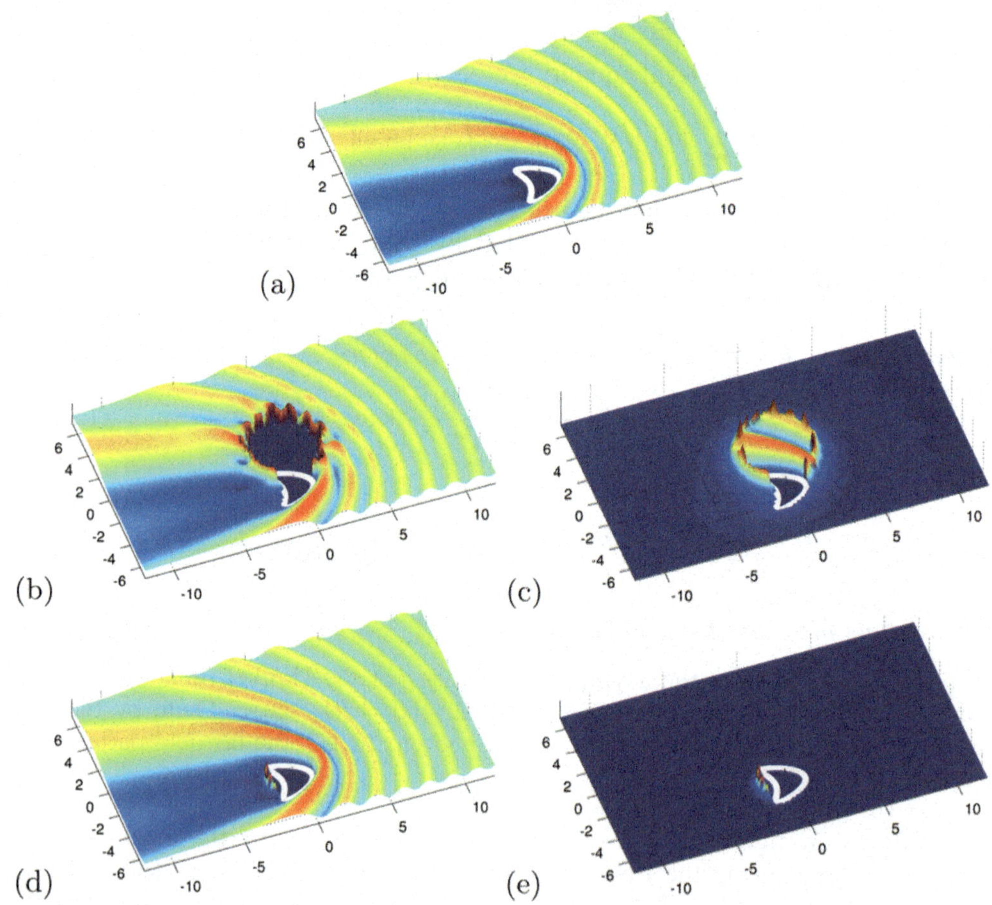

Figure 12.2. The original total field (a), the reconstructed total field via Fourier–Hankel series around the point $z = (0, 0)$ following code 12.1.5 (b) and the error for this reconstruction (c). In (d) and (e) the reconstruction and error are shown when the center is moved to $z = (-1, -3)$. We used $N_n = 30$, i.e. $2N_n + 1 = 61$ modes, and $N_\infty = 80$ far field points.

Here we work out some simple code to carry out the Fourier–Hankel reconstruction. Clearly, we first need to calculate the far field pattern u^{∞}, which is described in sections 8.2 and 8.3. Here, the corresponding calculations have been collected into a script `sim_12_1_5_a_preparations.m` and a solution of the integral equation by a collocation method in `sim_12_1_5_b_scattering.m`. The simulation is visualized by `sim_12_1_5_c_graphics_scattering.m`.

Code 12.1.5. *This script* `sim_12_1_5_d_FH_reconstruction.m` *demonstrates the reconstruction of a total field based on the Fourier–Hankel series (12.1.27). The graphical display of the reconstruction and its error is carried out by the script* `sim_12_1_5_e_graphics_FH_reconstruction.m`. *Run all scripts step-by-step using the control script* `sim_12_1_5_0_control.m`.

```
% I Preparations
Nn        = 20;                        % 2Nn+1 is number of FH coeffs
nvecF     = (1:Nn) - ceil(Nn/2); % indices for coefficients n

% far field shift factor
z1 = -1; z2 = -3;                      % center for series expansion
Mz = exp(1i*kappa*(z1*cos(tff)+z2*sin(tff))).'; % shift factor

% Calculate far field coeffs bn and Fourier-Hankel coeffs an from ff
FT    = 1/(2*pi)*exp(-1i*repmat(nvecF.',1,ffN).*repmat(tff,Nn,1))*hff;
bnF   = FT*(Mz.*ff);
anF   = sqrt(pi*kappa)/2*(1+1i)*((1i).^(abs(nvecF.')).*bnF);

% II Evaluation of Fourier-Hankel series via operator tFH
[tphimat,rvec] = cart2pol(pvec1-z1,pvec2-z2); % evaluation points
tFHn    = besselh(abs(nvecF),1,kappa*rvec); % setup matrix tFH
tenphi  = exp(1i*repmat(nvecF,M,1).*repmat(tphimat,1,Nn));
tFH     = tFHn.*tenphi;   % Fourier-Hankel matrix on evaluation domain

wsF = tFH*anF;             % evaluate Fourier-Hankel representation
wF = wi + wsF;             % total field

maskF = abs(wF)<2;         % mask set to zero for high values
wF = wF.*maskF;            % masking for better visibility
```

Field reconstructions can be obtained not only from far field measurements, but in the same way from measurements on some curve or surface Λ. The corresponding scripts are also available in the code repository, see `sim_12_1_6_0_control.m` and the scripts called from this control script.

12.2 Fourier plane-wave methods

Fourier plane-wave methods rely on the basic representation of a wave as a superposition of plane waves. They can be seen as a continuous version of the equations (12.1.1) and (12.1.2). The Fourier transform

$$(Ff)(k) := \int_{-\infty}^{\infty} e^{-ikt} f(t)\, ds(t), \quad k \in \mathbb{R} \tag{12.2.1}$$

is a universal tool in mathematics. It can also be used to solve inverse field reconstruction problems and we will need it to investigate time-dependent problems. We start with the observation that the terms

$$v_\kappa(x, k) := e^{i\left(kx_1 + \sqrt{\kappa^2 - k^2}\, x_2\right)}, \quad x \in \mathbb{R}^2 \tag{12.2.2}$$

for $k \in \mathbb{R}$ solve the Helmholtz equation with wave number $\kappa > 0$. For $|k| \leqslant \kappa$ the square root is positive and the field v corresponds to a plane wave with direction $(k, \sqrt{\kappa^2 - k^2})$ into the upper half-plane. For $|k| > \kappa$ we choose the branch of the square root such that we have

$$\sqrt{\kappa^2 - k^2} = i\sqrt{k^2 - \kappa^2}, \quad k > \kappa.$$

Then, v corresponds to a wave which is a plane wave along the x_1-axis, but is exponentially decaying in the x_2-direction. We are particularly interested in the upper half-space $U := \{x : x_2 \geqslant 0\}$. For any bounded integrable function φ defined on $\mathbb{R}$ the integral

$$\int_{\mathbb{R}} e^{i\left(kx_1 + \sqrt{\kappa^2 - k^2}\, x_2\right)} \varphi(k)\, dk, \quad x \in U \tag{12.2.3}$$

is well defined, which for $x_2 > 0$ can be seen by splitting the integration domain into $\{|k| \leqslant \kappa\}$ and $\{|k| > \kappa\}$, where for the second term the kernel of the integral is exponentially convergent to zero for $k \to \infty$.

Now, let u^s be a radiating field which is defined in U with boundary values $u^s|_{\Gamma_0}$ on $\Gamma_0 := \{x_2 = 0\}$. Then, we can calculate the Fourier transform Fu^s of u^s on $\mathbb{R}$ and represent u^s on U by

$$\begin{aligned} u^s(x) &= \int_{\mathbb{R}} e^{i\left(kx_1 + \sqrt{\kappa^2 - k^2}\, x_2\right)} Fu^s(k)\, dk, \\ &= \int_{\mathbb{R}} e^{ikx_1} e^{i\sqrt{\kappa^2 - k^2}\, x_2} Fu^s(k)\, dk, \quad x \in U. \end{aligned} \tag{12.2.4}$$

In general, the field $u^s|_{\{x_2 = 0\}}$ is not even integrable over an infinite plane surface or line. But it is possible to justify the application of the Fourier transform and the representation in the form (12.2.4).

Let us now discuss the *field reconstruction problem*. When measurements of u^s are given on line

$$\Gamma_c := \{x : x_2 = c\},$$

then we can reconstruct the field u^s on the straight line $\{x : x_2 = 0\}$ by a Fourier transform on Γ_c, which calculates the field

$$g(\kappa) = e^{i\sqrt{\kappa^2 - k^2}\, c} Fu^s(k), \quad k \in \mathbb{R}. \tag{12.2.5}$$

Multiplication with $e^{-i\sqrt{\kappa^2-k^2}}$ provides the field $Fu^s(k)$, which is the Fourier transform of the unknown field u^s on Γ_0. Now, $u^s|_{\Gamma_0}$ is calculated by an inverse Fourier transform. Finally, the Fourier plane-wave representation formula (12.2.4) provides a formula to calculate u^s in the whole half-space $\{x : x_2 \geqslant 0\}$.

Finally, we remark that this reconstruction is still exponentially ill-posed, although its Fourier transform steps are fully stable in $L^2(\Gamma_c)$ and $L^2(\Gamma_0)$. The multiplication by the term

$$\sigma(k) = e^{-i\sqrt{\kappa^2-k^2}}$$

for $k > \kappa$ is exponentially growing, since to obtain convergence in the representation (12.2.3) we had to choose the branch of the complex square root for which $i\sqrt{\kappa^2 - k^2}$ is negative for $|k| \geqslant |\kappa|$. This means that we need to employ some type of regularization by damping the factor. It can be understood as a version of *spectral damping* as introduced in section 3.13 and is known as a *low-pass filter* in electronics.

For more results on scattering by infinite rough surfaces and the use of Fourier plane-wave methods we refer the reader to, for example, the work of Chandler-Wilde *et al* [4–6].

For *numerical realization* of Fourier plane-wave methods we employ the *finite section method* and make use of the Fourier series

$$f(t) = \frac{1}{T} \sum_{m=-\infty}^{\infty} a_m e^{2\pi i m t/T}, \quad t \in [T_1, T_2] \tag{12.2.6}$$

with $T = T_2 - T_1$ and coefficients

$$a_m := \int_{T_1}^{T_2} e^{-2\pi i m t/T} f(t) dt, \quad m \in \mathbb{Z} \tag{12.2.7}$$

where it is well known that $\varphi_m(t) := \frac{1}{T} e^{2\pi i m t/T}$ for $m \in \mathbb{Z}$ form an orthonormal basis of $L^2([T_1, T_2])$. For the numerical realization of the Fourier transform we can use the implementation of the fast Fourier transform, which calculates

$$F_k := \sum_{n=1}^{N} e^{-i2\pi \frac{(k-1)(n-1)}{N}} f_n, \quad k = 1, \ldots, N \tag{12.2.8}$$

in an efficient way, see [7].

12.3 The potential or Kirsch–Kress method

The task of this section is to introduce the *potential method* or *Kirsch–Kress method* [3]. We will then review results on its convergence and present a convergence analysis for the method in a set-up suggested by Schulz and Potthast [8]. Part of the numerical realization of the method will provide detailed guidance on an easy numerical scheme including the use of masking operations.

The potential method was first suggested by Kirsch and Kress in 1986 as a scheme for shape reconstruction in acoustic scattering. The key idea is to seek a reconstruction of the scattered field $u^s(z)$ in the exterior $\mathbb{R}^m \backslash G$ of some auxiliary domain G with boundary $\Gamma = \partial G$ as shown in figure 12.3 (a) by a single-layer potential approach

$$u^s(x) = \int_\Gamma \Phi(x, y)\varphi(y)\, \mathrm{d}s(y), \quad x \in \mathbb{R}^m \backslash G \tag{12.3.1}$$

with far field pattern

$$u^\infty(\hat{x}) = \gamma \int_\Gamma \mathrm{e}^{-ik\hat{x}\cdot y}\varphi(y)\, \mathrm{d}s(y), \quad \hat{x} \in \mathbb{S}, \tag{12.3.2}$$

where γ is defined in (12.4.3). The density $\varphi \in L^2(\Gamma)$ is determined as a solution to the far field equation (12.3.2). Then, with the knowledge of the incident field u^i the total field $u = u^i + u^s$ can be calculated via (12.3.1) and, using the boundary condition $u = 0$, we can find the unknown boundary ∂D by a local minimization problem.

Equation (12.3.2) is an integral equation of the first kind with smooth kernel. The operator mapping φ onto u^∞ is a compact operator from $L^2(\Gamma)$ into $L^2(\mathbb{S})$ or in most other standard spaces. Thus, as shown in theorem 3.1.3, the equation (12.3.2) is *ill-posed* and needs *regularization* for its stable solution. However, in general the equation will not have a solution at all. Two different approaches have been suggested to treat this difficulty.

(1) In the original setting of the potential method the domain G was chosen as a subdomain of the unknown scatterer D, see figure 12.3(a). If the equation were be solvable, then the potential (12.3.1) would define a solution to the Helmholtz equation in $\mathbb{R}^m \backslash \bar{G}$. In general such a solution does not exist, for example when D has corners, then the field may not be extensible up to Γ. Kirsch and Kress suggested combining the far field equation (12.3.2) and the search for the zeros of u into one optimization problem

$$\mu(D, \varphi) := \left\| S^\infty\varphi - u^\infty \right\|^2_{L^2(\mathbb{S})} + \left\| u^i + S\varphi \right\|^2_{L^2(\partial D)} + \alpha\|\varphi\|^2, \tag{12.3.3}$$

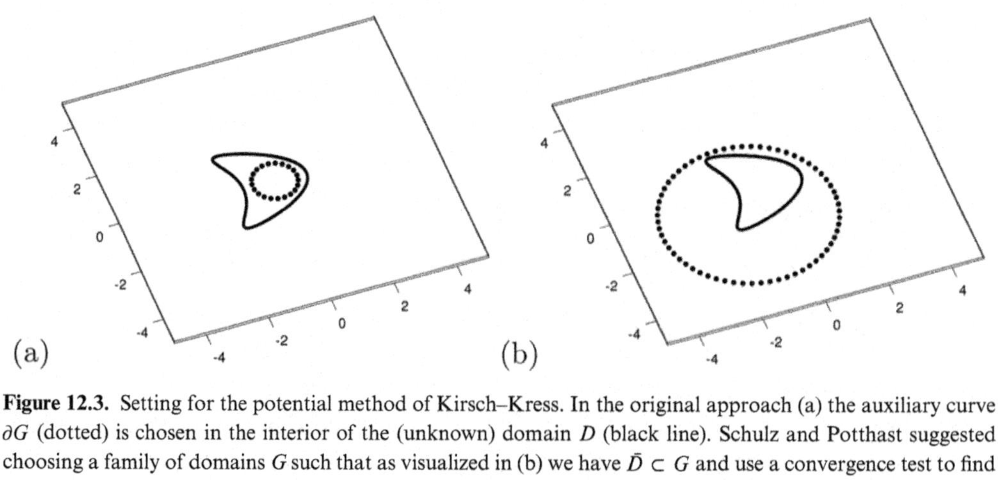

Figure 12.3. Setting for the potential method of Kirsch–Kress. In the original approach (a) the auxiliary curve ∂G (dotted) is chosen in the interior of the (unknown) domain D (black line). Schulz and Potthast suggested choosing a family of domains G such that as visualized in (b) we have $\bar{D} \subset G$ and use a convergence test to find *positive* domains for which in $\mathbb{R}^m \backslash G$ convergent field reconstructions can be obtained.

with regularization parameter α. One can prove (see [3]) that any optimal solution to this problem will represent an approximate solution to the inverse shape reconstruction problem and that for $\alpha \to 0$ under appropriate conditions convergence of a subsequence can be obtained.

(2) If we choose G to contain $\bar{D}$ in its interior as shown in figure 12.3(b), then we will see below that we obtain the solvability of the Kirsch–Kress equation (12.3.2). The search for the unknown boundary can be performed by using families of test domains G and using a *convergence test* for the solvability of the far field integral equation. This approach has been suggested by Schulz and Potthast [8].

Here, we will follow (2) and describe the realization of the potential method using a solvability test for (12.3.2) and *masking operations*. We use the set

$$\mathcal{T} := \left\{ G \subset \mathbb{R}^m : \ G \text{ is a non-vibrating domain} \right\}$$

of *admissible* test domains. Let T be a finite subset of $\mathcal{T}$ with N elements. As preparation we need the following general terminology.

Definition 12.3.1. *We call the domain* $G \in \mathcal{T}$ positive *with respect to* u^∞, *if the Kirsch–Kress equation* (*12.3.2*) *is solvable. We call it negative with respect to* u^∞, *if the equation* (*12.3.2*) *is not solvable.*

We will later study this terminology in more detail. A positive and a negative test domain are illustrated in figure 12.4. We will introduce methods for testing the solvability of (12.3.4) in section 15.1.

We are now prepared to formulate the Kirsch–Kress method based on *domain sampling* and *masking operations*.

Algorithm 12.3.2 (The Kirsch–Kress method via masking). *The* potential *or* Kirsch–Kress *method calculates an approximation to the unknown scatterer D from the far field pattern* u^∞ *on* $\mathbb{S}$ *for one or several incident waves* u^i. *For a selection M of test domains G in* $\mathcal{T}$ *we carry out the following steps.*

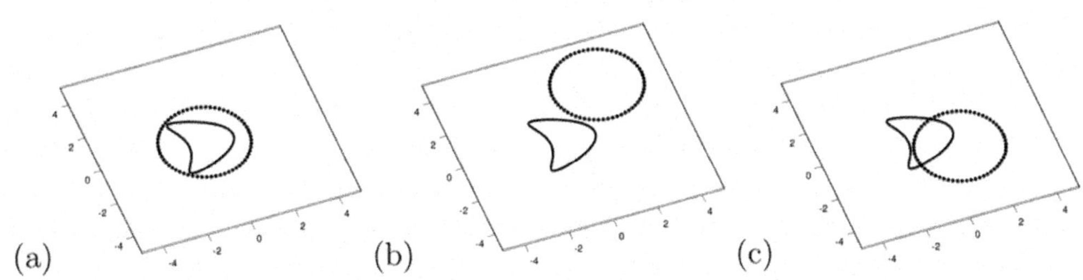

Figure 12.4. (a) A *positive* and (b) a *negative* test domain G (dotted) for scattering by some scatterer D (black line). If the test domain contains the scatterer in its interior and in its exterior, then the test domains are positive and negative, respectively. If (c) the boundary ∂G of the test domain G intersects the scattering domain D, the situation is more complicated and it depends on the extensibility of the field u^s into $(\mathbb{R}^m \backslash G) \cap D$ whether the test domain G is positive or negative. This topic leads to analytic continuation tests, which we treat in chapter 12.

A. *By using the notation* $\Gamma = \partial G$*, we first test the solvability of the equation*

$$\gamma H_\Gamma^* \varphi = u^\infty \quad \text{on } \mathbb{S} \tag{12.3.4}$$

for the test domains G under consideration, where H^* *is the adjoint of the Herglotz operator H which defines the Herglotz wave function. We obtain a set of positive test domains as defined in definition 12.3.1 and neglect negative test domains.*

An alternative is to work with arbitrary test domains *G and test convergence for the reconstruction of the* fields $u^s(x)$ *pointwise for each x and take those reconstructions for which convergence is observed, see figure 12.5, (d).*

B. *On some evaluation domain* Ω *as defined in (8.1.16) let* m_G *be a mask which is 1 on* $\Omega \backslash \bar{G}$*. We define*

$$s(x) := \sum_{G \in \mathcal{T},\ G \text{ positive}} m_G(x), \quad x \in \Omega. \tag{12.3.5}$$

The function s on Ω *counts the number of positive domains* $G \in \mathcal{T}$ *for which* $x \in \Omega$ *is in the exterior of G.*

An alternative is to define the mask by the convergence of $u_\alpha^s(x)$ *defined in C. With this mask, the following steps are identical.*

C. *For each positive domain G solve the equation (12.3.4) by calculating a regularized solution* $\varphi_{G,\alpha} \in L^2(\partial G)$ *for some* $\alpha > 0$*. Then evaluate the single-layer potential*

$$\left(S\varphi_{G,\alpha}\right)(x) = \int_\Gamma \Phi(x, y)\varphi_{G,\alpha}(y)\, \mathrm{d}s(y), \quad x \in \Omega. \tag{12.3.6}$$

An approximation of u^s *in* $\Omega \backslash \bar{G}$ *is given by*

$$u_{G,\alpha}^s(x) := \left(S\varphi_{G,\alpha}\right)(x) \cdot m_G(x), \quad x \in Q. \tag{12.3.7}$$

The function values $u_{G,\alpha}^s(x)$ *are zero for* $x \in \bar{G} \cap \mathcal{G}$*.*

D. *Calculate a full approximation of* u^s *by*

$$u_{\text{rec},\alpha}^s(x) := \frac{1}{\max\left\{s(x),\, 1\right\}} \sum_{G \in \mathcal{T},\ G \text{ positive}} u_{G,\alpha}^s(x) \tag{12.3.8}$$

for $x \in \Omega$*. Then* $u_{\text{rec},\alpha}^s$ *is zero on*

$$V := \left\{x \in \Omega :\ s(x) = 0\right\}. \tag{12.3.9}$$

E. *Finally, we search for the unknown boundary* ∂D *as the set of points where* $u_\alpha^s + u^i$ *is zero.*

For testing the solvability of the equation (12.3.4) or the convergence of the reconstruction of u^s we follow Erhard [9] and calculate

$$\mu(\alpha) := \|\varphi_{\alpha/2} - \varphi_\alpha\| \quad \text{or} \quad \mu_x^s(\alpha) := |u_{\alpha/2}^s(x) - u_\alpha^s(x)|, \tag{12.3.10}$$

which is a version of the *range test* and will be analyzed in section 15.1. Here, we follow some simple *thresholding approach*:

- If $\mu(\alpha) \leqslant c$ with some carefully chosen constant c we conclude that the equation is solvable,
- if $\mu(\alpha) > c$ we conclude that it is not solvable.

We need to solve equation (12.3.4) for many different test domains G. However, when translations of some reference domain G_0 are used, we can employ the following result.

Theorem 12.3.3. *Consider a domain $G = G(z) := G_0 + z$ arising from a reference domain G_0 by translation with translation vector $z \in \mathbb{R}^m$ and denote the far field operator for G_0 by H_0^*. Then the solution of (12.3.4) for G can be calculated by*

$$H_0^* \varphi = M_z u^\infty, \tag{12.3.11}$$

with the multiplication operator

$$M_z u^\infty(\hat{x}) = e^{ik\theta_k \cdot z} u^\infty(\hat{x}), \quad \hat{x} \in \mathbb{S}, \tag{12.3.12}$$

leading to the regularized solution

$$\varphi_{\alpha,z} = R_{0,\alpha} M_z u^\infty, \quad z \in Q \tag{12.3.13}$$

with $R_{0,\alpha} := (\alpha I + H_0 H_0^*)^{-1} H_0$.

Proof. A translation of the domain G_0 by $y = y_0 + z$ leads to

$$\begin{aligned} u^\infty(\hat{x}) &= \int_{\partial G} e^{-i\kappa \hat{x}\cdot y} \varphi(y) \, ds(y) \\ &= e^{-i\kappa \hat{x}\cdot z} \int_{\partial G_0} e^{-i\kappa \hat{x}\cdot y_0} \varphi(y_0 + z) \, ds(y_0), \end{aligned} \tag{12.3.14}$$

such that

$$e^{i\kappa \hat{x}\cdot z} u^\infty(\hat{x}) = \int_{\partial G_0} e^{-i\kappa \hat{x}\cdot y_0} \tilde{\varphi}(y_0) \, ds(y_0). \tag{12.3.15}$$

with $\tilde{\varphi}(y_0) = \varphi(y_0 + z)$ for $y_0 \in \partial G_0$. For simplicity write $A = H^*$ and denote A by A_0 if G is replaced by G_0. Then we have $A = M_z^* A_0$ with $M_z M_z^* = I$, such that $A^* A = A_0^* A_0$ and

$$\begin{aligned} R_\alpha &= (\alpha I + A^* A)^{-1} A^* \\ &= \left(\alpha I + A_0^* A_0\right)^{-1} A_0^* M_z \\ &= R_{0,\alpha} M_z, \end{aligned} \tag{12.3.16}$$

which proves (12.3.13).

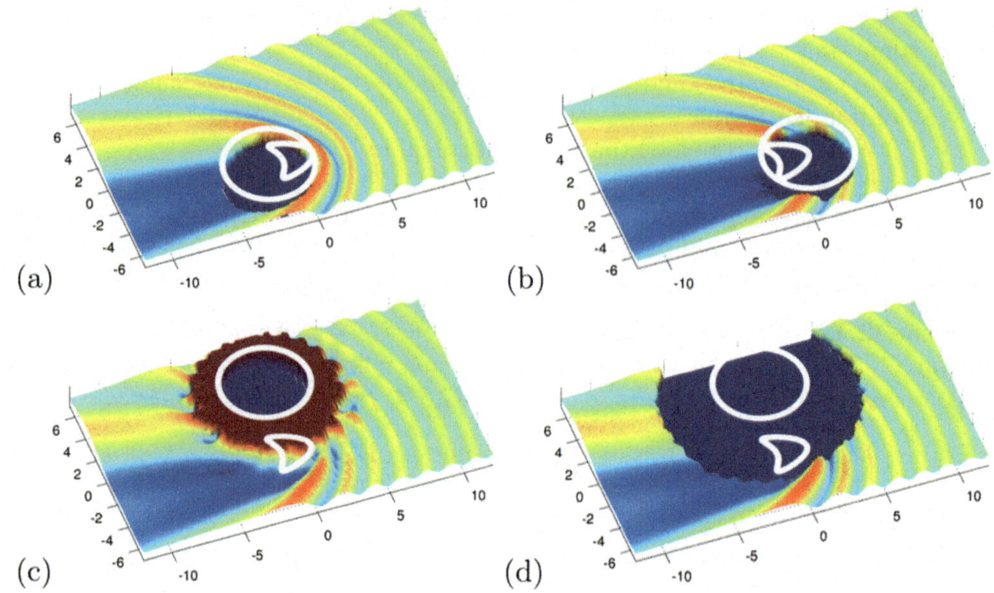

Figure 12.5. Reconstructions for different locations of a test domain for the Kirsch–Kress method. We have convergence in (a) and (b). (c) visualizes a negative test domain with large errors in the field reconstructions in some parts of the exterior $\mathbb{R}^m \backslash G$ of G. In (d) we apply a masking based on the convergence test for the reconstruction of u_α^s. The reconstructions are analogous to those of the point source methods, see figure 12.8.

If we employ a sufficiently chosen family of test domains such that $\mathbb{R}^m \backslash G$ covers the exterior of the scatterer D, we obtain $V \subset \bar{D}$ and

$$u_\alpha^s \to u^s \ \text{ on } Q \backslash \bar{D}. \tag{12.3.17}$$

Section 8.2 shows how to solve the scattering problem and section 8.3 how to calculate some far field pattern. We describe the boundary of circular test domains G by (8.1.6) with $z \in \mathbb{R}^m$. Of course, test domains G do not need to be circular, for example corresponding masks m_G are shown in figure 6.7. The construction of the discrete mask m_G has been described in (6.6.1) and (6.6.2).

We demonstrate the convergence test and the masking technique for the Kirsch–Kress method in a series of figures. First, we show four reconstructions with one fixed test domain each in figure 12.5. In panels (a) and (b) we show two reconstructions if $D \subset G$ and the Kirsch–Kress equation is solvable. Panels (c) and (d) show the results when the equation is not solvable. Clearly, the reconstructed field in (c) has large errors. In (d) we employ Erhard's convergence test for the reconstruction of u^s and construct a mask set to zero where we do not have convergence.

Code 12.3.4. *Script* `sim_12_3_4_c_KK.m` *to carry out the field reconstruction by the Kirsch–Kress method. Run script* `sim_12_3_4_a_scattering_ff.m` *first to generate* u^∞*, here named* `ff` and *some further variables.* `sim_12_3_4_b_graphics.m` *visualizes the total field on some evaluation domain* Ω*. The further* *parts of figure* 12.5 *with the field reconstructions are generated by* `sim_12_3_4_d_graphics.m`.

```
% I Preparations
NG = 120;                           % number of points on test dom G
z1 = 0;                             % center of test domain G, comp.1
z2 = 3;                             % center of test domain G, comp.2
hG = 2*pi/NG;                       % grid constant for test domain
tG = 0:hG:2*pi-hG;                  % parametrization grid for G
RG = 3;                             % radius of test domain
yG1 = RG*cos(tG)+z1;                % boundary test domain comp.1
yG2 = RG*sin(tG)+z2;                % boundary test domain comp.2

yGffmat1  = repmat(yff1.',1,NG);    % matrix of ff points comp.1
yGffmat2  = repmat(yff2.',1,NG);    % matrix of ff points comp.2
yGmat1    = repmat(yG1,ffN,1);      % matrix of points of G comp.1
yGmat2    = repmat(yG2,ffN,1);      % matrix of points of G comp.1

% II Potential Operators ffSG for potential on test domain G
ffSG  = fac*exp(-i*kappa*(yGffmat1.*yGmat1+yGffmat2.*yGmat2))*RG*hG;

% III reconstruct the density for far field representation
alphaG = 1e-11;                     % regularization parameter
varphiKK = (alphaG*eye(NG,NG) + ffSG'*ffSG)\ffSG'*ff; % density KK

% matrix of the norm differences of the grid to the curve points:
epsmat   = eps*ones(M,NG);          % matrix for cutting singularity
rGmat1   = repmat(pvec1,1,NG)-repmat(yG1,M,1);
rGmat2   = repmat(pvec2,1,NG)-repmat(yG2,M,1);
rGmat    = max(sqrt(rGmat1.^2 + rGmat2.^2),epsmat);

% IV Potential operator tSG - test domain boundary into far field
tSG  = i/4*besselh(0,1,kappa*rGmat)*RG*hG;

% V Calculation of reconstructed scattered, total fields and error
wsKK   = tSG*varphiKK;              % reconstructed scattered field
wKK    = wi + wsKK;                 % reconstructed total field
wKKerr = wSim-wKK;                  % error for reconstruction

% V convergence test for Kirsch-Kress method
alphaG = (1e-11)/2;                 % reg parameter for testing conv
varphiKK2 = (alphaG*eye(NG,NG) + ffSG'*ffSG)\ffSG'*ff; % density2

% VI Masking operations

 wsKK2  = tSG*varphiKK2;            % reconstructed scattered field 2
 maskG = sqrt( (pvec1-z1).^2 + (pvec2-z2).^2)>RG; % mask circle
 mask2 = abs(wsKK-wsKK2)<.02;     % mask via testing convergence
 mask3 = abs(wKKerr.*maskG)<.15; % mask via size of error
```

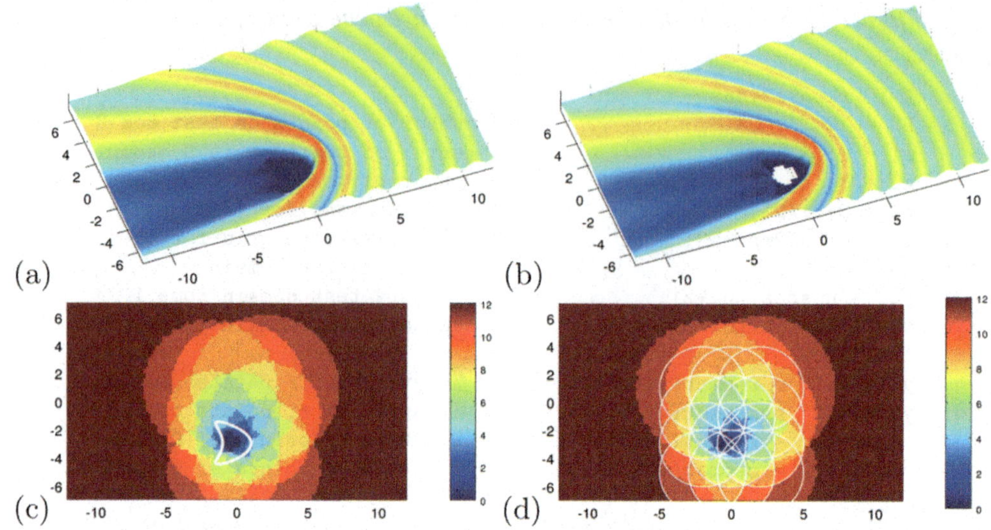

Figure 12.6. The simulated total field for scattering by some domain D (a) and its reconstruction by the Kirsch–Kress method with masking (b), for a setting with radius $r = 3$ of the test domains G as in figure 12.5, wave number $\kappa = 2$ and tests with a total of 12 test domains with center coordinates $x_1 \in \{-2, 0, 2\}$ and $x_2 \in \{-5, -3, -1, 1\}$. The test domains are displayed in (d), the function $s(x)$ as defined by the alternative to (12.3.5) in figure (c). The panels are analogous to figure 12.9.

Next, we study the combination of different reconstructions following the masking and summation technique (12.3.8), leading to a full field reconstruction in the exterior $\mathbb{R}^m \backslash D$ of D. The simulated and the reconstructed total field are shown in figure 12.6. The result shows that for a convex scatterer the methods provide a very satisfactory field reconstruction. The corresponding programs are given by `sim_12_3_5_a_KK_full.m`, based on `sim_12_3_4_a_scattering_ff.m`.

We end with two theoretical results on the convergence of the Kirsch–Kress method with the masking technique.

Lemma 12.3.5. *Consider a scatterer $D \subset Q$. We assume that the set M of test domains G is sufficiently rich such that*

$$\bar{D} = \bigcap_{G \in \mathcal{T},\, D \subset G} G. \tag{12.3.18}$$

Then the set V defined in (12.3.9) as the set of points where we do not obtain some convergent reconstruction of u^s is a subset of $\bar{D}$.

Proof. We will show that $x \notin \bar{D}$ implies $x \notin V$, which yields $V \subset \bar{D}$. Consider a point $x \in \Omega \backslash \bar{D}$. Then according to the condition (12.3.18) there is a domain $G \in \mathcal{T}$ with $\bar{D} \subset G$ and $x \notin G$. Since $\bar{D} \subset G$ as we will show in the framework of the range test lemma 15.1.1 the equation (12.3.4) is solvable and thus G is positive. But this yields $x \notin V$ and the proof is complete.

Theorem 12.3.6. *Under the conditions of lemma (12.3.5) we have the convergence*

$$u_\alpha^s(x) \to u^s(x), \quad \alpha \to 0, \tag{12.3.19}$$

for each $x \in \Omega \backslash V$, i.e. the Kirsch–Kress method with masking defines a convergent method for the reconstruction of u^s from u^∞.

Proof. First, we consider one fixed positive test domain $G \in M$. Then the equation (12.3.4) does have a solution which yields

$$u^s(x) = \int_\Gamma \Phi(x, y)\varphi(y)\, ds(y) \quad \text{in} \quad \mathbb{R}^m \backslash \bar{G} \tag{12.3.20}$$

by the Rellich uniqueness lemma. Now, the results of regularization theory apply and we obtain $\varphi_\alpha \to \varphi$ for $\alpha \to 0$ in $L^2(\Gamma)$ due to our choice of a convergent regularization scheme for the far field equation (12.3.4). This yields

$$\int_\Gamma \Phi(x, y)\varphi_\alpha(y)\, ds(y) \to \int_\Gamma \Phi(x, y)\varphi(y)\, ds(y) = u^s(x), \quad \alpha \to 0 \tag{12.3.21}$$

for each fixed $x \in \mathbb{R}^m \backslash \bar{G}$. In the second step consider the finite sum (12.3.8). Since we have convergence for each of the non-zero terms of $s(x)$ towards $u^s(x)$, the total sum converges towards

$$\begin{aligned} u_\alpha^s &\to \frac{1}{s(x)} \sum_{G \in \mathcal{T},\ G \text{ positive}} u^s(x) \cdot m_G(x) \\ &= u^s(x) \cdot \frac{1}{s(x)} \left(\sum_{G \in \mathcal{T},\ G \text{ positive}} m_G(x) \right) \\ &= u^s(x), \quad x \in \Omega \backslash V, \end{aligned} \tag{12.3.22}$$

for $\alpha \to 0$. □

12.4 The point source method

The point source method is a scheme based on *Green's formula* and a *point source approximation* to reconstruct a scattered wave u^s from measurements of its far field pattern u^∞ or from measurements $u^s|_\Lambda$ of the wave on some measurement surface Λ. It has been introduced since 1996 in a series of papers [1, 10, 11], see also [12]. We will see that it is a dual method with respect to the Kirsch–Kress scheme. To explain the basic idea we start with Green's formula

$$u^s(z) = \int_{\partial D} \left(\Phi(z, y)\frac{\partial u^s}{\partial \nu}(y) - \frac{\partial \Phi(z, y)}{\partial \nu(y)} u^s(y) \right) ds(y) \tag{12.4.1}$$

for $z \in \mathbb{R}^m \backslash \bar{D}$. Then, the far field pattern u^∞ of u^s is given by

$$u^\infty(\hat{x}) = \gamma \int_{\partial D} \left(e^{-i\kappa \hat{x} \cdot y} \frac{\partial u^s}{\partial \nu}(y) - \frac{\partial e^{-i\kappa \hat{x} \cdot y}}{\partial \nu(y)} u^s(y) \right) ds(y), \quad \hat{x} \in \mathbb{S}, \tag{12.4.2}$$

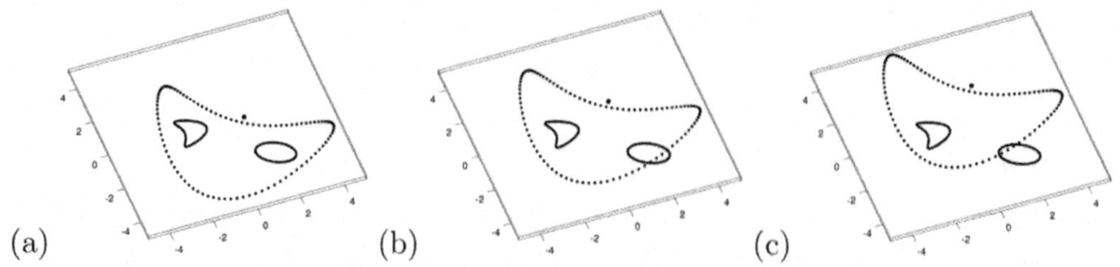

Figure 12.7. The setting for the point source method, which can be chosen in the same way as for the Kirsch–Kress method. Here we display a non-convex approximation domain G (dotted line). The figure shows a scatterer D (black line) with two separate components, also see figure 12.4. We will obtain convergence for an approximation domain G which contains the scatterer D in its interior as in (a). In the cases (b) and (c) we need convergence criteria and masking. The goal is to calculate the field $u(z)$ in a point $z \in \mathbb{R}^m \setminus \bar{G}$ (black dot).

with the constant

$$\gamma := \begin{cases} \dfrac{e^{i\pi/4}}{\sqrt{8\pi\kappa}}, & m = 2, \\ \dfrac{1}{4\pi}, & m = 3. \end{cases} \tag{12.4.3}$$

Let $G \subset \mathbb{R}^m$ denote some test domain with a boundary of class C^2 as visualized in figure 12.7. We assume that G is non-vibrating for the wave number κ, i.e. the homogeneous interior Dirichlet problem for G has only the trivial solution $u \equiv 0$. Further, let $z \in \mathbb{R}^m \setminus \bar{G}$ be some evaluation point.

Next, we study a superposition of plane waves, i.e. the Herglotz wave functions (8.3.12), on the domain G. Below we show that the set of superpositions Hg of plane waves with densities $g \in L^2(\mathbb{S})$ can approximate solutions to the Helmholtz equation on G up to an arbitrary precision. In particular, we can find a density $g \in L^2(\mathbb{S})$ such that

$$\Phi(\cdot, z) \approx Hg \tag{12.4.4}$$

in $L^2(\partial G)$ in the sense that given $\epsilon > 0$ there is $g \in L^2(\mathbb{S})$ such that

$$\left\| \Phi(\cdot, z) - Hg \right\|_{L^2(\partial G)} \leqslant \epsilon \tag{12.4.5}$$

(see theorem 17.7.3). The domain G is the *approximation domain* for the point source method and we say that g solves the *point source equation* (12.4.4) *with discrepancy* ϵ *on* ∂G if the estimate (12.4.5) is satisfied.

We now approximate the *point source* $\Phi(z, y) = \Phi(y, z)$ in Green's representation formula (12.4.1) by the superposition of plane waves Hg. This yields

$$\begin{aligned} u^s(z) &= \int_{\partial D} \left(\Phi(z, y) \frac{\partial u^s}{\partial \nu}(y) - \frac{\partial \Phi(z, y)}{\partial \nu(y)} u^s(y) \right) \mathrm{d}s(y) \\ &\approx \gamma \int_{\partial D} \int_{\mathbb{S}} \left(e^{i\kappa y \cdot d} \frac{\partial u^s}{\partial \nu}(y) - \frac{\partial e^{i\kappa y \cdot d}}{\partial \nu(y)} u^s(y) \right) g(d) \, \mathrm{d}s(d) \, \mathrm{d}s(y). \end{aligned} \tag{12.4.6}$$

An exchange of the order of integration and use of (12.4.2) with $-\hat{x} = d$ yields

$$u^s(z) \approx \int_{\mathbb{S}} u^\infty(-d)g(d)\,\mathrm{d}s(d) \tag{12.4.7}$$

$$= \int_{\mathbb{S}} u^\infty(d)g(-d)\,\mathrm{d}s(d). \tag{12.4.8}$$

This is a reconstruction formula or *back-projection formula* for the calculation of u^s from its far field pattern u^∞. We have presented the basic steps which lead to the following algorithm.

Algorithm 12.4.1 (Point source method). *The point source method calculates a reconstruction of the scattered field u^s for scattering by some scatterer D from its far field pattern u^∞ by the following steps.*

1. *Choose some domain G and a point $z \not\in \bar{G}$.*
2. *Calculate some density $g_\alpha \in L^2(\mathbb{S})$, $\alpha > 0$, such that (12.4.4) is satisfied.*
3. *Then use the* back-projection formula *(12.4.8) to calculate an approximation u_α^s to the scattered field u^s.*
4. *Test the convergence of the reconstruction u_α^s by an application of a convergence test as in (12.3.10).*
5. *For different test domains G, set the mask to one where the reconstruction of u^s is convergent and combine the reconstructions as in (12.3.8).*

We will see that the point source method is convergent for all settings of scatterers D, approximation domains G and points z where $\bar{D} \subset G$ and $z \not\in \bar{G}$ is satisfied. If D is not a subset of the approximation domain, in general the back-projection formula will *not* calculate the scattered field u^s, but convergence can appear under certain conditions. Here, the application of a convergence test in step 4 of the point source method is an important part of the algorithm.

Before we investigate the convergence of the point source method in more detail, let us carry out some reconstructions. Here, we choose the same set-up as for the Kirsch–Kress method in code 12.3.4. For the numerical solution we apply Nyström's method to the equation

$$(\alpha I + H^*H)g_\alpha = H^*f, \tag{12.4.9}$$

to calculate (7.3.11). The operator H is discretized by (8.3.15), its adjoint H^* by (8.3.16).

Code 12.4.2. *Script* `sim_12_4_2_c_PSM.m` *to carry out the field reconstruction by the point source method. Run script* `sim_12_4_2_a_scattering_ff.m` *first to generate u^∞, here named* `ff` *and some further variables.* `sim_12_4_2_b_graphics.m` *visualizes the total field on some evaluation domain Q. The further parts of figure 12.8 are generated by script* `sim_12_4_2_d_graphics.m`.

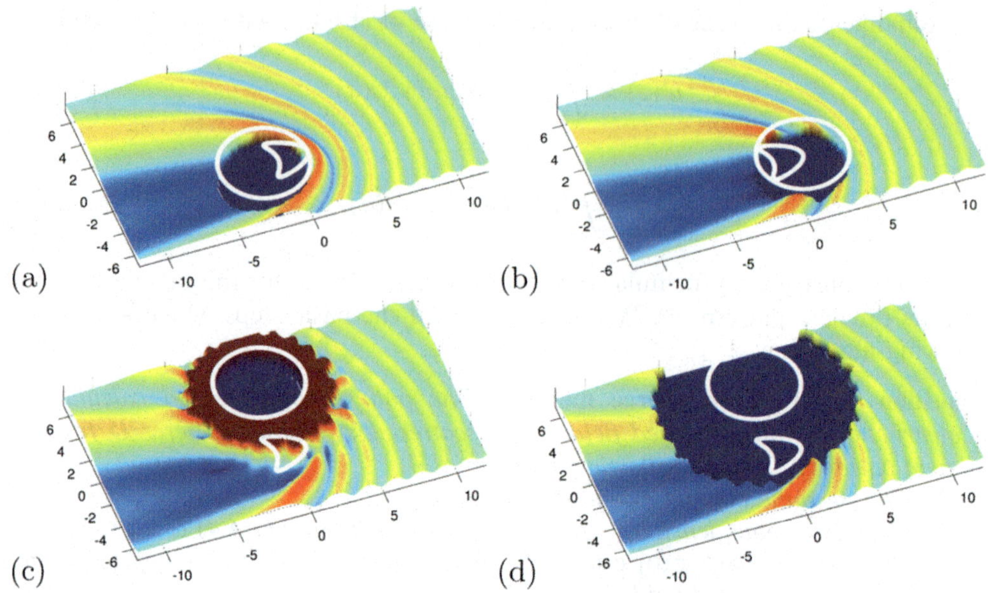

Figure 12.8. Reconstructions for different locations of a test domain for the point source method. We have convergence for $z \in \mathbb{R}^m \backslash \bar{G}$ in (a) and (b). (c) visualizes a test domain with large reconstruction errors in the field reconstructions in some parts of the exterior $\mathbb{R}^m \backslash G$ of G. In (d) we apply a masking based on the convergence test for the reconstruction of u_α^s. The reconstructions are analogous to those of the Kirsch–Kress method, see figure 12.5. In section 12.5 we prove the duality of both methods.

```
% I Preparations
NG = 120;                          % number of points on test dom G
z1 = 0;                            % center of test domain G, comp.1
z2 = 3;                            % center of test domain G, comp.2
hG = 2*pi/NG;                      % grid constant for test domain
tG = 0:hG:2*pi-hG;                 % parametrization grid for G
RG = 3;                            % radius of test domain
yG1 = RG*cos(tG)+z1;               % boundary of test domain comp.1
yG2 = RG*sin(tG)+z2;               % boundary of test domain comp.2

yGffmat1  = repmat(yff1,NG,1);    % matrix of ff points comp.1
yGffmat2  = repmat(yff2,NG,1);    % matrix of ff points comp.2
yGmat1    = repmat(yG1.',1,ffN);      % mat of points of G comp.1
yGmat2    = repmat(yG2.',1,ffN);      % mat of points of G comp.1

% II Herglotz operator for evaluation on test domain G
HG   = exp(i*kappa*(yGffmat1.*yGmat1+yGffmat2.*yGmat2))*hff;

% III Point Source Matrix for Evaluation
epsmat    = eps*ones(M,NG);        % matrix for cutting singularity
rGmat1    = repmat(pvec1,1,NG)-repmat(yG1,M,1);
rGmat2    = repmat(pvec2,1,NG)-repmat(yG2,M,1);
rGmat     = max(sqrt(rGmat1.^2 + rGmat2.^2),epsmat).';
PhiMatC   = conj(i/4*besselh(0,1,kappa*rGmat));
```

```
% IV Reconstruct the density for far field representation
alphaPSM = 1e-9;                    % regularization parameter
gPSM  = (alphaPSM*eye(ffN,ffN) + HG'*HG)\HG'*PhiMatC*hff;

% V Calculation of reconstructed scattered, total fields, error
wsPSM    = 1/fac*(gPSM')*ff;          % reconstructed scattered field
wPSM     = wi + wsPSM;                % reconstructed total field
wPSMerr = wSim-wPSM;                  % error for reconstruction

% V Convergence test for point source method
alphaPSM = (1e-9)/2;                 % reg parameter for testing conv
gPSM       = (alphaPSM*eye(ffN,ffN) + HG'*HG)\HG'*PhiMatC*hff;
           % density PSM for testing convergence

% VI Masking operations
hE      = 0.15; % mask cut parameter for error
hConv   = 0.02; % mask cut parameter for testing convergence
wsPSM2 = 1/fac*(gPSM')*ff;     % reconstructed scattered field
maskG   = sqrt( (pvec1-z1).^2 + (pvec2-z2).^2)>RG; % mask circle
mask2   = abs(wsPSM-wsPSM2)<hConv;  % mask via testing conv
mask3   = abs(wPSMerr.*maskG)<hE;   % mask via size of error
```

The point source method is based on the approximate solution of the *point source equation*

$$Hg = \Phi(\cdot, z) \quad \text{on} \quad \partial G \tag{12.4.10}$$

with the *Herglotz wave operator* H from $L^2(\mathbb{S})$ into $L^2(\partial G)$. We first show that this equation does not have a solution, but that the operator H has dense range from $L^2(\mathbb{S})$ into $L^2(\Gamma)$ and an approximate solution to the equation is obtained by *Tikhonov regularization*

$$g_\alpha = (\alpha I + H^*H)^{-1}H^*\Phi(\cdot, z), \quad \alpha > 0. \tag{12.4.11}$$

Here, $\alpha > 0$ is the *regularization parameter* for the equation. In (3.1.28) it has been shown that (12.4.11) is equivalent to the minimization of the functional

$$\mu_\alpha(g) := \|Hg - \Phi(\cdot, z)\|^2_{L^2(\partial G)} + \alpha\|g\|^2_{L^2(\Gamma)}. \tag{12.4.12}$$

Lemma 12.4.3 (Point source equation). *If the test domain G is non-vibrating, the operator $H : L^2(\mathbb{S}) \to L^2(\partial G)$ is injective and has a dense range. We have $\mu_\alpha(g_\alpha) \to 0$ for $\alpha \to 0$ and for each $\epsilon > 0$ equation* (12.4.10) *has an approximate solution g_α such that*

$$\|Hg_\alpha - \Phi(\cdot, z)\|^2_{L^2(\partial G)} \leqslant \epsilon, \tag{12.4.13}$$

where g_α is given by (12.4.11) with $\alpha > 0$ chosen sufficiently small.

Proof. Let $g \in L^2(\mathbb{S})$ be a function with $Hg = 0$ on ∂G. Then the Herglotz wave function v_g defined in (8.3.11) satisfies the Helmholtz equation in G and since G is non-vibrating it is identical to zero in G. This yields $g = 0$, and H is injective. Next, assume that for the adjoint H^* we have $H^*\varphi = 0$ with some $\varphi \in L^2(\partial G)$. Then the single-layer potential $u^s := S\varphi$ has far field $u^\infty = 0$, and by the Rellich lemma we obtain $S\varphi = 0$ in $\mathbb{R}^m \backslash G$. By continuity of the single-layer potential we have $S\varphi = 0$ on ∂G and since G is non-vibrating $\tilde{S}\varphi = 0$ in G. Now, the jump-relations for the normal derivative of the single-layer potential with L^2-densities yield $\varphi = 0$, i.e. H^* is injective and thus H has a dense range. We remark that there are various versions of this result, since to argue using the jump relations we need to be very careful about the regularity of the functions. There are nice arguments based on the Fredholm alternative in different spaces, we will provide this version in theorem 17.7.3 in the appendix.

If we show that for the minimizer g_α of (12.4.12) we have $\mu_\alpha(g_\alpha) \to 0$ for $\alpha \to 0$, we obtain the statement (12.4.13) as an immediate consequence. Given $\epsilon > 0$ by denseness of $R(H)$ we can choose $g^{(\epsilon)} \in L^2(\mathbb{S})$ such that

$$\|Hg^{(\epsilon)} - \Phi(\cdot, z)\|_{L^2}^2 \leqslant \frac{\epsilon}{2}.$$

Now, choosing $\alpha > 0$ such that $\alpha\|g^{(\epsilon)}\|^2 \leqslant \epsilon/2$ we obtain the estimate

$$\|Hg^{(\epsilon)} - \Phi(\cdot, z)\|_{L^2(\partial G)}^2 + \alpha\|g^{(\epsilon)}\|_{L^2(\mathbb{S})}^2 \leqslant \epsilon,$$

and the estimate is satisfied for all sufficiently small α. Then, the same estimate is also satisfied for the minimizer g_α of the functional μ_α, which yields $\mu_\alpha(g_\alpha) \to 0$ and thus (12.4.13). □

Convergence of the point source method is based on the following result.

Theorem 12.4.4 (Convergence of the point source method). *Assume that $\bar{D} \subset G$ and z in $\mathbb{R}^m \backslash \bar{G}$ with a test domain G which is non-vibrating. Then the field reconstruction of the point source methods is convergent in the sense that*

$$\left| u^s(z) - \int_{\mathbb{S}} u^\infty(d) g_{z,\alpha}(-d)\, \mathrm{d}s(d) \right| \to 0 \tag{12.4.14}$$

for $\alpha \to 0$, where $g_{z,\alpha}$ is the solution to equation (12.4.11).

Proof. By our previous theorem we know that $Hg_\alpha \to \Phi(\cdot, z)$ on ∂G in $L^2(\partial G)$. The solution to the interior Dirichlet problem for G depends continuously on the boundary data in $L^2(\partial G)$, such that we obtain convergence uniformly on compact subsets $\bar{D}$ of G.

We now argue as in (12.4.6), estimating the difference between u^s and u_α^s. Since the convergence is uniform on ∂D, we obtain (12.4.14) for $\alpha \to 0$.

Efficient speed-up of density calculations. The equation (12.4.10) needs to be solved for every domain G and every point z under consideration. The following *symmetry argument* can be employed to make this very efficient. A translation of the Herglotz wave function

$$v[g](x) = \int_{\mathbb{S}} e^{i\kappa x \cdot d} g(d)\, \mathrm{d}s(d) \tag{12.4.15}$$

by a vector $z \in \mathbb{R}^m$, $m = 2, 3$, can be carried out by multiplication of the density g with the complex function

$$f_z(d) := e^{-i\kappa z \cdot d}, \quad d \in \mathbb{S}. \tag{12.4.16}$$

Using the multiplication operator M_{f_z} defined in (6.6.12) we derive this from

$$\begin{aligned} v\left[M_{f_z} g\right](x) &= \int_{\mathbb{S}} e^{i\kappa x \cdot d}\left(f_z(d) \cdot g(d)\right) ds(d) \\ &= \int_{\mathbb{S}} e^{i\kappa x \cdot d} e^{-i\kappa z \cdot d} \cdot g(d)\, ds(d) \\ &= \int_{\mathbb{S}} e^{i\kappa (x-z) \cdot d} \cdot g(d)\, ds(d) \\ &= v[g](x - z) \end{aligned} \tag{12.4.17}$$

for $x, z \in \mathbb{R}^m$. We also translate a domain G by the vector z, from which we obtain a translated domain

$$G_z := G + z = \left\{x + z, \quad x \in G\right\} \subset \mathbb{R}^m. \tag{12.4.18}$$

Now we translate both the domain, the right-hand side $\Phi(\cdot, 0)$ and the Herglotz wave function Hg by a vector $z \in \mathbb{R}^m$. Then the translated Herglotz wave function approximates the translated right-hand side on the translated domain. Thus for any $z \in \mathbb{R}^m$ we can calculate solutions of (12.4.10) on ∂G_z by application of a multiplication operator. These results are summarized in the following lemma 12.4.5.

Lemma 12.4.5. *Consider the point $z = 0$, a non-vibrating domain G_0 for which $0 \notin \bar{G}_0$ and a solution $g^{(0)}$ to (12.4.10) with discrepancy ϵ on ∂G_0 in the sense of (12.4.5). Then for any $z \in \mathbb{R}^m$ the density $M_{f_z} g^{(0)} \in L^2(\mathbb{S})$ solves (12.4.10) with discrepancy ϵ on ∂G_z.*

With the above lemma we can now reformulate the calculation of the densities g_z and the evaluation of the Herglotz wave operator into the formula

$$u_\alpha^s(z) = \int_{\mathbb{S}} e^{i\kappa z \cdot \theta}\left(g_\alpha^{(0)}(\theta) \cdot u^\infty(-\theta)\right) ds(\theta), \quad z \in \mathbb{R}^m, \tag{12.4.19}$$

where $g^{(0)}$ solves the point source equation (12.4.10) for G_0 and $z = 0$.

The reconstruction of the *scatterer D* by the *point source method* is carried out by searching for the zero curve $\{x | \, u_\alpha(x) = 0\}$ of the reconstructed total field u. According to theorem 12.4.4, the field u_α is convergent to the true total field u on compact subsets of the exterior $\mathbb{R}^m \backslash \bar{D}$. We can use this convergence to derive the corresponding convergence of the domain reconstructions.

Theorem 12.4.6. *Assume that we are given a set of incident waves u_ξ^i for $\xi = 1, \ldots, L$ such that the sum of the absolute values of the corresponding total fields u_ξ is non-zero on $\mathbb{R}^m \backslash \bar{D}$. Then by*

$$I_\alpha(x) := \left(\sum_{\xi=1}^{L} |u_\alpha(x)|\right)^{-1}, \quad I(x) := \left(\sum_{\xi=1}^{L} |u(x)|\right)^{-1} \tag{12.4.20}$$

with the reconstruction u_α of the total field by the point source method an indicator function is defined. In this case the domain reconstruction by the point source method for the case of the Dirichlet boundary condition is locally convergent in the Hausdorff norm.

Proof. The theorem is a direct consequence of definition 10.6.3 together with theorem 10.6.4. □

12.5 Duality and equivalence for the potential method and the point source method

We have introduced the idea of reconstructing an unknown scattered field u^s by a *potential approach* in section 12.3 and by Green's formula and the point source method in section 12.4. At first sight both ideas seem to be independent, one approach involves the Kirsch–Kress equation (12.3.2), the other the point source equation (12.4.10). However, if we study the reconstruction results of figure 12.6 in comparison to figure 12.9, the similarity of the reconstructions is obvious.

Here we will present a duality principle for the point source method and the Kirsch–Kress method first proven in [1]. We will show that the two methods are *dual* to each other in the sense that they solve dual equations with respect to the L^2 scalar products on $\mathbb{S}$ and ∂G and can be transformed into each other by the classical operation of passing to the adjoint operator. As a consequence we prove that the

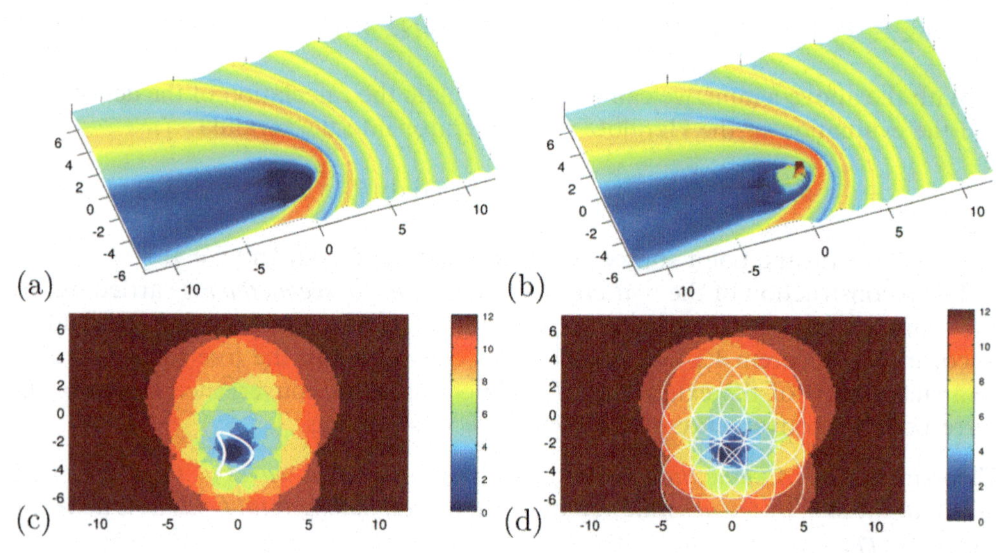

Figure 12.9. The simulated total field for scattering by some domain D (a) and its reconstruction by the point source method with masking (b), for a setting with radius $r = 3$ of the test domains G as in figure 12.8, wave number $\kappa = 2$ and tests with a total of 12 test domains with center coordinates $x_1 \in \{-2, 0, 2\}$ and $x_2 \in \{-5, -3, -1, 1\}$. The test domains are displayed in (d), the function $s(x)$ as defined by the alternative to (12.3.5) in figure (c). The images are analogous to figure 12.6, see section 12.5.

reconstruction quality of both methods is equivalent when carried out with the same geometrical set-up.

Theorem 12.5.1 (Duality and equivalence). *The Kirsch–Kress method and the point source method are dual to each other in the sense that their solution operators are dual with respect to the L^2scalar products on $L^2(\mathbb{S})$ and $L^2(\partial G)$ for a test domain G. In particular, for the regularized reconstructions we have*

$$u^s_{\mathrm{KK},\alpha}(z) = u^s_{\mathrm{PSM},\alpha}(z) \tag{12.5.1}$$

for $z \in \mathbb{R}^m \backslash \bar{G}$.

Proof. For one given test domain G we write the regularized version of the Kirsch–Kress method in the form

$$u^s_{\mathrm{KK},\alpha} = \tilde{S}(\alpha I + S^{\infty,*}S^{\infty})^{-1}S^{\infty,*}u^{\infty}. \tag{12.5.2}$$

The point source method (12.4.8) with the regularized back-projection density g_α given by (12.4.9) can be written as

$$\begin{aligned} u^s_{\mathrm{PSM},\alpha}(z) &= \int_{\mathbb{S}} \big((\alpha I + H^*H)^{-1}H^*\Phi(\cdot, z)\big)(\hat{x})u^{\infty}(\hat{x})\, \mathrm{d}s(\hat{x}) \\ &= \big\langle(\alpha I + H^*H)^{-1}H^*\Phi(\cdot, z), u^{\infty}\big\rangle_{L^2(\mathbb{S})}, \end{aligned} \tag{12.5.3}$$

where we recall the definition of the Herglotz wave operator (8.3.12). Writing the evaluation of the single-layer operator $\tilde{S}$ explicitly, we transform (12.5.2) into

$$\begin{aligned} u^s_{KK,\alpha}(z) &= \int_{\partial G} \Phi(z, y)\big((\alpha I + S^{\infty,*}S^{\infty})^{-1}S^{\infty,*}u^{\infty}\big)(y)\, \mathrm{d}s(y) \\ &= \big\langle \Phi(\cdot, z), (\alpha I + S^{\infty,*}S^{\infty})^{-1}S^{\infty,*}u^{\infty}\big\rangle_{L^2(\partial G)} \end{aligned} \tag{12.5.4}$$

using the symmetry $\Phi(y, z) = \Phi(z, y)$ for $z, y \in \mathbb{R}^m$ of the fundamental solution (8.1.3). We note the duality of the operators

$$(Hg)(y) = \int_{\mathbb{S}} \mathrm{e}^{\mathrm{i}\kappa y\cdot\hat{x}}g(\hat{x})\, \mathrm{d}s(\hat{x}), \quad y \in \partial G$$

and

$$(S^{\infty}\psi)(\hat{x}) = \int_{\partial G} \mathrm{e}^{-\mathrm{i}\kappa\hat{x}\cdot y}\psi(y)\, \mathrm{d}s(y), \quad \hat{x} \in \mathbb{S}$$

with respect to the scalar products of $L^2(\mathbb{S})$ and $L^2(\partial G)$, which are obtained from

$$\begin{aligned} \langle\psi, Hg\rangle_{L^2(\partial G)} &= \int_{\partial G} \psi(y)\overline{\left(\int_{\mathbb{S}} \mathrm{e}^{\mathrm{i}\kappa y\cdot\hat{x}}g(\hat{x})\, \mathrm{d}s(\hat{x})\right)}\, \mathrm{d}s(y) \\ &= \int_{\mathbb{S}} \left(\int_{\partial G} \mathrm{e}^{-\mathrm{i}\kappa y\cdot\hat{x}}\psi(y)\, \mathrm{d}s(\hat{y})\right)\overline{g(\hat{x})}\, \mathrm{d}s(\hat{x}) \\ &= \langle S^{\infty}\psi, g\rangle_{L^2(\mathbb{S})}, \end{aligned} \tag{12.5.5}$$

i.e. $H^* = S^\infty$ and $S^{\infty,*} = H$. Now, the duality of the Kirsch–Kress method and the point source method is a consequence of

$$\begin{aligned} \left((\alpha I + S^{\infty,*}S^{\infty})^{-1}S^{\infty,*}\right)^* &= H^*(\alpha I + HH^*)^{-1} \\ &= (\alpha I + HH^*)^{-1}H^* \end{aligned} \tag{12.5.6}$$

using (5.2.12) and the compactness of H and H^* together with the Riesz theorem 2.3.25, such that

$$\begin{aligned} u^s_{\mathrm{KK},\alpha}(z) &= \left\langle \Phi(\cdot, z), (\alpha I + S^{\infty,*}S^{\infty})^{-1}S^{\infty,*}u^\infty \right\rangle_{L^2(\partial G)} \\ &= \left\langle (\alpha I + H^*H)^{-1}H^*\Phi(\cdot, z), u^\infty \right\rangle_{L^2(\mathbb{S})} \\ &= u^s_{\mathrm{PSM},\alpha}(z) \end{aligned} \tag{12.5.7}$$

for $\alpha > 0$ and the proof is complete. □

Bibliography

[1] Liu J and Potthast R 2009 On the duality of the potential method and the point source method in inverse scattering problems *J. Integral Equ. Appl.* **21** 297–315

[2] Erdélyi A, Magnus W, Oberhettinger F and Tricomi F 1981 *Higher Transcendental Functions* vol 2 (Malabar, FL: Krieger)

[3] Colton D and Kress R 1998 *Inverse Acoustic and Electromagnetic Scattering Theory* (*Applied Mathematical Sciences* vol 93) 2nd edn (Berlin: Springer)

[4] Chandler-Wilde S N, Heinemeyer E and Potthast R 2006 Acoustic scattering by mildly rough unbounded surfaces in three dimensions *SIAM J. Appl. Math.* **66** 1002–6

[5] Chandler-Wilde S N, Heinemeyer E and Potthast R 2006 A well-posed integral equation formulation for three-dimensional rough surface scattering *Proc. R. Soc.* A **462** 3683–705

[6] Chandler-Wilde S N and Lindner M 2008 Boundary integral equations on unbounded rough surfaces: Fredholmness and the finite section method *J. Integral Equ. Appl.* **20** 13–48 2008.

[7] Eaton J W, Bateman D, Hauberg S and Wehbring R 2015 *GNU OCTAVE Version 4.0.0 Manual: a High-Level Interactive Language for Numerical Computations* (Boston, MA: Free Software Foundation)

[8] Potthast R and Schulz J 2005 From the Kirsch–Kress potential method via the range test to the singular sources method *J. Phys.: Conf. Ser* **12** 116–27

[9] Erhard K 2005 Point source approximation methods in inverse obstacle reconstruction problems *PhD Thesis* University of Göttingen

[10] Potthast R 1996 A fast new method to solve inverse scattering problems *Inverse Problems* **12** 731

[11] Potthast R 1998 A point-source method for inverse acoustic and electromagnetic obstacle scattering problems *IMA J. Appl. Math.* **61** 11940

[12] Potthast R 2001 *Point Sources and Multipoles in Inverse Scattering Theory* (*Chapman and Hall/CRC Research Notes in Mathematics* vol 427) (Boca Raton, FL: CRC)

Chapter 13

Sampling methods

Until the 1990s the basic approach to inverse scattering problems was led by the idea of minimizing some functional to adapt the far field pattern u^∞ of scattered fields to some measured far field pattern. This is an important idea which is very flexible and will remain of great importance. We have presented the Kirsch–Kress method in section 12.3 as an important example of this approach, and also the Newton and gradient methods introduced in section 9.4. An alternative approach has been suggested by Colton and Monk [1].

Some time in 1995 Andreas Kirsch, waiting in an airport on his way to Delaware, was playing with some numerical simulations of superpositions of plane waves such that the corresponding superposition of far field patterns would match the far field pattern of a point source. He observed that the norm of the solution in dependence on the source point tells us a lot about the unknown scattering object. This led to the *linear sampling method* (1996) suggested by Colton and Kirsch [2] and, in the case of an inhomogeneous medium, was given a theoretical foundation by Colton, Piana and Potthast [3]. We present this in section 13.2. It has inspired many further results since then, with the *factorization method* [4] as a prominent example.

In the same year, the idea of playing with point sources and the solution to integral equations in inverse scattering led to the *point source method* [5], which we have already introduced in section 12.4, as well as the *singular sources method* which we will present in section 14.1 (see [6]). Independently Ikehata [7] suggested the *probing method* which we introduce in section 14.2. This development leads to the class of *probe* or *probing methods*.

Years later, a very stable sampling method was suggested independently by Potthast and Ito, and Jin and Zou, with a convergence analysis carried out by Griesmaier, see [8–10]. It will be our starting point in this chapter. Before getting into the sections of this chapter, we have to note that for convenience of description from time to time we confine our argument to the three dimensional space, but the argument still works for the other dimensional space.

doi:10.1088/978-0-7503-1218-9ch13

13.1 Orthogonality or direct sampling

The *orthogonality sampling* or *direct sampling* method of Potthast [10], Griesmaier [8] and Ito, Jin and Zou [9] is based on the classical *Funck–Hecke formula* for spherical harmonics. The basic idea is to employ the standard scalar product on the unit circle or unit ball, respectively, to obtain information about an unknown scatterer. We will show that we can stably reconstruct the *reduced scattered field* or *Bessel potential*, which is an integral over the unknown boundary with a Bessel kernel. The Bessel potential carries key information about the location and shape of the scatterer, which can be evaluated in particular when several frequencies and directions of incidence are available.

We start with a representation of the scattered field u^s for a sound-soft scatterer D by

$$u^s(x) = \int_{\partial D} \Phi(x, y)\frac{\partial u}{\partial \nu}(y)\,\mathrm{d}s\,(y), \quad x \in \mathbb{R}^m\backslash\bar{D}. \tag{13.1.1}$$

The far field pattern of u^s is given by

$$u^\infty(\hat{x}) = \gamma \int_{\partial D} \mathrm{e}^{-\mathrm{i}\kappa\hat{x}\cdot y}\frac{\partial u}{\partial \nu}(y)\,\mathrm{d}s\,(y), \quad \hat{x} \in \mathbb{S}. \tag{13.1.2}$$

where γ is given in (12.4.3). Let us multiply $u^\infty(\hat{x})$ by $\overline{f_z}$ with $f_z(\hat{x}) := \mathrm{e}^{-\mathrm{i}\kappa\hat{x}\cdot z}$, $z \in \mathbb{R}^m$, and integrate over $\mathbb{S}$. We obtain

$$\begin{aligned}\int_{\mathbb{S}} u^\infty(\hat{x})\mathrm{e}^{\mathrm{i}\kappa\hat{x}\cdot z}\,\mathrm{d}s(\hat{x}) &= \gamma \int_{\mathbb{S}}\int_{\partial D} \mathrm{e}^{-\mathrm{i}\kappa\hat{x}\cdot(y-z)}\frac{\partial u}{\partial \nu}(y)\,\mathrm{d}s\,(y)\,\mathrm{d}s\,(\hat{x})\\ &= \gamma \int_{\partial D}\left(\int_{\mathbb{S}} \mathrm{e}^{-\mathrm{i}\kappa\hat{x}\cdot(y-z)}\,\mathrm{d}s(\hat{x})\right)\frac{\partial u}{\partial \nu}(y)\,\mathrm{d}s(y).\end{aligned}$$

Now we employ the *Funck–Hecke formula*

$$\int_{\mathbb{S}} \mathrm{e}^{-\mathrm{i}\kappa x\cdot\theta}\,Y_n(\theta)\,\mathrm{d}s(\theta) = \frac{4\pi}{\mathrm{i}^n}j_n(\kappa|x|)\,Y_n(\hat{x}), \quad x \in \mathbb{R}^m \tag{13.1.3}$$

for the three-dimensional case (taken from [1]) to derive

$$\int_{\mathbb{S}} u^\infty(\hat{x})\mathrm{e}^{\mathrm{i}\kappa\hat{x}\cdot z}\,\mathrm{d}s(\hat{x}) = 4\pi\gamma \int_{\partial D} j_0(\kappa|y - z|)\frac{\partial u}{\partial \nu}(y)\,\mathrm{d}s\,(y) \tag{13.1.4}$$

for $z \in \mathbb{R}^m$. The integral

$$u^s_{red}(z) := 4\pi\gamma \int_{\partial D} j_0(\kappa|z - y|)\frac{\partial u}{\partial \nu}(y)\,\mathrm{d}s(y) \tag{13.1.5}$$

is the *Bessel potential* for the dipole sources given by $\partial u/\partial \nu$ on ∂D, sometimes also called the *reduced* scattered field. We summarize the above derivation into the following lemma.

Lemma 13.1.1. *The far field pattern u^∞ of a scattered field u^s allows the stable reconstruction of the Bessel potential (13.1.5) by evaluation of the scalar product $\langle u^\infty, f_z\rangle$, $z \in \mathbb{R}^m$ as given by (13.1.4).*

If we know the Bessel potential for some scatterer for several wave numbers $\kappa \in [\kappa_1, \kappa_2]$ or for several incident waves with direction of incidence $d_j \in \mathbb{S}$, $j = 1, \ldots, J$, it is suggested in [10] to study the integral

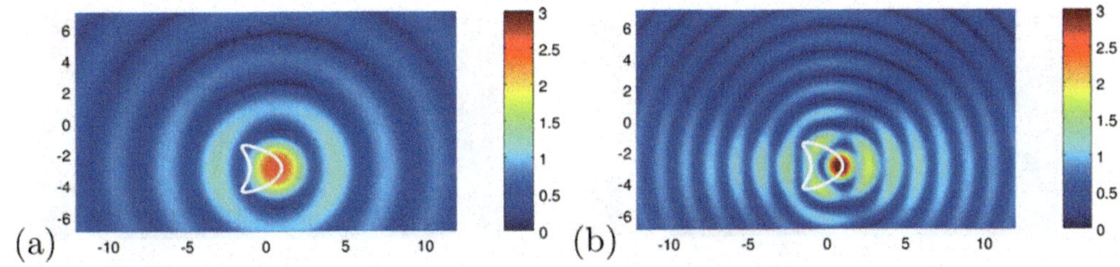

Figure 13.1. The reduced field for $\kappa = 1$ (a) and $\kappa = 2$ (b) for a single incident wave coming from the right. Clearly the field provides the location of the scatterer, but no information about the shape for one incident field and one frequency only.

$$\mu(z) := \sum_{j=1}^{J} \int_{\kappa_1}^{\kappa_2} \left| u^{s}_{\mathrm{red},\kappa}(z, d_j) \right| \mathrm{d}\kappa \qquad (13.1.6)$$

and use it as an *indicator function*, which is small inside and outside the scatterer D and has larger values at the boundary ∂D of D. The method provides very reasonable and stable reconstructions of complicated scatterers, but shows a resolution within the *diffraction limit*, i.e. the resolution is given by π/κ.

The evaluation of the scalar products $\langle u^\infty, f_z \rangle$ is not difficult and leads to a very fast method which provides significant insight into the geometry of the scatterer, in particular when it is not simple to parametrize it, i.e. when many separate components of different size are present. We show some examples in figure 13.1, 13.2 and 13.3 with different wave numbers and directions of incidence generated by code 13.1.2.

Code 13.1.2. *We show script* `sim_13_1_2_b_OrthoSampling.m` *to carry out the field reconstruction by orthogonality sampling or direct sampling. Run script* `sim_13_1_2_a_scattering_ff.m` *first to generate* u^∞*, here named* `ff`*, and some further variables. Script* `sim_13_1_2_c_graphics.m` *visualizes the simulated and reconstructed reduced scattered field on some evaluation domain* Q.

```
ossum = zeros(M,1);          % initialize summation vector for simulation
ossum2 = zeros(M,1);         % initialize summation vector for reconstruction
nc = 1;
for kappa=kappav             % loop over all wave numbers and directions
  disp(['kappa=' num2str(kappa)]);   % get wave number from stored vector

  % Setup and evaluate orthogonality scalar product = Herglotz wave operator
  hmat1    = repmat(pvec1,1,ffN).*repmat(yff1,M,1);  % setup f_z
  hmat2    = repmat(pvec2,1,ffN).*repmat(yff2,M,1);  % ~
  tH       = exp(i*kappa*(hmat1 + hmat2))*hff; % Herglotz wave operator

  tSred  = i/2*besselj(0,kappa*rmat).*drmat*ht;  % single-layer potential
  os     = tSred*varphiv(:,nc);            % reduced scattered field
  ossum = ossum + abs(os);                 % summation of reduced fields

  osff = -1/(4*pi*fac)*tH*ffv(:,nc);  % orthogonality sampling functional
  ossum2 = ossum2 + abs(osff);            % summation of reconstructed fields

  nc = nc+1;                 % counter
end
```

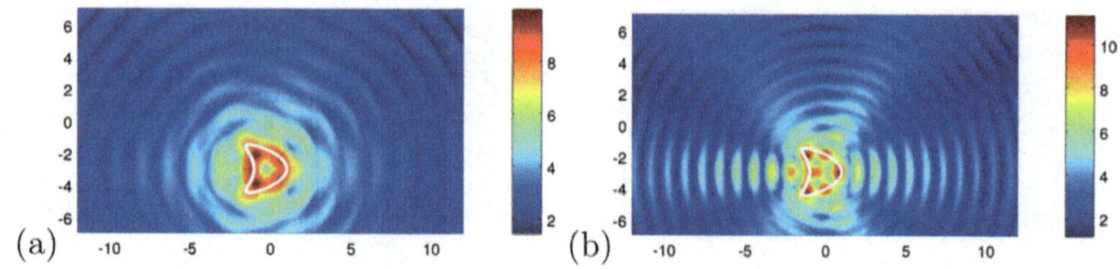

Figure 13.2. The reduced field for $\kappa = 2$ (a) and $\kappa = 3$ (b) for six different incident waves with the angle of incidence being multiples of $\pi/3$. This field already provides a lot of information about the shape of the scatterer, although it is still rather wavy.

13.2 The linear sampling method of Colton and Kirsch

The linear sampling method first suggested by Colton and Kirsch [2] investigates the *far field equation*

$$Fg = f_z \text{ on } \mathbb{S} \tag{13.2.1}$$

with $F : L^2(\mathbb{S}) \to L^2(\mathbb{S})$ defined by (8.3.18) and the right-hand side

$$f_z(\hat{x}) := \gamma e^{-i\kappa\hat{x}\cdot z}, \quad z \in \mathbb{R}^m \tag{13.2.2}$$

where γ is given in (12.4.3). The field f_z is the far field pattern of the fundamental solution $\Phi(\cdot, z)$ defined in (8.1.3). The basic idea is simple: the behavior of the solution g_z is different if $z \notin D$ or $z \in D$, and one can find the unknown scatterer D based on the norm $\|g_z\|$ of the solution. The norm of a regularized solution is larger if $z \notin D$ in comparison to $z \in D$. Before we develop a deeper understanding of the Colton–Kirsch equation (13.2.1), we first carry out reconstructions by a calculation of g_z. Results are shown in figure 13.4.

Code 13.2.1. *We show script* `sim_13_2_1_b_LinearSampling.m` *to carry out the shape reconstruction by the linear sampling method. Run script* `sim_13_2_1_a_F.m` *first to generate the simulated far field patterns and the far field operator F. Script* `sim_13_2_1_c_graphics.m` *visualizes the linear sampling indicator function on some evaluation domain Q. For an understanding of the following variables, in particular* `pvec1`, `pvec2`, *and* `yff1`, `yff2`, *have a look at codes 8.2.3 and 8.3.1 first.*

```
% Setup and evaluate far field pattern term f_z
hmat1   = repmat(pvec1,1,ffN).*repmat(yff1,M,1);
hmat2   = repmat(pvec2,1,ffN).*repmat(yff2,M,1);
Phiinf  = exp(-i*kappa*(hmat1 + hmat2)).';

alphaLS = 1e-8;                    % regularization parameter
gLS  = (alphaLS*eye(ffN,ffN) + F'*F)\(F'*Phiinf);  % solve equation

wLS = ( 1./sqrt(sum( abs(gLS).^2 )) ).';    % linear sampling functional
wmax = max(wLS); ca = 4; wLS = 2*ca*wLS / wmax;   % scaling for display
wLS = min(abs(wLS),ca);           % cut for better display
```

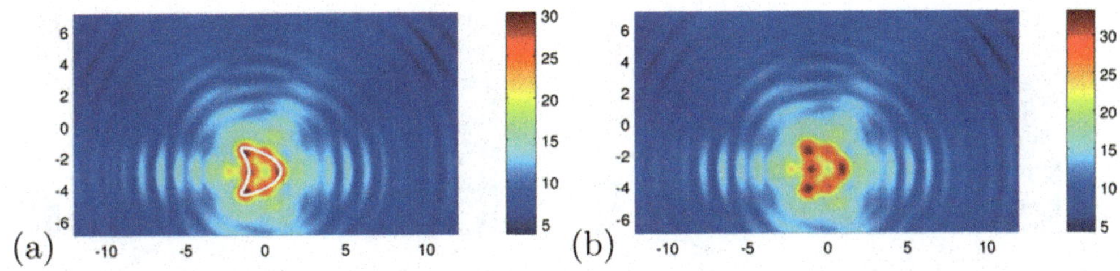

Figure 13.3. The sum of the modulus of the reconstructed reduced fields for $\kappa \in \{2, 2.5, 3\}$ with (a) and without (b) the scattering object for six different incident waves with the angle of incidence being multiples of $\pi/3$. Averaging the modulus of the field for several frequencies removes the oscillations and provides very stable and reliable shape reconstructions.

To understand the meaning of equation (13.2.1) we consider a simple scattering problem. That is scattering by an impenetrable scatterer D with a Dirichlet boundary condition. For acoustic wave scattering this means that D is sound soft. Then we can write F in the form

$$F = -G \circ H \tag{13.2.3}$$

with the Herglotz operator $H : L^2(\mathbb{S}) \to H^{1/2}(\partial D)$ defined in (8.3.12) modeling the superposition of plane waves and the mapping $G : H^{1/2}(\partial D) \to L^2(\mathbb{S})$ which maps the boundary values $u^s|_{\partial D}$ onto the far field pattern u^∞ of the radiating solution u^s to the Helmholtz equation in $\mathbb{R}^m \backslash \bar{D}$. Then, if $z \in D$, the equation

$$G\psi = f_z \tag{13.2.4}$$

has a solution. Note that equation (13.2.1) includes the additional condition that $\psi = Hg$ for some $g \in L^2(\mathbb{S})$. If $z \notin \bar{D}$, then (13.2.4) does not have a solution, since otherwise this solution would be bounded on compact subsets of $\mathbb{R}^m \backslash \bar{D}$, but it would also be equal to $\Phi(\cdot, z)$ and thus unbounded in a neighborhood of $z \in \mathbb{R}^m \backslash \bar{D}$—see lemma 13.2.4 below. This discriminating property, i.e. the solvability of (13.2.4) is simple, but when moving to the far field equation (13.2.1) the complication comes from $F = GH$ and the solvability issue with $\psi = Hg$ on ∂D.

For further argument we prepare the following lemma.

Lemma 13.2.2. *If $D \subset \mathbb{R}^m$ is non-vibrating, then the single-layer potential $S : H^{-1/2}(\partial D) \to H^{1/2}(\partial D)$ is an isomorphism.*

Proof. From the mapping property of $S : H^{-1/2}(\partial D) \to H^{1/2}(\partial D)$ is bounded (see chapter 6 of [11]). Also, it is well known that the single-layer potential operator $S_i : H^{-1/2}(\partial D) \to H^{1/2}(\partial D)$ associated to the meta-harmonic operator $\Delta - 1$ is invertible and $R = S - S_i : H^{-1/2}(\partial D) \to H^{1/2}(\partial D)$ is compact. Hence $S = S_i(I + T)$ with compact operator $T = S_i^{-1}R$ on $H^{-1/2}(\partial D)$. Hence by theorem 2.3.25 the injectivity of S implies its surjectivity and bounded invertibility which implies that S is an isomorphism. To see that S is injective, let $\phi \in H^{-1/2}(\partial D)$, $S\phi = 0$. Then $v(x) = \int_{\partial D} \Phi(x, y)\phi(y)\, ds(y)$ is the solution to the Dirichlet boundary value problem for the Helmholtz equation in $\mathbb{R}^m \backslash \bar{D}$ satisfying the Sommerfeld

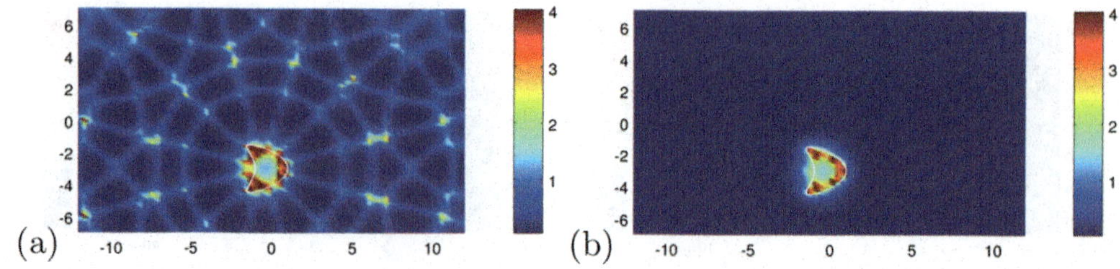

Figure 13.4. The inverse of the norm of g_z for $\kappa = 2$ for a dimension $F \in \mathbb{R}^{12\times 12}$ (a) and $F \in \mathbb{R}^{15\times 15}$ i.e. with 12 or 15 incident plane waves evaluated in the same 12 or 15 regularly distributed directions, solving equation (13.2.1) by Tikhonov regularization with $\alpha = 10^{-8}$. (a) still shows significant artifacts, but for (b) with some more waves we obtain a very good reconstruction.

radiation condition. By the uniqueness of this boundary value problem, we have $v = 0$ in $\mathbb{R}^m \backslash D$. Hence v also is the solution to the Dirichlet problem for the Helmholtz equation in D. Since D is non-vibrating, this implies $v = 0$ in $\overline{D}$. Then by the jump formula for the Neumann derivative of v at ∂D gives $\phi = 0$. Therefore S is injective. □

Note that for any $\varphi \in H^{-1/2}(\partial D)$ the far field pattern $G(S\varphi)$ of $S\varphi \in H^{1/2}(\partial D)$ with single-layer potential S defined by (8.1.31) is $\gamma H^*\varphi$ with the operators $H^* : H^{-1/2}(\partial D) \to L^2(\mathbb{S})$ given in (8.3.13). Hence we have $GS = \gamma H^*$. Since D is non-vibrating, $S : H^{-1/2}(\partial D) \to H^{1/2}(\partial D)$ is an isomorphism by lemma 13.2.2. Therefore we can write G in the form

$$G = \gamma H^* S^{-1}. \tag{13.2.5}$$

Now consider (13.2.3) for $z \in D$. Then, by Rellich's lemma, we can pass from the far field of equation (13.2.4) to the near field on ∂D and obtain an equivalence of (13.2.4) to

$$S^{-1}\psi = \Phi(\cdot, z). \tag{13.2.6}$$

Further, this is always solvable as far as $z \notin \partial D$, because $S^{-1} : H^{1/2}(\partial D) \to H^{-1/2}(\partial D)$ is an isomorphism because D is non-vibrating.

It is helpful for further insight to study the spectral representation of the Colton–Kirsch equation (13.2.1). Let (φ_n, g_n, μ_n) be a singular system of $F : L^2(\mathbb{S}) \to L^2(\mathbb{S})$. We will show in theorem 13.3.2 later that F has a dense range, this then implies that g_n is a complete orthonormal system of Y. Further, we denote the coefficients of f_z with respect to the complete orthonormal system $\{g_n\}$ by $\alpha_n^{(z)}$, i.e. we have

$$f_z = \sum_{n=1}^{\infty} \alpha_n^{(z)} g_n. \tag{13.2.7}$$

The solution $g = g^{(z)}$ of the equation

$$Fg = f_z \tag{13.2.8}$$

can then be written in spectral representation as

$$g = \sum_{n=1}^{\infty} \frac{\alpha_n^{(z)}}{\mu_n} \varphi_n. \tag{13.2.9}$$

Note that $\mu_n \to 0$ for $n \to \infty$, i.e. we divide by small numbers. Also, we expect $\alpha_n^{(z)} \to 0$ for $n \to \infty$. So the key question for the behavior of the solution g is which series tends to zero with which rate. First, we note the following basic result which is a consequence of the Picard's theorem 2.4.23.

Lemma 13.2.3. *The equation (13.2.8) does have a solution if and only if*

$$\sum_{n=1}^{\infty} \left| \frac{\alpha_n^{(z)}}{\mu_n} \right|^2 < \infty. \tag{13.2.10}$$

The condition (13.2.10) will be satisfied only in very particular situations, when the boundary values of $\Phi(\cdot, z)$ on ∂D can be represented as a Herglotz wave function Hg on ∂D. This is the consequence of the decomposition (13.2.3) of the far field operator F. In general the equation

$$\Phi(\cdot, z) = Hg \text{ on } \partial D \tag{13.2.11}$$

is not solvable. In particular, it will not be solvable for $z \notin \bar{D}$, since then the functions Hg and $\Phi(\cdot, z)$ would coincide in $\mathbb{R}^m \backslash \{z\}$, which cannot be the case since Hg is bounded in a neighborhood of z, but $\Phi(\cdot, z)$ is unbounded in z.

Traditionally, the convergence of *linear sampling* has been based on some special approximation results, which we will present next. For simplicity of description, we only give it for the three-dimensional case. As preparation we first prove the following two results.

Lemma 13.2.4. *If $z \notin D$, then $\Phi^\infty(\cdot, z) \notin R(G)$.*

Proof. There are two cases $z \notin \bar{D}$ and $z \in \partial D$. For both cases, we use the proof by contradiction. Suppose that we have $\Phi^\infty(\cdot, z) \in R(G)$. For the former case, there exists a unique and smooth outgoing solution u of the Helmholtz equation in $\mathbb{R}^m \backslash \bar{D}$ with Dirichlet data $-\Phi(\cdot, z)|_{\partial D}$. By Rellich's lemma, $u = \Phi(\cdot, z)$ in $\mathbb{R}^m \backslash (\bar{D} \cup \{z\})$. However, $\Phi(\cdot, z)$ is not smooth on $\mathbb{R}^m \backslash \bar{D}$, which is a contradiction.

For the latter case, take an open set $\tilde{D}$ with C^2 boundary such that $\bar{D} \subset \tilde{D}$ and $\mathbb{R}^m \backslash \bar{\tilde{D}}$, $\tilde{D} \backslash \bar{D}$ are connected. Then, likewise to the former case, we have $u = \Phi(\cdot, z)$ in $\mathbb{R}^m \backslash \bar{\tilde{D}}$ for the previously given u. Further, by the unique continuation of solutions to the Helmholtz equation, we have $u = \Phi(\cdot, z)$ in $\mathbb{R}^m \backslash \bar{D}$. The rest is just as the same as the one for the former case. Thus, we have proven lemma 13.2.4. □

Lemma 13.2.5. *$G : L^2(\partial D) \to L^2(\mathbb{S})$ is injective, compact and has a dense range.*

Proof. The injectivity of G immediately follows from Rellich's lemma. We have shown in sections 8.2 and 8.3 that by using the combined acoustic double- and single-layer potential according to Brackhage and Werner (8.2.7), the far field pattern $u^\infty(\hat{x})$ $(\hat{x} \in \mathbb{S})$ of radiating solution u^s in $\mathbb{R}^m \backslash \bar{D}$ can be given by

$$u^\infty(\hat{x}) = \frac{1}{2\pi} \int_{\partial D} \left(\frac{\partial\, e^{-i\kappa y \cdot \hat{x}}}{\partial \nu(y)} - ie^{-\kappa y \cdot \hat{x}} \right) \psi(y)\, ds\,(y)\ (\hat{x} \in \mathbb{S}), \tag{13.2.12}$$

with

$$\psi = (I + K - iS)^{-1} u^s|_{\partial D}, \tag{13.2.13}$$

where S and K are the single- and double-layer potential operators on ∂D, respectively. Since $(I + K - iS)^{-1} : L^2(\partial D) \to L^2(\partial D)$ is bounded, the compactness of G immediately follows from (13.2.12).

Finally, we will show that G has a dense range. Since the Herglotz operator $H : L^2(\mathbb{S}) \to L^2(\partial D)$ has a dense range, we can approximate $u^s|_{\partial D}$. Hence, it is enough to show that $\{u^\infty(\cdot, d) : d \in \mathbb{S}\}$ is dense in $L^2(\mathbb{S})$. That is if $h \in L^2(\mathbb{S})$ satisfies $\int_{\mathbb{S}} u^\infty(\hat{x}, d) h(\hat{x})\, ds(\hat{x}) = 0$ $(d \in \mathbb{S})$, then $h = 0$. By the far field reciprocity relation (8.4.1), the condition here for h is equivalent to

$$v_g^\infty(\hat{x}) = \int_{\mathbb{S}} u^\infty(\hat{x}, d)\, g(d)\, ds(d) = 0\ (\hat{x} \in \mathbb{S}), \tag{13.2.14}$$

where $g(d) = h(-d)$ and v_g^∞ is the far field pattern of the Herglotz wave functions $v_g^i = v_g = Hg$ with density g. Then, from Rellich's lemma, we have $v_g^s = 0$ in $\mathbb{R}^m \backslash \bar{D}$, where v_g^s is the scattered wave for the incident wave v_g^i. The boundary condition for v_g^s gives $v_g^s + v_g^i = 0$ on ∂D. Then, by the unique solvability of the exterior Dirichlet boundary value problem for the Helmholtz equation, we have $v_g^i = 0$ on ∂D. That is $(Hg)(x) = 0$ $(x \in \partial D)$. Then, the proof finishes if we recall the injectivity of H which is just a consequence of an assumption that D is non-vibrating. □

We are now prepared for deriving the traditional linear sampling approximation result. Note that the following theorem states the existence of particular solutions, which allow us to find the shape of D, but do not provide a method to calculate those particular solutions.

Theorem 13.2.6. *Let $D \subset \mathbb{R}^m$ be a non-vibrating domain.*

(i) *If $z \in D$, then for any $\varepsilon > 0$, there exists $g_z^\varepsilon \in L^2(\mathbb{S})$ locally bounded for $\varepsilon > 0$ such that*

$$\left\| g_z^\varepsilon \right\|_{L^2(\mathbb{S})} \to \infty, \qquad \left\| Hg_z^\varepsilon \right\|_{L^2(\partial D)} \to \infty$$

as $z \to \partial D$.

(ii) *If $z \notin D$, then for any $\varepsilon > 0$, $\delta > 0$, there exists $g_{\varepsilon,\delta}^z \in L^2(\mathbb{S})$ such that*

$$\| F g_{\varepsilon,\delta}^z - \Phi^\infty(\cdot, z) \|_{L^2(\mathbb{S})} < \varepsilon + \delta, \qquad \lim_{\delta \to 0} \left\| g_{\varepsilon,\delta}^z \right\|_{L^2(\mathbb{S})} = \infty. \tag{13.2.15}$$

Proof. We first prove (i). By lemma 13.2.5 and $H^{1/2}(\partial D)$ is dense in $L^2(\partial D)$, G can be naturally extended to an operator from $L^2(\partial D)$ to $L^2(\mathbb{S})$. We also denote this extension by G.

By using the denseness of the range of H, for $\varepsilon > 0$, $\delta > 0$, there exists $g_\varepsilon^z \in L^2(\mathbb{S})$ abbreviated by $g^z = g_z^\varepsilon$ such that

$$\|Hg^z - (-\Phi(\cdot, z))\|_{L^2(\partial D)} < \frac{\varepsilon}{\|G\|}. \tag{13.2.16}$$

Here, observe that the far field pattern $v_{g^z}^\infty(\hat{x})$ with $\hat{x} \in \mathbb{S}$ of $v_{g^z} = Hg^z$ is given by $(Fg^z)(\hat{x})$. Hence, by the definition of G, we have $Fg^z = G(Hg^z)$, and hence

$$\Phi^\infty(\cdot, z) = G(-\Phi(\cdot, z)). \tag{13.2.17}$$

Combining (13.2.16) and (13.2.17), we have

$$\|Fg^z - \Phi^\infty(\cdot, z)\|_{L^2(\mathbb{S})} = \|G(Hg^z - (-\Phi(\cdot, z))\|_{L^2(\mathbb{S})} < \varepsilon. \tag{13.2.18}$$

On the other hand, by the boundedness of the Herglotz operator H and (13.2.16), we have

$$\begin{aligned}\|g^z\|_{L^2(\mathbb{S})} &\geqslant C\|Hg^z\|_{L^2(\partial D)}\\ &\geqslant C(\| - \Phi(\cdot, z)\|_{L^2(\partial D)} - \|Hg^z - (-\Phi(\cdot, z))\|_{L^2(\partial D)})\\ &\geqslant C\|\Phi(\cdot, z)\|_{L^2(\partial D)} - \frac{C\varepsilon}{\|G\|} \to \infty(z \to \partial D)\end{aligned} \tag{13.2.19}$$

for some constant $C > 0$, because $\|\Phi(\cdot, z)\|_{L^2(\partial D)} \to \infty$ $(z \to \partial D)$. The local boundedness of g_z^ε is clear from how we can obtain $g^z = g_z^\varepsilon \in L^2(\mathbb{S})$ which satisfies (13.2.16) by using Tikhonov regularization. Hence we have shown (i).

Let us now prove (ii). Since G is injective and compact by lemma 13.2.5, we can define a singular system (μ_n, φ_n, g_n) of G. For any $\delta > 0$ and $\alpha > 0$, we further have by lemma 13.2.5 that there exists

$$\sigma_z^\alpha = -\sum_n \frac{\mu_n}{\alpha + \mu_n^2}\langle\Phi^\infty(\cdot, z), g_n\rangle\varphi_n \in L^2(\partial D)$$

which satisfies

$$\|G\sigma_z^\alpha + \Phi^\infty(\cdot, z)\|_{L^2(\mathbb{S})} < \delta. \tag{13.2.20}$$

Here, since $-\Phi^\infty(\cdot, z) \notin R(G)$, we have by Picard's theorem

$$\sum_n \frac{1}{\mu_n^2} |\langle\Phi^\infty(\cdot, z), g_n\rangle|^2 = \infty. \tag{13.2.21}$$

Hence, we have

$$\|\sigma_z^\alpha\|_{L^2(\partial D)}^2 = \sum_n \left(\frac{\mu_n}{\alpha + \mu_n}\right)^2 |\langle\Phi^\infty(\cdot, z), g_n\rangle|^2 \to \infty\ (\alpha \to 0). \tag{13.2.22}$$

Now, since $H : L^2(\mathbb{S}) \to L^2(\partial D)$ has a dense range, there exists $g_z^\alpha \in L^2(\mathbb{S})$ such that

$$\left\|\sigma_z^\alpha - Hg_z^\alpha\right\|_{L^2(\partial D)} < \frac{\varepsilon}{2\|G\|}, \quad \left\|G\sigma_z^\alpha - GHg_z^\alpha\right\|_{L^2(\mathbb{S})} < \frac{\varepsilon}{2}. \tag{13.2.23}$$

Recalling $F = -GH$ and combining the two inequalities (13.2.20), (13.2.23), we have

$$\left\|Fg_z^\alpha - \Phi^\infty(\cdot, z)\right\|_{L^2(\mathbb{S})} = \left\| - GHg_z^\alpha - \Phi^\infty(\cdot, z)\right\|_{L^2(\mathbb{S})} < \varepsilon + \delta. \tag{13.2.24}$$

By (13.2.22) and the first inequality of (13.2.23), we have $\|Hg_z^\alpha\|_{L^2(\partial D)} \to \infty$ as $\alpha \to 0$. Then $\|g_z^\alpha\|_{L^2(\mathbb{S})} \to \infty$ $(\alpha \to 0)$ immediately follows from the boundedness of H.

For further details on the convergence of the linear sampling method we refer the reader to the work of Arens [12] and to Cakoni and Colton [13] as well as Kirsch and Grinberg [4]. The method has been explored by a large number of researchers since its early beginnings and has experienced an astonishing popularity.

13.3 Kirsch's factorization method

The factorization method was suggested by Kirsch [14], see also [4], initially as a way to modify linear sampling to obtain a scheme with better convergence properties. The factorization method characterizes the range of the operator G (mapping the boundary values $u^s|_{\partial D}$ of some wave field onto its far field pattern u^∞) in terms of the far field operator F, where the operator F is given based on measurements.

Kirsch suggested replacing the Colton–Kirsch equation (13.2.1) by the equation

$$\left(F^*F\right)^{\frac{1}{4}}g = f_z \tag{13.3.1}$$

with $f_z = e^{-i\kappa \cdot z} \in L^2(\mathbb{S})$ for all points z on some evaluation domain Q. The equation is solvable if and only if $z \in D$, i.e. we obtain a constructive method to reconstruct the location and shape of a scattering object by calculating a regularized solution to (13.3.1) and testing the behavior of the solution.

Before we go into the theoretical foundation of the method, let us carry out some reconstructions. For the self-adjoint operator F^*F we can employ singular value decomposition to calculate the fourth root and then calculate a reconstruction by Tikhonov regularization.

Code 13.3.1. *We show script* `sim_13_3_1_b_Factorization.m` *to carry out the field reconstruction by the factorization method. Run script* `sim_13_3_1_a_F.m` *first to generate the simulated far field patterns and the far field operator F. Script* `sim_13_3_1_c_graphics.m` *visualizes the linear sampling indicator function on some evaluation domain Q. For an understanding of the following variables, in particular* `pvec1`, `pvec2`, *and* `yff1`, `yff2`, *have a look at codes 8.2.3 and 8.3.1 first. The result is displayed in figure 13.5.*

```
% Setup and evaluate far field pattern term f_z
hmat1    = repmat(pvec1,1,ffN).*repmat(yff1,M,1);
hmat2    = repmat(pvec2,1,ffN).*repmat(yff2,M,1);
Phiinf   = exp(-i*kappa*(hmat1 + hmat2)).';

alphaLS = 1e-8;                    % regularization parameter
[U,S,V]=svd(F);                    % singular value decomposition of F
A = U*sqrt(S)*V';                   % (F'*F)^(1/4)
gLS  = (alphaLS*eye(ffN,ffN) + A'*A)\\(A'*Phiinf);  % solve equation

wLS = ( 1./sqrt(sum( abs(gLS).^2 )) ).';    % linear sampling functional
wmax = max(wLS); ca = 4; wLS = ca*wLS / wmax;   % scaling for display
wLS = min(abs(wLS),ca);            % cut for better display
```

We now come to the analytic basis of the factorization method. It leads to a deeper understanding of the far field operator and its relationship to the scattering process. With the following results we follow Kirsch [14].

Theorem 13.3.2. *Assume that $D \subset \mathbb{R}^m$ is non-vibrating and let F be the far field operator, where $m = 2,3$ are the cases which have physical meaning.*

(i) *F has a factorization*

$$F = -\frac{1}{\gamma_m} G S^* G^*, \tag{13.3.2}$$

where S is the single-layer potential on ∂D and γ_m is given as in (12.4.3) by $\gamma_m = e^{i\pi/4}/\sqrt{8\pi\kappa}$ for $m = 2$ and $1/(4\pi)$ for $m = 3$.

(ii) *$F : L^2(\mathbb{S}) \to L^2(\mathbb{S})$ is an injective, compact and normal operator with a dense range. It has a complete set of orthonormal eigenvectors $\psi_j \in L^2(\mathbb{S})$ $(j = 1, 2, \cdots)$ with the corresponding eigenvalues $0 \neq \lambda_j \in \mathbb{C}$ on the circle*

$$\left\{ \lambda \in \mathbb{C} : |\lambda - (2\pi i)/\kappa| = (2\pi)/\kappa \right\}$$

and we have $\lambda_j \to 0$ as $j \to \infty$. Hence, the singular system of F is given by $(\sigma_j, \psi_j, \tilde{\psi}_j)$ with $\sigma_j = |\lambda_j|$, $\tilde{\psi}_j = \operatorname{sgn}(\lambda_j)\psi_j$. Furthermore, F has a dense range.

The convergence of the factorization method is established by the following characterization result.

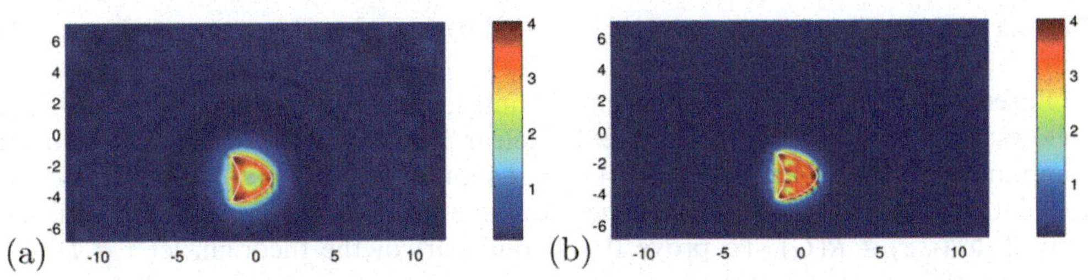

Figure 13.5. The inverse of the norm of g_z for a dimension $F \in \mathbb{R}^{15\times15}$ and $\kappa = 2$ (a) and $F \in \mathbb{R}^{50\times50}$ and $\kappa = 4$ (b), solving equation 13.3.1 by Tikhonov regularization with $\alpha = 10^{-8}$.

Theorem 13.3.3. *Let $D \subset \mathbb{R}^m$ $(m = 2, 3)$ be a non-vibrating domain. Then, based on F the scatterer D can be characterized by*

$$D = \left\{ z \in \mathbb{R}^m : \sum_{j=1}^{\infty} \frac{\left|\rho_j^{(z)}\right|^2}{\sigma_j} < \infty \right\}$$
$$= \left\{ z \in \mathbb{R}^m : \Phi^\infty(\cdot, z) \in (F^*F)^{\frac{1}{4}}(L^2(\mathbb{S})) \right\}, \qquad (13.3.3)$$

where $\rho_j^{(z)} = \langle \Phi^\infty(\cdot, z), \psi_j \rangle$. *Further, the behavior of* $\sum_{j=1}^{\infty} \frac{|\rho_j^{(z)}|^2}{\sigma_j}$ *as* $z \in D$ *tends to* ∂D *is given by*

$$\frac{1}{C^2} \|\Phi(\cdot, z)\|^2_{H^{1/2}(\partial D)} \leqslant \sum_{j=1}^{\infty} \frac{\left|\rho_j^{(z)}\right|^2}{\sigma_j} \leqslant C^2 \|\Phi(\cdot, z)\|^2_{H^{1/2}(\partial D)} \quad (z \in D) \qquad (13.3.4)$$

for some constant $C > 0$.

Proof. To keep the main line of reasoning we first collect some lemmas we need for the proof. These lemmas will be proven below, together with theorem 13.3.2.

Lemma 13.3.4. *For each* $j \in \mathbb{N}$, *define* φ_j *by*

$$G^* \psi_j = \sqrt{\lambda_j} \varphi_j \text{ with } \mathrm{Im} \sqrt{\lambda_j} > 0. \qquad (13.3.5)$$

Then, $\{\varphi_j\}_{j=1}^{\infty}$ *forms a Riesz basis in* $H^{-1/2}(\partial D)$. *That is there exists a bounded invertible linear operator T on* $H^{-1/2}(\partial D)$ *such that* $\{T\varphi_j\}_{j=1}^{\infty}$ *is an orthonormal system in* $H^{-1/2}(\partial D)$.

Lemma 13.3.5. *The range of* $G : H^{1/2}(\partial D) \to L^2(\mathbb{S})$ *can be given by*

$$G(H^{1/2}(\partial D)) = \left\{ \sum_{j=1}^{\infty} \rho_j \psi_j : \sum_{j=1}^{\infty} \frac{|\rho_j|^2}{|\lambda_j|} < \infty \right\}$$
$$= (F^*F)^{\frac{1}{4}}(L^2(\mathbb{S})). \qquad (13.3.6)$$

Lemma 13.3.6. *The mapping* $(F^*F)^{-\frac{1}{4}} G : H^{1/2}(\partial D) \to L^2(\mathbb{S})$ *is an isomorphism.*

Based on these properties of F and lemmas, we can prove theorem 13.3.3 as follows. The first part of the theorem is clear from lemma 13.2.4 for the linear sampling method and lemma 13.3.5. Here, it should be noted that if $z \in D$ it is easy to see that $\Phi^\infty(\cdot, z) \in R(G)$, and hence we have from lemma 13.2.4, $z \in D$ if and only if $\Phi^\infty(\cdot, z) \in R(G)$. To prove the second part of the theorem, let $z \in D$ and $g = \sum_{j=1}^{\infty} \frac{\rho_j^{(z)}}{\sqrt{\sigma_j}} \psi_j$. By lemma 13.3.2, we have

$$\left(F^*F\right)^{\frac{1}{4}}g = \sum_{j=1}^{\infty} \frac{\rho_j^{(z)}}{\sqrt{\sigma_j}} \sqrt{\sigma_j}\psi_j = \sum_{j=1}^{\infty} \rho_j^{(z)} \psi_j = \Phi^\infty(\cdot, z). \tag{13.3.7}$$

Now, for $f = \Phi(\cdot, z)|_{\partial D}$, observe that $Gf = \Phi^\infty(\cdot, z)$. Hence, $g = (F^*F)^{-1/4}Gf$, where we have used the injectivity of F. Here, by lemma 13.3.6, we have

$$\|g\|^2_{L^2(\mathbb{S})} = \sum_{j=1}^{\infty} \frac{\left|\rho_j^{(z)}\right|^2}{\sigma_j} \sim \|\Phi(\cdot, z)\|^2_{H^{1/2}(\partial D)},$$

which is nothing but (13.3.4). Here we used the notation '$\sim$' to denote the equivalence of the above two norms for G and $\Phi(\cdot, z)|_{\partial D}$. □

Derivation of the properties of F. We now work out the proof to theorem 13.3.2, starting with (i). By the assumption that D is non-vibrating, the single-layer potential operator $S : H^{-1/2}(\partial D) \to H^{1/2}(\partial D)$ is an isomorphism. Hence, the scattered wave u^{s} for the incident wave $u^{\mathrm{i}} = v_g$ given by the Herglotz wave function v_g can be expressed in the form

$$u^{\mathrm{s}} = S\varphi \text{ with } \varphi = S^{-1}(-u^{\mathrm{i}}|_{\partial D}). \tag{13.3.8}$$

This implies that the far field pattern u^∞ of u^{s} is given by

$$u^\infty(\hat{x}) = S^\infty\varphi = \int_{\partial D} \Phi^\infty(\hat{x}, y)\, \varphi(y)\, \mathrm{d}s(y)\ (\hat{x} \in \mathbb{S}), \tag{13.3.9}$$

and it is easy to see that $S^\infty = (\gamma_m H)^* = \overline{\gamma_m} H^*$. Hence, by the definition of G, we have

$$G(-u^{\mathrm{i}}|_{\partial D}) = u^\infty = S^\infty\varphi = S^\infty S^{-1}(-u^{\mathrm{i}}|_{\partial D}) = \overline{\gamma_m} H^* S^{-1}(-u^{\mathrm{i}}|_{\partial D}),$$

which means that

$$G = \overline{\gamma_m} H^* S^{-1}. \tag{13.3.10}$$

Further, by the definition of F, we have

$$Fg = u^\infty = -\overline{\gamma_m} H^* S^{-1} H g,$$

which means that $F = -\overline{\gamma_m} H^* S^{-1} H$ and by using (13.3.10), we have $F = -GH$. On the other hand, from (13.3.10) we obtain

$$GS = \overline{\gamma_m} H^*$$

and taking its dual

$$\gamma_m H = S^* G^*. \tag{13.3.11}$$

Therefore, combining this with $F = -GH$, we have

$$F = -G\left(\gamma_m^{-1} S^* G^*\right) = -\gamma_m^{-1} G S^* G^*,$$

which proves (i) of the theorem.

Next we will prove (ii). The key to the proof is the following identity given as

$$2\pi\{\langle Fg, h\rangle - \langle g, Fh\rangle\} = i\kappa\langle Fg, Fh\rangle \ (g, h \in L^2(\mathbb{S})). \tag{13.3.12}$$

This can be shown by taking the total fields $u = u_g^s + v_g$ and $w = u_h^s + v_h$ with incident waves given by Herglotz wave functions v_g and v_h, respectively, and consider the equality

$$0 = \int_{\partial D}\left(u\frac{\partial \bar{w}}{\partial \nu} - \bar{w}\frac{\partial u}{\partial \nu}\right)\mathrm{d}s = I_1 + I_2 + I_3 \tag{13.3.13}$$

with

$$I_1 = \int_{\partial D}\left(u_g^s\frac{\partial \overline{u_h^s}}{\partial \nu} - \overline{u_h^s}\frac{\partial u_g^s}{\partial \nu}\right)\mathrm{d}s, \quad I_2 = \int_{\partial D}\left(u_g^s\frac{\partial \overline{v_h}}{\partial \nu} - \overline{v_h}\frac{\partial u_g^s}{\partial \nu}\right)\mathrm{d}s$$

$$I_3 = \int_{\partial D}\left(v_g\frac{\partial \overline{u_h^s}}{\partial \nu} - \overline{u_h^s}\frac{\partial v_g}{\partial \nu}\right)\mathrm{d}s. \tag{13.3.14}$$

In fact, by using Green's formula and the asymptotic behaviors of the scattered waves u_g^s, u_k^s, we have $I_1 = -2i\kappa\langle Fg, Fh\rangle$, and for I_2, I_3, we further use the representation of far field pattern in terms of scattered field as an integral over ∂D to obtain $I_2 = 4\pi\langle Fg, h\rangle$, $I_3 = -4\pi\langle g, Fh\rangle$.

Now consider $\langle g, i\kappa F^*Fh\rangle$ for any $g, h \in L^2(\mathbb{S})$, then by (13.3.12), it can be easily seen to be equal to $\langle g, 2\pi(F - F^*)h\rangle$, and hence

$$i\kappa F^*F = 2\pi(F - F^*). \tag{13.3.15}$$

On the other hand from the reciprocity relation we know that

$$F^*g = \overline{RFR\bar{g}} \text{ with } (Rg)(\hat{x}) = g(-\hat{x}) \tag{13.3.16}$$

for any $g \in L^2(\mathbb{S})$. This yields

$$\langle g, -i\kappa FF^*h\rangle = \langle g, 2\pi(F - F^*)h\rangle$$

for any $g, h \in L^2(\mathbb{S})$ and hence

$$i\kappa FF^* = 2\pi(F - F^*). \tag{13.3.17}$$

Therefore, by (13.3.15), (13.3.17), we have $F^*F = FF^*$.

The location of any eigenvalue λ of F with eigenfunction g can be seen as follows. Let $f = g$ in (13.3.12). Then, this implies $2\pi\lambda\|g\|^2 = i\kappa\,|\lambda|^2\|g\|^2 + 2\pi\bar{\lambda}\|g\|^2$ and hence we have $i\kappa\,|\lambda|^2 + 2\pi(\bar{\lambda} - \lambda) = 0$ which is equivalent to $|\lambda - (2\pi i)/\kappa| = (2\pi)/\kappa$. As for $\lambda_j \to 0$ $(j \to \infty)$, it is just a general property for any compact operators.

By lemma 12.4.3, lemma 13.2.5 and 13.2.3, F is an injective, compact operator with a dense range. Once we have that F is an injective, compact and normal operator, then it is well known that F has a complete orthonormal system (see [15]). This finishes the proof. □

Finally, we present the proofs to the lemmas 13.3.4–13.3.6.

Proof of lemma 13.3.4. By theorem 13.3.2, we have

$$GS^*\varphi_j = -\gamma_m\sqrt{\lambda_j}\psi_j \quad (j \in \mathbb{N}). \tag{13.3.18}$$

Then, a simple calculation shows that

$$\langle S\varphi_j, \varphi_\ell\rangle = c_j\delta_{j\ell} \text{ with } c_j = -\overline{\gamma_m}\frac{\overline{\lambda_j}}{|\lambda_j|}. \tag{13.3.19}$$

Now, let S_i be the single-layer potential on ∂D for $\kappa = \text{i}$. Then, it is easy to see that $S_\text{i} : H^{-1}(\partial D) \to L^2(\partial D)$ is not only an isomorphism but also a positive operator, and $S - S_\text{i} : H^{-1/2}(\partial D) \to H^{1/2}(\partial D)$, $K := S_\text{i}^{-1/2}(S - S_\text{i})S_\text{i}^{-1/2} : L^2(\partial D) \to L^2(\partial D)$ are compact. Further, (13.3.19) becomes

$$\left\langle (I + K)\tilde{\varphi}_j, \tilde{\varphi}_\ell \right\rangle = c_j\delta_{j\ell} \text{ with } \tilde{\varphi}_j = S_\text{i}^{1/2}\varphi_j. \tag{13.3.20}$$

The rest of the proof relies on the following result from functional analysis.

Theorem 13.3.7. *Let X be a Hilbert space and K be a linear compact operator on X with the property*

$$\text{Im}\langle K\phi, \phi\rangle \neq 0 \quad (0 \neq \phi \in X). \tag{13.3.21}$$

Further, $\{\phi_n\}_{n=1}^\infty$ *is linearly independent and complete in X with the property*

$$\left\langle (I + K)\phi_j, \phi_\ell \right\rangle = c_j\delta_{j\ell} \quad (j, \ell \in \mathbb{N}), \tag{13.3.22}$$

where the sequence $\{c_j\}_{j=1}^\infty \subset \mathbb{C}$ *satisfies*

$$|c_j| = r(j \in \mathbb{N}), \quad \text{Im}(c_j) \to 0 \quad (j \to \infty). \tag{13.3.23}$$

Then, $\{\phi_j\}_{j=1}^\infty$ *form a Riesz basis in X.*

The proof of this theorem is very long and can be seen in the book by Cakoni and Colton [13], p 138, theorem 7.11. Let us check that this theorem can be applied to complete the proof if we take $X = L^2(\partial D)$, $\phi_j = \tilde{\varphi}_j$ and K as we have already defined. It is very easy to check that $\{c_j\}_{j=1}^\infty$ satisfies the properties. The linear independence and completeness of $\{\tilde{\varphi}_j\}_{j=1}^\infty$ in $L^2(\partial D)$ follows from $S_\text{i}^{1/2} : H^{-1/2}(\partial D) \to L^2(\partial D)$ is an isomorphism, $G^* : L^2(\mathbb{S}) \to H^{-1/2}(\partial D)$ is an injective operator having a dense range and $\{\psi_j\}_{j=1}^\infty$ is a complete orthonormal system in $L^2(\mathbb{S})$. So, we are left to check the property (13.3.21). Take $\varphi \in L^2(\partial D)$ such that $\varphi \neq 0$. Since S_i has a positive kernel, we have $\text{Im}(K\varphi, \varphi) = \text{Im}(S\psi, \psi)$ with $\psi = S_\text{i}^{-1/2}\varphi$. Hence, we only have to show that $\psi \in H^{-1/2}(\partial D)$, $\text{Im}(S\psi, \psi) \neq 0$ implies $\psi = 0$. For that let

$$v(x) = \int_{\partial D} \Phi(x, y)\,\psi(y)\,\text{d}s(y).$$

Then, by the jump formula of the normal derivative of the single-layer potential and asymptotic behavior of v, we have

$$(S\psi, \psi) = \langle v, \partial_\nu v_- - \partial_\nu v_+\rangle = \int_{B_R} (|\nabla v|^2 - \kappa^2 |v|^2)\,dx + i\kappa \int_{\partial B_R} |v|^2\,ds + o(1) \quad (R \to \infty),$$

where $\partial_\nu v_+$ and $\partial_\nu v_-$ are the traces of $\partial_\nu v$ with unit normal vector ν of ∂D directed outside D taken from outside D and inside D, respectively. Hence,

$$\kappa \lim_{R\to\infty} \int_{\partial B_R} |v|^2\,ds = \mathrm{Im}(S\psi, \psi) = 0$$

and by Rellich's lemma, we have $v = 0$ in $\mathbb{R}^m \backslash \bar{D}$. Then, by the continuity of the single-layer potential, we have $S\psi = 0$ and further using that $S : H^{-1/2}(\partial D) \to H^{1/2}(\partial D)$ is an isomorphism, we have $\psi = 0$. This completes the proof.

Proof of lemma 13.3.5. The second equality of (13.3.6) is easy to see. In fact $\psi \in (F^*F)^{1/4}(L^2(\mathbb{S}))$ is equivalent to

$$(F^*F)^{-1/4}\psi = \sum_{j=1}^{\infty} \sigma_j \psi_j \text{ with } \sum_{j=1}^{\infty} |\sigma_j|^2 < \infty$$

and this can be rewritten in the form

$$\psi = \sum_{j=1}^{\infty} \rho_j \psi_j \text{ with } \sum_{j=1}^{\infty} \frac{|\rho_j|^2}{|\lambda_j|} < \infty,$$

where $\rho_j = \sqrt{|\lambda_j|}\,\sigma_j$.

Now we proceed to prove the first equality of (13.3.6). To begin with, we first observe that $S^* : H^{-1/2}(\partial D) \to H^{1/2}(\partial D)$ is an isomorphism, because $S^*\varphi = \overline{S\bar{\varphi}}$ for $\varphi \in H^{-1//2}(\partial D)$ and $S : H^{-1/2}(\partial D) \to H^{1/2}(\partial D)$ is an isomorphism.

Let $\psi = G\varphi$ with $\varphi \in H^{1/2}(\partial D)$. Then, by lemma 13.3.4, $(S^*)^{-1}\varphi$ can be given in the form $(S^*)^{-1}\varphi = \sum_{j=1}^{\infty} \alpha_j \varphi_j$ with $\sum_{j=1}^{\infty} |\alpha_j|^2 < \infty$. Hence, by using (13.3.18), we have

$$\psi = (GS^*)\{(S^*)^{-1}\varphi\} = \sum_{j=1}^{\infty} \alpha_j (GS^*)\varphi_j = \sum_{j=1}^{\infty} \rho_j \psi_j$$

with $\rho_j = -\gamma_m \alpha_j \sqrt{\lambda_j}$. To derive our desired conclusion from this, we only have to note that

$$\sum_{j=1}^{\infty} \frac{|\rho_j|^2}{|\lambda_j|} = |\gamma_m|^2 \sum_{j=1}^{\infty} |\alpha_j|^2 < \infty.$$

Conversely, let

$$\psi = \sum_{j=1}^{\infty} \rho_j \psi_j \text{ with } \sum_{j=1}^{\infty} \frac{|\rho_j|^2}{|\lambda_j|} < \infty.$$

We need to find φ in $H^{-1/2}(\partial D)$ such that $G(S^*\varphi) = \psi$. For that take $\varphi = \sum_{j=1}^{\infty}\alpha_j\varphi_j$ with $\alpha_j = -\gamma_m^{-1}\rho_j/\sqrt{\lambda_j}$. Then, $\varphi \in H^{-1/2}(\partial D)$ because $\sum_{j=1}^{\infty}|\alpha_j|^2 < \infty$. Further, again by (13.3.18), $G(S^*\varphi) = \psi$. This completes the proof. □

Proof of lemma 13.3.6. By lemma 13.3.4, $\{\varphi_j\}_{j=1}^{\infty}$ is a Riesz basis in $H^{-1/2}(\partial D)$. Then, $\{S\varphi_j\}_{j=1}^{\infty}$ is a Riesz basis in $H^{1/2}(\partial D)$ because

$$S: H^{-1/2}(\partial D) \to H^{1/2}(\partial D)$$

is an isomorphism. Hence, it is enough to show that $\{(F^*F)^{-1/4}GS\varphi_j\}_{j=1}^{\infty}$ is a Riesz basis in $L^2(\mathbb{S})$. By recalling the definition of φ_j and (13.3.2), we have

$$\begin{aligned}(F^*F)^{-1/4}GS\varphi_j &= \frac{1}{\sqrt{\lambda_j}}(FF^*)^{-1/4}GSG^*\psi_j \\ &= -\gamma_m F\psi_j = -\gamma_m\lambda_j\psi_j = -\gamma_m\sqrt{\lambda_j/|\lambda_j|}\,\psi_j\end{aligned}$$

for any $j \in \mathbb{N}$ and this immediately shows that $\{(F^*F)^{-1/4}GS\varphi_j\}_{j=1}^{\infty}$ is a Riesz basis in $L^2(\mathbb{S})$ because $\{\psi_j\}_{j=1}^{\infty}$ is a complete orthonormal system in $L^2(\mathbb{S})$. This completes the proof. □

Bibliography

[1] Colton D and Kress R 1998 *Inverse Acoustic and Electromagnetic Scattering Theory* (*Applied Mathematical Sciences* vol 93) 2nd edn (Berlin: Springer)

[2] Colton D and Kirsch A 1996 A simple method for solving inverse scattering problems in the resonance region *Inverse Problems* **12** 383

[3] Colton D, Piana M and Potthast R 1997 A simple method using Morozov's discrepancy principle for solving inverse scattering problems *Inverse Problems* **13** 1477

[4] Kirsch A and Grinberg N 2008 *The Factorization Method for Inverse Problems* (Oxford: Oxford University Press)

[5] Potthast R 1996 A fast new method to solve inverse scattering problems *Inverse Problems* **12** 731

[6] Potthast R 2001 *Point Sources and Multipoles in Inverse Scattering Theory* (*Chapman and Hall/CRC Research Notes in Mathematics* vol 427) (Boca Raton, FL: CRC)

[7] Ikehata M 1998 Reconstruction of the shape of the inclusion by boundary measurements *Commun.* **23** 1459–74

[8] Griesmaier R 2011 Multi-frequency orthogonality sampling for inverse obstacle scattering problems *Inverse Problems* **27** 085005

[9] Ito K, Jin B and Zou J 2012 A direct sampling method to an inverse medium scattering problem *Inverse Problems* **28** 025003

[10] Potthast R 2010 A study on orthogonality sampling *Inverse Problems* **26** 074015

[11] McLean W 2000 *Strongly Elliptic Systems and Boundary Integral Equations* (Cambridge: Cambridge University Press)

[12] Arens T 2004 Why linear sampling works *Inverse Problems* **20** 163

[13] Cakoni F and Colton D 2006 *Qualitative Methods in Inverse Scattering Theory* (New York: Springer)

[14] Kirsch A 1998 Characterization of the shape of a scattering obstacle using the spectral data of the far field operator *Inverse Problems* **14** 1489

[15] Ringrose J R 1991 *Compact Non-Self-Adjoining Operators* (London: Van Nostrand Reinhold)

IOP Publishing

Inverse Modeling

An Introduction to the theory and methods of inverse problems and data assimilation

Gen Nakamura and Roland Potthast

Chapter 14

Probe methods

Probing the area where unknown scatterers are located is an old idea: *radar* sends a pulse into an unknown area and records the response. The time between sending the pulse and the answer tells you the distance of the object. Probe methods have seen a rapid development with many new ideas since the 1990s by authors such as Ikehata, Nakamura, Potthast, Luke, Sylvester, Kusiak, Liu, Schulz and many more.

Modern probe methods employ a range of particular ways to carry out probing. The idea is that with mathematical tools we can do more than just sending a physical pulse into an area. When we know the scattered field for a particular selection of incident waves (given by our physical device on which our measurements are based), we can construct the answer for other, very special waves. This provides a rich field of study, since the answer for special waves will tell us more about the location and physical nature of the scatterer under consideration.

In this chapter we provide an introduction to several probe methods compare figure 14.2. One basic approach is the construction of the scattered field or energy of *point sources* with source points sampling the area under consideration. This has already been employed by the *point source method* for field reconstruction, see section 12.4. Here, we use it to formulate probing procedures.

We will introduce and study the *singular sources method* (SSM) in section 14.1. It was suggested by Potthast in 1999 [1], based on ideas of Isakov, Kirsch and Kress for a uniqueness proof. It turned out to be the far field version of the *probe method* suggested by Ikehata in 1998 [2], which we describe and realize in section 14.2. The equivalence proof which was first given by Sini, Nakamura and Potthast can be found in section 14.2.2.

Using more general sampling functionals defined on domains has been investigated by Luke, Potthast, Schulz, Sini and others [3–5], for a review we refer the reader to [5]. The basic ideas are also visualized in figure 14.1. We describe a multi-wave version of the *no-response test* in section 14.3. The one-wave no-response test is a method to test analytic extensibility, see section 15.2. The multi-wave version as

doi:10.1088/978-0-7503-1218-9ch14

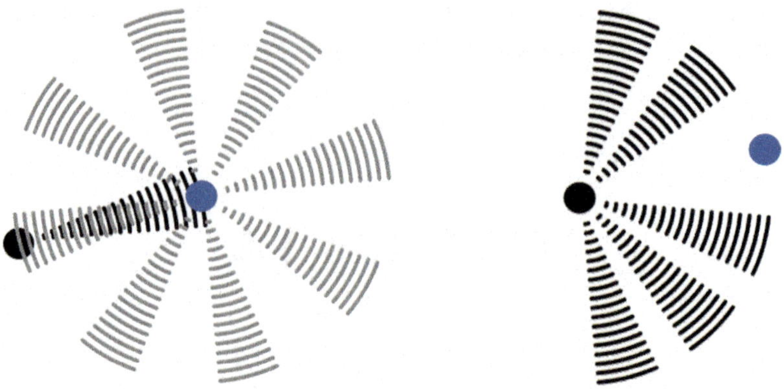

Figure 14.1. Two types of probe methods. In the *first* case, the black source sends pulses into different directions and records their response. Here, the *response* of the blue scatterer leads to its detection. We will study special *virtual* pulses, which can be constructed when the full scattered field for many different incident fields is measured. In the *second* example, pulses are sent everywhere except for one region. We can find the scatterer by searching for 'no response', the basic idea of the *no-response test*.

studied below will provide full shape reconstructions of objects. The method is equivalent to a multi-wave version of the *range test* which was introduced by Sylvester, Kusiak and Potthast (see section 15.1).

Recently probing methods have been successfully employed for solving scattering problems in the *time domain*. A *time-domain probe method* relying on the causality principle has been suggested by Burkard and Potthast [6] for rough surface reconstructions in 2009. We will describe the ideas and realization in chapter 16. Time-domain probe methods link back to the original ideas of travel time tomography. But they use modern construction tools whose range by far exceeds the linearized arguments based on pure travel time.

14.1 The SSM

The SSM employs point sources for probing some area under consideration to find the shape and physical properties of scatterers. Here, in subsection 14.1.1 we will first introduce the basic ideas of how to construct and employ the scattered field of point sources from the knowledge of the far field pattern for incident plane waves.

The convergence analysis is carried out in subsection 14.1.5. In subsection 14.1.4 we describe a very efficient *contraction scheme* for shape reconstructions when one or several objects are to be found. The *contraction scheme* is an *iterative method* for shape reconstruction based on indicator functions which can be evaluated only under special geometric assumptions. An alternative will be given by the *needle approach* of Ikehata and Nakamura in section 14.2.

14.1.1 Basic ideas and principles

We consider scattering by a singular source

$$u^{\mathrm{i}}(x) = \Phi(x, z), \quad x \in \mathbb{R}^m \tag{14.1.1}$$

with *source point* $z \in \mathbb{R}^m \backslash \bar{D}$. The scattered field for this singular source is denoted as $\Phi^s(x, z)$ for $x \in \mathbb{R}^m$. For scattering by a sound-soft obstacle from the boundary condition $\Phi^s(x, z) = -\Phi(x, z)$ for $x \in \partial D$, $z \in \mathbb{R}^m \backslash \bar{D}$ we derive

$$\left|\Phi^s(x, z)\right| \to \infty, \quad z \to x \in \partial D. \tag{14.1.2}$$

Moreover, in our convergence analysis section we will show that we have

$$\left|\Phi^s(z, z)\right| \to \infty, \quad z \to \partial D. \tag{14.1.3}$$

The behavior of the left-hand side of (14.1.3) is visualized in figure 14.3.

The idea of the SSM is to use the behavior (14.1.3) to detect the unknown shape ∂D. The indicator function

$$\mu(z) := \left|\Phi^s(z, z)\right| \tag{14.1.4}$$

for z in subsets of $\mathbb{R}^m \backslash \bar{D}$ can be constructed from the far field pattern or measured scattered field. We show the indicator function in figure 14.3.

In the simplest case we will now describe this reconstruction by the techniques of the point source method introduced in section 12.4. Later, we will also introduce an

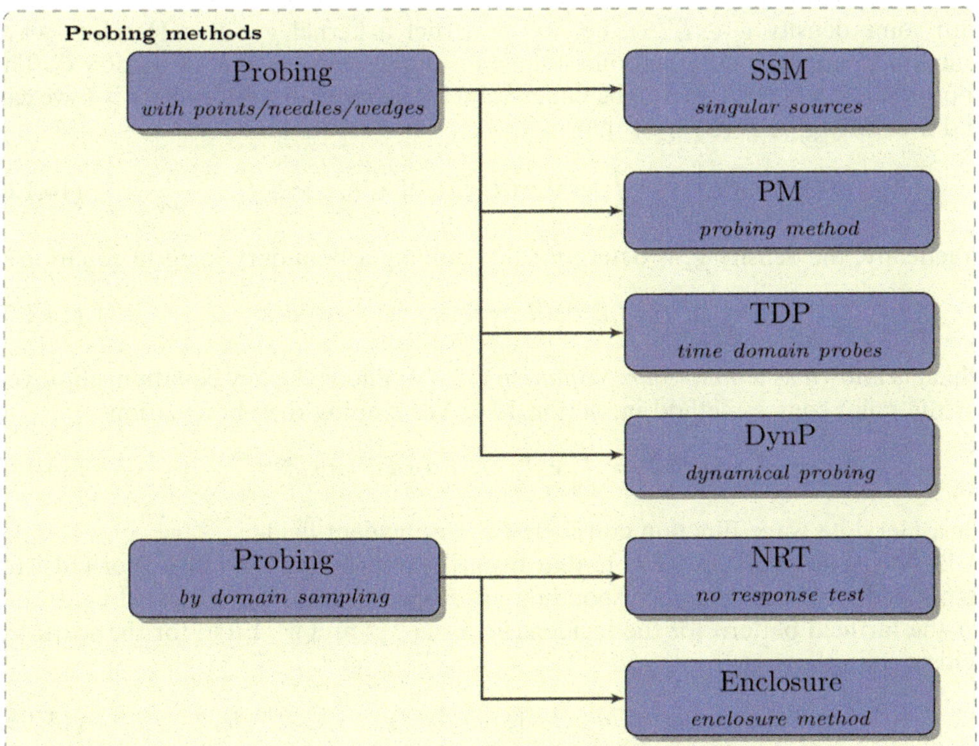

Figure 14.2. Different probing methods studied in this chapter.

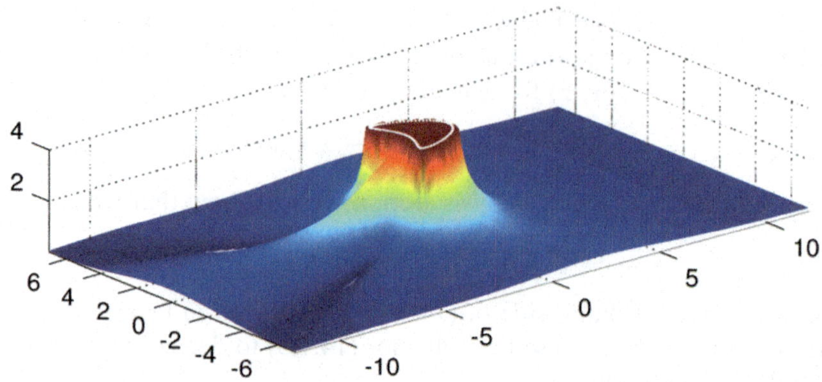

Figure 14.3. We show the indicator function (14.1.4) for the case of an impenetrable object with a Dirichlet boundary condition. The function has strong growth at the boundary of the scatterer.

alternative approach which relies on the potential approach of Kirsch and Kress introduced in section 12.3.

A. We first choose some approximation domain $G = G_z$ with $\bar{D} \subset G$ and $z \not\in \bar{G}$. Then we use the point source approximation

$$\Phi(x, z) \approx \int_{\mathbb{S}} e^{i\kappa x \cdot d} g_z(d)\, ds(d), \quad x \in \bar{G} \tag{14.1.5}$$

with some density $g_z \in L^2(\mathbb{S})$, i.e. we construct a kernel g_z for a Herglotz wave function to approximate the point source on the domain of approximation G. The approximation in (14.1.5) is to be understood in the sense that for every $\epsilon > 0$ we can find a density $g_z \in L^2(\mathbb{S})$ such that

$$\left| \Phi(x, z) - \int_{\mathbb{S}} e^{i\kappa x \cdot d} g_z(d)\, ds(d) \right| \leqslant \epsilon, \quad x \in \bar{G}. \tag{14.1.6}$$

Practically, the density g_z is determined by solving a boundary integral equation

$$\int_{\mathbb{S}} e^{i\kappa x \cdot d} g_z(d)\, ds(d) = \Phi(x, z), \quad x \in \partial G, \tag{14.1.7}$$

which is known as a *point source equation* (12.4.4) and is the key equation employed for the point source method in section 12.4. We employ the abbreviation

$$w^i[g_z](x) := \int_{\mathbb{S}} e^{i\kappa x \cdot d} g_z(d)\, ds(d), \quad x \in \mathbb{R}^m \tag{14.1.8}$$

for a Herglotz wave function considered as an incident field.

B. As the second step we note that from the approximation (14.1.5) or (14.1.6), respectively, we derive a corresponding approximation for the scattered fields and for the far field pattern for the incident fields $\Phi(\cdot, z)$ and w^i. First, for the scattered fields obtain the approximation

$$\Phi^s(x, z) \approx \underbrace{\int_{\mathbb{S}} u^s(x, d) g_z(d)\, ds(d)}_{=:\, w^s[g_z](x)}, \quad x \in \mathbb{R}^m \backslash D, \tag{14.1.9}$$

with some density $g_z \in L^2(\mathbb{S})$. Second, passing on both sides of (14.1.9) to the corresponding far field patterns we derive

$$\Phi^\infty(\hat{x}, z) \approx \underbrace{\int_{\mathbb{S}} u^\infty(\hat{x}, d) g_z(d)\, ds(d)}_{=:\, w^\infty[g_z](\hat{x})}, \quad \hat{x} \in \mathbb{S}. \tag{14.1.10}$$

The field $w^s[g_z]$ is the scattered field for the incident field $w^i[g_z]$ defined in (14.1.8). Its far field pattern is given by $w^\infty[g_z]$. Again, a rigorous convergence analysis is worked out in section 14.1.5.

C. Finally, we apply the point source method to the far field patterns $\Phi^\infty(\cdot, z)$ and w^∞ given in (14.1.10) to reconstruct the scattered field $\Phi^s(x, z)$ for $x \in \mathbb{R}^m \backslash \bar{D}$ and the field w^s on $x \in \mathbb{R}^m \backslash \bar{D}$. Then we obtain

$$\Phi^s(x, z) \approx \int_{\mathbb{S}} \int_{\mathbb{S}} u^\infty(-\hat{x}, d) g_z(d) g_x(\hat{x})\, ds(d)\, ds(\hat{x}) \tag{14.1.11}$$

$$= \int_{\mathbb{S}} \int_{\mathbb{S}} g_x(-\hat{x}) u^\infty(\hat{x}, d) g_z(d)\, ds(d)\, ds(\hat{x}) \tag{14.1.12}$$

for $x, z \in \mathbb{R}^m \backslash \bar{D}$.

Definition 14.1.1 (SSM). *The SSM in acoustics for impenetrable scatterers uses the indicator function $|\Phi^s(z, z)|$ as in (14.1.4) to find the shape of an unknown scattering obstacle. The indicator function is reconstructed using the point source method (14.1.5)–(14.1.11).*

The SSM can be formulated for penetrable homogeneous and for inhomogeneous scatterers and for cracks. In different settings the type of singularity under consideration needs to be adjusted to the properties of the scatterer under consideration.

We demonstrate the double application of the point source method in part IV of the following code 14.1.2. We assume the far field pattern $u^\infty(\theta, d)$ on a discrete set of points $\theta_\xi \in \mathbb{S}$, $\xi = 1, \ldots, m$ and $d_\xi \in \mathbb{S}$, $\xi = 1, \ldots, m$, to be given by the matrix $\mathbf{F} \in \mathbb{R}^{m \times m}$, containing a far field pattern $u^\infty(\cdot, d)$ for an incident plane wave of direction of incidence d in each of its columns. With the vectorial density $\mathbf{g}_1(z) \in \mathbb{R}^m$ solving (14.1.5) by Tikhonov regularization and $\mathbf{g}_2(z)$ analogously, we obtain the reconstruction in the form

$$\mathbf{w}^s(z, z) = \gamma\, \mathbf{g}_2^T(z) \circ \mathbf{F} \circ \mathbf{g}_1(z), \tag{14.1.13}$$

where $\circ$ denotes matrix multiplication, with γ given by (12.4.3).

In part IV of code 14.1.2 we realize the construction of the densities $\mathbf{g}_1, \mathbf{g}_2$ simultaneously for all points z in the grid of points (pvec1, pvec2), which can be carried out in matrix form by setting up the right-hand side $\mathbf{\Phi}(\cdot, z)$ as a matrix where each row corresponds to a different choice of z, see the definitions of `PhiMat` and `PhiMatC`.

Code 14.1.2. *The following script* `sim_14_1_2_b_SSM.m` *reconstructs the indicator function of the SSM in the exterior of some approximation domain G, it converges under the condition that* $\bar{D} \subset G$. *First run* `sim_14_1_2_a_F.m` *etc, to calculate the far field operator* `F` *and further variables. Finally, the graphical display is generated by* `sim_14_1_2_c_graphics.m` *The result is shown in figure 14.4.*

```
% I Setup of Test Domain
NG = 120;                            % number of points on test domain G
z1 = 3;                              % center of test domain G, comp.1
z2 = 0;                              % center of test domain G, comp.2
hG = 2*pi/NG;                        % grid constant for test domain
tG = 0:hG:2*pi-hG;                   % parametrization grid for G
RG = 4;                              % radius of test domain
yG1 = RG*cos(tG)+z1;                 % boundary of test domain comp.1
yG2 = RG*sin(tG)+z2;                 % boundary of test domain comp.2
yGffmat1  = repmat(yff1,NG,1);         % matrix of ff points comp.1
yGffmat2  = repmat(yff2,NG,1);         % matrix of ff points comp.2
yGmat1    = repmat(yG1.',1,ffN);       % matrix of points of G comp.1
yGmat2    = repmat(yG2.',1,ffN);       % matrix of points of G comp.1

% II Herglotz operator for evaluation on test domain G
HG    = exp(1i*kappa*(yGffmat1.*yGmat1+yGffmat2.*yGmat2))*hff;

% III Point Source Matrix for Evaluation
epsmat    = eps*ones(M,NG);         % matrix for cutting singularity
rGmat1    = repmat(pvec1,1,NG)-repmat(yG1,M,1);
rGmat2    = repmat(pvec2,1,NG)-repmat(yG2,M,1);
rGmat     = max(sqrt(rGmat1.^2 + rGmat2.^2),epsmat).';
PhiMat   = 1i/4*besselh(0,1,kappa*rGmat);
PhiMatC   = conj(1i/4*besselh(0,1,kappa*rGmat));

% IV Reconstruct the density for far field representation
alphaPSM = 1e-9;                    % regularization parameter
gPSM  = (alphaPSM*eye(ffN,ffN) + HG'*HG)\HG'*PhiMat;
gPSM2  = (alphaPSM*eye(ffN,ffN) + HG'*HG)\HG'*PhiMatC*hff;

for m=1:M  % reconstruction of SSM functional
    wRec(m,1) = 1/fac*(gPSM2(:,m)')*(F*gPSM(:,m));
end
Ampl = 20;     % amplification for better visualization
w = Ampl*wRec; % function for graphical representation
```

One important point for the reconstruction of the indicator function is the choice of the approximation domain G in (14.1.5). To obtain a proper reconstruction we need the unknown scatterer D to be in the interior of G. If we pick an arbitrary approximation domain G this will in general not be the case.

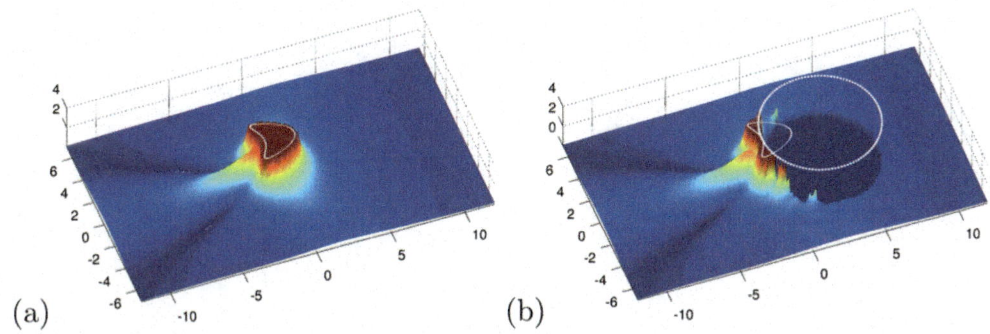

Figure 14.4. The simulated (a) and reconstructed (b) indicator function (14.1.4) for the case of an impenetrable object with Dirichlet boundary condition, following the elementary scheme (14.1.5)–(14.1.11) with one fixed test domain G which is indicated in (b) by a white circle. The code is given in code 14.1.2. The reconstruction can be valid only outside G, we have set the field to zero inside G. Three approaches to employ the elementary reconstruction tool are described in subsections 14.1.2–14.1.4.

There are three possible strategies to the geometrical question of how to choose the approximation domain G for implementing the SSM:

1. the *needle scheme*,
2. the *domain sampling method*, and
3. the *contraction scheme*.

All of them lead to convergent methods for shape reconstruction via the probe or singular sources functional. We will use the following subsections 14.1.2–14.1.4 to introduce these methods in detail and study their numerical realization in more depth.

14.1.2 The needle scheme for probe methods

The following geometrical idea can be applied to different probing methods. It was first suggested for the probe method of Ikehata, which we will explain later in section 14.2. His probe method uses the near field instead of the far field pattern. Here, we will develop a version of the probe method which uses the far field pattern. It becomes quite similar to the previous SSM based on the point source method, with equivalence studied in section 14.2.2.

To start with let $\Omega = B_R$ be an open circle or ball with radius $R > 0$ which contains $\bar{D}$. Let $c = \{c(t)\,;\, 0 \leqslant t \leqslant 1\}$ be a continuous curve called a *needle* which joins two points $c(0)$, $c(1) \in \partial B_R$, $c(t) \in B_R\,(0 < t < 1)$ and does not intersect to itself, see figure 14.5. For $z = c(t) \in \Omega$ with $0 < t < 1$, we define

$$c_z := \{c(s);\, 0 \leqslant s \leqslant t\}, \quad z = c(t), \tag{14.1.14}$$

and take a non-vibrating domain $G = G_z$ as an approximation domain which is strictly contained in $B_R \backslash \bar{c}_z$.

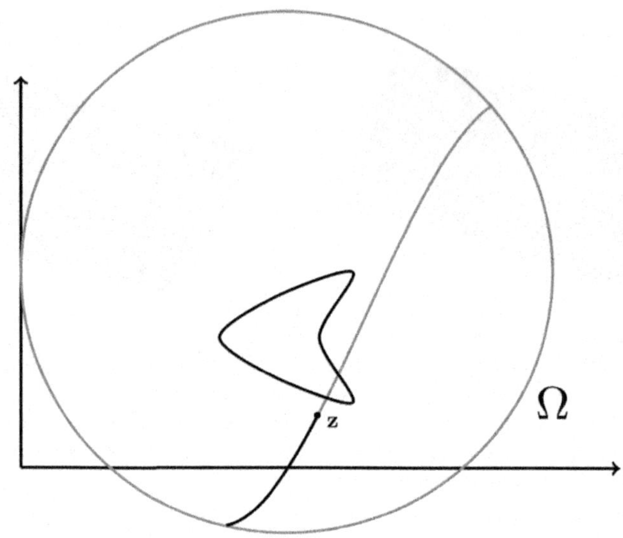

Figure 14.5. The key idea of the needle approach for probing an area with an unknown scatterer or inclusion. The point z moves along the curve which connects two points on the boundary $\partial\Omega$ of the domain Ω.

The idea is that in principle we use domains G_z such that $(B_R\backslash\bar{c}_z)\backslash\bar{G}_z$ is a very narrow domain. The specialty of the needle scheme as visualized in figure 14.5 is to take such an approximation domain attached to a needle and let $z = c(t)$ move along the needle c. We give an orientation to c such that $c(0)$ and $c(1)$ are the starting point and end point, respectively.

On ∂G_z, we approximate $\Phi(\cdot, z)$ by a sequence of Herglotz waves $v_{g_z^n}$. Then we compute the indicator function:

$$\mu(z, c) := \lim_{n\to\infty} \left| \int_{\mathbb{S}} \int_{\mathbb{S}} u^{\infty}(-\theta, d) g_z^n(\theta) g_z^n(d) \, \mathrm{d}s(\theta) \, \mathrm{d}s(d) \right|. \tag{14.1.15}$$

We state the following theorem.

Theorem 14.1.3 (Needle approach to probing). *For the needle approach the boundary ∂D of the unknown scatterer or inclusion is characterized by the following properties:*

(i) If a needle c *does not intersect with ∂D, then $\mu(z, c) < \infty$ $(z \in c)$.*

(ii) Let c *intersect with ∂D at $z = z(t)$ with some $0 < t < 1$ for the first time. Then, this t can be characterized as*

$$t = \sup\{0 < t' < 1; \mu(c(s), c) < \infty \; (0 < s < t')\}. \tag{14.1.16}$$

This means $\lim_{s\uparrow t} \mu(c(s), c) = \infty$ and we can say that the t is the smallest upper bound such that $\mu(c(s), c)$ stays bounded for $0 < s < t$.

The proof of this theorem will be worked out in section 14.1.5, where we will give the convergence for the SSM.

14.1.3 Domain sampling for probe methods

One of the disadvantages of the needle approach is the strong non-convexity of the approximation domains which follow the peaked geometry of the needle. This leads to strong numerical ill-conditioning of the point source approximation (14.1.5). It is much more stable if we use convex domains or domains which have a weak non-convexity.

One possible solution to the difficulty is what is known as *domain sampling*. the second possibility is the *contraction scheme* which will be introduced in the next section 14.1.4.

The key idea of domain sampling is shown in figure 14.6. We choose an approximation domain G_j, $j = 1, 2, 3, \ldots$, and a point $z \in \mathbb{R}^m \setminus G_j$. Then, the point source equation (14.1.7) is solved with z and G_j. If the scattered field Φ^s can be analytically extended from the exterior of some large ball into the exterior of G_j, then the reconstruction of $\Phi^s(x, z)$ by (14.1.11) is convergent for $x, z \in \mathbb{R}^m \setminus \bar{G}_j$. We call a domain G_j positive, if the reconstruction converges in the whole exterior of G. To test for positivity of a point z, we can carry out a simple *convergence test* as follows:

1. First, we solve the point source equation by Tikhonov regularization with two different positive small regularization parameters α_1 and $\alpha_2 := \alpha_1/2$. This leads to two different densities $g_{z,j}^{(\alpha_1)}$ and $g_{z,j}^{(\alpha_2)}$.
2. We then calculate reconstructions $\Phi^s_{\alpha_1,j}(z, z)$ and $\Phi^s_{\alpha_2,j}(z, z)$ for points $z \in \mathbb{R}^m \setminus \bar{G}_j$ using equation (14.1.11).
3. Finally, we mark the point z as positive for the sampling domain G_j if

$$\left|\Phi^s_{\alpha_1,j}(z, z) - \Phi^s_{\alpha_2,j}(z, z)\right| \leqslant \tau, \tag{14.1.17}$$

where τ is some small *discriminating* constant which has to be chosen appropriately depending on α_1, α_2. Positivity of points z indicates that the calculation of $\Phi^s(z, z)$ is convergent for the sampling domain G_j under consideration.

Domain sampling calculates reconstructions on sets $M_j := \mathbb{R}^m \setminus \bar{G}_j$ for positive domains G_j and then integrates the reconstructions on different domains by taking weighted averages. Here, we call M_j positive if the corresponding domain G_j is positive. For these averages we first need to count for every point $z \in \mathbb{R}^m$ by how many positive reconstructions we have. Assume that we have $\ell \in \mathbb{N}$ domains under consideration. Then, we define

$$L(z) := \{j: \ z \text{ positive for the reconstruction via } G_j\}. \tag{14.1.18}$$

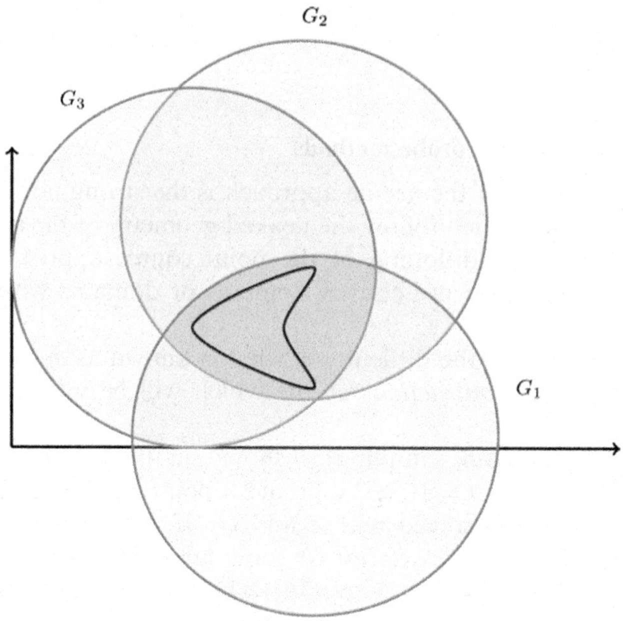

Figure 14.6. The key idea of domain sampling for the probe method. The dark gray area shows the intersection of positive domains, which approximates the unknown scatterer.

A reconstruction of $\Phi^s(z, z)$ is now calculated by

$$\Phi^s_{\text{rec}}(z, z) := \begin{cases} \dfrac{1}{\#L(z)} \sum\limits_{j \in L(z)} \Phi^s_{\alpha_1, j}(z, z), & \#L(z) > 0 \\ 0, & \text{otherwise.} \end{cases} \tag{14.1.19}$$

Here, we need to note that we calculate many reconstructions of $\Phi^s(z, z)$ arising from different approximation domains G_j, $j \in L(z)$. This is a very stable, but time consuming process. Of course, in principle one of these reconstructions is sufficient. Thus, for every point $z \in \mathbb{R}^m$ we could try to find one G_j with $z \notin \bar{G}_j$ which is positive. There is a lot of space for clever algorithms to choose the set of sampling domains G_j either by an automatic algorithm or by interactive choice.

We show the result of the reconstruction via (14.1.19) in figure 14.7. The sampling domains are shown in figure 14.8(b), the sum of the masks is displayed in figure 14.8(a). Here, we will skip the explicit representation of the algorithms, they can be found in the code repository, see `sim_14_1_2_h_SSM_domain_sampling.m` and `sim_14_1_2_i_graphics.m`.

14.1.4 The contraction scheme for probe methods

The *contraction scheme* suggested in [7] is an adaptive scheme for the choice of sampling domains for probe and sampling methods. It provides a set of rules for

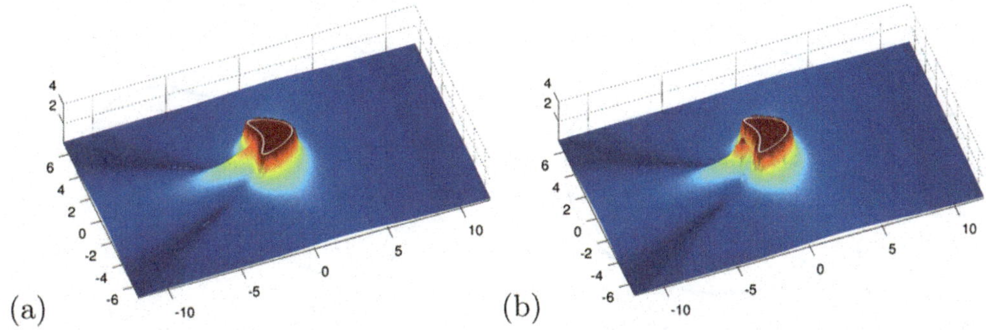

Figure 14.7. The simulated (a) and reconstructed (b) indicator function (14.1.4) for the case of an impenetrable object with a Dirichlet boundary condition, following the domain sampling scheme (14.1.19). Here, for better visibility, we set the reconstruction to a constant $c = 4$ where $\#L$ is zero or where $|\Phi^s_{\text{rec}}| > 4$. We have used $\kappa = 1$ with 80 far field measurement points and 15 sampling domains.

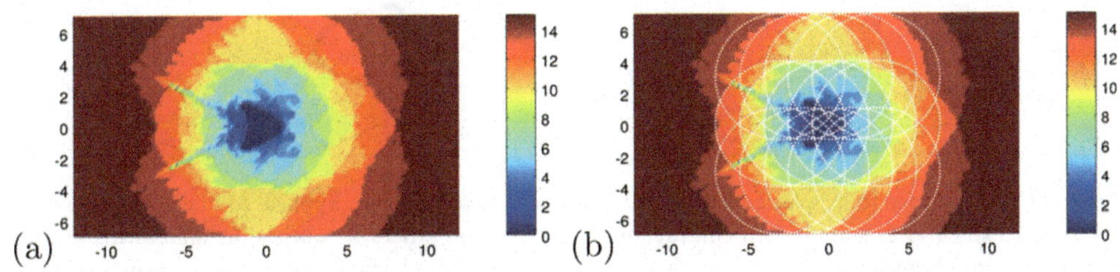

Figure 14.8. (a) The sum of the masks generated by the convergence test (14.1.17) applied to the reconstruction of $\Phi^s(z, z)$ by the SSM. In (b) we visualize the sampling domains which have been used for the reconstruction of figure 14.7. This is analogous to the masking techniques applied for field reconstructions, see figure 12.6 or figure 12.9.

step-by-step contraction of initially large sampling domains such that a successful reconstruction of the indicator function on the boundary of all sampling domains is possible.

The starting point for formulating the contraction scheme is the analytic result that the reconstruction of the probe functional by (14.1.11) outside a sampling domain G converges as long as the unknown scatterer D is inside G, i.e.

$$\bar{D} \subset G. \tag{14.1.20}$$

If we update G in an iterative procedure such that the *contraction condition* (14.1.20) is satisfied, we are always in the situation that we have convergence for the reconstruction of the indicator function outside our sampling domain. We will contract the test domain G in each step, where we stop contracting for points on the boundary of ∂G for which the indicator function reaches some threshold.

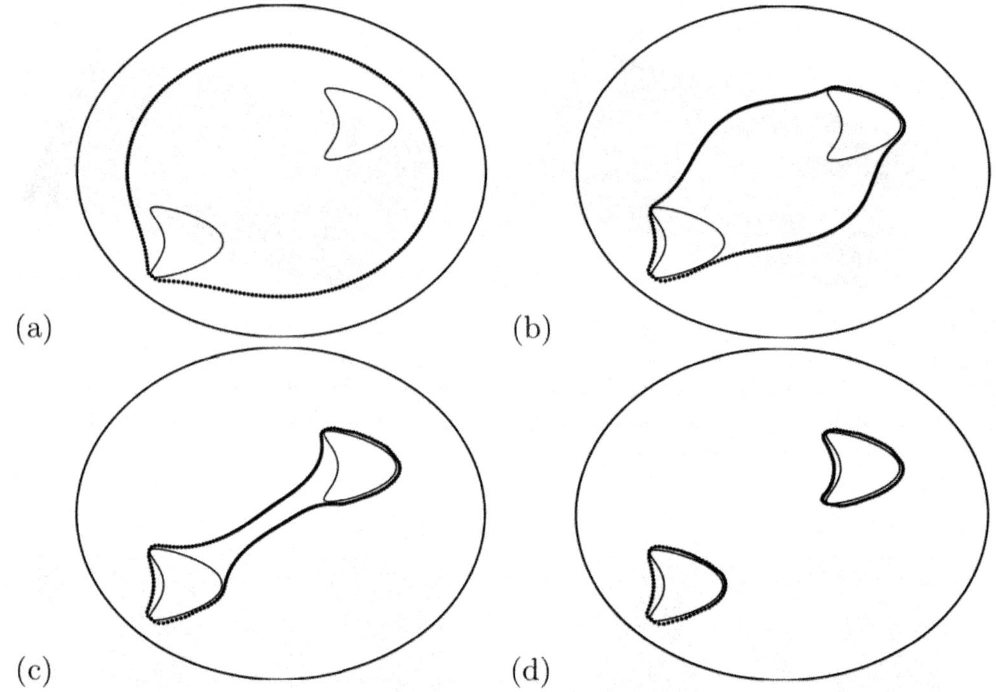

Figure 14.9. We show four selected steps from the *contraction scheme* applied to the *singular sources* functional. The curve is step-by-step contracting and only stops where it touches an unknown inclusion. Here, the measurement surface is indicated by the outer elipse. The original inclusions are shown as two kite-shaped curves.

In the following algorithm 14.1.4, we describe the contraction scheme for the reconstruction of a scatterer D which might consist of one or several separate components, see figure 14.9 with *two* such components.

Algorithm 14.1.4 (Contraction scheme for the probe method). *The contraction scheme for reconstructing one or several scatterers by the probe indicator function is an algorithm to iteratively update test domains* G *such that the unknown scatterers are kept in the interior of all test domains and the test domains converge towards an approximation of D. It consists of the following steps.*

1. *We start with some test domain G_0 such that the unknown scatterers D are inside G_0. Further, we define a constant c_{ssm} to discriminate* large *values of $\Phi^s(z, z)$ from* small *values. Then, we repeat the steps 2–5 until the stopping criterion is reached.*
2. *Given G_n, we reconstruct $\Phi^s(z, z)$ on the parallel surface*

$$G_n^{(h)} := \{x + h\nu(x) : \quad x \in \partial G_n\}, \tag{14.1.21}$$

where $\nu(x)$ is the outer unit normal of ∂G_n directed into the exterior of G_n. We call the reconstruction $w_{\text{rec}}(x)$, $x \in \partial G_n$.

3. *With some marching step size h_{m} we now define a marching operator*

$$\mathcal{M}(x) := \begin{cases} x - h_{\text{m}}\nu(x) & w_{\text{rec}}(x) < c_{\text{ssm}} \\ x & w_{\text{rec}}(x) \geqslant c_{\text{ssm}}. \end{cases} \tag{14.1.22}$$

An intermediate updated domain is now given by

$$\tilde{\Gamma}_{n+1} := \{\mathcal{M}(x) : \quad x \in \partial G_n\}. \tag{14.1.23}$$

This domain is no longer smooth due to the sharp discrimination between points for which $\Phi^{s}(z, z)$, $z = x + h\nu(x)$ is larger than c_{ssm} and those where the field is smaller. Further, by adding a multiple of the normal in each step, we lose one order of regularity in every update step. We need to employ the following smoothing step to restore regularity of the curve.

4. *We approximate $\tilde{\Gamma}_{n+1}$ by a sufficiently smooth surface $\Gamma_{n+1} = \partial G_{n+1}$. Here, we pose the further condition that points of $\tilde{\Gamma}_{n+1}$, where $w_{\text{rec}}(x)$ is larger than c_{ssm}, are not changed by the smoothing procedure.*
5. *If the boundary Γ_{n+1} comes close to intersecting itself, we split the domain G_{n+1} into two or more domains by removing the part of the boundary which comes close to itself and where w_{rec} is still small.*
6. *We check the stopping criterion. Here, we test whether there are still points $x \in \Gamma_{n+1}$ such that $w_{\text{rec}}(x) < c_{\text{ssm}}$. In this case we need further updates. But since we have the smoothing step which may partly revert the marching step, we might reach a stable state depending on our particular smoothing scheme. If we reach a stable state, we first reduce the amount of smoothing employed and try further updates. If the minimal smoothing level is reached, at a steady state we stop the iterations.*

The behavior of the contraction algorithm is visualized in figure 14.9. Starting with a large elipse, it shows a time step when the test domain touches the first unknown inclusion in (a). In figure 14.9(b) the test domain already fits half of the boundary of the two scatterers. Figure (c) shows how the curve contracts further, until it touches itself in the area between the two scatterers. Part of the curve is then removed and the scheme continues with a test domain G which consists of two separate components, which converge towards the unknown inclusion as shown in figure 14.9(d).

14.1.5 Convergence analysis for the SSM

There are two basic classes of convergence statements for probing methods. The first type of convergence is the *calculation of the indicator function* $\mu(z)$. Given the far field pattern $u^{\infty}(\cdot, d)$ for $d \in \mathbb{S}$ we calculate an approximation $\mu_{\alpha}(z)$ to $\mu(z)$, where α is some regularization parameter. The convergence

$$\mu_{\alpha}(z) \to \mu(z), \quad \alpha \to 0, \tag{14.1.24}$$

means that we can approximate the indicator function as well we wish under the condition that perfect measurements u^{∞} are available.

Second, there is convergence for the *shape reconstruction* problem. Given some approximate indicator function μ_α on Ω, we calculate an approximation D_α to the unknown scatterer D. Convergence of the shape reconstruction means that in some appropriate metric we have

$$D_\alpha \to D, \quad \alpha \to 0, \tag{14.1.25}$$

where $\alpha > 0$ denotes the regularization parameter of the problem.

For inverse problems with data error of size $\delta > 0$, the convergence statements are usually formulated in a slightly more general way. Given $\delta > 0$, there is $\alpha(\delta) > 0$ such that

$$\alpha(\delta) \to 0, \quad \delta \to 0, \tag{14.1.26}$$

and such that the convergence (14.1.24) or (14.1.25) is satisfied. We will see that by stability of the reconstruction of the indicator function, from (14.1.24) we can show both (14.1.25) and (14.1.26).

Before we go into the convergence of reconstructions of the indicator function, we need to study its behavior when $z \to \partial D$.

Lemma 14.1.5. *For scattering of an incident point source by a sound-soft scatterer D we have*

$$|\Phi^s(z, z)| \to \infty, \quad z \to \partial D, \tag{14.1.27}$$

uniformly for ∂D.

Proof. Close to the boundary ∂D with $z = x + h\nu(x)$, $x \in \partial D$, decompose the scattered field $\Phi^s(\cdot, z)$ into

$$\Phi^s(\cdot, x + h\nu(x)) = \Phi(\cdot, x - h\nu(x)) + u^s(\cdot, h) \tag{14.1.28}$$

where $u^s(\cdot, h)$ is the solution to the exterior Dirichlet problem for D with boundary values given by

$$f_h(y) := \Phi(y, x + h\nu(x)) - \Phi(y, x - h\nu(x)), \quad y \in \partial D$$

for $h > 0$ sufficiently small. For a boundary of class C^2 by elementary estimates for the singularity of type $1/|y - z|$ we can easily show that $f_h(y)$ is bounded by a constant for all sufficiently small $h > 0$, such that $u^s(\cdot, h)$ is bounded. Now, the behavior (14.1.27) is a consequence of the singularity of $\Phi(y, z)$ at $y = z$. □

Theorem 14.1.6 (Convergence of SSM indicator function). *For the SSM we have a convergence (14.1.24) of the indicator function defined by (14.1.4) or (14.1.15), for any set-up or strategy where the test domain G contains the unknown scatterer D in its interior and where* $z \in \mathbb{R}^3 \setminus \bar{G}$.

Proof. We proceed in several steps. In step (A) we show the convergence of the approximation of $\Phi^s(\cdot, z)$ by an integral over the far field pattern, construct a subsequent approximation of $\Phi^\infty(\cdot, z)$ in step (B) and then use the point source approximation again to obtain a reconstruction of $\Phi^s(z, z)$ in step (C).

(A) We first choose some approximation domain G with $\bar{D} \subset G$ and $z \not\in \bar{G}$. Then we use the point source approximation

$$\Phi(x, z) \approx \int_{\mathbb{S}} e^{i\kappa x \cdot d} g_z(d) \, ds(d), \quad x \in \bar{G} \tag{14.1.29}$$

with some density $g_z \in L^2(\mathbb{S})$, i.e. we construct a kernel g_z for a Herglotz wave function to approximate the point source on the domain of approximation G. The approximation in (14.1.29) is to be understood in the sense that for every $\epsilon > 0$ we can find a density $g_z \in L^2(\mathbb{S})$ such that

$$\left| \Phi(x, z) - \int_{\mathbb{S}} e^{i\kappa x \cdot d} g_z(d) \, ds(d) \right| \leqslant \epsilon, \quad x \in \bar{G}. \tag{14.1.30}$$

We recall the abbreviation $w^i[g_z]$ defined in (14.1.8) for the Herglotz wave function considered as an incident field.

(B) As the second step we note that from the approximation (14.1.29) or (14.1.30), respectively, we derive a corresponding approximation for the scattered fields and for the far field pattern for the incident fields $\Phi(\cdot, z)$ and w^i. First, for the scattered fields we estimate

$$\Phi^s(\hat{x}, z) \approx \int_{\mathbb{S}} u^s(\hat{x}, d) g_z(d) \, ds(d), \quad \hat{x} \in \mathbb{S}, \tag{14.1.31}$$

which means that given $\epsilon > 0$ there is a density $g_z \in L^2(\mathbb{S})$ such that

$$\left\| \Phi^s(\hat{x}, z) - \int_{\mathbb{S}} u^s(\hat{x}, d) g_z(d) \, ds(d) \right\|_\infty \leqslant c\epsilon, \tag{14.1.32}$$

with some constant c depending on the shape of the scatterer D and $L^\infty(\mathbb{S})$ norm $\| \cdot \|_\infty$. Second, the far field patterns can be estimated in the same way

$$\Phi^\infty(\hat{x}, z) \approx \int_{\mathbb{S}} u^\infty(\hat{x}, d) g_z(d) \, ds(d), \quad \hat{x} \in \mathbb{S}. \tag{14.1.33}$$

With the fields $w^s[g_z]$ and $w^\infty[g_z]$ defined in (14.1.9) and (14.1.10) the estimates can be written as

$$\left\| \Phi^s(\cdot, z) - w^s[g_z] \right\|_\infty \leqslant c\epsilon, \qquad \left\| \Phi^\infty(\cdot, z) - w^\infty[g_z] \right\|_\infty \leqslant \tilde{c}\epsilon. \tag{14.1.34}$$

(C) Finally, we apply the point source method to the far field patterns $\Phi^\infty(\cdot, z)$ and w^∞ to reconstruct the scattered field $\Phi^s(x, z)$ for $x \in \mathbb{R}^m \backslash \bar{D}$ and the field w^s on $x \in \mathbb{R}^m \backslash \bar{D}$. Then we obtain

$$\Phi^s(x, z) \approx \int_{\mathbb{S}} \int_{\mathbb{S}} u^\infty(\hat{x}, d) g_z(d) g_x(\hat{x}) \, \mathrm{d}s(d) \, \mathrm{d}s(\hat{x}) \tag{14.1.35}$$

for $x, z \in \mathbb{R}^m \backslash \bar{D}$ in the sense that given $\epsilon > 0$ there is g_z and g_x such that

$$\left| \Phi^s(x, z) - \int_{\mathbb{S}} \int_{\mathbb{S}} u^\infty(\hat{x}, d) g_z(d) g_x(\hat{x}) \, \mathrm{d}s(d) \, \mathrm{d}s(\hat{x}) \right| \leqslant \epsilon. \tag{14.1.36}$$

Here, we can consider ϵ as regularization parameter, where for $\epsilon \to 0$ the approximation converges towards the true indicator function $\Phi^s(z, z)$. This completes the proof. □

We have shown the convergence of the reconstruction of the indicator function of the SSM from measured far field patterns. This applies to any of its algorithmic realizations, i.e. to *domain sampling* as well as to the *needle scheme* or to the *contraction approach*, and it provides a proof of theorem 14.1.3 for the *needle approach*.

In contrast to the needle approach, which reconstructs single boundary points by testing individual needles, the domain sampling calculates a full approximation μ_α of the indicator function μ, and then calculates approximate shape reconstructions by taking level curves. To complete our analysis for this case as well, we finish with a convergence statement for shape reconstruction.

Theorem 14.1.7 (SSM convergence of shape reconstruction). *Assume that we construct the indicator function μ of the SSM where the algorithmic set-up of our reconstruction leads to a uniform convergence on compact subsets of $\mathbb{R}^m \backslash \bar{D}$. Then, we have the convergence $\Gamma_c \to \partial D$ for the level set Γ_c of size c for the indicator function μ.*

Proof. We have shown in theorem 10.6.4 that the convergence of an *indicator function* implies *convergence of the domain reconstructions* in the Hausdorff metric. Here, all assumptions are satisfied due to theorem 14.1.6 and theorem 14.1.5. □

14.2 The probing method for near field data by Ikehata

In this section we will explain the *probing method* as first suggested by Ikehata [2]. It employs the near field data based on the *Dirichlet-to-Neumann map*, i.e. the knowledge of the mapping which determines the normal derivatives of fields when the field values are prescribed on the boundary of some domain. The basic idea is to calculate an approximation of a point source $\Phi(\cdot, z)$ by a superposition of plane waves and then use the superposition on the measurement surface $\partial\Omega$ to define an indicator function which tends to infinity when the source point z tends to the boundary ∂D of the unknown scatterer.

Historically, Ikehata's method (1998) was developed independently of the point source method (1996)—although it uses the same type of point source approximation—and was published a year before the Potthast *SSM* (1999) which, based on far field data u^∞, evaluates a similar functional. Nakamura and Sini first realized that Ikehata's method is *equivalent* to the *SSM* in the sense that the probing functional can be bounded from below and above by the functional of the *SSM*. We will present the proof below.

14.2.1 Basic idea and principles

Let Ω be a bounded domain with C^2 smooth boundary $\partial\Omega$ which contains $\bar{D}$ and $\Omega\backslash\bar{D}$ is connected. For a given $f \in H^{1/2}(\partial\Omega)$, consider the boundary value problem

$$\Delta u + \kappa^2 u = 0 \quad \text{in} \quad \Omega\backslash\bar{D} \tag{14.2.1}$$

$$u = f \quad \text{on} \quad \partial\Omega \tag{14.2.2}$$

$$u = 0 \quad \text{on} \quad \partial D. \tag{14.2.3}$$

It is well known that there exists a unique solution $u = u^f \in H^1(\Omega\backslash\bar{D})$ to (14.2.1) if this boundary value problem only admits trivial solution when $f = 0$. Based on this we can define the so-called *Dirichlet-to-Neumann map* $\Lambda_D : H^{1/2}(\partial\Omega) \to H^{-1/2}(\partial\Omega)$ by

$$\Lambda_D f = \frac{\partial u^f}{\partial \nu} \quad \text{on} \quad \partial\Omega. \tag{14.2.4}$$

We also define $\Lambda_\varnothing$ as Λ_D with $D = \varnothing$ by assuming that κ^2 is not a Dirichlet eigenvalue for $-\Delta$ in Ω. This $\Lambda_\varnothing$ is the Dirichlet-to-Neumann map without any inclusion D in Ω.

Now, let us consider a needle c in $\bar{\Omega}$ as defined in equation (14.1.14) of section 14.1.2. Let $c_z \subset c$ be a connected subset of the needle c which contains $c(0)$ and $z \in c$. Further, we choose an approximation domain $G_z \subset \Omega\backslash c_z$ and the sequence of Herglotz wave functions

$$v_{g_z^n}(x) := w^{\mathrm{i}}\left[g_z^n\right](x) = \int_{\mathbb{S}} \mathrm{e}^{\mathrm{i}\kappa x\cdot y} g_z^n(y)\, \mathrm{d}s(y) \tag{14.2.5}$$

as defined by (14.1.8), with densities $g_z^n \in L^2(\mathbb{S})$, $n = 1, 2, 3, \ldots$, which approximates $\Phi(\cdot, z)$ on ∂G_z as in section 14.1.2. Integrating by parts we have

$$\int_{\partial\Omega} (\Lambda_D f)(x) \bar{f}(x)\, \mathrm{d}x = \int_{\Omega\backslash\bar{D}} \left(|\nabla u^f(x)|^2 - \kappa^2 |u^f(x)|^2\right) \mathrm{d}x \tag{14.2.6}$$

$$\int_{\partial\Omega} (\Lambda_\varnothing f)(x) \bar{f}(x)\, \mathrm{d}x = \int_{\Omega} \left(|\nabla v^f(x)|^2 - \kappa^2 |v^f(x)|^2\right) \mathrm{d}x \tag{14.2.7}$$

for any $f \in H^{1/2}(\partial\Omega)$, where $u^f \in H^1(\Omega\backslash\bar{D})$ is the solution to (14.2.1) and $v^f \in H^1(\Omega)$ is the solution to (14.2.1) for the case where there is no domain D in Ω.

The right-hand sides of (14.2.6) and (14.2.7) are the *steady state energies* of the fields u^f and v^f, respectively. Now, the *probe method* is to see the behavior of *the gap* $\mu_{\text{pb}}(z, c)$ of these energies for $f = v_{g_z^n}$ $(n = 1, 2, \dots)$ in the limit $n \to \infty$ as z moves along the needle c. The precise form of $\mu_{\text{pb}}(z, c)$ is given by

$$\mu_{\text{pb}}(z, c) = \lim_{n\to\infty} \int_{\partial\Omega} (\Lambda_D - \Lambda_\varnothing) v_{g_z^n}(x) \overline{v_{g_z^n}(x)} \, \mathrm{d}x. \tag{14.2.8}$$

Then, the behavior of $\mu_{\text{pb}}(z, c)$ as z moves along c is given in the next theorem. The behavior of μ_{pb} is used to identify the boundary ∂D of D.

Theorem 14.2.1 (Probe method). *The probe method is based on the behavior of the indicator function $\mu_{\text{pb}}(z)$ when z approaches the boundary ∂D of an unknown inclusion or scatterer, respectively.*

(i) *If a needle c does not intersect ∂D, then $\mu_{\text{pb}}(z, c) < \infty$ $(z \in c)$.*

(ii) *Let c intersect ∂D at $z = z(t)$ with some $0 < t < 1$ for the first time. Then, this t can be characterized as*

$$t = \sup\{0 < t' < 1;\ \mu(c(s), c) < \infty\ (0 < s < t')\}. \tag{14.2.9}$$

This means $\lim_{s\uparrow t} \mu_{\text{pb}}(c(s), c) = \infty$ and we can say that the t is the smallest upper bound such that $\mu(c(s), c)$ stays bounded for $0 < s < t$.

The proof of this theorem will be given in section 14.2.2, by proving the equivalence of the near field method with the far field needle approach as given in theorem 14.1.3.

Here, let us carry out a simple realization of the Dirichlet-to-Neumann map by single-layer potentials. The approach theoretically works for the case where κ is not an eigenvalue for D, Ω and $\Omega\setminus\bar{D}$. Practically it is stable for sufficiently small κ, see figure 14.10. We assume that $f \in H^1(\partial\Omega)$.

We search a solution to (14.2.1) by a single-layer potential over ∂D and $\partial\Omega$, i.e. we use the ansatz

$$u(x) = \int_{\partial D} \Phi(x, y)\varphi_1(y)\, \mathrm{d}s(y) + \int_{\partial\Omega} \Phi(x, y)\varphi_2(y)\, \mathrm{d}s(y), \quad x \in \mathbb{R}^m \tag{14.2.10}$$

with densities $\varphi_1 \in L^2(\partial D)$ and $\varphi_2 \in L^2(\partial\Omega)$. The boundary condition leads to the integral equation

$$\begin{pmatrix} S_{11} & \tilde{S}_{21} \\ \tilde{S}_{12} & S_{22} \end{pmatrix} \begin{pmatrix} \varphi_1 \\ \varphi_2 \end{pmatrix} = \begin{pmatrix} 0 \\ f \end{pmatrix}, \tag{14.2.11}$$

where we use the index 1 to denote integration or evaluation over ∂D and 2 for the corresponding operations on $\partial\Omega$. By multiplication with the operator matrix

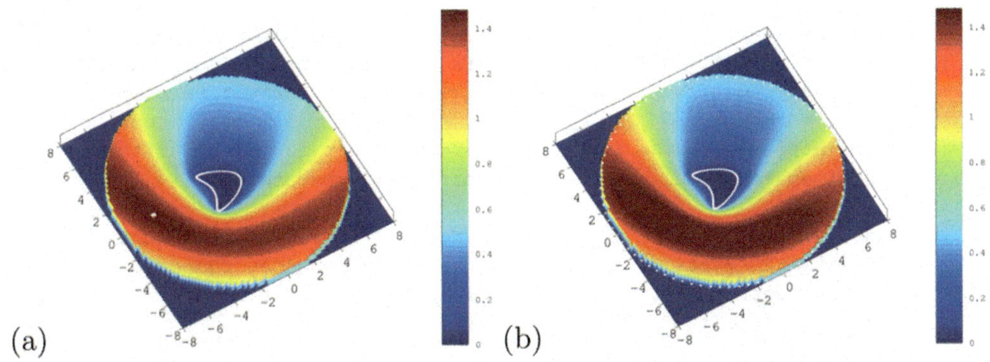

Figure 14.10. As a test case we take the total field $u = u^i + u^s|_{\partial\Omega}$ of a solution to the sound-soft scattering problem on the boundary $\partial\Omega$ of a domain Ω containing the scatterer D in its interior. Here, $\Omega = B_R$ is a ball with radius $R = 8$. Then, we solve the boundary value problem (14.2.1) using a single-layer approach. The resulting field is shown in (b). (a) is the simulation of the scattering problem. This is generated by `sim_14_2_2_0_control.m` calling several short OCTAVE scripts.

$$\begin{pmatrix} S_{11}^{-1} & 0 \\ 0 & S_{22}^{-1} \end{pmatrix},$$

where the inverse is taken for the mappings $L^2 \to H^1$ on ∂D or $\partial\Omega$, we obtain an operator equation of the form $(I + K)\varphi = b$ with a compact operator K on $L^2(\partial D) \times L^2(\partial\Omega)$. Injectivity of this equation can be obtained by standard arguments when the domains D, Ω and $\Omega\backslash\bar{D}$ are non-vibrating. In this case by the Riesz theory the operator has a bounded inverse.

Code 14.2.2. *We show a simple approach to the boundary value problem (14.2.1) for a sound-soft obstacle by a single-layer potential approach, which is valid if we avoid Dirichlet eigenvalues of D, Ω and $\Omega\backslash D$, i.e. if κ is sufficiently small. This is script* `sim_14_2_2_b_bvp.m`, *but note that it uses various variables and functions generated by* `sim_14_2_2_a_setup_rhs_for_test.m` *(not displayed here). In particular, the density* `varphi` *is calculated by solving the exterior scattering problem by a single-layer approach and here evaluating this on the boundary $\partial\Omega$ in line 24.*

```
1  % I Preparations for 2 domains D and Omega
2  N2=2*N;

3  % definition of domain D and tangential vectors
4  yD1  = cos(t)+0.65*cos(2*t)-0.65; yD2  = 1.5 * sin(t);
5  dyD1 = -sin(t)-0.65*2*sin(2*t);   dyD2 = 1.5 * cos(t);

6  % definition of domain G and tangential vectors
7  yG1     = R * cos(t);     yG2     = R * sin(t);
8  dyG1    = -R * sin(t);    dyG2    = R * cos(t);
9  y1 = [yD1 yG1];       y2  = [yD2 yG2];
10 dy1 = [dyD1 dyG1]; dy2 = [dyD2 dyG2];
```

```
ymat1  = repmat(y1,N2,1);  % matrix of domain points component 1
ymat2  = repmat(y2,N2,1);  % matrix of domain points component 2
dymat1 = repmat(dy1,N2,1); % matrix of domain derivative comp.1
dymat2 = repmat(dy2,N2,1); % matrix of domain derivative comp.2
eps = ht/3.6;             % set cut parameter for singularity
epsmat=eps*ones(N2,N2);     % matrix of cut parameter

rmat1=ymat1.'-ymat1;  % matrix of point differences compotent 1
rmat2=ymat2.'-ymat2;  % matrix of point differences component 2
rmat=max(sqrt(rmat1.^2 + rmat2.^2),epsmat); % matrix of ||x-y||
drmat=sqrt(dymat1.^2 + dymat2.^2);

% II Potential Operator S for D and Omega in a combined way
Sbvp = i/2*besselh(0,1,kappa*rmat).*drmat*ht;    % combined SLP Op
Spart = Sbvp((N+1):(2*N),1:N);    % part of SLP Op from D to Omega
usR = -1i*0.5*Spart*varphi(1:N); % evaluate SLP from scattering
uR = usR + exp(1i*kappa*(yG1*d1+yG2*d2)).'; % add incident field

rhs = [zeros(N,1); uR];  % setup right-hand side for BVP
alpha = 1e-8;            % regulariyation parameter
phi2 = (alpha*eye(N2,N2)+Sbvp'*Sbvp)\Sbvp'*(2*rhs); % solve Int.Eq

sim_14_2_2_b_Gbvp_test;
phi = [-1i*0.5*varphi(1:N); phiG];
```

Finally, the calculation of the normal derivative of the field u can be carried out as in code 9.4.1 and the Dirichlet-to-Neumann map is complete.

14.2.2 Convergence and equivalence of the probe and SSM

In this section, we will prove theorem 14.2.1 by showing that there is a link between the two indicator functions $\mu(z, c)$ and $\mu_{\text{pf}}(z, c)$ given as (14.1.15) and (14.2.8), respectively. We will see that this link will immediately give the equivalence of the SSM with the needle scheme and the near field probe method.

Assume $c_z \cap \bar{D} = \emptyset$ and let $v_{g_z^n} = w^{\text{i}}[g_z^n]$ $(n = 1, 2, \cdot)$ be the Herglotz wave functions which approximate $\Phi(\cdot, z)$ on $\Omega\backslash c_z$ using an approximation domain G_z. Also, let $v^{\text{s}}_{g_z^n} \in H^1(\Omega\backslash\bar{D})$ be the solution to (14.2.1) with $f = v_{g_z^n}|_{\partial\Omega}$. Then, $w_n := v^{\text{s}}_{g_z^n} - v_{g_z^n} \in H^1(\Omega\backslash\bar{D})$ satisfies

$$\begin{aligned} \Delta w_n + \kappa^2 w_n &= 0 \quad \text{in} \quad \Omega\backslash\bar{D} \\ w_n &= 0 \quad \text{on} \quad \partial\Omega \\ w_n &= -v_{g_z^n} \quad \text{on} \quad \partial D. \end{aligned} \tag{14.2.12}$$

Recall that we took the approximation domain $G_z \subset \Omega\backslash\bar{c}_z$ for which we approximated $\Phi(\cdot, z)$ on ∂G_z in terms of the $L^2(\partial G_z)$ norm as a non-vibrating domain such that $(\Omega\backslash\bar{c}_z)\backslash\bar{G}_z$ is very narrow. Hence, we can assume that $\bar{D} \subset G_z$.

By the regularity for the weak solutions of elliptic equations and its interior regularity the sequence $\{v_{g_z^n}|_{\partial D}\} \subset H^{1/2}(\partial D)$ is a Cauchy sequence. Then, by the well-posedness of the boundary value problem (14.2.12), the sequence $\{w_n\} \subset H^1(\Omega\backslash\bar{D})$ converges to $w = \Phi^s_\Omega(\cdot, z) \in H^1(\Omega\backslash\bar{D})$, where $w = \Phi^s_\Omega(\cdot, z) \in H^1(\Omega\backslash\bar{D})$ is the solution to

$$\begin{aligned} \Delta w + \kappa^2 w = 0 \quad &\text{in} \quad \Omega\backslash\bar{D} \\ w = 0 \quad &\text{on} \quad \partial\Omega \\ w = -\Phi(\cdot, z) \quad &\text{on } \partial D. \end{aligned} \tag{14.2.13}$$

Now, by integrating by parts, we have

$$\begin{aligned} \mu_{\text{pb}}(z, c) &= -\lim_{n\to\infty}\int_{\partial D}\left\{\frac{\partial v^s_{g_z^n}}{\partial\nu}(x)\overline{v_{g_z^n}(x)} - v^s_{g_z^n}(x)\overline{\frac{\partial v_{g_z^n}}{\partial\nu}(x)}\right\} ds(x) \\ &= -\int_{\partial D}\left\{\frac{\partial\Phi^s_\Omega}{\partial\nu}(x, z)\overline{\Phi(x, z)} - \Phi^s_\Omega(x, z)\overline{\frac{\partial\Phi}{\partial\nu}(x, z)}\right\} ds(x) \\ &= -\Phi^s_\Omega(z, z) + \int_{\partial\Omega}\frac{\partial\Phi^s_\Omega}{\partial\nu}(x, z)\overline{\Phi(x, z)}\, ds(x). \end{aligned} \tag{14.2.14}$$

Since $\Phi(\cdot, z) \in H^1(\Omega)$ stays bounded as $z \to \partial D$, the integral in the last line of this formula is also bounded as $z \to \partial D$. Further, it is easy to see that

$$|\Phi^s_\Omega(z, z) + \Phi^s(z, z)| = O(1) \tag{14.2.15}$$

as $z \to \partial D$, because $\Phi^s_\Omega(\cdot, z)$ and $\Phi^s(\cdot, z)$ satisfies the Helmholtz equation in $\Omega\backslash\bar{D}$ and the same boundary condition on ∂D with boundary data $-\Phi(\cdot, z)|_{\partial D}$. We summarize these results in the following theorem.

Theorem 14.2.3 (Equivalence of probing methods). *For the two indicator functions $\mu(z, c)$ and $\mu_{\text{pf}}(z, c)$ given as (14.1.15) and (14.2.8) we have*

$$\left|\mu_{\text{pb}}(z, c) - \mu(z, c)\right| = O(1) \tag{14.2.16}$$

as $z \to \partial D$ for any needle c and z with $c_z \cap \bar{D} = \varnothing$.

We have shown the *equivalence* of the *SSM* of defined by (14.1.4) or (14.1.15), and the *probe method* as given by (14.2.8). This also proves the convergence of the probe method when we employ a *needle approach*, when we use *domain sampling* or the *contraction scheme* based on the functional (14.2.8).

14.3 The multi-wave no-response and range test of Schulz and Potthast

So far we have based our probing approaches on approximations to the fundamental solution $\Phi(\cdot, z)$ with some source point z, where we could show that for $z \to \partial D$ some indicator function would have a particular behavior which can be used to identify the location and shape of the unknown object D.

We have already introduced the concept of *domain sampling*, where we employed a particular set of *test domains* as a tool to reconstruct fields by the *Kirsch–Kress* or the *point source method*, or to construct an *indicator function* for the SSM.

Here, we will fully bring the *domain sampling* idea to life as a key ingredient of the multi-wave version of the *no-response test* originally suggested by Schulz and Potthast [8] in 2005. The key point here is to fully treat domains as domains and carry out sampling on the set of test domains.

The basic idea of the no-response test is to make a pulse small on a test domain G, but leave it arbitrary outside G. It is used to probe the region of interest. As long as the unknown scatterer is inside G, then its response should also be small. If D is not included in G a key point to show is that there are pulses which will lead to a large response. This is used to probe the region by testing various domains.

Let G be any non-vibrating domain. We define the indicator function for the *multi-wave no-response test* by

$$\mu(G) := \lim_{\varepsilon \downarrow 0} \tilde{\mu}_\varepsilon(G), \tag{14.3.1}$$

where with the operator H defined by (8.3.12)

$$\tilde{\mu}_\epsilon(G) := \sup\left\{ I_\epsilon(f, g) : \|Hf\|_{L^2(\partial G)} < \epsilon,\ \|Hg\|_{L^2(\partial G)} < \epsilon \right\} \tag{14.3.2}$$

with

$$I_\epsilon(f, g) := \left| \int_{\mathbb{S}} \int_{\mathbb{S}} u^\infty(-\theta, d) f(\theta) g(d)\, \mathrm{d}s(\theta)\, \mathrm{d}s(d) \right| \tag{14.3.3}$$

and the Herglotz wave functions $v_f = Hf$, $v_g = Hg$, see (8.3.11). For the set $\mathcal{G}$ of all the non-vibrating domains G the *no-response test* calculates the indicator function $\mu(G)$ and builds the intersection

$$D_{\mathrm{rec}} := \bigcap_{G \in \mathbf{G}} G \tag{14.3.4}$$

where

$$\mathbf{G} := \{ G \in \mathcal{G} : \mu(G) = 0 \}. \tag{14.3.5}$$

Now, with these preparations, we can prove the following characterization of D from the far field pattern.

Theorem 14.3.1 (Convergence of the multi-wave no-reponse test). *Given the far field pattern u^∞ for all incident plane waves, i.e. given the far field operator F, we define an indicator function μ on non-vibrating test domains G by (14.3.1) and a domain reconstruction by (14.3.4). Then,*

(i) *if $\bar{D} \subset G$, then $\mu(G) = 0$.*

(ii) *if $\bar{D} \not\subset G$,* then $\mu(G) = \infty$.

Thus, the unknown scatterer D is given by the intersection of all non-vibrating domains G for which $\mu(G)$ is zero, that is

$$\bar{D} = D_{\mathrm{rec}}. \tag{14.3.6}$$

Proof. First, we consider the case where $\bar{D} \subset G$. Recall that v_g is the Herglotz wave function with kernel $g \in L^2(\mathbb{S})$. For $\|v_g\|_{L^2(\partial G)} < \epsilon$ then from the regularity theory of the very weak solutions of the elliptic equations and its interior estimate, we have

$$\left\|v_g^s\right\|_{C(\partial D)} < c\epsilon, \quad \left\|\frac{\partial v_g^s}{\partial \nu}\right\|_{C(\partial D)} < c\epsilon \tag{14.3.7}$$

with some constant $c > 0$. Using (14.3.3) and the fact that $\|v_f\|_{C^1(\bar{D})} < \tilde{c}\epsilon$, we obtain

$$|I_\epsilon(f, g)| \leqslant C\epsilon^2 \tag{14.3.8}$$

with some constant $C > 0$ and thus

$$\mu(G) = \lim_{\epsilon\to 0} \sup\left\{I_\epsilon(f, g) : \left\|v_g\right\|_{L^2(\partial G)} < \epsilon, \left\|v_f\right\|_{L^2(\partial G)} < \epsilon\right\} = 0. \tag{14.3.9}$$

Next, for simplicity we confine to the case $n = 3$, let $\bar{D} \not\subset G$. Let $z \in \partial D$ such that z is on the boundary of the unbounded component of $\mathbb{R}^3\backslash\overline{D \cup G}$. Then, there exists a sequence of points

$$(z_p)_{p\in\mathbb{N}} \subset \mathbb{R}^3\backslash(\overline{G \cup D}) \tag{14.3.10}$$

such that z_p tends to z. We consider the sequence of point sources $\Phi(\cdot, z_p)$. We set G_p as a sequence of non-vibrating domains such that $\overline{G \cup D} \subset G_p$ and $z_p \in \mathbb{R}^m\backslash\bar{G}_p$. In this case, due to the denseness property of the Herglotz wave operator (see [9], lemma 3.1.3) we take $g^p{}_n$ as a sequence such that for every p fixed

$$\left\|v_{g_n^p} - \frac{\epsilon}{2}\alpha_p\Phi(\cdot, z_p)\right\|_{L^2(\partial G_p)} \to 0, \quad n \to \infty, \tag{14.3.11}$$

where

$$\alpha_p := \left\|\Phi(\cdot, z_p)\right\|_{L^2(\partial G)}^{-1}. \tag{14.3.12}$$

Hence, by a combination of (14.3.11) and (14.3.12) and the well-posedness of the interior Dirichlet problem in G_p we derive that for every p fixed we have

$$\left\|v_{g_n^p}\right\|_{L^2(\partial G)} < \epsilon, \tag{14.3.13}$$

for n large enough. On the other hand, from (14.1.35), replacing (f, g) by (g_n^p, g_n^p), we deduce that

$$\begin{aligned}&\lim_{n\to\infty} \int_{\mathbb{S}}\int_{\mathbb{S}} u^\infty(-\theta, d)g_n^p(\theta)g_n^p(d)\, ds(\theta)\, ds(d)\\ &= \frac{\epsilon^2\alpha_p^2}{16\pi}\Phi^s(z_p, z_p).\end{aligned} \tag{14.3.14}$$

Hence using the property

$$\left|\Phi^s(z_p, z_p)\right| \geqslant c_1\left[d\left(z_p, \partial D\right)\right]^{-1} \tag{14.3.15}$$

as shown in theorem 2.1.15 of [9] and the fact that

$$\alpha_p^2 := \left\|\Phi(\cdot, z_p)\right\|_{L^2(\partial G)}^{-2} \geqslant c_2\left[\ln\left(d\left(z_p, \partial B\right)\right)\right]^{-1} \tag{14.3.16}$$

for some positive constants c_1, c_2, we deduce that

$$\lim_{p\to\infty}\lim_{n\to\infty}\int_{\mathbb{S}}\int_{\mathbb{S}} u^{\infty}(-\theta, d)g_n^p(\theta)g_n^p(d)\,\mathrm{d}\theta\mathrm{d}s(d) = \infty. \tag{14.3.17}$$

Then $\mu(G) = \infty$.

From points (i) and (ii), μ may have only the values 0 and $+\infty$. In addition, μ, as a set function, is monotonically decreasing. From such properties, we obtain $\bar{D} = D_{\mathrm{rec}}$. □

For the numerical realization of the *no-response test* it is of course a key question how to generate the test densities g such that v_g is smaller than ϵ on some test domain G and has appropriate variations outside G.

From our proofs we know that, when we employ singular sources with a singularity outside $\bar{G}$, approximating such functions leads to the desired behavior. Thus, it has been suggested by Schulz to take a selection of rescaled singular sources $c_z\Phi(\cdot, z)$ with scaling factor c_z and then solve $Hg = c_z\Phi(\cdot, z)$ on the boundary ∂G of a test domain G by a regularization approach to generate proper densities g. For a sufficiently rich set of such functions we obtain convergence as worked out in the proof of theorem 14.3.1. We show a simple realization of the *no-response test* in code 14.3.2.

Code 14.3.2. *The following script* `sim_14_3_2_b_NRT.m` *calculates the domain-based indicator function of the* no-response test *for a selection of test domains and then builds the intersection of all domains for which the indicator is smaller than a constant* η. *First run* `sim_14_3_2_a_F.m` *etc to calculate the far field operator* `F` *and further variables. Finally, the graphical display is generated by* `sim_14_3_2_c_graphics.m`. *The result is shown in figure 14.11.*

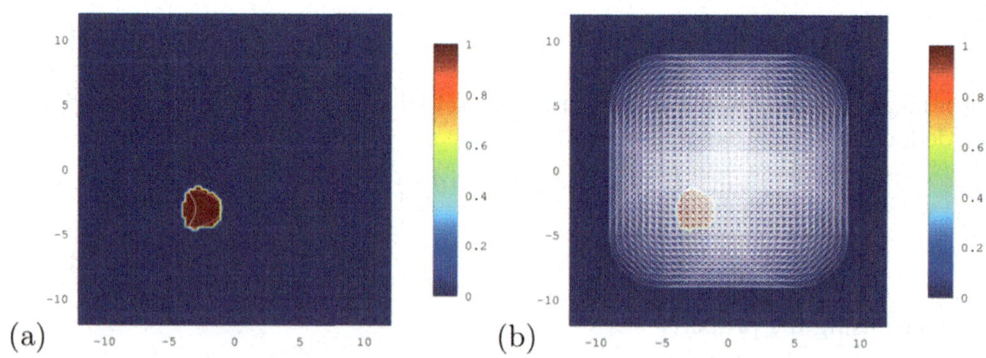

Figure 14.11. For the multi-wave *no-response test* of Schulz and Potthast we show the results of the intersection of all test domains with an indicator function smaller than 1 (a), based on the functional (14.3.3), here again for the case of an impenetrable object with Dirichlet boundary condition. We employ $441 = 11 \times 11$ test domains with centers on a grid with grid spacing 0.5 on $[-5,5]$ (b). Since we use circular test domains, we cannot expect to find the non-convex part of D, but obtain the circular hull of D where circles with radius $R = 4$ have been used.

```
% Loop over sampling domains
z1vec = -5:0.5:5;          % setup centers of sampling domains
z2vec = -5:0.5:5;          % ~
maskSum = ones(M,1);       % initialize summation variable for mask
pc = 1;                    % counter
yG1all = [];yG2all = []; % initialize vector to store test domains
for z1= z1vec
   for z2 = z2vec
      % I Setup of test domain G
      NG = 120;                          % number of points on test domain G
      hG = 2*pi/NG;                      % grid constant for test domain
      tG = 0:hG:2*pi-hG;                 % parametrization grid for G
      RG = 4;                            % radius of test domain
      Rs = 4.2;                          % radius of sources boundary
      yG1 = RG*cos(tG)+z1;               % boundary of test domain comp.1
      yG2 = RG*sin(tG)+z2;               % boundary of test domain comp.2
      ys1 = (Rs*cos(tG)+z1)';               % boundary of test domain comp.1
      ys2 = (Rs*sin(tG)+z2)';               % boundary of test domain comp.2
      yG1all = [yG1all; yG1];               % collect test domains
      yG2all = [yG2all; yG2];               %    for later display
      yGffmat1  = repmat(yff1,NG,1);         % matrix of ff points comp.1
      yGffmat2  = repmat(yff2,NG,1);         % matrix of ff points comp.2
      yGmat1    = repmat(yG1.',1,ffN);      % matrix of points of G comp.1
      yGmat2    = repmat(yG2.',1,ffN);      % matrix of points of G comp.1

      % II Herglotz operator for evaluation on test domain G
      HG    = exp(1i*kappa*(yGffmat1.*yGmat1+yGffmat2.*yGmat2))*hff;

      % III Point Source Matrix for Evaluation
      epsmat   = eps*ones(NG,NG);       % matrix for cutting singularity
      rGmat1   = repmat(ys1,1,NG)-repmat(yG1,NG,1);   % coordinate
      rGmat2   = repmat(ys2,1,NG)-repmat(yG2,NG,1);   %   difference
      rGmat    = max(sqrt(rGmat1.^2 + rGmat2.^2),epsmat).';
      PhiMat   = 1i/4*besselh(0,1,kappa*rGmat);            % right-hand side
      PhiMatC  = conj(1i/4*besselh(0,1,kappa*rGmat)); % ~

      % IV construct the density g for calculating the NRT functional
      alphaPSM = 1e-9;                   % regularization parameter
      gPSM  = (alphaPSM*eye(ffN,ffN) + HG'*HG)\HG'*PhiMat;
      gPSM2  = (alphaPSM*eye(ffN,ffN) + HG'*HG)\HG'*PhiMatC*hff;
      maskG = sqrt( (pvec1-z1).^2 + (pvec2-z2).^2)<RG; % mask for circle
      Ampl = 20;
      for m=1:NG                         % evaluate functional
         wRecVec(m)   = Ampl/fac*(gPSM2(:,m)')*(F*gPSM(:,m));
      end
      maxRecVec = max(abs(wRecVec)); % maximal response for all g
      if(maxRecVec<4)
```

```
            maskSum = maskSum.*maskG;  % intersection of domains
        end
        pc = pc + 1;                    % increase counter
    end
end
```

It is possible to see the *no-response test* as a version of the SSM, in particular when the reconstruction of the densities g make use of the point source approximations (14.3.11) as carried out by code 14.3.2.

However, from a *conceptional* point of view its *domain sampling approach* is a big step forward. It is neither an iterative method nor is it reconstructing a scattered field, nor is it calculating an indicator function in space, but its indicator function lives on a set of test domains.

Multi-wave range test. Finally, let us briefly introduce the *range test* originally introduced by Kusiak, Sylvester and Potthast for one wave [3] and developed by Schulz and Potthast [8] for the multi-wave case. The idea of the range test will be worked out in more detail in section 15.1.

If the far field pattern for all incident plane waves is given, we obtain the far field pattern of a superposition of plane waves v_g with superposition kernel $g \in L^2(\mathbb{S})$ by the corresponding superposition of the far field patterns, i.e. by

$$v_g^\infty(\hat{x}) = \int_{\mathbb{S}} u^\infty(\hat{x}, d)\, ds(d), \quad \hat{x} \in \mathbb{S}. \tag{14.3.18}$$

Following theorem 3.1.10 we can now test the solvability of the far field equation of the single-layer potential $S^\infty \psi = v_g^\infty$ for all Herglotz wave functions for which the L^2-norm of v_g is bounded by 1 on ∂B. Testing the solvability corresponds to testing the norm ψ_g^α of a regularized solution, leading to the indicator function

$$\mu_{\mathrm{rt}}(B) = \lim_{\alpha \to 0} \sup_{\|v_g\|_{L^2(\partial B)} \leqslant 1} \left\| \psi_g^\alpha \right\|_{L^2(\partial B)} \tag{14.3.19}$$

with $\psi_g^\alpha \in L^2(\partial B)$ given by

$$\psi_g^\alpha = (\alpha + (S^\infty)^* S^\infty)^{-1} (S^\infty)^* v_g^\infty. \tag{14.3.20}$$

We will see below that this approach is equivalent to the *no-response test* and thus, with an identical set-up and appropriately chosen constants, leads to identical reconstructions as shown in figure 14.11.

14.4 Equivalence results

The goal of this section is to investigate the equivalences between the different probe methods under consideration.

We first need to formulate precisely what we mean when we speak about *equivalence* of methods. It cannot be the fact that two methods calculate the same quantity, nor can it be the case that both methods converge under the same assumptions. Here we will use the term *equivalence* for probe methods if we can prove estimates for the corresponding indicator functions in the sense that there are constants c_1 and c_2 such that

$$\mu_2(G) \leqslant c_1\mu_1(G) \tag{14.4.1}$$

$$\mu_1(G) \leqslant c_2\mu_2(G) \tag{14.4.2}$$

for all test domains $G \in \mathcal{G}$ with the set $\mathcal{G}$ of test domains under consideration. This approach to *equivalence* is used for example when talking about the equivalence of norms, see definition 2.1.6. In the section 12.5 on *field reconstruction* we have used a stronger *equivalence concept*, calling methods equivalent if under an identical geometrical set-up they provide identical approximations to the fields under consideration.

14.4.1 Equivalence of SSM and the no-response test

The indicator function $I(z)$ of the SSM is defined for each point z outside the scatterer D, where its reconstruction employs an approximation domain B in which the fundamental solution $\Phi(\cdot, z)$ is approximated by the Herglotz wave function, i.e. we have the geometrical restrictions $z \not\in B$, $\bar{D} \subset B$.

In contrast to the singular sources indicator function, the *no-response test* indicator function is defined on the set of sampling domains B (where often for domain sampling we use the letter G). Here, to relate the *SSM* to the *no-response test*, we transfer the indicator function of the SSM to a test domain B by taking the supremum of the singular sources indicator over a parallel surface of ∂B.

The indicator function $\mu_{\text{SSM}}(B)$ for a non-vibrating domain B is defined as follows.

Definition 14.4.1. *For* ε_1, ε_2 $(0 < \varepsilon_1 < \varepsilon_2 \ll 1)$, *let*

$$B_{\varepsilon_j} = \left\{x \in \mathbb{R}^3 : \ \text{dist}(x, B) < \varepsilon_j\right\} \quad (j = 1, 2)$$

be non-vibrating domains. Then, with $g_z, g_\zeta \in L^2(\mathbb{S})$ *such that* $v_{g_z} \approx \alpha_z\Phi(\cdot, z)$, $v_{g_\zeta} \approx \alpha_\zeta\Phi(\cdot, \zeta)$ *with error smaller than* δ *in* $L^2(\partial B)$, *where we employ the* normalization $\alpha_z = (1 - \delta)\|\Phi(\cdot, z)\|^{-1}_{L^2(\partial B)}$, $\alpha_\zeta = (1 - \delta)\|\Phi(\cdot, \zeta)\|^{-1}_{L^2(\partial B)}$, *we define* $\mu_{\text{SSM}}(B)$ *by*

$$\mu_{\text{SSM}}(B) = \lim_{\varepsilon_1\to 0}\lim_{\delta\to 0}\sup_{z,\zeta\in\partial B_{\varepsilon_1}}\left|\int_{\mathbb{S}}\int_{\mathbb{S}} u^\infty(-\hat{x}, d)g_z(d)g_\zeta(\hat{x})\, \mathrm{d}s(d)\, \mathrm{d}s(\hat{x})\right|. \tag{14.4.3}$$

We will call this $\mu_{\text{SSM}}(B)$ the *domain related indicator function* of the SSM. As we are taking the supremum with respect to $z, \zeta \in \partial B_{\varepsilon_1}$, we fully stay within the restriction $z \not\in \bar{B}$ of the original SSM.

We define a slightly modified indicator function $\mu_{\text{NRT}}(B)$ of the no-response test defined by

$$\mu_{\mathrm{NRT}}(B) = \sup\left\{\left|\int_{\mathbb{S}}\int_{\mathbb{S}} u^{\infty}(-\hat{x}, d)g(d)\tilde{g}(\hat{x})\,\mathrm{d}s(d)\,\mathrm{d}s(\hat{x})\right| : \|v_g\|_{L^2(\partial B)} \leqslant 1, \|v_{\tilde{g}}\|_{L^2(\partial B)} \leqslant 1\right\}, \tag{14.4.4}$$

which no longer takes the limit in terms of the smallness of the $L^2(\partial B)$ norms of the Herglotz wave functions. We will see that restricting the approach to the bound 1 is sufficient. We are now prepared to show equivalence.

Theorem 14.4.2 (Equivalence of SSM and the no-response test). *For any non-vibrating domain B, we have*

$$\mu_{\mathrm{SSM}} \sim \mu_{\mathrm{NRT}}. \tag{14.4.5}$$

Proof. By the definitions (14.4.3) and (14.4.4) all admissible densities g, $\tilde{g}$ for the SSM are admissible for the no-response test. Thus we have for any non-vibrating domain B,

$$\mu_{\mathrm{SSM}}(B) \leqslant C_1\,\mu_{\mathrm{NRT}}(B), \tag{14.4.6}$$

for some constant $C_1 > 0$.

Next we will show that there exists a constant $C_2 > 0$ such that the estimate

$$\mu_{\mathrm{NRT}}(B) \leqslant C_2\,\mu_{\mathrm{SSM}}(B) \tag{14.4.7}$$

holds for any non-vibrating domain B. Let

$$\|v_h\|_{L^2(\partial B)} \leqslant 1, \ \|v_{\tilde{h}}\|_{L^2(\partial B)} \leqslant 1. \tag{14.4.8}$$

Since B_{ε_1} is non-vibrating, the single-layer potential operator S from $L^2(\partial B_{\varepsilon_1})$ into $H^1(\partial B_{\varepsilon_1})$ is an isomorphism. Hence, there exist $\varphi, \psi \in L^2(\partial B_{\varepsilon_1})$ such that

$$S\varphi = v_h, \quad S\psi = v_{\tilde{h}} \tag{14.4.9}$$

is satisfied. Then, for $y \in \partial B$ we have

$$\begin{aligned} v_h(y) = (S\varphi)(y) &= \int_{\partial B_{\varepsilon_1}} \varphi(z)\Phi(y, z)\,\mathrm{d}s(z) \\ &\approx \int_{\partial B_{\varepsilon_1}} \varphi(z)\left(\int_{\mathbb{S}} \mathrm{e}^{\mathrm{i}\kappa y\cdot d} g_z(d)\,\mathrm{d}s(d)\right)\mathrm{d}s(z) \\ &= \int_{\mathbb{S}} \mathrm{e}^{\mathrm{i}\kappa y\cdot z} h'(z)\,\mathrm{d}s(z), \end{aligned} \tag{14.4.10}$$

with

$$h'(z) = \int_{\partial B_{\varepsilon_1}} g_z(d)\varphi(z)\,\mathrm{d}s(z) \tag{14.4.11}$$

for some $g_z \in L^2(\mathbb{S})$. Here we changed the order of integrations. For this we need to know that we can take $g_z(d)$ as a measurable function of (z, d). This can be justified by taking $\varphi(z)$ as a continuous function on ∂B_{ϵ_1} and approximating $\int_{\partial B_{\epsilon_1}} \varphi(z)\Phi(y, z)\, ds(z)$ by a Riemann sum. Similarly, we have for

$$\tilde{h}'(\hat{x}) = \int_{\partial B_{\epsilon_1}} g_\zeta(\hat{x})\varphi(\zeta)\, ds(\zeta) \tag{14.4.12}$$

with some $g_\zeta \in L^2(\mathbb{S})$,

$$v_{\tilde{h}}(y) = (S\psi)(y) \approx \int_{\mathbb{S}} e^{i\kappa y\cdot\hat{x}}\tilde{h}'(\hat{x})\, ds(\hat{x}). \tag{14.4.13}$$

By using that $S : L^2(\partial B_{\epsilon_1}) \to H^1(\partial B_{\epsilon_1})$ is an isomorphism and the interior estimate of solutions of the Helmholtz equation, there exists a constant $C_0 > 0$ such that

$$\|\varphi\|_{L^2(\partial B_{\epsilon_1})} \leqslant C_0,\ \|\psi\|_{L^2(\partial B_{\epsilon_1})} \leqslant C_0. \tag{14.4.14}$$

By the continuity with respect to ϵ_1 and invertibility of S on ∂B these bounds are uniform in ϵ_1. Hence, for $D \subset B$ we have

$$\begin{aligned}
&\left|\int_{\mathbb{S}}\int_{\mathbb{S}} u^\infty(-\hat{x}, d)h(d)\tilde{h}(\hat{x})\, ds(d)\, ds(\hat{x})\right| \\
&\approx\left|\int_{\mathbb{S}}\int_{\mathbb{S}} u^\infty(-\hat{x}, d)h'(d)\tilde{h}'(\hat{x})\, ds(d)\, ds(\hat{x})\right| \\
&=\left|\int_{\partial B_{\epsilon_1}}\int_{\partial B_{\epsilon_1}}\left(\int_{\mathbb{S}}\int_{\mathbb{S}} u^\infty(-\hat{x}, d)g_z(d)g_\zeta(\hat{x})\, ds(d)\, ds(\hat{x})\right)\right. \\
&\qquad\qquad \left.\varphi(z)\psi(\zeta)\, ds(z)\, ds(\zeta)\right| \\
&\leqslant C_0'(\mu_{\mathrm{SSM}}(B) + \eta)\left(\int_{\partial B_{\epsilon_1}}|\varphi(z)|\, ds(z)\right)\left(\int_{\partial B_{\epsilon_1}}|\psi(\zeta)|\, ds(\zeta)\right) \\
&\leqslant C_0'|\partial B_{\epsilon_1}|\,\|\varphi\|_{L^2(\partial B_{\epsilon_1})}\|\psi\|_{L^2(\partial B_{\epsilon_1})}(\mu_{\mathrm{SSM}}(B) + \eta) \\
&\leqslant C_0^2C_0'|\partial B_{\epsilon_1}|(\mu_{\mathrm{SSM}}(B) + \eta)
\end{aligned} \tag{14.4.15}$$

for some constant $C_0' > 0$ and with a parameter η with $\eta \to 0$ for $\epsilon_1 \to 0$, where $|\partial B_{\epsilon_1}|$ denotes the area of ∂B_{ϵ_1}. If $D\not\subset B$, we have $\mu_{\mathrm{SSM}}(B) = \infty$ according to lemma 14.1.5 and (14.1.11), such that (14.4.7) is also satisfied in this case. Collecting all estimates we obtain the inequality (14.4.7). □

14.4.2 Equivalence of the no-response test and the range test

Let $B \subset \mathbb{R}^m$ $(m = 2, 3)$ be a non-vibrating domain and recall the definitions of the indicator functions $\mu_{\mathrm{NRT}}(B)$ and $\mu_{\mathrm{RT}}(B)$ of the no-response test (14.3.1) in its form (14.4.4) and multi-wave range test (14.3.19), respectively. We have

$$\mu_{\mathrm{NRT}}(B) = \sup\left\{\left|\int_{\mathbb{S}}\int_{\mathbb{S}} u^\infty(-\theta, d)f(\theta)g(d)\, ds(\theta)\, ds(d)\right| : \|v_g\|_{L^2(\partial B)} \leqslant 1,\ \|v_f\|_{L^2(\partial B)} \leqslant 1\right\} \tag{14.4.16}$$

and

$$\mu_{\mathrm{RT}}(B) = \lim_{\alpha\to 0} \sup_{\|v_g\|_{L^2(\partial B)}\leqslant 1} \left\|\psi_g^\alpha\right\|_{L^2(\partial B)} \tag{14.4.17}$$

with $\psi_g^\alpha \in L^2(\partial B)$ given by

$$\psi_g^\alpha = (\alpha + (S^\infty)^* S^\infty)^{-1}(S^\infty)^* v_g^\infty, \tag{14.4.18}$$

.

Theorem 14.4.3. *The indicator functions of the no-response test $\mu_{\mathrm{NRT}}(B)$ and range test $\mu_{\mathrm{RT}}(B)$ are the same up to a positive constant which is independent of B and depends only on m.*

Proof. We calculate

$$\begin{aligned}\mu_{\mathrm{RT}}(B) &= \lim_{\alpha\to 0}\left\|\psi_g^\alpha\right\|_{L^2(\partial B)} = \lim_{\alpha\to 0}\sup_{\|v\|_{L^2(\partial B)}\leqslant 1}\left|\left\langle v, \psi_g^\alpha\right\rangle_{L^2(\partial B)}\right| \\ &= \lim_{\alpha\to 0}\sup_{\|v\|_{L^2(\partial B)}\leqslant 1}\left|\left\langle v, (\alpha + (S^\infty)^* S^\infty)^{-1}(S^\infty)^* v_g^\infty\right\rangle_{L^2(\partial B)}\right| \\ &= \lim_{\alpha\to 0}\sup_{\|v\|_{L^2(\partial B)}\leqslant 1}\left|\left\langle \left((\alpha + (S^\infty)^* S^\infty)^{-1}(S^\infty)^*\right)^* v, v_g^\infty\right\rangle_{L^2(\partial B)}\right|.\end{aligned}$$

Here, by noticing that $H : L^2(\mathbb{S}) \to L^2(\partial B)$ has a dense range and $(S^\infty)^* = \gamma H$ with $\gamma = \frac{e^{i\pi/4}}{\sqrt{8\pi\kappa}}(m = 2)$, $1/(4\pi)$ $(m = 3)$, we can continue our computation to have

$$\begin{aligned}\mu_{rt}(B) &= \lim_{\alpha\to}\sup_{\|Hf\|_{L^2(\partial B)}\leqslant 1}\left|\left\langle\left((\alpha + (\gamma H)(\gamma H)^*)^{-1}(\gamma H)\right)^* Hf, v_g^\infty\right\rangle_{L^2(\mathbb{S})}\right| \\ &= \lim_{\alpha\to 0}\sup_{\| Hf\|_{L^2(\partial B)\leqslant 1}}\left|\gamma^{-1}\left\langle\left((\alpha\gamma^{-2} + HH^*)^{-1}H\right)^* Hf, v_g^\infty\right\rangle_{L^2(\mathbb{S})}\right|.\end{aligned}$$

To proceed further, let $K = H^* : L^2(\partial B) \to L^2(\mathbb{S})$, $\beta = \alpha\gamma^{-2}$. Then, we have

$$\left(\left(\alpha\gamma^{-2} + HH^*\right)^{-1} H\right)^* Hf = K(\beta + K^*K)^{-1}K^*f. \tag{14.4.19}$$

This right-hand side can be computed using the singular system (μ_n, φ_n, g_n) of K as follows.

$$(\beta + K^*K)^{-1}K^*f = \sum_n \frac{\mu}{\beta + \mu_n^2}\langle f, g_n\rangle\varphi_n. \tag{14.4.20}$$

Hence, we have

$$K(\beta + K^*K)^{-1}K^* = \sum_n \frac{\mu_n^2}{\beta + \mu_n^2}\langle f, g_n\rangle g_n \to \sum_n \langle f, g_n\rangle g_n \ (\beta \to 0). \tag{14.4.21}$$

Since $K^* = (H^*)^* = H$ is injective, $K : L^2(\partial B) \to L^2(\mathbb{S})$ has a dense range. Then, taking account that $\varphi_n \in L^2(\partial B)$ $(n = 1, 2, \ldots)$ forms a complete orthonormal system, $g_n = 1/\mu_n K\varphi_n$ $(n = 1, 2, \ldots)$ is also a complete orthonormal system in $L^2(\mathbb{S})$. Hence, $\sum_n \langle f, g_n\rangle g_n = f$. The equivalence of the two indicators $\mu_{\mathrm{NRT}}(B)$ and $\mu_{\mathrm{RT}}(B)$ for any non-vibrating domain B can be seen the following final computation:

$$\begin{aligned}\mu_{\mathrm{RT}}(B) &= \gamma^{-1} \sup_{\|Hf\|_{L^2(\partial B)} \leqslant 1} \left| \left\langle f, v_g^\infty \right\rangle_{L^2(\mathbb{S})} \right| \\ &= \gamma^{-1} \sup_{\|v_f\|_{L^2(\partial B)} \leqslant 1} \left| \int_{\mathbb{S}} \int_{\mathbb{S}} \overline{u^\infty(\theta, d)} f(\theta) \overline{g(d)}\, \mathrm{d}s(\theta)\, \mathrm{d}s(d) \right| \\ &= \gamma^{-1} \mu_{\mathrm{NRT}}(B)\end{aligned} \tag{14.4.22}$$

and the proof is complete. □

14.5 The multi-wave enclosure method of Ikehata

The point source $\Phi(\cdot, z)$ is not the only special solution which can be used as a basis to probe an unknown area and identify a scatterer. It is well known from uniqueness proofs in inverse scattering that *geometrical optics solutions* have very useful properties. Ikehata's multi-wave enclosure method for inverse scattering problems was given in two steps via near field measurements [10]. Here we will give a reconstruction method directly using the far field patterns, see also [13] and [14].

Definition 14.5.1. *Let $\tau > \kappa$ and $\omega, \omega^\perp \in S^2$ with $\omega \perp \omega^\perp$. The complex geometric optic solution w of the Helmholtz equation* (8.2.1) *is defined by*

$$w(x, \tau, \omega) = \mathrm{e}^{x\cdot\left(\mathrm{i}\tau\omega^\perp + \sqrt{\tau^2 - \kappa^2}\,\omega\right)} \tag{14.5.1}$$

for $x \in \mathbb{R}^m$.

The complex geometric optics solution is a plane wave with wave number τ along the direction $\omega^\perp$. In the direction ω it is exponentially increasing, i.e. in $-\omega$ it is exponentially decreasing with the exponent $\sqrt{\tau^2 - \kappa^2} \approx \tau$ for $\tau \gg \kappa$.

Let D be a sound-soft obstacle with boundary ∂D of class C^2. We define the *support function* h_D of D by

$$h_D(\omega) := \sup_{x \in D} x \cdot \omega \tag{14.5.2}$$

as the maximum of the projection of vectors $x \in D$ onto the direction ω. We call ω a *regular direction* with respect to D if $\{x \cdot \omega = h_D(\omega) : x \in D\}$ consists of just one point.

Let Ω be some non-vibrating domain with a boundary of class C^2 and $\bar{D} \subset \Omega$. In $L^2(\partial\Omega)$ we approximate $w(\cdot, \tau, \omega)$ by a Herglotz wave function (8.3.11) with kernel $g \in L^2(\mathbb{S})$ with error smaller than δ. Then, we define the *indicator function* η based on

$$\tilde{\eta}(\tau, t, \omega) := \lim_{\delta \to 0} e^{-2t\tau} \left| \int_{\mathbb{S}} \int_{\mathbb{S}} u^{\infty}(\hat{x}, \theta) \overline{g(\hat{x})} g(\theta) \, ds(x) \, ds(\theta) \right| \tag{14.5.3}$$

by

$$\eta(t, \omega) := \lim_{\tau \to \infty} \tilde{\eta}(\tau, t, \omega). \tag{14.5.4}$$

Here, for numerical tests we also use the *shifted* geometrical optics solution

$$w(x, \tau, t, \omega) = e^{-t\tau} \cdot e^{x \cdot \left(i\tau\omega^{\perp} + \sqrt{\tau^2 - \kappa^2}\omega\right)}, \quad x \in \mathbb{R}^m, \tag{14.5.5}$$

for shift parameter t along the ω direction. In $\tilde{\eta}$ defined in (14.5.3) it is reflected by the factor $e^{-2t\tau}$. In figure 14.12, we employ $t = x \cdot \omega$ and in each point show the result of $\tilde{\eta}(\tau, t, \omega)$ for two different directions of ω.

The function η or $\tilde{\eta}$ allows us to recover the functional $h_D(\omega)$ and the convex hull of the unknown scatterer D as follows.

Theorem 14.5.2. *Let $\omega \in \mathbb{S}$ be a regular direction for the unknown scatterer D. Then we have*

$$\begin{aligned} \eta(t, \omega) &= 0 \ (t > h_D(\omega)) \\ \eta(t, \omega) &= \infty \ (t = h_D(\omega)) \\ \eta(t, \omega) &= \infty \ (t < h_D(\omega)). \end{aligned} \tag{14.5.6}$$

and the convergence

$$\lim_{\tau \to \infty} (2\tau)^{-1} \log|\tilde{\eta}(\tau, t, \omega)| = h_D(\omega) - t \tag{14.5.7}$$

which allows us to reconstruct $h_D(\omega)$ for each regular direction ω.

The functional $\tilde{\eta}$ can be used to probe some area with affine half-planes and test whether the unknown scatterer D is inside some half-plane or not by evaluating the indicator function $\tilde{\eta}$. Before we work out the convergence proof, let us study the numerical realization of the method.

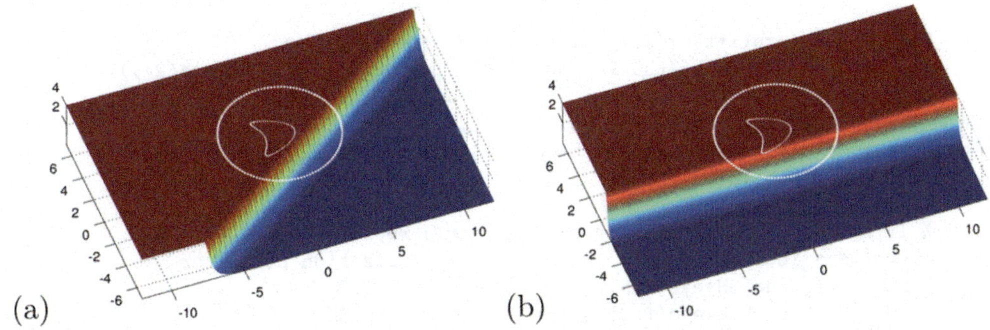

Figure 14.12. The result of code 14.5.3, where for each point x we define $t = x \cdot \omega$ and display the behavior of $\tilde{\eta}(\tau, t_x, \omega)$. The method easily detects the convex hull of the scatterer.

Code 14.5.3. *The following script* `sim_14_5_3_b_MWEM.m` *reconstructs the indicator function* $\tilde{\eta}$ *for some* $\tau > 0$ *of the multi-wave enclosure method on some evaluation area and with a fixed approximation domain G under the condition that* $\bar{D} \subset G$. *First run* `sim_14_4_3_a_F.m` *etc to calculate the far field operator* `F` *and further variables. Finally, the graphical display is generated by* `sim_14_5_3_c_graphics.m`. *The result is shown in figure 14.12.*

```
% I Setup of Approximation Domain
NG = 200;                          % number: points on test domain G
z1 = 0;                           % center of test domain G, comp.1
z2 = 0;                            % center of test domain G, comp.2
hG = 2*pi/NG;                      % grid constant for test domain
tG = 0:hG:2*pi-hG;                 % parametrization grid for G
RG = 6;                            % radius of test domain
yG1 = RG*cos(tG)+z1;               % boundary of test domain comp.1
yG2 = RG*sin(tG)+z2;               % boundary of test domain comp.2
yGffmat1  = repmat(yff1,NG,1);         % matrix of ff points comp.1
yGffmat2  = repmat(yff2,NG,1);         % matrix of ff points comp.2
yGmat1    = repmat(yG1.',1,ffN);       % matrix of points of G comp.1
yGmat2    = repmat(yG2.',1,ffN);       % matrix of points of G comp.2

% II Herglotz operator for evaluation on test domain G
HG    = exp(1i*kappa*(yGffmat1.*yGmat1+yGffmat2.*yGmat2))*hff;

% III Complex Geometric Optics Matrix for Evaluation
tau        = 4; tau2 = tau+1;            % exponential parameters
beta       = 1.5*pi;                   % angle for direction omega
omega      = [cos(beta);sin(beta)];    % setup omega
p1mat      = repmat(pvec1',NG,1);      % prepare evaluation
p2mat      = repmat(pvec2',NG,1);      % on all points of the
yGmat1     = repmat(yG1.',1,M);        % matrix of points of G comp.1
yGmat2     = repmat(yG2.',1,M);        % matrix of points of G comp.2
omat1      = repmat(omega(1),NG,M);    % evaluation domain Q
omat2      = repmat(omega(2),NG,M);    % ~
```

```
CGO       = (exp(i*tau*(yG1.*omega(2)-yG2.*omega(1)))...
           .*exp(sqrt(tau^2-kappa^2)*(yG1.*omega(1)+yG2.*omega(2)))).';
CGOC      = conj(CGO);                    % conjugation
CGO2      = (exp(i*tau2*(yG1.*omega(2)-yG2.*omega(1)))...
           .*exp(sqrt(tau2^2-kappa^2)*(yG1.*omega(1)+yG2.*omega(2)))).';
CGOC2     = conj(CGO2);                    % conjugation

% IV Reconstruct the density for far field representation
alphaPSM = 1e-9;                          % regularization parameter
gPSM    = (alphaPSM*eye(ffN,ffN) + HG'*HG)\HG'*CGO;
gPSMb   = (alphaPSM*eye(ffN,ffN) + HG'*HG)\HG'*CGOC;
gPSM2    = (alphaPSM*eye(ffN,ffN) + HG'*HG)\HG'*CGO2;
gPSM2b  = (alphaPSM*eye(ffN,ffN) + HG'*HG)\HG'*CGOC2;

rfac    = 1/fac*(gPSMb')*(F*gPSM)*(hff)^2;
rfac2   = 1/fac*(gPSM2b')*(F*gPSM2)*(hff)^2;
hD      = log(abs(rfac2/rfac))/(sqrt(tau2^2-kappa^2)-sqrt(tau^2-kappa^2))/2;
for m=1:M  % reconstruction of MW EN functional
    wRec(m,1) = exp( -2*(pvec1(m)*omega(1)+pvec2(m)*omega(2)) ...
                    *tau)*rfac;
end
Ampl = 1;       % amplification for better visualization
w = Ampl*wRec; % function for graphical representation
```

Proof of theorem 14.5.2. The basic idea of the proof of convergence is to use the formula

$$\int_{\mathbb{S}}\int_{\mathbb{S}} u^{\infty}(\hat{x}, \theta)\overline{g(\hat{x})}g(\theta)\ \mathrm{d}s(\hat{x})\ \mathrm{d}s(\theta)$$
$$= -\gamma\int_{\partial D}\left(\frac{\partial v_g^{\mathrm{s}}}{\partial \nu}\overline{v_g^{\mathrm{i}}} - v_g^{\mathrm{s}}\frac{\overline{\partial v_g^{\mathrm{i}}}}{\partial \nu}\right)\mathrm{d}s, \tag{14.5.8}$$

where we take $v_g^{\mathrm{i}} = Hg$ which approximates

$$w = w_\tau(x;\, \omega) = \mathrm{e}^{x\cdot\left(\mathrm{i}\tau\omega^{\perp}+\sqrt{\tau^2-\kappa^2}\,\omega\right)} \tag{14.5.9}$$

in a neighborhood of $\bar{D}$ together with their gradient with respect to the L^2 norm and v_g^{s} is the scattered wave of v_g^{i}.

Let x_0 be the point given by $\{x \cdot \omega = h_D(\omega)\} \cap \partial D$. The first formula of (14.5.6) is clear since v_g^{i} is exponentially decaying with τ. Also, the third formula of (14.5.6) immediately follows from the second formula of (14.5.6) since for $s < t$ we have $\mathrm{e}^{-2s\tau} > \mathrm{e}^{-2t\tau}$.

We need to prove the second formula of (14.5.6). For the simplicity of description, we only show this for $n = 2$. The basic idea for proving this is to extract a dominant part of v_g^{s} as $\tau \to \infty$.

It is easy to see that $w = e^{\tau x\cdot(\omega+i\omega^{\perp})}(1 + O(1/\tau))\ (\tau \to \infty)$. Then, by the well-posedness of the exterior Dirichlet problem, the dominant part is generated by $w_0 = e^{\tau x\cdot(\omega+i\omega^{\perp})}$. Hence, $v = v_g^s$ is dominated by the solution v_0 to

$$\begin{aligned}(\Delta + \kappa^2)v_0 &= 0 \quad \text{in} \quad \mathbb{R}^2\backslash\bar{D}\\ v_0 &= -w_0 \quad \text{on} \quad \partial D\\ &\text{radiation condition.}\end{aligned} \tag{14.5.10}$$

Due to the invariance of the Helmholtz equation under translation and rotation, we can assume that $x^0 = (0, 0)$, $\omega = (0, 1)$, $\omega^{\perp} = (1, 0)$. Then, w_0 becomes $w_0 = e^{\tau(x_2+ix_1)}$. One of the key observations for the coming argument is that w_0 is $O(1)\ (\tau \to \infty)$ near the origin and exponentially decaying except at the origin as $\tau \to \infty$.

Let $\mathcal{B}_1$, $\mathcal{B}_2$ be two small open discs centered at the origin such that $\bar{\mathcal{B}}_1 \subset \mathcal{B}_2$. We look for z which satisfies

$$\begin{aligned}(\Delta + \kappa^2)z &= 0 \quad \text{in} \quad \mathcal{B}_2\backslash\bar{D}\\ z &= -w_0 \quad \text{on} \quad \mathcal{B}_2 \cap \partial D \text{ near the origin}\end{aligned} \tag{14.5.11}$$

and the following decaying properties

$$\begin{aligned}&z = O(\tau^{-3/4}),\ \nabla z = O(\tau^{1/4})\ (\tau \to \infty) \text{ w.r.t. the } L^2 \text{ norm in } \mathcal{B}_2\backslash\bar{D}\\ &z = O(\tau^{-7/4}),\ \nabla z = O(\tau^{-5/4})(\tau \to \infty)\\ &\qquad \text{w.r.t. the } L^2 \text{ norm in } \left(\mathcal{B}_2\backslash\mathcal{B}_1\right) \cap \left(\mathbb{R}^2\backslash\bar{D}\right),\end{aligned}$$

where the abbreviation 'w.r.t.' is used to denote 'with respect to'. Then, χz can be the dominant part of v_0, where $\chi \in C_0^{\infty}(\mathcal{B}_2)$ and it satisfies $\chi = 1$ in B_1. In fact, $r = v_0 - \chi z$ satisfies

$$\begin{aligned}(\Delta + \kappa^2)r &= -2\nabla\chi \cdot \nabla z - (\Delta\chi)z \quad \text{in} \quad \mathbb{R}^2\backslash\bar{D}\\ r &= (\chi - 1)w_0 \quad \text{on} \quad \partial D\\ &\text{radiation condition.}\end{aligned} \tag{14.5.12}$$

Both of the right-hand sides of (14.5.12) are $O(\tau^{-5/4})\ (\tau \to \infty)$ with respect to the L^2 norms. Hence, by the well-posedness of the exterior Dirichlet problem, $r = O(\tau^{-5/4})\ (\tau \to \infty)$ with respect to the $L^2(\mathbb{R}^2\backslash\bar{D})$ norm and $H^{3/2}(\partial D)$ norm, respectively.

The construction of r is analogous to the construction of the so-called oscillating decaying solution (see subsection 15.4.1). However, we must be careful about the nonlinear phase of the Dirichlet data of (14.5.12). To start with let ∂D and D be given as $x_2 = \varphi(x_1)$ and $x_2 < \varphi(x_1)$ near the origin, respectively, where $\varphi(x_1)$ satisfies $\varphi(0) = \varphi'(0) = 0$ and $\varphi(x_1) < 0\ (x_1 \neq 0)$. In terms of the local coordinates $y_1 = x_1$, $y_2 = x_2 - \varphi(x_1)$, $(\Delta + \kappa^2)z = 0$ and $z = -w_0$ become

$$Lz := \sum_{j,k=1}^{2} g^{jk}\partial_j\partial_k z + \sum_{k=1}^{2} b_k\partial_k z + \kappa^2 z = 0 \quad \text{locally in} \quad y_2 > 0 \tag{14.5.13}$$

and

$$z = -w_0 \quad \text{locally on} \quad y_2 = 0, \tag{14.5.14}$$

respectively, where $\partial_j = \frac{\partial}{\partial y_j}$ and g^{jk}, b^k are given as

$$g^{jk} = \sum_{i=1}^{2} \frac{\partial y_j}{\partial x_i}\frac{\partial y_k}{\partial x_i}, \qquad b_k = \sum_{i,j=1}^{2} \frac{\partial y_j}{\partial x_i}\partial_j\left(\frac{\partial y_k}{\partial x_i}\right) \tag{14.5.15}$$

for j, k $(1 \leqslant j, k \leqslant 2)$.

Let us look for z in the form $z = e^{i\tau y_1}\zeta$. Then, ζ has to satisfy $M\zeta = 0$ locally in $y_2 > 0$ and $\zeta = -w_0$ locally on $y_2 = 0$ near the origin, where $M\cdot = e^{-i\tau y_1}L(e^{i\tau y_1}\cdot\,)$. Further, in order to satisfy the *decaying property*, we expect ζ to satisfy $\partial_1^j\partial_2^k\zeta = O(\tau^{(-3+2j+4k)/4)})$ $(\tau \to \infty)$ with respect to the local L^2 norm.

Due to the fact that $t^j e^{-t}$ $(t \geqslant 0)$ is bounded for each $j = 0, 1, 2, \ldots$, we have $y_2^j\zeta = O(\tau^{-3/4-j})$ for each $j \in \mathbb{N}$, the previous expected estimate for $\partial_1^j\partial_2^k\zeta$, $\partial_2^j\zeta = O(\tau^{-3/4+j})$ $(\tau \to \infty)$ with respect to the local L^2 norm. This brings us to introduce the concept of order. That is, for $j = 0, 1, 2, \ldots,$ we consider applying ∂_1^j and ∂_2^j as order $j/2$ and j operators, respectively, and multiplying y_2^j as an order $-j$ operator. Hence, by Taylor's expansion theorem, we can grade M and D as follows:

$$M = M_2 + M_1 + M_0 + \cdots, \tag{14.5.16}$$

where M_{2-j} s are operators of order $2 - j/2$. We only need the explicit forms of M_2, D_1, D_0. These are given as

$$M_2 = \overset{\circ}{g}{}^{22}\partial_2^2 + 2i\tau\,\overset{\circ}{g}{}^{12}\partial_2 - \tau^2\,\overset{\circ}{g}{}^{11} \tag{14.5.17}$$

with $\overset{\circ}{g}{}^{jk} = g^{jk}|_{y_2=0}$ $(1 \leqslant j, k \leqslant 2)$.

Adapting to the decomposition (14.5.16), we look for ζ in the form:

$$\zeta = \zeta_0 + \zeta_{-1} + \cdots, \tag{14.5.18}$$

where each ζ_{-j} has to satisfy

$$\begin{cases} M_2\zeta_0 = 0, \quad \zeta_0 = -w_0\big|_{y_2=0} \\ \begin{cases} M_2\zeta_j = -\sum_{k=1}^{j} M_{2-k}\zeta_{-j+k} \\ \zeta_{-j} = 0 \\ \qquad\qquad (j = 1, 2, \cdots). \end{cases} \end{cases} \tag{14.5.19}$$

The explicit form of ζ_0 is given as

$$\zeta_0 = -e^{\tau(\phi(y_1)-\lambda y_2)} \tag{14.5.20}$$

with

$$\lambda = \left(\overset{\circ 22}{g}\right)^{-1}\left(i\overset{\circ 21}{g} + h\right), \qquad h = \sqrt{\overset{\circ 11}{g}\overset{\circ 22}{g} - \left(\overset{\circ 12}{g}\right)^2}. \tag{14.5.21}$$

By using lemma 14.5.4 given below, it is easy to show that ζ_{-j} satisfy the estimates

$$\begin{cases} \partial_1^k \partial_2^\ell \zeta_{-j} = O(\tau^{(-3-2j+2k+4\ell)/4}) \\ \partial_1^k \partial_2^\ell \left(\left(\sum_{i=0}^{I} M_{2-i}\right)\left(\sum_{j=0}^{J} \zeta_j\right)\right) = O(\tau^{(3-2K+2k+4\ell)/4}) \end{cases} \tag{14.5.22}$$

as $\tau \to \infty$ with respect to the local L^2 norm, where $K = 2\min\{I, J\}$. Since these estimates (14.5.22) do not change even we replace ζ_{-j} by $\chi\zeta_{-j}$ with another $\chi \in C_0^\infty(\mathcal{B}_2)$, by solving a Dirichlet boundary value problem for $e^{i\tau y_1}s$ with $s = \zeta - \sum_{j=0}^{4} \zeta_{-j}$, we have the desired z and its dominant part is of course $z_0 := e^{i\tau y_1}\zeta_0$.

Some useful lemma for the proof.

Lemma 14.5.4. *Let $P(\tau)$ be a polynomial in τ of degree 2 with real constant coefficients. Also, let the equation $P(\tau) = 0$ have two roots $\tau_\pm$ such that $\pm\mathrm{Re}\tau_\pm > 0$. For simplicity, we denote $\lambda = \tau_-$. For a given polynomial $f(t)$ in t of degree d, consider the solution $v = v(t)(t \geqslant 0)$ to*

$$P(d/dt)v = f(t)e^{\lambda t} \quad \text{in} \quad t > 0, \qquad v = 0 \quad \text{on} \quad t = 0 \tag{14.5.23}$$

which decays exponentially as $t \to \infty$. Then, it is given by

$$v(t) = \ell(t)e^{\lambda t}, \tag{14.5.24}$$

where $\ell(t)$ is a polynomial in t of degree $d + 1$ such that $\ell(0) = 0$.

Now we are ready to analyze the behavior of our indicator function as $\tau \to \infty$. It is enough to analyze the behavior of the integral

$$I(\tau) := \int_{\partial D} \left(\frac{\partial(\chi z_0)}{\partial \nu}\overline{w_0} - (\chi z_0)\frac{\overline{\partial w_0}}{\partial \nu} \right) ds. \tag{14.5.25}$$

By using the explicit form of w_0 and z_0, we have

$$I(\tau) = \tau \int_{-R}^{R} e^{2\tau\varphi(y_1)}\overset{\circ 22}{g}(\lambda + 1)\left|\left(\nabla_y x\right)^T e_2\right|^{-1} \sqrt{1 + \varphi'(y_1)^2}\, dy_1 + O(e^{-\delta'\tau}) \tag{14.5.26}$$

for some $\delta' > 0$ as $\tau \to \infty$, where we took $R > 0$ such that $\chi(y_1) = 1$ on $|y_1| \leqslant R$. Since $\varphi(y_1) = -\mu y_1^2 + O(y_1^3)$ $(y_1 \to 0)$ with $\mu = -\frac{1}{2}\varphi''(0) < 0$, $e^{2\tau\varphi(y_1)} \leqslant e^{-\tau\mu y_1^2}$ for $|y_1| \leqslant R$ and large τ by taking $R > 0$ small. Also, $\int_{-R}^{R} e^{-\mu\tau y_1^2} a(y_1)dy_1 = \frac{\sqrt{\pi}}{\sqrt{\mu\tau}}a(0) + O(1/\tau)$ as $\tau \to \infty$ for any bounded function $a(y_1)$ on $\mathbb{R}$ of C^1 class with bounded derivative. This immediately implies that $|I(\tau)|$ blows up by the order of $\tau^{1/2}$ as $\tau \to \infty$.

Finally, we prove (ii). By

$$e^{2\tau(h_D(\omega)-t)}|\tilde{\eta}(\tau, h_D(\omega), \omega)| = |\tilde{\eta}(\tau, t, \omega)|, \tag{14.5.27}$$

we have

$$h_D(\omega) - t = \frac{\log|\tilde{\eta}(\tau, t, \omega)|}{2\tau} - \frac{\log|\tilde{\eta}(\tau, h_D(\omega), \omega)|}{2\tau}. \tag{14.5.28}$$

Then, since the second term in the right-hand side of the above formula tends to 0 as $\tau \to \infty$, we immediately have (ii). □

Remark 14.5.5. *The method given here has the possibility not only to obtain $h_D(\omega)$ in the regular direction, but also to obtain a further geometric information on ∂D for regular direction ω. To remove having the regular direction, there is an important work by Sini and Yoshida* [11], *and further having one step multi-wave enclosure is given in* [12].

Bibliography

[1] Potthast R 1999 Point sources and multipoles in inverse scattering theory *Habilitation Thesis* University of Göttingen

[2] Ikehata M 1998 Reconstruction of the shape of the inclusion by boundary measurements *Commun. PDE* **23** 1459–74

[3] Ikehata M 1998 Reconstruction of an obstacle from the scattering amplitude at a fixed frequency *Inverse Problems* **14** 949–54

[4] Kar M and Sini M 2014 Reconstructing obstacles by the enclosure method using the far field measurements in one step *Appl. Anal.* **93** 1327–36

[5] Potthast R 2001 *Point Sources and Multipoles in Inverse Scattering Theory* (*Chapman & Hall/CRC Research Notes in Mathematics* vol 427) (Boca Raton, FL: CRC)

[6] Sini M and Yoshida K 2012 On the reconstruction of interfaces using complex geometrical optics solutions for the acoustic case *Inverse Problems* **28** 055013

[7] Potthast R and Schulz J 2005 From the Kirsch–Kress potential method via the range test to the singular sources method *J. Phys.: Conf. Ser.* **12** 116–27

[8] Potthast R 2006 A survey on sampling and probe methods for inverse problems *Inverse Problems* **22** R1–47

[9] Kusiak S, Potthast R and Sylvester J 2003 A range test for determining scatterers with unknown physical properties *Inverse Problems* **19** 533–47

[10] Burkard C and Potthast R 2009 A time-domain probe method for three-dimensional rough surface reconstructions *Inverse Probl. Imaging.* **3** 259–74

[11] Potthast R 2011 An iterative contractive framework for probe methods: LASSO *Radio Sci.* **46** RS0E14

[12] Russell Luke D and Potthast R 2003 The no response test—a sampling method for inverse scattering problems *SIAM J. Appl. Math.* **63** 1292–312

[13] Nakamura G 2004 Applications of the oscillating-decaying solutions to inverse problems *New Analytic and Geometric Methods in Inverse Problems* (New York: Springer) pp 353–65

[14] Nakamura G, Uhlmann G and Wang J-N 2005 Oscillating–decaying solutions, runge approximation property for the anisotropic elasticity system and their applications to inverse problems *J. Math. Pure. Appl.* **84** 21–54

IOP Publishing

Inverse Modeling

An introduction to the theory and methods of inverse problems and data assimilation

Gen Nakamura and Roland Potthast

Chapter 15

Analytic continuation tests

In our previous chapters we have introduced sampling and probe methods to reconstruct the location, shape and potentially further properties of scatterers. Most of these schemes are based on data for incident waves with many different directions of incidence, as given by the far field operator F.

As a parallel line of development, methods have been explored which can be carried out with fewer data, for example using the far field pattern u^∞ for scattering of one single plane wave only. Methods have been formulated by several researchers and groups. We will see that they can be understood as methods testing the analytic extensibility of the scattered field into areas of space, thus finding either the boundary where singularities of the field arise, or a subset of the interior of the scatterer.

15.1 The range test

The *range test* is a method using a test for the solvability of integral equations of the first kind to find the location and shape of unknown scatterers in acoustic or electromagnetic scattering theory. It was first suggested by Potthast, Kusiak and Sylvester [1].

The range test relies on one or any given number of far field patterns. It does not need to know the boundary condition of a scatterer to provide rough estimates for the location and shape from the far field pattern for scattering of one time-harmonic wave. It is used as a tool in other methods. We will also see that the ideas here can be used to establish convergence of other schemes such as the *no-response test* via duality arguments.

Consider the inverse shape reconstruction problem in acoustics with some scatterer D and a time-harmonic incident field u^i, its scattered field u^s and far field pattern u^∞. For some test domain $G \subset \mathbb{R}^m$ the range test investigates the relation between the domain D and the test domain G.

To understand the set-up of the range test we study the single-layer potential

$$w(x) := \int_{\partial G} \Phi(x, y)\varphi(y)\, \mathrm{d}s(y), \quad x \in \mathbb{R}^m, \tag{15.1.1}$$

doi:10.1088/978-0-7503-1218-9ch15

with density $\varphi \in L^2(\partial G)$. The field w solves the Helmholtz equation in $\mathbb{R}^m \backslash \bar{G}$ and satisfies the Sommerfeld radiation condition. According to (8.3.2) the far field pattern of the field w is given by

$$w^{\infty}(\hat{x}) = \gamma \int_{\partial G} e^{-i\kappa \hat{x} \cdot y} \varphi(y) \, ds(y), \quad \hat{x} \in \mathbb{S}, \tag{15.1.2}$$

with constant γ defined in (12.4.3).

If the closure $\bar{D}$ of the scatterer D is contained in the test domain G, then we will show in theorem 15.1.1 that the scattered field u^s can be represented as a single-layer potential (15.1.1) in $\mathbb{R}^m \backslash \bar{G}$. In this case the far field pattern u^{∞} is equal to w^{∞} and the equation

$$H^* \varphi = u^{\infty} \quad \text{on } \mathbb{S} \tag{15.1.3}$$

with H^* given in (8.3.13) has a solution $\varphi \in L^2(\partial G)$. If the representation of u^s in $\mathbb{R}^m \backslash \bar{G}$ is not possible, then (15.1.3) does not have a solution. Thus, we can use the solvability of equation (15.1.3) to test whether the scattered field u^s can be analytically extended into $\mathbb{R}^m \backslash \bar{G}$.

Finally, we need a simple criterion to judge whether (15.1.3) has a solution. According to theorems 3.1.8 and 3.1.10 for the Tikhonov regularization the regularized solution

$$\varphi_{\alpha} := (\alpha I + HH^*)^{-1} H^{*\prime} u^{\infty} \tag{15.1.4}$$

of (15.1.3) converges for $\alpha \to 0$ if and only if the equation has a solution. If the equation (15.1.3) does not have a solution, then the norm $\|\varphi_{\alpha}\|$ tends to infinity for $\alpha \to 0$. Thus, the norm of a regularized solution of (15.1.3) can be used as indicator function for the sampling domain G in the sense of (10.6.26).

Alternatively, we can use the following simple test for the convergence of φ_{α} suggested by Erhard [2]. In the case of solvability we have

$$\|\varphi_{\alpha} - \varphi_{\alpha/2}\| \to 0, \quad \alpha \to 0 \tag{15.1.5}$$

and if the equation (15.1.3) is not solvable, then

$$\|\varphi_{\alpha} - \varphi_{\alpha/2}\| \not\to 0, \quad \alpha \to 0. \tag{15.1.6}$$

This leads to the criterion: the equation (15.1.3) is solvable if and only if for each $c_{\mathrm{RT}} > 0$ there exists $\alpha_0 > 0$ such that

$$\left\| \varphi_{\alpha_1} - \varphi_{\alpha_2} \right\| \leqslant c_{\mathrm{RT}} \tag{15.1.7}$$

for all $\alpha_1, \alpha_2 \leqslant \alpha_0$. Thus, the criterion (15.1.7) can be employed to decide whether a test domain G is *positive* (if the criterion is satisfied) or *negative* (if the criterion is not satisfied) in the sense of definition 12.3.1.

We have argued that the criterion (15.1.7) can be used to test whether the field u^s can be analytically extended into $R^m \backslash \bar{G}$ for some test domain G, i.e. G is a positive test domain. This is the basic *range test technique* which is used as a tool in many other probe or sampling methods. Here, we will now describe a direct method to obtain shape reconstructions via a *domain sampling* process.

Consider two positive and convex test domains G_1 and G_2. Then, the scattered field can be analytically extended into the larger domain $\mathbb{R}^m \backslash \overline{G_1 \cap G_2}$. More generally, we can analytically extend u^s into $\mathbb{R}^m \backslash M$ with

$$M := \bigcap_{G \text{ positive}} \bar{G}. \tag{15.1.8}$$

If the set of test domains is sufficiently rich, we will show in our convergence analysis that the set M is a subset of the scatterer D. Thus, it can be used to provide an estimate for the location and shape of D.

We would like to complete this section with some remarks on the flexibility of the constants c_{RT} in the estimate (15.1.7). Consider a ball-like scatterer $D = B_R(z_0)$. It is well known that the scattered field for an incident plane wave can be analytically extended up to the center point z_0, see [3]. In this case we would not expect the range test to find more than the center point. However, with an appropriate choice of the constant c_{RT} it is possible to find approximation to the full shape of D. This is demonstrated in figure 15.1, where reconstructions of a ball of radius $R = 2$ with a Dirichlet boundary condition and a ball of radius $R = 3$ with a Neumann boundary condition with $c_{\mathrm{RT}} = 0.1$ are shown.

The numerical set-up for the range test is the same as for the Kirsch–Kress or the *point source method*, see figure 12.4. We need to define an appropriate set of test domains and for each test domain G a discretized version of the integral equation (15.1.3) needs to be solved via Tikhonov regularization such that the test (15.1.7) can be carried out.

In the simplest case the test domains can be realized as circles or balls $B_R(z)$, where z is chosen from a regular grid of points, see (8.1.16). The operator $\mathbf{H}^*$ has been worked out in (8.3.16). The discretized version of equation (15.1.3) is then given by

$$\mathbf{H}^* \circ \boldsymbol{\varphi} = \mathbf{u}^\infty, \tag{15.1.9}$$

which is carried out in code 12.3.4 for the Kirsch–Kress method. Its Tikhonov solution is

$$\varphi_\alpha := (\alpha \boldsymbol{I} + \mathbf{H}\mathbf{H}^*)^{-1}\mathbf{H}^{*\prime}\mathbf{u}^\infty, \tag{15.1.10}$$

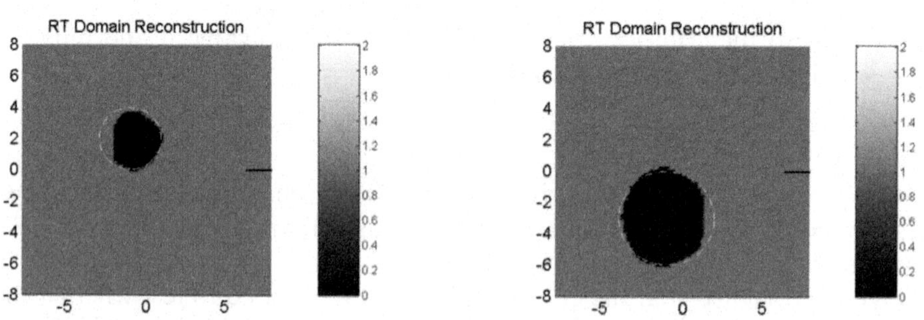

(a) (b)

Figure 15.1. Range test reconstructions for scattering by a ball like obstacle with a Dirichlet (a) or Neumann (b) boundary condition. Here, the constant c_{RT} is chosen to be equal to 0.1 or 0.06, respectively, and we use balls of radius $r = 6$ as sampling domains. The regularization parameter is $\alpha = 10^{-9}$.

see also figure 12.5 where the convergence of solution to the Kirsch–Kress equation has been tested for different locations of the test domain G. The operator $H^* = \gamma_m^{-1} S^\infty$ is defined in line 15, the regularized solution is calculated in lines 18 and 32, and a convergence test is carried out in line 36 of code 12.3.4. The solution is calculated for two choices of α, depending on the size of the data error under consideration. For data with only numerical errors we used $\alpha_1 = 10^{-9}$ and $\alpha_2 = \alpha_1/2$. Then, the criterion

$$\left\| \varphi_{\alpha_1} - \varphi_{\alpha_2} \right\| \leqslant c_{\mathrm{RT}} \tag{15.1.11}$$

is used to judge whether the equation is solvable or unsolvable. The choice of this constant is strongly linked to the overall numerical set-up, i.e. to the number of discretization points for the densities φ and the choice of the wave number κ. For our numerical experiments with a parameter set as shown in table 15.1 we usually used a fixed constant around $c_{\mathrm{RT}} = 0.1$, which has been successful both for smaller and larger scatterers and for Dirichlet or Neumann boundary conditions.

Figure 15.2 demonstrates the reconstruction of two boat-like scatterers, one with a Dirichlet and the other with a Neumann boundary condition.

The calculation of the intersections (15.1.8) can be realized via masking operations for the masks representing the different test domains G. Let m_{G_j} be the mask for the test domain G_j, $j = 1, 2$, then the mask for the intersection of G_1 and G_2 is given by

$$m_{G_1 \cap G_2} = m_{G_1} \cdot m_{G_2}, \tag{15.1.12}$$

where the dot $\cdot$ denotes pointwise multiplication of the matrices m_{G_1} and m_{G_2}.

Numerical results for the range test are shown in figure 15.1 for circular obstacles and in figure 15.2 for boat-shaped scatterers. To calculate the reconstructions we took the intersection of all positive test domains, with the result shown in black. Here, the behavior of the algorithm under different boundary conditions on ∂D has been tested as well, also its dependence on the constants when scatterers of different size are to be reconstructed. The results are quite satisfactory, in particular since we

Table 15.1. The typical parameter set which has often been used in reconstructions for two-dimensional sample problems. With these parameters all algorithms can be carried out within seconds or minutes with a precision of the results in the range of 1%–2%. This allows extensive numerical studies and testing as is necessary and useful for educational purposes.

Functionality	Variable	Value
Wave number κ	kappa	1
Approximate radius of scatterer D	scale	1, 2 or 3
Number of far field $u^\infty(\hat{x}_k)$ points	nff	100
Boundary discretization $\Gamma(t_j)$	n	100
Evaluation domain size $[-a, a] \times [-a, a]$	a	8
Evaluation domain discretization	np × np	80 × 80
Radius of test domains $G = B_r(z)$	r	6

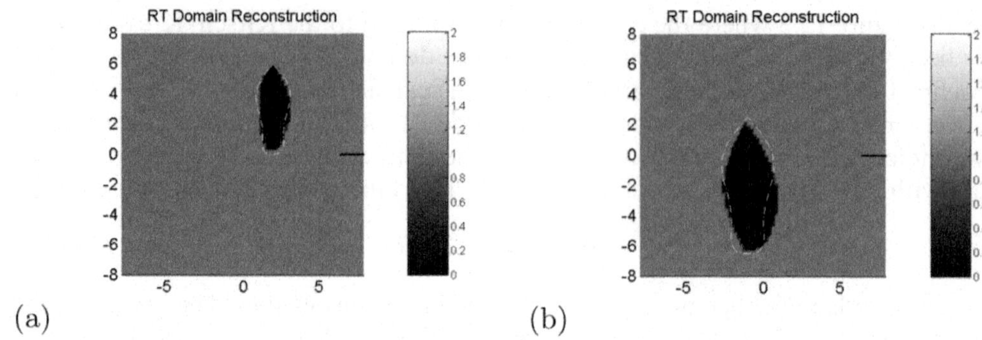

Figure 15.2. Numerical reconstructions of a smaller (a) and larger (b) boat like scatterer via the range test. Here, we used a Neumann boundary condition on the scatterer in (a) and a Dirichlet boundary condition for the scatterer in (b), both with the parameters given in table 15.1.

employ the scattered field for one incident plane wave only and we do not use the boundary condition for reconstructions.

Let us now carry out the convergence analysis of the *range test*.

Theorem 15.1.1. *We assume that the domain G is non-vibrating. The equation (15.1.3) is solvable in* $L^2(\partial G)$ *if and only if* u^s *with far field pattern* u^∞ *can be analytically extended into the exterior* $\mathbb{R}^m \backslash \bar{G}$ *of G with boundary value* $u_+^s \in H^1(\partial G)$.

Proof. If the equation is solvable in $L^2(\partial G)$, then there is $\phi \in L^2(\partial G)$ such that $H^* \phi = u^\infty$ on $\mathbb{S}$. Let

$$v(x) = \left(S_{\partial G}\left(\gamma_m^{-1}\phi\right)\right)(x) := \int_{\partial G} \Phi(x, y)\gamma_m^{-1}\phi(y)\, \mathrm{d}s(y) \tag{15.1.13}$$

for $x \in \mathbb{R}^m \backslash G$. That is, v is the single-layer potential with density γ_m^{-1} which is analytic in $\mathbb{R}^m \backslash \bar{G}$. Since v satisfies the Helmholtz equation in $\mathbb{R}^m \backslash \bar{G}$ and its far field pattern is u^∞, we have $u^s = v$ in $\mathbb{R}^m \backslash \overline{(D \cup M)}$ by the Rellich lemma. Hence u^s can be analytically extended into $\mathbb{R}^m \backslash \bar{G}$ and by the property that $S_{\partial G}$ maps $L^2(\partial G)$ bijectively into $H^1(\partial G)$, we have $u_+^s \in H^1(\partial G)$.

Now, we assume that u^s can be analytically extended into $R^m \backslash \bar{G}$ and u^s has an analytic extension into $\mathbb{R}^m \backslash \bar{G}$ and $u_+^s \in H^1(\partial G)$. Let $\phi \in L^2(\partial G)$ be the unique solution of the equation

$$S_{\partial G}\phi = u_+^s \quad \text{on } \partial G, \tag{15.1.14}$$

which is solvable due to the bijectivity of $S_{\partial G} : L^2(\partial G) \to H^1(\partial G)$. Also, let $w = S_{\partial G}\phi$ be the single-layer potential with density ϕ. Then, w is the unique solution to the boundary value problem for the Helmholtz equation in $\mathbb{R}^m \backslash G$ with radiation condition and Dirichlet condition $w = u^s$ on ∂G. By the uniqueness of this boundary value problem, we have $w = u^s$ in $\mathbb{R}^m \backslash G$ and hence

$$H^*(\gamma_m \phi) = w^\infty = u^\infty \quad \text{on } \mathbb{S}, \tag{15.1.15}$$

where w^∞ is the far field pattern of w. □

As an immediate consequence of the previous theorem combined with theorem 3.1.8, we obtain the convergence of the range test for the analytic continuation of scattered field and the estimate for the location of the scatterer.

Theorem 15.1.2. *For a scatterer D let u^s be a scattered field with far field pattern u^∞. Then*

- *(i) for any non-vibrating positive test domain G the field u^s can be analytically extended into $\mathbb{R}^m \setminus \bar{G}$ with boundary values in $H^1(\partial G)$,*
- *(ii) if the set $\mathcal{G}$ of test domains G is sufficiently rich such that*

$$\bar{D} \subset \bigcap_{G \in \mathcal{G},\ \bar{D} \subset G} G \tag{15.1.16}$$

the set M defined by (15.1.8) is a subset of $\bar{D}$.

Remark 15.1.3. *Finally, we note that there has been significant further work on the range test, for example by Kusiak and Sylvester [4]. The minimal set M which supports the scattered field has been called the* scattering support. *When intersections of convex test domains are considered, the term convex* scattering support *is used.*

For circles, the solvability of the Kirsch–Kress integral equation can be expressed in terms of its Fourier–Hankel series (12.1.5), and the behavior of the coefficients determines the analytic extensibility of the scattered field into the exterior of some ball B_R around the center point, see our test in figure 12.2.

15.2 The no-response test of Luke–Potthast

The *no-response test* [5, 6] is a general scheme probing a region with incident waves which are constructed to be small on some test domain. If the unknown scatterer is inside the test domain, then the corresponding scattered field is small. The idea comes from the fact that it is easy to create an incident field which is small on some particular region and large everywhere else, but it is difficult to have it large on some region but small everywhere else.

The idea of the *no-response test* is similar to the general idea to shoot into the darkness and if you hear a bang, then you have hit a target. But here, we instead search for the case where no hit is obtained, since we do not probe with large, but with small fields. So the idea is to some extent in the spirit of the singular sources or probe methods. But note that here we formulate a scheme which works for *one* incident time-harmonic wave and, thus, can be applied to settings where the singular sources or probe methods are not applicable.

To introduce the no-response test we start with a Herglotz wave function

$$v_g(x) = \int_{\mathbb{S}} e^{i\kappa x \cdot d} g(d)\, ds(d), \quad x \in \mathbb{R}^m,$$

as defined in (8.3.11) and we study this wave field on some test domain G. Consider a density $g \in L^2(\mathbb{S})$ such that

$$\left| v_g(x) \right| \leqslant \epsilon, \quad x \in \bar{G}, \tag{15.2.1}$$

i.e. the Herglotz wave function is smaller than ϵ on the test domain G. We consider v_g as incident field for scattering by a scatterer D. Then, since the scattering problem is bounded and linear with respect to the incident wave, the scattered field is given by

$$v_g^s(x) = \int_{\mathbb{S}} u^s(x, d)g(d)\,\mathrm{d}s(d), \quad x \in \mathbb{R}^m \backslash D, \tag{15.2.2}$$

with the scattered field $u^s(\cdot, d)$ for the incident field $\mathrm{e}^{i\kappa x \cdot d}$, $x \in \mathbb{R}^m$, $d \in \mathbb{S}$. Its far field pattern is

$$v_g^\infty(\hat{x}) = \int_{\mathbb{S}} u^\infty(\hat{x}, d)g(d)\,\mathrm{d}s(d), \quad \hat{x} \in \mathbb{S}, \tag{15.2.3}$$

with the far field pattern $u^\infty(\cdot, d)$ of $u^s(\cdot, d)$, $d \in \mathbb{S}$. By reciprocity (18.4.1) we obtain that

$$v_g^\infty(\hat{x}) = \int_{\mathbb{S}} u^\infty(-d, -\hat{x})g(d)\,\mathrm{d}s(d), \quad \hat{x} \in \mathbb{S}. \tag{15.2.4}$$

This means that the far field pattern $v_g^\infty(\hat{x})$ at $\hat{x} \in \mathbb{S}$ can be calculated from the far field pattern $u^\infty(d, -\hat{x})$ for $d \in \mathbb{S}$, i.e. from the far field pattern for *one* incident plane wave of direction of incidence $-\hat{x}$.

The no-response test is based on the following observation. If the unknown scatterer D is a subset of G, then the incident field is bounded by ϵ on the domain of the scatterer and by the linearity and boundedness of the scattering problem the scattered field must be bounded by $c\epsilon$ with some constant c. The same holds for its far field pattern, which is continuously depending on the scattered field. If the scatterer D is in the exterior of the test domain G (in fact we need some positive distance), then we can show that in general the scattered field will not be bounded by $c\epsilon$.

This now leads to a method to detect the location of a scatterer. We investigate the far field pattern (15.2.3) for many incident fields v_g with density $g \in L^2(\mathbb{S})$ chosen such that (15.2.1) is satisfied. The set of such densities is denoted by $\mathcal{M}_\epsilon(G)$. The far field pattern is observed in direction $-\hat{x}$. If for any $C > 0$ there is $\epsilon > 0$ such that the maximal modulus

$$\tilde{\mu}_\epsilon(G) := \sup_{g \in \mathcal{M}_\epsilon(G)} \left| v_g^\infty(-\hat{x}) \right| \tag{15.2.5}$$

is smaller than C, we mark the test domain as *positive*. If this is not the case, we mark it as *negative*.

In a second step we calculate the intersection of all positive test domains

$$M := \bigcap_{G \text{ positive}} G. \tag{15.2.6}$$

The one way no-response test of Luke–Potthast is to test a non-vibrating domain G in $\mathbb{R}^m$ ($m = 2, 3$) with $I(g)$ when g satisfies the property that $\|Hg\|_{L^2(\partial G)}$ is small. Multiplying the well-known formula

$$u^\infty(-d, -\hat{x}) = \gamma \int_{\partial D} \left\{ e^{i\kappa d\cdot y} \frac{\partial u^s(y, -\hat{x})}{\partial \nu}(y) - \frac{\partial e^{i\kappa d\cdot y}}{\partial \nu(y)} u^s(y, -\hat{x}) \right\} ds(y),$$

by $g(d)$ and integrating over $\mathbb{S}$ using reciprocity we obtain

$$v_g^\infty(\hat{x}) = \gamma \int_{\partial D} \left\{ \frac{\partial u^s}{\partial \nu} v_g - u^s \frac{\partial v_g}{\partial \nu} \right\} ds. \tag{15.2.7}$$

Hence, if the scatterer D is contained in G, then v_g is small, which means that there is almost no response coming from the scatterer. A more precise formulation of the one wave no-response test is as follows. For $\varepsilon > 0$, a non-vibrating domain G and $g_\varepsilon \in L^2(\mathbb{S})$ with $\|Hg_\varepsilon\|_{L^2(\partial G)} < \varepsilon$, let

$$\mu(G) = \lim_{\varepsilon \downarrow 0} \tilde{\mu}_\varepsilon(G), \tag{15.2.8}$$

where $\tilde{\mu}_\varepsilon(G)$ is defined by (15.2.5).

Please note that for a domain G the definitions that G is positive if $\mu(G) = 0$, and G is positive if for $C > 0$ there is $\epsilon > 0$ such that $\tilde{\mu}_\epsilon(G) \leqslant C$, are equivalent.

We note that if for $g \in L^2(\mathbb{S})$ we have $\|v_g\|_{L^2(\partial G)} \leqslant 1$, then ϵg satisfies

$$\|v_{\epsilon g}\|_{L^2(\partial G)} \leqslant \epsilon \tag{15.2.9}$$

and vice versa. This means that $\tilde{\mu}_\epsilon(G) = \epsilon \tilde{\mu}_1(G)$. Thus, $\mu(G) = 0$ is equivalent to $\tilde{\mu}_1(G)$ being bounded. In the next section we show by duality arguments that the *no-response test* tests for analytic extensibility into $\mathbb{R}^m \backslash \bar{G}$. Here, we formulate this result before we move towards numerical calculations.

Theorem 15.2.1. *First, if the scatterer D is subset of the test domain G, then G is positive, i.e. (i) if $D \subset G$ then $\mu(G) = 0$. The condition $D \subset G$ is not necessary for $\mu(G) = 0$. We have (ii) $\mu(G) = 0$ if the scattered field u^s can be analytically extended into $\mathbb{R}^m \backslash \bar{G}$, and $\mu(G) = \infty$ if u^s cannot be analytically extended into $\mathbb{R}^m \backslash \bar{G}$.*

Let us introduce a simple numerical realization of the *no-response test*. Here, we actually use an implementation based on the equivalence equation (15.3.6), i.e. we calculate the maximal response by solving the *range test* equation (15.1.3). We remark that this approach is an alternative to the construction of a set of densities as carried out in code 14.3.2.

Code 15.2.2. *Script* `sim_15_2_2_c_nrt_full.m` *to carry out the field reconstruction by the no-response test. Run script* `sim_15_2_2_a_scattering_ff.m` *first to generate u^∞, here named* `ff`*, and some further variables.* `sim_15_2_2_b_graphics.m` *visualizes the total field on some evaluation domain Ω. The other*

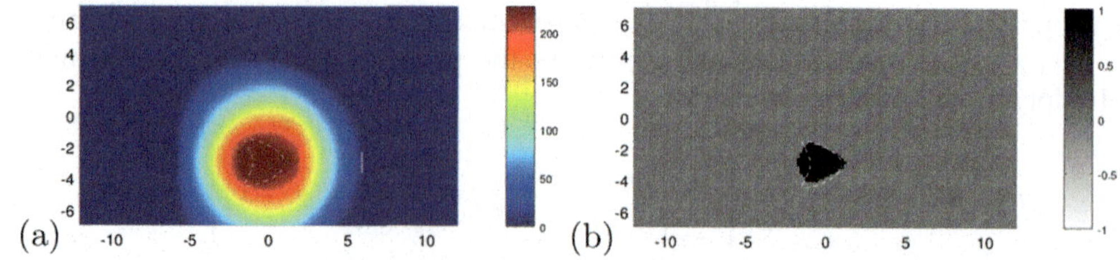

Figure 15.3. The result of the *no-response test* for reconstructing some domain D using the far field pattern for one incident plane wave coming from the right. (a) The sum of all positive masks, which should be largest on their intersection M. (b) The intersection of all positive test domains. The images are analogous to figure 12.9.

parts of figure 15.3 with the field reconstructions are generated by `sim_12_3_4_d_graphics.m`.

```
masksum  = ones(M,1);  % initialization of mask sums for total fields
masksum2 = zeros(M,1);  % initialization of mask sums for total fields
yG1all = []; yG2all = []; % initialization of the test domain collector
% loop over different domains using their center [z1 z2]
for z1 = -2:0.3:2
  for z2 = -6:0.3:1
    % I Preparations
    NG = 120;                          % number of points on test domain G
    hG = 2*pi/NG;                      % grid constant for test domain
    tG = 0:hG:2*pi-hG;                 % parametrization grid for G
    RG = 4;                            % radius of test domain
    yG1 = RG*cos(tG)+z1;               % boundary test domain comp.1
    yG2 = RG*sin(tG)+z2;               % boundary test domain comp.2

    yGffmat1  = repmat(yff1.',1,NG);     % matrix of far field points comp.1
    yGffmat2  = repmat(yff2.',1,NG);     % matrix of far field points comp.2
    yGmat1    = repmat(yG1,ffN,1);       % matrix of points of G comp.1
    yGmat2    = repmat(yG2,ffN,1);       % matrix of points of G comp.1

    % II far field of single-layer potential
    ffSG  = fac*exp(-i*kappa*(yGffmat1.*yGmat1+yGffmat2.*yGmat2))*RG*hG;

    % III regularized solution of far field equation
    alphaG = 1e-11;                   % regularization parameter
    varphiKK = (alphaG*eye(NG,NG) + ffSG'*ffSG)\ffSG'*ff; % density KK method
    alphaG = (1e-11)/2;               % reg parameter for testing convergence
    varphiKK2 = (alphaG*eye(NG,NG) + ffSG'*ffSG)\ffSG'*ff; % 2nd density

    % VI Masking operations
    maskG = sqrt( (pvec1-z1).^2 + (pvec2-z2).^2)<RG; % mask for circle

    if( norm(varphiKK - varphiKK2)/NG < 0.002 )
       masksum = masksum.*maskG;
       masksum2 = masksum2+maskG;
    end
    yG1all = [yG1all; yG1];   % for display of test domains
    yG2all = [yG2all; yG2];   % for display of test domains
  end
end
```

15.3 Duality and equivalence for the range test and no-response test

One important task in the mathematical investigation of inversion algorithms is to study the relations between the different methods. Here, we prove a *duality principle* for the range test and the no-response test. In particular, we will show that when the set-up of both methods is identical, then both schemes are *dual* to each other in the sense that dual equations are solved and show *equivalence* of the method for the case when a Tikhonov regularization scheme is employed.

Let us start with a definition of the *duality* of two methods.

Definition 15.3.1 (Duality). *Consider operators $A : X \to Y$ and $B : Y \to X$ with two Hilbert spaces X and Y. We call two equations $A\varphi = a$ and $B\psi = b$ dual to each other, if A and B are adjoint operators in the sense that*

$$\langle A\varphi, \psi\rangle_Y = \langle \varphi, B\psi\rangle_X \tag{15.3.1}$$

with the scalar products $\langle\cdot,\cdot\rangle_X$ and $\langle\cdot,\cdot\rangle_Y$ of X and Y, respectively. We call two algorithms dual, if they solve dual equations.

In the acoustic case we immediately see that for the operators S^∞ and H equation (15.3.1) can be verified by simple exchange of the integration variables according to Fubini's theorem, i.e. $S^\infty = \gamma_m H^*$.

The basic step of the *range test* for some domain G is to evaluate the indicator function $\mu(G)$ for G. The function is given by

$$\mu_{\mathrm{RT}}(G) := \lim_{\alpha\to 0} \|\varphi_\alpha\|^2 \tag{15.3.2}$$

where

$$\varphi_\alpha := (\alpha I + (S^\infty)^* S^\infty)^{-1}(S^\infty)^* u^\infty, \quad \alpha > 0, \tag{15.3.3}$$

with the operator

$$(S^\infty\varphi)(\hat{x}) = \int_{\partial G} \mathrm{e}^{-\mathrm{i}\kappa\hat{x}\cdot y}\varphi(y)\,\mathrm{d}s(y), \quad \hat{x}\in\mathbb{S}. \tag{15.3.4}$$

The definition of the *no-response test* relies on the indicator function μ_{NRT} defined by

$$\mu_{\mathrm{NRT}}(G) := \sup_{\|Hg\|_{L^2(\partial G)}\leqslant 1} \left|\langle g, u^\infty\rangle_{L^2(\mathbb{S})}\right|. \tag{15.3.5}$$

Here, in the condition $\|Hg\| \leqslant \epsilon$ we used $\epsilon = 1$, which can be understood as a simple scaling of the indicator function by $1/\epsilon$. We remark that with arguments as in (3.2.10) the inverse

$$(\alpha I + (S^\infty)^* S^\infty)^{-1} = (\alpha I + HH^*)^{-1}$$

is the adjoint of $(\alpha I + (S^{\infty})^{*}S^{\infty})^{-1}$. Under the condition that G is non-vibrating, the operator H has a dense range in $L^2(\mathbb{S})$. Then, we can rewrite the *range test* as

$$
\begin{aligned}
\mu_{\mathrm{RT}}(G) &= \lim_{\alpha\to 0} \sup_{\|v\|\leqslant 1} \left| \langle v, \varphi_\alpha \rangle_{L^2(\partial G)} \right| \\
&= \lim_{\alpha\to 0} \sup_{\|v\|\leqslant 1, v\in R(H)} \left| \langle v, (\alpha I + (S^{\infty})^{*}S^{\infty})^{-1}(S^{\infty})^{*}u^{\infty} \rangle_{L^2(\partial G)} \right| \\
&= \gamma_m^{-1} \lim_{\alpha\to 0} \sup_{\|v\|\leqslant 1, v\in R(H)} \left| \langle H^{*}(\alpha I + HH^{*})^{-1}v, u^{\infty} \rangle_{L^2(\mathbb{S})} \right| \\
&= \gamma_m^{-1} \lim_{\alpha\to 0} \sup_{\|v\|\leqslant 1, v\in R(H)} \left| \langle (\alpha I + H^{*}H)^{-1}H^{*}v, u^{\infty} \rangle_{L^2(\mathbb{S})} \right| \\
&= \gamma_m^{-1} \sup_{\|Hg\|\leqslant 1} \left| \langle g, u^{\infty} \rangle_{L^2(\mathbb{S})} \right| \\
&= \gamma_m^{-1} \mu_{\mathrm{NRT}}(G).
\end{aligned}
\tag{15.3.6}
$$

This proves that

(a) The *range test* solves the *dual equation* with respect to the *no-response test*.

(b) Both methods have the *same convergence properties* and, indeed, provide even the same numerical reconstructions when used with an equivalent set-up for the one-wave case.

(c) We can use the range test as an efficient way to calculate the no-response test function $\mu(G)$. This has been carried out in code 15.2.2.

We summarize these arguments in the following basic result.

Theorem 15.3.2. *The range test and the no-response test are dual in the sense of definition 15.3.1.*

15.4 Ikehata's enclosure method

We have already talked about the multi-wave version of Ikehata's enclosure method in section 14.5. Here we introduce, study and extend its one-wave version. The task is to reconstruct the location or shape of a scatterer from the far field pattern of a scattered wave.

The basic idea of the enclosure method is to employ the scaled *complex geometrical optics solution*

$$
w = \mathrm{e}^{-t\tau}\mathrm{e}^{x\cdot\left(i\tau\omega^{\perp}+\sqrt{\tau^2-\kappa^2}\,\omega\right)}
\tag{15.4.1}
$$

to the Helmholtz equation (8.2.1), where $\omega \in \mathbb{S}$, $\omega^{\perp} \perp \omega$ and $\tau > \kappa$ is a decay parameter and $t \in \mathbb{R}$ controls the shift of the function along the ω direction. Recall that this solution is a plane wave along $\omega^{\perp}$, an exponentially increasing solution to (8.2.1) along ω and exponentially decaying in the $-\omega$-direction, see figure 15.4. The parameter τ controls the wave number and the rate of divergence or decay at the same time.

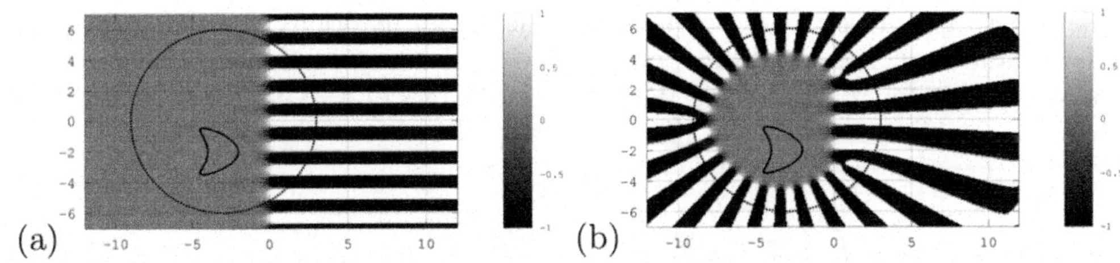

Figure 15.4. The real part of a complex geometric optics solution with $t = 0$ and $\omega = (1, 0)$ (a) and its approximation by a Herglotz wave function (b). We remark that the approximation is numerically challenging—the exponentially increasing parts for $x_1 > 0$ lead to significant errors in a neighborhood of the approximating circle.

We have already discussed the approximation of solutions to the Helmholtz equation by the point source method (12.4.13), i.e. for a non-vibrating domain G there is a density $g \in L^2(\mathbb{S})$ such that

$$Hg \approx w \quad \text{in } L^2(\partial G). \tag{15.4.2}$$

Then, up to the constant we obtain approximations to w and its derivatives on compact subsets of G.

We assume that we measure the far field pattern $u^\infty(\hat{x})$ in all directions $\hat{x} \in \mathbb{S}$. We remark that the *limited-aperture case*, i.e. where measurements are taken only on some limited angle $\beta < 2\pi$, in principle can be treated in exactly the same way as the full-aperture case, since the denseness of the Herglotz wave functions in $L^2(\partial G)$ is also satisfied when we integrate over an open subset of the unit sphere $\mathbb{S}$, such that we can restrict our attention to measurements on the full angle for an introductory presentation. This remark applied to all sampling and probe methods and to the field reconstruction methods of chapter 12 as well.

For the original enclosure method, the key ingredient is the use of the family (15.4.1) of the complex geometric optic solutions which are complex plane-wave type solutions with a large parameter. They have the special property that each of them is exponentially growing and decaying on each side of a line or hyperplane depending on the space dimension, which we name the *threshold level set* from now on, and oscillating along this level set as the parameter tends to infinity. This threshold level set is attached to each complex geometric optic solution to form a one parameter family of threshold level sets $\{\Sigma_t(\omega) : t \in \mathbb{R}\}$ with

$$\Sigma_t(\omega) = \{x |\ x \cdot \omega = t\}, \quad t \in \mathbb{R}, \tag{15.4.3}$$

for the complex geometrical optics solution (15.4.1).

An *indicator function* is defined which tends to infinity when the level set $\Sigma_t(\omega)$ touches a point p of the scatterer at which the solution u of the Helmholtz equation outside D cannot be continued into D for the one wave enclosure method. We will refer this point p as the *singular point* of the solution u.

Before we go into deeper analytical arguments, let us briefly describe a heuristic understanding of the enclosure method with the complex geometrical optics solution, see [7] for more details. We remark that by (12.4.2) we have

$$\int_{\mathbb{S}} u^{\infty}(\hat{x}, d)\overline{g(\hat{x})}\,\mathrm{d}s(\hat{x}) = \gamma \int_{\partial B} \left(\overline{Hg}(y)\frac{\partial u^s}{\partial \nu}(y, d) - \frac{\partial \overline{Hg}}{\partial \nu}(y)u^s(y, d) \right) \mathrm{d}s(y) \tag{15.4.4}$$

for any domain B for which u^s can be analytically extended into $\mathbb{R}^m \backslash \bar{B}$ with sufficiently smooth boundary values. Let ∂B be in the half-space where the complex geometrical optics solution decays exponentially, with a part Λ of the boundary ∂B which coincides with the line along $\omega^{\perp}$ where the solution w oscillates. Then, it has been argued in [7] that we can understand the functional (15.4.4) as an approximation to the Fourier transform of some function which has the same smoothness level as $u^s|_{\Lambda}$.

- If the field u^s is analytic in a neighborhood of ∂B and thus on $\Lambda \subset \partial B$, then the limit of the functional for $\tau \to \infty$ is zero.
- If the field u^s has a singularity in some point z and if we approach z with the line defined by ω and the shift parameter t, then the functional (15.4.4) will diverge for $\tau \to \infty$, since the derivatives of u^s cannot remain bounded.

The complex geometrical optics solution is bounded to the geometry of half-planes, thus we can only find the convex hull of the *scattering support* (see remark 15.1.3) of the far field u^{∞}. In the following sections we will introduce a method to go beyond half-spaces. For this extension we need to pay a price: we will no longer have an explicit representation of the probing function w.

15.4.1 Oscillating-decaying solutions

The disadvantage of the complex geometric optic solution is that it cannot be localized. In the following presentation we will introduce the *oscillating-decaying solution* given by Nakamura [8] as a substitute for the complex geometric optic solution. It can be explicitly defined on one side of a threshold level set. And it can be localized, which is an important step forward.

To use a locally defined oscillating-decaying function to define an indicator function as the substitute of the complex geometric optic solution, we further extend this complex geometric solution to the other side of the threshold level set. At this point we lose complete control of the behavior of this extended oscillating-decaying solution on the other side of the threshold level set. As a consequence we can only approach the scatterer up to some value $t_0 \in \mathbb{R}$ at which $\{\Sigma_t(\omega) : t \in \mathbb{R}\}$ touches the scatterer for the first time. Here we note that we are moving $t \in \mathbb{R}$ from $t = \infty$.

In order to simplify the analysis of the behavior of solution u at p, we only consider the two space dimensional case. To begin, we first define the oscillating-decaying solution. Let Ω, Ω_1 and $\tilde{\Omega}$ be bounded domains in $\mathbb{R}^2$ with smooth boundaries $\partial\Omega$, $\partial\Omega_1$ and $\partial\tilde{\Omega}$, respectively. We assume that these domains satisfy the relations

$$\bar{\Omega} \subset \Omega_1, \quad \overline{\Omega_1} \subset \tilde{\Omega}. \tag{15.4.5}$$

Let $\omega,\ \omega^\perp \in S^1 := \{\xi \in \mathbb{R}^2;\ |\xi| = 1\}$. For $t \in \mathbb{R}$ such that the line $x \cdot \omega = t$ intersects Ω, we define $\tilde{\Omega}_t(\omega) \subset \{x \cdot \omega < t\}$ as a slight modification of $\tilde{\Omega} \cap \{x \cdot \omega < t\}$ such that it has a smooth boundary $\partial\tilde{\Omega}_t(\omega)$ which coincides with $\Sigma_t(\omega) := \Omega_1 \cap \{x \cdot \omega = t\}$ in Ω_1.

The oscillating-decaying solution $z \in C^\infty(\overline{\tilde{\Omega}_t(\omega)})$ is a function with large parameter $\tau \geqslant 1$ which satisfies

$$(\Delta + \kappa^2)z = 0 \quad \text{in } \tilde{\Omega}_t(\omega) \tag{15.4.6}$$

and behaves like $\chi(x \cdot \omega^\perp)\exp[\tau\{\mathrm{i}x \cdot (\omega^\perp - \mathrm{i}\omega) - t\}]$ in $\overline{\tilde{\Omega}_t(\omega)}$, where $\chi(\eta) \in C_0^\infty(\mathbb{R})$ such that $\operatorname{supp}\chi(x \cdot \omega^\perp) \cap \{x \cdot \omega = t\} \subset \Sigma_t(\omega)$ and $\chi(x_0 \cdot \omega^\perp) = 1$ for a given $x_0 \in \Sigma_t(\omega)$.

For $N \in \mathbb{N}$ we will seek z in the form:

$$z = \mathrm{e}^{\mathrm{i}\tau x\cdot\omega^\perp} w_N + r. \tag{15.4.7}$$

Here $w_N \in C^\infty(\overline{\tilde{\Omega}_t(\omega)})$ has to satisfy $\operatorname{supp} w_N \subset \operatorname{supp}\chi(x \cdot \omega^\perp)$ and the following estimates. For any multi-index α, there exists a constant $C_\alpha > 0$ such that

$$\left|\partial^\alpha\left(w_N - \chi(x \cdot \omega^\perp)\mathrm{e}^{\tau(x\cdot\omega - t)}\right)\right| \leqslant C_\alpha \tau^{|\alpha|-1} \mathrm{e}^{\tau(x\cdot\omega-t)} \tag{15.4.8}$$

for $x \in \overline{\tilde{\Omega}_t(\omega)}$, $\tau \geqslant 1$, where $\partial^\alpha := (\partial/\partial x_1)^{\alpha_1}(\partial/\partial x_2)^{\alpha_2}$ for $x = (x_1, x_2)$, $\alpha = (\alpha_1, \alpha_2)$. Defining Q by

$$Q(\cdot) := \mathrm{e}^{-\mathrm{i}\tau x\cdot\omega^\perp}(\Delta + \kappa^2)\,(\mathrm{e}^{\mathrm{i}\tau x\cdot\omega^\perp}\cdot), \tag{15.4.9}$$

we have

$$Qw_N = O\left(\tau^{-N}\mathrm{e}^{\tau(x\cdot\omega-t)}\right) \quad (\tau \to \infty). \tag{15.4.10}$$

Moreover, $r \in C^\infty(\overline{\tilde{\Omega}_t(\omega)})$ which depends on N satisfies

$$(\Delta + \kappa^2)r = f \quad \text{in } \tilde{\Omega}_t(\omega) \quad \text{with } f = -\mathrm{e}^{\mathrm{i}\tau x\cdot\omega^\perp} Qw_N \tag{15.4.11}$$

and the estimate (15.4.19) given later.

Finally, we will give a brief illustration of the construction of w_N. For simplicity we assume $\omega = (0, 1)$, $\omega^\perp = (1, 0)$ and $x_0 = (0, 0)$. First we introduce the concept of order in the following manner. That is, we consider that $\partial/\partial x_2$, τ are of order 1 and $\partial/\partial x_1$ is of order 0. Then, decompose w_N in the form

$$w_N = \sum_{j=0}^{N} w_N^{(-j)} \tag{15.4.12}$$

with $w_N^{(-j)}$ is of order $-j$. The estimate we have in mind for $w_N^{(-j)}$ is

$$w_N^{(-j)} = O(\tau^{-j})(\tau \to \infty) \tag{15.4.13}$$

which turns out to be true if we estimate each $w_N^{(-j)}$ given below in (15.4.18). We look for w such that

$$\begin{aligned} Qw_N &= O(\tau^{-N+1})(\tau \to \infty) \\ w_N|_{x_2=t} &= \chi(x_1)\mathrm{e}^{\mathrm{i}\tau x_1}. \end{aligned} \tag{15.4.14}$$

Then, grading Q

$$\begin{aligned} Q &= Q_2 + Q_1 + \cdots \\ Q_2 &= \frac{\partial^2}{\partial x_2^2} - \tau^2 \\ Q_1 &= 2\mathrm{i}\tau\frac{\partial}{\partial x_1} \\ &\cdots \end{aligned} \tag{15.4.15}$$

with respect to the order and collecting terms with the same order by expanding Qw, $w_N^{(-j)}$ $(j = 0, 1, 2, \cdots)$ have to satisfy

$$\begin{aligned} &\left(\frac{\partial^2}{\partial x_2^2} - \tau^2\right)w_N^{(0)} = 0, \; w_N^{(0)}\Big|_{x_2=t} = \chi(x_1) \\ &\left(\frac{\partial^2}{\partial x_2^2} - \tau^2\right)w_N^{(-1)} = -2\mathrm{i}\tau\frac{\partial}{\partial x_1}w_N^{(0)}, \; w_N^{(-1)}\Big|_{x_2=t} = 0 \end{aligned} \tag{15.4.16}$$

$$\cdots \tag{15.4.17}$$

and $w_N^{(-j)}$ $(j = 0, 1, 2, \cdots)$ are given by

$$\begin{aligned} w_N^{(0)} &= \chi(x_1)\mathrm{e}^{\tau(x_2-t)} \\ w_N^{(-1)} &= -\mathrm{i}(x_2 - t)\chi'(x_1)\mathrm{e}^{\tau(x_2-t)}. \\ &\cdots \end{aligned} \tag{15.4.18}$$

Estimate of the remainder r. We will work out a proof for the following estimate for r.

Theorem 15.4.1. *There exists* $r \in C^\infty(\overline{\tilde{\Omega}_t(\omega)})$ *which satisfies (15.4.11) and the estimate*

$$\|\partial^\alpha r\|_{L^2(\tilde{\Omega}_t(\omega))} = O(\tau^{|\alpha|+1-N}\mathrm{e}^{\tau(x\cdot\omega-t)}), \quad \tau \to \infty \tag{15.4.19}$$

for each multi-index α $(|\alpha| \leqslant 2)$.

Proof. In order to simplify the notations in the proof, we give the proof for the case

$$x_0 = 0, \quad t = 0$$
$$\omega = (0, 1), \quad \omega^{\perp} = (1, 0)$$
$$\overline{\tilde{\Omega}_t(\omega)} \subset \left\{ |x_1| < \frac{R}{2}, \ |x_2| < \delta/2 \right\}. \tag{15.4.20}$$

For a small $\delta > 0$, let

$$\begin{aligned} C_{\pm}(\delta, R) &:= \{|x_1| < R, \quad 0 < \pm x_2 < \delta\} \\ C(\delta, R) &:= \{|x_1| < R, \quad |x_2| < \delta\} \end{aligned} \tag{15.4.21}$$

and define the even extension f_1 of f across $x_2 = 0$ by

$$H^1_{\mathrm{loc}}(\mathbb{R}^2) \ni f_1(x_1, x_2; \tau) = \begin{cases} f(x_1, x_2; \tau) & \text{in } C_-(\delta, R) \\ f(x_1, -x_2; \tau) & \text{in } C_+(\delta, R). \end{cases} \tag{15.4.22}$$

By (15.4.10) and (15.4.22), it is easy to see that

$$\tilde{f}_1(x; \tau) := \mathrm{e}^{-\tau x_2} f_1(x; \tau) \tag{15.4.23}$$

satisfies the estimate:

$$\left\| \tilde{f}_1 \right\| = O(\tau^{-N})(\tau \to \infty), \tag{15.4.24}$$

where $\| \ \|$ denotes the $L^2(C(\delta, R))$ norm. Now consider

$$(\Delta + \kappa^2) r_1 = f_1 \quad \text{in } C(\delta, R). \tag{15.4.25}$$

Then in terms of

$$\tilde{r}_1 := \mathrm{e}^{-\tau x_2} r_1, \tag{15.4.26}$$

(15.4.25) is equivalent to

$$\Delta \tilde{r}_1 + 2\tau \partial_2 \tilde{r}_1 + (\tau^2 + \kappa^2) \tilde{r}_1 = \tilde{f}_1, \tag{15.4.27}$$

where $\partial_j := \partial / \partial x_j (j = 1, 2)$. In order to solve (15.4.27) with a good estimate for $\tilde{r}_1$, we will follow Hähner's argument [9]. By defining the grid in terms of

$$\Gamma := \left\{ \alpha = (\alpha_1, \alpha_2) \in \mathbb{R}^2; \ \frac{R\alpha_1}{\pi} \in \mathbb{Z}, \ \frac{\delta\left(\alpha_2 - \dfrac{1}{2}\right)}{\pi} \in \mathbb{Z} \right\} \tag{15.4.28}$$

we can seek the solution $\tilde{r}_1$ in the form:

$$\tilde{r}_1 = \sum_{\alpha \in \Gamma} \tilde{r}_{1\alpha} e_\alpha, \tag{15.4.29}$$

where each $e_\alpha(x)$ and $\tilde{r}_1$ are given by

$$e_\alpha(x) := \frac{1}{2\sqrt{\delta R}} e^{i\alpha\cdot x} \ (x \in C(\delta, R), \ \alpha \in \Gamma)$$

$$\tilde{r}_{1\alpha} = \langle \tilde{r}_{1\alpha}, e_\alpha \rangle, \tag{15.4.30}$$

with the $L^2(C(\delta, R))$ inner product $\langle , \rangle$. Substituting (15.4.29) into (15.4.27), each $\tilde{r}_{1\alpha}$ has to satisfy

$$(-\alpha \cdot \alpha + 2i\tau\alpha_2 + (\tau^2 + \kappa^2))\, \tilde{r}_{1\alpha} = \tilde{f}_{1\alpha}, \tag{15.4.31}$$

where $\tilde{f}_1 = \sum_{\alpha\in\Gamma} \tilde{f}_{1\alpha} e_\alpha$, $\tilde{f}_{1\alpha} = \langle \tilde{f}_1, e_\alpha \rangle$. Then, since

$$\frac{\delta\left(\alpha_2 - \frac{1}{2}\right)}{\pi} \in \mathbb{Z} \Longrightarrow |\alpha_2| \geqslant \frac{1}{2}, \tag{15.4.32}$$

$$|-\alpha \cdot \alpha + 2i\tau\alpha_2 + \tau^2| \geqslant |\mathrm{Im}(-\alpha \cdot \alpha + 2i\tau\alpha_2 + (\tau^2 + \kappa^2)| = \tau. \tag{15.4.33}$$

Hence, each $\tilde{r}_{1\alpha}$ is given by

$$\tilde{r}_{1\alpha} = \frac{\tilde{f}_{1\alpha}}{-\alpha \cdot \alpha + 2i\tau\alpha_2 + (\tau^2 + \kappa^2)}. \tag{15.4.34}$$

By (15.4.33), we have

$$\|\tilde{r}_1\| \leqslant \frac{1}{\tau}\left\|\tilde{f}_1\right\|. \tag{15.4.35}$$

Now, for estimating $\nabla\tilde{r}_1$, we use the estimates:

$$\begin{aligned}
&\left|-\alpha \cdot \alpha + 2i\tau\alpha_2 + \left(\tau^2 + \kappa^2\right)\right| \geqslant \tau \\
&\geqslant \frac{-\sqrt{\tau^2 + \kappa^2} + \sqrt{\tau^2 + \kappa^2 + 4\tau}}{2}|\alpha| \\
&\text{if} \quad |\alpha| \leqslant \frac{\sqrt{\tau^2 + \kappa^2} + \sqrt{\tau^2 + \kappa^2 + 4\tau}}{2} \\
&\left|-\alpha \cdot \alpha + 2i\tau\alpha_2 + \left(\tau^2 + \kappa^2\right)\right| \\
&\geqslant \left(|\alpha| + \sqrt{\tau^2 + \kappa^2}\right)\left(|\alpha| - \sqrt{\tau^2 + \kappa^2}\right) \\
&\geqslant \frac{-\sqrt{\tau^2 + \kappa^2} + \sqrt{\tau^2 + \kappa^2 + 4\tau}}{2}|\alpha| \\
&\text{if} \quad |\alpha| \geqslant \frac{\sqrt{\tau^2 + \kappa^2} + \sqrt{\tau^2 + \kappa^2 + 4\tau}}{2}.
\end{aligned} \tag{15.4.36}$$

This immediately implies

$$\|\nabla \tilde{r}_1\| \leqslant \frac{\sqrt{\tau^2 + \kappa^2} + \sqrt{\tau^2 + \kappa^2 + 4\tau}}{2\tau} \left\| \tilde{f}_1 \right\|. \tag{15.4.37}$$

Using the equation $\Delta \tilde{r}_1 = -2\tau \partial_2 \tilde{r}_1 - (\tau^2 + \kappa^2)\tilde{r}_1 + \tilde{f}_1$, (15.4.35) and (15.4.37), we have

$$\|\Delta \tilde{r}_1\| \leqslant \left(2\sqrt{\tau^2 + \kappa^2} + \sqrt{\tau^2 + 4\tau} + 1\right) \left\| \tilde{f}_1 \right\|. \tag{15.4.38}$$

Since any second derivative of $\tilde{r}_1$ can be estimated using (15.4.35), (15.4.38), we have (15.4.19) for any $\alpha(|\alpha| \leqslant 2)$. Hence in order to finish the proof we only need to define r as the restriction of $\tilde{r}_1$ to $\overline{\Omega_t(\omega)}$. □

Remark 15.4.2. *For constructing r with the estimate (15.4.19), we could have used a Carleman estimate. See [10, 11] for details.*

15.4.2 Identification of the singular points

In this part we will build a one wave enclosure method based on the oscillating-decaying solution for calculating information about the singular points of the scattered field or the shape of an unknown polygonal cavity D, from the far field pattern $u^\infty(\cdot, d)$ of the scattered field $u^s = u^s(\cdot, d)$ which satisfies the scattering problem introduced in section 8.2, i.e.

$$\begin{aligned} &\left(\Delta + \kappa^2\right)u^s = 0 \quad \text{in } \mathbb{R}^2 \backslash \bar{D} \\ &\left.\frac{\partial u}{\partial \nu}\right|_{\partial D} = 0 \\ &\lim_{|x| \to \infty} \sqrt{|x|}\left(\frac{\partial u^s}{\partial |x|} - \mathrm{i}\kappa u^s\right)(x, d) = 0, \end{aligned} \tag{15.4.39}$$

where $u = u^s + u^i$ and $u^i(x, d)$ denote the total wave field and an incident wave $u^i(x, d) = e^{i\kappa x \cdot d}$ with incident direction d, respectively. Then, our inverse problem is as follows.

Definition 15.4.3 (Identification of singular points). *The task of the singular point identification problem is to find the singular points of the field u^s at the boundary ∂D or inside D from the far field pattern $u^\infty(\hat{x}, d)(\hat{x} \in \mathbb{S})$ of the scattered wave $u^s(\cdot, d)$ with direction of incidence $d \in \mathbb{S}$.*

The singular points of u at the boundary ∂D carry the geometric information of D which will be useful for identifying D. This might only be insufficient information about the shape of the scatterer D.

However, if D is a *convex polygon*, then the possible singular points of u are the *vertices* of D. Hence, in this case, the identification of the singular points will lead to full information of the location of D and also fully reconstruct its shape.

Let Ω_1, $\tilde{\Omega}, \Sigma_t(\omega)$ with $\omega \in \mathbb{S}$ be those given above and $V := \{y_k\,;\ 1 \leqslant k \leqslant K\}$ be a set of some vertices of a polygon D. We assume the following assumptions (A1) to (A3):

(A1) The points in V are well separated. That is we assume that we know the smallest distance d between any two points in V.

(A2) $\Sigma_t(\omega) \cap \partial D = \emptyset$ for $t > t_0$ and $\Sigma_{t_0}(\omega) \cap \partial D = V$.

(A3) u satisfying (15.4.39) has the strongest singularity at each y_k $(1 \leqslant k \leqslant K)$. That is $u \notin H^2_{\mathrm{loc}}$ near each y_k.

Let $z = z_{\chi,N,t,\omega}(x, \tau)$ be the oscillating-decaying solution in $\overline{\tilde{\Omega}_t(\omega)}$ constructed in section 15.4.1. Here, we take the diameter of suppχ to be less than d. Then, by the *point source method* or the Runge approximation theorem, for any ϵ $(0 < \epsilon \ll 1)$, there exist Herglotz wave functions $\tilde{z}_{\epsilon,j} = Hg_{\epsilon,j}$ with densities $g_{\epsilon,j} \in L^2(\mathbb{S})$ $(j = 1, 2, \cdots)$ such that $\tilde{z}_{\epsilon,j} \to z_{\chi,N,t+\epsilon,\omega}(x, \tau)$ in $H^2(\Omega_t(\omega))$ $(j \to \infty)$, where $\Omega_t(\omega) := \{x \in \Omega\,;\ x \cdot \omega < t\}$. Then, define the indicator function $I(\tau, t, \chi, \omega)$ by

$$I(\tau, t, \omega) = \sup|J(\tau, t, \chi, \omega)|, \tag{15.4.40}$$

where the supremum is taken for all $\chi \in C_0^\infty(\mathbb{R})$ which satisfy

$$\text{diameter of } \operatorname{supp}(\chi) < d \tag{15.4.41}$$

and

$$J(\tau, t, \chi, \omega) = \lim_{\epsilon \to 0} \lim_{j \to \infty} \int_{\mathbb{S}} u^\infty(\theta, d) g_{\epsilon,j}(\theta)\, \mathrm{d}s(\theta) \tag{15.4.42}$$

which is equal to

$$-\int_{\partial D}\left(\frac{\partial u^s}{\partial \nu}(y, d) z(y) - \frac{\partial z}{\partial \nu}(y) u^s(y, d)\right) \mathrm{d}s(y)$$
$$= \int_{\partial D} \frac{\partial z}{\partial \nu}(y)(u(y, d) - \lambda)\, \mathrm{d}s(y)$$

with $z = z_{\chi,N,t,\omega}$ for any constant λ due to

$$\int_{\mathbb{S}} u^\infty(\theta, d) g(\theta)\, \mathrm{d}s(\theta)$$
$$= -\int_{\partial D}\left(\frac{\partial u^s}{\partial \nu}(y, d)(Hg)(y) - \frac{\partial (Hg)}{\partial \nu}(y) u^s(y, d)\right) \mathrm{d}s(y)$$

and

$$\int_{\partial D} \lambda \frac{\partial (Hg)}{\partial \nu}(y)\, \mathrm{d}s(y) = 0$$

for any $g \in L^2(\mathbb{S})$.

Now let y_0 be one of $y_k (1 \leqslant k \leqslant K)$ and the local behavior of u which satisfy (A3) be $u = u_0 + O(|x - y_0|^\sigma)$ $(|x - y_0| \to 0)$, where u_0 is a constant and $\sigma := \pi/\Theta$ with exterior angle Θ of the polygon D at the vertex y_0. Then, take $\lambda = u_0$.

Then, we can use the asymptotic analysis of the indicator function given in Ikehata [12], because the dominant part of $v_{\chi,N,t,\omega}$ is the same as the exponentially growing solution which was used to define the indicator function in [12] provided that $\chi(x \cdot \omega^{\perp}) = 1$ at y_0. More precisely, the indicator function in [12] is (15.4.42) with $v_{\epsilon,j}$ replaced by the exponentially growing solution $e^{\tau(x\cdot\omega - t + ix\cdot\omega^{\perp})}$ and deleting the limits with respect to j and ϵ.

Then, we have the following lemma.

Lemma 15.4.4. *For the indicator function I defined by (15.4.40) we have*

(i) *If* $t \in (t_0, \infty)$, *then* $I(\tau, t, \omega) = O(e^{-(t-t_0)\tau})$ $(\tau \to \infty)$.

(ii) *There exists a constant* $c \neq 0$ *such that*

$$I(\tau, t_0, \omega) = c\tau^{-\sigma}(1 + o(1)), \quad \tau \to \infty. \tag{15.4.43}$$

(iii) *We have the following Polya type identity:*

$$t_0 - t = \lim_{\tau \to \infty} \frac{|\log I(\tau, t, \omega)|}{\tau} \tag{15.4.44}$$

for $t \in [t_0, 1]$.

The above lemma can be used to predict some information about the location of V. In particular, the *Polya type identity* may be useful to search t_0, because we can expect that $\log|I(\tau, \omega, t)|/\tau + t$ for a large fixed τ will only variate around some fixed value as we move $t \in (t_0, \infty)$ toward t_0 and this value can possibly be very close to t_0.

Bibliography

[1] Kusiak S, Potthast R and Sylvester J 2003 A 'range test' for determining scatterers with unknown physical properties *Inverse Problems* **19** 533–47

[2] Erhard K 2005 Point source approximation methods in inverse obstacle reconstruction problems *PhD Thesis* University of Göttingen

[3] Colton D and Kress R 1998 *Inverse Acoustic and Electromagnetic Scattering Theory* (*Applied Mathematical Sciences* vol 93) 2nd edn (Berlin: Springer)

[4] Kusiak S and Sylvester J 2003 The scattering support *Commun. Pure Appl. Math.* **56** 1525–48

[5] Luke D R and Potthast R 2003 The no response test—a sampling method for inverse scattering problems *SIAM J. Appl. Math.* **63** 1292–312

[6] Potthast R 2007 On the convergence of the no response test *SIAM J. Math. Anal.* **38** 1808–24

[7] Potthast R 2006 A survey on sampling and probe methods for inverse problems *Inverse Problems* **22** R1–47

[8] Nakamura G 2004 Applications of the oscillating-decaying solutions to inverse problems *New Analytic and Geometric Methods in Inverse Problems* (Berlin: Springer) pp 353–65

[9] Hähner P 1996 A periodic Faddeev-type solution operator *J. Differ. Equ.* **128** 300–8

[10] Kenig C, Sjöstrand J and Uhlmann G 2007 The Calderón problem with partial data *Ann. Math.* **165** 567–91

[11] Nakamura G and Yoshida K 2007 Identification of a non-convex obstacle for acoustical scattering *J. Inverse Ill-Posed Probl.* **15** 611–24

[12] Ikehata M 1999 Enclosing a polygonal cavity in a two-dimensional bounded domain from Cauchy data *Inverse Problems* **15** 1231–41

IOP Publishing

Inverse Modeling
An introduction to the theory and methods of inverse problems and data assimilation
Gen Nakamura and Roland Potthast

Chapter 16

Dynamical sampling and probe methods

In the previous chapters we have introduced the main ideas of sampling and probing methods. This chapter can be viewed as an *advanced topics* section, where we treat *dynamical* cases, i.e. where a *time dimension* and the behavior of the field in time are taken into account.

In the first section 16.1 we treat the heat equation, which describes the development of the heat field over time. With the corresponding fundamental solution which lives in a four-dimensional space, we will develop the basic theory of the *linear sampling method.*

In section 16.2 we describe the *dynamical probe method* of Nakamura. It is concerned with thermographic non-destructive testing to identify some cavity D based on the heat equation. This testing is to identify D from the measurements which apply the *heat flux* (sometime called thermal load) to $\partial\Omega$ many times and measure the corresponding *temperature* on $\partial\Omega$.

Section 16.3 introduces a way to employ probing in the time domain. Based on the work of Burkard and Potthast [7] we use a time-dependent pulse to probe some unknown area and find the boundary as the set of points where a scattered field arises at the first arrival time. Reconstructions of the time-dependent scattered field $U^s(x, t)$ are carried out based on the *point source method* based on [40].

If we can measure waves over some time interval at the boundary of the medium, we can use the *boundary control method* (BC method) by Belishev–Kurylev, introduced in section 16.4. It is a very powerful method for identifying the *coefficients* of the second order scalar hyperbolic equation which describes the propagation of waves inside the medium. The measured data of the BC method are usually taken as the *Neumann-to-Dirichlet map*, which is a set of infinitely many Cauchy data at the boundary of solutions to the initial boundary value problem for the hyperbolic equation without a source term and with a homogeneous initial condition.

doi:10.1088/978-0-7503-1218-9ch16

16.1 Linear sampling method for identifying cavities in a heat conductor

There are many kinds of non-destructive tests. They depend on what kind of physical phenomena are used. For example electric impedance tomography (EIT), magnetic tomography, ultrasonic inspection and thermography use electric current, magnetic field, ultrasound waves and heat, respectively. The reconstruction methods given for inverse acoustic wave scattering problems in chapters 13–15 can be adapted for ultrasonic inspection.

In this section we will provide some mathematical basis for active thermography ([9, 22, 38]). It is a non-destructive test to detect anomalies such as cracks, cavities and inclusions inside a heat conductor. The method involves injecting a heat flux using a flash lamp or heater into a heat conductor and measuring the corresponding temperature distribution at the boundary of the conductor using an infrared light camera. This is a very fast non-contact measurement which can be repeated many times. The principle of active thermography is based on the following observation. That is, if there is an anomaly, then it will influence the heat flow inside the conductor and change the corresponding temperature distribution. The model equations which describe the phenomenon are the heat equation for cracks and cavities, and the heat equation with discontinuous coefficients for inclusions. We note that these equations are not formally self-adjoint which makes us very curious to study the reconstruction method for thermography. We first look for a linear sampling type method for active thermography.

First we have to note that there is a large difference between the measurements for inverse scattering and active thermography. That is, inverse scattering uses far field measurements and active thermography uses near field measurements. EIT is a typical inverse boundary value problem which uses near field measurement. For the inverse boundary value problem for EIT, Somersalo gave a linear sampling method in his unpublished paper [41] to identify unknown inclusions. We will basically follow his argument to provide a linear sampling type method for active thermography to identify unknown cavities inside a heat conductor.

To begin with let $\Omega \subset \mathbb{R}^m (m = 1, 2, 3)$ be a heat conductor with unknown cavities $D \subset \Omega$ which consist of several disjoint domains such that $\bar{D} \subset \Omega$ and $\Omega \backslash \bar{D}$ are connected. Each of these domains can be considered as a cavity. We assume that the boundaries $\partial\Omega$ and ∂D of Ω and D, respectively, are of the C^2 class. We assume that the initial temperature distribution of Ω with cavities D is zero. A single measurement of active thermography measures the corresponding temperature distribution on $\partial\Omega$ over $(0, T)$ for a given heat flux f on $\partial\Omega$ over the time interval $(0, T)$. This measurement can be quickly repeated many times because the effect of the heat flux injection will die out very quickly. The reason for this is because the heat flux is only injected into part of the boundary and the rest of the boundary has a uniform constant temperature distribution equal to that of the surrounding heat conductor. In fact it can be proven mathematically that the temperature generated by the heat flux returns to the surrounding temperature exponentially fast in time after the heat flux is stopped. However, to make our argument simple, we will restrict ourselves to the case where the heat flux is injected on the whole $\partial\Omega$ and D consists of a single cavity, i.e. D is a domain. Also, based on the above observation we assume that we can have as our measurement the set of all possible sets of the aforementioned single measurement which will be a graph of the so-called Neumann-to-Dirichlet map, defined below, between some function spaces.

Before introducing the precise definition of the Neumann-to-Dirichlet map, we prepare some notation for the anisotropic Sobolev spaces which we are going to use for analyzing solutions of the heat equation in this section and the heat equation with discontinuous coefficients in the next section.

For $p, q \geqslant 0$ let

$$H^{p,q}(\mathbb{R}^m \times \mathbb{R}) := L^2(\mathbb{R}; H^p(\mathbb{R}^m)) \bigcap H^q(\mathbb{R}; L^2(\mathbb{R}^m)).$$

and for $p, q \leqslant 0$ define by duality

$$H^{p,q}(\mathbb{R}^m \times \mathbb{R}) := H^{-p,-q}(\mathbb{R}^m \times \mathbb{R}).$$

In terms of the Fourier transform a tempered distribution f defined on $\mathbb{R} \times \mathbb{R}^m$ is in $H^{p,q}(\mathbb{R}^m \times \mathbb{R})$ if and only if we have

$$\{(1 + |\xi|^2)^{p/2} + (1 + |\tau|^2)^{q/2}\}\hat{f}(\xi, \tau) \in L^2(\mathbb{R}^m \times \mathbb{R}).$$

For any set $E \subset \mathbb{R}^m$, we use the notation E_T to denote the cylindrical set $E \times (0, T)$. Take X to be an open set in $\mathbb{R}^m$ and denote by $H^{p,q}(X_T)$ the set of all the restriction of elements of $H^{p,q}(\mathbb{R}^m \times \mathbb{R})$ to $X \times (0, T)$. By assuming ∂X has the structure of a manifold with some regularity, $H^{p,q}((\partial X)_T)$ analogously by using a partition of unity and transforming ∂X locally to an Euclidean space. Here X and ∂X are a bounded domain in $\mathbb{R}^m$ and its boundary, respectively. We also need the following spaces:

$$\tilde{H}^{1,\frac{1}{2}}(X_T) := \left\{u \in H^{1,\frac{1}{2}}(X \times (-\infty, T)): \; u(x, t) = 0 \text{ for } t < 0\right\}$$

and

$$H^{1,\frac{1}{2}}(X_T; \partial_t - \Delta) := \left\{u \in H^{1,\frac{1}{2}}(X_T): (\partial_t - \Delta)u \in L^2(X_T)\right\}.$$

We note that for each $f \in H^{-\frac{1}{2},-\frac{1}{4}}((\partial\Omega)_T)$, there exists a unique solution $u = u^f = u(f) \in \tilde{H}^{1,\frac{1}{2}}((\Omega\backslash\bar{D})_T)$ to the initial boundary value problem

$$\begin{cases} (\partial_t - \Delta)u = 0 \text{ in } (\Omega\backslash\bar{D}) \times (0, T) =: (\Omega\backslash\bar{D})_T, \\ \partial_\nu u = f \text{ on } \partial\Omega \times (0, T) =: (\partial\Omega)_T, \\ \partial_\nu u = 0 \text{ on } \partial D \times (0, \mathrm{T}) =: (\partial D)_T, \\ u = 0 \text{ at } t = 0 \end{cases} \tag{16.1.1}$$

and it depends continuously on f, where ν is the unit normal vector to $\partial\Omega$(or ∂D) directed outside Ω(or D) (see corollary 3.17 of [13]). Based on this we define the Neumann–Dirichlet Λ_D by

$$\Lambda_D : H^{-\frac{1}{2},-\frac{1}{4}}\left((\partial\Omega)_T\right) \to H^{\frac{1}{2},\frac{1}{4}}\left((\partial\Omega)_T\right), \quad f \mapsto u^f|_{(\partial\Omega)_T}. \tag{16.1.2}$$

We will use the notation $\Lambda_\emptyset$ to denote the Neumann-to-Dirichlet map when there is not any cavity, i.e. $D = \emptyset$.

By taking the Neumann-to-Dirichlet map Λ_D as the measured data, we formulate our inverse problem as follows:

Inverse problem: reconstruct D from Λ_D.

The rest of our arguments are as follows. We will first give the linear sampling method for the cases $y \in D$ and $y \notin D$. Some properties of the heat potential used to give the linear sampling will be given after that.

16.1.1 Tools and theoretical foundation

Let

$$\Gamma_{(y,s)}(x, t) := \Gamma(x, t; y, s) = \begin{cases} \dfrac{1}{\left(4\pi(t-s)\right)^{m/2}} \exp\left(-\dfrac{|x-y|^2}{4(t-s)}\right), & t > s, \\ 0, & t \leqslant s \end{cases}$$

be the fundamental solution of the heat operator $\partial_t - \Delta$ and denote the Green function of the heat operator in Ω_T with a Neumann boundary condition on $(\partial\Omega)_T$ by $\Gamma^0_{(y,s)}(x, t)$. Also, we define the operators R, Q, A, F as follows:

(i)

$$R : H^{-\frac{1}{2},-\frac{1}{4}}\left((\partial\Omega)_T\right) \to \tilde{H}^{1,\frac{1}{2}}\left((\Omega\backslash\bar{D})_T\right)$$

and

$$Q : H^{-\frac{1}{2},-\frac{1}{4}}\left((\partial\Omega)_T\right) \to H^{-\frac{1}{2},-\frac{1}{4}}\left((\partial D)_T\right)$$

are defined by $Rf := u^f$, $Qf := \partial_\nu Qf|_{(\partial D)_T}$ with the solution u^f for

$$\begin{cases} (\partial_t - \Delta)u^f = 0 \text{ in } \Omega_T, \\ \partial_\nu u^f = f \text{ on } (\partial\Omega)_T, \\ u^f = 0 \text{ at } t = 0; \end{cases} \tag{16.1.1}$$

(ii)

$$A : H^{-\frac{1}{2},-\frac{1}{4}}\left((\partial D)_T\right) \to H^{\frac{1}{2},\frac{1}{4}}\left((\partial\Omega)_T\right)$$

is defined by $Ag := z^g|_{(\partial\Omega)_T}$ with the solution z^g for

$$\begin{cases} (\partial_t - \Delta)z^g = 0 \text{ in } (\Omega\backslash\bar{D})_T, \\ \partial_\nu z^g = g \text{ on } (\partial D)_T, \\ \partial_\nu z^g = 0 \text{ on } (\partial\Omega)_T, \\ z^g = 0 \text{ at } t = 0; \end{cases} \tag{16.1.2}$$

(iii)

$$F: H^{-\frac{1}{2},-\frac{1}{4}}\big((\partial\Omega)_T\big) \to H^{\frac{1}{2},\frac{1}{4}}\big((\partial\Omega)_T\big)$$

is defined by

$$F := \Lambda_D - \Lambda_\varnothing.$$

Then, we have the following two lemmas.

Lemma 16.1.1. $F = -AQ$.

Proof. Let $v = Rf$ with $f \in H^{-\frac{1}{2},-\frac{1}{4}}((\partial\Omega)_T)$. Consider the solutions w and u for

$$\begin{cases} (\partial_t - \Delta)w = 0 \text{ in } (\Omega\backslash\bar{D})_T, \\ \partial_\nu w = -\partial_\nu v \text{ on } (\partial D)_T, \\ \partial_\nu w = 0 \text{ on } (\partial\Omega)_T, \\ w = 0 \text{ at } t = 0 \end{cases}$$

and

$$\begin{cases} (\partial_t - \Delta)u = 0 \text{ in } (\Omega\backslash\bar{D})_T, \\ \partial_\nu u = 0 \text{ on } (\partial D)_T, \\ \partial_\nu u = \partial_\nu v \text{ on } (\partial\Omega)_T, \\ u = 0 \text{ at } t = 0, \end{cases}$$

respectively. By $(\partial_t - \Delta)v = 0$ in $(\Omega\backslash\bar{D})_T$, we have

$$\begin{cases} (\partial_t - \Delta)(u - v) = 0 \text{ in } (\Omega\backslash\bar{D})_T, \\ \partial_\nu(u - v) = \partial_\nu w \text{ on } (\partial D)_T, \\ \partial_\nu(u - v) = 0 \text{ on } (\partial\Omega)_T, \\ u - v = 0 \text{ at } t = 0. \end{cases}$$

Then by the unique solvability of (16.1.1), we have $w = u - v$. Hence, by $\partial_\nu v|_{(\partial D)_T} = Qf$, we have

$$\begin{aligned} A(-Qf) = w|_{(\partial\Omega)_T} &= (u - v)\big|(\partial\Omega)_T \\ &= (\Lambda_D - \Lambda_\varnothing)\big(\partial_\nu v|_{(\partial\Omega)_T}\big) \\ &= (\Lambda_D - \Lambda_\varnothing)f = Ff. \end{aligned}$$

Thus we have obtained $F = -AQ$. □

Lemma 16.1.2. *The operator*

$$Q: H^{-\frac{1}{2},-\frac{1}{4}}\big((\partial\Omega)_T\big) \to H^{-\frac{1}{2},-\frac{1}{4}}\big((\partial D)_T\big)$$

is a continuous operator with a dense range.

Proof. For the proof we will use the layer potential approach for the heat equation. Let $u := u^f \in \tilde{H}^{1,\frac{1}{2}}(\Omega_T)$ with $f \in H^{-\frac{1}{2},-\frac{1}{4}}((\partial\Omega)_T)$ satisfy

$$\begin{cases} (\partial_t - \Delta)u = 0 \text{ in } \Omega_T, \\ \partial_\nu u = f \text{ on } (\partial\Omega)_T, \\ u = 0 \text{ at } t = 0. \end{cases} \tag{16.1.3}$$

Then by corollary 3.17 in [13], the solution u to (16.1.3) can be expressed as a single-layer heat potential:

$$u(x, t) = K_0\phi := \int_0^t \int_{\partial\Omega} \Gamma(x, t; y, s)\phi(y, s)\,\mathrm{d}s(y)\,\mathrm{d}s, \quad (x, t) \in \Omega_T \tag{16.1.4}$$

with a density $\phi \in H^{-\frac{1}{2},-\frac{1}{4}}((\partial\Omega)_T)$.

Now we define the integral operator N by

$$N\phi := \frac{1}{2}\left[\gamma_1(K_0\phi)|_{\Omega\times(0,T)} + \gamma_1(K_0\phi)|_{\Omega^c\times(0,T)}\right],$$

where $\Omega^c := B_R \backslash \bar{\Omega}$ with R large enough such that $\bar{\Omega} \subset B_R$ and $\gamma_1 : H^{1,\frac{1}{2}}(B_T; \partial_t - \Delta) \to H^{-\frac{1}{2},-\frac{1}{4}}((\partial B)_T)$ (here B is taken as Ω or Ω^c) is a continuous linear map defined as follows [13]. Consider the bilinear form $b(u, v)$ given by

$$b(u, v) := \int_0^T \int_\Omega [\nabla u \cdot \nabla v - (\partial_t - \Delta)u\, v]\mathrm{d}x\mathrm{d}t + \int_{\Omega\times\mathbb{R}} \partial_t u\, v\mathrm{d}x\mathrm{d}t.$$

Define $\gamma_1 u$ as the continuous linear form defined by $\gamma_1 u : \varphi \mapsto b(u, \gamma^-\varphi)$, where γ^- is a continuous right inverse of the trace map $\gamma : u \mapsto u|_{(\partial B)_T}$. We note that if $u \in C^2(\bar{B}_T)$, then we have $\gamma_1 u = \partial_\nu u|_{(\partial B)_T}$. Using the jump relations of layer potentials in theorem 3.7 of [13], we have

$$\frac{1}{2}\phi + N\phi = f. \tag{16.1.5}$$

Further by corollary 3.14 in [13], we know that

$$\frac{1}{2}I + N : H^{-\frac{1}{2},-\frac{1}{4}}\left((\partial\Omega)_T\right) \to H^{-\frac{1}{2},-\frac{1}{4}}\left((\partial\Omega)_T\right) \tag{16.1.6}$$

is an isomorphism. Hence defining

$$\mathcal{K}\psi := \partial_\nu(K_0\psi)\Big|_{(\partial D)_T}$$

for $\psi \in H^{-\frac{1}{2},-\frac{1}{4}}((\partial\Omega)_T)$ with a normal derivative from the exterior of D, we have

$$Q = \mathcal{K}\left(\frac{1}{2}I + N\right)^{-1}.$$

Therefore in order to show the denseness of H, it is enough to show that the operator

$$\mathcal{K}: H^{-\frac{1}{2},-\frac{1}{4}}\big((\partial\Omega)_T\big) \to H^{-\frac{1}{2},-\frac{1}{4}}\big((\partial D)_T\big)$$

has a dense range and this can be shown as follows. First, direct calculations give

$$\mathcal{K}\psi(x, t) = \int_0^t \int_{\partial\Omega} M(x, t; y, s)\psi(y, s)\mathrm{d}s(y)\mathrm{d}s, \quad (x, t) \in (\partial D)_T \tag{16.1.7}$$

with

$$M(x, t; y, s) := \partial_{\nu(x)}\Gamma(x, t; y, s).$$

Then the adjoint

$$\mathcal{K}^*: H^{\frac{1}{2},\frac{1}{4}}\big((\partial D)_T\big) \to H^{\frac{1}{2},\frac{1}{4}}\big((\partial\Omega)_T\big)$$

of $\mathcal{K}$ is given as

$$\mathcal{K}^*\eta(y, s) = \int_s^T \int_{\partial D} M(x, t; y, s)\eta(x, t)\mathrm{d}s(x)\mathrm{d}t, \, (y, s) \in (\partial\Omega)_T. \tag{16.1.8}$$

For proving the denseness of $\mathcal{K}$, it suffices to show that $\eta = 0$ in $H^{\frac{1}{2},\frac{1}{4}}((\partial D)_T)$ if we have

$$(\mathcal{K}\psi, \eta) = (\psi, \mathcal{K}^*\eta) = 0$$

for all $\psi \in H^{-\frac{1}{2},-\frac{1}{4}}((\partial\Omega)_T)$. By the property of potential given later in this subsection, we have

$$w(y, s) := \int_s^T \int_{\partial D} M(x, t; y, s)\eta(x, t)\mathrm{d}s(x)\mathrm{d}t, \, (y, s) \in (\mathbb{R}^m \backslash \partial D)_T \tag{16.1.9}$$

satisfies

$$w = 0 \quad \text{in } (\mathbb{R}^m \backslash \bar{D})_T \tag{16.1.10}$$

and it clearly satisfies

$$\begin{cases} \partial_s w + \Delta w = 0 \text{ in } (\mathbb{R}^m \backslash \bar{D}) \times (0, T), \\ w = 0 \text{ at } s = T. \end{cases}$$

From (16.1.10) this implies

$$\frac{\partial w^+}{\partial \nu} = 0 \text{ on } (\partial D)_T, \tag{16.1.11}$$

where '+' means that we take the normal derivative from the exterior of D. Further combining (16.1.11) with (16.1.2) given later in this subsection, we have

$$\frac{\partial w^-}{\partial \nu} = 0 \text{ on } (\partial D)_T, \tag{16.1.12}$$

where '–' means that the normal derivative is taken from the interior of D.

To finish the proof observe that w given by (16.1.9) also satisfies

$$\begin{cases} \partial_s w + \Delta w = 0 \text{ in } D_T, \\ w = 0 \text{ at } s = T. \end{cases}$$

Then from the uniqueness result for the above initial boundary value problem, we have

$$w = 0 \quad \text{in } D_T.$$

Consequently, combining this with (16.1.10), we conclude that $\eta = 0$ from the jump formula (16.1.25) given later. Hence the proof is complete. □

Now we can give the foundation for the linear sampling for the case $y \in D$.

Theorem 16.1.3. *Fix $s \in (0, T)$ and let $y \in D$. Then, we can find a function $g^y \in H^{-\frac{1}{2},-\frac{1}{4}}((\partial\Omega)_T)$ which satisfies the inequality*

$$\left\| Fg^y - \Gamma^0_{(y,s)} \right\|_{H^{\frac{1}{2},\frac{1}{4}}\left((\partial\Omega)_T\right)} < \varepsilon \tag{16.1.13}$$

and has the behaviors $y \to \partial D$:

$$\lim_{y\to\partial D} \|g^y\|_{H^{-\frac{1}{2},-\frac{1}{4}}\left((\partial\Omega)_T\right)} = \infty \tag{16.1.14}$$

and

$$\lim_{y\to\partial D} \|Qg^y\|_{H^{-\frac{1}{2},-\frac{1}{4}}\left((\partial D)_T\right)} = \infty. \tag{16.1.15}$$

Proof. We will basically follow the proof for the inverse scattering problem given in [8] and [39]. But it has to be adapted to our situation. Denote the norm of the operator

$$A : H^{-\frac{1}{2},-\frac{1}{4}}\left((\partial D)_T\right) \to H^{\frac{1}{2},\frac{1}{4}}\left((\partial\Omega)_T\right)$$

by $\|A\|$. For given $\varepsilon > 0$. By lemma 16.1.2, there exists a function $g^y \in H^{-\frac{1}{2},-\frac{1}{4}}((\partial\Omega)_T)$ such that

$$\left\| Qg^y - \left(-\partial_\nu \Gamma^0_{(y,s)}\right) \right\|_{H^{-\frac{1}{2},-\frac{1}{4}}\left((\partial D)_T\right)} < \frac{\varepsilon}{\|A\|}.$$

By $\partial_\nu \Gamma^0_{(y,s)}|_{(\partial\Omega)_T} = 0$, we have

$$A\left(\partial_\nu \Gamma^0_{(y,s)}|_{(\partial D)_T}\right) = \Gamma^0_{(y,s)}\Big|_{(\partial\Omega)_T}.$$

Hence, using lemma 16.1.1, we have

$$\begin{aligned} &\left\| Fg^y - \Gamma^0_{(y,s)} \right\|_{H^{\frac{1}{2},\frac{1}{4}}\left((\partial\Omega)_T\right)} \\ &= \left\| -AQg^y - A\left(\partial_\nu \Gamma^0_{(y,s)}|_{(\partial D)_T}\right) \right\|_{H^{\frac{1}{2},\frac{1}{4}}\left((\partial\Omega)_T\right)} \\ &\leqslant \|A\| \left\| Qg^y - \left(-\partial_\nu \Gamma^0_{(y,s)}|_{(\partial D)_T}\right) \right\|_{H^{-\frac{1}{2},-\frac{1}{4}}\left((\partial D)_T\right)} < \varepsilon. \end{aligned}$$

On the other hand, by the boundedness of H, we have

$$\begin{aligned}
&\|g^y\|_{H^{-\frac{1}{2},-\frac{1}{4}}((\partial\Omega)_T)} \\
&\geqslant c\|Qg^y\|_{H^{-\frac{1}{2},-\frac{1}{4}}((\partial D)_T)} \\
&\geqslant c\left(\left\|\partial_\nu\Gamma^0_{(y,s)}\right\|_{H^{-\frac{1}{2},-\frac{1}{4}}((\partial D)_T)} - \left\|Qg^y - \left(-\partial_\nu\Gamma^0_{(y,s)}\right)\right\|_{H^{-\frac{1}{2},-\frac{1}{4}}((\partial D)_T)}\right) \\
&\geqslant c\left\|\partial_\nu\Gamma^0_{(y,s)}\right\|_{H^{-\frac{1}{2},-\frac{1}{4}}((\partial D)_T)} - \frac{c\varepsilon}{\|A\|}
\end{aligned}$$

with some constant $c > 0$.

Our next task is to prove

$$\left\|\partial_\nu\Gamma^0_{(y,s)}\right\|_{H^{-\frac{1}{2},-\frac{1}{4}}((\partial D)_T)} \to \infty \ (y \to \partial D).$$

Since the $m = 2$ case is easier to prove, we only give the proof for the case $m = 3$. By the assumption that ∂D is of the C^2 class, for any point $x_0 \in \partial D$, there exists a C^2-function f such that

$$D \cap B(x_0, r) = \{x \in B(x_0, r) : \ x_3 > f(x_1, x_2)\}.$$

Take a new orthonormal basis $\{e_j\}, j = 1, \cdots, 3$, centered at x_0 with $e_3 = -\nu$ with the unit outward normal vector ν to the boundary at x_0 and also the vectors e_1, e_2 lying in the tangent plane to ∂D at x_0. Let η be the local coordinates defined by the basis $\{e_j\}$ and define the local transformation of coordinates $\eta = F(x)$ as follows:

$$\eta' = x', \ \eta_3 = x_3 - f(x') \text{ with } x' = (x_1, x_2), \ \eta' = (\eta_1, \eta_2).$$

Hence, in terms of the new coordinates system, we can let $\xi = (0, 0, \xi_3)$, $\xi_3 > 0$ and $\eta' \in [-l, l] \times [-l, l] := D_2$ with a sufficiently small constant l. Assuming $s = 0$ without loss of generality, we have

$$\begin{aligned}
&\left\|\partial_\nu\Gamma^0_{(y,s)}\right\|_{H^{-\frac{1}{2},-\frac{1}{4}}((\partial D)_T)} \\
&\geqslant c_1\left\|\partial_\nu\Gamma^0_{(\xi,s)}\right\|_{H^{-\frac{1}{2},-\frac{1}{4}}((D_2)_T)} \\
&= c_1 \sup_{\|\varphi\|_{H^{\frac{1}{2},\frac{1}{4}}((D_2)_T)} \leqslant 1} \left|\int_0^T \int_{D_2} \partial_\nu\Gamma^0_{(\xi,s)}(\eta', t)\varphi(\eta')\mathrm{d}\eta'\mathrm{d}t\right|,
\end{aligned}$$

where c_1 is a positive constant.

To estimate the last term in the above inequality, we introduce an auxiliary function $\varphi(\eta') = c\ h(t)t^{-\alpha}\mathrm{e}^{-\frac{|\eta'|^2}{4t}}$, where $0 < \alpha \ll 1$ and $h(t) \in C_0^\infty((0, T))$ with

$0 \leqslant h \leqslant 1$. It is easy to see that $\varphi \in H^{\frac{1}{2},\frac{1}{4}}((D_2)_T)$ and $\|\varphi\|_{H^{\frac{1}{2},\frac{1}{4}}((D_2)_T)} \leqslant 1$ for sufficiently small positive constant c. By observing

$$\nu(\eta) \cdot (\eta - \xi) = (0, 0, -1) \cdot (\eta', -\xi_3) = \xi_3, \ |\eta - \xi|^2 = |\eta'|^2 + \xi_3^2,$$

we have

$$\begin{aligned}\partial_\nu \Gamma^0_{(\xi,s)}(\eta, t) &= \frac{1}{\left(\sqrt{4\pi t}\right)^3} \frac{-\nu(\eta) \cdot (\eta - \xi)}{2t} \exp\left(-\frac{|\eta - \xi|^2}{4t}\right) \\ &= -\frac{1}{16\pi^{3/2}} t^{-5/2} \xi_3 \exp\left(-\frac{|\eta'|^2 + \xi_3^2}{4t}\right).\end{aligned}$$

Then taking $\varphi = ch(t)t^{-\alpha}\mathrm{e}^{-\frac{|\eta'|^2}{4t}}$ we have

$$\begin{aligned}&\sup_{\|\varphi\|_{H^{\frac{1}{2},\frac{1}{4}}\left((D_2)_T\right)} \leqslant 1} \left| \int_0^T \int_{D_2} \partial_\nu \Gamma^0_{(\xi,s)}(\eta', t)\varphi(\eta')\mathrm{d}\eta'\mathrm{d}t \right| \\ &\geqslant \frac{c}{16\pi^{3/2}} \xi_3 \int_0^T t^{-5/2} \mathrm{e}^{-\frac{\xi_3^2}{4t}} h(t) t^{-\alpha} \int_{D_2} \mathrm{e}^{-\frac{|\eta'|^2}{2t}} \mathrm{d}\eta'\mathrm{d}t \\ &= \frac{c}{8\pi^{3/2}} \xi_3 \int_0^T t^{-3/2} \mathrm{e}^{-\frac{\xi_3^2}{4t}} h(t) t^{-\alpha} \int_{D_t} \mathrm{e}^{-|\gamma'|^2} \mathrm{d}\gamma'\mathrm{d}t\end{aligned}$$

with

$$D_t = \left[-\frac{l}{\sqrt{2t}}, \frac{l}{\sqrt{2t}}\right] \times \left[-\frac{l}{\sqrt{2t}}, \frac{l}{\sqrt{2t}}\right].$$

Since

$$I(t) := \int_{D_t} \mathrm{e}^{-|\gamma'|^2} \mathrm{d}\gamma'$$

is monotonically decreasing in $[0, T]$, we have

$$\begin{aligned}&\left\| \partial_\nu \Gamma^0_{(y,s)} \right\|_{H^{-\frac{1}{2},-\frac{1}{4}}\left((\partial D)_T\right)} \\ &\geqslant \frac{cI(T)}{8\pi^{3/2}} \xi_3 \int_0^T t^{-3/2} \mathrm{e}^{-\frac{\xi_3^2}{4t}} h(t) t^{-\alpha} \mathrm{d}t \\ &= \frac{cI(T)}{8\pi^{3/2}} \xi_3^{-2\alpha} \int_{\frac{\xi_3^2}{T}}^{\infty} \tau^{\alpha - \frac{1}{2}} \mathrm{e}^{-\frac{\tau}{4}} \tilde{h}(\tau) \mathrm{d}\tau \to \infty \quad (\xi_3 \to 0 \text{ as } y \to \partial D),\end{aligned}$$

where $\tilde{h} \in C_0^\infty(\xi_3^2/T, +\infty)$. This completes the proof. □

For further investigation of the behavior of the density g^y when $y \not\in D$, we need the following two lemmas.

Lemma 16.1.4. *The operator*

$$A : H^{-\frac{1}{2},-\frac{1}{4}}\big((\partial D)_T\big) \to H^{\frac{1}{2},\frac{1}{4}}\big((\partial\Omega)_T\big)$$

is injective, compact and has a dense range.

Proof. The injectivity is a direct consequence of the unique continuation property (UCP) for $\partial_t - \Delta$. To show that A has a dense range, we will adopt the proof of lemma 4.1 given in [11]. Let $g_j \in H^{-\frac{1}{2},-\frac{1}{4}}((\partial D)_T)$ ($j \in \mathbb{N}$) be such that the linear hull $[\{g_j\}]$ of $\{g_j\}$ is dense in $H^{-\frac{1}{2},-\frac{1}{4}}((\partial D)_T)$. Then, by corollary 2.4.2, it is enough to prove the following:

$$f \in H^{-\frac{1}{2},-\frac{1}{4}}\big((\partial\Omega)_T\big) : \int_{(\partial\Omega)_T} \varphi_j f \,\mathrm{d}s(x)\mathrm{d}t = 0 \text{ for all } j \in \mathbb{N} \text{ with } \varphi_j := z^{g_j}|_{(\partial\Omega)_T}$$

implies $f = 0$.

For this consider the solution v in $\tilde{H}^{1,\frac{1}{2}}((\Omega\backslash\bar{D})_T)$ to the following well-posed initial boundary value problem

$$\begin{cases} (\partial_t + \Delta)v = 0 \text{ in } (\Omega\backslash\bar{D})_T, \\ \partial_\nu v = 0 \text{ on } (\partial D)_T, \\ \partial_\nu v = f \text{ on } (\partial\Omega)_T, \\ v = 0 \text{ at } t = T, \end{cases}$$

and set $z_j = z^{g_j}$. Then, we have

$$\begin{aligned} 0 &= \int_{(\Omega\backslash\bar{D})_T} \big(v\Delta z_j - z_j\Delta v\big)\mathrm{d}x\mathrm{d}t \\ &= \int_{(\partial\Omega)_T} \big(\partial_\nu z_j v - \partial_\nu v z_j\big)\mathrm{d}s\mathrm{d}t - \int_{(\partial D)_T} \big(\partial_\nu z_j v - \partial_\nu v z_j\big)\mathrm{d}s\mathrm{d}t \\ &= -\int_{(\partial D)_T} g_j v \mathrm{d}s\mathrm{d}t, \end{aligned}$$

which implies $v = 0$ on $(\partial D)_T$. Combining this with $\partial_\nu v|_{(\partial D)_T} = 0$, we have $v = 0$ in $(\Omega\backslash\bar{D})_T$ by the unique continuation for $\partial_t + \Delta$. Hence $f = \partial_\nu v|_{(\partial\Omega)_T} = 0$.

Finally, to see the compactness of A, note that there is a unique density $\varphi \in H^{-\frac{1}{2},-\frac{1}{4}}((\partial D)_T)$ such that the solution z^g to (16.1.2) is given by

$$z^g(x, t) = \int_0^t \int_{\partial D} \Gamma^0(x, t; y, s)\varphi(y, s)\,\mathrm{d}s(y)\,\mathrm{d}s.$$

Then, we can immediately see that A is compact due to the fact $\Gamma^0(x, t; y, s)$ is smooth for $x \in \partial\Omega$, $y \in \partial D$ and $0 \leqslant s \leqslant t \leqslant T$, $Ag = z^g|_{(\partial\Omega)_T} \in C^\infty(\partial\Omega \times [0, T])$. □

Lemma 16.1.5. *For any fixed* $s \in (0, T)$, *if* $y \in \Omega \backslash D$, *then* $\Gamma^0_{(y,s)} \notin R(A) :=$ *range of* A.

Proof. Suppose that we have $\Gamma^0_{(y,s)} \in R(A)$. Then there is a function $f \in H^{-\frac{1}{2},-\frac{1}{4}}((\partial D)_T)$ and the solution $w^f \in \tilde{H}^{1,\frac{1}{2}}((\Omega \backslash \bar{D})_T)$ to

$$\begin{cases} (\partial_t - \Delta) w^f = 0 \text{ in } (\Omega \backslash \bar{D})_T, \\ \partial_\nu w^f = f \text{ on } (\partial D)_T, \\ \partial_\nu w^f = 0 \text{ on } (\partial \Omega)_T, \\ w^f = 0 \text{ at } t = 0 \end{cases}$$

such that

$$w^f|_{(\partial\Omega)_T} = \Gamma^0_{(y,s)}|_{(\partial\Omega)_T}.$$

Since $\partial_\nu w^f|_{(\partial\Omega)_T} = \partial_\nu \Gamma^0_{(y,s)}|_{(\partial\Omega)_T} = 0$, we have

$$w^f = \Gamma^0_{(y,s)} \quad \text{in} \quad \left(\Omega \backslash \left(\bar{D} \cup \{y\}\right)\right)_T$$

by the unique continuation principle and hence

$$\left\| \Gamma^0_{(y,s)} \right\|_{\tilde{H}^{1,\frac{1}{2}}\left((\Omega \backslash \bar{D})_T\right)} = \| w^f \|_{\tilde{H}^{1,\frac{1}{2}}\left((\Omega \backslash \bar{D})_T\right)} < \infty.$$

Here we can show that for the case $y \in \partial D$ and the case $y \in \Omega \backslash \bar{D}$, it holds that

$$\left\| \Gamma^0_{(y,s)} \right\|_{\tilde{H}^{1,\frac{1}{2}}\left((\Omega \backslash \bar{D})_T\right)} = \infty. \tag{16.1.16}$$

Since the other cases are easier to demonstrate, we consider the case such that $m = 3$, $s = 0$ and $y \in \partial D$. Then, for some $\delta (0 < \delta < \pi/2)$, $\gamma (|\gamma| < \pi)$, R, $\tau (0 < R, \tau \ll 1)$ and any $\varepsilon (0 < \epsilon \ll 1)$, we have

$$\begin{aligned} &\left(\frac{1}{8\pi^{3/2}}\right)^2 \int_0^T \int_{\Omega \backslash \bar{D}} t^{-3} \exp\left(-\frac{|x - y|^2}{2t}\right) \mathrm{d}x \mathrm{d}t \\ &\geqslant \left(\frac{1}{8\pi^{3/2}}\right)^2 \int_\varepsilon^\tau \mathrm{d}t \int_\delta^{\pi-\delta} \mathrm{d}\varphi \int_{-\gamma}^\gamma \mathrm{d}\theta \int_0^R t^{-3} \exp\left(-\frac{r^2}{2t}\right) r^2 \sin \varphi \mathrm{d}r \\ &= \frac{1}{64\pi^3} [-\cos \varphi]_\delta^{\pi-\delta} 2\gamma \int_\varepsilon^\tau t^{-3} \mathrm{d}t \int_0^R r^2 \exp\left(-\frac{r^2}{2t}\right) \mathrm{d}r \\ &\geqslant \frac{1}{16\pi^3} \gamma \cos \delta \int_\varepsilon^\tau t^{-1} \mathrm{d}t \to \infty (\varepsilon \to 0) \end{aligned}$$

and hence

$$\left\| \Gamma^0_{(y,s)} \right\|_{L^2((0,T);L^2(\Omega \backslash \bar{D}))} = \infty$$

which implies

$$\left\| \Gamma^0_{(y,s)} \right\|_{\tilde{H}^{1,\frac{1}{2}}\left((\Omega\setminus\bar{D})_T\right)} = \infty.$$

This is a contradiction. □

In contrast to proposition 16.1.3 for $y \in D$, we now establish the following unstable property of a solution g^y of the far-field equation with respect to the perturbation of discrepancy for the case $y \notin D$ as follows.

Theorem 16.1.6. *Fix $s \in (0, T)$ and let $y \in \Omega\setminus D$. Then, for every $\varepsilon > 0$ and $\delta > 0$, there exists $g^y = g^y_{\varepsilon,\delta} \in H^{-\frac{1}{2},-\frac{1}{4}}((\partial\Omega)_T)$ which satisfies the inequality*

$$\left\| Fg^y - \Gamma^0_{(y,s)} \right\|_{H^{\frac{1}{2},\frac{1}{4}}\left((\partial\Omega)_T\right)} < \varepsilon + \delta \tag{16.1.17}$$

and has the following behaviors:

$$\lim_{\delta\to 0} \| g^y \|_{H^{-\frac{1}{2},-\frac{1}{4}}\left((\partial\Omega)_T\right)} = \infty \tag{16.1.18}$$

and

$$\lim_{\delta\to 0} \| Qg^y \|_{H^{-\frac{1}{2},-\frac{1}{4}}\left((\partial D)_T\right)} = \infty. \tag{16.1.19}$$

Proof. We will adopt the argument given in [8] and [39]. Let (μ_n, φ_n, g_n) be the singular system of A. Then since A is injective and has dense range, for arbitrary $\delta > 0$ there exists $\alpha = \alpha(\delta) > 0$ such that a function

$$f^y_\alpha = \sum_{n=1}^{+\infty} \frac{\mu_n}{\alpha + \mu_n^2} (\Gamma^0_{(y,s)}, g_n)\varphi_n$$

satisfies

$$\left\| Af^y_\alpha - \Gamma^0_{(y,s)} \right\|_{H^{\frac{1}{2},\frac{1}{4}}\left((\partial\Omega)_T\right)} < \delta.$$

By lemma 16.1.5, we have $\Gamma^0_{(y,s)} \notin R(A)$. Hence taking account of $\Gamma^0_{(y,s)} \in H^{\frac{1}{2},\frac{1}{4}}((\partial\Omega)_T) = \overline{R(A)}$, we have

$$\left\| f^y_\alpha \right\|_{H^{-\frac{1}{2},-\frac{1}{4}}\left((\partial D)_T\right)} \to \infty \ (\alpha \to 0)$$

by Picard's theorem. From the denseness of $R(Q)$ in $H^{-\frac{1}{2},-\frac{1}{4}}((\partial D)_T)$, there exists

$$g^y_\alpha \in H^{-\frac{1}{2},-\frac{1}{4}}\left((\partial\Omega)_T\right)$$

such that

$$\left\| Qg^y_\alpha + f^y_\alpha \right\|_{H^{-\frac{1}{2},-\frac{1}{4}}\left((\partial D)_T\right)} \leqslant \frac{\varepsilon}{\|A\| + 1} \tag{16.1.20}$$

and then

$$\left\|Af_\alpha^y + AQg_\alpha^y\right\|_{H^{\frac{1}{2},\frac{1}{4}}\left((\partial\Omega)_T\right)} < \varepsilon.$$

Hence, we have

$$\begin{aligned}
&\left\|Fg_\alpha^y - \Gamma^0_{(y,s)}\right\|_{H^{\frac{1}{2},\frac{1}{4}}\left((\partial\Omega)_T\right)}\\
&\quad = \left\|-AQg_\alpha^y - \Gamma^0_{(y,s)}\right\|_{H^{\frac{1}{2},\frac{1}{4}}\left((\partial\Omega)_T\right)}\\
&\quad \leqslant \left\|-AQg_\alpha^y - Af_\alpha^y\right\|_{H^{\frac{1}{2},\frac{1}{4}}\left((\partial\Omega)_T\right)} + \left\|Af_\alpha^y - \Gamma^0_{(y,s)}\right\|_{H^{\frac{1}{2},\frac{1}{4}}\left((\partial\Omega)_T\right)}\\
&\quad < \varepsilon + \delta.
\end{aligned}$$

To derive the behaviors of g^y and Hg^y, recall (16.1.20) and $\|f_\alpha^y\|_{H^{-\frac{1}{2},-\frac{1}{4}}((\partial D)_T)} \to \infty(\alpha \to 0)$. These imply

$$\left\|Qg_\alpha^y\right\|_{H^{-\frac{1}{2},-\frac{1}{4}}\left((\partial D)_T\right)} \to \infty\ (\alpha \to 0)$$

and hence we have

$$\left\|g_\alpha^y\right\|_{H^{-\frac{1}{2},-\frac{1}{4}}\left((\partial\Omega)_T\right)} \to \infty\ (\alpha \to 0).$$

Then, the proof can be achieved by noting that $\alpha = \alpha(\delta) \to 0\ (\delta \to 0)$. □

16.1.2 Property of potential

The aim here is to show that for $\eta \in H^{\frac{1}{2},\frac{1}{4}}((\partial\Omega)_T)$,

$$w(x, t) := \int_t^T\int_{\partial D} M(y, s; x, t)\eta(y, s)\, \mathrm{d}s(y)\mathrm{d}s = 0 \text{ in } (\mathbb{R}^m\backslash\bar{D}) \times (0, T)$$

if $w = 0$ on $(\partial\Omega)_T$. It can easily be shown that

$$\partial_t w + \Delta w = 0 \quad \text{in} \quad (\mathbb{R}^m\backslash\bar{D}) \times (0, T)$$

and

$$w = 0 \quad \text{at} \quad t = T.$$

Let $N(\mathcal{K}^*)$ be the null space of the operator $\mathcal{K}^*$ and take $\eta = \eta(x, t) \in N(\mathcal{K}^*)$ from $C^0((\partial D)_T)$ such that $\eta = 0$ at $t = 0$. Note that such η are dense in $N(\mathcal{K}^*)$. We use the same notation to denote any continuous extension of η to $t < 0$.

We first show the uniqueness of solutions to the following problem:

$$\begin{cases}\partial_t w + \Delta w = 0 \text{ in } (\mathbb{R}^m\backslash\bar{\Omega}) \times (-\infty, T),\\ w = 0 \text{ on } \partial\Omega \times (-\infty, T),\\ w = 0 \text{ at } t = T.\end{cases} \tag{16.1.21}$$

For $s > t$ and $x \neq y$, using theorem 6.15 in [30] and the inequality

$$r^{\beta}e^{-r} \leqslant \beta^{\beta}e^{-\beta} \quad \text{for } 0 < \mathrm{r},\ \beta < +\infty,$$

we have

$$\begin{aligned} |M(y, s; x, t)| &= \left| \frac{1}{2\sqrt{4\pi(s-t)}^{m}} \frac{(\nu(y), x-y)}{s-t} \exp\left(-\frac{|x-y|^2}{4(s-t)}\right) \right| \\ &\leqslant \frac{C_1}{\sqrt{(s-t)}^{m}} \frac{|x-y|^2}{s-t} \exp\left(-\frac{|x-y|^2}{4(s-t)}\right) \\ &\leqslant \frac{C_2}{(s-t)^{\alpha_1}|x-y|^{m-2\alpha_1}} \end{aligned} \tag{16.1.22}$$

for $0 < \alpha_1 < 1/2$, where C_1, C_2 are positive constants. Similarly, we have

$$\begin{aligned} &|\partial_{\nu(x)}M(y, s; x, t)| \\ &= \left| \frac{1}{4\sqrt{4\pi(s-t)}^{m}} \frac{(\nu(y), x-y)}{s-t} \frac{(\nu(x), y-x)}{s-t} \exp\left(-\frac{|x-y|^2}{4(s-t)}\right) \right. \\ &\quad \left. + \frac{1}{2\sqrt{4\pi(s-t)}^{m}} \frac{(\nu(y), \nu(x))}{s-t} \exp\left(-\frac{|x-y|^2}{4(s-t)}\right) \right| \\ &\leqslant \frac{C_3}{\sqrt{(s-t)}^{m}} \left(\frac{|x-y|^2}{s-t}\right)^2 \exp\left(-\frac{|x-y|^2}{4(s-t)}\right) \\ &\quad + \frac{C_4}{\sqrt{(s-t)}^{m}} \frac{1}{s-t} \exp\left(-\frac{|x-y|^2}{4(s-t)}\right) \\ &\leqslant \frac{C_5}{(s-t)^{\alpha_2}|x-y|^{m-2\alpha_2}} + \frac{C_6}{(s-t)^{\alpha_3}|x-y|^{2+m-2\alpha_3}} \end{aligned} \tag{16.1.23}$$

for $0 < \alpha_2, \alpha_3 < 1/2$, $s > t$ and $x \neq y$, where $C_j (j = 3, \dots, 6)$ are positive constants. Then, by the estimate (16.1.22), we have

$$\|w(\cdot, -\infty)\|_{L^2(\Omega_R)} = 0$$

for large enough $R > 0$, where $\Omega_R := (\mathbb{R}^m \backslash \bar{\Omega}) \cap B_R$. Combining this with $\|w(\cdot, T)\|_{L^2(\Omega_R)} = 0$ and $w = 0$ on $\partial\Omega \times (-\infty, T)$, we have

$$\begin{aligned} 0 &= \int_{\Omega_R \times (-\infty, T)} w(\partial_t + \Delta) w \mathrm{d}s(x) \mathrm{d}t \\ &= -\int_{\Omega_R \times (-\infty, T)} |\nabla_x w|^2 \mathrm{d}x \mathrm{d}t + \int_{\partial B_R \times (-\infty, T)} w\, \partial_\nu w\, \mathrm{d}s(x) \mathrm{d}t. \end{aligned} \tag{16.1.24}$$

Using the estimates (16.1.22) and (16.1.23), we can easily show that

$$\int_{\partial B_R \times (-\infty, T)} w\, \partial_\nu w\, \mathrm{d}s(x)\, \mathrm{d}t \to 0 (R \to +\infty).$$

Then from (16.1.2), we have

$$\lim_{R \to \infty} \int_{\Omega_R \times (-\infty, T)} |\nabla_x w|^2 \mathrm{d}x \mathrm{d}t = 0,$$

and hence

$$w = 0 \text{ in } (\mathbb{R}^m \backslash \bar{\Omega}) \times (-\infty, T)$$

by observing $w = 0$ on $\partial\Omega \times (-\infty, T)$. Therefore by the unique continuation principle (see [26]), we have

$$w = 0 \text{ in } (\mathbb{R}^m \backslash \bar{D}) \times (-\infty, T).$$

The proof is complete.

16.1.3 The jump relations of $\mathcal{K}^*$

The aim here is to show the jump relations of the adjoint $\mathcal{K}^*$ of $\mathcal{K}$ given as follows:

$$\begin{aligned} &\lim_{h \to 0+} (\mathcal{K}^* \eta)(x + h\nu(x), t) \\ &\quad = \lim_{h \to 0+} (\mathcal{K}^* \eta)(x - h\nu(x), t) + \eta(x, t), \quad x \in \partial D, t \in (0, T), \end{aligned} \tag{16.1.25}$$

$$\begin{aligned} &\lim_{h \to 0+} \frac{\partial}{\partial \nu(x)} (\mathcal{K}^* \eta)(x + h\nu(x), t) \\ &= \lim_{h \to 0+} \frac{\partial}{\partial \nu(x)} (\mathcal{K}^* \eta)(x - h\nu(x), t), \quad x \in \partial D, t \in (0, T) \end{aligned} \tag{16.1.26}$$

for all $\eta \in H^{\frac{1}{2}, \frac{1}{4}}((\partial D)_T)$. By interchanging the variables (x, t) and (y, s) in (16.1.8), we have

$$\mathcal{K}^* \eta(x, t) = \int_t^T \int_{\partial D} M(y, s; x, t) \eta(y, s)\, \mathrm{d}s(y) \mathrm{d}s \tag{16.1.27}$$

with

$$M(y, s; x, t) = \frac{1}{[4\pi(s-t)]^{m/2}} \partial_{\nu(y)} \exp\left(-\frac{|y-x|^2}{4(s-t)}\right).$$

Then by writing this in the form

$$\mathcal{K}^*\eta(x, t) = \int_0^{T-t} \int_{\partial D} \partial_{\nu(y)} \Gamma(x, T-t; y, \tau)\eta(y, T-\tau)\, ds(y) d\tau,$$

(16.1.25) and (16.1.26) follow from the classical jump relations of the double-layer heat potential which are shown in theorem 3.4 in [13].

Remark 16.1.7. *Two new features which we can observe here are as follows. The first is that we do not need to assume that 0 is not an eigenvalue of Δ in D with a Dirichlet boundary condition. The other is that there is a freedom to choose s, which suggests that we have more data than the linear sampling methods for the inverse boundary value problem for EIT and inverse scattering problem with near field data. Note that the theoretical results obtained in this paper do not provide a way of choosing s. But if we relate s to the sampling point y of $\Gamma^0_{(y,s)}$, we may have enough sampling points by conducting just one measurement to generate an approximate identification of D.*

16.2 Nakamura's dynamical probe method

16.2.1 Inverse boundary value problem for heat conductors with inclusions

Let a heat conductor Ω be a bounded domain in $\mathbb{R}^2$ with C^2 boundary $\partial\Omega$ for simplicity. We consider the case that a heat conductor Ω has an unknown inclusion D such that D is an open set with the properties $\bar{D} \subset \Omega$, $\Omega \backslash \bar{D}$ is connected and the boundary ∂D of D is of class $C^{1,\alpha}$ $(0 < \alpha \leqslant 1)$. Let the heat conductivity $\gamma(x) \in L^\infty(\Omega)$ of Ω be given as follows.

$$\gamma(x) = \begin{cases} 1 \text{ for } & x \in \Omega \backslash \bar{D}, \\ k \text{ for } & x \in D \end{cases} \tag{16.2.1}$$

with a positive constant $k \neq 1$. Using the characteristic function χ_D of D, $\gamma(x)$ can be given as $\gamma(x) = 1 + (k-1)\chi_D$.

The forward problem which describes a single measurement of active thermography is that for given heat flux $f \in L^2((0, T); H^{-1/2}(\partial\Omega))$ measures the corresponding temperature distribution $u(f)|_{(\partial\Omega)_T}$, where $u = u(f)$ in

$$W(\Omega_T) := \left\{ v \in H^{1,0}(\Omega_T) \colon \partial_t v \in L^2\left((0, T); \left(H^1(\Omega)\right)^*\right)\right\}$$

is a unique solution to the following initial boundary value problem

$$\begin{cases} \mathcal{P}_D u(x, t) := \partial_t u(x, t) - \operatorname{div}_x(\gamma(x)\nabla_x u(x, t)) = 0 \quad \text{in } \Omega_T \\ \partial_\nu u(x, t) = f(x, t) \quad \text{in } \partial\Omega_T, \quad u(x, 0) = 0 \quad \text{for } x \in \Omega. \end{cases} \tag{16.2.2}$$

It is well known that this initial boundary value problem (16.2.2) is well-posed (see [43]), which means that is there exists a unique solution $u = u(f) \in W(\Omega_T)$

to (16.2.2) and $u(f)$ depends continuously on $f \in L^2((0, T); (H^{1/2}(\partial\Omega))^*)$. Based on this and the same reason as before for the linear sampling method applied to the active thermography, we mathematically idealize many measurements of the active thermography at $\partial\Omega$ over the time interval $(0, T)$ to be considered as the *Neumann-to-Dirichlet map* $\Lambda_D : L^2((0, T); (H^{1/2}(\partial\Omega))^*) \to L^2((0, T); H^{1/2}(\partial\Omega))$ defined by $\Lambda_D(f) = u(f)|_{\partial\Omega_T}$.

Then, our *inverse problem* is given to reconstruct the unknown inclusion D from Λ_D. We will give a reconstruction method called the *dynamical probe method* which is an analogue of the probe method for identifying the shape of an unknown sound-soft obstacle for the inverse acoustic scattering problem.

We will close this subsection by giving a brief history on the inverse problem identifying unknown inclusions in a heat conductor by using the Neumann–Dirichlet map. The uniqueness was shown by Elayyan and Isakov [16] even for the time-dependent inclusions using the localized Neumann–Dirichlet map. A log type stability estimate was given by Di Cristo and Vessella [15] for the piece-wise homogeneous isotropic heat conductor with time-dependent unknown inclusions. As for the reconstruction methods, Daido, Kang and Nakamura first initiated a reconstruction method called the dynamical probing method, which was later called the dynamical probe method [14]. After that several other reconstruction methods appeared. They are the linear sampling type method [36] and the two different kinds of enclosure type methods given for example in [25] and in [19], [20]. It should be noted here that the former enclosure type method can extract some information on the unknown inclusions by one measurement. There are advantages and disadvantages to these reconstruction methods which have to be examined further. In particular, what kind of data and indicator functions should be used.

16.2.2 Tools and theoretical foundation

For $(y, s), (y, s') \in \mathbb{R}^2_T$, let $\Gamma(x, t; y, s)$ and $\Gamma^*(x, t; y, s')$ for $(x, t) \in \mathbb{R}^2_T$ be the fundamental solutions for the heat operator $\mathcal{P}_\emptyset$ and its dual operator $\mathcal{P}^*_\emptyset$ given by

$$\Gamma(x, t; y, s) = \begin{cases} \dfrac{1}{4\pi(t-s)}\exp\left[-\dfrac{|x-y|^2}{4(t-s)}\right], & t > s, \\ 0, & t \leqslant s, \end{cases} \tag{16.2.3}$$

and

$$\Gamma^*(x, t; y, s') = \begin{cases} 0, & t \geqslant s', \\ \dfrac{1}{4\pi(s'-t)}\exp\left[-\dfrac{|x-y|^2}{4(s'-t)}\right], & t < s', \end{cases} \tag{16.2.4}$$

respectively.

From what is provided in subsection 16.2.4 and the interior regularity estimate (see, for example [18]), we have the following.

Lemma 16.2.1. *There are two sequences of functions* $\{v^j_{(y,s)}\}$ *and* $\{\varphi^j_{(y,s')}\}$ *in* $H^{2,1}(\Omega_{(-\eta,T+\eta)})$ *called* Runge's *approximation functions for arbitrary constant* $\eta > 0$ *such that*

$$\begin{cases} \mathcal{P}_{\emptyset} v^j_{(y,s)} = 0 & \text{in } \Omega_{(-\eta,T+\eta)}, \\ v^j_{(y,s)}(x,\, t) = 0 & \text{if } -\eta < t \leqslant 0, \\ v^j_{(y,s)} \to \Gamma(\cdot,\, \cdot;\, y,\, s) & \text{in } H^{2,1}(U_T) \quad \text{as} \quad j \to \infty, \end{cases} \tag{16.2.5}$$

and

$$\begin{cases} \mathcal{P}^*_{\emptyset} \varphi^j_{(y,s')} = 0 & \text{in } \Omega_{(-\eta,T+\eta)}, \\ \varphi^j_{(y,s')}(x,\, t) = 0 & \text{if } \ T \leqslant t < T + \eta, \\ \varphi^j_{(y,s')} \to \Gamma^*(\cdot,\, \cdot;\, y,\, s') & \text{in } H^{2,1}(U_T) \quad \text{as} \quad j \to \infty, \end{cases} \tag{16.2.6}$$

respectively, for each $s,\, s' \in (0,\, T)$ *and open set* U *with a Lipschitz boundary such that* $\bar{D} \subset U$, $\bar{U} \subset \Omega$, $\Omega \backslash \bar{U}$ *is connected and* $\bar{U}$ *does not contain* y.

We now define the pre-indicator function which will lead us to define our mathematical testing machine called the indicator function to identify the unknown inclusion.

Definition 16.2.2. *Let* $(y,\, s),\, (y,\, s') \in \Omega_T$ *and* $\{v^j_{(y,s)}\}$, $\{\varphi^j_{(y,s')}\}$ *be Runge's approximation functions in* $H^{2,1}(\Omega_{(-\eta,T+\eta)})$. *Then, define the pre-indicator function* $I(y,\, s';\, y,\, s)$ *by*

$$I(y,\, s';\, y,\, s) = \lim_{j\to\infty} \int_{\partial\Omega_T} \left[\partial_\nu v^j_{(y,s)}\Big|_{\partial\Omega_T} \varphi^j_{(y,s')}\Big|_{\partial\Omega_T} - \Lambda_D\left(\partial_\nu v^j_{(y,s)}\Big|_{\partial\Omega_T}\right) \partial_\nu \varphi^j_{(y,s')}\Big|_{\partial\Omega_T} \right] \tag{16.2.7}$$

whenever the limit exists.

For Runge's approximation functions $\{v^j_{(y,s)}\} \subset H^{2,1}(\Omega_{(-\eta,T+\eta)})$, let $u^j_{(y,s)} := u(\partial_\nu v^j_{(y,s)}|_{\partial\Omega_T})$ and $w^j_{(y,s)} := u^j_{(y,s)} - v^j_{(y,s)}$. Then, $w^j_{(y,s)} \in W(\Omega_T)$ is the solution to the following initial boundary value problem

$$\begin{cases} \mathcal{P}_D w^j_{(y,s)} = (k-1)\operatorname{div}_x\!\left(\chi_D \nabla_x v^j_{(y,s)}\right) \text{ in } \Omega_T, \\ \partial_\nu w^j_{(y,s)} = 0 \quad \text{on } \partial\Omega_T,\ w^j_{(y,s)}(x,\, 0) = 0 \quad \text{for } x \in \Omega. \end{cases} \tag{16.2.8}$$

Take $y \notin \bar{D}$. Then, since $\{v^j_{(y,s)}\}$ converges in $H^{2,1}(D_T)$ and this initial boundary value problem is well-posed, $\{w^j_{(y,s)}\}$ has a limit $w_{(y,s)}$ in $W(\Omega_T)$ and it is the solution to the initial boundary value problem:

$$\begin{cases} \mathcal{P}_D w_{(y,s)} = (k-1)\mathrm{div}_x(\chi_D \nabla_x \Gamma(\cdot, \cdot; y, s)) \text{ in } \Omega_T, \\ \partial_\nu w_{(y,s)} = 0 \quad \text{on } \partial\Omega_T,\ w_{(y,s)}(x, 0) = 0 \quad \text{for } x \in \Omega. \end{cases} \tag{16.2.9}$$

We call $w_{(y,s)}$ the *reflected solution.*

By using the Green formula, we can have the following representation formula for the pre-indicator function in terms of the reflected solution.

Theorem 16.2.3. *For $(y, s), (y', s') \in \Omega_T$ such that $y, y' \notin \bar{D}$, $(y, s) \neq (y', s')$, we have*

$$I(y, s'; y, s) = -w_{(y,s)}(y, s') - \int_{\partial\Omega_T} w_{(y,s)}(x, t)\partial_\nu \Gamma^*(x, t; y, s')ds(x)dt. \tag{16.2.10}$$

Remark 16.2.4. *Fixing $s \in (0, T)$ and taking $s' \in (0, T)$ close to s, if we let $y = y' \in \Omega \backslash \bar{D}$ tend to ∂D, then* (16.2.10) *tells us that the behavior of the pre-indicator function is controlled by that of the reflected solution.*

Now we are ready to define the indicator function as follows.

Definition 16.2.5. *For a needle $C := \{c(\lambda); 0 \leqslant \lambda \leqslant 1\}$ in $\bar{\Omega}$ joining two points $c(0), c(1) \in \partial\Omega$, define for each $s \in (0, T)$ and $c(\lambda)$ with $\lambda \in [0, 1]$, define the indicator function $J(c(\lambda), s)$ by*

$$J(c(\lambda), s) := \lim_{\epsilon \to 0} \liminf_{\delta \downarrow 0} |I(c(\lambda - \delta), s + \epsilon^2; c(\lambda - \delta), s)| \tag{16.2.11}$$

whenever the limit exists.

By using this indicator function, we can recover D as follows.

Theorem 16.2.6. *For a fixed $s \in (0, T)$, we have the following.*

(i) *If C never touches $\bar{D}$, then $J(c(\lambda), s) < \infty$ for all λ $(0 \leqslant \lambda \leqslant 1)$.*

(ii) *If C touches $\bar{D}$ and λ_0 is the smallest λ such that $c(\lambda) \in \partial D$, then $J(c(\lambda), s)$ becomes infinity for the first time at λ_0 as we observe the behavior of $J(c(\lambda), s)$ starting from $\lambda = 0$.*

16.2.3 Proof of theorem 16.2.6

Clearly it is enough to show (ii). By remark 16.2.4, we only need to analyze the behavior of $\limsup_{\delta\downarrow 0}|w_{(y,s)}(y, s + \varepsilon)|$ with $y = y(\delta) := c(\lambda_0 - \delta)$ as $\varepsilon \to 0$.

Since ∂D is $C^{1,\alpha}$ $(0 < \alpha \leqslant 1)$, there is a $C^{1,\alpha}$ diffeomorphism $\Phi : \mathbb{R}^2 \to \mathbb{R}^2$ which transforms $P := c(\lambda_0)$ to the origin 0 in $\mathbb{R}^2$, $\Phi(D) \subset \mathbb{R}^2_- = \{x = (x_1, x_2) \in \mathbb{R}^2; x_2 < 0\}$ and $D\Phi(P) = I$ with the 2×2 identity matrix (see [1] for the details).

By observing that $E(x, t; y, s) := w_{(y,s)}(x, t) + \Gamma(x, t; y, s)$ is the fundamental solution of $\mathcal{P}_D$, decompose $w_{(y,s)}$ as follows.

$$\begin{aligned} w_{(y,s)}(x, t) &= E(x, t; y, s) - \Gamma(x, t; y, s) \\ &= \{E(x, t; y, s) - \Gamma_-(\Phi(x), t; \Phi(y), s)\} \\ &\quad + \{\Gamma_-(\Phi(x), t; \Phi(y), s) - \Gamma(\Phi(x), t; \Phi(y), s)\} \\ &\quad + \{\Gamma(\Phi(x), t; \Phi(y), s) - \Gamma(x, t; y, s)\}, \end{aligned} \tag{16.2.12}$$

where Γ_- is the fundamental solution of $\partial_t - \operatorname{div}(1 + (k-1)\chi_-)\nabla)$ with the characteristic function χ_- of $\mathbb{R}^2_-$.

Let $\varepsilon > 0$. Put $\xi = \Phi(x)$, $\eta = \Phi(y)$. Then concerning (16.2.12) as $\delta \downarrow 0$, we are in the situation $\xi = \eta \to \Phi(P) = 0$. By the definition of $\Gamma(x, t; y, s)$ the last term on the right-hand side of (16.2.12) is zero.

We will show that the second term on the right-hand side of (16.2.12) is the dominant term. To begin with we have the following lemma.

Lemma 16.2.7.

$$\limsup_{\delta \downarrow 0} (\tilde{E} - \Gamma_-)(\eta, s + \varepsilon^2; \eta, s) = O(\varepsilon^{\alpha - 2}) \quad as \quad \varepsilon \to 0. \tag{16.2.13}$$

Proof. We first note that $\tilde{E}$ satisfies

$$\left[\partial_t - \nabla_\xi \cdot \left((1 + (k-1)\chi_-)M(\xi)\nabla_\xi\right)\right]\tilde{E}(\xi, t; \eta, s) = \delta(\xi - \eta)\delta(t - s) \tag{16.2.14}$$

in $R^2 \times R^1$, where $M(\xi) = JJ^T$ with $J = \frac{\partial \xi}{\partial x}(\Phi^{-1}(\xi))$ and $\tilde{E}(\xi, t; \eta, s) = 0$ for $t \leqslant s$. Then $\tilde{R}(\xi, t; \eta, s) := \tilde{E}(\xi, t; \eta, s) - \Gamma_-(\xi, t; \eta, s)$ satisfies

$$\begin{aligned} &\left[\partial_t - \nabla_\xi \cdot \left((1 + (k-1)\chi_-)\nabla_\xi\right)\right]\tilde{R}(\xi, t; \eta, s) \\ &\quad = \nabla_\xi \cdot \left((1 + (k-1)\chi_-)(M - I)\nabla_\xi\right)\tilde{E}(\xi, t; \eta, s) \end{aligned} \tag{16.2.15}$$

in $R^2 \times R^1$.

Let Γ^*_- be the fundamental solution for $-\partial_t - \operatorname{div}_\xi((1 + (k-1)\chi_-)\nabla_\xi)$ such that $\Gamma^*_-(\xi, t; z, \tau) = 0$ for $t \geqslant \tau$ and choose a ball B_r such that $\overline{\Phi(\Omega)} \subset B_r$. Then, we can represent $\tilde{R}(\xi, t; \eta, s)$ in the form

$$\begin{aligned} &\tilde{R}(\xi, t; \eta, s) \\ &= \int_s^t \int_{B_r} (1 + (k-1)\chi_-)(I - M)\nabla_z \tilde{E}(z, \tau; \eta, s) \cdot \nabla_z \Gamma^*_-(z, \tau; \xi, t) \mathrm{d}z \mathrm{d}\tau \\ &+ \int_s^t \int_{\partial B_r} (1 + (k-1)\chi_-)\left[\frac{\partial}{\partial \nu_z}\tilde{R}(z, \tau; \eta, s)\Gamma^*_-(z, \tau; \xi, t)\right. \\ &\left. - \tilde{R}(z, \tau; \eta, s)\frac{\partial}{\partial \nu_z}\Gamma^*_-(z, \tau; \xi, t)\right] \mathrm{d}s(z)\mathrm{d}\tau. \end{aligned}$$

Clearly the integration we have here on $\partial B_r \times (s, t)$ with $t = s + \varepsilon^2$ is bounded as $\xi = \eta \to 0$. For the other integration we use the following estimates ([17, 31–33])

$$\begin{cases} |\nabla_z \tilde{E}(z, \tau; \eta, s)| \leqslant c_1(\tau - s)^{-\frac{3}{2}} \exp\left(-\dfrac{|z-\eta|^2}{c_2(\tau - s)}\right), \\ |\nabla_z \Gamma^*_-(z, \tau; \xi, t)| \leqslant c_3(t - \tau)^{-\frac{3}{2}} \exp\left(-\dfrac{|z-\xi|^2}{c_4(t-\tau)}\right) \end{cases}$$

for some positive constants c_i, $1 \leqslant i \leqslant 4$. To proceed further, we use the estimate $|M(z) - I| \leqslant c_5 |z|^\alpha$ ($|z| < r$) for some constant $c_5 > 0$, because ∂D is $C^{1,\alpha}$. From now on, we use c for constants which do not depend on ε. By using the Fatou lemma, the $\limsup_{\delta\downarrow 0}$ of the absolute value of the integration can be bounded from above by a constant multiple of the following:

$$F := \int_s^{s+\varepsilon^2} \int_{|z|<r} |z|^\alpha |\nabla_z \tilde{E}(z, \tau; O, s)||\nabla_z \Gamma^*_-(z, \tau; O, s+\varepsilon^2)| \mathrm{d}z\mathrm{d}\tau. \quad (16.2.16)$$

We have from (16.2.16)

$$\begin{aligned} F &\leqslant c \int_s^{s+\varepsilon^2} \int_{|z|<r} \frac{|z|^\alpha}{(\tau - s)^{3/2}(s + \varepsilon^2 - \tau)^{3/2}} \exp\left[-\frac{|z|^2}{c(\tau - s)} - \frac{|z|^2}{c(s + \varepsilon^2 - \tau)}\right] \mathrm{d}z\mathrm{d}\tau \\ &= c\varepsilon^{\alpha-2} \int_0^1 \int_{|\zeta|<r\varepsilon^{-1}} |\zeta|^\alpha (\mu(1-\mu))^{-\frac{3}{2}} \exp\left[-\frac{|\zeta|^2}{c\mu} - \frac{|\zeta|^2}{c(1-\mu)}\right] \mathrm{d}\zeta\mathrm{d}\mu \\ &\leqslant c\varepsilon^{\alpha-2} \int_0^1 \int_{R^2} |\zeta|^\alpha (\mu(1-\mu))^{-\frac{3}{2}} \exp\left[-\frac{|\zeta|^2}{c\mu(1-\mu)}\right] \mathrm{d}\zeta\mathrm{d}\mu. \end{aligned}$$

Further, using polar coordinates, we have

$$\begin{aligned} F &\leqslant c\varepsilon^{\alpha-2} \int_0^1 \int_0^\infty r^{\alpha+1} (\mu(1-\mu))^{-\frac{3}{2}} \exp\left[-\frac{r^2}{c\mu(1-\mu)}\right] \mathrm{d}r\mathrm{d}\mu \\ &\leqslant c\varepsilon^{\alpha-2} \int_0^1 [\mu(1-\mu)]^{\frac{\alpha-1}{2}} \mathrm{d}\mu \int_0^\infty s^{\alpha+1} \mathrm{e}^{-s^2} \mathrm{d}s \\ &\leqslant c\varepsilon^{\alpha-2}. \end{aligned}$$

Thus we have

$$\limsup_{\delta\downarrow 0} |\tilde{R}(\eta, s+\varepsilon^2; \eta, s)| \leqslant c\varepsilon^{\alpha-2}, \quad (16.2.17)$$

□

Now suppressing (η, s) define

$$W(\xi, t) := \Gamma_{-}(\xi, t; \eta, s) - \Gamma(\xi, t; \eta, s) \tag{16.2.18}$$

and denote $W(\xi, t)$ for $\pm\xi_2 > 0$ by $W^{\pm}(\xi, t)$. Then, we have the following lemma.

Lemma 16.2.8. *If y approaches the boundary of D along C to P, then there exists a constant C_0 independent of ε such that*

$$\lim_{\delta\downarrow 0} W^{+}(\eta, s + \varepsilon^2) = C_0\varepsilon^{-2} \quad as\ \varepsilon \to 0. \tag{16.2.19}$$

Proof. Since the case $k < 1$ can be handled in the same way as the case $k > 1$, we show the proof for the case $k > 1$. From

$$\begin{cases} \partial_t\Gamma_{-}(\xi, t; \eta, s) - \nabla_\xi \cdot (1 + (k - 1)\chi_{-})\nabla_\xi\Gamma_{-}(\xi, t; \eta, s) = \delta(\xi - \eta)\delta(t - s), \\ \partial_t\Gamma(\xi, t; \eta, s) - \Delta_\xi\Gamma(\xi, t; \eta, s) = \delta(\xi - \eta)\delta(t - s) \end{cases}$$

in $R^2 \times R^1$, W satisfies

$$\partial_t W(\xi, t) - \nabla_\xi \cdot \big((1 + (k - 1)\chi_{-})\nabla_\xi W(\xi, t)\big) = (k - 1)\nabla_\xi \cdot \big(\chi_{-}\nabla_\xi\Gamma(\xi, t; \eta, s)\big) \tag{16.2.20}$$

in $\mathbb{R}^2 \times \mathbb{R}^1$.

By using the Laplace transform in t and Fourier transform in ξ_1 to reduce (16.2.20) to two ordinary differential equations for the Laplace and Fourier transformed $W^{\pm}$ with transmission boundary at $\xi_2 = 0$ and then transforming back using their inverses, we can show $W^{+}(\xi, t)$ with $\xi = \eta$, $t = s + \varepsilon^2$ has the asymptotic expression of the form:

$$W^{+}(\xi, t) = \frac{\sqrt{k-1}}{8\pi^2}\int_0^1 \frac{\sqrt{r} + \mathrm{i}\sqrt{k(1-r)}}{\sqrt{r}\left(\sqrt{r} - \mathrm{i}\sqrt{k(1-r)}\right)}F(r)\mathrm{d}r + O(\varepsilon^{-1}),\ 0 < \varepsilon \ll 1 \tag{16.2.21}$$

uniformly with respect to small $\delta > 0$, where

$$F(r) = \int_R |\zeta| \mathrm{e}^{\mathrm{i}(\xi_1 - \eta_1)\zeta}\Big[\exp\Big\{-(t - s)\zeta^2(kr - r + 1) - \mathrm{i}(\xi_2 + \eta_2)\sqrt{k-1}\,|\zeta|\sqrt{r}\Big\} + \exp\Big\{-(t - s)\zeta^2(kr - r + 1) + \mathrm{i}(\xi_2 + \eta_2)\sqrt{k-1}\,|\zeta|\sqrt{r}\Big\}\Big]\mathrm{d}\zeta.$$

The details for the case $m = 3$ were worked out in [27]. But the error term such as $O(\varepsilon^{-1})$ is missing which comes from the non-cancellation of the two integrals along $\overrightarrow{CD}$ and $\overrightarrow{IJ}$ in figure 3, appendix of [27]. By letting $\delta \downarrow 0$, we have

$$\begin{aligned}\lim_{\delta\downarrow 0} F(r) &= 2\int_R |\zeta|\exp[-\varepsilon^2(kr - r + 1)\zeta^2]\mathrm{d}\zeta \\ &= 4\int_0^\infty \zeta \exp[-\varepsilon^2(kr - r + 1)\zeta^2]\mathrm{d}\zeta \\ &= \frac{2\varepsilon^{-2}}{kr - r + 1}\end{aligned}$$

which implies

$$\begin{aligned}\lim_{\delta\downarrow 0} W^+\left(\xi, s + \varepsilon^2\right) &= \frac{\sqrt{k-1}}{4\pi^2}\varepsilon^{-2}\int_0^1 \frac{\sqrt{r} + \mathrm{i}\sqrt{k(1-r)}}{\sqrt{r}\left(\sqrt{r} - \mathrm{i}\sqrt{k(1-r)}\right)}\frac{1}{kr - r + 1}\mathrm{d}r \\ &=: \frac{\sqrt{k-1}}{4\pi^2}\varepsilon^{-2}H.\end{aligned}$$

Here note that

$$\left|\frac{\sqrt{r} + \mathrm{i}\sqrt{k(1-r)}}{\sqrt{r} - \mathrm{i}\sqrt{k(1-r)}}\right| = 1,\ 1 \leqslant kr - r + 1 \leqslant k,\ 1 \leqslant r + k(1-r) \leqslant k. \tag{16.2.22}$$

Then we have

$$|H| \leqslant \int_0^1 \frac{1}{\sqrt{r}}\mathrm{d}r = 2 < \infty, \tag{16.2.23}$$

and the imaginary part $\mathrm{Im}H$ of H has the estimate

$$\mathrm{Im}H = \int_0^1 \frac{2\sqrt{k}\sqrt{1-r}}{(kr - r + 1)(r + k(1-r))}\mathrm{d}r \geqslant \frac{2}{k\sqrt{k}}\int_0^1 \sqrt{1-r}\,\mathrm{d}r = \frac{4}{3k\sqrt{k}} > 0. \tag{16.2.24}$$

This completes the proof.

Therefore summing up what we have obtained so far, we have proved

$$\liminf_{\delta\downarrow 0}\left|w_{(y(\delta),s)}\left(y(\delta), s + \varepsilon^2\right)\right| \geqslant C_0'\,\varepsilon^{-2} \tag{16.2.25}$$

as $\varepsilon \to 0$ for some positive constant C_0' independent of ε. □

16.2.4 Existence of Runge's approximation functions

In this subsection, we prove the existence of Runge's approximation functions. We will only show the existence of $\{v^j_{(y,s)}\}$, because that of $\{\psi^j_{(y',s')}\}$ can be shown in the same way. Let U be the same as in lemma 16.2.1. Consider $\tilde{S} : L^2((\partial\Omega)_T \to L^2((\partial U)_T)$ defined by

$$\left(\tilde{S}\varphi\right)(x, t) = \int_{(\partial\Omega)_T} \Gamma(x, t; y, s)\varphi(y, s)\, \mathrm{d}s(y)\mathrm{d}s \tag{16.2.26}$$

for $\varphi \in L^2((\partial\Omega)_T)$. Since $L^2((\partial U)_T) = \overline{R(\tilde{S})} \oplus N(\tilde{S}^*)$ with the null space $N(\tilde{S}^*)$ of $\tilde{S}^*$ and the interior regularity estimate of solutions of the heat equation, it is enough to prove $\Gamma(x, t; \xi, \tau) \in N(\tilde{S}^*)^\perp$.

First, we note that $\tilde{S}^*$ is given by

$$\Psi(y, s) := \tilde{S}^*(\psi)(y, s) = \int_{(\partial U)\times(s,T)} \Gamma(x, t; y, s)\psi(x, t)\mathrm{d}s(x)\mathrm{d}t \tag{16.2.27}$$

for $\psi \in L^2((\partial U)_T)$. Take $\psi = \psi(x, t) \in N(\tilde{S}^*)$ to be such that $\psi \in C(\overline{(\partial U)_T})$ and $\psi|_{t=T} = 0$. Since the set of such ψ is dense in $N(\tilde{S}^*)$, we only need to prove $\Psi = 0$. We further continuously extend ψ to $t < 0$ with compact support with respect to t and denote the extended ψ by the same notation. Then, the associated Ψ satisfies

$$\begin{cases} \mathcal{P}_\phi^*\Psi = 0 & \text{in } (\mathbb{R}^2\backslash\bar{\Omega})_{(-\infty,T)} \\ \Psi = 0 & \text{on } \partial\Omega_{(-\infty,T)}, \quad \Psi|_{\tau=T} = 0 \quad \text{on } \mathbb{R}^2\backslash\bar{\Omega}. \end{cases}$$

Let $R > 0$ large enough such that $\bar{\Omega} \subset B_R$ and set $\Omega_R = (\mathbb{R}^2\backslash\bar{\Omega}) \cap B_R$. Then, since we can easily show that

$$\|\Psi(\cdot, T)\|_{L^2(\Omega_R)} = \|\Psi(\cdot, -\infty)\|_{L^2(\Omega_R)} = 0 \tag{16.2.28}$$

and

$$\int_{(\partial B_R)\times(-\infty,T)} \Psi\partial_\nu\Psi\, \mathrm{d}\sigma(\xi)\mathrm{d}\tau \to 0, \quad R \to \infty \tag{16.2.29}$$

by using $\mathrm{dist}(U, \partial\Omega) > 0$ and $r^\beta \mathrm{e}^{-r} \leqslant \beta^\beta \mathrm{e}^{-\beta}$ for any $r, \beta > 0$, we have

$$\begin{aligned} 0 &= \int_{\Omega_R\times(-\infty,T)} \Psi\mathcal{P}_\phi^*\Psi\, \mathrm{d}\xi\, \mathrm{d}\tau \\ &= \int_{\Omega_R\times(-\infty,T)} \left|\nabla_\xi\Psi\right|^2 \mathrm{d}\xi\, \mathrm{d}\tau - \int_{(\partial B_R)\times(-\infty,T)} \Psi\partial_\nu\Psi\mathrm{d}\sigma(\xi)\mathrm{d}\tau \\ &\to \int_{\Omega_R\times(-\infty,T)} \left|\nabla_\xi\Psi\right|^2 \mathrm{d}\xi\, \mathrm{d}\tau, \quad R \to \infty. \end{aligned} \tag{16.2.30}$$

Therefore by $\Psi(y, T) = 0$, we have

$$\Psi = 0 \quad \text{in } (\mathbb{R}^2\backslash\bar{\Omega}) \times (-\infty, T). \tag{16.2.31}$$

Therefore, by the unique continuation,

$$\Psi(\xi, \tau) = \int_{(\partial U)_T} \Gamma(x, t; \xi, \tau)\psi(x, t)\mathrm{d}s(x)\mathrm{d}t = 0 \tag{16.2.32}$$

for $\psi \in N(\tilde{S}^*)$. This completes the proof.

16.3 The time-domain probe method

The use of waves to carry out probing in the time domain was suggested by Burkard and Potthast [7]. The basic idea is to send a pulse into some region of space and to use the time-dependent scattered field $U^s(x, t)$ as the indicator function.

- As long as an incident pulse $U^i(x, t)$ has not reached a point $x \in \mathbb{R}^3 \backslash D$, there will not be an incident field.
- When the incident field U^i reaches a point $x \in \mathbb{R}^m$ with $d(x, D) > \rho > 0$ at time $T > 0$, then the scattered field $U^s(x, t)$ will be zero in a neighborhood of T, since the scattered field arises at the boundary ∂D of D and needs some time to travel to the point x.
- When $x \in \partial D$, we know that there is a scattered field $U^s(x, t)$ at the time T where $U^i(x, t)$ first touches x.

The above phenomena can be used to distinguish between boundary points $x \in \partial D$ and points x with $d(x, D) > \rho > 0$, if we can reconstruct the field $U^s(x, t)$ from measurements. It was suggested by [7] to use the *time-domain point source method* developed in [35, 40] for this reconstruction. Practical tests on real data have been carried out in [40] by Fazi, Nelson and Potthast, realizing a wave splitting base on an array of microphones which allows the identification of a particular speaker from a crowd of different people.

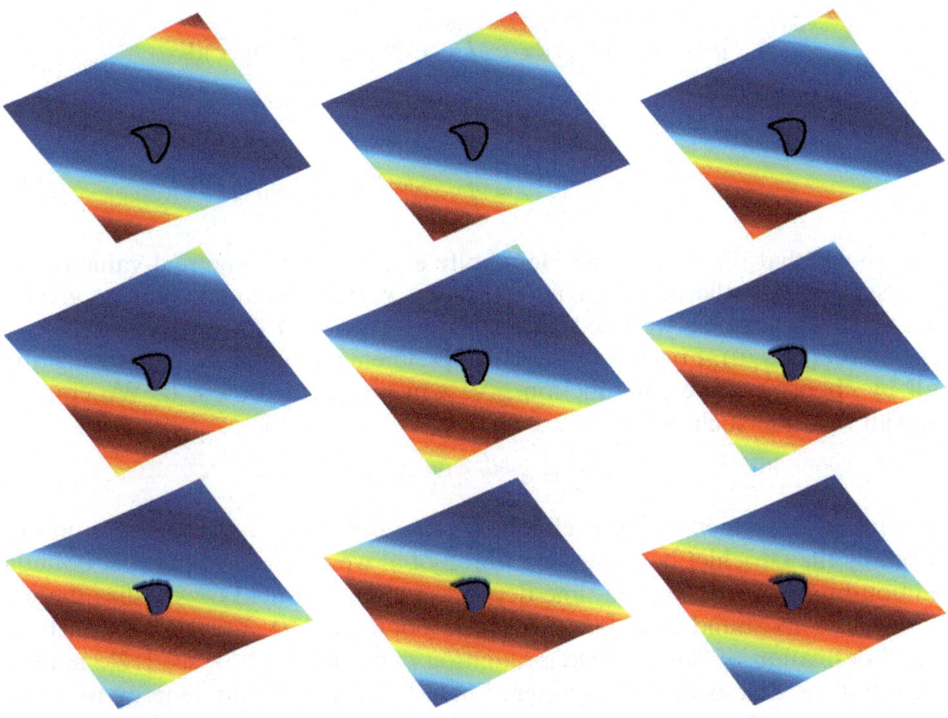

Figure 16.1. The time-dependent incident pulse as it progresses and hits the object D. The first hitting time is displayed in the third image of the top row; then, step-by-step, the pulse hits the back of the object in the second image in the third row. For the scattered fields see figure 16.2.

Let us study an incident pulse as shown in figure 16.1, which is a superposition of incident plane waves with frequency $\kappa \in \mathbb{R}$ given by

$$U^{\mathrm{i}}(x, t, d) = \int_{\mathbb{R}} \mathrm{e}^{\mathrm{i}\kappa ct} \mathrm{e}^{\mathrm{i}\kappa x \cdot d} g(\kappa)\, \mathrm{d}\kappa, \quad x \in \mathbb{R}^m,\ t \in \mathbb{R}, \tag{16.3.1}$$

with spectral density $g \in L^2(\mathbb{R})$ and wave speed c, which is the *Fourier transform*

$$F(u^{\mathrm{i}}(\cdot, \kappa, d) g(\kappa))(ct), \quad t \in \mathbb{R}. \tag{16.3.2}$$

Since $u^{\mathrm{i}}(x, \kappa, d) = \mathrm{e}^{\mathrm{i}\kappa x \cdot d}$ satisfies $\Delta u^{\mathrm{i}} + \kappa^2 u^{\mathrm{i}} = 0$, the field U^{i} satisfies

$$\begin{aligned} \frac{\partial^2 U^{\mathrm{i}}}{\partial t^2}(x, t, d) &= c^2 \int_{\mathbb{R}} (-\kappa^2) \mathrm{e}^{\mathrm{i}\kappa ct} \mathrm{e}^{\mathrm{i}\kappa x \cdot d} g(\kappa)\, \mathrm{d}\kappa, \\ &= c^2 \Delta U^{\mathrm{i}}(x, t, d), \end{aligned} \tag{16.3.3}$$

i.e. U^{i} solves the wave equation with wave speed c. We define a scattered field $U^{\mathrm{s}}(x, t, d)$ in the time domain by a corresponding superposition of scattered fields $u^{\mathrm{s}}(x, \kappa, d)$ with Dirichlet boundary condition (8.2.3) on ∂D, i.e.

$$U^{\mathrm{s}}(x, t, d) := \int_{\mathbb{R}} \mathrm{e}^{\mathrm{i}\kappa ct} u^{\mathrm{s}}(x, \kappa, d) g(\kappa)\, \mathrm{d}\kappa, \quad x \in \mathbb{R}^m, \quad t \in \mathbb{R}. \tag{16.3.4}$$

An example is shown in figure 16.2. The case where $\kappa < 0$ is treated using the following equality

$$\begin{aligned} \overline{\int_{-\infty}^{0} \mathrm{e}^{\mathrm{i}\kappa ct} u^{\mathrm{s}}(x, \kappa, d) g(\kappa)\, \mathrm{d}\kappa} &= \int_{-\infty}^{0} \mathrm{e}^{-\mathrm{i}\kappa ct} \overline{u^{\mathrm{s}}(x, \kappa, d) g(\kappa)} \mathrm{d}\kappa \\ &= \int_{0}^{\infty} \mathrm{e}^{\mathrm{i}\kappa ct} \overline{u^{\mathrm{s}}(x, -\kappa, d) g(-\kappa)} \mathrm{d}\kappa \\ &= \int_{0}^{\infty} \mathrm{e}^{\mathrm{i}\kappa ct} \overline{u^{\mathrm{s}}(x, \kappa, -d) g(-\kappa)} \mathrm{d}\kappa. \end{aligned} \tag{16.3.5}$$

where we use that if u^{s} solves the Helmholtz equation for some real-valued $\kappa \in \mathbb{R}$, then also $\overline{u^{\mathrm{s}}}$ solves the equation for the same κ, with boundary values given by $\overline{u^{\mathrm{i}}} = \mathrm{e}^{-\mathrm{i}\kappa x \cdot d}$. As a result, we need to evaluate $u^{\mathrm{s}}(\cdot, \kappa, -d)$ for $\kappa > 0$ to obtain the integral over $(-\infty, 0)$. For our numerical experiments we will avoid using negative κ at all by choosing a weight function $g(\kappa)$ which is zero for $\kappa < 0$. We employ a Gaussian function such as

$$g(\kappa) = \mathrm{e}^{-\sigma(\kappa - \kappa_0)^2}, \quad \kappa \in \mathbb{R}$$

with some parameter σ, which is effectively zero outside some neighborhood of κ_0.

Then by the Dirichlet boundary condition of u^{s} we derive the Dirichlet boundary condition of U^{s} on ∂D, i.e. the time-dependent scattered field U^{s} satisfies the wave equation and its sound field is zero on the boundary ∂D. Further, following the original derivation of Sommerfeld's radiation condition (8.2.6), see [12] chapter 3, we see that the time-dependent scattered field U^{s} is transporting energy to infinity, i.e. the superposition of the radiating scattered fields u^{s} is the correct radiating scattered field in the time domain.

Clearly, we cannot work with far field patterns in the time domain, since pulses need an infinite time to reach infinity. However, using the far field pattern to describe an approximation to the field u^s on the boundary of some ball $B_R(0)$, they are very useful. In this sense, we can assume that we are measuring the far field pattern

$$U^\infty(\hat{x}, t, d) := \int_{\mathbb{R}} e^{i\kappa c t} u^\infty(\hat{x}, \kappa, d) g(\kappa) ds(\kappa), \quad \hat{x} \in \mathbb{S}, \; t \in \mathbb{R}. \tag{16.3.6}$$

The suggestion of Luke and Potthast [35] is to use the inverse Fourier transform on the measurement data $U^\infty(\hat{x}, \cdot, d)$ to recover $u^\infty(\hat{x}, \kappa, d)g(\kappa)$ for $\kappa \in \mathbb{R}$. This approach has been tested successfully by Fazi *et al* [40] for real measured acoustic data. Then, a reconstruction of $u^s(x, \kappa, d)$ from $u^\infty(\cdot, \kappa, d)$ for $x \in \mathbb{R}^m \backslash \bar{D}$ can be carried out using the *point source method* of section 12.4. The scattered field $U^s(\cdot, t, d)$ is reconstructed calculating the integral (16.3.4).

The convergence of the time-domain field reconstructions can be shown by the same arguments as the convergence of the frequency-domain point source method in section 12.4, see [35] for more details. Here, we will show that the approach can easily be realized based on the codes introduced in sections 8.2 and 12.4.

Code 16.3.1. *Script* `sim_16_3_1_c_frequency_loop.m` *to calculate the scattering in the frequency domain and* `sim_16_3_1_d_FT.m` to *calculate its Fourier transform and obtain the incident, scattered and total fields in the time domain.*

```
tic;                          % for recording calculation times
hk = 0.3;                     % grid spacing for Fourier integral
kappav = 0.1:hk:4;            % vector of wave numbers
Nk = size(kappav,2);          % number of wave numbers
jj=1;                         % counter
for kappa = kappav            % loop over wave numbers
   sim_16_3_1_a_total_field; % solve scattering problem for kappa
   wv(:,jj)=w;                % total field in frequency domain
   wiv(:,jj)=wi;              % incident field in frequency domain
   wsv(:,jj)=ws;              % scattered field in frequency domain
   jj = jj+1;                 % counter update
end                           %
t = toc;                      % enquire evaluation time
disp(['time needed t=' num2str(t) ' sec for Nk=' num2str(Nk) ' wave numbers']);
```

```
tv   = (-10:0.5:-2).';             % setup vector with times
Nt   = size(tv,1);                 % number of points in time
tmat = repmat(tv,1,Nk);            % matrix of times
kmat = repmat(kappav,Nt,1);        % matrix of wave numbers
F    = exp(-1i*tmat.*kmat)*hk;     % Fourier matrix, no constant(!)
g    = exp(-6*(kappav-1.5).^2);    % weighting function
gmat = repmat(g.',1,M);            % weighting matrix
U    = (F*((wv.').*gmat)).';       % total field in time
Ui   = (F*((wiv.').*gmat)).';      % incident field in time
Us   = (F*((wsv.').*gmat)).';      % scattered field in time
```

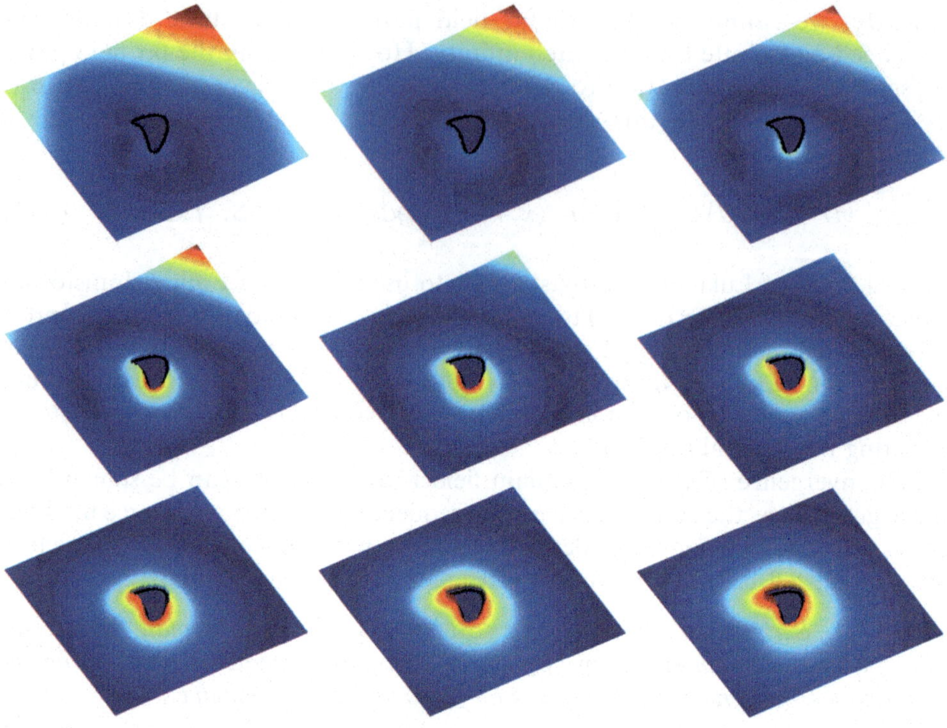

Figure 16.2. The time-dependent scattered field as it arises at the same times as in figure 16.1, when the object is hit by the incident pulse in the third image of the top row. In the second image of the last row, the incident pulse has reached the back of the object and the scattered field has appeared everywhere on the boundary ∂D of D. Note that it is straightforward to reconstruct the scattered field from the far field pattern based on the *point source method*. A simple reconstruction is shown in figure 16.3.

Based on the above calculations, we can now mask the scattered field with the first arrival time mask of the incident field and reconstruct the boundary of the scatterer D. An easy example of such a reconstruction is shown by the black points in figure 16.3.

16.4 The BC method of Belishev for the wave equation

As we have already seen in the inverse scattering problem for the Helmholtz equation, waves are used for non-destructive testing such as to detect unknown cavities inside a medium. These waves are time-harmonic waves. If we can measure waves over some time interval at the boundary of the medium we can use the BC method, originated by Belishev [2] and Kurylev also played an important role in developing this method [5, 6]. This is a very powerful method for identifying the coefficients of the second order scalar hyperbolic equation which describes the propagation of waves inside the medium. The measured data of the BC method are usually taken as the Neumann-to-Dirichlet map which is a set of infinitely many Cauchy data at the boundary of solutions to the initial boundary value problem for the hyperbolic equation without a source term and

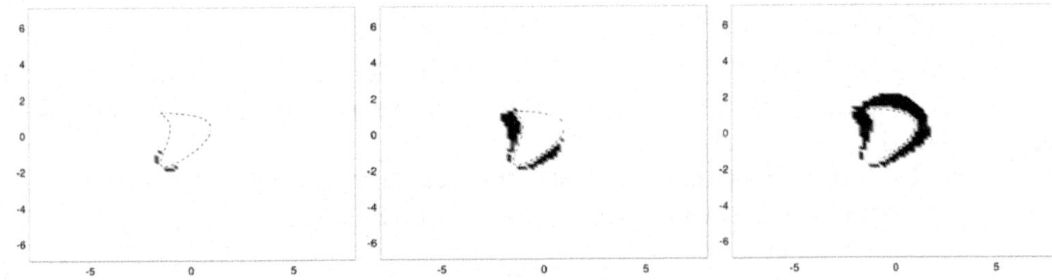

Figure 16.3. A simple example for reconstructions generated from the time-dependent scattered field U^s, where we employ a mask which is one only at points at which the incident field U^i is large the first time only, as shown in figure 16.1, and where U^s is large as well, see figure 16.2. The method first finds the lightened part of the object and then in the following time steps the other parts. This is generated by `sim_16_3_1_h_probing.m` found in the code repository.

with a homogeneous initial condition. For the acoustic wave equation, the physical meaning of the measured data called the Neumann-to-Dirichlet map is to measure all the displacements at the boundary for any given pressures at the boundary.

In this section, by adapting the idea of Belishev in [3], we will show how the BC method can be used to identify an unknown inclusion inside the medium. This idea is quite similar to the probing type argument for the Helmholtz equation and heat equation which we explained previously. For the details of the BC method, see [4, 28].

Let $\Omega \subset \mathbb{R}^m$ $(m \geqslant 2)$ be a bounded domain with C^2 boundary $\partial\Omega$ and D be a domain with C^2 boundary ∂D such that $\Omega\backslash\bar{D}$ is connected. Physically, we consider Ω as an acoustic medium with an inclusion D inside. We assume the density ρ of Ω is piece-wise constant and it only changes across ∂D. More precisely, we assume that $\rho = 1 + k\chi_D$ with a constant $k > -1$, where χ_D denotes the characteristic function of D. For simplicity we consider the case $k > 0$, whi ch means that the speed of the acoustic wave inside D is slower than that outside. If an acoustical pressure $f \in L^2(\partial\Omega \times (0, T))$ is given at $\partial\Omega$ over the time interval $(0, T)$ and the medium Ω is at rest in the past $t < 0$, then an acoustic wave will propagate inside the medium and its displacement $u(x, t) \in L^2((0, T) \times \Omega)$ is given as a weak unique solution to the initial boundary value problem

$$\begin{cases} Lu = \rho\dfrac{\partial^2 u}{\partial t^2} - \Delta u = 0 \quad \text{in } \Omega_T = \Omega \times (0, T) \\ \dfrac{\partial u}{\partial \nu} = f \text{ on } (\partial\Omega)_T = \partial\Omega \times (0, T) \\ u(\cdot, 0) = \dfrac{\partial u}{\partial t}(\cdot, 0) = 0 \quad \text{in } \Omega, \end{cases} \tag{16.4.1}$$

where ν denotes the outer unit normals to $\partial\Omega$ and $u = u^f$ is called the weak solution to (16.4.1) if it satisfies

$$\int_Q u\, Lv \,\mathrm{d}x\,\mathrm{d}t - \int_{\Sigma_T} fv \,\mathrm{d}s\,\mathrm{d}t = 0 \tag{16.4.2}$$

for any $v \in H^2(\Omega_T)$ with $\frac{\partial v}{\partial \nu} = 0$ on $(\partial\Omega)_T$ and $v|_{t=T} = \frac{\partial v}{\partial t}|_{t=T} = 0$. It can be shown that $u^f \in L^2((0, T); H^1(\Omega)) \cap C^0([0, T]; L^2(\Omega))$ depends continuously on $f \in L^2((\partial\Omega)_T)$ in the same way as in [34]. Hereafter, we assume that all the functions are real valued.

For any fixed proper open set $\Sigma \subset \partial\Omega$, this regularity of solutions allows us to define the Neumann-to-Dirichlet map $\Lambda_T : F^T \to L^2(\Sigma_T)$ by

$$\Lambda_T f = u^f|_{\Sigma T}, \tag{16.4.3}$$

where $F^T = \{f \in L^2((\partial\Omega)_T)\text{supp} f \subset \Sigma\}$ and $\Sigma_T = \Sigma \times (0, T)$. The further consequences of the regularity of solutions are that the operators defined by $W^T : F^T \ni f \mapsto u^f(\cdot, T) \in \mathcal{H} = L^2_\rho(\Omega)$ and $C^T = (W^T)^* W^T : F^T \mapsto F^T$ are a bounded operator and a self-adjoint operator, respectively, where $L^2_\rho(\Omega)$ denotes L^2 space in Ω with measure ρdx.

The inverse problem which we are interested in is to reconstruct D from Λ_{2T} if we do not know D and k.

To study this inverse problem, we first prepare some preliminary facts we need. Let $f, g \in F^T$ and $u^f, u^g \in L^2((0, 2T); H^1(\Omega))$ be the corresponding unique solutions to (16.4.1) with Neumann data f, g. Then, by the definition of Λ_T, the real inner product of $u^f(\cdot, t)$ and $u^g(\cdot, s)$ in $\mathcal{H}$ denoted by $q(t, s) = (u^f(\cdot, t), u^g(\cdot, s))_{\mathcal{H}} \in C([0, 2T] \times [0, 2T])$ is a weak solution of

$$\left(\frac{\partial^2}{\partial t^2} - \frac{\partial^2}{\partial s^2}\right) q(t, s) = \int_\Sigma (f(x, t)(\Lambda_{2T} g)(x, s) - g(x, s)(\Lambda_{2T} f)(x, t))\, ds(x) \tag{16.4.4}$$

in $(0, 2T) \times (0, 2T)$, where the right-hand side is a known bounded measurable function in $(0, 2T) \times (0, 2T)$ and by the initial conditions for u^f and u^g, $q(t, s) = 0$ at $t = 0$ and $s = 0$. Hence, by integrating this equation and setting $t = s = T$, we conclude that C^T is determined by Λ_{2T}.

The so-called UCP is known for our operator L. That is the continuation of the zero set of weak solutions of $Lu = 0$ in $C^0([0, T]; H^1(\Omega)) \cap C^1([0, T]; L^2(\Omega))$ is known. For its proof we refer the reader to consult the papers [37, 42]. Based on this, we define our domain of influence as follows.

Definition 16.4.1. *We define the domain of influence $\tilde{E}^T$ as the maximal subdomain of Ω in which all the weak solutions of $Lu = 0$ in $C^0([0, 2T]; H^1(\Omega)) \cap C^1([0, 2T]; L^2(\Omega))$ will become zero at $t = T$ if their Cauchy data on Σ_{2T} are zero.*

Remark 16.4.2. *Let $E^T = \{x \in \bar{\Omega} : \text{dist}(x, \Sigma) < T\}$. If $E^T \cap D = \emptyset$, then $\tilde{E}^T = E^T$ by the global Holmgren–John–Tataru uniqueness theorem which says that all the solutions with zero Cauchy data on Σ_T vanish in the double cone $\{(x, t) : dist(x, \Sigma) \leqslant T - |t - T|\}$ (see* [28]*). Further, by the finite speed of propagation and the result given in* [37] *or* [42] *on UCP for a wave equation with a jump in density, we have $\tilde{E}^T \subset E^T$.*

Then, we have the following two very important theorems for the BC method.

Theorem 16.4.3 (Approximate controllability). *The closure $cl(W^T F^T)$ of $W^T F^T$ in $\mathcal{H}$ is equal to $L^2(\tilde{E}^T) \subset \mathcal{H}$. That is*

$$cl\left(W^T F^T\right) = L^2\left(\tilde{E}^T\right).$$

Proof. By corolllary 2.4.2, it is enough to show that $\psi \in L^2(\tilde{E}^T)$ satisfies $u^f(\cdot, T), \psi)_{\mathcal{H}} = 0$ for any $f \in C_0^\infty(\Sigma_T)$ implies $\psi = 0$. In order to do that, let $e \in C^0([0, T]; H^1(\Omega)) \cap C^1([0, T]; L^2(\Omega))$ be the weak solution to

$$\begin{cases} Le = 0 & \text{in } \Omega_T \\ \dfrac{\partial e}{\partial \nu} = 0 & \text{on } \Sigma_T \\ e = 0, \ \dfrac{\partial e}{\partial t} = \psi & \text{at } t = T. \end{cases} \tag{16.4.5}$$

Then, integrating by parts, we have

$$\begin{aligned} 0 &= \int_{\Omega_T} \left\{u^f(Le) - (Lu^f)e\right\} \mathrm{d}x \, \mathrm{d}t \\ &= \int_\Omega u^f(\cdot, T)\psi\rho \, \mathrm{d}x + \int_{(\partial\Omega)_T} fe \, \mathrm{d}s \, \mathrm{d}t = \int_{\Sigma_T} fe \, \mathrm{d}s \, \mathrm{d}t, \end{aligned}$$

for any $f \in C_0^\infty(\Sigma_T)$. Hence, the Cauchy data of e on Σ_T are zero.

Now let $E(x, t) = e(x, t)(t \leqslant T), -e(x, 2T - t)\,(t > T)$. Then, due to $e = 0$ at $t = T, E \in C^0([0, 2T]; H^1(\Omega)) \cap C^1([0, 2T]; L^2(\Omega))$, and it satisfies $LE = 0$ in Ω_{2T} and

$$E = \frac{\partial E}{\partial \nu} = 0 \quad \text{on } \Sigma_{2T}.$$

By applying the global Holmgren–John–Tataru uniqueness theorem, we have $\psi = E(\cdot, T) = 0$. □

Theorem 16.4.4. *Let $T < T_* := \inf\{T{:}\ E^T = \Omega\}$. Then, the kernel* Ker C^T *of C^T just consists of* 0.

Proof. Let $f \in F^T$ satisfy $C^{Tf} = 0$ which immediately implies $u^f(\cdot, T) = 0$ in Ω. Then consider an extension $u_0(\cdot, t)$ of $u^f(\cdot, t)$ to $\Omega \times \mathbb{R}$ defined by

$$u_0(\cdot, t) = \begin{cases} 0 & (t < 0) \\ u^f(\cdot, t) & (0 \leqslant t \leqslant T) \\ -u^f(\cdot, 2T - t) & (T < t \leqslant 2T) \\ 0 & (2T < t). \end{cases}$$

It is easy to see that $u_0 \in L^2(\mathbb{R}; H^1(\Omega)$ satisfies

$$\begin{cases} Lu_0 = 0 & \text{in } \Omega \times \mathbb{R} \\ \dfrac{\partial u_0}{\partial \nu} = f_0 & \text{on } \partial\Omega \times \mathbb{R}, \end{cases} \tag{16.4.6}$$

where f_0 is the extension of f by a similar extension as u_0. Then, the Fourier transform $\hat{u}_0(\cdot, \omega)$ of $u_0(\cdot, t)$ with respect to t satisfies

$$\begin{cases} \left(\Delta + \rho\omega^2\right)\hat{u}_0 = 0 & \text{in } \Omega \\ \dfrac{\partial \hat{u}_0}{\partial \nu} = \hat{f}_0 & \text{on } \partial\Omega \end{cases} \tag{16.4.7}$$

for any $\omega \in \mathbb{R}$, where $\hat{f}_0(\cdot, \omega)$ is the Fourier transform of $f(\cdot, t)$ with respect to t. Since by assumption $T < T_*$ and theorem 16.4.3, $\hat{u}_0(\cdot, \omega)$ is zero in the non-empty set $\Omega \backslash E^T$ for any $\omega \in \mathbb{R}$. Hence, by a UCP for $\Delta + \rho\omega^2$ similar to the above-mentioned UCP for L, $\hat{u}_0(\cdot, \omega) = 0$ in Ω for any $\omega \in \mathbb{R}$. Therefore, $\hat{f}_0(\cdot, \omega) = 0$ for any $\omega \in \mathbb{R}$ and hence $f = 0$.

For any fixed $\in (0, T_*)$, let Φ^T be a Hilbert space obtained by completing F^T with respect to the new inner product $\langle f, g\rangle_{\Phi^T} = \langle C^T f, g\rangle_{F^T} (f, g \in F^T)$. Also, by theorem 16.4.3, let $\mathcal{W}^T$ be the continuous extension of W^T to Φ^T and $C^T = (\mathcal{W}^T)^* \mathcal{W}^T$. Then, $\mathcal{W}^T : F^T \to L^2(\tilde{E}^T)$ becomes an isometry and $C^T : \Phi^T \to \Phi^T$ becomes an isomorphism.

Now we will make the following observation. That is for $f \in F^T$ and $V \in H^1(\Omega)$ with $\frac{\partial V}{\partial \nu}|_{\partial\Omega} \in L^2(\partial\Omega)$ and $\Delta V = 0$ in Ω, we have by the Green formula

$$\begin{aligned} \left\langle u^f(\cdot, T), V \right\rangle_{\mathcal{H}} &= \int_\Omega V(x) \left\{ \int_0^T (T - t) \frac{\partial^2 u^f}{\partial t^2}(x, t) \mathrm{d}t \right\} \mathrm{d}x \\ &= \int_{\Sigma_T} \left\{ (T - t) V(x) - (\Lambda_T)^* \left((T - t) \frac{\partial V}{\partial \nu}(x) \right) \right\} f(x, t)\, \mathrm{d}s(x)\, \mathrm{d}t, \end{aligned} \tag{16.4.8}$$

with the adjoint Λ_T^* of Λ_T. Based on this observation, we define q_V^T by

$$q_V^T = (C^T)^{-1} Y \text{ with } Y = (T - t)V - (\Lambda_T)^* \left((T - t) \frac{\partial V}{\partial \nu} \right) \text{ on } \Sigma_T. \tag{16.4.9}$$

Then, q_V^T satisfies

$$\mathcal{W}^T q_v^T = V \text{ in } \tilde{E}^T. \tag{16.4.10}$$

In fact, for any $f \in F^T$, we have

$$\langle Y, f\rangle_{F^T} = \langle (C^T)^{-1} Y, f\rangle_{\Phi T} = \left\langle \mathcal{W}^T \left((C^T)^{-1} Y \right), \mathcal{W}^T f \right\rangle_{\mathcal{H}}.$$

On the other hand, from (16.4.8),

$$\langle Y, f\rangle_{F^T} = \left\langle W^T f, V \right\rangle_{\mathcal{H}} = \langle V, \rangle \mathcal{W}^T f_{\mathcal{H}}.$$

Then, (16.4.10) follows from these two equations and recalling that $\mathcal{W}^T f$ $(f \in F^T)$ is dense in $C_0^\infty(\tilde{E}^T)$ with respect to the $L^2(\tilde{E}^T)$ norm.

We will build our reconstruction scheme to reconstruct D from Λ_{2T} on these preliminaries. In order to simplify our description we restrict to the case $m = 3$ in the rest of this section. Let $T < T_*$, $x_0 \in \Omega \backslash \overline{\tilde{E}^T}$, linearly independent vectors $a_j \in \mathbb{R}^3 (j = 1, 2, 3)$ and $\varepsilon_{x_0}(x)$ be the fundamental solution of Δ with singularity at x_0, namely $\varepsilon_{x_0}(x) = \Phi(x, x_0)$ with wave number $\kappa = 0$, where $\Phi(x, x_0)$ is the fundamental solution of the Helmholtz equation with wave number κ. Further, let $G_{a_j, x_0} = a_j \cdot \nabla \varepsilon_{x_0}$. By the point source method or Runge approximation theorem, for $x_0 \notin \overline{\tilde{E}^T}$ and each $j (j = 1, 2, 3)$, there exists a sequence $\{h_{j,\ell}\}_{\ell=1}^{\infty} \subset H^2(\Omega)$ such that

$$\Delta h_{j,\ell} = 0 \text{ in } \Omega,\ h_{j,\ell} \to G_{a_j, x_0}\ (\ell \to \infty) \text{ in } L^2\left(\tilde{E}^T\right).$$

Then, by taking $q_{j,\ell} = q_V^T$ of (16.4.9) with $V = h_{j,\ell}$, we define our indicator function $\sigma(x_0, T)$ by

$$\sigma(x_0, T) = \sum_{j=1}^{3} \left(C^T q_{j,\ell_0}, q_{j,\ell_0}\right)_{F^T}, \tag{16.4.11}$$

where $\ell_0 = \inf\{\ell : \|h_{j,\ell} - G_{a_j, x_0}\|_{L^2(\tilde{E}^T)} \leqslant 1\ (j = 1, 2, 3)\}$. Then, clearly $\sigma(x_0, T)$ is finite. To see the behavior of $\sigma(x_0, T)$ as $x_0 \to \tilde{E}^T$, observe the estimate

$$\begin{aligned} \sigma(x_0, T) &= \sum_{j=1}^{3} \int_{\Omega} |u^{q_{j,\ell_0}}(x, T)|^2 \rho(x) \mathrm{d}x \geqslant \sum_{j=1}^{3} \int_{\tilde{E}^T} |u^{q_{j,\ell_0}}(x, T)|^2 \rho(x) \mathrm{d}x \\ &\geqslant \sum_{j=1}^{3} \int_{\tilde{E}^T} \left|G_{a_j, x_0}(x)\right|^2 \rho(x) \mathrm{d}x - K \end{aligned} \tag{16.4.12}$$

with some constant $K > 0$ which is independent of x_0 and T. We will use this indicator function to detect ∂D as follows. Let $C = \{c(\alpha) : \alpha \in [0, 1]\}$ be a needle which connects $c(0) \in \Sigma$ to $c(1) \in \partial\Omega \backslash \bar{\Sigma}$ and $\gamma(T)$ be the unique touching point of C to $\partial \tilde{E}^T$ when we traces C from $c(1)$. By the estimate (16.4.12), $\gamma(T)$ is given by

$$\gamma(T) = c(\alpha_T) \text{with } \alpha_T = \inf\{\alpha' \in (0, 1) : \sigma(c(\alpha), T) < \infty\ (\alpha' < \alpha \leqslant 1)\}. \tag{16.4.13}$$

Further, by remark 16.4.2, we have

$$\begin{aligned} T &= \operatorname{dist}(\gamma(T), \Sigma) \text{ if } \tilde{E}^T \cap \bar{D} = \varnothing \\ T &> \operatorname{dist}(\gamma(T), \Sigma) \text{ if } \tilde{E}^T \cap D \neq \varnothing. \end{aligned} \tag{16.4.14}$$

Hence, by these (16.4.13), (16.4.14), we can in principle recover the point at which C touches ∂D at the first time by tracing C from its end point $c(1) \notin \Sigma$. By taking several needles, we can recover ∂D. Of course $\partial \tilde{E}^T$ may become complicated as T becomes large and Σ has a complicated shape. These cases have to be avoided.

Remark 16.4.5. *The idea given here works even in the case where the density ρ in $\Omega \backslash \bar{D}$ is unknown and C^∞ smooth. For this case we can even reconstruct ρ there.*

Bibliography

[1] Alessandrini G and Di Cristo M 2005 Stable determination of an inclusion by boundary measurements *SIAM J. Math. Anal.* **37** 200–17

[2] Belishev M 1987 An approach to multidimensional inverse problems for wave equation *Dokl. Akad. Nauk SSSR* **297** 524–7
Belishev M 1988 *Sov. Math. Dokl.* **36** 481–4 (Engl. transl.)

[3] Belishev M 1987 Equations of the Gelfand-Levitan type in a multidimensional inverse problem for the wave equations *Zap. Nauchn. Semin. LOMI* **173** 30–41
Belishev M 1991 *J. Sov. Math.* **55** 1663 (Engl. transl.)

[4] Belishev M 1997 Boundary control in reconstruction of manifolds and metrics (the BC method) *Inverse Problems* **13** R1–R45

[5] Belishev M and Kurylev Ya 1986 An inverse problem of acoustic scattering in a space with local inhomogeneity *Zap. Nauchn. Sem. Leningrad Otdel. Mat. Inst. Steklov. (LOMI)* 156 (Mat. Voprosy Teor. Rasprostr. Voln. 16):24–34 184

[6] Belishev M and Kurylev Ya 1987 A nonstationary inverse problem for the multidimensional wave equation 'in the large' *Zap. Nauchn. Sem. Leningrad. Otdel. Mat. Inst. Steklov. (LOMI)* 165

[7] Burkard C and Potthast R 2009 A time-domain probe method for three-dimensional rough surface reconstructions *Inverse Probl. Imaging* **3** 259–74

[8] Cakoni F and Colton D 2006 *Qualitative Methods in Inverse Scattering Theory* (New York: Springer)

[9] Cantwell W J and Morton J 1992 The significance of damage and defects and their detection in composite materials: a review *J. Strain Anal. Eng. Des.* **27** 29–42

[10] Chen Q, Haddar H, Lechleiter A and Monk P 2010 A sampling method for inverse scattering in the time domain *Inverse Problems* **26** 085001

[11] Cheng J, Liu J and Nakamura G 2003 Recoverly of the shape of an obstacle and the boundary impedance from the far-field pattern *J. Math. Kyoto Univ.* **43** 165–86

[12] Colton D and Kress R 1983 *Integral Equation Methods in Scattering Theory* (New York: Wiley)

[13] Costabel M 1990 Boundary integral operators for the heat equation *Integral Equ. Oper. Theor.* **13** 498–552

[14] Daido Y, Kang H and Nakamura G 2007 A probe method for the inverse boundary value problem of non-stationary heat equations *Inverse Problems* **23** 1787–800

[15] Di Cristo M and Vessella S 2010 Stable determination of the discontinuous conductivity coefficient of a parabolic equation *SIAM J. Math. Anal.* **42** 183–217

[16] Elayyan A and Isakov V 1997 On uniqueness of recovery of the discontinuous conductivity coefficient of a parabolic equation *SIAM J. Math. Anal.* **28** 49–59

[17] Fan J, Kim K, Nagayasu S and Nakamura G 2013 A gradient estimate for solutions to parabolic equations with discontinuous coefficients *Electron. J. Differ. Equ.* **153** 91–151

[18] Friedman A 1964 *Partial Differential Equations of Parabolic Type* (Englewood Cliffs, NJ: Prentice Hall)

[19] Gaitan P, Isozaki H, Poisson O, Siltanen S and Tamminen J P 2012 Probing for inclusions in heat conductive bodies *Inverse Probl. Imaging* **6** 423–46

[20] Gaitan P, Isozaki H, Poisson O, Siltanen S and Tamminen J P 2015 Inverse problems for time-dependent singular heat conductivities: multi-dimensional case *Comm. Partial Differ. Equ.* **40** 837–77

[21] Haddar H, Lechleiter A and Marmorat S 2014 An improved time domain linear sampling method for Robin and Neumann obstacles *Appl. Anal.* **93** 369–90

[22] Ibarra-Castanedo C, Piau J and Gulberte S *et al* 2009 Comparative study of active thermography techniques for the nondestructive evaluation of honeycomb structure *Res. Nondestr. Eval.* **20** 1–31

[23] Ikehata M 2013 The enclosure method for inverse obstacle scattering problems with dynamical data over a finite time interval: III. Sound-soft obstacle and bistatic data *Inverse Problems* **29** 085013

[24] Ikehata M 2013 The enclosure method for inverse obstacle scattering problems with dynamical data over a finite time interval: III. Sound-soft obstacle and bistatic data *Inverse Problems* **29** 085013

[25] Ikehata M and Kawashita M 2010 On the reconstruction of inclusions in a heat conductive body from dynamical boundary data over a finite time interval *Inverse Problems* **26** 095004

[26] Isakov V 1998 *Inverse Problems for Partial Differential Equations Springer Series in Applied Mathematical Science* vol 127 (Berlin: Springer)

[27] Isakov V, Kim K and Nakamura G 2010 Reconstruction of an unknown inclusion by thermography *Ann. Sc. Norm. Super. Pisa Cl. Sci.* **9** 725–58

[28] Katchalov A, Kurylev Y and Lassas M 2001 *Inverse Boundary Spectral Problems* (*Monographs and Surveys in Pure and Applied Mathematics* vol 123) (Boca Raton, FL: CRC)

[29] Kirpichnikova A and Kurylev Y 2012 Inverse boundary spectral problem for Riemannian polyhedra *Math. Ann.* **354** 1003–28

[30] Kress R 1999 *Linear Integral Equations* (*Applied Mathematical Sciences* vol 82) 2nd edn (New York: Springer)

[31] Ladyzenskaja O, Solonnikov V and Uralceva N 1968 *Linear and Quasilinear Equations of Parabolic Type* (*Translations of Mathematical Monographs* vol 23) (Providence, RI: American Mathematical Society)

[32] Ladyzenskaja O A, Ja Rivkind V and Uralceva N N 1966 Solvability of diffraction problems in the classical sense *Trudy Mat. Inst. Steklov* **92** 116–46

[33] Li H and Li Y 2011 Gradient estimates for parabolic systems from composite material arXiv:1105.1437v1

[34] Lions J 1971 *Optimal Control of Systems Governed by Partial Differential Equations* (Berlin: Springer)

[35] Luke D R and Potthast R 2006 The point source method for inverse scattering in the time domain *Math. Methods Appl. Sci.* **29** 1501–21

[36] Nakamura G and Wang H 2013 Linear sampling method for the heat equation with inclusions *Inverse Problems* **29** 104015 23

[37] Oksanen L 2013 Inverse obstacle problem for the non-stationary wave equation with an unknown background *Commun. PDE* **38** 1492–518

[38] Patel P M, Lau S K and Almond D P 1992 A review of image analysis techniques applied in transient thermographic nondestructive testing *Nondestruct. Test Eval.* **6** 343–64

[39] Potthast R 2001 *Point Sources and Multipoles in Inverse Scattering Theory* (*Chapman and Hall/CRC Research Notes in Mathematics* vol 427) (Boca Raton, FL: CRC)

[40] Potthast R, Fazi F and Nelson P 2010 Source splitting via the point source method *Inverse Problems* **26** 045002

[41] Somersalo E 2001 Locating anisotropy in electrical impedance tomography arXiv:math/0110299v1

[42] Stefanov P and Uhlmann G 2011 Thermoacoustic tomography arising in brain imaging *Inverse Problems* **27** 045004

[43] Wloka J 1987 *Partial Differential Equations* (Cambridge: Cambridge University Press)

IOP Publishing

Inverse Modeling

An introduction to the theory and methods of inverse problems and data assimilation

Gen Nakamura and Roland Potthast

Chapter 17

Targeted observations and meta-inverse problems

The *direct* problem simulates some natural phenomenon, given all its constituents and parts. The *inverse* problem usually measures some function simulated by the *direct* problem and aims to *reconstruct* other parts or constituents of the direct problem, which are not known for the particular application under consideration.

Often, there is a lot of freedom in the choice of measurement positions or frequencies. We are able to plan experiments, or we are able to modify our set-up depending on our findings in a first measurement step. We call the questions of *how* an inverse problem should be designed and *where* measurements should be taken *meta* inverse problems, which include the problems of *targeted observations* and *experimental design*.

The overall strategy to design an experiment tries to find the best set-up to reconstruct the unknown quantity. In this sense we can view it as an *optimization* problem. It includes the direct and inverse problems under consideration. This means that we need to take particular care to include the tools which are used to solve the *inverse* problem. In particular, we need to include *regularization theory* into the optimization task. This will be carried out in section 17.1, where we develop a conceptual framework for solving meta-inverse problems.

When measurement points (or in general the set-up of the measurements) are chosen to view more details of a particular inverse solution, we denote the meta-inverse algorithm as *zoom* or *framework adaption*, see section 17.2. An application to a discrete acoustic source problem is treated in section 17.3.

17.1 A framework for meta-inverse problems

The goal of this section is to formulate the *meta-inverse problem* based on formulations of the direct and inverse problems as developed by Udosen and Potthast [1, 2]. Here, we will use an example of an acoustic source problem to illustrate the key ideas and, in parallel, work out a generic framework for a broad class of problems.

In the general case we have a linear or nonlinear mapping

$$A : X \to Y, \quad \varphi \mapsto A(\varphi) \tag{17.1.1}$$

from a normed space X into a normed space Y. This set-up corresponds to a system with a state $\varphi \in X$ and measured values $\psi \in Y$ where A is the measurement operator which assigns the measurements $\psi = A(\varphi)$ to φ. Usually, the measurements depend strongly on the overall set-up of the system. Here, we extend the model to incorporate the set-up. We assume that the set-up is given by a variable q in a set of possible set-ups Q. Then we have a general mapping

$$A : Q \times X \to Y, \quad (q, \varphi) \mapsto A[q](\varphi). \tag{17.1.2}$$

Here, the model (17.1.2) implies that the system state φ is independent of the set-up. The same is true for the data space Y, i.e. we assume that for all choices $q \in Q$ we have the same fixed data space Y for our measurements. Usually, this can be achieved by appropriate mappings from different data spaces into one reference space. Here, we will focus on a set-up which is limited by the framework (17.1.2) and consider the solution of

$$A[q](\varphi) = \psi + \xi \tag{17.1.3}$$

with some measurement error ξ. For the further arguments, we assume that we have a family of regularization operator $R_\alpha[q]$ such that we obtain a regularized solution

$$\varphi_\alpha := R_\alpha[q](\psi + \xi), \quad \alpha > 0 \tag{17.1.4}$$

of the problem (17.1.2) finding φ from $\psi + \xi$ when $q \in Q$ is given. Here, α is known as the *regularization parameter* and we assume that

$$\varphi_\alpha \to \varphi, \quad \alpha \to 0 \tag{17.1.5}$$

if we use true data $\psi = A[q](\varphi)$.

As an *example*, we consider the radiation of acoustic pressure from a vibrating bounded planar surface Γ into a homogeneous isotropic medium. We denote the acoustic field by u, it solves the Helmholtz equation

$$\Delta u + \kappa^2 u = 0 \quad \text{in } \mathbb{R}^3 \backslash \Gamma \tag{17.1.6}$$

with wave number κ which satisfies $\mathrm{Re}(\kappa) > 0$ and $\mathrm{Im}(\kappa) \geqslant 0$. Further, we assume that u satisfies the Sommerfeld radiation condition

$$\frac{\partial u}{\partial r} - iku(x) \to 0, \quad r = |x| \to \infty \tag{17.1.7}$$

uniformly for all directions, see [3].

1. Here, the *direct problem* is to calculate the radiated field u given sources on Γ such that (17.1.6) and (17.1.7) are satisfied. We will place receivers on some second planar surface Λ which is parallel to Γ. Then, the mapping A maps

the source strength φ into the measurements $\psi_j = u(x_j)$ with $x_j \in \Lambda$ for $j = 1, \ldots, M$, $M \in \mathbb{N}$.

2. The *inverse problem* is, given measurements ψ_j, $j = 1, \ldots, M$ of u in points x_j on Λ for $j = 1, \ldots, M$, to reconstruct the strength $\varphi \in X$ of the acoustic sources on Γ.

In practice, the above inverse source problem is strongly ill-posed and requires regularization algorithms of the form (17.1.4) to calculate plausible solutions. For example, we may use *Tikhonov regularization* as introduced in section 3.1.4

$$R_\alpha := \left(\alpha I + A^* A\right)^{-1} A^*; \qquad \|R_\alpha\| \leqslant 1/2\sqrt{\alpha} \tag{17.1.8}$$

with *regularization parameter* $\alpha > 0$ employed and we assume that there is a strategy for making a choice of α where

$$\alpha = \alpha(\psi) = \alpha(A[q]). \tag{17.1.9}$$

Particular strategies for our study will be

1. a constant $\alpha > 0$ and
2. α values chosen to be optimal for the given problem (using the knowledge of the true solutions in the simulated case).

Consider the general set-up (17.1.2). We now formulate what we call the *meta-inverse problem*, which is an optimal parameter problem involving the solution of the inverse problem.

Definition 17.1.1 (The meta-inverse problem). *Given a set of systems Q, let the data space Y be fixed and assume we can measure $\psi = A[q]\varphi$ depending on the particular set-up $q \in Q$. The meta-inverse problem is finding an optimal element $q \in Q$ such that for measurements $\psi = A[q]\varphi + \xi$ with some data error ξ the reconstruction error*

$$E(q) := \|R_\alpha[q]\psi - \varphi\|_X \tag{17.1.10}$$

is minimized, where $\alpha = \alpha(A[q]\varphi)$ is chosen according to some prescribed strategy (17.1.9).

In the case of the *acoustic source reconstruction problem* q describes the spatial positions of the receivers employed to measure the acoustic pressure in the radiated field. For other applications, q could represent the frequency of an incident field or other physical input parameters such as input directions and waveforms.

The *meta* inverse problem is to modify the inverse problem in a way such that better reconstructions are obtained, expressed by minimizing the reconstruction error (17.1.10). Clearly, there are many ingredients of this meta-inverse problem and many different possible questions. Here, we discuss some basic points.

1. **Regularization strategy.** The choice, or strategy, for selection of the regularization parameter α might strongly influence the reconstructions and also the optimal measurement set-up. We need to take this choice into account when we formulate a solution to the meta-inverse problem.
2. **Measurement error.** Usually the measurements

$$\psi_\delta = \psi + \chi \tag{17.1.11}$$

contain some measurement error χ, which in applications is usually modeled by some stochastic distribution or by some overall error bound $\delta > 0$ such that

$$\|\psi_\delta - \psi\|_Y \leqslant \delta. \tag{17.1.12}$$

The size of the errors have a crucial influence on the quality of reconstructions, and the size and distribution of errors will be important ingredients for the meta-inverse problem.

3. **Classes of systems.** It is crucial whether we are studying the reconstruction of one element $\varphi \in U$ from one measurement $\psi \in Y$ only, of whether we consider the best possible set-up for a class of experiments where $\varphi \in U$ might be chosen in a particular way. Here, different stochastic and deterministic approaches are possible with different norms or estimators to be selected. We visualize a tree with the options in figure 17.1.

In the following steps we describe a framework for the solution of the meta-inverse problem, working out the different components of the problem which need to be studied to obtain *optimal* solutions. We will then develop an approach to use this meta-inverse problem for improving the reconstruction quality for an actual inverse problem by an *adaptive procedure* which involves the solution of meta-inverse problems. The example shows that the approach indeed can be solved and clearly improves the quality of the reconstructions. But there are many choices for the set-up, among other questions we need to clarify

- the type and role of errors under consideration,
- how we deal with regularization,
- what set U of possible solutions we need to take into account, and
- whether we are interested in stochastic estimation of optimal parameters or deterministic optimization.

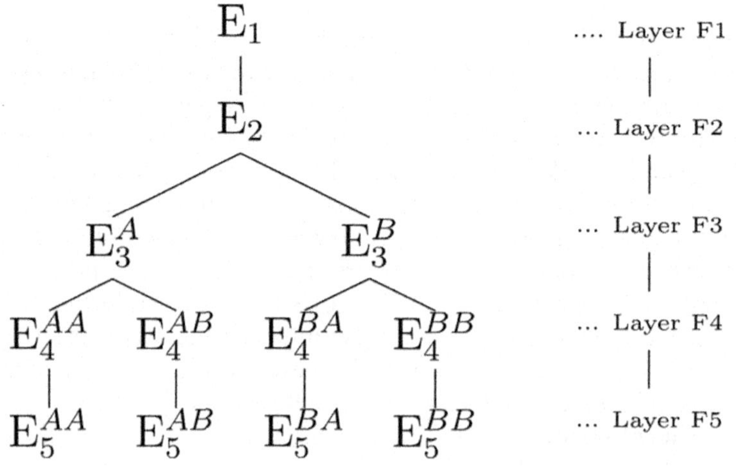

Figure 17.1. Framework for solving meta-inverse problems following [1].

F1. Reconstruction error. We begin by defining the total approximation error (17.1.13) in dependence of all its input parameters. We set

$$E_1(\alpha, \varphi, \xi, q) = \left\|\varphi_\alpha^\xi - \varphi\right\| \tag{17.1.13}$$

where

$$\varphi_\alpha^\xi = R_\alpha(A[q]\varphi + \xi). \tag{17.1.14}$$

Here we have $\varphi \in U$ for the true solution, ξ is the imposed random error within the range described by our error level $\delta \geqslant 0$ or governed by a distribution which is controlled by δ; α is the regularization parameter, and $q \in Q$ is an element describing the system set-up. The function E_1 basically describes the pointwise reconstruction error for one single quantity $\varphi \in U$ with one set of measurement data given by $y \in Y$.

In (17.1.13) the regularization parameter α can be chosen in dependence of the data y via some adequate strategy as in (17.1.9). Here we will study either a constant $\alpha = \alpha_0$ or the optimal setting, where we choose the best possible α using the knowledge of the true solution. In fact, the second approach is not an academic exercise, but relevant for applications since it will tell us which set-up can provide optimal reconstructions and how to calculate the minimal error which can be achieved with this optimal measurement set-up.

F2. Regularization. To find the optimal regularization parameter, we minimize E_1 over α. This is given by the function

$$E_2(\varphi, \xi, q) = \min_{\alpha \in [\alpha_1, \alpha_2]} E_1(\alpha, \varphi, \xi, q) \tag{17.1.15}$$

with some interval $[\alpha_1, \alpha_2] \subset \mathbb{R}$, which provides the error for the optimal α. Clearly, we have $E_2 \geqslant 0$, since E_1 is positive or zero. The case $\alpha_1 = \alpha_2 = \alpha_0$ corresponds to a constant choice of the regularization parameter.

F3. Error estimate. We next need to discuss the treatment of the error ξ. We will study stochastic errors below, here we first consider some error which is known to be bounded by

$$||\xi|| \leqslant \delta \tag{17.1.16}$$

for some parameter $\delta > 0$. A supremum estimate is then calculated by the supremum of (17.1.15) over all ξ with $||\xi|| \leqslant \delta$

$$E_3^A(\delta, \varphi, q) = \sup_{\xi \in Y,\ ||\xi|| \leqslant \delta} E_2(\varphi, \xi, q). \tag{17.1.17}$$

This computation generates the maximal reconstruction error for data with a maximal data error of size δ and, hence, simulates the worst-case scenario.

A second possibility is to work with error distributions and optimize the expectation of the function, i.e.

$$E_3^B(\delta, \varphi, q) = \mathbb{E}[E_2(\varphi, \cdot, q)] = \int E_2(\varphi, \xi, q) \mathrm{d}p_\delta(\xi) \tag{17.1.18}$$

where we assume that some appropriate distribution $\mathrm{d}p_\delta$ depending on the parameter δ of the error ξ is given. Here, δ could for example correspond to the variance of a Gaussian distribution.

F4. Range of solutions. To find the optimal system set-up for several $\varphi \in U$ with data controlled by $\delta > 0$, we take the following steps. We compute the maximum estimate of function E_3 over all $\varphi \in U$ to generate maximal error for the optimal true solution $\varphi \in U$. Here, we can calculate a supremum estimate if we consider the deterministic set-up.

Alternatively, if some probability distribution π on $U \subset X$ is given, we can calculate an estimate for the reconstruction error. This leads to the following functions:

$$E_4^{AA}(\delta, U, q) = \sup_{\varphi \in U} E_3^A(\delta, \varphi, q) \quad (17.1.19)$$

$$E_4^{AB}(\delta, U, q) = \int_U E_3^A(\delta, \varphi, q)\, \mathrm{d}\pi(\varphi) \quad (17.1.20)$$

$$E_4^{BA}(\delta, U, q) = \sup_{\varphi \in U} E_3^B(\delta, \varphi, q) \quad (17.1.21)$$

$$E_4^{BB}(\delta, U, q) = \int_U E_3^B(\delta, \varphi, q) \mathrm{d}\pi(\varphi) \quad (17.1.22)$$

where we might follow the branch A or B from above. Here, the case where we prescribe a particular $\varphi \in X$ is included by the possibility of defining $U = \{\varphi\}$ to consist out of one element only.

F5. Optimal set-up. We are now prepared for studying the minimization with respect to the set-up of the problem. Recall that we summarize the set-up parameters or parameter functions into the variable $q \in Q$ where the set Q reflects the full range of systems which are under consideration. The several options are visualized in the tree structure of figure 17.1.

The optimal system set-up with data error controlled by $\delta > 0$ and data in the set U is found by minimizing $\mathrm{E}_4{}^Z$ with Z given by one of the pairs AA, AB, BA or BB over all $q \in Q$ such that

$$q_{\mathrm{opt}}^Z := \arg\min_{q \in Q} E_4^Z(\delta, U, q), \quad (17.1.23)$$

$$E_5^Z(\delta, \varphi) := \min_{q \in Q} E_3(\delta, U, q), \quad (17.1.24)$$

where $Z \in \{AA, AB, BA, BB\}$. The function $\mathrm{E}_5{}^Z$ describes the minimum reconstruction error with respect to the system set-up $q \in Q$ for true solutions $\varphi \in U$. The minimization (17.1.23) calculates a system $q_{\mathrm{opt}}^A \in Q$ such that for measurements $\psi = A[q]\varphi + \xi$ the reconstruction error (17.1.10) or its expectation value is minimized.

Condition number criteria. A natural choice would be to use the condition number $\mathrm{cond}(A[q])$, i.e. the measure of the extent of ill-posedness of a linear problem

$A[q]\varphi = \psi$, to search for an optimal set-up. We recall that cond($A[q]$) measures the sensitivity of a given linear system (17.1.3) for which in the presence of some data error ξ bounded by some $\delta > 0$ is rewritten as

$$A(\varphi + \delta\varphi) = \psi + \xi. \tag{17.1.25}$$

The relative error is therefore estimated by

$$\frac{\|\delta\varphi\|}{\|\varphi\|} \leqslant \|A\| \cdot \|A^{-1}\| \frac{\|\xi\|}{\|\psi\|} \tag{17.1.26}$$

where $\|A\| \cdot \|A^{-1}\|$ denotes the condition number of matrix A.

Several authors have considered the use of the condition number as an objective function for minimization. In engineering fields [4, 5], work has been carried out on the determination of optimal locations for balancing planes via minimization of the condition number.

The use of condition numbers for judging a set-up has strong advantages and disadvantages.

1. First, the calculation of a condition number is carried out much more quickly than the processing of several nested minimization and maximization routines required to solve the meta-inverse problem. This is an advantage of condition number estimates as indicators.
2. However, condition numbers are usually large and show strong fluctuations when the set-up q is modified. To find minima might be a large challenge for numerical optimization methods, even though clear and stable minima are usually present.
3. The condition number as a target function provides a criterion to find a set-up which is optimal with respect to the complete number of possible solutions of the system $A[q]\varphi = \psi$ under consideration. It is not specific with respect to a particular subset of solutions and does not provide the possibility of including results carried out from measurements.
4. Condition number estimates do not take care of the need for regularization and are not able to allow sophisticated algorithms for the choice of the regularization parameter.
5. Target functions based on condition numbers do not provide the possibility of including more general stochastic distributions of the errors or the probability in our solution space.

We have carried out numerical experiments using the condition number as an alternative to the meta-inverse problem as a measure to determine the optimal set-up $q \in Q$ in [2]. For the full meta-inverse problem we show numerical examples in section 17.3.

17.2 Framework adaption or zoom

In this section we approach our second main task, the use of meta-inverse problems as part of an adaptive inversion scheme following [1, 2]. Here, in contrast to classical adaptive schemes where the input data are fixed, we study the adaption of the framework or set-up of our inverse problem after initial measurements.

The key idea is simple: you measure something and carry out an initial inversion. This inversion tells you something about the unknown quantity. You then use the initial knowledge to change your measurement set-up slightly or completely, such you can obtain a better 'view' of the unknown quantity. In terms of a microscope you might zoom in where you want to see something better. For some general inverse problem this *zooming in* is expressed in terms of a change of the framework or set-up of the inverse problem. We can determine the new set-up by solving a meta-inverse problem.

To be more formal, we call the initial set-up $q_0 \in Q$. The initial reconstruction is denoted by $\varphi^{(0)}$. If the old set-up is described by some value $q_n \in Q$, $n \in \mathbb{N}_0$, with reconstruction $\varphi^{(n)}$, the new set-up will be a new value $q_{n+1} \in Q$, determined by the meta-inverse problem to find an optimal framework $q_{n+1} \in Q$ for the reconstruction of $\varphi^{(n)}$. We can now carry out a new reconstruction from new measurements with the set-up q_{n+1}. This is then iterated until a desired result or precision is obtained.

We formulate this iterative adaption of the framework into the following algorithmic form.

Algorithm 17.2.1 (Inversion with framework adaption/zoom). *Assume that the general framework (17.1.1), (17.1.2) is provided and for given $q \in Q$ we can measure the function $y \in Y$. Further, we assume that from y and q we can calculate a reconstruction $\varphi_\alpha \in X$ using a regularization operator $R_\alpha[q]$ as described in (17.1.4).*

1. *Given an initial set-up $q_0 \in Q$ and data $y_0 \in Y$ we calculate a reconstruction $\varphi_0 = R_\alpha[q_0](y)$. Then, for $n = 0, 1, 2, \ldots$ we iteratively carry out the following steps (2) and (3).*
2. *Given the current reconstruction φ_n, solve the meta-inverse problem (17.1.23) based on (17.1.13)–(17.1.22) to determine the optimal set-up q_{n+1}.*
3. *Carry out new measurements for the set-up q_{n+1}, leading to data values $y_{n+1} \in Y$.*
4. *Then solve the inverse problem to determine a reconstruction*

$$\varphi_{n+1} = R_\alpha[q_{n+1}]y_{n+1} \tag{17.2.1}$$

from y_{n+1} and q_{n+1}.

We will study the applicability of the above framework adaption as an iterative algorithm to the following inverse source problem and show by numerical results the great potential of the above ideas.

17.3 Inverse source problems

The task of this section is to solve a meta-inverse problem for the inverse source problem of acoustic radiation. We will show that framework adaption or zoom can be realized algorithmically and leads to strong improvements of reconstructions.

Let us study a set-up with a number of acoustic sources which are spatially located at

$$y_l := \left(c_1 + h_1 \cdot \left(l - \frac{m_1}{2}\right), H\right), \quad l = 1, \dots, m_1, \tag{17.3.1}$$

where m_1 denotes the number of sources, c_1 is the center of the source locations, h_1 denotes the source grid constant and H the height of sources above the receivers. For simplicity, here we employ a uniform grid to illustrate and demonstrate the main effects. We also assume that acoustic pressure is measured at a number of receivers with spatial positions

$$x_k := \left(c_2 + h_2 \cdot \left(k - \frac{m_2}{2}\right), 0\right), \quad k = 1, \dots, m_2, \tag{17.3.2}$$

where m_2 denotes number of receivers, c_2 is the center of the receiver locations and h_2 the receiver grid constant (see figure 17.2). For $x, y \in \mathbb{R}^m$, $x \neq y$, the fundamental solution of the Helmholtz equation

$$\Phi(x, y) := \begin{cases} \frac{\mathrm{i}}{4} H_0^{(1)}(\kappa|x - y|), & m = 2, \\ \frac{1}{4\pi} \frac{e^{\mathrm{i}\kappa|x-y|}}{|x - y|}, & m = 3, \end{cases} \tag{17.3.3}$$

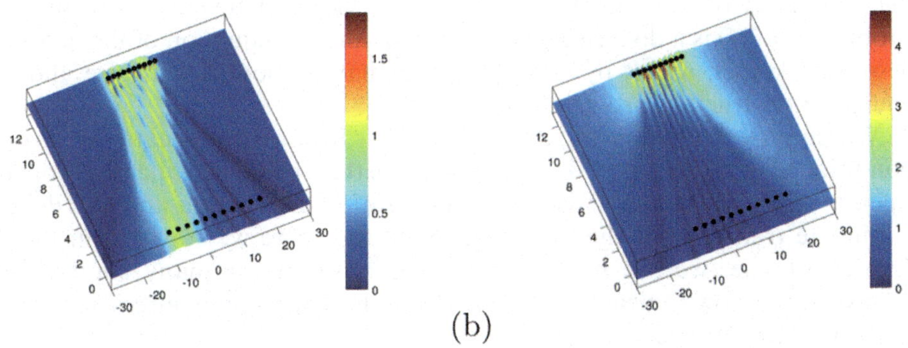

(a) (b)

Figure 17.2. In (a) and (b) we show the modulus of the acoustic fields (17.3.4) of two different source distributions on some evaluation domain Q. The sources are located at $x_2 = H = 12$, the receivers on the line $x_2 = 0$. Here, we have chosen a wave number $\kappa = 2$. Clearly, to reconstruct the sources we need different measurement locations in each of the cases, since when there is no field, we cannot expect to extract information from it. The simulation and images are generated by code 17.3.2.

where $H_0^{(1)}$ denotes the *Hankel function* of the first kind and of order zero, represents the field of a point source located at y in two or three dimensions. We model their superposition by

$$u(x_k) = \sum_{l=1}^{m_1} \Phi(x_k, y_l)\varphi_l, \quad k = 1, \ldots, m_2, \tag{17.3.4}$$

where $\varphi_l \in \mathbb{R}$ denotes the strength of the acoustic source at y_l.

Definition 17.3.1 (Acoustic source problem). *The task of the acoustic source problem is to reconstruct the modeled acoustic source strengths φ_l, $l = 1, \ldots, m_1$ from measurements of the pressure field $u(x_k)$, $k = 1, \ldots, m_2$, obtained at discrete receiver locations x_k, $k = 1, \ldots, m_2$.*

Here, the inversion problem is equivalent to solving the linear system

$$A\varphi = \psi \tag{17.3.5}$$

with

$$A := \left(\Phi(x_k, y_l)\right)_{l=1,\ldots,m_1;k=1,\ldots,m_2} \quad \psi := \begin{pmatrix} u(x_1) \\ \vdots \\ u(x_{m_2}) \end{pmatrix}, \tag{17.3.6}$$

where A represents the point-to-point response function matrix, vector ψ represents the acoustic pressures measured at the receivers and $\varphi \in \mathbb{C}^{m_1}$ denotes the strengths of the acoustic sources, where their phase is modeled by the complex phase of the components of φ.

In the previous chapters we have seen that the inversion of A to obtain φ from measurements of ψ is an ill-conditioned or ill-posed inverse problem, since the problem (17.3.4) is a discretized version of an integral equation of the first kind with a smooth kernel, which by (2.3.71) cannot have a bounded inverse and thus the source reconstruction problem is ill-posed.

We employ Tikhonov regularization (17.1.8) to approximate the strongly ill-conditioned inverse $A^{-1} : Y \mapsto X$ by a better behaved linear reconstruction operator R_α, where $\alpha > 0$ is the regularization parameter and A^* denotes the conjugate transpose of the matrix A. Note that here we stay within the discrete framework, i.e. we are not interested in the limit for many sources or many receivers, but how to deal with the inversion task when m_2 sources are present and when we have m_1 receivers.

Code 17.3.2. *Script* `sim_17_3_2_c_meta_demo.m` *to show the variability in reconstruction quality for the acoustic source problem when the parameters are modified. Run script* `sim_17_3_2_a_source_fields.m` *first to generate the measurements u and some further variables.* `sim_17_3_2_b_graphics.m` *visualizes the total field on some evaluation domain Ω. The other parts of figures 17.3 and 17.4 are generated by* `sim_17_3_2_d_graphics.m`.

```
m1    = 10;         % number of sources
H     = 12;         % define line where sources are located
m2    = 11;         % number of receivers
c1    = 0;          % center of sources
h1    = 5*pi/m1;    % distance of sources
kappa = 2;          % wave number
x1    = [(c1-h1*(m1-1)/2):h1:(c1+h1*(m1-1)/2)]; % source locations
beta  = 0.6*pi; %-1;         % density frequency shift
varphi = (exp(i*beta*x1)).';       % true source density
%varphi = (zeros(size(x1))).';     % different options
%varphi(1) = 1;
alpha  = 1e-1;         % fixed regularization parameter

j      = 1;            % counter for error evaluation
c2v    = -20:0.5:20;   % vector of centers
h2v    = 0.01:0.09:10; % vector of h2 measurement point distances
for c2 = c2v           % center of receiver array
  for h2 = h2v         % distance of receiver points
     x2    = [(c2-h2*(m2-1)/2):h2:(c2+h2*(m2-1)/2)];  % receiver locations
     rmat  = sqrt( (repmat(x2',1,m1)-repmat(x1,m2,1)).^2 ...
             + H^2);                         % matrix of distances
     A     = besselh(0,1,kappa*rmat);    % acoustic transfer matrix
     u     = A*varphi;
     % now reconstruct the sources
     varphi_rec  = (alpha*eye(m1,m1)+A'*A)\A'*u;
     % calculate reconstruction error
     ev(j)       = norm(varphi_rec - varphi);
     j = j + 1;
  end
end
```

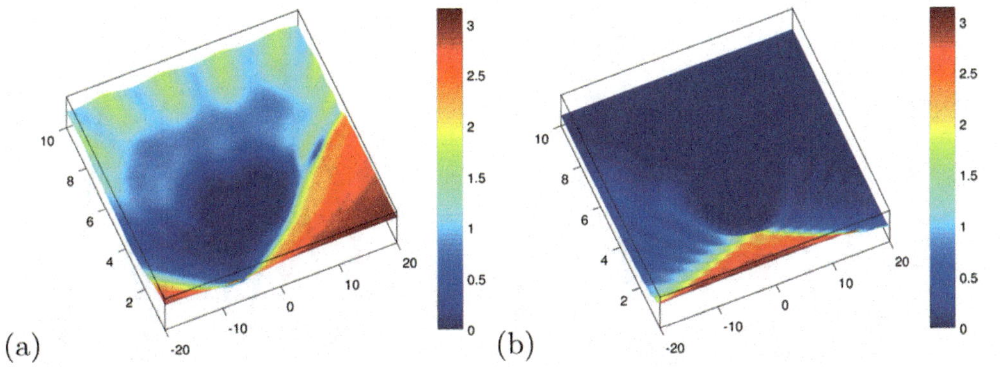

Figure 17.3. We evaluate the reconstruction error when reconstructing a source distribution φ with $m_1 = 10$ with a fixed regularization parameter $\alpha = 10^{-5}$ using code 17.3.2. (a) The result when the sources are given as in figure 17.2(a). (b) The situation when the fields behave as in figure 17.2(b). We display the errors for center points $c_2 \in [-20, 20]$ and the spacing of the measurement array $h_2 \in [0, 10]$.

Code 17.3.2 generates figure 17.3, which displays the reconstruction error in two different situations as visualized in figures 17.2(a) and (b). For (a) good reconstructions can be achieved when the measurements hit the field ray which is located around $x_1 = -8$. In the case (b) the best reconstructions are obtained when the spacing is rather large, reflecting the fact that in this case the field is merely reaching the array location plane.

Clearly, in both situations displayed in figure 17.3 the parameter sets for which good reconstructions are obtained are quite different, with some overlap in a region around $c_2 = 0$ and $h_2 = 4$. Here we need to note that the geometrical set-up for both situations is the same, the results are fully *solution-dependent*.

The dependence of the best approximation on the regularization parameter α is shown in figure 17.4, where we display $\alpha = 10^{-2}$ and $\alpha = 1$. When we have data with significant error, a good choice of the set-up becomes crucial.

We complete this part with a demonstration that *zoom* or *framework adaption* works and can significantly improve reconstructions even in simple situations.

In figure 17.5 we display the result of the iterative algorithm 17.2.1 with two iterations. We have applied a set-up where we have no measurement error and a

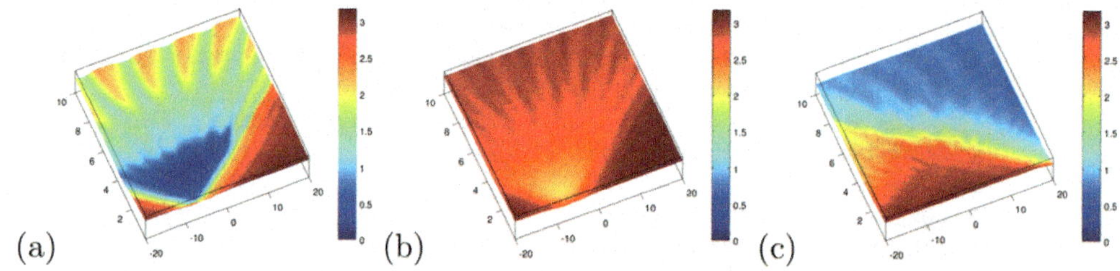

Figure 17.4. The same result as in figure 17.3(a), but with $\alpha = 10^{-2}$ in (a) and $\alpha = 1$ in (b). (c) shows the situation for the other source density, i.e. figure 17.3(b), now with $\alpha = 10^{-1}$. Here, we employ perfect data, but regularize for different data errors, showing the best possible situation for that case. If the data error is larger, we need to regularize more strongly and the dependence on the set-up increases.

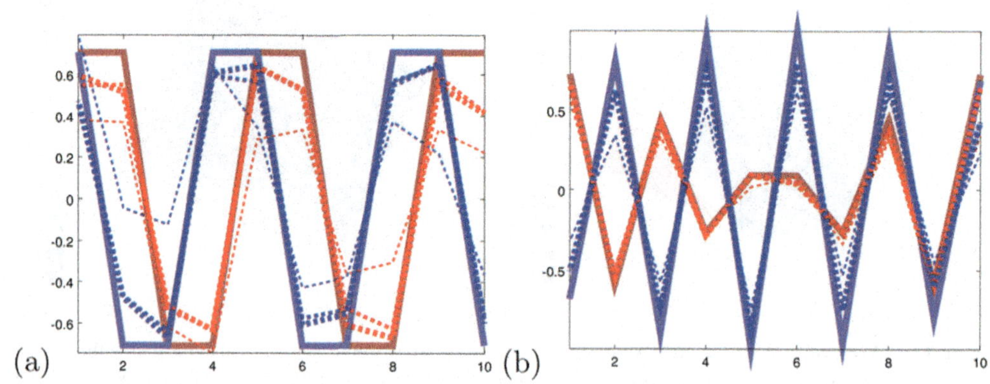

Figure 17.5. Reconstructions of the real and imaginary part of the source density φ before and after the first and the second step of the adaptive method defined in algorithm 17.2.1. In (a) we reconstruct the density leading to the field shown in figure 17.2(a), in (b) the corresponding density leading to (b), in both cases with a fixed regularization parameter $\alpha = 10^{-1}$. The thin dashed lines are the two iterations, with iteration two shown by a slightly thicker dashed line. The true density is displayed by the full thick lines, the real part in red and the imaginary part in blue.

fixed regularization parameter $\alpha = 10^{-1}$. This reflects the situation where we have medium measurement error and need to clearly regularize, but then study the case where perfect data are measured as a best case situation.

We start both iterations with $c_2 = 0$ and $h_2 = 5$. For the first density φ in (a), the optimal values are $c_2 = -6.5$ and $h_2 = 1.6$ after the second iteration step. For the second density φ in (b) we obtain $c_2 = 20$ and $h_2 = 4.8$ as parameter settings where best reconstructions are achieved, which is consistent with the error surfaces displayed in figures 17.3 and 17.4. We observe that the iterations strongly decrease the reconstruction error. Also, we note that the optimal set-up for reconstructing source strength $\varphi^{(a)}$ leads to large errors when reconstructing source strength $\varphi^{(b)}$ and vice versa.

Practically, the two iterations mean that we measure three times, with the initial set-up, then with the set-up calculated in the first iteration and a second time with the set-up from the second iteration based on the reconstruction with the previous (second) measurements. The *final* reconstruction is then based on the second optimal measurement set-up for the unknown source distribution.

Here, we have solved the optimization problem by brute force, calculating the functional for many values and selecting the minimizing arguments, see `sim_17_3_2_e_zoom.m` in the code repository. Clearly, there is the strong need for developing efficient schemes for large-scale problems in the future.

Bibliography

[1] Udosen N 2013 Algorithms for selecting optimal measurement locations in electrical resistivity tomography *PhD Thesis* University of Reading, UK

[2] Udosen N and Potthast R 2015 Algorithms for selecting optimal measurement locations in electrical resistivity tomography *at press*

[3] Colton D and Kress R 1992 *Inverse Acoustic and Electromagnetic Scattering Theory* (*Applied Mathematical Sciences* vol 93) 2nd edn (Berlin: Springer)

[4] Kang Y, Lin T-W, Chang Y-J, Chang Y-P and Wang C-C 2008 Optimal balancing of flexible rotors by minimizing the condition number of influence coefficients *Mech. Mach. Theory* **43** 891–908

[5] Lin T-W, Kang Y, Wang C-C, Chang C-W and Chang C-P 2005 An investigation in optimal locations by genetic algorithm for balancing flexible rotor-bearing systems *Proc. GT2005 ASME Turbo Expo 2005: Power for Land, Sea and Air* (*Reno-Tahoe, NV, June 6–9*) paper GT2005-69048 in preparation

IOP Publishing

Inverse Modeling

An introduction to the theory and methods of inverse problems and data assimilation

Gen Nakamura and Roland Potthast

Appendix A

The appendix collects important formulas and definitions which are needed throughout the set-up and analysis of our tools and methods. We also provide some reasoning of how and why the notation was chosen.

A.1 Basic notation and abbreviations

It is usually important to employ a homogeneous notation in a coherent introductory book. However, working with very diverse communities such as the *mathematical analysis* community, the *stochastics* community, the *data assimilation* community, diverse *engineering* communities and the language developed by *functional analysis*, a homogeneous representation of results in one joint language is difficult to achieve. Also, the different communities usually present their results in their own language, such that it is important to learn to speak that language. We have, thus, made the decision to keep different notations active throughout the book, reflecting the interfaces and links of our methods.

- $\mathbb{N}$ is the set of natural numbers $\mathbb{N} = \{1, 2, 3, \ldots\}$. Also $\mathbb{N}_0 := \mathbb{N} \cup \{0\}$. $\mathbb{R}$ is the set of real numbers. $\mathbb{C}$ is the set of complex numbers.
- We use the notation $B_\rho(x)$ or $B(x, \rho)$ for an open ball with center x and radius ρ either in $\mathbb{R}^m$ for $m \in \mathbb{N}$ or in some Banach space X.
- We use the notation $B(x, \rho)$ for an open ball with center x and radius ρ either in $\mathbb{R}^m$ for $m \in \mathbb{N}$ or in some normed space X and the abbreviation $B_\rho = B(0, \rho)$.
- For any set $E \subset \mathbb{R}^m$, we use the notation ∂E to denotes its boundary and E_T to denotes $E \times (0, T)$.

We will often work both with functions and with their discretized versions, i.e. with vectors in a space $\mathbb{R}^m$ or $\mathbb{C}^n$.

- We use the notation $\varphi(x)$ for a function φ with argument x. But x itself can denote a vector or function, in particular when we talk about *data assimilation* or *probability distributions*. Then, we might use $p \in \mathbb{R}^m$ as the function argument or evaluation point.
- We sometimes use bold letters for discretized versions of functions, i.e. $\boldsymbol{\varphi} \in \mathbb{R}^m$ can be a vector of discretized values of a function φ. This approach is used when we explicitly talk about the discretized version of some continuous problem.

doi:10.1088/978-0-7503-1218-9ch18

- But we also employ normal letters φ or ψ, a, b, f, y to denote vectors $\varphi \in \mathbb{R}^m$. The notation a_k can either denote a scalar $a_k \in \mathbb{R}$, or it can denote a vector $a_k \in \mathbb{R}^m$ with index k, which can for example be a time index. In this case we have

$$a_k = \begin{pmatrix} a_{k,1} \\ a_{k,2} \\ \vdots \\ a_{k,n} \end{pmatrix} \in \mathbb{R}^m.$$

  Usually, the context clarifies quickly which notation is used. These conventions are also used for the case where $\mathbb{R}$ is replaced by $\mathbb{C}$.
- Equations can be written in the form $Ax = y$ or $A\varphi = f$ or $A\varphi = \psi$.

We gave preference to having a nice form of formulas and presentation over a completely consistent way to distinguish finite and infinite sets, vectors and functions.

- Operators or mappings are often denoted by capital letters, $A: X \to Y$ with linear spaces X and Y, but we take the freedom to use any letter to denote a mapping, in particular in the analytic sections.
- Sets of functions, vector spaces, normed spaces and Banach or Hilbert spaces are usually denoted by capital letters such as X, Z, U, M, Y etc.

Let $G \subset \mathbb{R}^m$ be some open set. Our basic notation includes

- the use of $C(G)$ or $BC(G)$ for the space of continuous functions on G, usually equipped with the supremum norm.
- The notation $C^n(G)$ or $BC^n(G)$ is for the functions which have $n \in \mathbb{N}$ continuous derivatives on G which are bounded. The canonical norm here is the supremum of all n derivatives on G up to order n including order 0.
- We use $L^2(G)$ for the space of square integrable functions on G.
- $H^1(G)$ is the space of functions in $L^2(G)$ which are weakly differentiable with a derivative in $L^2(G)$. To work with the exterior part of domains, one usually employs the space $H^1_{\text{loc}}(\mathbb{R}^m \backslash G)$, denoting the space which is $H^1(M)$ for any compact subset M of $\mathbb{R}^m \backslash G$.
- The notation $H^s(G)$ is used for the Sobolev space of order $s \in \mathbb{R}$. For further details about Sobolev spaces see [1].

When working with normed spaces, and Banach and Hilbert spaces

- the scalar product is written in the form $\langle \varphi, \psi \rangle$, but sometimes the common notation (φ, ψ) might still be used.
- Norms are written as $\|\varphi\|$, where we note that in general this is a norm in a Banach space environment, not a particular choice of such a norm.
- For some operator $A: X \to Y$, the adjoint operator is denoted by A^* or A'. We use A' in particular when the L^2 or ℓ^2 scalar products are used, A^* is then the adjoint with respect to some other more general scalar products.

Arguments and results should apply to any Banach or Hilbert space and corresponding norm, if not explicitly stated otherwise.

A.2 Important integration formulas

Our chapters on waves and magnetic fields make heavy use of integration formulas which are valid for a domain G with sufficiently smooth boundary ∂G and outward unit normal ν. We note the *divergence theorem*

$$\int_G \nabla \cdot A \, dx = \int_{\partial G} \nu \cdot A \, ds$$
$$\int_G \nabla \times A \, dx = \int_{\partial G} n \times A \, ds$$
$$\int_G \nabla \varphi \, dx = \int_{\partial G} \varphi \, ds.$$

For working with the Helmholtz equation *Green's formula* is very important. We have *Green's first identity*

$$\int_G (\nabla u \cdot \nabla v + u\Delta v) \, dx = \int_{\partial G} u\frac{\partial v}{\partial \nu} \, ds \qquad (A.2.1)$$

as long as we have sufficient regularity up to the boundary ∂G of G. For functions u which satisfy the Helmholtz equation $\Delta u + \kappa^2 u = 0$, this yields

$$\int_G (|\nabla u|^2 - \kappa^2 |u|^2) \, dx = \int_{\partial G} \bar{u}\frac{\partial u}{\partial \nu} \, ds. \qquad (A.2.2)$$

We also note *Green's second identity*

$$\int_G (v\Delta u - u\Delta v) \, dx = \int_{\partial G} \left(v\frac{\partial u}{\partial \nu} - \frac{\partial v}{\partial \nu}u \right) ds.$$

If u and v are two functions which satisfy the Helmholtz equation in G with sufficiently regularity on ∂G, Green's second identity implies

$$\int_{\partial G} \left(v\frac{\partial u}{\partial \nu} - \frac{\partial v}{\partial \nu}u \right) ds = 0. \qquad (A.2.3)$$

The derivation of the mesh rules employs *Stokes' theorem* which is valid for a surface patch Λ in $\mathbb{R}^m$, $m = 2, 3$, with boundary curve $\partial\Lambda$ with tangential vector γ. We have

$$\int_\Lambda \nabla \times A \, ds = \int_{\partial \Lambda} A \cdot d\gamma,$$

where we need to be careful with the orientation of the curve and the normal vector.

A.3 Vector calculus

It is very useful to have standard formulas of vector calculus at hand for various derivations and transformations. Here, we collect formulas which are heavily used throughout this book. We start with various product rules. Let φ, ψ denote scalar functions and A, B, C denote vector functions in $\mathbb{R}^3$. Then, we have the basic identities

$$(A \times B) \times C = (A \cdot C)B - (A \cdot B)C$$
$$(A \times B) \cdot (C \times D) = (A \cdot C)(B \cdot D) - (B \cdot C)(A \cdot D).$$

Many other formulas can be derived from these using $A \times B = -B \times A$. Further, we are often working with the ∇ operator, which is the vector of partial derivatives, such that $\nabla \cdot \varphi = \mathrm{div}(\varphi)$ and $\nabla \times A = \mathrm{curl}\ A$. We have

$$\nabla \cdot (\varphi A) = A \cdot \nabla\varphi + \varphi\nabla \cdot A$$
$$\nabla \times (\varphi A) = \varphi\nabla \times A + (\nabla\varphi) \times A.$$

We also need the vector cross product

$$\nabla \cdot (A \times B) = (\nabla \times A) \cdot B - A \cdot (\nabla \times B)$$
$$\nabla \times (A \times B) = A(\nabla \cdot B) - B(\nabla \cdot A) + (B \cdot \nabla)A - (A \times \nabla)B.$$

Very often, the following vector calculus formulas

$$\nabla \times (\nabla\varphi) = 0$$
$$\nabla \cdot (\nabla \times A) = 0$$

are used, and we note

$$\Delta\varphi = \nabla \cdot (\nabla\varphi)$$
$$\nabla \times (\nabla \times A) = \nabla(\nabla \cdot A) - \Delta A.$$

Today, these formulas are also readily available on-line in many places.

A.4 Some further helpful analytic results

Here, we collect some further helpful analytic results which are used in particular in the sections on inverse scattering problems.

First, many arguments are based on Rellich's lemma, showing that the far field pattern u^∞ or the asymptotic behavior of the scattered field u^s, uniquely determines the whole scattered field u^s outside the scatterer D.

Theorem A.4.1 (Rellich's lemma). *Let $D \subset \mathbb{R}^m$ with $m = 2, 3$ be a bounded domain such that $\mathbb{R}^m \backslash \bar{D}$ is connected. If $u \in H^1_{\mathrm{loc}}(\mathbb{R}^m \backslash \bar{D})$ satisfies*

$$(\Delta + \kappa^2)u = 0 \quad \text{in} \quad \mathbb{R}^m \backslash \bar{D} \tag{A.4.1}$$

and

$$\lim_{R\to\infty} \int_{|x|=R} |u(x)|^2 \, \mathrm{d}\sigma = 0, \tag{A.4.2}$$

then $u = 0$ in $\mathbb{R}^m \backslash \bar{D}$.

Proof. We will only show the outline of the proof for the case $m = 3$. Expand $u(x)$ for $|x| \gg 1$ in the form

$$u(x) = \sum_{n=0}^{\infty} \sum_{|\ell| \leqslant n} a_n^\ell(r) Y_n^\ell(\hat{x}), \tag{A.4.3}$$

where $r = |x|$ and $Y_n^\ell(\hat{x})$ ($|\ell| \leqslant n$) are the spherical harmonics of order n for $\hat{x} = (\sin\theta\cos\varphi, \sin\theta\sin\varphi, \cos\theta)$ with $\theta \in [0, \pi]$, $\varphi \in [0, 2\pi)$ given by

$$Y_n^\ell(\hat{x}) = \sqrt{\frac{2n+1}{4\pi}\frac{(n-|\ell|)!}{(n+|\ell|)!}}\,P_n^{|\ell|}(\cos\theta)\mathrm{e}^{\mathrm{i}\ell\varphi} \tag{A.4.4}$$

with $P_n^{|\ell|}(t) = (1-t^2)^{\ell/2}\frac{\mathrm{d}^{|\ell|}P_n(t)}{\mathrm{d}t^{|\ell|}}$ and the Legendre polynomial $P_n(t)$ whose generating function is

$$\frac{1}{\sqrt{1-2tz+z^2}} = \sum_{n=0}^{\infty} P_n(t)z^n, \quad z \in \mathbb{R}, \quad |z| < 1. \tag{A.4.5}$$

It is well known that $\{Y_n^\ell\}$ is a complete orthonormal system of $L^2(\mathbb{S})$ and hence each a_n^ℓ is given by

$$a_n^\ell(r) = \int_{\mathbb{S}} u(r\hat{x})\overline{Y_n^\ell(\hat{x})}\,\mathrm{d}s(\hat{x}) \tag{A.4.6}$$

and it can be clearly estimated as

$$|a_n^\ell(r)| \leqslant \left(\int_{\mathbb{S}} |u(r\hat{x})|^2\,\mathrm{d}s(\hat{x})\right)^{1/2}, \quad r \gg 1. \tag{A.4.7}$$

Since $u(r\hat{x}) \in C^\infty(\mathbb{S})$ for each fixed $r \gg 1$ by the interior regularity of solutions for the Helmholtz equation, the infinite series (17.7) of functions and its term-wise derivatives of any order are absolutely and uniformly convergent on $\mathbb{S}$ by using the fact $r^2\Delta Y_n^\ell = -n(n+1)Y_n^\ell$ for $|\ell| \leqslant n$. Further based on this, it is known that each $a_n^\ell(r)$ satisfies the spherical Bessel equation given by

$$\frac{\mathrm{d}^2 a_n^\ell}{\mathrm{d}r^2} + \frac{2}{r}\frac{\mathrm{d}a_n^\ell}{\mathrm{d}r} + \left(\kappa^2 - \frac{n(n+1)}{r^2}\right)a_n^\ell = 0, \quad r \gg 1. \tag{A.4.8}$$

The fundamental solutions of (17.7) are the spherical Hankel functions $h_n^{(1)}(\kappa r)$ and $h_n^{(2)}(\kappa r)$ of the first and second kind, respectively. Hence, $a_n^\ell(r)$ can be expressed as

$$a_n^\ell(r) = \alpha_n^\ell h_n^{(1)}(\kappa r) + \beta_n^\ell h_n^{(2)}(\kappa r) \tag{A.4.9}$$

with some constants $\alpha_n^\ell, \beta_n^\ell \in \mathbb{C}$. By (A.4.2), (A.4.7) and the asymptotic behaviors of $h_n^{(j)}(z)$ as $z \in \mathbb{R}$, $z \to \infty$:

$$h_n^{(j)}(z) = \frac{1}{z}\exp\left\{(-1)^{j-1}i\left(z - \frac{n+1}{2}\right)\right\}\left(1 + O\left(\frac{1}{z}\right)\right), \quad j = 1, 2 \tag{A.4.10}$$

each $\alpha_n^\ell = \beta_n^\ell = 0$. Hence $u(x) = 0$ for $|x| \gg 1$. Then by the unique continuation property (UCP) of solutions for the Helmholtz equation, we have $u = 0$ in $\mathbb{R}^3\backslash\bar{D}$. □

Corollary A.4.2. *Let $D \subset \mathbb{R}^m$ with $m = 2, 3$ be a bounded domain such that $\mathbb{R}^m\backslash\bar{D}$ is connected. Also, let $u \in H^1_{\mathrm{loc}}(\mathbb{R}^m\backslash\bar{D})$ be a solution of the Helmholtz equation in $\mathbb{R}^m\backslash\bar{D}$*

and satisfy the Sommerfeld radiation condition. Then, if the far field pattern $u^{\infty}(\hat{x}) = 0$ *for any* $\hat{x} \in \mathbb{S}$, *then* $u = 0$ *in* $\mathbb{R}^m \setminus \bar{D}$.

Proof. Since the proof is similar for the case $m = 2$, we only prove for the case $m = 3$. By 8.3.1

$$\int_{|x|=R} |u(x)|^2 \, \mathrm{d}s(x) = O\left(\frac{1}{R}\right), \quad R \to \infty. \tag{A.4.11}$$

Hence by Rellich's lemma, $u = 0$ in $\mathbb{R}^3 \setminus \bar{D}$. □

Next we provide a variant of a basic denseness result on which many of our field reconstruction methods and probing methods are based, see lemma 12.4.3. The main ideas can be found in [2].

Theorem A.4.3. *Let* $G \subset \mathbb{R}^m$ *be a non-vibrating domain. Then, the map* $H : L^2(\mathbb{S}) \longrightarrow L^2(\partial G)$ *has a dense range.*

Proof. It is enough to show that $H^* : L^2(\partial G) \longrightarrow L^2(\mathbb{S})$ is injective. In order to show this let $\phi \in L^2(\partial G)$ satisfy $(H^*\phi)(\hat{x}) = 0 (\hat{x} \in \mathbb{S})$. Here $(H^*\phi)(\hat{x})$ is given as

$$(H^*\phi)(\hat{x}) = \int_{\partial G} \mathrm{e}^{-\mathrm{i}\kappa\hat{x}\cdot y} \phi(y) \, \mathrm{d}s(y), \quad \hat{x} \in \mathbb{S}, \tag{A.4.12}$$

$\gamma(H^*\phi)(\hat{x})(\hat{x} \in \mathbb{S})$ is the far field pattern of the single-layer potential

$$(S\phi)(x) = \int_{\partial G} \Phi(x, y)\phi(y) \, \mathrm{d}s(y), \quad x \in \mathbb{R}^m. \tag{A.4.13}$$

Hence, by Rellich's lemma,

$$(S\phi)(x) = 0, \quad x \in \mathbb{R}^m \setminus \bar{G}. \tag{A.4.14}$$

Further, by the jump formula of the single-layer potential, $(I - K^*)\phi = 0$ on ∂G, where K^* is the L^2 dual of the double-layer potential.

Now, by applying the first Fredholm alternative (theorem 1.28 in [3]) to $I - K^*$ for the dual systems $\langle C(\partial G), L^2(\partial G)\rangle$ and $\langle L^2(\partial G), L^2(\partial G)\rangle$, the null space of $I - K^*$ in $L^2(\partial G)$ and $C(\partial G)$ are the same. Hence, $\phi \in C(\partial G)$.

By the continuity of a single-layer potential across ∂G with continuous density, we have $(S\phi)(x) = 0 (x \in \partial G)$. Since $S\phi$ solves the Dirichlet problem for $\Delta + \kappa^2$ in G and G is a non-vibrating domain, we have $S\phi = 0$ in $\mathbb{R}^m$. Then, by the jump formula for the single-layer potential, we have $\phi = 0$. □

We have used the UCP of solutions for the heat equation. Since it is slightly different from that for scalar strongly elliptic equations, we will state it precisely for general second order scalar parabolic equations. To begin with let $\Omega \subset \mathbb{R}^m$ be a domain and $P = \partial_t - L$ with the second order elliptic operator L of the form:

$$L = \nabla \cdot A\nabla + b \cdot \nabla + c, \tag{A.4.15}$$

where $A = A(x)$ is positive and continuous matrix in Ω, vector valued function $b = b(x)$ and scalar function $c = c(x)$ are continuous in Ω. Note that b and c do not have to be real vector valued nor real valued. Then we have the following UCP for P.

Theorem A.4.4. *Let $u = u(x, t) \in L^2((0, T); H^2_{\text{loc}}(\Omega))$ satisfy $Pu = 0$ in $\Omega_T = \Omega \times (0, T)$. Suppose $u = 0$ in an open subset of Ω_T. Then u vanishes in the horizontal component of ω in Ω_T. That is, it vanishes in $\{(x, t) \in \Omega_T : (y, t) \in \omega$ for some $y\}$.*

For the proof of this theorem, see theorem 3.3.6 of [4].

Family of domains are used as approximation domains for point source method, singular sources method and domain sampling for no response test, range test, enclosure method. Here for a given domain compactly embedded in a ball, we will give a continuous deformation of the ball to touch the domain at a given point on its boundary. In order to state this more precisely, let B be an open ball and A be a domain with C^m smooth boundary ∂A for $m \in \mathcal{N} \cup \{0\}$ compactly embedded in B. We assume that $B\backslash\bar{A}$ is connected. We call a bounded domain whose boundary is C^m diffeomorphic to a unit sphere a C^m spherical domain.

Theorem A.4.5. *Let $q \in \partial A$. Then, there is a C^m spherical domain $G(q)$ satisfying the following properties (i) and (ii).*

- (i) *$A \subset G(q) \subset B$ and $q \in \partial G(q)$.*
- (ii) *There exists a family of C^m spherical domain $G_s(q)$ $(0 \leqslant s \leqslant 1)$ such that $G_0(q) = G(q)$, $B_1(q) = \Gamma$, $\overline{G_s(q)} \subset G_{s'}(q)$ $(s < s')$ and the dependency of $G_s(q)$ on s is of C^m smooth. We call such a family $\{G_s(q)\}$ strict deformation family of q and Γ.*

Proof. Although the argument which will be given here works on any dimension, we restrict to the three dimensional case. Take $p \in \partial B$. By $B\backslash\bar{A}$ is connected, there is a C^{m+1} curve $\ell := \{\ell(t) : 0 \leqslant t \leqslant 1\}$ such that $\ell(0) = q, \ell(1) = p, \ell(t) \in B\backslash\bar{A}(t \in (0, 1))$. Moreover we can assume ℓ does not intersect with itself and transversal to ∂A, ∂B. Extend ℓ a little bit beyond p and q so that we can consider that ℓ is defined by $\ell := \{\ell(t) : -\varepsilon < t < 1 + \varepsilon\}$ with small $\varepsilon > 0$. Consider a subbundle E of the tangent bundle $T\mathcal{R}$ over ℓ with fiber $\{\frac{d\ell}{dt}(t)\}^\perp \subset \mathbb{R}^3$ at $\ell(t)$. This E is a trivial bundle, because ℓ is retractable. Hence, there exist linearly independent sections $v_j(t) \in C^m((-\varepsilon, 1 + \varepsilon), E)$ $(j = 1, 2)$ such that a tubular neighborhood T of ℓ is given by $T = \{\ell(t) + xv_1(t) + yv_2(t) : t \in (-\varepsilon, 1 + \varepsilon), (x, y) \in V\}$ with an open neighborhood $V \subset \mathbb{R}^2$ of $(0, 0) \in \mathbb{R}^2$. This representation of T gives a C^m diffeomorphism h:

$$h : U \ni (t, x, y) \to \ell(t) + xv_1(t) + yv_2(t) \in T,$$

where $U := (-\varepsilon, 1 + \varepsilon) \times V$.

Let $B_\delta := \{x \in B : \mathrm{dist}(x, \partial B) > \delta\}$ with a sufficiently small $\delta > 0$. By the transversality of ∂A and ∂B to ℓ, for any $0 \leqslant \delta' \leqslant \delta$, $h^{-1}(\partial A)$ and $h^{-1}(\partial B_{\delta'})$ are given by $t = \varphi(x, y)$ and $t = \psi_{\delta'}(x, y)$ in V with $\varphi, \psi_{\delta'} \in C^m(V)$, $\varphi(x, y) > \psi_{\delta'}(x, y)$, respectively.

Now, let $\chi(x, y) \in C_0^m(V)$, $0 \leqslant \chi(x, y) \leqslant 1$, $\chi(0, 0) = 1$ and define a family of C^m surfaces $\{Z_s\}$ in U by

$$Z_s := \left\{t = s\chi(x, y)\big(\varphi(x, y) - \psi_{s\delta}(x, y)\big) + \psi_{s\delta}(x, y), (x, y) \in V\right\}.$$

Moreover, associated to the family of C^m surfaces $\{h(Z_s)\}$ defined in T, define a family $\{X_s\}$ of C^m surfaces in $\mathbb{R}^3$ by

$$X_s = \begin{cases} h(Z_{1-s}) \text{ in } T \\ \partial\big(B_{(1-s)\delta}\big) \text{ outside } T. \end{cases}$$

Since X_s is isomorphic to a sphere, $\mathcal{R}^3 \backslash X_s$ has two connected components. We choose the bounded component of $\mathcal{R}^3 \backslash X_s$ as $G_s(q)$. Then, this gives the desired strict deformation family of q and B.

Remark. Since it is well known that the eigenvalues of the Laplacian in a bounded domain with Dirichlet boundary condition strictly decreases as the domain increases and their dependence on domain is continuous due to computing eigenvalues using the mini-max principle (see [5]), we can assume that $G_s(q)$ $(0 < s < 1)$ in Theorem 1.1 can be assumed non-vibrating.

Bibliography

[1] Adams R A and Fournier J J F 2003 Sobolev spaces, volume 140 of Pure and Applied Mathematics (Amsterdam). Elsevier/Academic Press, Amsterdam, second edition, 2003

[2] Colton D and Kress R 1998 *Inverse Acoustic and Electromagnetic Scattering Theory* (*Applied Mathematical Sciences* vol 93) 2nd edn (Berlin: Springer)

[3] Colton D and Monk P 1998 The simple method for solving the electromagnetic inverse scattering problem: the case of TE polarized waves *Inverse Problems* **14** 597–614

[4] Isakov V 1998 Inverse Problems for Partial Differential Equations *Springer Series in Applied Math. Science* vol 127 (Berlin: Springer)

[5] Courant R and Hilbert D 1953 *Methods of Mathematical Physics*, (New York: Interscience)

CPSIA information can be obtained
at www.ICGtesting.com
Printed in the USA
BVOW07*1124281017
498664BV00002B/30/P

9 780750 312196